Condensed Systems of Low Dimensionality

NATO ASI Series

Advanced Science Institutes Series

A series presenting the results of activities sponsored by the NATO Science Committee, which aims at the dissemination of advanced scientific and technological knowledge, with a view to strengthening links between scientific communities.

The series is published by an international board of publishers in conjunction with the NATO Scientific Affairs Division

A	**Life Sciences**	Plenum Publishing Corporation
B	**Physics**	New York and London
C	**Mathematical and Physical Sciences**	Kluwer Academic Publishers
D	**Behavioral and Social Sciences**	Dordrecht, Boston, and London
E	**Applied Sciences**	
F	**Computer and Systems Sciences**	Springer-Verlag
G	**Ecological Sciences**	Berlin, Heidelberg, New York, London,
H	**Cell Biology**	Paris, Tokyo, Hong Kong, and Barcelona
I	**Global Environmental Change**	

Recent Volumes in this Series

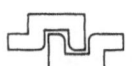

Series B: Physics

Condensed Systems of Low Dimensionality

Edited by

J. L. Beeby

University of Leicester
Leicester, United Kingdom

Associate Editors

P. K. Bhattacharya

University of Michigan
Ann Arbor, Michigan

P. Ch. Gravelle

CNRS-PIRSEM
Paris, France

F. Koch

Technical University of Munich
Garching, Germany

and

D. J. Lockwood

National Research Council
Ottawa, Canada

Plenum Press
New York and London
Published in cooperation with NATO Scientific Affairs Division

Proceedings of a NATO Advanced Research Workshop on
Condensed Systems of Low Dimensionality,
held April 23–27, 1990,
in Marmaris, Turkey

Library of Congress Cataloging-in-Publication Data

NATO Advanced Research Workshop on Condensed Systems of Low
   Dimensionality (1990 : Marmaris, Turkey)
    Condensed systems of low dimensionality / edited by J.L. Beeby ...
[et al.].
        p.   cm. -- (NATO ASI series. Series B, Physics ; v. 253)
    "Published in cooperation with NATO Scientific Affairs Division."
    "Proceedings of a NATO Advanced Research Workshop on Condensed
Systems of Low Dimensionality, held April 23-27, 1990, in Marmaris,
Turkey"--T.p. verso.
    Includes bibliographical references and index.

    ISBN-13: 978-1-4684-1350-2   e-ISBN-13: 978-1-4684-1348-9

    DOI: 10.1007/978-1-4684-1348-9

    1. Condensed matter--Congresses. 2. Dimensional analysis-
-Congresses.   I. Beeby, J. L. II. North Atlantic Treaty
Organization. Scientific Affairs Division. III. Title.
IV. Series.
QC173.4.C65N3713   1990
530.4'1--dc20   91-2721
                                                             CIP

© 1991 Plenum Press, New York
Softcover reprint of the hardcover 1st edition 1991

A Division of Plenum Publishing Corporation
233 Spring Street, New York, N.Y. 10013

SPECIAL PROGRAM ON CONDENSED SYSTEMS OF LOW DIMENSIONALITY

This book contains the proceedings of a NATO Advanced Research Workshop held within the program of activities of the NATO Special Program on Condensed Systems of Low Dimensionality, running from 1985 to 1990 as part of the activities of the NATO Science Committee.

Other books previously published as a result of the activities of the Special Program are:

SPECIAL PROGRAM ON CONDENSED SYSTEMS OF LOW DIMENSIONALITY

PREFACE

The NATO Special Programme Panel on Condensed Systems of Low Dimensionality began its work in 1985 at a time of considerable activity in the field. The Panel has since funded many Advanced Research Workshops, Advanced Study Institutes, Cooperative Research Grants and Research Visits across the breadth of its remit, which stretches from self-organizing organic molecules to semiconductor structures having two, one and zero dimensions. The funded activities, especially the workshops, have allowed researchers from within NATO countries to exchange ideas and work together at a period of development of the field when such interactions are most valuable. Such timely support has undoubtedly assisted the development of national programs, particularly in the countries of the alliance wishing to strengthen their science base.

A closing Workshop to mark the end of the Panel's activities was organized in Marmaris, Turkey from April 23-27, 1990, with the same title as the Panel: Condensed Systems of Low Dimensionality. This volume contains papers presented at that meeting, which sought to bring together chemists, physicists and engineers from across the spectrum of the Panel's activities to discuss topics of current interest in their special fields and to exchange ideas about the effects of low dimensionality. As the following pages show, this is a topic of extraordinary interest and challenge which produces entirely new scientific phenomena, and at the same time offers the possibility of novel technological applications.

I would like to thank the members of the Special Programme Panel and Dr. G. Venturi of the NATO Scientific Affairs Division for their continuing support during the organization of the workshop. I would also like to acknowledge the help and advice of my co-directors and associate editors, Pallab Bhattacharya, Pierre Gravelle, Fred Koch and David Lockwood, who contributed so effectively to the high quality of the program. Finally, all participants would wish me to thank Salim Ciraci and his Turkish colleagues for the excellence of the local arrangements which contributed so much to the enjoyment of the meeting.

J.L. Beeby
University of Leicester
UK

CONTENTS

MESOSCOPIC

SEMICONDUCTOR STRAINED LAYERS

GROWTH AND RELAXATION

GROUP IV

ELECTRONIC AND OPTICAL PROPERTIES

DEVICES

INTRODUCTION

The influence of dimensionality on the properties of
materials has been evident to scientists for generations, even
centuries, but the ability to explore the microscopic detail of
that influence had to await the developments of techniques to
produce well-characterized samples in which the low dimensional
properties became manifest. Over the last twenty or so years
chemical and physical methods for the production of such samples
have improved dramatically, beginning with layered, two-
dimensional samples and progressing to complex structures
including those with effectively zero dimensions. As the
samples have improved so has the science become more varied,
with entirely unexpected novel phenomena such as the fractional
quantum Hall effect being investigated alongside quantum well
lasers for use in mass-produced electronic equipment. Thus the
subject of low dimensionality has developed into an exciting,
inter-disciplinary research and development activity. This
volume covers only a representative selection of the work which
is currently underway worldwide, but should give a clear picture
of many recent developments.

Semiconductors form an overwhelming majority of the
materials studied because of the precision with which layered
structures can be formed from them using techniques such as
Molecular Beam Epitaxy (MBE) and Metal-Organic Chemical Vapour
Deposition (MOCVD). So the first three chapters of this volume
are devoted to optical and transport properties of
semiconductors and to strained layers formed from them. The
fourth and fifth chapters are concerned with structures not
formed from semiconductors; although less well known, the
effects of low dimensionality also influence many of the
properties of these materials.

Chapter 1, on semiconductor optical properties, reflects
the remarkable advances in the experimental and theoretical
investigations of layered structures including superlattices.
Several papers document the detailed studies which have been
made of plasmons and excitons. Optical techniques also allow
the behavior of charge carriers and the influence of magnetic
fields upon them to be investigated in many different systems.
This includes the possibility of investigating the decay times
of excited carriers. In addition to a number of papers on these
topics there are several which document detailed comparisons
between theoretical predictions of electronic band structures in
these systems and the optical experiments which probe them.

Electronic transport in semiconductors, discussed in Chapter 2, presents many interesting and challenging problems. One aspect of this is the need to refine the basic understanding of conductivity when quantum phenomena become important. The influence of contacts on the measurement process and the changes which occur when the electrons must pass through a narrow gate are but two of the microscopic effects which must be taken into account. The very nature of scattering processes and the way in which dimensionality and disorder interact are also probed by experiments on very small structures. Together with papers on these topics, the chapter contains papers concerned with the effects on conductivity of individual electrons in traps or many electrons between the barriers of resonant tunneling structures.

Strained layers, the topic of Chapter 3, have begun to present several exciting technological possibilities in addition to the challenges involved in creating and characterizing them. Their growth can now be accomplished in a number of ways, of varying effectiveness, but the study of their structures and, for example, the ways in which those structures relax to equilibrium, is still of great importance. Once created, the electronic, vibrational and optical properties of the strained layer systems present interesting new results. Most importantly, strained layer systems allow the device designer much greater flexibility in the choice of materials, which has opened up interesting prospects for both electronic and optoelectronic devices. All these features are represented amongst the papers of this section.

Studies of layered compounds have a much longer history than the equivalent semiconductor structures, since they sometimes appear in nature and have been used in the laboratory and in applications for many years. The quality of such compounds requires careful control and much effort goes into characterization. However, particularly with the appearance of high temperature superconductors, novel methods of preparation are beginning to develop and the influence of dimensionality is becoming more apparent. The papers in Chapter 4 reflect a number of research lines within this topic.

The possiblity that organic materials might substitute for metals or semiconductors in electronic devices has been of interest for many years. Indeed, layered systems can be assembled and interesting transport properties observed, though without the high quality of semiconductors. There is also the possibility that the considerable variety of organic molecules will allow the fine tuning of the system properties for given applications. These topics are discussed in the papers in Chapter 5, as is the possibility that molecules can be made to interact in organized ways.

This book covers a wide range of materials and properties, linked by their common low dimensionality. The topic still has many exciting phenomena to be discovered and applications to be made, but has now reached a point at which a text of this kind can be useful as a general summary of the present state of the art.

SEMICONDUCTOR OPTICAL PROPERTIES

OPTICAL SPECTROSCOPY OF THE TWO-DIMENSIONAL ELECTRON GAS IN GaAs QUANTUM WELLS

A. Pinczuk, B.B. Goldberg,† D. Heiman,‡
L.N. Pfeiffer and K.W.West

AT & T Bell Laboratories
Murray Hill, NJ 07974, USA
†Physics Department
Boston University, Boston MA02215, USA
‡MIT National Magnet Laboratory
Cambridge MA 02139, USA

ABSTRACT

This paper presents a short review of recent optical investigations of the 2D electron gas in GaAs quantum wells. Absorption, emission and resonant inelastic light scattering have revealed intriguing behaviors related to electron-electron interactions. Anomalies in band gap absorption and emission in the magnetic quantum limit and in the regime of the fractional quantum Hall effect are interpreted as the screening response of the electron gas to the positively charged valence hole. Light scattering by intersubband excitations of ultra-high electron mobility single quantum wells shows that exchange interactions are larger than previously anticipated. The unexpected observations of light scattering by large wavevector inter-Landau-level excitations display the excitonic binding and roton minima in the mode dispersions that are predicted by Hartree-Fock theories.

1.   INTRODUCTION

During the last decade there has been great interest in two-dimensional electron systems at the interfaces of semi-conductor microstructures and nanostructures. Such ultra-thin semiconductor layers, smooth on an atomic scale, are obtained by means of sophisticated epitaxial growth methods that allow excellent control of composition and doping. In these structures discontinuities in the band gaps at the interface of the two materials and space charge electric fields act to quantize the electron motion along the normal to the interface into discrete energy levels. When the electrons remain free to move in a plane parallel to the interface each of the discrete energy levels gives rise to a two-dimensional subband. A population of free electrons in conduction subband states of semiconductor heterojunctions and quantum wells have behaviors in common with those of an idealized two-dimensional (2D) electron gas[1].

Well defined discrete energy levels occur when electron wavefunctions are confined within physical dimensions comparable to characteristic lengths that determine electron behavior, like the de Broglie wavelength and the mean free path. In typical semiconductors these lengths range between 10 Å and 1000 Å. The greatly enhanced electron mobility achieved in modulation doped semiconductor heterojunctions and quantum wells have stimulated much interest in the physics of the 2D electron gas. A perpendicular magnetic field reduces further the dimensionality by quantizing the continuum of kinetic energy states into discrete Landau levels[1]. One of the highlights of fundamental research of the 2D electron gas in semiconductor heterojunctions has been the discovery of the fractional quantum Hall effect (FQHE) in 1982[2]. The FQHE is a magnetotransport manifestation of condensation of the electron gas into a quantum liquid with behavior determined by electron-electron interactions[3-5]. The discovery of the fractional quantum Hall effect shows that in electron systems of high perfection, reduced dimensionality and many-particle interactions are associated with a novel class of surprising and intriguing collective phenomena.

It is now well established that the many-body interactions of free carriers introduce fundamental changes in the optical properties of semiconductor quantum wells[6]. The earliest studies of optical emission and absorption processes in modulation doped GaAs-AlGaAs quantum wells had already shown large band gap renormalization and final state, or excitonic, interactions[7]. Further experimental and theoretical studies, in GaAs and InGaAs quantum wells, gave a more complete picture of these phenomena[8-16]. Of great current interest are the optical investigations of the fractional quantum Hall effect in Si-mosfets[17,18] and in GaAs-AlGaAs heterostructures[19-24].

Inelastic light scattering is a powerful method to study the excitations of 2D electron gases in semiconductor heterojunctions and quantum wells[25,26]. The impact of the light scattering method goes beyond its flexibility as a spectroscopic tool that gives the energies of the electronic excitations. By means of polarization selection rules it is possible to obtain separate spectra of spin-density and charge-density excitations. This feature is important in the quantitative determinations of energy levels and electron-electron interactions[27]. Recent work has also shown that for a 2D electron gas in a perpendicular magnetic field the light scattering method gives access to elementary excitations in the wavevector range that is most relevant to studies of many-body interactions[28].

This paper presents a brief review of studies of fundamental optical processes, band gap absorption and emission, and inelastic light scattering of the 2D electron gas in high mobility modulation doped GaAs quantum wells. In the following section we consider optical absorption and emission with emphasis on recent work in the regime of the fractional quantum Hall effect. The next section considers recent inelastic light scattering studies of electron-electron interactions in the 2D electron gas. In the last section we present our concluding remarks.

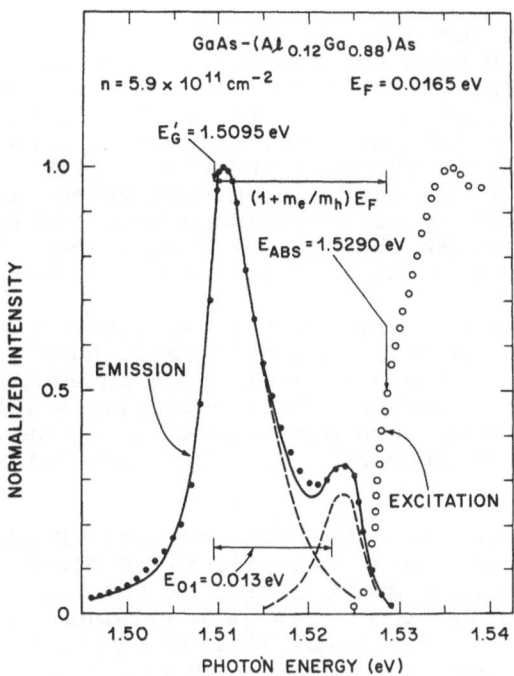

Fig. 1. Optical emission (full dots) and luminescence
excitation (open dots) measured in modulation
doped multiple quantum wells. The lines are
fits to the emission spectra. Dashed lines are
the contributions due to each of the two
conduction subbands occupied by the electrons.
The full line is the total fit. After Ref. 7.

2.   OPTICAL ABSORPTION AND EMISSION PROCESSES OF THE 2D ELECTRON GAS

Figure 1 compares spectra of photoluminescence (PL) and of
photoluminescence excitation (PLE) from a modulation doped GaAs-
AlGaAs multiple quantum well sample[7]. These results reveal
the energy separation between absorption (in PLE) and emission
(in PL) that is the signature of optical processes of the 2D
electron gas. The difference between the optical emission band
gap E'_G and the onset of absorption, E_{ABS}, has been represented
in the form that corresponds to a 2D electron gas with Fermi
energy E_F embedded in a semiconductor quantum well with
parabolic conduction and valence subbands having the GaAs
effective masses $m_e = 0.68m_0$ and $m_h = 0.39m_0$. The results in
Fig. 1 show that the optical emission gap E'_G is considerably
smaller than that in bulk GaAs. This observation was interpreted as evidence of large effects due to band gap renormalization. Another striking feature in Fig. 1 is the peak in the
PLE spectrum above E_{ABS} that indicates substantial final state,
or excitonic, interactions.

In modulation doped semiconductor quantum wells the
energies of optical transitions can be written as[6,9,29]

$$E = E_G + E_C^e + E_C^h + E_{ex} + E_{CH}^e + E_{CH}^h + E_B \qquad (1)$$

5

where E_G is gap of the bulk, E_C^e and E_C^h are the electron and hole energies associated with the confinement potential of the quantum well. These terms include the direct, or Hartree, part of the electron-electron interactions. E_{ex} is the self-energy due to the exchange interactions. E_{CH}^h and E_{CH}^e are the Coulomb-hole self-energies for the hole and the electron. These terms describe the effects of correlation that change the electron gas density in the vicinity of the valence hole and conduction electron taking part in the optical transitions. E_B represents the energy associated with excitonic interactions.

Studies of optical processes in modulation doped and photoexcited GaAs and InGaAs quantum wells have revealed large contributions from the electron-electron interactions considered in Eqn. 1. This research has been extensively reviewed in Ref. 6. In this section we turn our attention to the optical studies of the fractional quantum Hall effect in modulation doped GaAs quantum wells.

The FQHE occurs when the high mobility 2D electron gas is under a large perpendicular magnetic field B[2-5]. It appears at low temperature as a quantization of the Hall resistivity to values $\rho_{xy} = e^2/hi$, where i = p/q is a rational fraction that coincides with the Landau level filling factor $\nu = 2\pi n l_0^2$. n is the free electron density and $l_0 = (\hbar c/eB)^{1/2}$ is the magnetic length. In the FQHE there is also vanishingly small longitudinal magnetoresistivity $\rho_{xx} \sim \exp(-\xi/2kT)$, where ξ is the activation energy. Quantization at fractional values of ν, when the Fermi energy is within a partially populated spin-split Landau level, is evidence of unexpected phenomena related to electron-electron interactions. The FQHE is associated with the formation of an energy gap that separates new highly correlated many-body states of a quantum liquid from its excited states. The higher lying states are characterized by quasi-particle excitations with fractional charge.

Near band gap optical recombination in the regime of the FQHE was reported in silicon metal-insulator-semiconductor transistors and in modulation doped GaAs-AlAs quantum wells. The silicon work was carried out at T = 1.5 K and at filling factors 7/3 and 8/3[17,18]. The measurements in GaAs are at much lower temperatures and filling factors[20-24]. We consider here magneto-optical experiments in GaAs quantum wells under conditions where simultaneous transport measurements display the characteristic signatures of the FQHE[17-22,30]. Magneto-optic experiments at larger filling factors, $\nu > 2$, have been reported in Refs. 31-35.

The samples used in the initial work are multiple quantum wells with free electron areal densities in the range 1.4 x 10^{11} < n < 6 x 10^{11} cm^{-2} and mobilities 10^5 < μ < 5 x 10^5 cm^2/Vsec. GaAs substrates were removed for optical transmission measurements and the ultra-thin samples were cemented on a thin AlGaAs epilayer[35]. The more recent work was carried out in one-side modulation doped single quantum wells with electron mobilities in the range 2 x 10^6 < μ < 5 x 10^6 cm^{-1}/Vsec. A fiber optic apparatus[36,37] was used to carry the excitation light to the sample and return the luminescence or transmitted light from the sample to a spectrometer equipped with multichannel detection. The apparatus was placed either in ^{3}He or dilution cryostats where temperatures as low as 120 mK could be reached. Constant

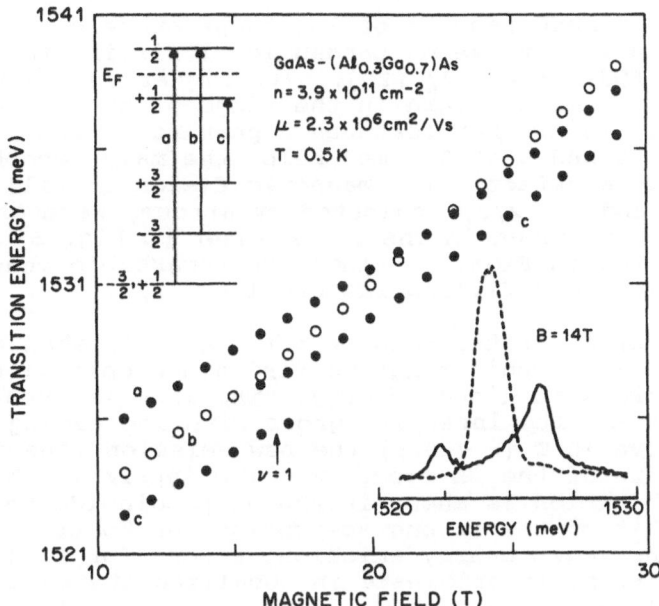

Fig. 2. Energies of the three lowest optical absorption
bands in modulation doped quantum wells as
functions of magnetic fields in the range $\nu < 2$.
Lower inset: dashed (solid) lines are absorption
spectra in σ^+ (σ^-) polarization. Upper inset:
initial and final states. After Ref. 30.

magnetic fields of up to 30 T could be achieved in hybrid
(superconductor plus Bitter solenoid) magnets. The power of
incident light was kept below 1 mW/cm^2.

The upper inset to Fig. 2 shows the three lowest optical
transitions between states in the valence and conduction
subbands. The states are Landau levels labeled by their
components of angular momenta along the normal to the plane[35].
The diagram shows the two spin-split states of the lowest Landau
level and the three higher valence states. At $\nu = 1$ the Fermi
level is in the middle of the gap between the lowest spin-split
conduction Landau levels. The lower of these levels has spin $+\frac{1}{2}$
because the g-factor is negative. The valence subbands have
extensive mixing of different terms of angular momentum
components m_j[38-40]. At fields above 16 T the highest valence
state is mostly $m_j = \frac{3}{2}$. The second state has $m_j = -\frac{3}{2}$ and the
third state has a considerable admixture of $+\frac{1}{2}$ and $-\frac{3}{2}$ m_j
components[35].

The data points in Fig. 2 show the three lowest absorption
transitions in the range $1.5 > \nu > 0.56$[30]. The lower inset
shows typical absorption spectra obtained at $\nu = 1.16$. The
measured circular polarizations are consistent with the assign-
ments in the upper inset. Transitions a and c are measured in
σ^- polarization and transition b in σ^+ polarization. The observation of transitions c for $\nu > 1$ is unexpected at 0.5 K since
the lowest Landau level is fully occupied. The appearance at
this low energy absorption in σ^+ polarization is explained as
evidence that residual disorder and localization promote states
with spin $+\frac{1}{2}$ above the Fermi level.

7

Figures 3-5 show results of low temperature optical emission over a wide range of Landau level filling factors, including the FQHE at $\nu = 2/3$[20]. The higher energy features in these emission spectra are in the range of absorption due to transitions c. The weaker features appearing in the PL spectra shown in Figs. 3 and 5 at 3-5 meV below the main recombination are assigned to a defect. The magnetic fields of filling factors $\nu = 1$ and $\nu = 2/3$, indicated by arrows, were obtained from ρ_{xx} traces as shown in the lower inset to Fig. 5. The results presented in Figs. 3-5 indicate remarkable changes in optical emission near filling factors 1 and 2/3.

We consider first the range $\nu < 1$. As indicated by the arrows in Fig. 3, in this range we find a new optical emission line which emerges at higher energy. The line is absent below 16.2 T ($\nu = 1$) and its intensity grows with increasing field. For fields above 18 T ($\nu < 0.9$) the new emission line is the dominant feature of the PL spectrum. The intrinsic character of the new recombination is shown in the near coincidence of its energy with that of the strong absorption due to transitions c in Fig. 2. The lower energy emission, strong for fields below 17 T ($\nu > 0.95$), could originate in localized states in the lower Landau level.

The remarkable growth of the strength of intrinisic recombination in the field range $\nu < 1$ is seen when the Fermi level shifts from the region of localized states into that of the extended states near the center of the lowest, spin-split, Landau level. To interpret this "anomalous" behavior we proposed that electron-electron interactions, like those considered in Eqn. 1, play a fundamental role in optical recombination in the magnetic quantum limit ($\nu < 1$)[20]. We pointed out that the response of the electron gas to the Coulomb potential enhances the electron density near the positive hole in the valence subband. Such a "screening response" is possible when the Fermi level moves into the region of extended states and there is partial occupation of the Landau level[41-43]. The enhanced electron hole overlap leads to a larger optical matrix element that explains the observed increasing strength of optical recombination for $\nu < 1$. The interactions responsible for this response of the electron gas are similar to those involved in the Coulomb hole term for the valence hole.

Our recent results for much higher mobility single quantum wells has revealed further features of optical recombination in the magnetic quantum limit[22]. In these spectra the emission from localized states is much weaker and the intrinsic recombination could be measured as the field was swept through the region about $\nu = 1$. We found that the intensity of intrinsic optical recombination has a minimum at $\nu = 1$, and at this filling factor we also observe a blue-shift of about 0.5 meV in the recombination energy. These observations are consistent with the picture of optical recombination proposed in Ref. 20. The intensity minimum at $\nu = 1$ is explained by the reduction in the response of the electron gas to the presence of the hole in the valence subband when the lowest, spin-split, Landau level is full and the others are empty. The blue-shift of the recombination energy is interpreted as a minimum in the absolute value of the negative Coulomb-hole self-energy terms associated with the reduced response of the electron gas at $\nu = 1$.

8

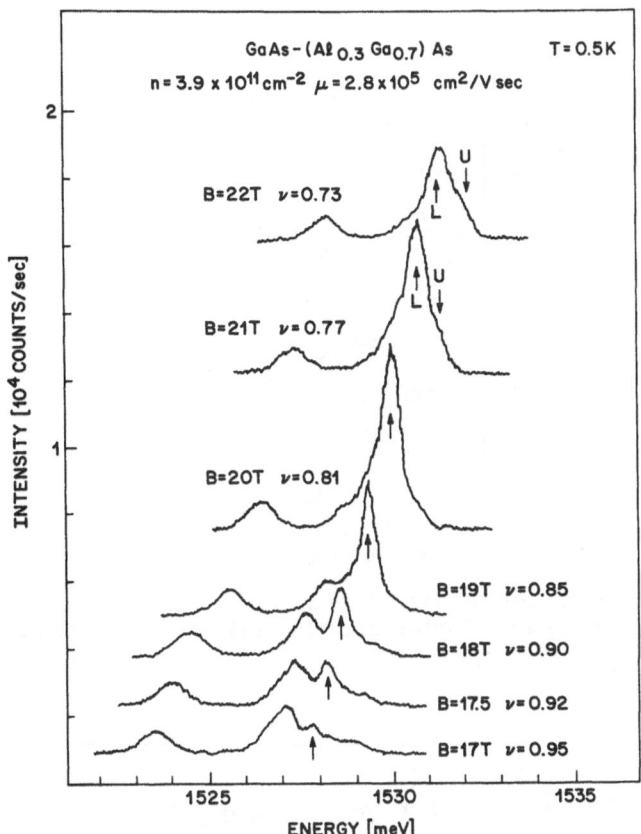

Fig. 3. Low temperature photoluminescence spectra at
several magnetic fields in the range
$0.95 < \nu < 0.73$. The arrows indicate the new
peak appearing for $\nu < 1$. After Ref. 20.

Optical recombination in the regime of the FQHE also has
remarkable "anomalous" behavior[22]. The results in Fig. 3 show
that when $\nu < 0.8(B > 20$ T) a new emission band labeled U
appears in the spectra. Initially it is seen as a weak high
energy shoulder on the main emission (labeled L). The evolution
of the low temperature emission in the regime of the FQHE is
shown in Figs. 4 and 5. With increasing field the intensity of
the new high energy emission increases until the recombination
becomes a well defined doublet for $\nu = 2/3$. At smaller filling
factors the high energy component of the doublet becomes
dominant.

The temperature dependence of optical emission in the
regime of the FQHE is very different from that observed in the
vicinity of $\nu = 1$. The spectral lineshapes and intensities at
$\nu \lesssim 1$ undergo only minor changes in the temperature range
$0.4 < T < 3$ K. On the other hand, in the regime of the FQHE the
temperature dependence of optical emission is striking[20,22].
Figure 6 shows results for filling factors 0.63 and 0.65. We
see that at these filling factors close to 2/3 small variations
in sample temperature result in large changes in the relative
intensities of the two components of the emission doublet. This

9

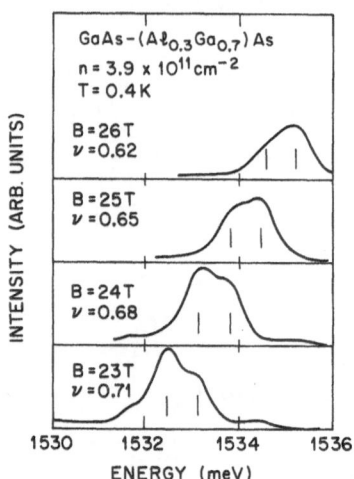

Fig. 4. Low temperature emission spectra in the
regime of the fractional quantum Hall
effect at $\nu = 2/3$. After Ref. 20.

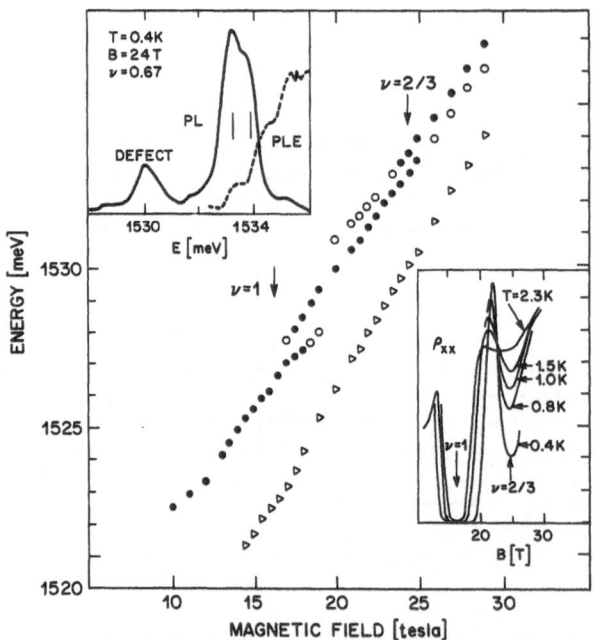

Fig. 5. Peak energies in low temperature optical
emission spectra. The instrinsic peaks
are represented by full (open) circles for
strong (weak) intensities. The triangles
are for extrinsic or defect peaks. After.
Ref. 20.

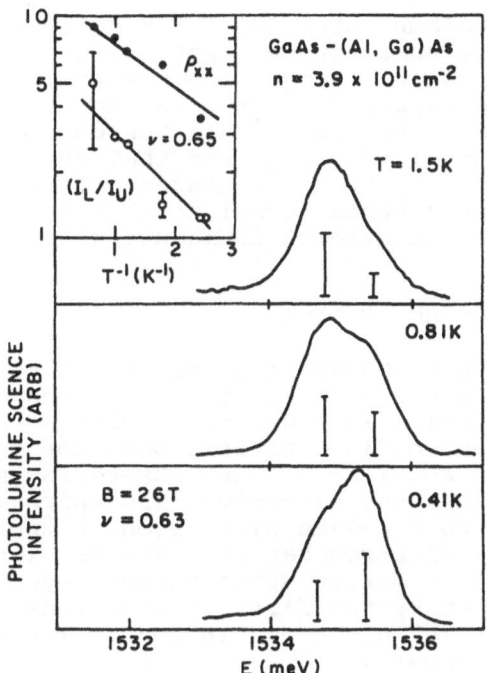

Fig. 6. Temperature dependence of photoluminescence in
the regime of the fractional quantum Hall
effect. The inset shows $\log(I_L/I_U)$ as function
of 1/T. After Ref. 20.

temperature dependence is remarkable because the overall
temperature variation is much smaller than the separation of the
peaks.

The inset to Fig. 6 shows the temperature dependence of the
intensities ratio (I_L/I_U) of the two components of the doublet.
The inset also shows the activated behavior of the longitudinal
magnetoresistivity in the regime of the FQHE at $\nu = 2/3$. The
results for ρ_{xx} show an activated behavior with $(\xi/2) = 0.44$ K =
0.04 meV. The temperature dependence of (I_L/I_U) is similar. In
fact, at temperatures above 1.5 K, where the FQHE minimum in ρ_{xx}
is very weak, the optical emission develops into a singlet. The
temperature dependence suggests that the "anomalous" behavior of
optical emission at $\nu \sim 2/3$ is related to the FQHE. However,
the splitting of the two components of the emission doublet
(~ 0.5 meV) is considerably larger than the activation of the
FQHE in this sample. For this reason we have not identified the
energy spacing between the two components of the emission
doublet with the gap of quasi-particle excitations of the FQHE.

To interpret optical recombination in the regime of the
FQHE we need to understand the response of the electron gas to
the sudden appearance of the hole in the valence subband. At
the present time there is no theoretical framework for such
phenomena. However, it is conceivable that the response of the
states of the FQHE to the presence of the hole has features in
common with the response to charged impurities and disorder[44-
46]. In this approach the hole would be represented by its long

range attractive Coulomb potential. Finite system numerical
calculations suggest that for such potential there is a large
accumulation of screening charge within a radius $r < 2 \, l_0$ about
the position of the impurity[45,46]. We conjecture that in the
case of an extended state hole such a response could lead to the
formation of exciton-like states, and that the emission doublets
could be associated with two such states. In this framework the
optical emission experiments reveal novel behavior of the states
of the FQHE that is not evident in magnetotransport.

3.   INELASTIC LIGHT SCATTERING

        A communication by Burstein et al. presented at the 14th
Internation Conference on the Physics of Semiconductors[47],
proposed resonant inelastic light scattering as a method to
study free electrons confined to semiconductor space charge
layers. This paper considered light scattering mechanisms and
selection rules within the effective mass approximation[48] and
emphasized that in bulk n-GaAs with resonant enhancements it is
possible to measure electron gas excitations within a volume
comparable to those of narrow space charge layers[49,50]. The
proposal was followed by the first observations of light
scattering by the 2D electron gas in modulation doped GaAs-
AlGaAs heterostructures[51,52].

        During the last decade there has been extensive inelastic
light scattering research of the elementary excitations of free
electrons in quantum well and superlattices. This work has
shown that the light scattering method is very effective in
studies of quantum confinement and electron-electron
interactions of electron systems with reduced dimensionality
[25,26,53]. In this section we consider recent results from
ultra-high mobility 2D electron systems in GaAs quantum wells
[27,28,54].

        Figure 7 shows the possible single particle transitions of
a 2D electron system in the absence of a magnetic field. Figure
7a shows a vertical intersubband transition between the two
lowest subbands. If the two subbands are parallel, the
transition energy is $E_{01} = E_1 - E_0$ and independent of the
wavevector $\vec{K}$. Figure 7b represents an intrasubband transition
with an in-plane wavevector transfer q. These transitions have

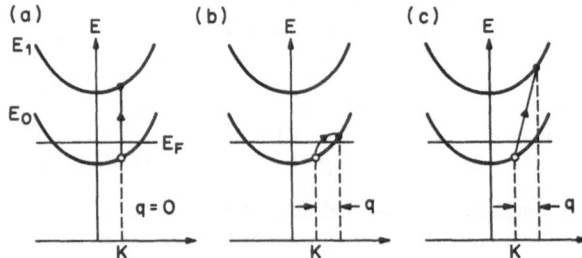

Fig. 7. Single-particle transitions of the 2D electron
        gas in a quantum well. The two lowest subbands
        are shown: (a) Vertical intersubband transition;
        (b) Intrasubband transition.   (c) Nonvertical
        intersubband transition. After Ref. 26.

a continuum of energies $\hbar^2(\vec{K}\cdot\vec{q}+\vec{q}'/2)m^*$, where m^* is the effective mass of the carriers. Figure 7c shows a nonvertical intersubband transition. These excitations have a continuum of energies centered at E_{01}. In a large perpendicular magnetic field all the single particle transitions are discrete. They are intersubband transitions, inter-Landau-level transitions and combined transitions (with simultaneous change of Landau level and subband).

In the case of GaAs the orbital angular momentum of the free electrons is zero and the electron hole-pairs in the transitions shown in Fig. 7 can be classified according to their total spin angular momentum as singlet ($J = 0$) or triplet ($J = 1$) states. The singlet states describe charge-density excitations and the triplets correspond to spin-density excitations. Only singlet states are active in optical absorption[1]. In the case of inelastic light scattering the two types of excitations are active[25,26,47]. Charge-density excitations are measured in polarized spectra in which polarization of incident and scattered light are parallel. Spin-density excitations are active in depolarized spectra where the two polarizations are orthogonal. These features of light scattering spectra makes possible unique applications in studies of electron-electron interactions.

Charge-density and spin-density excitations are collective modes of the electron gas. Their energies are shifted from those of the single-particle transitions by the strong Coulomb interactions. The spin-density modes are shifted downwards by the exchange terms. These are the final state, or excitonic, interactions also known as vertex corrections[1]. The charge-density modes are shifted upwards from spin-density excitations by the direct (or Hartree) terms. The Hartree interactions include coupling to polar optical phonons[55,56]. They are the origin of the well known depolarization-field effect in optical absorption[1]. The energies of these collective modes of the electron gas are calculated by means of charge-density and spin-density response functions[1,57-62].

The capability to obtain separate spectra of spin-density and charge-density excitations has been a major factor in the impact of the light scattering method in studies of the 2D electron gas. The selection rules for polarized and depolarized spectra served as a basis for the interpretation of light scattering measurements in GaAs quantum wells and hetero-junctions[25,26]. Because exchange interactions were expected to be small in GaAs[1,47,48,58,60,61], the spin-density excitations were assumed to give single particle energies [25,26,63]. However, recent work in ultra-high mobility single GaAs quantum wells has shown that for intersubband excitations exchange interactions are more important than had been anticipated[27]. It is also possible that large exchange interactions also account for anomalies encountered in resonant light scattering spectra of intrasubband excitations[26,64,65].

3.1 Intersubband Excitations

We consider next the light scattering studies of intersubband excitations of high mobility single GaAs quantum wells[27]. In this work the spectra are measured in a conventional backscattering geometry such that the in-plane component

13

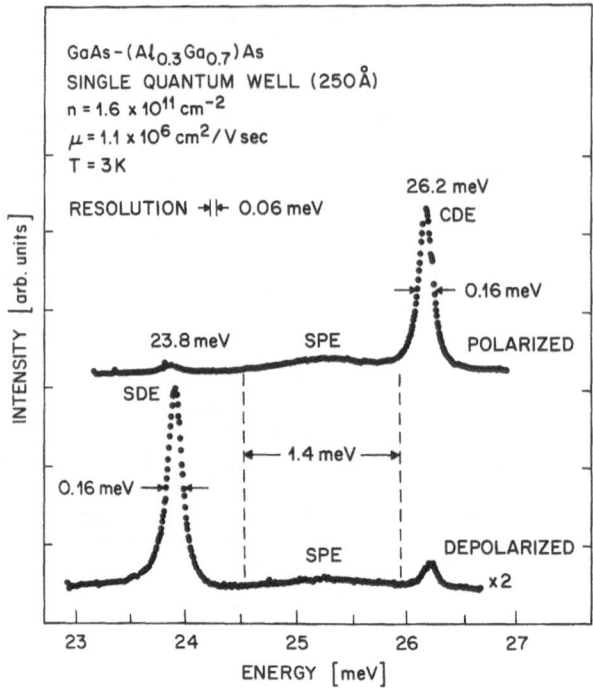

Fig. 8. Light scattering spectra of vertical
intersubband excitations. The peaks of spin-
density excitations (SDE), charge-density
excitations (CDE) and single-particle
excitations (SPE) are shown. After Ref. 27.

of the scattering wavevector is small. For this experimental
configuration we expect that the excitations derived from the
vertical intersubband transitions shown in Fig. 7a will be
active. In the work considered here the incident photon
energies are resonant with sharp excitonic features of the GaAs
quantum wells[66].

Figure 8 shows spectra of intersubband excitations of a
high mobility single quantum well[27]. The illuminated area of
the sample is a circle of about 50 μm in radius. Optical
multichannel detection was used to acquire spectra of such high
resolution. The sharp peaks in Fig. 8 are, to the best of our
knowledge, the narrowest excitations reported for the 2D
electron gas in GaAs. Because of the well defined polarization
selection rules they have been assigned to the collective spin-
density (SDE) and charge-density (CDE) intersubband excitations.

The spectra in Fig. 8 also show unexpected bands at
energies between those of SDE and CDE. These bands have been
identified as arising from single particle intersubband
excitations (SPE)[27]. The assignment was made on the basis of
studies of their dependence on the in-plane scattering
wavevector k. Results from a higher density sample are
presented in Fig. 9 and 10. The inset to Fig. 9 shows the
scattering geometry of the experiments and the expression for
the in-plane scattering wavevector. The spectra in Fig. 9

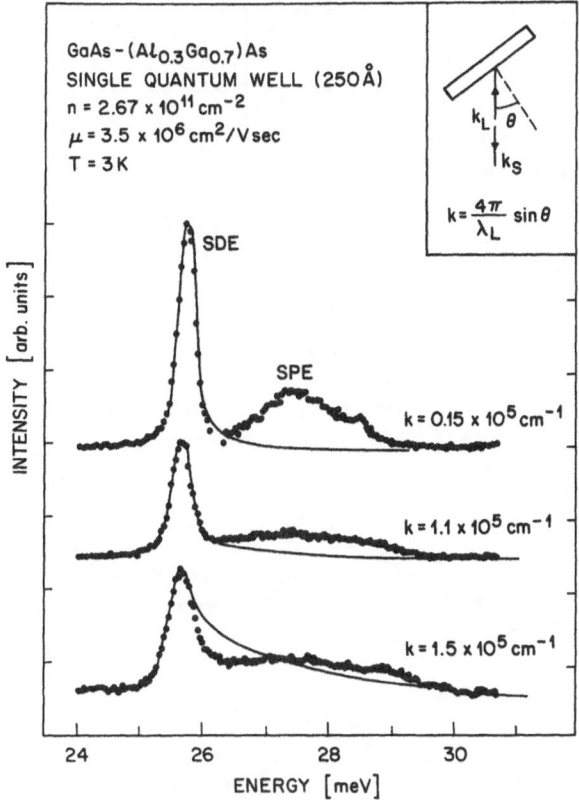

Fig. 9. Depolarized light scattering spectra of
intersubband excitations for three values of the
scattering wavevector. The lines are fits of
SDE spectra with Eqns. 2-7. The inset shows the
scattering geometry and the expression for k.
After Ref. 27.

reveal that the width of the SPE band has a pronounced dependence on k. Under experimental conditions that allow for wavevector conservation (k = q), such behavior is expected for the
"nonvertical" single particle intersubband excitations of Fig.
7c. This is evident in the inset to Fig. 10, where we show that
the energies of these transitions are within the continuum
defined by $E_{01} \pm kv_F$, where v_F is the Fermi velocity. The
assignment of the SPE bands follows from the fact that at the
largest values of k their widths are about $2kv_F$.

The remarkable changes in the lineshapes of CDE seen in the
polarized spectra of Fig. 10 is also explained by the assignment
of the SPE bands to single particle intersubband excitations.
When k is increased the energies of the CDE overlap with the
continuum of single particle excitations, as indicated in the
inset to Fig. 10. At these values of k Landau damping, i.e. the
decay of the collective charge-density modes into single
particle intersubband excitations, causes the observed
broadening of the spectra. The effect of Landau damping is less
pronounced in the case of spin-density excitations because the
k ~ 0 SDE is further away from E_{01}.

15

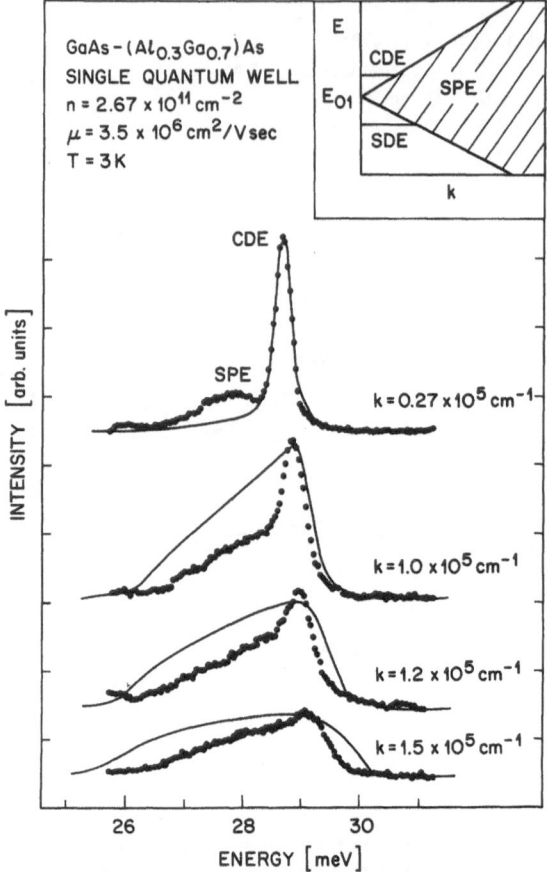

Fig. 10. Polarized light scattering spectra of intersubband excitations at four values of the scattering wavevector. The lines are fits of CDE spectra with Eqns. 2-8. The inset is a sketch of the k-dependence of long wavelength intersubband excitations. After Ref. 27.

The measurements of the energies of the two collective modes, and of the spacing between the subbands E_{01}, from the position of the SPE band, allows a determination of Coulomb interactions in intersubband transitions based solely on inelastic light scattering spectra. In the quantitative analysis the light scattering intensities are written as[27]

$$I_j(k,\omega) \sim Im\chi_j(k,\omega) \qquad (2)$$

where $\chi_j(k,\omega)$ are the response functions for the collective excitations. Within a generalized random-phase approximation they are[58,60-63]

$$\chi_j(k,\omega) = \frac{\chi_0(k,\omega)}{1 - \gamma_j(k)\chi_0(k,\omega)} \qquad (3)$$

where $\chi_0(k,\omega)$ is the intersubband susceptibility[57-62,67,68].

Equations 2 and 3 predict collective modes at energies

$$\omega_j^2 = E_{01}^2 + 2n\,\gamma_j(0)\,E_{01} \; .$$ (4)

For spin-density excitations

$$\gamma_j \equiv \gamma_{SD} = -\,\beta_{01}$$ (5)

where β_{01} is a positive parameter that accounts for the exchange term of the Coulomb interaction. It describes the excitonic, or final-state, effect in intersubband transitions. In the case of the charge-density excitations we have

$$\gamma_j = \gamma_{CD} = \left[\alpha_{01}/\epsilon(\omega)\right] - \beta_{01}$$ (6)

where α_{01} corresponds to the direct term and $\epsilon(\omega)$ is the dielectric function of the polar lattice. The first term on the right side of Eqn. 6 describes the depolarization field effect and coupling to longitudinal optical phonons.

Equations 4–6 yield the following expressions for β_{01} and α_{01}

$$2n\beta_{01} = \frac{E_{01}^2 - \omega_{SD}^2}{E_{01}} \; ,$$ (7)

$$2n\alpha_{01} = \frac{\omega_{CD}^2 - \omega_{SD}^2}{E_{01}} \; .$$ (8)

β_{01} and α_{01} have been considered as parameters that are obtained by means of Eqns. 7 and 8 and the energies of the features in the light scattering spectra[27,69]. For free electron densities $n < 3 \times 10^{11}$ cm^{-2} the ratio of the strengths of the exchange and direct terms of the Coulomb interaction is approximately independent of electron density and has the large value $(\beta_{01}/\alpha_{01}) \cong 0.4[27]$. This is an unexpected result. It suggests that exchange interactions are larger than the predictions of local density-functional theory[58].

The full lines in Figs. 9 and 10 are theoretical spectra calculated with Eqns. 2 and 3. The calculations reproduce the overall behavior of the collective excitations but fail to predict the SPE features. The deviations at large values of k could be due to several experimental factors. To describe the weak dispersion of spin-density excitations, less than 0.3 meV in Fig. 9, the exchange terms β_{01} were assumed to decrease with k. This behavior represents a departure from the predictions of local spin-density-functional theory[58,60]. Better agreement with local density-functional theory was recently reported by Gammon et al.[69]. At the present time there is no definitive interpretation for the unexpected presence of the SPE bands. The effects associated with residual disorder in these high-mobility electron systems, in conjunction with large resonant enhancements of the light scattering cross-section, are the most likely explanations[27].

17

3.2 Inter-Landau-Level Excitations

The remarkable behaviors due to electron-electron interactions within the 2D electron gas in a perpendicular magnetic
field also manifest in the dispersion of elementary excitations
based on transitions between Landau levels[70-74]. These inter-
Landau-level excitations, magnetoplasmons and spin-density
excitations, have energies

$$\omega(q) = \omega_c + \Delta(q,\omega) \tag{9}$$

where ω_c is the cyclotron frequency. Hartree-Fock calculations
of the dispersions $\Delta(q,\omega)$ display characteristic roton minima at
finite wavevectors $q > q_0 = 1/l_0$. The roton is due to the
reduction at large wavevectors of the excitonic binding between
the electron in the excited Landau level and the hole left in
the lower level[71-76]. These interactions also play leading
roles in the theory of collective excitations of the fractional
quantum Hall effect. The magnetoroton minimum associated with
the gap of the FQHE is related to the excitonic attraction
between fractionally charged quasi-particles[77,78].

The range of wavevectors that is most relevant to these
electron-electron interactions is $q \geq q_0 \sim 10^6$ cm^{-1}.
Unfortunately, this wavevector range is not easily accessible in
inelastic light scattering or infrared absorption experiments
[25,26,79]. It has been pointed out that with the breakdown of
translational invariance due to impurities and residual
disorder, the $q \sim 0$ Landau level transitions are expected to
couple to large wavevector excitations[80]. These are several
reports of cyclotron resonance that display effects of electron-
electron interactions[81-83].

The observation of roton structure was recently reported in
resonant inelastic light scattering spectra of inter-Landau-
level excitations[28,54]. These measurements were carried out
at integral values of the Landau level filling factor in
modulation doped GaAs-AlGaAs quantum wells and single
heterojunctions of extremely high electron mobility. The
spectra were interpreted in terms of the critical points of the
mode dispersions, where ($\partial\omega/\partial q$) = 0. The rotons manifest in
these spectra as characteristic structure associated with
multiple critical points. In these experiments there is massive
breakdown of wavevector conservation that allows the observation
of modes with $q > q_0 \sim 10^6$ cm^{-1}. We have attributed the loss of
translational invariance implied by these measurements to the
well known[1,86-88] reduction in screening of the disorder
potential at integral values of the Landau level filling factor.
In this framework light scattering processes by large wavevector
excitations take place with intermediate virtual transitions
between levels broadened by disorder.

The initial results of light scattering by large wavevector
inter-Landau-level excitations were obtained in multiple quantum
wells and in single heterojunctions. In the more interesting
case of the ultra-high mobility single heterojunctions
comparison with theory is complicated by the energy overlap
between inter-Landau-level and intersubband excitations[28,85].
To overcome this difficulty the more recent results are obtained
in single GaAs quantum wells with electron mobilities comparable
to those of the low-disorder single heterojunctions[54]. In the

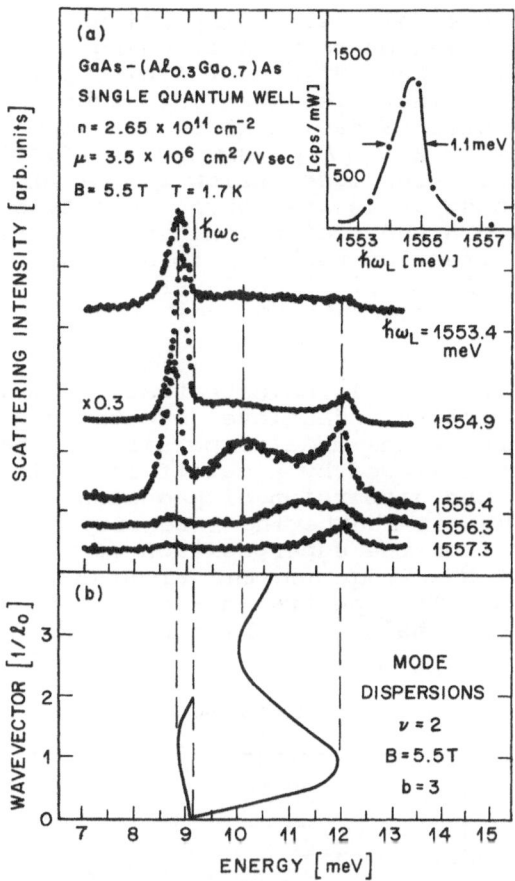

Fig. 11. (a) Light scattering spectra at four incident
photon energies and $\nu = 2$. The inset shows the
profile of resonant enhancement of intensities
at an energy of 8.8 meV. L is a weak luminescence. (b) Calculated mode dispersions.
After Ref. 54.

single quantum wells the subband spacings can be adjusted so as
not to overlap the energies of the Landau level excitations.
This allowed a comparison between measured spectra and the
predictions of calculated mode dispersions.

Figure 11a presents results obtained at $\nu = 2$. The
incident photon energies were in resonance with optical
transitions between the Landau levels of the ground heavy
valence subband h_0 and the first excited conduction subband c_1,
as shown in the lower inset to Fig. 12. The inset to Fig. 11
shows the variation of the light scattering intensity as a
function of incident photon energy $\hbar\omega_L$. The spectra are
continua in which we can identify several clearly defined peaks.
The relative intensities of the structures in the spectra have a
marked dependence on $\hbar\omega_L$. This behavior of the intensities is
characteristic of spectra measured under the strong and sharp
resonant enhancements like that shown in the inset to Fig. 11a.

Figure 11b shows calculated dispersions of inter-Landau-level excitations based on results of Cheng[89]. The calculations consider the effects of finite thickness in the 2D electron gas[1,90]. They also take into account the couplings to higher energy inter-Landau-level excitations that yield higher-order terms in E_c/ω_c[72], where $E_c = e^2/\epsilon_0 l_0$ and ϵ_0 is the dielectric constant. The modes below ω_c are spin-density excitations and those above are charge-density excitations (or magnetoplasmons). In the excitations shown in Fig. 11b the finite thickness effect corresponds to a value of the inverse of the average width of the electron layer of b = 3 (in units of $1/3l_0$). Coupling to intersubband excitations, at energies above 24 meV can be ignored.

The results in Fig. 11 indicate that the multiple structure in the spectra of inter-Landau-level excitations is related to the critical points in the mode dispersions. The high energy scattering maximum followed by a cutoff is explained by the critical point in the dispersion of magnetoplasmons near $q = q_0 = 1/l_0$. The maximum in scattering intensity seen near 10 meV appears to result from the superposition of the roton minimum of magnetoplasmons at $q \sim 2.5q_0$ and the large density of states for larger values of q. The spectra in Fig. 11a have an onset and maximum that are well below ω_c. This is evidence of the

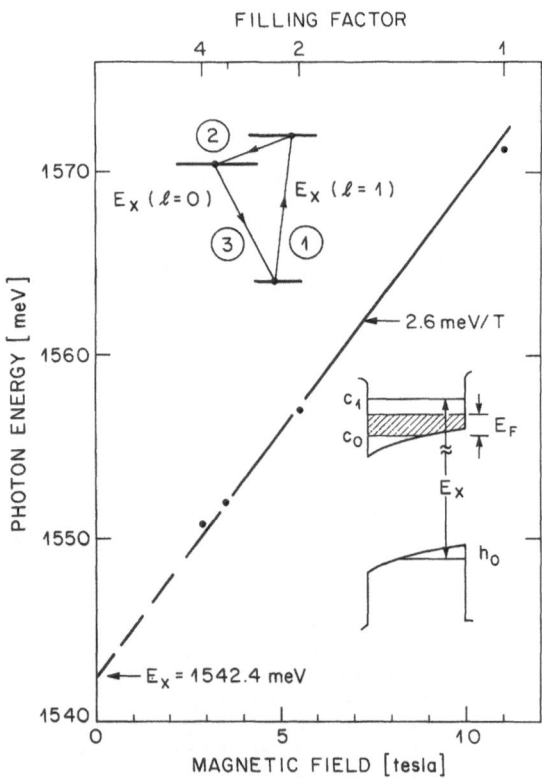

Fig. 12. Photon energy of the maximum resonant enhancement as a function of magnetic field. The
lower inset shows B = 0 quantum well energy
levels. The upper inset represents a light
scattering process in third-order perturbation
theory. After. Ref. 54.

excitonic binding and roton minimum of spin-density inter-Landau-level excitations. Similar agreement between measured structures in the spectra and calculated mode dispersions has been found at filling factor $\nu = 1$[54].

The profile of resonant enhancement of the intensities gives an insight into the light scattering mechanisms by the large wavevector inter-Landau-level excitations. Figure 12 shows the energies of incident photons that correspond to the maxima in intensities as function of magnetic field. The four points shown correspond to filling factors $\nu = 1,2,3,4$. The zero field extrapolation gives the energy of the excitonic transition shown in the lower inset. The slope of the line, 2.6 meV/T, can be written as $(1 + 1/2)\omega'_c$, and ω'_c is the cyclotron energy for the optical reduced effective mass ($\mu* \simeq 0.064m_0$). Good quantitative agreement with the measured slope is obtained for $l = 1$. This result points to a light scattering mechanism as shown in the upper inset to Fig. 12. In such third order processes the intermediate transitions take place in the time sequence indicated by the numbers. Steps 1 and 3 are the optical transitions. The inter-Landau-level excitations are created, or annihilated, in step 2. The requirement of wavevector conservation is relaxed for optical transitions between Landau levels broadened by residual disorder.

CONCLUDING REMARKS

In this short review we have considered recent research on band gap optical emission and absorption processes and inelastic light scattering by the 2D electron gas in semiconductor quantum wells and heterojunctions. In the high mobility systems currently available in GaAs-AlGaAs heterostructures the optical spectroscopies reveal the effects of confinement of electron wavefunctions and the remarkable phenomena associated with strong electron-electron interactions. Emission and absorption processes of the 2D electron gas in a perpendicular magnetic field are very sensitive to Landau level filling factors and localization. We have proposed that in the extreme quantum limit ($\nu < 1$) the fascinating changes observed in optical emission spectra are related to the response of the electron gas to the Coulomb potential of the hole in the valence subband. At this time we require a theory that describes exchange-correlation effects and excitonic interactions in the magnetic quantum limit and in the regime of the fractional quantum Hall effect. However, it is by now clear that studies of optical emission reveal aspects of the states of the FQHE, like the screening response to a strong Coulomb potential, that are not evident in magnetotransport experiments. Measurements of resonant inelastic light scattering spectra in ultra-high electron mobility systems give unexpected insights into many-body interactions, including the large wavevector excitations in a perpendicular magnetic field. The method can now be applied to studies of the intriguing physics of electron-electron interactions of the lowest disorder systems. One of the most exciting areas for current light scattering work is the investigation of the elementary excitations, magnetorotons and spin-density excitations, in the regime of the fractional quantum Hall effect. It is conceivable that formation of gaps of the FQHE could greatly reduce screening of the disorder potential by the electron gas. Such residual disorder is followed by loss of translational invariance that may give access to light scattering by the large wavevector excitations relevant to the FQHE[4,5,73,74,77,78].

Acknowledgments

L. Rubin and B. Brandt provided valuable assistance at the Francis Bitter National Magnet Laboratory. Work supported in part by National Science Foundation grants DMR-8807682 and DMR-8813164.

REFERENCES

1. T. Ando, A.B. Fowler and F. Stern, **Rev. Mod. Phys.**, 54:437 (1982)
2. D.C. Tsui, H.L. Stormer and A.C. Gossard, **Phys. Rev. Lett.**, 48:1559 (1982)
3. R.B. Laughlin, **Phys. Rev. Lett.**, 50:1395 (1983)
4. "The Quantum Hall Effect", R.Prange and S.M. Girvin, eds., Springer-Verlag, New York (1987); T.Chakraborty and P. Pietilainen, "The Fractional Quantum Hall Effect", Springer-Verlag, Berlin (1988)
5. J.P. Eisenstein and H.L. Stormer, **Science**, 248:1510 (1990)
6. S. Schmitt-Rink, D.S. Chemla and D.A.B. Miller, **Adv. Phys.**, 38:89 (1989)
7. A. Pinczuk, J. Shah, R.C. Miller, A.C. Gossard and W. Wiegmann, **Solid State Commun.**, 50:735 (1984)
8. R.C. Miller and D.A. Kleinman, **J. Lumin.**, 30:520 (1985)
9. G.E.W. Bauer and T. Ando, **Phys. Rev.B**, 31:8321 (1985)
10. M.H. Meynadier, J. Orgonasi, C. Delalande, J.A. Brum, G. Bastard and M. Voos, **Phys. Rev.B**, 34:2482 (1986)
11. M.S. Skolnick, J.M. Rorison, K.J. Nash, D.J. Mowbray, P.R. Tapster, S.J. Bass and A.D. Pitt, **Phys. Rev. Lett.**, 58:1264 (1987)
12. A.E. Ruckenstein, S. Schmitt-Rink and R.C. Miller, **Phys. Rev. Lett.**, 56:504 (1987)
13. Y.C. Chang and G.D. Sanders, **Phys. Rev. B**, 32:5521 (1985)
14. R. Sooryakumar, A. Pinczuk, A.C. Gossard, D.S. Chemla and L.J. Sham, **Phys. Rev. Lett.**, 58:1150 (1987)
15. C. Delalande, G. Bastard, J. Orgonasi, J.A. Brum, H.W. Liu, M. Voss, G. Weimann and W. Schlapp, **Phys. Rev. Lett.**, 59:2690 (1987)
16. G. Livescu, D.A.B. Miller, D.S. Chemla, M. Ramaswamy, T.Y. Chang, N. Sauer, A.C. Gossard and J.H. English, **IEEE J. Quantum Electron.**, 24:1677 (1988)
17. I.V. Kukushkin and V.B. Timofeev, **JETP Lett.**, 44:228 (1986)
18. I.V. Kukushkin and V.B. Timofeev, **Surface Sci.**, 196:196 (1988)
19. B.B. Goldberg, D. Heiman, A. Pinczuk, C.W. Tu, A.C. Gossard and J.H. English, **Surface Sci.**, 196:209 (1988)
20. D. Heiman, B.B. Goldberg, A. Pinczuk, C.W. Tu, A.C. Gossard and J.H. English, **Phys. Rev. Lett.**, 61:605 (1988)
21. D. Heiman, B.B. Goldberg, A. Pinczuk, C.W. Tu, J.H. English, A.C. Gossard, M. Santos and M. Shayegan, in: "High Magnetic Fields in Semiconductor Physics II", G. Landwehr, ed., Springer-Verlag, New York (1989)
22. B.B. Goldberg, D. Heiman, A. Pinczuk, L.N. Pfeiffer and K.W. West, **Phys. Rev. Lett.**, 65:641 (1990)
23. A.J. Turberfield, S.R. Haynes, P.A. Wright, R.A. Ford, R.G. Clark, J.F. Ryan, J.J. Harris and C.T. Foxon, **Phys. Rev. Lett.**, 65:637 (1990)

24. H. Buhmann, W. Joss, K. von Klitzing, I.V. Kukushkin, G. Martinez, A.J. Plant, K. Ploog and V.B. Timofeev, **Phys. Rev. Lett.**, 65:1056 (1990)

25. G. Abstreiter, R. Merlin and A. Pinczuk, **IEEE J. Quantum Electron.**, 22:1771 (1986)

26. A. Pinczuk and G. Abstreiter, **in**: "Light Scattering in Solids V", M. Cardona and G. Guentherodt, eds., Springer-Verlag, Berlin (1989)

27. A. Pinczuk, S. Schmitt-Rink, G. Danan, J.P. Valladares, L.N. Pfeiffer and R.W. West, **Phys. Rev. Lett.**, 63:1633 (1989)

28. A. Pinczuk, J.P. Valladares, D. Heiman, A.C. Gossard, J.H. English, C.W. Tu, L. Pfeiffer and K. West, **Phys. Rev. Lett.**, 61:2701 (1988)

29. A.B. Ruckenstein and S. Scmitt-Rink, **Phys. Rev. B**, 35:5655 (1987)

30. B.B. Goldberg, D. Heiman and A. Pinczuk, **Phys. Rev. Lett.**, 63:1102 (1989)

31. M.C. Smith, A. Petrou, C.H. Perry, J.M. Worlock and R.L. Aggrawal, **in**: "Proc. of the 17th Int. Conf. on the Physics of Semiconductors", D.J. Chadi and W.A. Harrison, eds., Springer-Verlag, New York (1985)

32. C.H. Perry, J.M. Worlock, M.C. Smith and A. Petrou, **in**: "High Magnetic Fields in Semiconductor Physics", G. Landwehr, ed., Springer-Verlag, New York (1987)

33. F. Meseguer, J.C. Maan and K. Ploog, **Phys. Rev. B**, 35:2505 (1987)

34. J. Sanhez-Dehesa, F. Meseguer, F. Borondo and J.C. Maan, **Phys. Rev. B**, 36:5070 (1987)

35. B.B. Goldberg, D. Heiman, M.J. Graf, D.A. Broido, A. Pinczuk, C.W. Tu, J.H. English and A.C. Gossard, **Phys. Rev. B**, 38:10131 (1988)

36. D. Heiman, **Rev. Sci. Instrum.**, 56:684 (1985)

37. E.D. Isaacs and D. Heiman, ibid. 58:1672 (1987)

38. T. Ando, **J. Phys. Soc. Japan**, 54:1528 (1985)

39. G. Bastard and J.A. Brum, **IEEE J. Quantum Electron.**, 22:1625 (1986)

40. D.A. Broido and L.J. Sham, **Phys Rev. B**, 31:888 (1985)

41. S. Das Sarma, **Solid State Commun.**, 36:357 (1980)

42. T. Uenoyama and L.J. Sham, **Phys. Rev. B**, 39:11044 (1989)

43. S. Katayama and T. Ando, **Solid State Commun.**, 70:97 (1989)

44. F.C. Zhang, V.Z. Vulovic, Y. Guo and S. Das Sarma, **Phys. Rev. B**, 32:6920 (1985)

45. E.H. Rezayi and F.D.M. Haldane, **Phys. Rev. B**, 32:6924 (1985)

46. A.H. MacDonald, K.L. Liiu, S.M. Girvin and P.M. Platzman, **Phys. Rev. B**, 33:4014 (1986)

47. E. Burstein, A. Pinczuk and S. Buchner, **in**: "Physics of Semiconductors 1978", B.L.H. Wilson, ed., The Institute of Physics, London (1979)

48. D. Hamilton and A.L. McWhorter, **in**: "Light Scattering Spectra of Solids", G.B. Wright, ed., Springer-Verlag, New York (1969)

49. A. Pinczuk, L. Brillson, E. Anastassakis and E. Burstein, **Phys. Rev. Lett.**, 27:317 (1971)

50. A. Pinczuk, G. Abstreiter, R. Trommer and M. Cardona, **Solid State Commun.**, 30:429 (1979)

51. G. Abstreiter and K. Ploog, **Phys. Rev. Lett.**, 42:1308 (1979)

52. A. Pinczuk, H.L. Stormer, R. Dingle, J.M. Worlock, W. Wiegmann and A.C. Gossard, **Solid State Commun.**, 32:1001 (1979)

53. G. Fasol, D. Richards and K. Ploog, p.35 of this publication
54. A. Pinczuk, J.P. Valladares, D. Heiman, L.N. Pfeiffer and K.W. West, **Surface Sci.**, 229:384 (1990)
55. E. Burstein, A. Pinczuk and D.L. Mills, **Surface Sci.**, 95:451 (1980)
56. A. Pinczuk, J.M. Worlock, H.L. Stormer, R. Dingle, W. Wiegmann and A.C. Gossard, **Solid State Commun.**, 36:341 (1980)
57. D.A. Dahl and L.J. Sham, **Phys. Rev.B**, 16:651 (1977)
58. T. Ando, **J. Phys. Soc. Japan**, 51:3893 (1982)
59. S. Das Sarma, **Appl. Surface Sci.**
60. A.C. Tsellis and J.J. Quinn, **Phys. Rev. B**, 29:3318 (1984)
61. S. Katayama and T. Ando, **J. Phys. Soc. Japan**, 54:1615 (1985)
62. G. Eliasson, P. Hawrilak and J.J. Quinn, **Phys. Rev. B**, 35:5569 (1987)
63. A. Pinczuk and J.M. Worlock, **Surface Sci.**, 113:69 (1982)
64. G. Fasol, N. Mestres, H.P. Hughes, A. Fischer and K. Ploog, **Phys. Rev. Lett.**, 56:2517 (1986)
65. A. Pinczuk, J.P. Valladares, C.W. Tu, A.C. Gossard and J.H. English, **Bull. Am. Phys. Soc.**, 32:756 (1987)
66. G. Danan, A. Pinczuk, J.P. Valladares, L.N. Pfeiffer, K.W. West and C.W. Tu, **Phys. Rev. B**, 39:55121 (1989)
67. L. Wendler and R. Pechstedt, **Phys. Rev. B**, 35:5887 (1987)
68. D.H. Ehlers, **Phys. Rev. B**, 38:9706 (1988)
69. D. Gammon, B.V. Shanabrook, J.C. Ryan and D.S. Katzer, **Phys. Rev. B**, in press.
70. K.W. Chiu and J.J. Quinn, **Phys. Rev. B**, 9:4724 (1974)
71. C. Kallin and B.I. Halperin, **Phys. Rev. B**, 30:5655 (1984)
72. A.H. MacDonald, **J. Phys. C**, 18:1003 (1987)
73. A.H. MacDonald, H.C.A. Oji and S.M. Girvin, **Phys. Rev. Lett.**, 55:2208 (1985)
74. P. Pietilainen and T. Chakraborty, **Europhys. Lett.**, 5:157 (1988)
75. I.V. Lerner and Yu. E. Lozovik, **Sov. Phys. JETP**, 51:588 (1980)
76. Yu. A. Bychkov, S.V. Iordanskii and G.M. Eliashberg, **JETP Lett.**, 33:143 (1981)
77. S.M. Girvin, A.H. MacDonald and P.M. Platzman, **Phys. Rev. Lett.**, 54:481 (1985)
78. S.M. Girvin, A.H. MacDonald and P.M. Platzman, **Phys. Rev. B**, 33:7481 (1986)
79. W. Hansen, W. Horst, J.P. Kotthaus, U. Merkt, Ch. Sikorski and K. Ploog, **Phys. Rev. Lett.**, 58:2586 (1987)
80. C. Kallin and B.I. Halperin, **Phys. Rev. B**, 31:3635 (1985)
81. Z. Schlesinger, W.I. Wang and A.H. MacDonald, **Phys. Rev. Lett.**, 58:73 (1987)
82. E. Battke and C.W. Tu, **Phys. Rev. Lett.**, 58:2474 (1987)
83. R.J. Nicholas, D.J. Barnes, R.G. Clark, S.R. Haynes, J.R. Mallet, A.M. Suckling, A. Usher, J.J. Harris, C.T. Foxon and R.J. Willett, **in**: "High Magnetic Fields in Semiconductor Physics II", G. Landwehr, ed., Springer-Verlag, New York (1989)
84. S. Huant, G. Martinez and B. Etienne, **Europhys. Lett.**, 9:397 (1989)
85 . A. Pinczuk, J.P. Valladares, C.W. Tu, D. Heiman, A.C. Gossard and J.H. English, **Bull. Am. Phys. Soc.**, 33:573 (1988)

86. R. Lassnig and E. Gornik, **Solid State Commun.**, 47:9591 (1983)
87. T. Ando and Y. Murayama, **J. Phys. Soc. Japan**, 54:1519 (1985)
88. S. Das Sarma and X.C. Xie, **Phys. Rev. Lett.**, 61:7381 (1988)
89. S.C. Cheng, PhD Thesis, Yale University (1989)
90. C. Kallin, **in**: "Interface Quantum Wells and Superlattices", C.R. Leavens and R. Taylor, eds., Plenum Press, New York (1988)

OPTICAL DETECTION OF THE INTEGER AND FRACTIONAL QUANTUM HALL

EFFECTS IN GaAs AT MILLIKELVIN TEMPERATURES

A.J. Turberfield, S.R. Haynes, P.A. Wright, R.A. Ford,
R.G. Clark, J.F. Ryan, J.J. Harris,* and C.T. Foxon*

Clarendon Laboratory, Parks Road
Oxford OX1 3PU, UK
*Philips Research Laboratories, Redhill
Surrey RH1 5HA, UK

ABSTRACT

We describe a definitive optical measurement, using band
gap photoluminescence, of the integer and fractional quantum
Hall effects in GaAs by a comprehensive study of integer states
from ν = 1 to 10 and fractional states of the ν = 2/3 hierarchy
out to the 5/9 daughter state, in an ultra-high mobility single
heterojunction at 120 mK.

1.   INTRODUCTION

The integer and fractional quantum Hall effects[1,2] (IQHE
and FQHE) occur in two-dimensional electron systems in a
perpendicular magnetic field when the Landau level filling
factor ν is an integer or a fraction p/q (where p is integer and
q is odd integer in the l = 0 level). They are characterized in
transport measurements by plateaus in the Hall resistivity ρ_{xy}
at quantized values $h/\nu e^2$, with corresponding minima in the
longitudinal resistivity ρ_{xx}. The IQHE is observed at
relatively high temperatures (< 50 K), and can be explained
theoretically within a single particle framework. In the FQHE
electron-electron interactions play a crucial role: it is
believed to result from the condensation of electrons into an
incompressible liquid[3], with an energy gap ~ 1 K to excited
states. Elementary excitations of the electron liquid are
fractionally charged quasi-electrons and quasi-holes (e* =
±e/q)[3-5], which recondense to form successive daughter states
in a hierarchical scheme[4] that are only clearly observed at
milliKelvin temperatures. To date information about the IQHE
and FQHE has been obtained almost exclusively from transport
studies. For example, energy gaps occurring at integer and
fractional ν have been determined from the dependence of ρ_{xx} on
temperature; the transport energy gap Δ_ν, twice the thermal
activation energy, is interpreted as being the sum of quasi-
electron and quasi-hole energies. Optical spectroscopy is a
potentially important technique for probing these electron

states: for example, recombination of electrons and holes is considered to involve the creation or annihilation of q quasi-particles, so that in principle quasi-particle gaps can be measured. However, only a few optical results have been obtained so far. Whilst there has been a photoluminescence measurement[6] of the FQHE at ν = 7/3 and 8/3 in a Si MOSFET at 1.5 K, optical experiments that probe the QHE in GaAs have remained a major challenge. Photoluminescence intensity minima have been reported for GaAs/GaAlAs modulation-doped multiple quantum wells (MQW) at 1.8K, in the IQHE regime, but there was no correlation with ρ_{xx} minima[7]. Weak anomalies in the energy of photoluminescence have been observed[8] at fields corresponding to integer ν at relatively high ν. The first indication of the FQHE in the optical spectrum was the observation[9] of energy anomalies near ν = 2/3 in the luminescence spectrum of GaAs MQWS at T $\approx$ 0.4 K. In this paper we describe optical measurements[10] of the integer and fractional QHEs in GaAs by a comprehensive study of integer states from ν = 1 to 10 and fractional states within the ν = 2/3 hierarchy in an ultra-high mobility single GaAs/GaAlAs heterojunction at 120 mK.

2. EXPERIMENTAL

Our experiment measures band gap photoluminescence from electrons in n = 0 and n = 1 subbands of the two-dimensional electron gas (2DEG) confined at a single GaAs/Ga$_{.68}$Al$_{.32}$As heterojunction, with a density n$_s$ = 9.7 x 10^{10} cm^{-2} and mobility μ = 9 x 10^6 cm^{2}V^{-1}s^{-1}. The sample structure is shown schematically in the Fig. 1 inset: the GaAs layer (5000 Å) was very weakly n-type (~ 10^{14} cm^{-3}) in order to produce flat bands to give sharp luminescence lines. The incorporation of a GaAs/GaAlAs superlattice (SL) buffer acts to prevent carbon acceptors reaching the 2DEG. The unetched sample (with electrical contacts) was mounted in the dilute phase of a dilution refrigerator with a base temperature of 25 mK. The optical system consisted of an optical fibre to deliver light from a tunable dye laser at wavelength 740 nm (i.e. below the GaAlAs band gap), and a fibre bundle to transmit the luminescence to the detection system. The latter consisted of a 1m spectrograph, with a resolution of 0.05 meV, and a liquid nitrogen cooled charge-coupled-device detector. We estimate that the laser intensity at the sample was < 10^{-4} Wcm^{-2}. A sample temperature of 40 mK was achieved with the laser off; this increased to 120 mK under continuous laser illumination, as determined by comparison with transport measurements at elevated refrigerator temperatures with the laser off. In this experiment the maximum magnetic field strength available was 9 T; in a forthcoming paper[11] we will report measurements made at lower electron densities for fields < 15 T.

3. RESULTS

Luminescence spectra obtained at 120 mK are shown in Fig. 1 for several fields. They show two lines E$_0$ and E$_1$, whose B = 0 energies are 1.509 eV and 1.511 eV respectively which we assign to recombination of electrons in n = 0 and n = 1 subbands of the confining potential. The much higher intensity of E$_1$ compared to E$_0$ is due to the greater E$_1$ electron wavefunction penetration

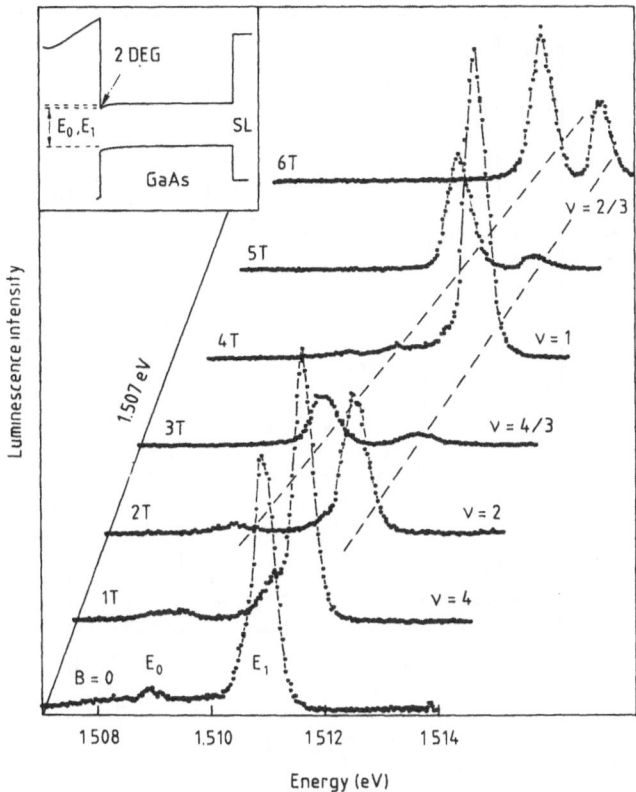

Fig. 1. Selected photoluminescence spectra in the field
range B = 0 to 6 T at 120 mK.

into the GaAs layer where there is overlap with the photoexcited
hole wavefunctions. On increasing the magnetic field, both
lines shift to higher energy as expected for $l_e = 0 \rightarrow l_h = 0$
inter-Landau level transitions, although a weak nonlinear
variation of energy with field at low fields suggests that there
is excitonic character to the states, especially E_1. The
intensity of E_1 generally decreases with increasing field and
that of E_0 increases. For B > 2.5 T their intensities are
reversed from the zero field situation, with E_0 now being the
more intense. At B = 4 T ($\nu = 1$) the E_1 intensity undergoes a
dramatic enhancement, whereas E_0 decreases substantially.
Closer inspection of the spectra reveals a remarkable field-
induced modulation of the intensities.

3.1 Low Field ($\nu > 1$): IQHE

The peak intensities of E_0 and E_1 at 120 mK are shown in
Fig. 2a, together with ρ_{xx} (Fig. 2b) which was measured
simultaneously. The ρ_{xx} data show well-developed minima at
integer filling $\nu = 1$ to 10 and at fractional states, most
notably 4/3 and 5/3. The luminescence data reveal that the E_1
line has sharp intensity maxima at integer ν, stronger at ν even
that at ν odd (with the exception of $\nu = 1$). Underlying this
behavior there is a general decrease in E_1 intensity with
increasing field, with step-like drops at $\nu = 5$, 3 and 5/3; the

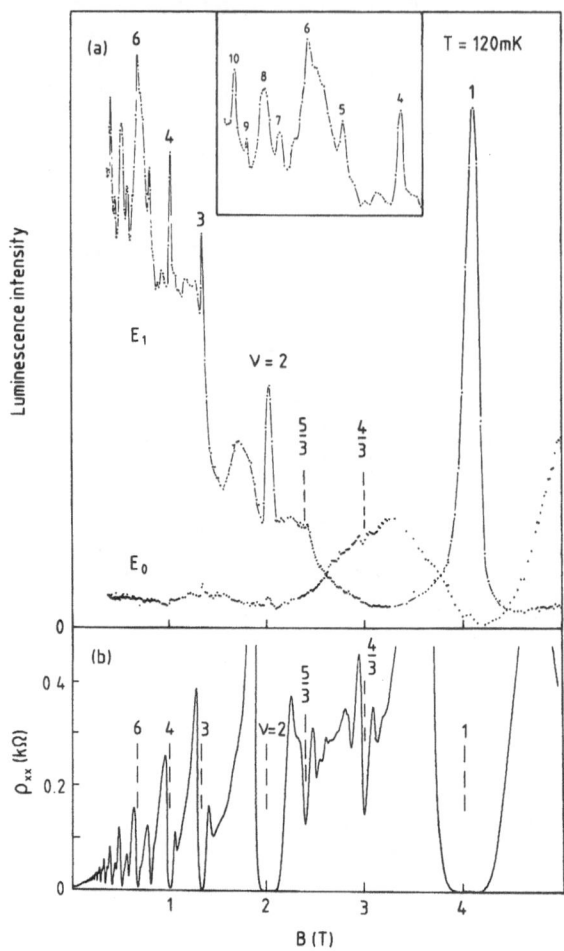

Fig. 2. (a) Intensities of the E_0 and E_1 luminescence lines at 120 mK as a function of magnetic field for $\nu \geq 1$; the inset shows the region $\nu = 10$ to 4 on an expanded scale. (b) Simultaneous ρ_{xx} data.

E_1 intensity recovers at fields intermediate between $\nu = 3$ and 4, and more strongly between $\nu = 2$ and 3. In contrast to this behavior the intensity of E_0 shows very little structure for fields up to $\nu = 2$. For $\nu < 2$, however, the E_0 intensity first increases, shows a broad maximum, with a small but distinct dip at $\nu = 4/3$, and then decreases to a small value at $\nu = 1$. At this occupancy E_1 undergoes a dramatic enhancement by a factor of ~ 35. In fact, the peak occurs at 4.08 T whereas the field for $\nu = 1$ filling estimated from fractional states is 4.0 T. At substantially higher sample temperatures (T > 1 K) the sharp structure in the E_1 intensity disappears and instead we observe[11] broad intensity oscillations with maxima occurring near ν odd[12].

In this experiment the electron ground and excited states are probed by the photoexcited hole. The field dependence of the E_0 intensity depends on a number of factors, the most

important of which may be the screening response of the
electrons. An electron can scatter between states in a
partially filled Landau level in the n = 0 subband to screen the
potential of a photoexcited hole, thereby increasing the
electron-hole wavefunction overlap; for a completely filled
level screening is suppressed. Since the l = 0 level is full
for $\nu > 2$, the E_0 signal which arises from $l_e = 0 \rightarrow l_h = 0$
recombination might be expected to be weak and approximately
field-independent in this range, as observed. For $1 < \nu < 2$ the
$l_e = 0$ level is partly filled and screening becomes active,
consistent with the observed increase in E_0 intensity. At $\nu = 1$
screening is strongly suppressed and the E_0 intensity falls very
close to zero. It should be recalled that the electronic g-
factor is exceptionally enhanced at $\nu = 1$ due to unscreened
electron-electron interactions[13]. For $\nu < 1$ the E_0 intensity
is expected to increase again, as electron-hole screening
increases; this is confirmed by the experiment, as described
below, but there are important anomalies at $\nu = p/q$ fractional
filling.

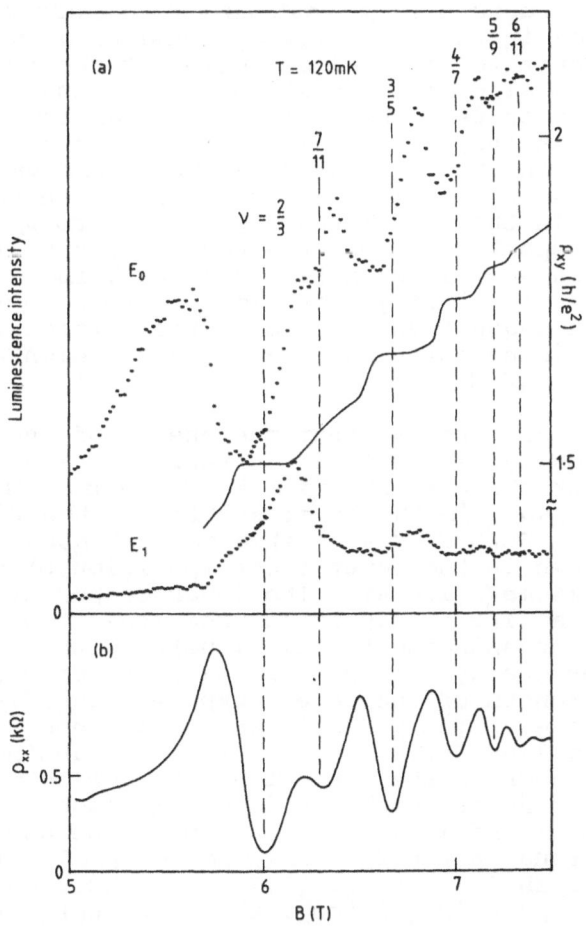

Fig. 3. (a) Intensities of the E_0 and E_1 luminescence
lines at 120 mK for $\nu < 1$; (b) simultaneous ρ_{xx}
data.

The enhanced intensity of E_1 at integer ν can arise both
from increased occupancy of the E_1 subband and from enhanced
optical transition rates. The former depends on the density of
vacant n = 0 states, and on the relaxation rate from E_1 to E_0,
which can be reduced dramatically when the Fermi level
approaches the l = 0 level of E_1 due to the sharp cut-off of the
electron-phonon interaction at low energy (< 1 meV)[14]. The
recombination rate also depends on screening which is strongly
affected by state occupancy; in this case we expect the
excitonic character of the E_1 state to be weakened due to
screening by n = 0 electrons, so that the E_1 intensity will be
enhanced at integral and fractional filling where screening is
suppressed.

3.2 High Field ($\nu < 1$): FQHE

In Fig. 3 we compare E_0 and E_1 luminescence intensities
with ρ_{xx} date for fields at $\nu < 1$. Also shown are ρ_{xy}
measurements of a Hall bar sample from the same wafer and with
the same electron density, at the same temperature but in the
absence of laser illumination. The ρ_{xx} data show well defined
FQHE states at ν = 2/3, 7/11, 3/5, 4/7, 5/9 and 6/11. The
resolution and structure in ρ_{xx} out to ν = 5/9 are mirrored in
remarkable detail in the E_0 intensity. The results provide
definitive evidence for optical detection of the FQHE. The E_1
luminescence line also responds strongly to changes in the
electronic ground state. Whereas the E_0 intensity shows *minima*
at the above fractions, E_1 intensity develops *maxima* (as for the
IQHE). In addition, the strengths of the E_0 minima and E_1
maxima decrease as the filling factor is swept through higher
order fractions in the 2/3 hierarchy, similar to ρ_{xx}. The
optical and transport structure are not in precise field
alignment (see dashed lines in Fig. 3): E_1 maxima occur at
fields above ν = p/q filling, whereas E_0 minima occur at lower
fields. This is suggestive of localization effects. In fact
the maxima and minima correlate well with the high and low field
extents of the ρ_{xy} plateaus.

The spectra also reveal that the energy of the E_0 line
changes dramatically as the field is varied. Figure 4 shows
this behavior for the ν = 2/3 and ν = 1 regions. The effect
around ν = 2/3, shown in the lower section of the figure, is
striking. At B = 6.25 T (ν = 0.64) a doublet structure is
resolved, as shown in the lower inset (position b) with a
splitting of 0.16 meV (1.8 K). The lower component of the
doublet lies on a line extrapolated from lower fields (arrows
mark energies corresponding to this line). With decreasing
field the higher energy component becomes the more intense, and
the relative strength of the lower component rapidly diminishes
(inset spectra a-e). The peak of the luminescence remains above
the extrapolated line for a range of fields spanning ν = 2/3,
before shifting down to the line at B = 5.75 T (ν = 0.7),
although a weaker upper component is still discernible at this
(inset spectrum g) and lower fields. In the neighborhood of ν =
3/5 a similar trend in the E_0 energy is observed, although it is
less distinct: a shift 0.06 meV (0.7 K) is observed. At ν = 1
the luminescence peak is shifted upwards in energy by 0.2 meV
(2.4 K) and an upper component becomes resolved again (inset
spectra h, i, j), a further 0.2 meV higher in energy. The
doublet is not well resolved at lower fields, but the shift to
higher energy is maintained. In contrast, E_1 remains unshifted

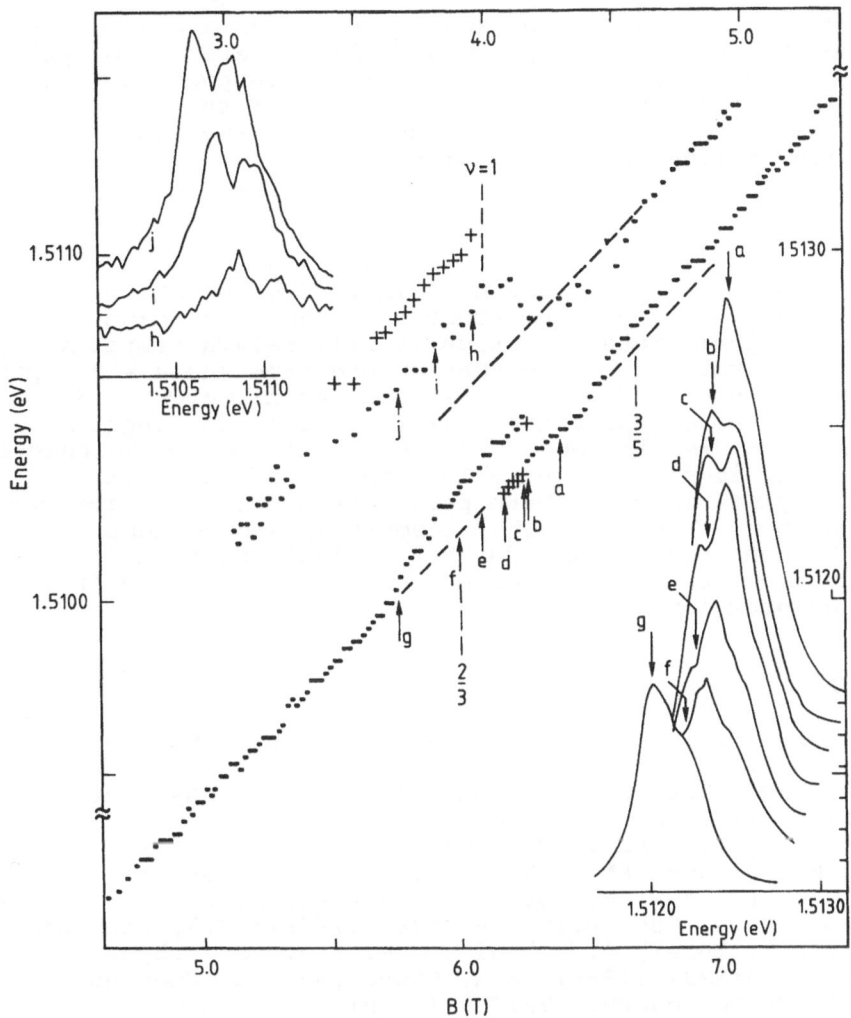

Fig. 4. Magnetic field dependence of the energy of the
E_0 luminescence line. The solid circles denote
the more intense line.

from the normal linear B-dependent behavior throughout this
range of magnetic field.

The energy of the E_0 line is a measure of the electronic
ground state energy and so it may in principle give information
about quasi-particle energies. Kukushkin and Timofeev[6]
observed optical shifts of +0.34 meV (4 K) and -0.26 meV (3 K)
above and below $\nu = 7/3$ (in filling factor) in a Si MOSFET;
these shifts were interpreted to be $3\epsilon_e$ and $3\epsilon_h$ respectively
where ϵ_e, ϵ_h are the quasi-electron, quasi-hole creation
energies, which agreed quantitatively with measurements of the
$\Delta_{7/3}$ transport gap. On the other hand, Heiman et al.[9]
observed a shift of 0.7 meV (8.1 K) in the luminescence spectrum
at $\nu \approx 2/3$ which could not be identified as a quasi-particle
energy. In the present case, our observed doublet separation of
0.16 meV (1.8 K) at $\nu = 2/3$ differs in magnitude from the GaAs
quantum well measurements[9], and is smaller than the transport

gap $\Delta_{2/3}$ = 4.6 K measured for this sample. It is clear that the splitting observed at ν = 2/3 is not simply related to the quasi-particle energies. A quantitative interpretation of our observations will require a detailed model of the recombination of photoexcited holes with electrons in the many-body ground state which includes excitonic effects.

CONCLUSIONS

In summary, we have made a definitive observation of the FQHE and IQHE by optical spectroscopy of an ultra-high mobility GaAs/GaAlAs single heterojunction at milliKelvin temperatures. Our main observations of the FQHE extend from the ν = 2/3 parent state out to the 5/9 daughter state in the hierarchy. A splitting of the ground state luminescence in the region of ν = 1, 2/3, 3/5 is also observed. At ν = 2/3 the magnitude of the splitting is significantly less than the transport gap. In the IQHE regime, intense sharp peaks are observed at integer filling from ν = 1 to 10, with exceptional enhancement at ν = 1. This work opens the way for further optical spectroscopy of the many-body electron liquid states and the Wigner solid in the extreme quantum limit.

REFERENCES

1. K. von Klitzing, G. Dorda and M. Pepper, **Phys. Rev. Lett.**, 45:494 (1980)
2. D.C. Tsui, H.L. Störmer and A.C. Gossard, **Phys. Rev. Lett.**, 48:1559 (1982)
3. R.B. Laughlin, **Phys. Rev. Lett.**, 50:1395 (1983)
4. F.D.M. Haldane, **Phys. Rev. Lett.**, 51:605 (1983)
5. R.G. Clark, J.R. Mallett, S.R. Haynes, J.J. Harris and C.T. Foxon, **Phys. Rev. Lett.**, 60:1747 (1988); J.A. Simmons, H.P. Wei, L.W. Engel, D.C. Tsui and M. Shayegan, **Phys. Rev. Lett.**, 63:1731 (1989); A.M. Chang and J.E. Cunningham, **Solid State Commun.**, 72:651 (1989)
6. I.V. Kukushkin and V.B. Timofeev, **Pis'ma Zh. Eksp. Teor. Fiz.**, 44:179 (1986) [**JETP Lett.** 44:228 (1986)]
7. C.H. Perry, J.M. Worlock, M.C. Smith and A. Petrou, p.202 **in**: "High Magnetic Fields in Semiconductor Physics", G. Landwehr, ed., Springer-Verlag, New York (1987)
8. R. Stepniewski, W. Knap, A. Raymond, G. Martinez, T. Rötger, J.C. Maan and J.P. André, p.62 **in**: "High Magnetic Fields in Semiconductor Physics", G. Landwehr, ed., Springer-Verlag, New York (1989)
9. D. Heiman, B.B. Goldberg, A. Pinczuk, C.W. Tu, A.C. Gossard and J.H. English, **Phys. Rev. Lett.**, 61:605 (1988)
10. A.J. Turberfield, S.R. Haynes, P.A. Wright, R.A. Ford, R.G. Clark, J.F. Ryan, J.J. Harris and C.T. Foxon, **Phys. Rev. Lett.**, 65:637 (1990)
11. R.G. Clark, R.A. Ford, S.R. Haynes, J.F. Ryan, A.J. Turberfield, P.A. Wright, C.T. Foxon and J.J. Harris, **in**: "High Magnetic Fields in Semiconductor Physics", G. Landwehr, ed., Springer-Verlag, New York (1991), to be published.
12. W. Chen, M. Fritze, A.V. Nurmikko, D. Ackley, C. Colvard and H. Lee, **Phys. Rev. Lett.**, 64:2434 (1990)
13. Th. Englert, D.C. Tsui, A.C. Gossard and Ch. Uihlein, **Surf. Sci.**, 113:295 (1982)
14. P. Maksym, private communication.

RAMAN LIGHT SCATTERING FROM PLASMONS IN MODULATION-DOPED QUANTUM

WELLS

G. Fasol, D. Richards and K. Ploog*

Cavendish Laboratory, Madingley Road
Cambridge CB3 0HE, England
*Max-Planck-Institut für Festkörperforschung
D-7000 Stuttgart 80, Fed. Rep. of Germany

ABSTRACT

Electronic Raman scattering is a very powerful technique for the contactless optical characterization of quantum wells and heterojunctions and has much promise for the spatially resolved characterization of quantum wire structures. We show that electronic Raman scattering measurements of the plasmon dispersion is a contactless optical alternative to measurements of Shubnikov-de Haas oscillations for the determination of carrier concentractions in modulation-doped single- and multi-quantum wells and heterojunctions. We demonstrate the optical control of carrier density in a single heterojunction with an adjacent δ-layer of acceptors: we use Raman measurements of the plasmon by a probe laser beam to directly determine the change in carrier concentration induced by illumination with a pump laser beam of varying intensity. Using Raman measurements in combination with calculations of the random phase approximation (RPA) dielectric response and self-consistent electronic subband calculations we determine the subband structure, populations, potential and wavefunctions of modulation-doped $GaAs/Al_xGa_{1-x}As$ multi-quantum wells with multiple subband occupancy.

1.   INTRODUCTION: ELECTRONIC RAMAN SCATTERING IN QUANTUM WELLS
     AND HETEROJUNCTIONS

Since the suggestion of resonance electronic Raman scattering from single 2D electron layers[1], and its experimental discovery[2-4], much work has been done on excitations of equilibrium electrons in modulation-doped semiconductor structures, on optically excited carriers in undoped structures (for reviews see Ref.5 and 6), and recently also in quantum wire structures. Almost all Raman experiments on low dimensional semiconductor systems exploit the resonant enhancement of the Raman scattering efficiency when the incident laser beam, the scattered light beam, or both, are close to resonance with an electronic resonance of the semiconductor structure. In IV-IV and in III-V semiconductor structures, usually the E_0, the $E_0 + \Delta_0$ or the E_1 resonance enhancement are employed. The scattering volume of a single 50 Å or 100 Å electron layer is very small, and therefore the resonance enhancement is usually crucial to observe any signal.

The following excitations of quasi-two-dimensional electrons have been found in Raman spectroscopy:

- Single particle intersubband excitations[5-8]

- intersubband collective excitations[5,6]

- intersubband spin density excitations[5-8]

- intrasubband collective excitations (plasmons)[5,6,9,10,11,12]

- intrasubband single particle excitations[10,13].

Intersubband excitations (i.e. carrier excitations from a lower subband to a higher bound subband) have frequently been used for the characterisation of modulation-doped quantum wells. The recent experimental discovery[8] of the difference between single particle excitations and spin density fluctuations has clarified a puzzle which now make the determination of the subband spacing from electronic Raman scattering possible.

Collective intrasubband excitations (plasmons) in a modulation-doped quantum well system have first been observed by Olego et al.[9]. In this case a sample structure with a large number of periods was studied, and a single plasmon peak was observed, reflecting the conservation of $k_\perp$ in the scattering process. The situation is different in the case of a small number of layers: Fasol et al.[10,12] showed, that in complex modulation-doped quantum well systems the *inter*-subband and the *intra*-subband plasmon eigenmodes show the effects of coupling via the *intra*-well and the *inter*-well Coulomb interactions - and that these are observable in the case of a small number of layers (in practice around 5 wells). A multiple well system with N wells and M filled subbands per quantum well has M x N plasmon branches[12]. These have strong $k_\parallel$ dependence. Therefore Raman measurements of the full $k_\parallel$ dependence of the plasmon resonances yield a very detailed fingerprint of the electronic structure of a multiple quantum well structure. We have demonstrated recently that Raman measurements of plasmons represent a very powerful method for the determination of the subband structure and populations[14]. A multiple quantum wire structure is expected to have an equivalent pattern of multiple coupled wire plasmon modes - therefore this Raman measurements in combination with electronic structure calculations should also be able to yield the detailed electronic structure of multiple quantum wires. In the present paper we demonstrate how to determine the electronic structure of complex modulation-doped structures (e.g. multiple modulation-doped structures, single heterostructures incorporating δ-function doping, pump light induced carrier density changes), by combining selfconsistent subband structure calculations, calculations of the dielectric response in full Random Phase Approximation (RPA), and electronic Raman measurements. Thus we show how electronic Raman scattering may be used for detailed electronic structure determination complementing Shubnikov-de Haas measurements or Hall measurements. The Raman techniques introduced here have certain advantages:

1.   No contacts are necessary.

2.   Electron subbands with similar carrier densities can clearly be distinguished, since dispersions of the coupled plasmon bands are split by the intrawell Coulomb interaction.

3.   Micro-Raman techniques allow spatial resolution.

2.   PLASMONS IN QUASI-2D SYSTEMS

The plasmons of an electron gas are self-sustaining carrier oscillations of the electrons. In a single two-dimensional electron gas (2DEG), the plasmon energy goes as $k_\parallel^{1/2}$ as a function of in-plane wavevector $k_\parallel$ in the long-wavelength high frequency approximation[15]. When there are more than one parallel two-dimensional electron gas layers, the electron sheets are coupled together by the Coulomb interaction, which causes the plasmon branches for the electrons in each well to be coupled. So a modulation-doped quantum well sample with N periods and a single occupied subband in each well will have a set of N intrasubband plasmon modes[10,11,12,16] – collective excitations in the plane of the electron sheets which set up charge distributions along the layer. An accurate description of plasma properties must go beyond the 2DEG approach and take into account quantum size effects from the carriers being confined within wells of finite width[17,18]. For the modulation-doped quantum well system studied here, the GaAs wells are sufficiently far apart for there to be no overlap between the electron states in neighbouring wells, although they are of course still coupled by the relatively weaker long range Coulomb interaction. With M occupied subbands per well there is a further strong intra-well Coulomb interaction due to overlap of electron wavefunctions which splits the plasmon modes into M groups. Thus there will be N x M intrasubband plasmon eigenmodes[12]. In addition, intersubband plasmons, which are collective intersubband excitations with an induced charge density normal to the plane of the layer, will also exist for wells with more than one bound subband. These are not dispersive at very long wavelength and are depolarization shifted to a frequency higher than the intersubband transition energy. Das Sarma has noted that there is a coupling between the intersubband and intrasubband plasmons[17]. The inter-subband plasmon is only affected by mode coupling effects as $O(k_\parallel^2)$ but the intrasubband plasmon frequency is affected in the leading order and if the two modes cross they will split. We show clear evidence of the coupling of the longitudinal and the transverse plasmon excitations both in the frequencies as well as in the intensities of the Raman spectra.

The dispersion of plasmons in 2DEG systems can be probed by electronic resonant Raman scattering measurements at liquid helium temperatures[9-11]. There is no conservation of $k_\perp$ in the case of light scattering from a sample of only a few quantum wells, since there is no translational symmetry pependicular to the sample growth direction. So light coupling to all in-plane plasmons is possible. The plasma dispersion is measured as a

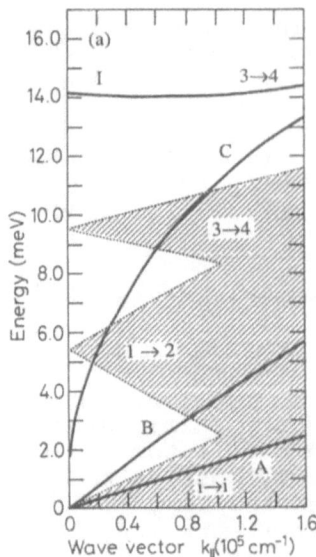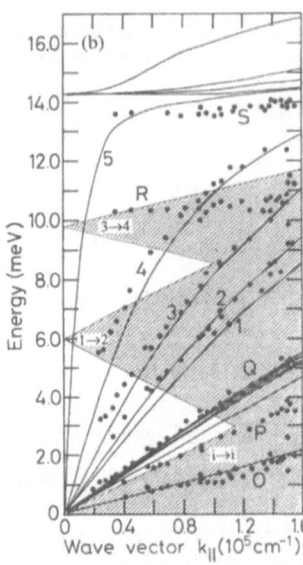

Fig. 1. (a) Results of a full RPA calculation of the
plasmon modes of a single modulation-doped
quantum well. As there are three filled
subbands in the quantum well, there are three
coupled *intra*subband plasmon modes A,B,C. There
is also one *inter*subband plasmon I from the 3 →
4 transition. The shaded areas show the
intrasubband and the 1 → 2 and 3 → 4 single
particle excitation regimes where Landau damping
occurs. (b) Calculated plasmon dispersion (thin
solid lines) for a multi quantum well structure
with five periods. The interwell interaction
splits mode C of the single well into modes
1,2,3,4,5. Mode B is split into group Q, mode P
being the asymmetric 'acoustic' plasmon mode of
the first well. Group O is a degenerate set of
plasmons corresponding to mode A. Experimental
points are also given for Raman scattering
measurements.

function of $k_{\parallel}$ by varying the angles of the incident and
scattered light with respect to the plane of the quantum wells.

3. CALCULATION OF THE DIELECTRIC RESPONSE OF A SINGLE QUANTUM
WELL

In the present Section we illustrate the calculation of the
dielectric response to obtain the plasmon dispersion for the
case of a single quantum well, modulation-doped on both
sides[11]. We demonstrate the calculations here for a structure
similar to those investigated experimentally in Section 5. The
well is 500 Å wide and has spacer layers of 50 Å AlAs at the
inverted interface and 100Å $Al_{0.3}Ga_{0.7}As$ at the normal interface
and we assume has a total carrier concentration of 1.8×10^{12}
cm^{-2}. Self-consistent electron subband calculations give three
occupied subbands for this structure. We show in Fig. 1a the
calculated plasmon dispersion of such a structure. As there are

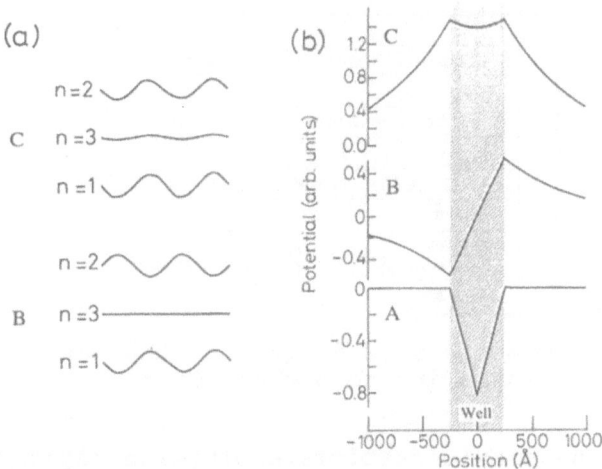

Fig. 2. (a) Charge-density modulation parallel to the
quantum well plane for the plasmon modes B and
C, in Fig. 1. We do not show the result for the
Landau damped mode A. (b) The electrical
potential associated with plasmon modes A,B,C of
the single quantum well. Here we have taken the
approximation that the charge densities are
infinitely thin layers, the two lowest subbands
located at the "inverted" and "normal"
interfaces of the well with equal populations n_1
= n_2 and the third subband located in the middle
of the well with population $n_3 \ll n_1$. The
shaded area shows the thickness of the quantum
well.

three occupied subbands in the quantum well, there are three
intrasubband plasmon modes resulting from the strong coupling
between the different subbands. For the highest frequency
modes, the dispersion goes like $k_{\parallel}^{1/2}$, electrons in all the
subbands moving in phase. The other two modes are so-called
acoustic plasmons[19] and disperse more like $k_{\parallel}$, electrons in
different subbands moving in antiphase to one another. We show
in Fig. 2a the carrier density modulations parallel to the plane
of the well for the two highest frequency modes and in Fig. 2b
the electrical potential profiles corresponding to the coupled
plasmon modes. As all subbands are in phase for the highest
frequency plasmon C, the electric potential outside the well is
greatest for this mode. Thus, if we bring several such quantum
wells together, the coupling of plasmons between wells due to
the long range Coulomb potential is strongest for this mode,
resulting in a strong splitting of plasmon modes. This effect
is less for the lower frequency plasmons where the electric
field outside the well is correspondingly smaller. Note in Fig.
2b that for the lowest energy mode A, the carriers move such
that the potential outside the well is zero. Thus there can be
no Coulomb coupling between wells for mode A. We show in Fig.
1b the calculated plasmon dispersion for a five-quantum well
sample[12] together with experimental points from Raman
scattering measurements. In these calculations we have taken
the first well to have half the carrier density of the others
and to only have two filled subbands. The shaded regions in

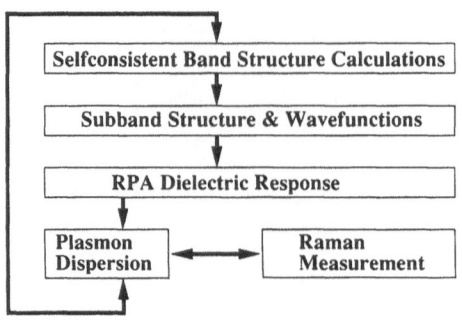

Fig. 3. Schematic flow diagram of the technique
employed in the present work.

Fig. 1 denote the single particle excitation regimes where
Landau damping can occur. It is remarkable that Landau damping
appears to be very weak as seen for modes O and P. We have
recently investigated this Landau damping in detail and found
evidence for a Fano resonance effect.

4.   DETERMINATION OF THE ELECTRONIC STRUCTURE OF MODULATION-
     DOPED QUANTUM WELLS BY RAMAN MEASUREMENTS OF THE PLASMON
     DISPERSION

     We show in the present Section that electronic Raman
scattering measurements of the plasmon dispersion is a valuable
tool for the complete determination of the electronic structure
of modulation-doped quantum wells. A layered electron gas comprising a total of N 2D electron subbands will yield N coupled
intrasubband plasmons[12]. Thus, if all these plasmons are
measured, a complete characterization of the multi-quantum well
structure is possible. Alternative characterization methods,
such as Shubnikov de Haas measurements, are able to determine
the carrier densities of multiple subbands in a *single* modulation doped quantum well if the densities are sufficiently
different. However, for *multiple* quantum wells, signals from
subbands with similar densities in the same or in different
wells are very difficult to separate.

     Figure 3 illustrates our technique schematically. Plasmon
energies are measured from electronic Raman scattering measurements for varying in-plane wavevectors $k_\parallel$. The plasmon
dispersion is then calculated from the RPA dielectric response,
including self-consistent electron subband calculations.
Experimental and calculated plasmon dispersions are then
compared and parameters varied until agreement between theory
and experiment is obtained, yielding carrier concentrations and
subband wavefunctions and energies. When analyzing an experimental dispersion, the calculated plasmon energies are not very
sensitive to small changes in the form of the electron wavefunctions, although such quantum size effects are crucial in
fitting the data. However, the dispersion of the intrasubband
plasmon modes is *very* sensitive to subband carrier concentrations. Hence an initial 'guess' at the subband structure can
be made from self-consistent calculations. Then the carrier
densities and intersubband energies can be 'fine tuned' to fit
the experimental results. Note that if we know the carrier

densities of two subbands in the same quantum well, then we know
their intersubband energy (assuming a parabolic conduction
band).

$$E_f - E^\nu = \frac{\hbar^2 \pi N^\nu}{m^*} \quad \text{hence} \quad E^\nu - E^{\nu'} = \frac{\hbar^2 \pi}{m^*} (N^{\nu'} - N^\nu) \tag{1}$$

where E_f is the Fermi energy and E^ν and N^ν are the confinement
energy and carrier density respectively for subband ν.

When we start tending towards a reasonable fit, self-
consistent calculations can be performed again, varying
parameters of large uncertainty, such as dopant concentrations,
to obtain a good agreement between calculated and experimentally
obtained plasmon dispersion.

5.   RAMAN DETERMINATION OF THE ELECTRONIC STRUCTURE OF A DOUBLE
     QUANTUM WELL

We demonstrate the above technique by considering a
representative GaAs/$Al_{0.3}Ga_{0.7}$As sample grown by MBE consisting
of two 500 Å wide modulation-doped quantum wells. The two wells
have different environments, the well nearest the substrate
(hereafter the 'substrate well') being only doped on one side
and thus having an electronic structure similar to that of a
heterojunction. The second well (hereafter the 'surface well')
is doped on both sides, although surface states may completely
deplete the doped layer on the surface side of the well, thus
affecting the electronic structure of the quasi-2DEG in the
well.

In Fig. 4 are shown the Raman spectra of the plasmon modes
of this sample, measured for different in-plane wavevectors $k_\parallel$
at a laser energy close to the E_0 resonance of the quantum well.
There are four dispersive plasmons (labelled P,Q,1,2) and two
less dispersive modes (R and S). The plasmon energies are shown
as a function of $k_\parallel$ in Fig. 5 - the point size indicating the
Raman signal intensity (this display technique is due to J.C.
Maan). The four dispersive $intrasubband$ plasmons indicate the
presence of four 2D electron subbands in the structure. As the
asymmetric substrate well is likely to have only one occupied
subband, we can infer that there are three occupied subbands in
the surface well. Modes 1,2 are the plasmons coupled by the
interwell Coulomb interaction, where all charges within each
well are moving in phase. From the dispersion of these modes we
can determine the total carrier concentrations for the surface
and substrate wells to be 1.6×10^{12} cm^{-2} and 6.5×10^{11} cm^{-2}
respectively. These values can be compared with a total carrier
concentration for the structure, obtained from a Hall
measurement with the sample illuminated, of 2.0×10^{12} cm^{-2}.
The intrasubband modes P,Q arise from splitting of the intrasubband collective excitations in the surface well due to the
strong intrawell interaction. Analysis of these modes thus
enables one to determine the occupancies for the different
subbands in the surface well. The nondispersive modes R,S are
$intersubband$ plasmons due to collective excitations in the
surface well. Note that mode P shows very weak Landau damping.
Weak Landau damping of coupled plasmons has first been reported

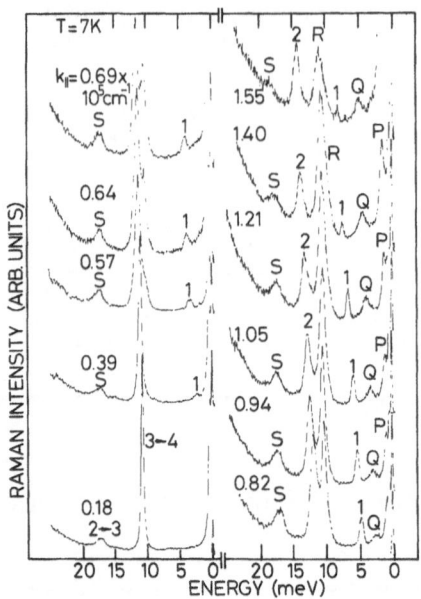

Fig. 4. Raman spectra of the plasmon modes
of two modulation-doped quantum
wells, for different in-plane
wavevectors $k_\parallel$. Letters (P,Q,R,S)
and numbers (1,2) refer to the
different types of plasmon mode
discussed in the text.

by Fasol et al.[10]. We have recently investigated Landau
damping in modulation-doped quantum wells extensively within the
RPA approximation and predicted the possibility of a Fano
resonance effect.

Taking the total concentrations of 1.6×10^{12} cm^{-2} and
6.5×10^{11} cm^{-2}, the subband wavefunctions and energies for the
two wells were calculated, the probability density distributions
$|\psi|^2$ and the potential for the structure are shown in Fig. 6. A
band bending offset Δ was used as a fitting parameter to
describe the effect of surface depletion on the first well. The
plasmon dispersion could then be determined from calculations of
the full RPA dielectric response, utilizing the self-consistently determined wavefunctions and energies with $\Delta = 75$ meV.
The calculated plasmon dispersion (dotted lines) is shown in
Fig. 5, together with that obtained by experiment (full
experimental points), and we can assign the intersubband
plasmons R and S as the collective excitations between the third
and fourth and the second and third subbands respectively. The
intersubband plasmon between the first and second subbands does
not occur as the spatial separation between these two states is
too great, since they are localized at opposite interfaces. The
electron subband energies and carrier concentrations determined
by this technique are given in Table I.

Of particular interest is the dispersion of the $2 \rightarrow 3$
intersubband mode S at large in-plane wavevector. The
intersubband plasmon frequency $\omega_{\nu\nu'}$ for a transition from

Table 1. Subband energies and subband occupations for the present double quantum well sample under saturated illumination. (Energies are with respect to the first subband.) ([a] The energy of the first excited state ($\nu = 4$) in the surface well was calculated to be 39.3 meV. The value given was chosen to give a $3 \rightarrow 4$ transition energy of 7.7 meV.)

	Subband ν	Electron density x 10^{11} cm^{-2}	Energy meV
1. Surface Well: 1.5×10^{12} cm^{-2}	1	9.27	0.0
	2	5.48	13.65
	3	1.25	28.9
	4	-	36.6[a]
2. Substrate Well 6.5×10^{11} cm^{-2}	1	6.5	0.0
	2	-	32.1

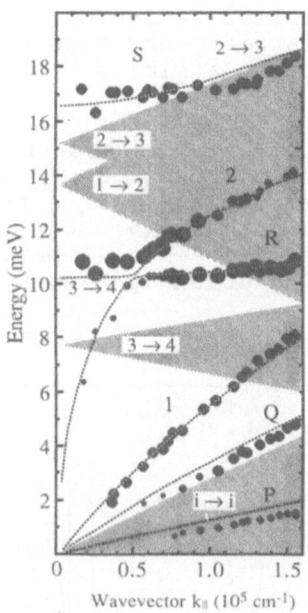

Fig. 5. Experimental points and full RPA calculation (dashed lines) of plasmon modes, as a function of the in-plane wavevector $k_\parallel$, of the two quantum well sample discussed in Section 6. Point size indicates relative Raman signal strength for each plasmon resonance (displaying signal strength as area of points is a graphical method due to J.C. Maan). Modes P,Q,1,2 are *intra*subband plasmons. R,S are *inter*subband collective excitations. The shaded areas show the intrasubband and the $3 \rightarrow 4$, $1 \rightarrow 2$ and $2 \rightarrow 3$ single particle excitation continua where Landau damping occurs.

subband ν to ν' can be approximately given for small in-plane wavevector by,

$$\hbar\omega_{\nu\nu'} = \left[(E^{\nu} - E^{\nu'})^2 + W_p^{\,2}\right]^{1/2} \tag{2}$$

where the depolarization shift W_p is given by

$$W_p^{\,2} = \frac{4\pi e^2}{\epsilon} (E^{\nu} - E^{\nu'})(N^{\nu'} - N^{\nu})\left[vk_{(\parallel)}\right]_{k_{\parallel} \to 0} , \tag{3}$$

$v(k_{\parallel})$ is a matrix element term.

Thus there are two influences which will tend to reduce the depolarization shift of the $2 \to 3$ plasmon (S) compared to the $3 \to 4$ plasmon (R):

a) The wavefunction overlap is smaller for $2 \to 3$ as the second subband is localized at the normal interface, whereas 3 and 4 are localized over the whole well.

b) As $2 \to 3$ is from one occupied subband to another occupied subband, and the depolarization shift goes as the difference in the carrier densities of each subband, it is reduced with respect to a similar transition to an empty subband.

The net result is that the $2 \to 3$ intersubband collective excitation approaches its corresponding single particle excitation regime at low in-plane wavevector, at which point it appears to be repelled and thus gives the mode some dispersion. It then follows the edge of the single particle regime before finally entering and being strongly Landau damped.

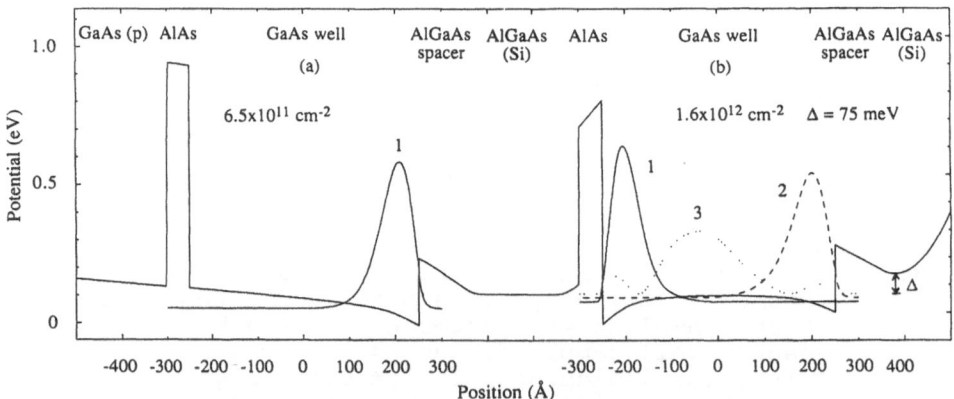

Fig. 6. Self-consistently determined potential and probability density distributions $|\Psi|^2$ for the modulation-doped two quantum well sample. We show in the 'substrate well' (a) the probability density for the ground state (1). In the 'surface well' (b), probability densities are shown for the three occupied subbands (1,2,3).

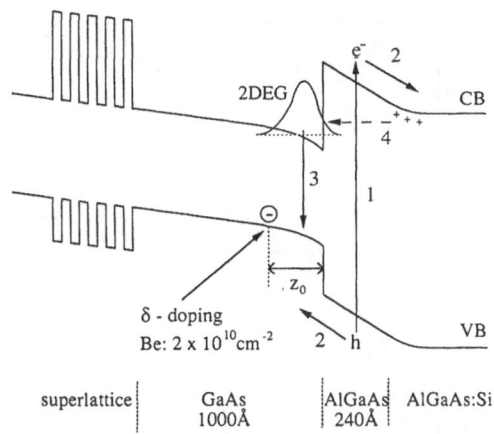

superlattice | GaAs | AlGaAs | AlGaAs:Si
              | 1000Å | 240Å |

Fig. 7. Schematic diagram of conduction and valence
        bands of the heterojunction, illustrating the
        mechanism for the depletion of electrons from
        the 2D channel under illumination. (1)
        Electron-hole pairs are created across the
        AlGaAs band gap. (2) Electrons and holes
        separated by in-built electric field. (3)
        Electrons from the 2D channel recombine with
        holes. (4) In the steady state, electrons
        tunnel through the barrier to the 2D channel.

6.   RAMAN DETERMINATION OF THE CARRIER DENSITY IN A
     HETEROJUNCTION FROM THE INTRASUBBAND PLASMON

     Electronic Raman scattering from plasmons in modulation-
doped *multi*-quantum well structures has been extensively studied
in the past years[9-12,14]. In such structures there are a
large number of adjacent 2DEG sheets, leading to complicated
plasmon dispersions, as described in the previous Section. For
a structure containing just one sheet of electrons, there will
be just one intrasubband plasmon mode with a dispersion that
goes like $k_\parallel^{1/2}$. Raman spectra of intrasubband excitations have
previously been obtained for a one side doped asymmetric quantum
well[20], where the presence of an undoped AlGaAs barrier at the
inverted interface ensures the confinement of photocreated holes
in the same region as the confined electrons, thus enabling
resonant Raman processes to occur. Here we present Raman
scattering measurements on a new type of heterojunction[21], the
structure of which we illustrate schematically in Fig. 7. A
δ-doped layer of Be acceptors a well defined distance z_0 from
the interface enables the enchancement of resonant Raman
scattering.

     We show in Fig. 8 the dispersion of the plasmon for two
samples with Be doping δ layers at $z_0 = 250$ Å (#6639) and $z_0 =$
300 Å (#6505). To these experimentally determined dispersions
we have fit calculated plasmon modes for carrier densities of
2.6×10^{11} cm^{-2} (#6639) and 3.4×10^{11} cm^{-2} (#6505). These can
be compared with respective carrier densities obtained from Hall
measurements of 3.8×10^{11} cm^{-2} (#6639 at 300 K) and 5.2×10^{11}
cm^{-2} (#6505 at 77 K). The discrepancies between Hall and
plasmon measurements of carrier concentration are most probably
due to differences in temperature and illumination conditions.

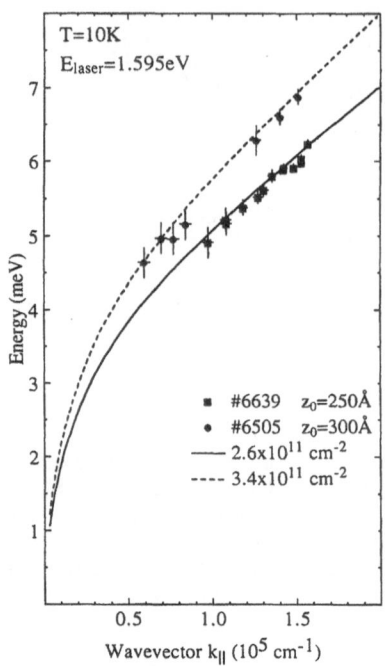

Fig. 8. Experimental plasmon dispersion for two heterojunctions (No. 6639 and No. 6505). Plasmon
modes calculated to fit the experimental points
enable a determination of sheet carrier density.

As the dispersions determined from Raman scattering match
perfectly the expected square-root-like behavior, we can infer
that there is little or no parallel conduction present in the
AlGaAs barrier. The plasmon dispersion obtained from an
asymmetric quantum well which we studied separately (results not
shown here) was at a much lower energy than expected and did not
follow the expected behavior for a single 2DEG. This was found
to be due to the presence of a second quasi-2DEG of large
carrier density in the AlGaAs barrier which, due to the Coulomb
interaction, resulted in the plasmon mode being reduced in
energy and having a more acoustic-like behavior. Thus Raman
determination of the plasmon is a sensitive probe for parallel
conduction.

7. OPTICAL CONTROL OF ELECTRON DENSITY (- *Not Persistent
Photoconductivity*)

It has been shown[22,23], that the carrier concentration of
the 2DEG at a modulation-doped heterointerface can be reduced
with increasing intensity of illumination above the AlGaAs
barrier band gap. This effect, as explained by Chaves et al.
[22], is due to the photoexcitation of electron hole pairs
across the AlGaAs band gap. The mechanism is shown schematically in Fig. 7. The electrons and holes are then separated by
the built-in electric field, the electrons residing in the Si
doped AlGaAs whereas the holes move to the more favorable
potential in the GaAs where they recombine with electrons from
the quasi-2DEG, thus reducing the 2DEG concentration. The

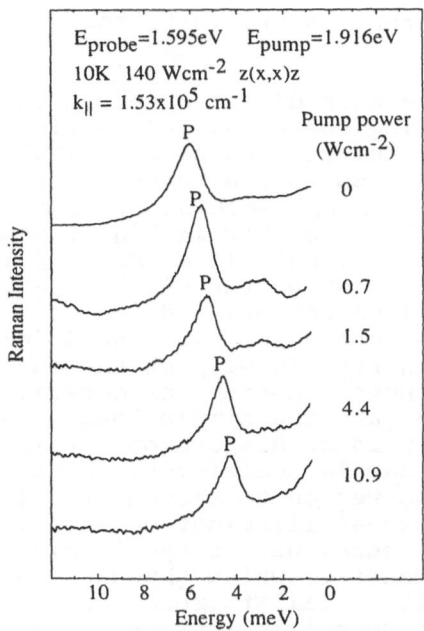

Fig. 9. Raman spectra of the plasmon mode P in the
heterojunction for different pump power
densities. In-plane wavevector $k_\parallel$ =
1.53×10^5 cm^{-1}. There is a strong shift to
lower energy for increasing pump power density,
corresponding to a reduction in carrier
concentration.

effective transfer of electrons from the 2D channel to the Si
donors shifts the quasi-Fermi levels in the different regions
relative to one another. In the steady state the electrons then
tunnel through the barrier to the 2D channel. This is a dynamic

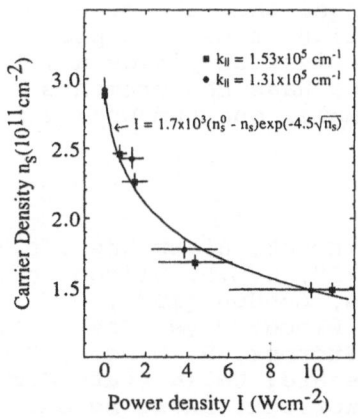

Fig. 10. Carrier concentration, determined from plasmon
energy, plot versus pump power density I. This
has been fit using the simple model proposed in
Ref. 22.

effect entirely different in origin to the well known persistent photoconductivity.

We present here a more direct observation of this phenomenon, by measuring the change of carrier concentration from the variation of the plasmon energy with varying illumination. In addition to the "probe" beam, of energy close to the E_0 resonance, a "pump" spot of 647.1 nm illumination was focused onto the sample. We show in Fig. 9 Raman sepctra for varying power densities of multiline red pump illumination. A clear shift to lower energy of the plasmon peak can be observed for increasing pump illumination, and indeed this effect is in no way persistent, in accordance with the model proposed by Chaves et al.[22]. In Fig. 10 we plot the change in carrier concentration versus power density, as determined from the plasmon energy, using just the simple long wavelength $k_\parallel^{1/2}$ approximation to the plasmon dispersion - this will overestimate the determination of carrier density by up to 10%. Power densities have been corrected to account for the change of sample angle to the excess illumination when varying $k_\parallel$. Thus we show that Raman measurements of the in-plane plasmon directly measure the carrier density, while Shubnikov-de Haas experiments have been reported[22] to fail to recognize the change in carrier concentration upon illumination.

SUMMARY

We have shown that Raman measurements in combination with RPA calculations of the dielectric response (including self-consistently determined wavefunctions) are a very versatile technique to determine carrier concentrations, population of subbands, and wavefunctions in quantum wells. We demonstrated this technique by determining the electronic structure of a double quantum well structure and by investigating single heterojunctions with a δ function p-type doping close to the heterointerface. We show the influence of interwell interactions and intrawell interactions on the dispersion of the coupled well plasmons.

It should be stressed that the power of the present method comes from the combination of microscopic calculations and Raman measurements. This combination yields a method which complements Shubnikov-de Haas measurements for the characterization of complex quantum well structures.

REFERENCES

1.   E. Burstein, A. Pinczuk, S. Buchner, p.1231 **in**: "Physics of Semiconductors 1978", B.L.H. Wilson, ed., The Inst. of Phys. Conf. Series, London (1979)
2.   G. Abstreiter, K. Ploog, **Phys. Rev. Lett.**, 42:1308 (1979)
3.   A. Pinczuk, H.L. Störmer, R. Dingle, J.M. Worlock, W. Wiegmann, A.C. Gossard, **Solid State Commun.**, 32:1001 (1979)
4.   A. Pinczuk, J.M. Worlock, **Surf. Science**, 113:69 (1982)
5.   G. Abstreiter, M. Cardona and A Pinczuk, p.5 **in**: "Light Scattering in Solids IV", M. Cardona and G. Güntherodt, eds., Springer-Verlag, Berlin, Heidelberg, New York. (1984)
6.   A. Pinczuk, G. Abstreiter, **in**: "Light Scattering in Solids V", M. Cardona and G. Güntherodt, eds., Springer-Verlag,

Berlin, Heidelberg, New York. (1989) p.153

7. G. Danan, A. Pinczuk, J.P. Valladares, L.N. Pfeiffer, K.W. West, C.W. Tu, **Phys. Rev. B**, 39:5512 (1989)

8. A. Pinczuk, S. Schmitt-Rink, G. Danan, J.P. Valladares, L.N. Pfeiffer, K.W. West, **Phys. Rev. Lett.**, 63:1633 (1989)

9. D. Olego, A. Pinczuk, A.C. Gossard, W. Wiegmann, **Phys. Rev. B**, 25:7867 (1982)

10. G. Fasol, N. Mestres, H.P. Hughes, A. Fischer, K. Ploog, **Phys. Rev. Lett.**, 56:2517 (1986)

11. A. Pinczuk, M.G. Lamont, A.C. Gossard, **Phys. Rev. Lett.**, 56:2092 (1986)

12. G. Fasol, R.D. King-Smith, D. Richards, U. Ekenberg, N. Mestres and K. Ploog, **Phys Rev. B**, 39:12695 (1989)

13. G. Fasol, N. Mestres, M. Dobers, A. Fischer, K. Ploog, **Phys. Rev B**, 36:1536 (1987)

14. D. Richards, G. Fasol and K. Ploog, **Appl. Phys. Lett.**, 56:1649 (1990)

15. F. Stern, **Phys. Rev. Lett.**, 18:546 (1967)

16. J.K. Jain, P.B. Allen, **Phys. Rev. Lett.**, 54:2437 (1985)

17. S. Das Sarma, **Phys. Rev. B**, 29:2334 (1984)

18. J.K. Jain, S. Das Sarma, **Surf. Science**, 196:466 (1988)

19. S. Das Sarma and A. Madhukar, **Phys. Rev. B**, 23:805 (1981)

20. B. Jusserand, D.R. Richards, G. Fasol, G. Weimann and W. Schlapp, **Surf. Sci.**, 229:394 (1990)

21. I.V. Kukushkin, K. von Klitzing, K. Ploog and V.B. Timofeev, **Phys. Rev. B**, 40:7788 (1989)

22. A.S. Chaves, A.F.S. Penna, J.M. Worlock, G. Weimann and W. Schlapp, **Surf. Sience**, 170:618 (1986)

23. B. Jusserand, J.A. Brum, D. Gardin, H.W. Liu, G. Weimann and W. Schlapp, **Phys. Rev. B**, 40:4220 (1989)

CONDENSATION OF CONFINED ELECTRON-HOLE PLASMAS

Gerrit E.W. Bauer

Philips Research Laboratories
5600 J A Eindhoven, The Netherlands

ABSTRACT

The mean-field theory of electron-hole plasma condensation is reviewed. The predicted optical properties of the excitonic insulator state in confined, neutral plasmas suggest that electron-hole pairing phenomena should be observable in the optical spectra of excited quantum wells.

1. INTRODUCTION

Bardeen, Cooper and Schrieffer (BCS)[1] were the first to show that at sufficiently low temperatures a two component fermion system with an attractive interparticle interaction is unstable with respect to a transformation to a symmetry broken ground state. The instability is caused by a Bose-like condensation of weakly bound ("Cooper") pairs, which is accompanied by the formation of an energy gap at the chemical potential. After BCS it was soon realized that an analogous phenomenon should occur in electron-hole plasmas[2,3], where the attractive interaction is simply provided by the Coulomb potential. But while the Cooper pairs in the BCS super-conducting ground state carry charge, the electron-hole pairs in plasmas are neutral. The originally metallic plasma thus becomes insulating because a threshold electric field is required to create free carriers, i.e. an "excitonic insulator". This term was initially introduced for the transition in semimetals or semiconductors in which the exciton binding energy exceeds the energy gap. Highly excited semiconductors provide a very similar situation, however. The ground state of a photoexcited electron-hole plasma has been discussed extensively by Comte and Nozieres for direct-gap bulk semiconductors[4,5]. In the BCS mean-field approximation they find a smooth transition from a Bose-condensed gas of excitons (i.e. tightly bound electrons and holes) at low densities to the excitonic insulator phase in the high density limit. The absence of an abrupt transition as a function of carrier density has attracted interest also in the context of superconductivity[6]. To the best of my knowledge no clear experimental evidence of electron-hole plasma condensation in optically pumped semiconductors or in semimetals has been published to date. It is hoped that the present study will contribute to change this unsatisfactory situation.

The performance of semiconductor lasers can be improved by inserting quantum wells into the active layer[7]. The first study of electron-hole plasmas in quantum wells[8] was motivated by the interest in the band gap renormalization, which is an important parameter for laser applications. Subsequent research on the band gap renormalization in excited quantum wells is briefly reviewed in Ref. 9. Another line of research is concerned with the optical nonlinearities of quantum wells at low excitation densities and the dynamical Stark effect[10]. Only very recently experimental activity has been devoted to the search for plasma condensation in quasi-two-dimensional systems. Motivated by predictions of Lerner and Lozovik[11] and Pacquet et al.[12] about Bose condensation of strictly two-dimensional excitons in the limit of infinitely high magnetic fields, Potemski and Maan[13] studied the magnetoluminescence of highly excited quantum wells. They arrived at the surprising result that excitonic effects survive in high density, high temperature plasmas. This effect has been tentatively ascribed to short-lived, localized excitonic correlations[14], but has probably no relation with the plasma condensation. Golub et al.[15] demonstrated that electrons and holes may be separately confined in asymmetrically coupled quantum wells by applying an electric field, without drastically reducing the exciton binding energy. Besides facilitating condensation due to the increased carrier life-time, such a system might also be suited to study superconductivity as predicted by Lozovik and Yudson[16] in a condensed plasma of spatially separated electron and holes. In this context GaSb/InAs type II quantum wells should be mentioned too, which are semimetals containing spatially separated electrons and holes even in the ground state[17].

It is noted that the optical properties of neutral electron-hole plasmas discussed in the following are drastically different from charged plasmas in modulation-doped, weakly excited quantum wells. Though excitonic effects are important for both systems[18,19], the electron-hole pair condensation vanishes when the difference between the individual carrier densities becomes large. The main prediction of mean field theory for the optical properties of doped quantum wells is an abrupt exciton unbinding transition observable in photo-luminescence spectra[19].

In recent communications[9,18,19] a Hartree-Fock theory of electron-hole plasmas in quantum wells has been introduced, where the term Hartree-Fock is used for the mean-field approximation with unscreened interaction potentials. It is an extension of formal theories[20,21] to take into account the effects of realistic confinement potentials and magnetic fields. Here the relevant equations are given explicitly in a form which is easily solved numerically (Section 2). It is hoped that the results of the computations (Section 3) will be helpful in the interpretation of experiments. Hartree-Fock theory and the reliability of its predictions is discussed critically in Section 4.

2. OPTICAL PROPERTIES OF THE CONFINED EXCITONIC INSULATOR

Following Refs. 20,21, the equations for the optical properties of the condensed state as obtained in the Hartree-Fock approximation are summarized below. The corresponding expressions for the optical properties of the normal state have been given in Ref. 19.

The single-particle properties of the excitonic insulator state are easily obtained by a straightforward transcription of BCS theory. Assuming thermal equilibrium ($\beta = 1/k_B T$) at an applied magnetic field normal to the interfaces, disregarding couplings to higher subbands, and denoting Landau level indices of free electron-hole pair states by m,n, the gap equation reads:

$$\Delta(m) = \sum_n V_{eh}(m,n) \frac{F(n)\Delta(n)}{\overline{E}(n)} . \tag{1}$$

The functions appearing in Eqn. 1 are defined by the Eqns. 2-12:

$$F(n) = f(\epsilon_1(n)) - f(\epsilon_2(n)); \quad f(\epsilon) = \frac{1}{e^{\beta\epsilon} + 1} \tag{2}$$

$$\overline{E}(n) = \sqrt{E(n)^2 + \Delta(n)^2}; \quad E(n) = \epsilon_e(n) + \epsilon_h(n) \tag{3}$$

$$\epsilon_{1,2}(n) = \frac{1}{2} [\epsilon_e(n) - \epsilon_h(n) \pm \overline{E}(n)] \tag{4}$$

$$\epsilon_e(n) = \epsilon_e^{sub} + \frac{eB\hbar}{m_e} (n+\frac{1}{2}) - \mu_e - \sum_m \{f(\epsilon_1(n))u_n^2 + f(\epsilon_2(n))v_n^2\}V_{ee}(n,m) \tag{5}$$

$$\epsilon_h(n) = \epsilon_h^{sub} + \frac{eB\hbar}{m_h} (n+\frac{1}{2}) - \mu_h - \sum_m \{f(-\epsilon_1(n))v_n^2 + f(-\epsilon_2(n))u_n^2\}V_{hh}(n,m) \tag{6}$$

$$u_n^2 = 1 - v_n^2 = \frac{1}{2} (1 + \frac{E(n)}{\overline{E}(n)}) \tag{7}$$

$$\rho_e = \sum_m \{f(\epsilon_1(n))u_n^2 + f(\epsilon_2(n))v_n^2\} \tag{8}$$

$$\rho_h = \sum_m \{f(-\epsilon_2(n))u_n^2 + f(-\epsilon_1(n))v_n^2\} . \tag{9}$$

The (quasi) chemical potentials for electrons and holes μ_e and μ_h are defined implicitly by Eqns. 8 and 9, assuming that the densities ρ_e and ρ_h are fixed.

The Coulomb matrix elements read ($1 = (c\hbar/eH)^{\frac{1}{2}}$ is the magnetic length and k the dielectic constant):

$$V_{ab}(n,m) = \frac{e^2}{k1} \int dq |J_{nm}(q)|^2 F_{ab}(q).$$

(10)

If $M = \max(m,n)$ and $N = \min(m,n)$,

$$J_{nm}(q) = \sqrt{\frac{N!}{M!}} \; e^{-1^2 q^2/4} (1q/\sqrt{2})^{N-M} L_M^{(N-M)} (1^2 q^2/2)$$

(11)

in terms of the associated Laguerre polymonials $L_n^{(\alpha)}$. The form factors of the subbands reflect the weakening of the Coulomb interactions as compared with the strictly two-dimensional limit:

$$F_{ab}(q) = \int dz dz' |\varsigma_a(z)|^2 |\varsigma_b(z')|^2 e^{-q|z-z'|},$$

(12)

where the ς denote subband envelope functions which are chosen to be Gaussians with half-widths which minimize the total Hartree energy[22]. Explicit expressions for the integral in Eqn. 12 have been given by Chu and Chang[23].

The optical properties, i.e. the linear response to a weak phonon probe, can be expressed in terms of the two-particle electron-hole Green function[24]. The gain/absorption spectra are proportional to ($s,t = \pm 1$)

$$I^{abs}(\hbar\omega) = \frac{1}{2\pi 1^2} \frac{1}{\pi} \; \text{Im} \sum_{msnt} G_{eh}(ms,nt;\hbar\omega - \mu_e - \mu_h)$$

(13)

and the photoluminescence intensity is obtained as:

$$I^{lum}(\hbar\omega) = I^{abs}(\hbar\omega)/(e^{\beta(\hbar\omega-\mu_e-\mu_h)} - 1).$$

(14)

The trace of the Green function can be expressed as

$$\sum_{msnt} G_{eh}(ms,nt;E) = \sum_i \frac{1}{E-\lambda_i-i\delta} \frac{\left| \sum_{ns} \phi_i(ns) \sqrt{F(n)} \{u_n^2 \delta_{s+} + v_n^2 \delta_{s-}\} \right|^2}{\sum_{ks'} \text{sgn}\{s'\} |\phi_i(ks')|^2},$$

(15)

where the ϕ_i's and the λ_i's are the solution of the eigenvalue problem

$$\sum_{ns'} \left[E(m)\delta_{nm}\delta_{ss'} - \{\gamma_{mn}^2\delta_{ss'} + \bar{\gamma}_{mn}^2(1-\delta_{ss'})\}\sqrt{F(m)}\,V(m,n)\sqrt{F(n)} \right]\phi_i(ns') \tag{16}$$

$$= \text{sgn}\{s\}\lambda_i\phi_i(ms).$$

The coherence factors γ, $\bar{\gamma}$ are defined as

$$\gamma_{mn} = u_m u_n + v_n v_m, \quad \bar{\gamma}_{mn} = u_m v_n - v_m u_n. \tag{17}$$

The oscillator strengths from Eqn. 15 are singular at the chemical potential. Via a separate treatment we obtain

$$I^{abs}(\mu_e+\mu_h) = \frac{1}{2\pi l^2}\frac{1}{\pi}\frac{\displaystyle\sum_m \Delta(m)F(m)/\bar{E}(m)\sum_n (u_n^2-v_n^2)\Delta(n)F(n)/\bar{E}(n)}{\displaystyle\sum_o \left[\Delta(o)F(o)/\bar{E}(o)\right]^2}, \tag{18}$$

$$I^{lum}(\mu_e+\mu_h) = \frac{1}{2\pi l^2}\frac{1}{\pi}\frac{\displaystyle\sum_m \Delta(m)F(m)/\bar{E}(m)\sum_n v_n^2\Delta(n)F(n)/\bar{E}(n)}{\displaystyle\sum_o \left[\Delta(o)F(o)/\bar{E}(o)\right]^2}. \tag{19}$$

The above expressions have been evaluated numerically and the results are presented in the following Section.

3.   NUMERICAL RESULTS

The effects of the electron-hole interaction can be separated in three parts. First, the single-particle energies are renormalized (Eqns. 5 and 6). Next the condensation into the excitonic insulator state occurs and an energy gap opens up in the one-particle density of states. Finally, the optical properties are strongly influenced by the excitonic vertex correction. The results presented here have been chosen to shed light on this distinction.

The formalism can be used to study the effects of magnetic fields. Here attention is focused on zero-field results, however. By applying a small field of 0.3 T singular integral equations are converted into easily solvable matrix equations. The calculations are carried out for a representative 100 Å quantum well with a basis of 350 Landau levels which is accurate enough for the present purposes. The matrix to be diagonalized has a dimension of 700, in contrast to 350 in the normal state. The parameters for GaAs $m_e = 0.067$, $m_h = 0.3$, $k = 12.5$ for well and barrier material and a valence/conduction band offset ratio of 0.3/0.7 are used.

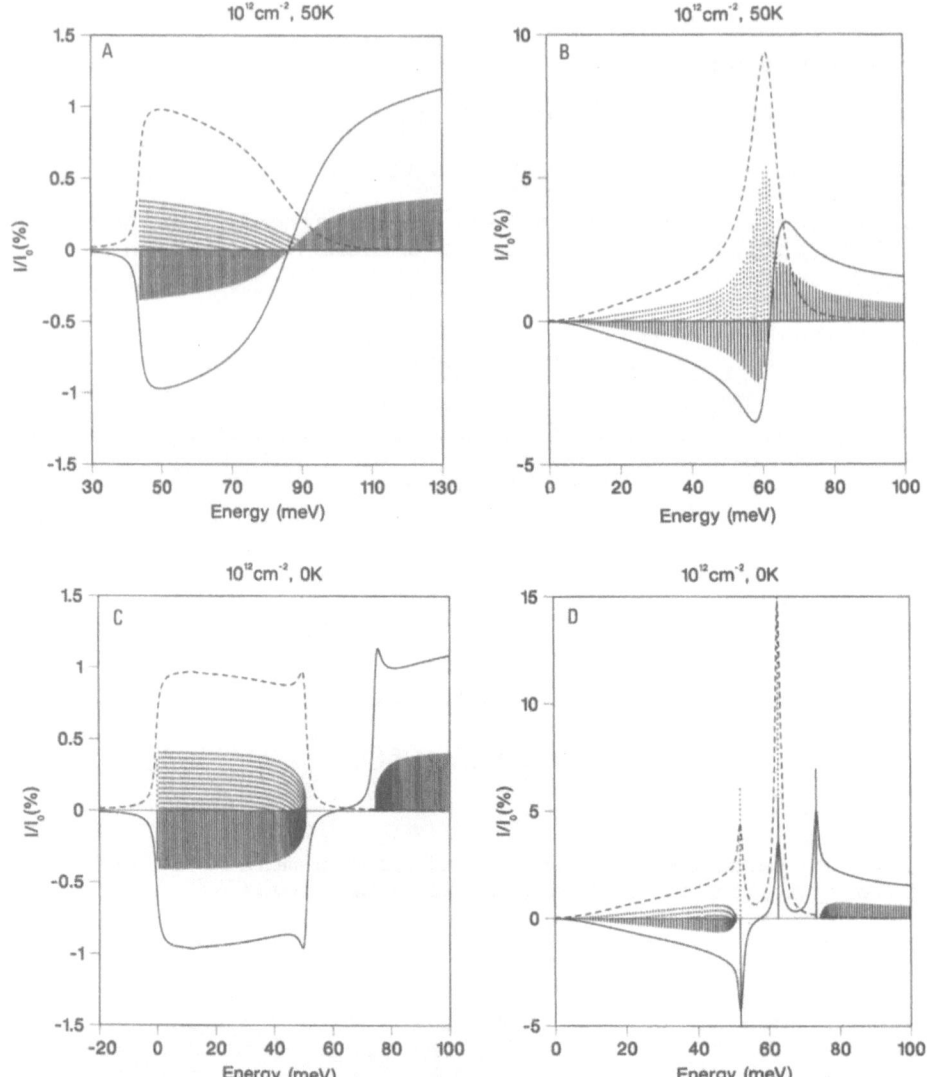

Fig. 1. Results of Hartree-Fock theory for the
photoluminescence (dashed lines) and gain/
absorption (continuous lines) spectra of a 100 Å
GaAs/Al$_{.3}$Ga$_{.7}$As intrinsic quantum well containing an electron-hole plasma with a pair-density
of 10^{12} cm^{-2}.   Results for two temperatures
(50 K: upper two figures, 0 K: lower two
figures) are given, where the excitonic vertex
correction is neglected in the left two figures.
$I_0 = 8/(\pi a_B^2)$, where a_B is the exciton Bohr
radius.   The vertical lines are the calculated
line spectra which are convoluted by a
Lorentzian with FWHM of 2 meV to obtain the
continuous curves.

In Fig. 1 the absorption and luminescence spectra are plotted with and without the excitonic vertex correction and at temperatures of 0 and 50 K, i.e. above and below the critical temperature T_c, which for the present density is between 34 and 35 K. Switching on the Coulomb potential in the normal state results in a strong enhancement of the oscillator strength and a modified line shape. Note that this enhancement is not restricted to the energy region close to the chemical potential as in the case of statically screened interactions[25,26]. The condensation gap is clearly seen in the lower two panels of Figs. 1. When the vertex correction is switched on, a 1s bound state is created in the center of the gap, i.e. at the chemical potential, and excited exciton states develop at both band edges.

In Fig. 2 the gap function Δ (Eqn. 1) is plotted as a function of temperature and Landau level index. As expected, the gap shrinks to zero at the transition temperature. The gap function is peaked at the uppermost occupied Landau levels and only slightly changes its shape with temperature. The transition can therefore be described by a single order parameter. The temperature dependent spectra are not shown here, but they clearly reflect the shrinkage of the condensation gap. The normal state spectra diverge at the chemical potential when the critical temperature is approached. Close to the transition temperature the spectra are strongly dominated by the oscillator strength from the chemical potential.

The density dependence of the gain/absorption spectra is plotted in Fig. 3 as a function of pair density. In order to

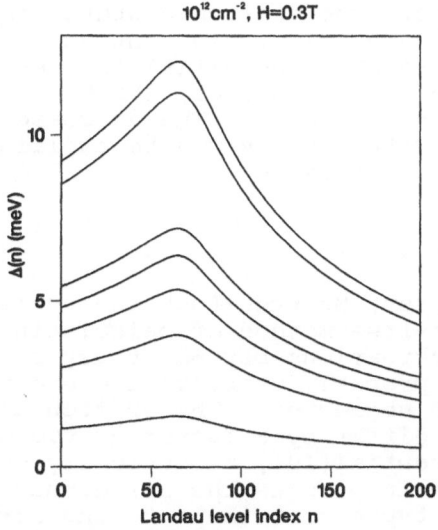

Fig. 2. Results of Hartree-Fock theory for the gap function Δ (Eqn. 1) as a function of temperature of a 100 Å GaAs/Al$_{.3}$Ga$_{.7}$As intrinsic quantum well containing an electron-hole plasma of a pair-density of 10^{12} cm^{-2}. The gap is shown to decrease with increasing temperatures of 0,20,30,31,32,33,34 K.

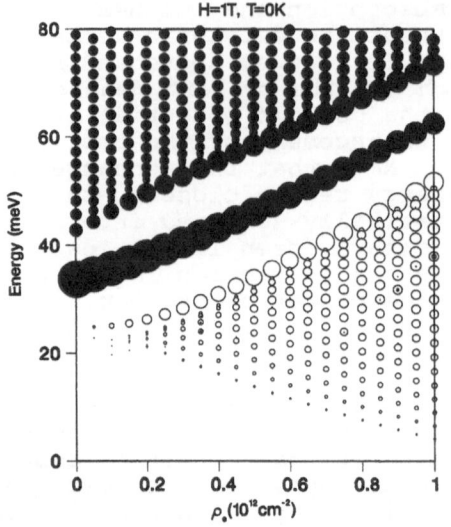

Fig. 3. Results of Hartree-Fock theory for the
absorption/gain spectra of a 100 Å
GaAs/Al$_{.3}$Ga$_{.7}$As intrinsic quantum well as
a function of plasma pair-density (T = 0
K, H = 1 T). The oscillator strengths are
proportional to the dot area. The
chemical potential coincides with the
exciton in the center of the gap.

resolve the oscillator strengths a slightly higher field of 1 T
has been applied here. The excitonic states in the gap are
clearly seen for all densities. In contrast to the charged
plasma in modulation-doped samples[19], we see that the exciton
remains stable up to high densities, which is closely related to
its approximate Boson character. The decrease of the gap and
oscillator strength with density can be explained in terms of
the Pauli-blocking of states[24].

4.   DISCUSSION

The self-consistent Hartree-Fock theory is appealing as a
simple and parameter-free method of calculating many-body
effects, but the fluctuations beyond it may be significant.
The present results probably overestimate excitonic effects
because screening is neglected. The exciton binding energies in
doped quantum wells differ by a factor of two when, instead of
the bare Coulomb potential[19], a statically screened one[27] is
used. In electron-hole plasmas the difference is even larger,
because two carrier types contribute to the screening. The
critical temperature calculated here is therefore too large.
Another complication is the asphericity of the hole Fermi
surface due to valence band mixing[28], which might decrease the
transition temperature in wide quantum wells. For low magnetic
fields the present theory could be adapted to a statically
screened interaction, but it is not from the outset clear that
the results would be more reliable than the present ones.

Formally, in a two-dimensional system at nonzero
temperature long range order will be destroyed by the coupling

of the condensate to long wavelength collective excitations, from which it can be concluded that mean-field theory cannot be employed to estimate transition temperatures at all[21,22]. This criticism of mean-field theory might be rather academic, however, since it applies only to infinite systems, where a topological (Kosterlitz-Thouless) phase transition[29] takes over, with transition temperatures which at high densities are not expected to differ much from the mean-field predictions.

Optical nonresonant high excitation or electrical injection via a PIN structure necessarily creates hot carriers and the rather short radiative lifetime of the electrons and holes in direct-gap semiconductors can prevent cooling of the plasma below the transition temperature. To observe condensation in conventional quantum wells, careful resonant excitation is therefore required. Highly excited type II quantum wells are interesting candidates because of the long recombination lifetimes, as are asymmetrically coupled quantum wells with an externally applied field which separates excited carriers spatially[15]. The excitonic interaction is reduced, but so is the screening, and sizable transition temperatures could possibly be achieved.

Noted added: The recent observation of a phase transition of excitons around 10 K in coupled GaAs quantum wells with an applied electric field by T. Fukuzawa, E.E. Mendez et al. was communicated by L. Esaki at the workshop[30].

Acknowledgment

The support of Prof. M.F.H. Schuurmans is gratefully acknowledged.

REFERENCES

1. See e.g. J.R. Schrieffer, "Theory of Superconductivity", Benjamin, New York (1964)
2. B.I. Halperin and T.M. Rice, **Solid State Physics**, 21:115 (1968)
3. L.V. Keldysh and A.N. Koslov, **Zh. Eksp. Teor. Fiz.**, 54:978 (1968) [**Soviet Physics JETP**, 27:521 (1968)]
4. C. Comte and P. Nozieres, **J. Phys.**, 43:1069 (1982)
5. P. Nozieres and C. Comte, **J. Phys.**, 43:1083 (1982)
6. P. Nozieres and S. Schmitt-Rink, **J. Low Temp. Phys.**, 59:195 (1985)
7. See e.g. the special issue on semiconductor lasers, **IEEE J. Quant. Electr.**, QE-23 (1987)
8. S. Tarucha, H. Kobayashi, Y. Horikoshi and H. Okamoto, **Jpn. J. Appl. Phys.**, 23:874 (1984)
9. G.E.W. Bauer, "Proceedings of the 8th EP2DS", Grenoble, 4-8 September 1989, **Surf. Science**, 229:374 (1990)
10. S. Schmitt-Rink, D.S. Chemla and D.A.B. Miller, **Adv. Physics**, 38:89 (1989)
11. I.V. Lerner and Yu.E. Lozovik, **Zh. Eksp. Teor. Fiz.**, 80:1488 (1981) [**Soviet Physics JETP**, 53:763 (1981)]
12. D. Paquet, T.M. Rice and K. Ueda, **Phys. Rev. B**, 32:5208 (1985)
13. M. Potemski, J.C. Maan, K. Ploog and G. Weimann, **in:** "Proceedings of the 19th Int. Conf. on the Physics of Semiconductors", W.Zawadzki, ed., Institute of Physics, Polish Academy of Science, Warsaw (1988)

14. J.C. Maan, **in**: "Localization and Confinement of Electrons in Semiconductors", G. Bauer, F. Kuchar, H. Heinrich, eds., Springer, Berlin (1990)

15. J.E. Golub, P.F. Liao, D.J. Eilenberger, J.P. Harbison and L.T. Florez, **Solid State Commun.**, 72:735 (1989)

16. Yu.E. Lozovik and V.I. Yudson, **Zh. Eksp. Teor. Fiz.**, 71:738 (1976) [**Soviet Physics JETP**, 44:389 (1976)]

17. L.L. Chang and L. Esaki, **Surf. Sci.**, 98:70 (1980)

18. G.E.W. Bauer, **Phys. Rev. Lett.**, 64:60 (1990)

19. G.E.W. Bauer **in**: "Localization and Confinement of Electrons in Semiconductors", G. Bauer, F. Kuchar, H. Heinrich, eds., Springer, Berlin (1990)

20. D. Jerome, T.M. Rice and W. Kohn, **Phys. Rev.**, 158:462 (1967)

21. R. Cote and A. Griffin, **Phys. Rev. B**, 37:4539 (1988)

22. G.E.W. Bauer and T. Ando, **Phys. Rev. B**, 31:8321 (1985); **J. Phys.C**, 19:1553 (1986)

23. H. Chu and Y.-C. Chang, **Phys. Rev.B**, 40:5497 (1989)

24. H. Haug and S. Schmitt-Rink, **Prog. Quantum Electron.**, 9:3 (1984)

25. S. Schmitt-Rink, C. Ell and H. Haug, **Phys. Rev.B**, 33:1183 (1986)

26. G.E.W. Bauer **in**: "Proceedings of the 19th Int. Conf. on the Physics of Semiconductors", W.Zawadzki, ed., Institute of Physics, Polish Academy of Science, Warsaw (1988)

27. D.A. Kleinman, **Phys. Rev. B**, 32:3766 (1985)

28. T. Ando, **J. Phys. Soc. Jpn.**, 54:1528 (1985)

29. J.M. Kosterlitz and D.J. Thouless, **J. Phys. C**, 6:1181 (1973)

30. T. Fukuzawa, E.E. Mendez and J.M. Hong, **Phys. Rev. Lett.**, 64:3066 (1990)

COOLING OF AN ELECTRON-HOLE PLASMA IN GaAs

W.W. Rühle, J. Collet,† M. Pugnet† and K. Leo*

Max-Planck-Institut für Festkörperforschung
Heisenbergstrasse 1, 7000 Stuttgart 80, FRG
†Laboratoire de Physique des Solides
Institut National des Sciences Appliquées
Avenue de Rangueil, 31077 Toulouse Cedex, France

ABSTRACT

Several difficulties involved in the analysis of optical experiments on the energy relaxation of carriers are reported. Carrier temperatures must be evaluated by lineshape fits with a many-body theory rather than by simple exponential fits to the high energy tail of the band-to-band transitions. The latter procedure always delivers too high values. Lifetimes must be determined from the density decay as obtained from lineshape fits to the transient spectra rather than from the decay of luminescence intensity. Electron and hole temperatures are for pulsed excitation even for hundreds of picoseconds considerably different at low excitation intensities. We get a good agreement between experimental cooling data and a theoretical cooling curve without any adjustable parameter if we use a correct analysis.

1. INTRODUCTION

Cooling of an electron-hole plasma has been studied by optical methods in the last years by several groups[1-17]. Such experiments actually have several advantages in comparison with electrical measurements: i) direct information on electron-phonon interaction is obtained, free of distortions by imperfections, such as crystal defects or rough interfaces which limit the mobility in electrical measurements, ii) the time resolution of optical experiments is clearly superior to electrical experiments making a direct investigation of elementary scattering processes such as electron-electron or electron-phonon interactions possible[11,15].

A direct comparison of cooling in the ps regime showed that the electron-polar-optical phonon interaction in two-dimensional

---

*Present address: AT & T Bell Laboratories, Holmdel, NJ 07733, USA.

Condensed Systems of Low Dimensionality
Edited by J. L. Beeby *et al.*, Plenum Press, New York, 1991

quantum well systems[8,9] and in GaAs/AlAs superlattices[17] is
very similar to bulk GaAs. These comparative experiments proved
that the influence of dimensionality is not very large.
Quantitatively, however, one problem remained up to now
unresolved: experimentally, the cooling is strongly reduced at
high densities[1,2]. Initially it was thought that screening of
the carrier-polar-optical phonon interaction causes this
reduction. However, it was shown later theoretically[18] and
experimentally[16], that a "hot" phonon distribution is
responsible for the slower energy loss rates at high densities.
A nonthermal phonon population builds up since many phonons are
emitted by a high density plasma, and these phonons have a long
lifetime of about 10 ps. The reabsorption of these hot phonons
then leads to a reduction of the energy loss rate. It was
essential to distinguish between these two effects. The effect
which sets in at lower densities will also strongly dominate the
cooling behavior at all higher densities[16]. Quantitatively,
however, the experimentally observed reduction was always up to
more than an order of magnitude stronger than expected
theoretically.

     In this contribution, we want to address for the first time
not only the merits but also the difficulties which are inherent
as well in time-resolved as in dc optical investigations. In
particular, three effects complicate considerably the interpretation of the results: i) holes, always present in an optical
experiment, have at lower excitation densities a quite different
temperature than the electrons. ii) Many-body effects strongly
influence the optical transitions at high densities, and a
simple evaluation of the experimental data is obviously wrong.
iii) Errors can also be made if lifetimes are simply deduced
from the decay of the luminescence intensity.

     A quite reasonable agreement between experiment and theory
of the cooling of an electron-hole plasma, without any adjustable parameter, is obtained if all these effects are taken into
account. However, the theories involved are only approximations, and in particular no many-body theory exists for
the lineshape analysis in two-dimensional systems. Therefore a
quantitative, not only comparative, but absolute evaluation of
the data meets serious difficulties.

Table 1. Material parameters used for the
theoretical cooling curve.

Electron effective mass[a]	$0.067 \, m_0$
Hole effective mass[a]	$0.50 \, m_0$
Static dielectric constant[a]	12.75
Optical dielectric constant[a]	10.94
LO-phonon energy[a]	36.4 meV
Electron deformation potential[b]	7.0 eV
Hole deformation potential[c]	4.8 eV
Density of GaAs[a]	$5.32 \, \mathrm{g \, cm^{-3}}$
Optical phonon lifetime[d]	10 ps

[a] Ref. 31. [b] From D.L. Rode[32]. [c] Calculated
according to J.D. Wiley[33] with the recent values
for elastic moduli given in Ref. 31. [d] From Ref. 34.

2. EXPERIMENTAL

The photoluminescence is excited with a synchronously pumped dye laser (5 ps pulse width, 80 MHz repetition rate, λ = 746 nm corresponding to an excess energy of 150 meV). The luminescence is dispersed in a 0.32 m spectrometer and temporarily resolved by a two-dimensional streak camera with a time resolution of 12 ps (FWHM of laser pulse on streak camera). The excitation density is varied by changing the focus spot size of the laser, the minimum spot having a diameter of 15 μm.

The sample is a 0.2 μm thick GaAs layer imbedded in a few micrometer thick $Al_{0.4}Ga_{0.6}As$ barriers which prevent inhomogeneous excitation and vertical diffusion. The samples are mounted on the cold finger of a temperature regulated cryostat.

3. RESULTS AND DISCUSSION

3.1 Theoretical Cooling Curves

The theoretical cooling curves are calculated taking into account[14,18-20]:

i) Fermi-Dirac statistics, i.e. the Pauli exclusion principle. This is especially important for high densities and short lifetimes, where effects like "recombination heating" or "degeneracy cooling" become important - for details see Refs. 14 and 20.

ii) Polar-optical scattering of electrons and holes via the Fröhlich interaction. We use an unscreened potential. Static screening overestimates the effect of screening and dynamical screening is equivalent to no screening for the densities investigated[14,21].

iii) Finite lifetime of the optical phonons generated during the cooling process. We calculate the complete, coupled electron-phonon system, i.e. we take into account the reabsorption of phonons via the carriers. The optical phonons have a finite lifetime before they decay via anharmonic decay into acoustical phonons. A "hot" or "nonthermal" phonon population builds up in a certain range of k space and strongly influences the cooling process as pointed out by Pötz and Kocevar[18].

iv) Acoustical deformation potential scattering of electrons and, in particular, holes. The inclusion of deformation potential scattering is important as soon as polar scattering is strongly reduced by nonthermal phonons.

v)   Optical deformation potential scattering of holes. The inclusion of the TO phonon scattering becomes important if the energy loss rate via emission of LO phonons is reduced by the buildup of a nonthermal phonon population.

vi) Piezoelectric scattering. This scattering becomes important at lower temperatures than investigated here and its inclusion is rather for sake of completeness than of practical importance.

vii) Different electron and hole temperatures. Differences in electron and hole temperatures were already discussed theoretically[22-24]. However, degeneracy of holes was not taken into account and until now it was not realized that at low densities differences between electron and hole temperatures might be sustained for hundreds of ps as will be discussed in the next section.

The parameters used for the theoretical cooling curves are taken from literature and are compiled in Table 1. For a more detailed description of the procedure for calculating the cooling see Refs. 14 and 18-20.

3.2 Differences in Electron and Hole Temperatures

First, we discuss the case of low excitation, where a single particle fit to the lineshape $I(h\nu)$ is still a good approximation. For $h\nu \gg E_g$ and E_F, the energy gap and the difference in Fermi levels of electrons and holes, we obtain:

$$I(h\nu) \sim \exp(-h\nu/k_B T_{eff}), \tag{1}$$

where the effective temperature T_{eff} is a composite of electron and hole temperatures, T_e and T_h, weighted with the respective effective masses m_e and m_h (isotropic heavy hole mass) according to:

$$(m_e + m_h)/T_{eff} = m_h/T_e + m_e/T_h. \tag{2}$$

The effective temperature thus is close to T_e. The much lighter electrons obtain initially most of the excess energy $\Delta E = h\nu - E_g$. Assuming an instant accomplishment of a different T_e and T_h via electron-electron and hole-hole scattering we calculate the cooling curve of electrons and holes separately. Two transfer mechanisms of energy from the electron to the hole system are considered (for details see appendix in Ref. 19), a statically screened electron-hole scattering and a transfer of energy via emission and absorption of nonthermal phonons, the latter process being the most important at high densities. Theoretical cooling curves are shown in Fig. 1. The difference $T_e - T_h$ amounts at the lowest densities to about 10 K even up to 100 ps after the laser pulse. Now, carrier temperature enters exponentially in the energy loss rate via emission of optical phonons, i.e. at low densities essentially only the electrons dissipate energy. However, both particles contribute with $3/2 \, k_B$ to the specific heat of the carrier system. This elucidates why in recent publications[8,16,17,25,26] always a discrepancy of about a factor of two was obtained between experimental and theoretical cooling at low densities. The agreement between experiment and theory becomes satisfactory at low densities as shown by the experimental crosses and the dotted line in Fig. 2, if the difference $T_e \neq T_h$ is taken into account. At higher density (compare experimental squares and theoretical solid line in Fig. 2), the discrepancy is still enormous although the nonthermal phonons are correctly included in theory. At the highest densities of about 10^{18} cm^{-3} the theoretical energy loss rate is more than an order of magnitude faster than the experimental one. The theoretical cooling is even at higher densities still relatively fast for the following reason: holes emit phonons in a larger volume in k space than electrons. In this larger volume in k space the nonthermal

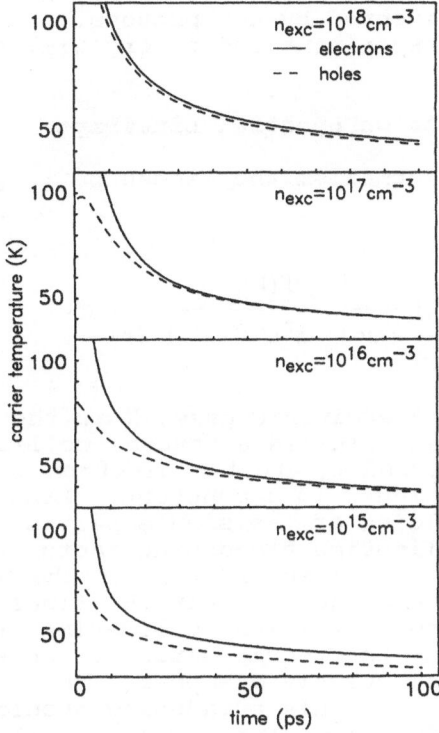

Fig. 1. Theoretical cooling curves for electron
and holes for different excitation
densities.

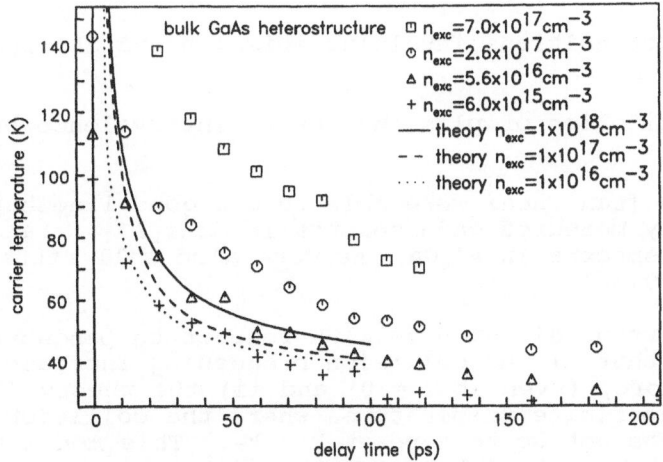

Fig. 2. Theoretical cooling curves $T_e(t)$ and
experimental cooling data for different
excitation densities.

65

phonon population then stays smaller. Therefore, the electron-hole plasma still cools relatively fast via emission of phonons by holes although the dissipation of energy by electrons is almost completely prevented by hot phonons. This strong discrepancy between experiment and theory will be solved in the next sections.

3.3 Many Body Effects on Spectral Lineshape

In a many body formalism the lineshape is given by[27,19]:

$$I(h\nu) = C \int_0^\infty f_e(k) f_h(k) \frac{\Gamma(k)}{(h\nu - E_g - \hbar^2 k^2/2\mu)^2 + \Gamma^2(k)} \, dk \qquad (3)$$

where μ is the reduced excitonic mass, k is the quasi-momentum of electrons and holes, $\Gamma(k)$ is a general collision broadening or damping factor, f_e and f_h are the electron and hole Fermi-Dirac distributions, and C is a constant. Landsberg first introduced such a folding of the single particle lineshape with a Lorentzian with a lifetime broadening factor Γ in order to explain the low-energy tail for $h\nu < E_g$ of the band-to-band recombination[28]. He suggested that the final state of a radiative band-to-band transition is lifetime broadened: the hole left behind after the recombination of an electron out of the electron Fermi sea quickly disappears due to scattering of other electrons into it. This broadening should go to zero for $T = 0$ at the Fermi edge, i.e. for $k = k_F$.

Haug and Tran Thoai treated this collision broadening factor Γ in a more general, correct way by treating optical transitions including plasmons in the plasmon pole approximation and taking into account[27]:

i) Lifetime broadening of the states by plasmon collisions.

ii) Optical transitions involving emission and absorption of plasmons.

iii) Finite lifetime of plasmons due to intervalence band absorption.

Haug and Tran Thoai were able to get good lineshape fits to experimentally measured gain spectra in GaAs[27]. Later, luminescence spectra in $Al_xGa_{1-x}As$ were also well fitted using this theory[29,30].

Two important differences with respect to Landsberg's approach are that i) the collision broadening increases for $k > k_F$ once more, (even at $T = 0$) and ii) the theory also treats the problem at finite temperatures where the collision broadening does not go to zero at $k = k_F$. This means that not only the low-energy tail but also the high-energy tail of the recombination is strongly affected by collision broadening. The exponential of the single particle approach in Eqn. 1 must actually be folded with Lorentzians (being different for electrons and holes). This folding, however, always causes a transfer from luminescence intensity from regions of higher

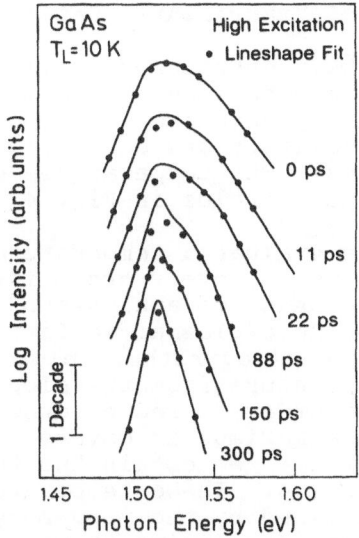

Fig. 3. Many-body lineshape fits (dots) to the
semilogarithmic transient spectra (lines)
as obtained for high excitation of n_{exc} =
7 x 10^{17} cm^{-3}.

intensity to regions of lower intensity, i.e. the slope of the
high-energy tail is generally decreased. A simple exponential
fit will therefore always deliver too high temperatures. This
effect becomes especially important if the width of collision
broadening Γ is comparable to the carrier temperature T_{eff}.
Figure 3 actually reveals that this is the case in our cooling
experiment: in the semilogarithmic plot of $I(h\nu)$ the low-energy
tail, influenced only by collision broadening, shows very
similar slopes to the high energy tail revealing that under our
conditions $\Gamma \approx k_B T_{eff}$.

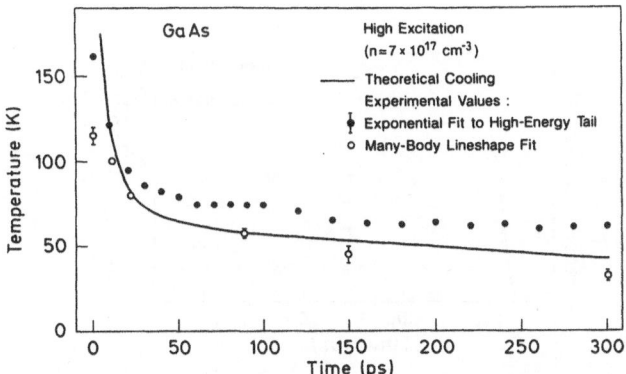

Fig. 4. Comparison of the temperature decay as obtained
by an exponential fit to the high-energy tail
(filled dots) or by the correct many-body
lineshape fit (circules). The solid line is the
theoretical cooling $T_e(t)$ calculated with the
parameters given in Table 1 and an electron-hole
lifetime of 110 ps.

3.4 Correct Analysis of Experimental Data

We now evaluate the temperatures at different times after the laser pulse (t = 0 at the laser pulse maximum) using the many-body lineshape fit by Haug and Tran Thoai (see Fig. 3). The only adjustable parameters are density and temperature of the electron hole plasma. The temperatures obtained from these fits are depicted as open circles in Fig. 4.

For comparison, we evaluated temperatures also according to a procedure used up to now by every group involved in time resolved or dc photoluminescence analysis of cooling behavior of a photoexcited electron hole plasma: we fitted the high-energy tail ($h\nu \gg E_g, E_F$) in a semilogarithmic plot by a straight line and calculated the temperature from the slope of the line. (An experimental detail: in order to reduce cross talk in the streak camera, the luminescence maximum is covered and only the high-energy tails are analyzed.) We obtain the temperatures depicted as closed circles in Fig. 4. These temperatures are always higher than those determined by the many-body lineshape fits, as expected according to the discussion above. The difference amounts to 25 K at long times. The experimental errors, however, are also quite large.

We discuss now another error which is normally introduced by a too simple analysis of the experimental data: usually the lifetime of the electron-hole plasma is obtained from the decay of the wavelength integrated luminescence. This time behavior is depicted in Fig. 5 by the solid line. An initial slow increase is observed before the luminescence actually decays with a time constant of 525 ps. The slow increase of luminescence is possibly due to an increase in the radiative recombination coefficient when the plasma cools down. We would conclude from these data that the lifetime is initially very long and then decreases to about 525 ps. The many-body lineshape fits, however, reveal that the plasma density actually drops much faster. The dots in Fig. 5 depict the density variation as obtained from this analysis. A 1/e lifetime of

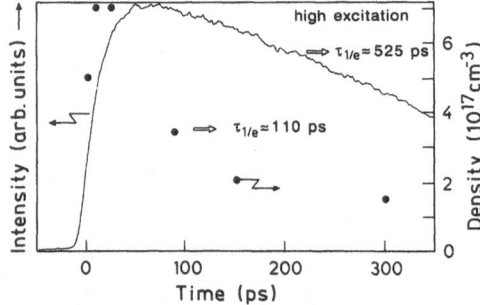

Fig. 5. Time evolution of the wavelength-integrated luminescence (solid line, left, arbitrary but linear scale) for an excitation density of 2 x 10^{17} cm^{-3} in comparison with the density evolution (dots, right scale) as obtained from many-body lineshape fits to the transient spectra. Lifetimes (1/e decay) of 525 ps and 110 ps are obtained respectively.

only 100 ps approximates this density decay. The origin for
this discrepancy is not unambiguously clear: ambipolar lateral
diffusion should be negligible on this time scale. It could be
that the density is quickly reduced via stimulated emission in
the lateral direction. In any case, the experimentally observed
fast decay of density has serious influence on the theoretical
cooling curve since such a short lifetime leads to a strong
recombination heating effect (for a detailed explanation of this
effect see Refs. 14 and 20). Finally, it turns out that the
correct experimental values (open circles) and the theoretical
cooling curve calculated with the correct lifetime (solid line)
are rather close together in Fig. 4.

Figure 6 is a compilation of the results for four different
excitation densities. The agreement between theoretical cooling
and experimental cooling as obtained from a many-body lineshape
fit is quite good, except for the lowest density. It should be
mentioned that at the lowest densities the many-body lineshape
fit to the spectra is difficult. For the spectra at t = 0 also

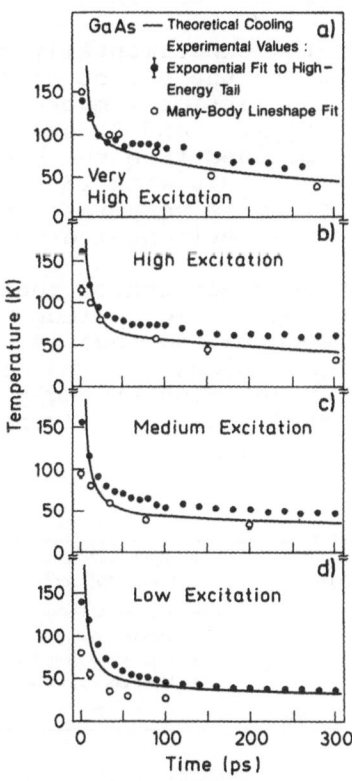

Fig. 6. Comparison of experimental cooling (dots: wrong
procedure, taking the temperature from a fit to
the high-energy tail; circles: correct temperatures as obtained from many-body lineshape
fits) with theory (solid line). The initial
densities and the lifetimes as obtained from the
many-body line-shape fits are: a) 1.7×10^{18}
cm^{-3}, 91 ps; b) 7×10^{17} cm^{-3}, 110 ps; c) $2 \times$
10^{17} cm^{-3}, 723 ps; d) 6×10^{16} cm^{-3}, 450 ps.

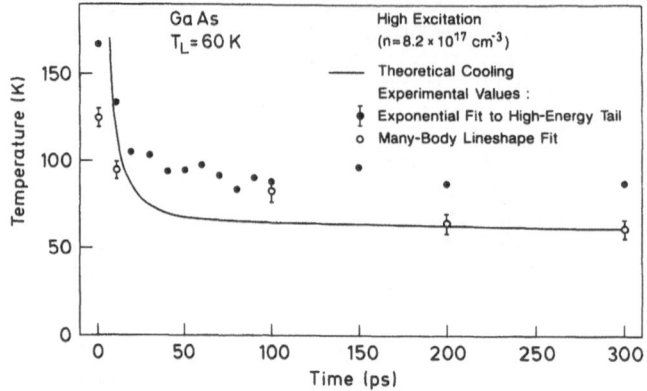

Fig. 7. Comparison of experimental and theoretical
cooling for a lattice temperature T_L = 60 K.

a deviation between experiment and theory is observed.   This is
probably simply an artefact of our limited time resolution.

     We finally investigated experimentally the temperature
limit up to which temperature the effect of collision broadening
is important by setting the lattice temperature to 60 and 100 K.
The results are shown in Figs. 7 and 8: at 60 K the discrepancy
between the temperature obtained from the high-energy tail and
from a many-body lineshape fit is with 25 K still considerable.
However, at 100 K the cooling is very fast and the difference
lies already almost within experimental error.

     In conclusion, we have demonstrated that a correct investi-
gation of the plasma cooling should include many-body effects in
the analysis of optical spectra and thus is much more difficult
than generally assumed. Additionally, the difference in
electron and hole temperatures also strongly influences the
analysis of the experimental data. A reasonable agreement

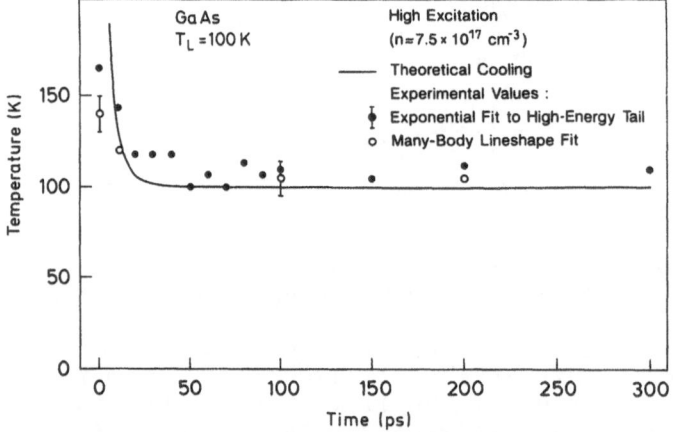

Fig. 8. Comparison of experimental and theoretical
cooling for lattice temperature T_L = 100 K.

70

between experimental cooling data and theory is obtained if all these effects are correctly taken into account.

Acknowledgment

We acknowledge many fruitful discussions with P. Kocevar, help on the computer by M.G.W. Alexander, sample growth by liquid phase epitaxy in the group of E. Bauser, and the expert technical assistance by K. Rother and H. Klann. The project was financially supported by the Bundesministerium für Forschung und Technologie.

REFERENCES

1.   J. Shah, **Solid State Electron.**, 32:1051 (1989)
2.   S. Lyon, **J. Lumin.**, 35:121 (1986)
3.   J. Shah, **IEEE J. Quantum Electr.**, QE22:1728 (1986)
4.   C.V. Shank, R.L. Fork, R. Yen, J. Shah, B.I. Greene, A.C. Gossard and C. Weisbuch, **Solid State Commun.**, 47:981 (1982)
5.   R.J. Seymour, M.R. Junnarkar and R.R. Alfano, **Solid State Commun.**, 41:657 (1982)
6.   W. Graudszus and E.O. Göbel, **Physica**, 117 & 118B:555 (1983)
7.   J.F. Ryan, R.A. Taylor, A.J. Turberfield, A. Maciel, J.J. Worlock, A.C. Gossard and W. Wiegmann, **Phys. Rev. Lett.**, 53:1841 (1984)
8.   K. Leo, W.W. Rühle and K. Ploog, **Phys. Rev. B**, 38:1947 (1988)
9.   J. Shah, A. Pinczuk, A.C. Gossard and W. Weigmann, **Phys. Rev. Lett.**, 54:2045 (1985)
10. T. Elsässer, R.J. Bäuerle and W. Kaiser, **Phys. Rev.**, 40:2976 (1989)
11. C.W.W. Bradley, R.A. Taylor and J.F. Ryan, **Solid State Electron.**, 32:1173 (1989)
12. Z.Y. Xu and C.L. Tang, **Appl. Phys. Lett.**, 44:692 (1984)
13. H. Uchiki, T. Kobayashi and H. Sakaki, **Solid State Commun.**, 55:311 (1985)
14. K. Leo, Ph.D Thesis, University of Stuttgart (1988)
15. W.H. Knox, **Solid State Electron.**, 32:1057 (1989)
16. W.W. Rühle, K. Leo and E. Bauser, **Phys. Rev. B**, 40:1756 (1989)
17. K. Leo, W.W. Rühle and K. Ploog, **Solid State Commun.**, 71:101 (1989)
18. W. Pötz and P. Kocevar, **Phys. Rev. B**, 28:7040 (1983)
19. J.H. Collet, W.W. Rühle, M. Pugnet, K. Leo and A. Million, **Phys. Rev. B**, 40:12296 (1989)
20. K. Leo and W.W. Rühle, **Solid State Commun.**, 62:659 (1987)
21. J. Collet, **Phys. Rev.**, 39:7659 (1989)
22. M. Asche and O.G. Sarbei, **Phys. Stat. Sol.**, B126:607 (1984)
23. W. Pötz, **Phys. Rev. B**, 36:5016 (1987)
24. M. Asche and O.G. Sarbei, **Phys. Stat. Sol.**, B141:487 (1987)
25. W.W. Rühle and H.J. Polland, **Phys. Rev. B**, 36:1683 (1987)
26. J.H. Polland, W.W. Rühle, K. Ploog and C.W. Tu, **Phys. Rev. B**, 36:7722 (1987)
27. H. Haug and D.B. Tran Thoai, **Phys. Stat. Sol.**, B98:581 (1980)
28. P.T. Landsberg, **Phys. Stat. Sol.**, 15:623 (1966)
29. A. Selloni, S. Modesti and M. Capizzi, **Phys. Rev. B**, 30:821 (1984)
30. M. Capizzi, S. Modesti, A. Frova, J.L. Staehli, M. Guzzi and R.A. Logan, **Phys. Rev. B**, 29:2028 (1984)

31. "Semiconductors", vol. 17 of Landolt-Börnstein, New Series, K.H. Hellwege, ed., Springer, Berlin (1982)
32. D.L. Rode, **Phys. Rev. B**, 2:1012 (1970)
33. J.D. Wiley, Mobility of Holes in III-V Compounds, vol. 10 in "Semiconductors and Semimetals", R.K. Willardson and A.C. Beer, eds., Academic Press, New York, San Francisco, London (1978)
34. J.A. Kash, J.C. Tsang and J.M. Hvam, **Phys. Rev. Lett.**, 54:2151 (1985)

MAGNETO-OPTICS OF [111] GaAs/GaAlAs QUANTUM WELLS

L. Viña, F. Calle, C. López, J.M. Calleja and
W.I. Wang*

Instituto de Ciencia de Materiales-CSIC & Departamento
de Fisica Aplicada, Universidad Autónoma
E-28049 Madrid, Spain
*Electrical Engineering Department and Center for
Telecommunication Research, Columbia University
New York, NY 10027, USA

1. INTRODUCTION

The electronic and optical properties of GaAs/GaAlAs
quantum wells (QWs) have been profusely studied in the last
twenty years, since its concept was introduced by Esaki and
Tsu[1]. Their optical properties are dominated by strong
excitonic effects, i.e. the Coulomb interaction between
electrons and holes[2]. The main attention has been devoted to
samples grown on substrates oriented along the [100] axis,
mainly due to the superior crystal quality which can be attained
for this orientation. However, recent Molecular Beam Epitaxy
(MBE) studies have shown that GaAs/GaAlAs QWs can be fabricated
with the quantization axis along a variety of crystallographic
directions[3-6]. The complexity of the valence band, especially
its anisotropy, offers an extra flexibility for tuning the
optical properties and for the design of electro-optical
devices. Although it is difficult to prepare high quality QWs
on non-(100)-oriented substrates, it has been shown that by
slightly misorienting (2°) the (111) substrates towards the
(100) orientation, dramatic improvements in the electrical and
optical properties can be obtained[5]. The photoluminescence
(PL) intensity of GaAs films grown on misoriented substrates is
comparable to that of (100) GaAs[5,7].

Several works have appeared recently in the literature
dedicated to the optical properties of GaAs/GaAlAs QWs grown on
different crystallographic directions[8-16]. These studies find
differences in the band structure depending on the growth
orientation and suggest possible improvements of the electronic
properties of QWs with the quantization axis along the [111]
direction. An effective mass for the heavy hole band along the
[111] direction amounting to 0.7-0.9 has been determined
[8,9,11]. This implies an enlargement by a factor of 2-2.6
compared with the (100) orientation. As a consequence of the
heavier mass, the subbands are more closely spaced and the

Condensed Systems of Low Dimensionality
Edited by J. L. Beeby *et al.*, Plenum Press, New York, 1991

relative ordering of the heavy and light hole subbands changes. Therefore the band mixing also changes and the subband dispersion is modified. Hayakawa et al. reported on an enhancement of the optical transition rates and a decrease in the threshold current densities for (111) oriented QW lasers relative to (100) oriented ones[10]. An improvement of the peak-to-valley ratio by a factor of two for AlAs/GaAs/AlAs double barrier heterostructures grown along the (111) crystal axis compared to samples grown in the (100) orientation has also been reported[17]. From a theoretical study of the behavior of deep impurity levels in [111] superlattices (Sls) it has been suggested that these SLs could be a better material for electronic applications[18]. It should be also mentioned that in the case of strained layer SLs grown along the [111] axis, as for example GaInAs/AlInAs, new features not present in [100]-growth-axis strained layer SLs are found. The most striking one is the appearance of strain-induced polarization fields in the layers grown along the [111] direction, which causes nonlinear optical response of these SLs[19,20].

We have used high resolution photoluminescence excitation spectroscopy (PLE) to study the pseudo-absorption of [111] oriented GaAs/GaAlAs QWs in the presence of an external magnetic field. The application of the magnetic field provides a supplementary possibility of tuning the electronic states in addition to the usual confinement effects. It also enhances oscillator strengths and increases the energy differences between exciton states, therefore allowing the study of excited states in not very high quality samples. Additionally, the magnetic-field-induced breaking of the Kramers degeneracy, combined with the use of circularly polarized light, allows us to select excitons associated with different spin orientation of the holes. An analysis of the polarization of the emitted light gives further information on the spin relaxation of photo-generated carriers[21], and/or on the mixing of different excitonic states. Although the magneto-absorption of semi-conductors has received great attention in the past, because it provides very useful information about band structure parameters[22], to the best of our knowledge, no magneto-optical study of QWs grown on the [111] direction has been performed.

2. EXPERIMENTAL DETAILS

The GaAs/GaAlAs QWs used in our experiments were grown by MBE on (111)B GaAs substrates, with crystal orientation tilted 3° towards the (100) direction to improve their optical quality[5]. A 1 μm GaAs buffer layer was followed by 475 Å of $Ga_{1-x}Al_xAs$ and five quantum wells separated by 200 Å $Ga_{1-x}Al_xAs$ barriers with x = 0.30. The samples were capped by 500 Å of undoped $Ga_{1-x}Al_xAs$. Two samples have been investigated. From growth parameters, their nominal QW thicknesses were 90 Å and 120 Å. However, by fitting the excitonic energies observed in PLE experiments at 0 T, using a Kronig-Penney model, QW widths of 75 Å and 100 Å were obtained, respectively. For the calculations, we have used the measured binding energies (see below) and the effective mass parameters from Ref. 11.

The light from a LD700 dye laser, pumped by a Kr^+-ion laser, was focused onto the sample with a spherical lens and the power density was kept below 0.1 Wcm^{-2}. The emitted light was

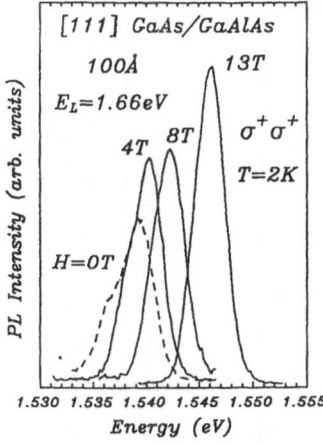

Fig. 1. PL spectra of a 100 Å QW at several
magnetic fields. The excitation was
performed in the σ^+ polarization, and only
the σ^+ component of the emission was
detected.

analysed with a Jarrel-Ash 1 m double-grating monochromator and
detected with standard photon counting techniques. The incident
light was circularly polarized by means of an achromatic $\lambda/4$
plate. The emission was analysed into its σ^+ and σ^- components
using a combination of a second achromatic $\lambda/4$ plate and a
linear polarizer. The sample was immersed in a liquid He bath-
cryostat within a standard-coil superconducting magnet, which
enabled us to apply external magnetic fields, in the Faraday
configuration, up to 13.5 T.

3.   RESULTS AND DISCUSSION

    Photoluminescence spectra of the 100 Å sample, recorded
with σ^+ polarized light and analysed into their σ^+ component,
are depicted in Fig. 1 for four different magnetic fields. The
PL was excited at 1.66 eV, below the GaAlAs band gap. The
typical quadratic diamagnetic shift of the heavy hole exciton is
clearly observed. It amounts to 6 meV at 13 T. The FWHM
diminishes from 4.6 meV at 0 T to 3.3 meV at 13 T. This
decrease is due to the quenching of the low energy band of the
PL, seen at 0 T, with increasing magnetic field. For the 75 Å
QW a FWHM at 0 T of 5.4 meV was measured. The values of the
FWHM are a factor ~1.5 better than previously reported
widths[12]. The nonmonotonic behavior of the intensities with H
can be related to the changes in the absorption coefficient at
1.66 eV for different magnetic fields. Furthermore, the PL
strength also depends on the polarization configuration, not
only because of the absorption changes but also due to the
relaxation processes.

    Additional information can be obtained by analysing the
polarization of the emission, as demonstrated in Fig. 2. The
left panel shows PLE spectra of the 75 Å QW at 9 T for σ^-
excitation, where the two lowest traces have been analysed into
left- and right-handed polarizations, respectively. The most

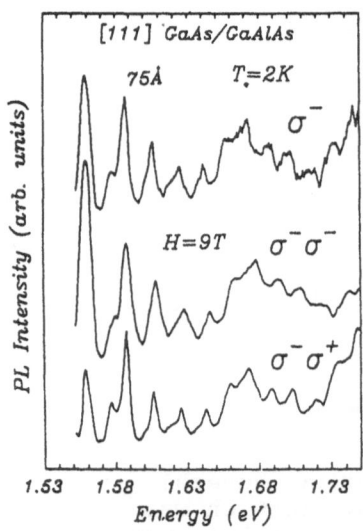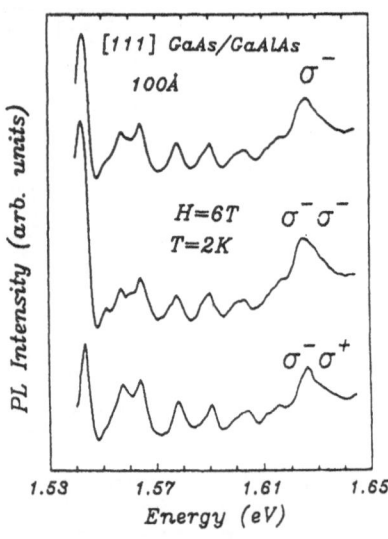

Fig. 2. PLE spectra performed with σ^- incident light for
the 75 Å QW at 9 T (left) and the 100 Å at 6 T
(right). The lower traces correspond to the
analysis of the emission into its circular
polarizations.

conspicuous difference between them is the smaller intensity of
the ground state heavy hole exciton (1.56 eV) observed in the
$\sigma^-\sigma^+$ configuration. This is expected from selection rules,
since the heavy hole, derived mainly from $|\tfrac{3}{2}\rangle$ states, should be
ideally observed only in parallel polarizations (parallel refers
either to $\sigma^-\sigma^-$ or $\sigma^+\sigma^+$). The rest of the excitonic transitions
show approximately the same strength for the three spectra.
This indicates that in the process of energy relaxation of the
photocreated carriers (for excitons above the ground heavy hole
exciton), in [111] QWs in the presence of a magnetic field, the
electron magnetic moment is not conserved. This result is at
variance with recent experiments by Potemski et al.[21];
however, it agrees with the fast carrier spin-relaxation found
in GaAs/GaAlAs multi-quantum wells and CdTe/CdMnTe heterostructures by Freeman and co-workers[23]. A similar effect is
observed for the 100 Å QW at 6 T depicted in the right panel of
Fig. 2. Besides the lower intensity of the ground state heavy
hole exciton in $\sigma^-\sigma^+$ configuration, this figure also illustrates
the additional resolution of fine structure in the spectra when
the emitted light is analysed into its circular polarizations.
The triplet resolved only in $\sigma^-\sigma^-$ originates from $h_1(2p)$, $l_1(1s)$
and $h_1(2s)$ excitons, as will be shown below.

Let us turn now to the dependence of the excitation spectra
with magnetic field. Very useful information about band
structure parameters[24] and ground state exciton binding
energies[25-28] can be obtained from magneto-optical data.
Figure 3 shows PLE spectra of the 100 Å QW at four different
magnetic fields. The richest spectra are generally obtained in
$\sigma^-\sigma^+$ configuration. The peaks at 6 T are assigned to the
following excitons: $h_1(1s)$, 1.542 eV; $l_1(1s)$, 1.557 eV; $h_1(2s)$,
1.564 eV; $h_1(3s)$, 1.577 eV; higher excited states and $h_2(1s)$,
1.625 eV. Besides the diamagnetic shift of the excitons,

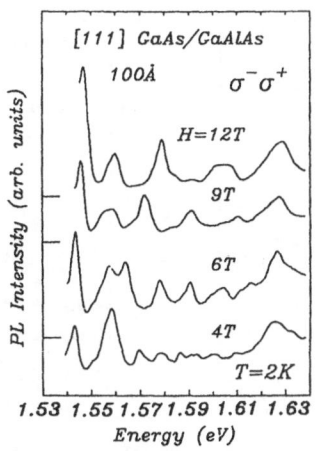

Fig. 3. PLE spectra of the 100 Å QW recorded at
different magnetic fields perpendicular to
the layers, in the $\sigma^-\sigma^+$ configuration.

an enhancement of the oscillator strength of the transitions is
seen with increasing magnetic field. The latter effect is due
to the reduction of the exciton radius and a corresponding
increase in the binding energy, and has been already observed in
[100] GaAs/GaAlAs QWs[29]. At 4 T the ground state light hole
exciton is hidden below $h_1(2s)$. Both states belong to different
symmetries and therefore they do not couple[29,30], as borne out
by our experiments.

Figure 4 shows the energies of the ground state heavy and
light hole excitons and those of the first two excited states of
h_1 as a function of magnetic field for the 75 Å and 100 Å QW in
the left and right panel, respectively. The solid (open)
symbols correspond to data obtained from PLE recorded with σ^-
(σ^+) polarization. The dashed (solid) lines represent the best
fit to the data also for $\sigma^-(\sigma^+)$ polarization. In the case of
the 100 Å QW a transition is observed up to 6 T (9 T) midway
between $h_1(1s)$ and $l_1(1s)$ for right-(left-)handed polarized
light. We tentatively assign this structure to the forbidden
$h_1(2p)$ exciton. Its zero-field energy coincides with the
extrapolation of the excited "ns" states of h_1. The Zeeman
splitting of $h_1(1s)$ is relatively small in the field range of
our measurements. The spin-up ground state excitons lie above
their spin-down counterparts, while the reverse situation holds
for the excited states, for which the Zeeman splitting is more
noticeable.

Recently, Bauer has studied theoretically the anisotropy of
the magneto-optical properties of GaAs/GaAlAs QWs, taking into
account exciton mixing interactions[16]. His results can be
compared directly with our experiments, since he has considered
a 76 Å QW with an aluminum composition of 30%. The theoretically obtained shifts are shown in Fig. 5 for spin-down (σ^+)
excitons in the left panel and for spin-up (σ^-) excitons in the
right panel. The transition oscillator strength at integer
magnetic fields are proportional to the area of the points. The
agreement between theory and experiment is excellent. The spin
splitting at 10 T for $h_1(1s)$ amount to 1.4 meV in the

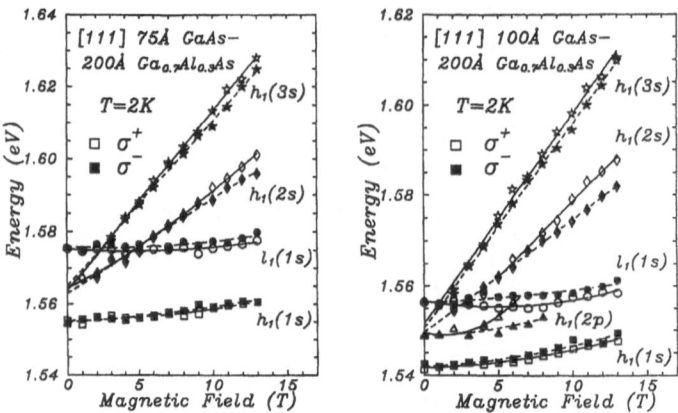

Fig. 4. Experimental energies of the ground state heavy
and light hole excitons and the first excited
states of h_1 versus magnetic field, for the 75 Å
(left) and 100 Å (right) QW. For the sake of
clarity, other transitions have been ignored.

theory[16], while 1.6 meV is obtained at this field from a
quadratic fitting of our data. The agreement deteriorates
somewhat for the l_1(1s) exciton, whose experimental splitting is
2 meV, while the theory obtains 0.7 meV. The spin averaged
diamagnetic shift at 10 T is 4.6 meV (1.5 meV) in the theory for
h_1 (l_1) in accordance with the experimental value of 4.2 meV (1.1
meV). The splitting of the excited states is larger for the
h_1(2s) states than for h_1(3s) states, both in theory and
experiment. The crossing between l_1(1s) and h_1(2s) takes place
between 5 T and 6 T, a finding also reproduced by the theory.
Finally, the oscillator strength of h_1(3s) at high magnetic

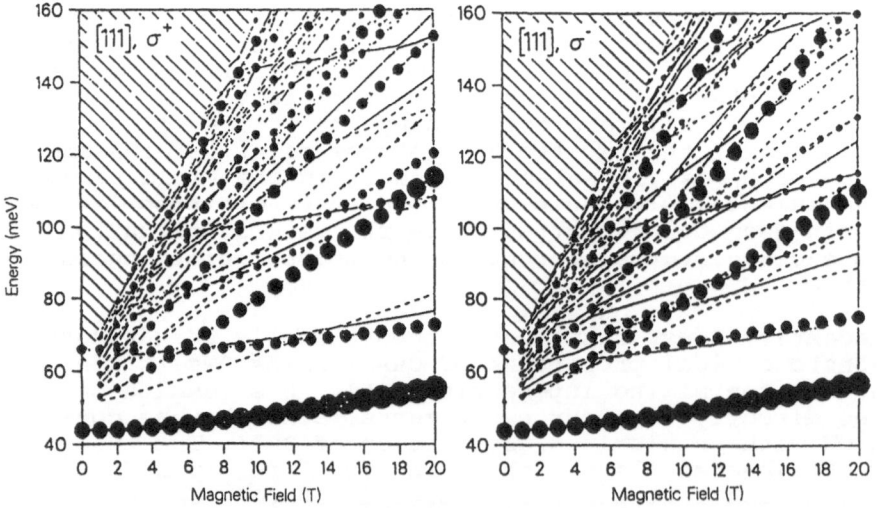

Fig. 5. Theoretical energies of excitons and transition
oscillator strengths of 76 Å QW for σ^+ (left
panel) and σ^- (right panel), as a function of
magnetic field. (After Ref. 16).

fields is ~ 30% larger for spin-up (σ^- excitation) excitons, roughly the same amount that can be extracted from the area of the dots depicted in the theoretical curves. We are not able to observe the $h_1(2p)$ exciton in 75 Å QW, which according to theory should have a very small oscillator strength. However, the diamagnetic shift and the Zeeman splitting of this exciton resolved in the 100 Å QW, are in qualitative agreement with the theoretical results[16].

Hayakawa and co-workers[9] have found an increase by about 10% in the ground state exciton binding energy of [111] GaAs/GaAlAs QWs compared with [100] QWs. They report a term splitting $E[h_1(2s)] - E[h_1(1s)]$ of 9 meV for a 100 Å [111] QW. For this thickness, our value, obtained from the extrapolation to zero field of the energies of the $h_1(ns)$ states, is only 7.7 meV. It should be mentioned that this extrapolation agrees, within the experimental error, with the energy of the $h_1(2p)$ exciton, resolved at 0 T (the difference between the binding energies of $h_1(2p)$ and $h_1(2s)$ excitons amounts to 0.4 meV for a 160 Å [110] QW[31], and it is too small to be measured in our [111] samples, of worse quality). Assuming a binding energy of 1.5 meV for the $h_1(2s)$ exciton[9,13], we obtain 9.2 meV for the binding energy of $h_1(1s)$. This value is somehow larger than the theoretical result of Broido[15] (8 meV), and it is slightly smaller than the 9.5 meV binding energy of [100] QWs of the same thickness[32]. This tendency is also found in the calculations which predict smaller binding energies for [111] QWs compared to [100] oriented ones[13,16].

The agreement between the zero-field extrapolation of the "s" excited states and the energy of the $h_1(2p)$ exciton, mentioned above, allows us to extract a reliable binding energy of h_1 also for the 75 Å QW from the magneto-excitation spectra. The discrepancies for the binding energies obtained from magneto-optical experiments and from direct observation of excited states reported in the literature[32], are not expected in our case, due to the observation of excited states at fields as low as 1 T. With the same assumption as above for the binding energy of $h_1(2s)$, we obtain for $h_1(1s)$ a value of 10.5 meV. This is larger than the theoretical values of 9 meV (Ref. 15) and 9.4 meV (Ref. 16), and again slightly smaller than the experimental values 11.3 meV (Ref. 9) and 11 meV (Ref. 32) for [100] QWs.

From the theoretical results of Ref. 16 the effective mass parallel to the well, and thus the band gap density of states, is expected to decrease in [111] QWs relative to [100] oriented QWs (this also explains the smaller values of the binding energies for [111] orientation). However, a selective enhancement of the strength of heavy hole transitions in [111] oriented QWs relatively to the light hole ones has been reported[10]. We present in Fig. 6 a comparison between the pseudo-absorption of two 100 Å QW at zero field. The left (right) panel depicts, for the four possible polarization configurations, the PLE spectra for [111] ([100]) oriented QWs, in the region of the ground state heavy hole and light hole excitons (note that $h_1(2p)$ is also seen for parallel polarizations). The dashed lines represent the photoluminescence. The quality of the sample, as determined from the FWHM of the peaks and the Stokes shift between PL and PLE, is in this case slightly better for the [111] QW. One should notice that the

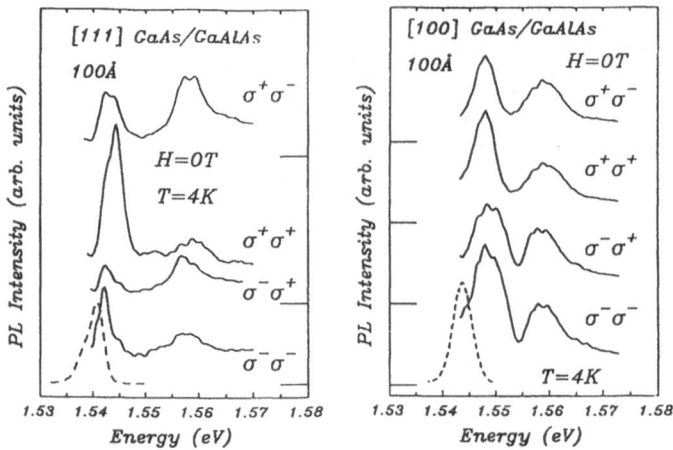

Fig. 6. PLE spectra recorded at 0 T in the four
polarization configurations for 100 Å QWs grown
along [111] (left panel) and [100] (right panel)
crystallographic directions. The horizontal
ticks mark the offset of the spectra. Dashed
lines depict the PL emission.

energy position of l_1(1s) is approximately the same for both
orientations (1.56 eV), while h_1(1s) for [111] lies ~ 6 meV
below the corresponding exciton of the [100] QW. This fact is
due to the large anisotropy of the heavy hole subband, whereas
the light hole subband is almost isotropic[11], and has been
observed previously[9]. In unpolarized PLE experiments no
enhancement of the heavy hole relative to the light hole exciton
for [111] samples, as compared with [100] QWs, was found. The
spectra are very similar in the case of the [100] QW for all the
polarization configurations, and no noticeable changes in the
relative intensities of h_1(1s) and l_1(1s) are observed. Similar
results were obtained for a 130 Å [100] QW of higher quality
(FWHM of the peaks ~ 1.5 meV instead of the ~ 5 meV value of the
100 Å QW). However, striking differences are seen in the [111]
QW when light of different helicity is utilized to excite the
sample, and the polarization of the emission is analysed. A
larger strength for h_1(1s) relative to l_1(1s) is obtained in
parallel configurations, while the situation is reversed for
orthogonal polarizations. This result is expected from the
selection rules of the absorption, which predict that the heavy
hole (light hole) exciton should be only observable in parallel
(orthogonal) polarizations. More intriguing is the fact that,
at zero field, marked differences are seen between the $\sigma^+\sigma^+$
polarization and the $\sigma^-\sigma^-$ one (the same applies for crossed
polarizations). They should be equivalent in the absence of any
magnetic field. However, a much larger oscillator strength for
h_1 is obtained exciting with right-handed polarization. This
finding seems to be specific for the [111] oriented QWs and
could be related to the details of the valence band structure.
Changes in the spin-relaxation of the excitons and/or excitonic
mixing could be also responsible for the behavior of the light
hole exciton.

The ratio of the intensity of h_1(1s) to l_1(1s) seen in the
different polarizations is strongly field-dependent. As an

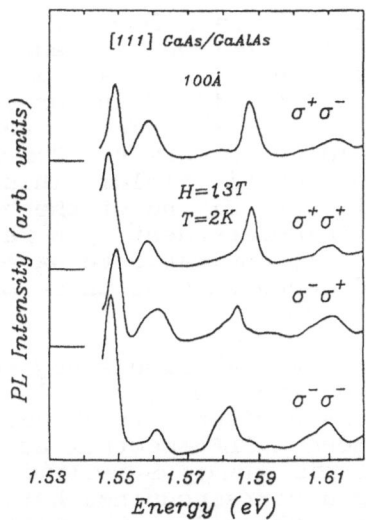

Fig. 7. PLE spectra for the 100 Å QW at 13 T in
the different polarization configurations.
The ticks in the vertical axis show the
offset of the spectra.

example, we show in Fig. 7 PLE spectra of the 100 Å QW at 13 T
for the four polarizations. At this field, the ground state
heavy hole exciton is always larger than l_1(1s). A greater
oscillator strength is still observed for l_1(1s) in orthogonal
polarizations compared with parallel ones. In this figure is
also clearly demonstrated the Zeeman spin-splitting of the
excitons and the resolution of further structure when both
exciting and analysed light are polarized (see, for example, the
broad peak at 1.58 eV in $\sigma^+\sigma^-$ also observed as a shoulder in the
$\sigma^-\sigma^-$ configuration).

Figure 8 depicts the ratio of the oscillator strength of
the ground state heavy hole exciton to that of the ground state

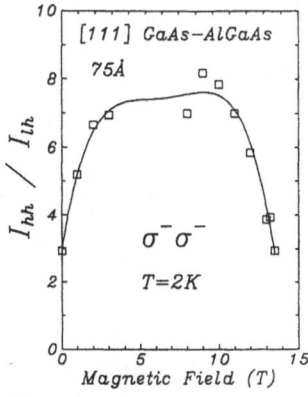

Fig. 8. Dependence on magnetic field of the
intensity ratio of the heavy hole exciton
to the light hole one for a 75 Å QW and σ^-
σ^-. The line is a guide to the eye.

light hole exciton, for $\sigma^-\sigma^-$, as a function of magnetic field for the 75 Å QW. The values between 4 T and 7 T have been omitted since in this region $l_1(1s)$ overlaps with $h_1(2s)$. A ratio of 3 is obtained at zero field in agreement with the polarization selection rules. A large increase is observed with increasing magnetic field up to ~ 10 T. The ratio drops with a further increase of the magnetic field. This behavior is maybe due to magnetic field induced mixing of the heavy hole and light hole excitons or to a field-dependent spin-relaxation of the excitons. It would be very desirable to have a theoretical estimation of the field dependence of these processes in order to distinguish between them.

From our experiments we can also study the degree of polarization of the pseudo-absorption, defined as $p = (I^+ - I^-)/(I^+ + I^-)$, where I^+ means excited with σ^+ polarization[33]. The left panel of Fig. 9 shows a PLE spectrum at 2 T of the 100 Å QW for $\sigma^+\sigma^-$, together with the energy dependence of the degree of polarization. There is a correspondence between the peaks of the PLE and the features observed in the polarization spectrum. p has a value of ~ 40% for the $h_1(1s)$. A strong drop of p is observed at the energy of the light hole exciton, reaching a negative value. A new positive peak is observed in the spectral region of the $h_2(1s)$ exciton, as it corresponds to a heavy hole state. Similar spectra are depicted in the right panel for a field of 5 T. Very useful information on the degree of polarization can still be obtained at this magnetic field (a peak for $h_1(1s)$, a negative peak for $l_1(1s)$ and a positive peak for $h_2(1s)$), which can be used to extract the heavy or light character of the excitons. However, the analysis becomes more complicated for higher excitons, since the Zeeman splitting between spin-up and spin-down excitons, especially important for excited states, causes shifts between the features observed in

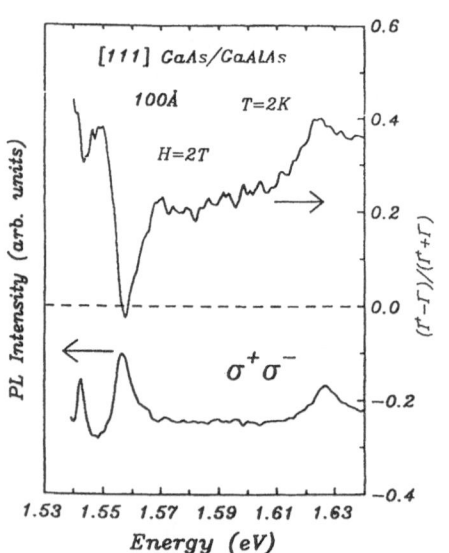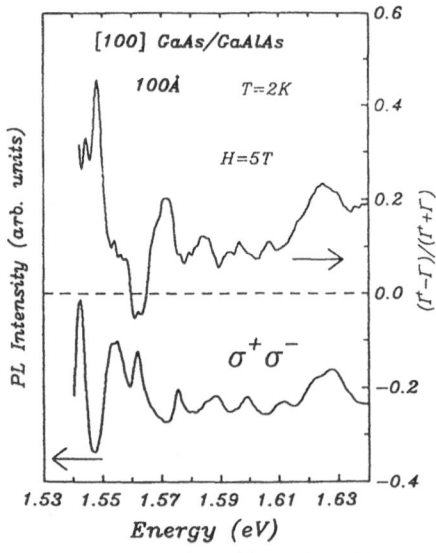

Fig. 9. Energy dependence of the degree of polarization and PLE for a 100 Å QW at 2 T (left panel) and 5 T (right panel).

PLE and p, and it is no longer possible to perform an unequivocal assignment of the structures seen in both spectra.

SUMMARY

We have studied the magneto-optical properties of [111] oriented GaAs/GaAlAs QWs. From the observation of excited excitonic states at fields as low as 1 T, we were able to extract the binding energy of the ground state heavy hole exciton for two different well widths. Our results show that the binding energies are very similar, and possibly slightly smaller, to those of [100] oriented QWs, in agreement with recent theoretical calculations. The magnitude of the Zeeman spin-splittings and diamagnetic shifts are also in good agreement with full calculations of magneto-excitons in [111] QWs. From a comparison between measurements in [100] and [111] QWs, we do not find any appreciable enhancement of the oscillator strength of the heavy hole exciton in the latter case. A complete analysis of the polarization of the emitted light, using circularly polarized excitation, opens new possibilities to the study of spin-relaxation of excitons and excitonic mixing in quasi-two-dimensional systems.

Acknowledgment

We want to thank G.E.W. Bauer for the availability of Fig. 5. One of the authors (WIW) wants to acknowledge financial support from ARO. This work was sponsored in part by CICYT Grant No. MAT-88-0116-C02.

REFERENCES

1. L. Esaki and R. Tsu, **IBM J. Res. Develop.**, 14:61 (1970)
2. See, for example, Excitons in Confined Systems, **in** Vol.25 "Springer Proceedings in Physics", R. Del Sole, A.D. Andrea and A. Lapiccirella, eds., Springer-Verlag, Berlin (1988)
3. W.I. Wang, **J. Vac. Sci. Technol. B**, 1:630 (1983); **Surf. Sci.**, 174:31 (1986)
4. T. Fukanaga, T. Takamori and H. Nakashima, **J. Cryst. Growth**, 81:85 (1987)
5. L. Viña and W.I. Wang, **Appl. Phys. Lett.**, 48:36 (1986)
6. S. Subbanna, H. Kroemer and J.L. Merz, **J. Appl. Phys.**, 59:488 (1986)
7. T. Hayakawa, M. Kondo, T. Suyama, K. Takahashi, S. Yamamoto and T. Hijikata, **Jpn. J. Appl. Phys.**, 26:L302 (1987)
8. B.V. Shanabrook, O.J. Glembocki, D.A. Broido, L. Viña and W.I. Wang, **J. Phys. (Paris) Colloq.**, 48:C5-235 (1987)
9. T. Hayakawa, K. Takahashi, M. Kondo, T. Suyama, S. Yamamoto and T. Hijikata, **Phys. Rev. B**, 38:1526 (1988)
10. T. Hayakawa, K. Takahashi, M. Kondo, T. Suyama, S. Yamamoto and T. Hijikata, **Phys. Rev. Lett.**, 60:349 (1988)
11. L.W. Molenkamp, R. Eppenga, G.W. 't Hooft, P. Dawson, C.T. Foxon and K.J. Moore, **Phys. Rev. B**, 38:4314 (1988)
12. L.W. Molenhamp, G.E.W. Bauer, R. Eppenga and C.T. Foxon, **Phys. Rev. B.**, 38:6147 (1988)
13. L.W. Molenhamp, G.E.W. Bauer, R. Eppenga, G.W. 't Hooft, and C.T. Foxon, **Superl. and Micros.**, 5:355 (1989)
14. M. Nakayama, J. Kimura, H. Nishimura, T. Komatsu and Y. Kaifu, **Solid State Commun.**, 71:1137 (1989)

15. D.A. Broido, **Superl. and Micros.**, 5:471 (1989)
16. G.E.W. Bauer, **in**: "Spectroscopy of Semiconductor Microstructures", NATO ASI Series B, G. Fasol, A. Fasolino and P. Lugli, eds., Plenum Press, NY. 206:381 (1990)
17. L.F. Luo, R. Beresford, W.I. Wang and E.E. Mendez, **Appl. Phys. Lett.**, 54:2133 (1989)
18. S.Y. Ren and J.D. Dow, **J. Appl. Phys.**, 65:1987 (1989)
19. C. Mailhiot and D.L. Smith, **Phys. Rev. B**, 35:1242 (1987); D.L. Smith and C. Mailhiot, **Phys. Rev. Lett.**, 58:1264 (1987)
20. J.G. Beery, B.K. Laurich, C.J. Maggine, D.L. Smith, K. Elcess, C.G. Fonstad and C. Mailhiot, **Appl. Phys. Lett.**, 54:233 (1989)
21. M. Potemski, J.C. Maan, A. Fasolino, K. Ploog and G. Weimann, **Phys. Rev. Lett.**, 63:2409 (1989)
22. See, for example, L. Viña, **in**: "Spectroscopy of Semiconductor Microstructures", NATO ASI Series B, G. Fasol, A. Fasolino and P. Lugli, eds., Plenum Press. NY. 206:367 (1990) and references therein
23. M.R. Freeman, D.D. Awschalom, J.M. Hong and L.L. Chang (preprint); M.R. Freeman, **Bull. Am. Phys. Soc.**, 35:457 (1990)
24. See, for example, J.C. Maan, **Surf. Sci.**, 196:518 (1988)
25. J.C. Maan, G. Belle, A. Fasolino, M. Altarelli and K. Ploog, **Phys. Rev. B**, 30:2253 (1984)
26. D.C. Rogers, J. Singleton, R.J. Nicholas, C.T. Foxon and K. Woodbridge, **Phys. Rev. B**, 34:4002 (1986)
27. S. Tarucha, H. Okamoto, Y. Iwasa and M. Miura, **Solid State Commun.**, 52:815 (1984)
28. W. Ossau, B. Jäkel, E. Bangert, G. Landwehr and G. Weiman, **Surf. Sci.**, 174:188 (1986)
29. L. Viña, G.E.W. Bauer, M. Potemski, J.C. Maan, E.E. Mendez and W.I. Wang, **Phys. Rev. B**, 41:10767 (1990)
30. G.E.W. Bauer and T. Ando. **Phys. Rev. B**, 38:6015 (1988)
31. L. Viña, R.T. Collins, E.E. Mendez and W.I. Wang, **Phys. Rev. Lett.**, 58:832 (1987)
32. E.S. Koteles and J.Y. Chi, **Phys. Rev. B**, 37:6332 (1988)
33. C. Weisbuch, R.C. Miller, R. Dingle, A.C. Gossard and W. Wiegmann, **Solid State Commun.**, 37:219 (1981)

SPIN MEMORY OF PHOTOCREATED CARRIERS IN QUANTUM WELLS IN HIGH

MAGNETIC FIELDS

J.C. Maan, M. Potemski, A. Fasolino,*
K. Ploog,† and G. Weimann‡

Max-Planck-Institut für Festkörperforschung
HML, 166X F-38042 Grenoble Cedex, France
*SISSA Strada Costiera 11, I-34014 Trieste, Italy
†Max-Planck-Institut für Festkörperforschung
 Heisenbergstrasse 1, D-7000 Stuttgart 80, FRG
‡Walter Schottky Institut, TU München, D-8046
 Garching, FRG

ABSTRACT

Spin relaxation of photocreated carriers in GaAs/GaAlAs quantum wells in high magnetic fields is investigated by luminescence experiments under selective optical excitation. The complete quantization of the two-dimensional energy structure in high magnetic fields implies a simultaneous exchange of energy and momentum for spin relaxation and makes spin conserving relaxation processes, both in thermalization and recombination, much faster than spin-flip processes. This results in a new spectroscopic tool to identify magneto excitonic states with the same electronic spin orientation. We apply this technique to study the zero field splitting of the heavy hole exciton ground state often observed in high quality GaAs/GaAlAs quantum wells. We find that each peak of this doublet splits in a magnetic field into two components, and using the selective relaxation we can attribute the conduction band spin character to each of these. It turns out that the ordering of the electron spins of the Zeeman splitting of the upper component of the doublet is inverted with respect to that of the lower component. This observation strongly suggests a magnetic mechanism, a possibility would be exchange, as the origin of the zero field splitting.

1.   INTRODUCTION

The energy spectrum of a two-dimensional (2D) system, like a quantum well, in high magnetic fields perpendicular to the layers, is fully quantized, i.e. characterized by discrete states (Landau levels with Zeeman splitting), separated by gaps. In three dimensions there is always the residual degree of freedom for movement in the direction of the field for which, with the exception of the ground state, there can be states with different spin orientation and/or different Landau level number at the same energy. Since in 2D in high fields this is no longer true, the relaxation of electron spin between Zeeman split conduction band states can only take place by a process which carries away simultaneously both the full Zeeman energy and the magnetic moment. Elastic processes, in which the spin relaxation takes place in two steps, namely, first, a simultaneous change of momentum *and* magnetic moment, but no change in energy and, secondly, a relaxation of energy within the same band and with the same magnetic moment, are therefore not possible anymore. Such elastic processes are usually invoked[1-3] to explain the experimentally observed rapid spin relaxation in bulk materials[4-6] or in 2D systems without magnetic field[7-9]. This inhibition of the elastic spin-flip processes may thus lead to a substantial increase of the spin relaxation time[10,11].

These effects can conveniently be studied by means of interband photoluminescence and photoluminescence-excitation experiments on GaAs/GaAlAs quantum wells in high magnetic fields. In such experiments a photon at a given energy can promote an electron from a particular full valence band Landau level into another particular empty conduction band Landau level with a well defined electron spin magnetic moment. Both electron and hole relax to the lowest, respectively highest, available states and recombine under emission of a photon. In a magnetic field, due to Zeeman splitting, there are in fact two "lowest" energies, namely one for each component of the electron spin orientation, one being slightly higher than the other. Observing the intensity of each of these two components as a function of the energy of the exciting light (the usual photoluminescence excitation experiment) gives therefore information not only about the intensity of the absorption in some particular higher excited state, but also to which of the two ground states with different spin orientation these excited particles relax. That excitation spectroscopy can be used to study not only the absorption in excited states but also the subsequent relaxation to the ground state is, in fact, the main issue of the present work. It will be demonstrated that from excitation spectra in high magnetic fields it can be concluded that: i) the relaxation time of the magnetic moment of the electrons is much longer than the recombination time and ii) the magnetic moment is conserved both in recombination and during thermalization from the excited states.

Spin conserving phenomena of this type are very well known for relaxation between states in free atoms - states which are both localized and discrete - but they are rarely observed for band states in solids. As we mentioned before, in a two-dimensional system in a magnetic field the band states are fully quantized and in this sense the spectrum is very similar to an atomic spectrum. On the other hand these discrete states describe free particles in a solid which are delocalized contrary to those in atoms. Our experimental results demonstrating the inhibition of spin relaxation show therefore that this full quantization alone is already enough to produce such an atom-like behavior.

A direct consequence of these long spin relaxation times is the possibility to perform optical pumping experiments, i.e. to populate electronic states with preferential spin orientation

using selective optical excitation. This technique is known to
be a powerful spectroscopic tool in atomic physics[10] or, in
general, for localized centers which possess spin relaxation
times longer than the radiative lifetime. Optical pumping
experiments on band states in bulk semiconductors are instead
less easy because of the rather rapid spin relaxation of
photocreated carriers[4]; typical values of the spin relaxation
time at low temperatures are often shorter than the recombination time[5,6]. We will use this optical pumping technique to
identify the magnetic signature of the fine structure of the
heavy hole ground state. The interpretation of this fine
structure is a subject of controversy[12-15]. The optical
pumping experiments show that the interaction leading to this
splitting must be of magnetic origin, a possibility being
exchange interaction[15].

2.   EXPERIMENTAL RESULTS

     We have performed the experiments on high quality GaAs
undoped quantum wells (thickness between 4.5 nm and 20 nm)
sandwiched between $Ga_{1-x}Al_xAs$ barriers (x = 0.26 - 0.43). The
photoluminescence and photoluminescence-excitation spectra have
been measured at 1.7 K and in magnetic fields up to 20 T in the
Faraday configuration. Both exciting and emitted light was
analyzed with the different helicity (σ^+ and σ^-) of the circular
polarization. Dye lasers (with LD700 or DCM as dyes) pumped by
ion lasers (Kr^+ or Ar^+) have been used as excitation sources.

     Due to size quantization, the energy spectrum of GaAs
quantum wells consists of several subbands and, at zero magnetic
field, the optical spectra are characterized by a series of
excitonic lines associated with each of these subbands. The
lowest state, i.e. the heavy hole exciton ground state, is
observed in low temperature luminescence experiments. In
excitation spectroscopy the intensity of this luminescence is
measured as a function of the energy of the exciting light.
Only if i) all excited carriers relax to the ground state and
ii) all of them then recombine radiatively, this excitation
spectrum reflects the absorption. Usually these conditions are
fulfulled and quantities like transition matrix elements and
oscillator strengths can be inferred from the data. Under the
influence of a magnetic field, the heavy hole ground state
splits into two components[16-18]. Interestingly enough, at
sufficiently high magnetic fields, both these components,
designated A and B, are seen in luminescence spectra; as
expected from the selection rules they have opposite circular
polarization. The interesting fact is that also the higher
energy component B is visible in emission, even though the
thermal energy (kT = 0.13 meV) is much less than the splitting
between A and B (e.g. 1.5 meV for 9 nm QW at 18 T) and even if
the excitation power is very low (less than 0.1 mW/cm^2), which
excludes heating as an explanation. This experimental fact
shows that carriers do not relax from the higher lying state of
the doublet to the lower one but rather recombine directly with
emission of light; i.e. the relaxation time B to A is comparable
to the radiative recombination time, which is typically of the
order of 10^{-9} s.

     We have measured excitation spectra on each of these two
peaks A and B and the results are shown in Fig. 1. It is useful

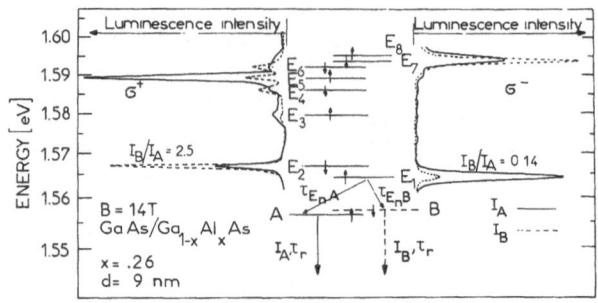

Fig. 1. Luminescence intensity as a function of the
energy of the σ^- (left) and σ^+ (right) polarized
exciting light in a field of 14 T. A and B
designate the energy position of the two components of the heavy hole spin split ground state.
The luminescence of A (B) is σ^- (σ^+) polarized.
The solid lines show the excitation spectra
taken by measuring the intensity of A, and the
dashed lines by measuring the intensity of B.
The relevant relaxation processes are indicated
in the central part. The arrows label the
orientation of the conduction band magnetic
moment involved in the transition as extracted
from the comparison between theory and experiments shown in Fig. 2.

to bear in mind that these spectra in fact reflect the intensity
of the luminescence A or B for excitation in a particular state
with a particular polarization. As can be seen from Fig. 2, the
intensity ratio between these two luminescence peaks is found to
depend strongly on the energy and helicity of the exciting light
- i.e. on which excited state is actually pumped. For a given
polarization of the exciting light, say σ^+, the peak positions
in the excitation spectra are the same because they involve the
absorption in the same state, and it does not matter whether the
luminescence intensity (I_A) from the lower component (A) or that
(I_B) from the higher component (B) of the ground state is
measured. However, the relative intensity of the peaks in the
two excitation spectra is drastically different. For σ^- polarization a similar behavior but, of course, involving other
excited states, is observed. Since, for a fixed polarization,
the exciting light is absorbed in the same states, the difference in the excitation spectra for A and B must be attributed to
different relaxation rates. Therefore we conclude that carriers
excited in some higher magnetoexcitonic level, designated E_n,
relax preferentially to either one or to the other component (A
or B) of the ground state. Considering for example the states
shown in Fig. 1, we conclude that E_1, E_3, E_5 and E_8 relax
preferentially to A, and E_2, E_4, E_6 and E_7 to B. All the states
among which the relaxation is efficient must have a certain
common property: in the following, we will show that all these
states possess the same orientation of the electron spin.

3. ANALYSIS

The energy levels of magnetoexcitons in quantum wells have
been studied theoretically[17,18] and experimentally[16,19] in

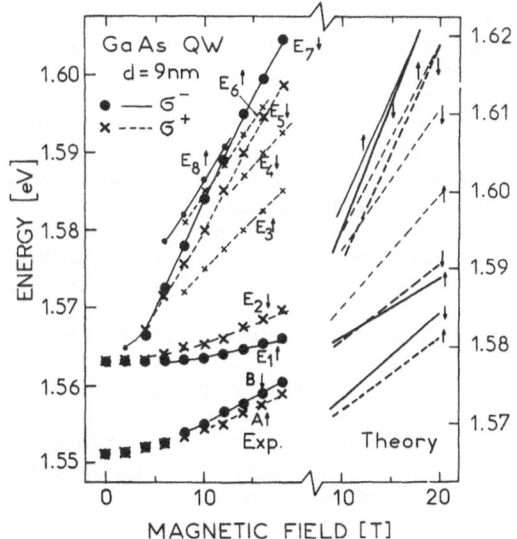

Fig. 2. Magnetic field dependence of the peaks in the
excitation spectra in the two circular
polarizations. The size of the symbols indicate
the strength of the transition (left panel).
Calculated field dependence of interband
transitions in both polarizations; the thickness
of the lines is proportional to the calculated
intensity of the transitions (right panel). The
spin magnetic moment of the conduction band
involved in the transitions (indicated by
arrows) is assigned by comparing theory and
experiments.

the recent past. Although the resulting optical spectra (see
e.g. Figs. 1, 2) are rather complicated, it has been shown that,
at high fields, the main features of experimental data can be
well described with calculated transitions between the free
electron and hole Landau levels[20]. The energy level structure
of the electrons in a magnetic field can be described by a
simple Landau level ladder, with each Landau level split into
two well defined spin components. The hole Landau levels are,
however, more complicated[20] and, due to the effect of spin-
orbit coupling, do not have a simple spin-up or spin-down
character. In the left panel of Fig. 2 we show the field
dependence of the same magnetoexcitonic states which, for
B = 14 T, are shown in Fig. 1. In the right panel of this
figure the calculated free electron hole interband transition
energies at high fields are shown. The theoretical curves are,
of course, not identical to the experimental ones because in the
calculations excitonic effects are not included. The effect of
the Coulomb interaction is a shift to lower energies of the
heavy and light hole states and a change of slope of the
transitions as a function of magnetic field; it affects more the
lower states than the higher ones. However, the qualitative
agreement between theory and experiment is good enough to allow
the identification of the transitions, also thanks to the
calculated intensities[18]. We can therefore attribute to each
level E_n an electron spin orientation, corresponding to the

conduction band Landau level involved (see Fig. 2). In particular, the peaks A and B, observed in luminescence, are seen to possess the opposite spin orientation. In Fig. 1, we have labeled all the observed states with their electron spin orientation, and it can be seen that, in a given excitation spectrum, those states are enhanced which possess the same conduction band spin orientation both in absorption and emission. Therefore, the experimental results, together with the theoretical assignment, show that the inhibition of spin-flip processes is responsible both for the nonthermalization between the spin split ground states (peaks A and B) as well as for the preferential relaxation from a higher excited state to that ground state which has the same electron spin.

It is useful to bear in mind that the description of the experimental data and their analysis is entirely based on an interband transition picture without reference to excitonic effects. Strictly speaking the ground state of the system is the full valence and the empty conduction band. The first excited state is the excitonic ground state and all higher states are excitonic states which, by their very nature, are always some combined electron hole state. The relaxation from the excited states toward the two different Zeeman split components of the excitonic ground state can therefore not simply be described in terms of the individual relaxation of electrons and holes, but rather as a relaxation through different excited states of the exciton. All these states have simultaneously a valence band and a conduction band character, i.e. the wavefunctions of these states are made from some combination of valence band and conduction band wavefunctions belonging to some band states. The selective relaxation of the electron spin should therefore more properly be seen as a preferential relaxation between those excitonic excited states which are mainly made from conduction band states having the same electron spin. The underlying assumption in describing the relaxation as was done previously in terms of band states only, neglecting excitonic effects, is that at high fields the cyclotron energy dominates the Coulomb energy. At the high fields where the effects discussed in this paper are observed this condition is fulfilled and the simple band picture may be used.

To estimate the relevant relaxation times we analyze the intensity ratio $R = I_B/I_A$ between the luminescence peaks A and B using rate equations. A simplified scheme of the processes involved is shown in Fig. 1. The relevant processes can be described in the following way. Carriers are created optically in some arbitrary state E_n. From this state they relax toward the states A and B, with characteristic relaxation times τ_{EnA} and τ_{EnB}. The deviation from unity of the ratio between these two relaxation times indicates preferential relaxation, i.e. spin conserving processes. If spin-flip processes would be completely forbidden then one of these times would be infinitely long and the relaxation would exclusively take place from E_n to A or from E_n to B, depending on which pair of states has the same electron spin. When the carriers have relaxed down to the state A, which is the lowest in energy, they can only recombine radiatively, giving rise to the luminescence I_A. Carriers which relax down to state B can instead either relax further to A, which implies an electron spin-flip, or recombine directly, giving rise to the luminescence I_B. The intensity ratio

R = I_B/I_A will therefore depend also on the ratio of the recombination lifetime τ_r (assumed to be the same for A and B) to the relaxation time between A and B, τ_{BA}. With this model we can write:

$$R = \frac{I_B}{I_A} = \left\{ \frac{\tau_{BA}}{\tau_r} \right\} \bigg/ \left\{ \left[\frac{\tau_{EnB}}{\tau_{EnA}} \right]\left[1 + \frac{\tau_{BA}}{\tau_r} \right] + 1 \right\} \tag{1}$$

Since there are two ratios of characteristic times which determine R, it is impossible to evaluate these characteristic times directly from a single measurement of the intensity ratio which involves only one particular excited state. However, the same equation is valid for different excited states E_n, with, of course, different values of τ_{EnB}/τ_{EnA}, but with the same value for τ_{BA}/τ_r since this latter ratio has no relation with the excited state. To estimate the relaxation times we note that as follows from Eqn. 1, for any E_n:

$$\tau_{BA}/\tau_r \geq R \tag{2}$$

the equal sign occurring when τ_{EnB}/τ_{EnA} is zero, i.e. when all carriers excited in E_n relax to B only.

In the data shown in Fig. 1, we observe R = 2.5 for the excited state E_2; we thus conclude that $\tau_{BA}/\tau_r \geq 2.5$, implying that the spin relaxation time between the two components of the spin split heavy hole ground states is about three times longer than the recombination lifetime. Taking τ_{BA}/τ_r equal to 2.5, also in the case when R = 0.14 - which corresponds to the excitation to the E_1 state (see Fig. 1) - we obtain $\tau_{E1B}/\tau_{E1A} = 5$, which means that the spin conserving relaxation for this state is at least five times more efficient than the spin-flip process. In the rate equation model presented here, an infinite density of states is assumed for each level. This assumption is only fulfilled when the number of excited particles is less than the degeneracy of the levels (low excitation power). Upon increasing power the values of R are changed, as we have indeed observed. The results we presented here were therefore all taken at low excitation power (< 0.1 W/cm^2).

It is important to note that the inhibition of spin-flip processes is strong enough to create a population inversion between spin split levels using selective optical excitation (optical pumping) as shown in Fig. 1. More intense luminescence from the higher energy component of the ground state is observed when one of the E_2, E_4, E_6 or E_7 excited state is pumped, which is clear evidence of population inversion. From a study of this effect as a function of magnetic field, exciting energy and for different samples, it is found that inhibition of spin-flip is more pronounced in the following conditions: i) at higher fields (for the same sample of Fig. 1 but at 18 T, we have found $\tau_{AB}/\tau_r \geq 4$ and $\tau_{E2B}/\tau_{E2A} = 8$ while no spin conserving processes are observed at vanishing magnetic field); ii) for lower lying excited states and iii) for better samples (i.e. with sharper peaks). The common denominator of these trends is the reduced overlap in the density of states between spin split levels. We therefore conclude that the conservation of magnetic moment is the result of the discrete nature of the levels in 2D systems in a magnetic field.

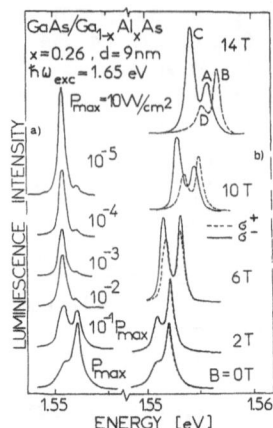

Fig. 3. Evolution of the luminescence spectrum
with the excitation intensity at zero
magnetic field. The scale is not the same
for all spectra (a). Magnetic field
dependence of the σ^- (solid) and σ^+
(dashed) luminescence at P = 10 W/cm^{-2}.

4.   USING THE TOOL

     The possibility to distinguish between transitions allowed
in either left or right circularly polarized light is an
important tool for the identification of transitions in magneto-
optical experiments. Our results demonstrate an additional
spectroscopical possibility, namely to distinguish between
states with different electron spin orientation. It is
interesting to note that this is indeed a complementary tool to
polarization, because there are cases where light is absorbed in
one polarization and gives rise to a strong luminescence of
opposite polarization; for instance, as shown in Fig. 1,
absorption in the state E_2 of σ^+ radiation gives rise to a
strong luminescence of B in σ^- radiation, and similarly for E_1
and A. As an example of the possibilities of this new
spectroscopic tool we will study a fine splitting of the heavy
hole exciton ground state which is often observed in high
quality GaAs/GaAlAs quantum wells and for which there is not yet
a satisfactory explanation[12-15]. We will study in more detail
the same 9 nm sample which was used before, but all seven other
samples which we studied showed the same behavior.

     At zero magnetic field the luminescence shows already a
doublet structure which is the subject of discussion here. In
Fig. 3 we show this luminescence at B = 0 as a function of
excitation power, and as a function of field at fixed power.
With varying excitation power it can be seen that the higher
intensity peak grows faster than the lower one and as a function
of the magnetic field each peak of the doublet is split into two
peaks with opposite circular polarization. Figure 4 shows that
the doublet structure is visible both in excitation spectra as
well as in reflectance spectra. There is a slight Stokes shift
between luminescence and excitation (0.3 meV) and both in
excitation as in reflectance the high energy peak is much more
pronounced. The experiments on spin relaxation were all done on

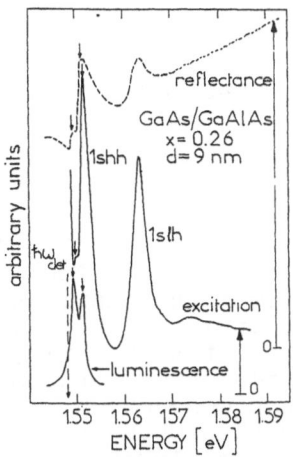

Fig. 4. Comparison of luminescence, luminescence
excitation and reflectance spectra at
$B = 0$ T, at power density $P = 1$ W/cm^2.
All the spectra show the double structure
of the heavy hole exciton ground state,
which is more pronounced in luminescence,
however.

this high energy peak and on its two components A and B when
there is a magnetic field. In any case, it was found that in
all cases with or without a magnetic field, the energy positions
of the peaks in the excitation spectra were always the same
independent of whether these spectra were taken recording the
intensity of A, B, C or D (see Fig. 3). The intensities were,
of course, different as was explained before. On the basis of
these results we conclude that: i) the observed fine structure
cannot be due to well width fluctuations since we do not observe
any variations in the position excited states; ii) since the
lowest peak is observed also in excitation and in reflection it
is improbable that it can be attributed to bound excitons; iii)
the evolution of power excludes the formation of biexcitons
because then the lowest peak (the biexciton) would be observed
at high intensities.

In high magnetic fields the luminescence consists of four
components A, B, C and D (Fig. 3). Using the selective electron
spin relaxation effect discussed before, we can attribute the
electron-spin character to each of the components. In Fig. 5 we
show, therefore, a part of the excitation spectrum recorded on
each of these peaks. By inspection of Fig. 5 we can then divide
the four components into two groups {A,D} and {B,C}. The
relative intensities of the excitation spectra within each group
are very similar, whereas those belonging to different groups
are very different. As before the differences cannot be
explained by different oscillator strengths because in a given
polarization light is absorbed in the same excited states.
Using the previous analysis of the selective relaxation we can
conclude that the thing which the states A and D have in common
and which distinguishes them from B and C is that {A,D} is
associated with one conduction band spin, whereas {B,C} is
associated with the opposite electron spin.

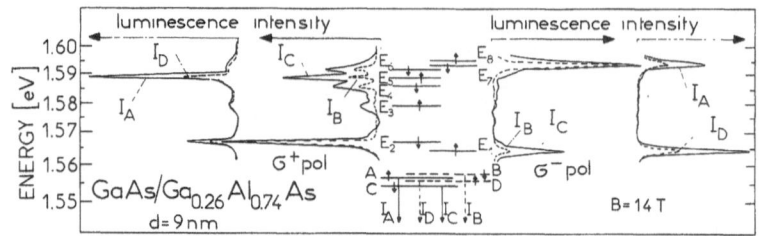

Fig. 5. Excitation spectra of the four components of the
ground state for both polarizations of the
exciting light at 14 T. Different amplitude
ratios of the excitation maxima reflect the
preferential relaxation between states which
involve the same conduction band spin
orientation (marked with the arrows). The
highest (B) and the lowest (C) component involve
the same conduction band spin orientation (↓),
similarly the two intermediate components (A and
D) show spin up character (↑). This distinction
cannot be deduced from the polarization rules
since A and C are σ^+ and B,D σ^- polarized.

In Fig. 6 the field dependence of A,B, C and D is shown.
Having assigned an electron spin to each of the peaks it can be
seen that the highest state B and the lowest state C have the
same spin orientation which is opposite to that of the two
intermediate sates A and D. This fact clearly shows that the
origin of the zero field splitting must be due to a magnetic
mechanism, since the magnetic order of the doublet A,B is
different from C,D. Previous work [15,21,22] has attributed
this zero field splitting to exchange interaction; i.e. the
fourfold degeneracy of the heavy hole ground state is lifted by
exchange interaction which leads to one doubly degenerate
electron-hole spin polarized and another electron-hole spin
unpolarized state at a slightly lower energy due to
exchange[21]. This mechanism would explain the present data
very well[22], but we can not prove that this is the only
possible explanation. In fact all theoretical estimates seem to
indicate that[21,23] the observed zero field splitting is a few

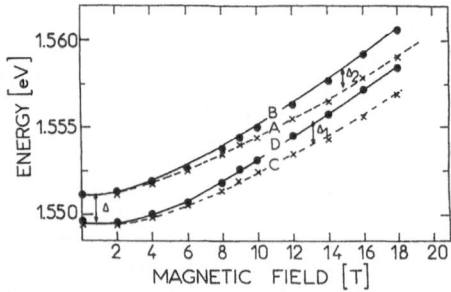

Fig. 6. Energy of the four components of the ground
state as a function of magnetic field for a
9 nm GaAs quantum well. Lines are a guide to
the eye.

orders of magnitude larger than would be expected[23] from the known values for exchange interaction in bulk GaAs. However, the experimental tool which we developed and which allows the determination of the electron spin character of each of the states does clearly show the magnetic origin of the splitting, which limits severely the domain of possible explanations.

CONCLUSIONS

In conclusion, we have shown experimentally that energy relaxation of photo-created carriers in quantum wells in quantizing magnetic fields takes place preferentially with conservation of the electron magnetic moment. We believe that these long spin relaxation times are consequence of the discrete nature of the spin-split Landau levels (zero density of states between the levels), which makes these states analogous to energy levels in atoms where similar effects are observed. In combination with excitation spectroscopy, these long spin relaxation lifetimes can be used as a tool to distinguish between states which have the same and states which have the opposite electron spin magnetic moment with respect to each of the two components of the spin-split ground state. We have used this tool to show that the mechanism underlying the fine structure of the heavy hole exciton must be of magnetic origin. Exchange interaction would be a plausible, although not the only, candidate. In any case the experimental proof that there must be a magnetic mechanism underlying the zero field splitting strongly limits possible explanations.

As a consequence of the specific relaxation, our results show that excitation spectra at high fields may not necessarily reflect the absorption intensity (oscillator strengths) of a transition. The proper oscillator strength can be obtained from the sum of the luminescence intensity from both spin-split components of the ground state. The strong tendency of conservation of the electron magnetic moment is shown to be sufficient to create a population inversion between the two spin split components of the ground state. We think that our experimental studies contribute to a better understanding of spin dependent properties of 2D systems and open the way to many interesting experiments such as for instance optically detected resonance (ESR and NMR) at high magnetic fields, the conception of a spin flip laser etc., etc. In this connection our work closes in on the recent renewed interest of the magnetic properties of 2D systems[24-27] and we hope it will stimulate further research on this subject.

Acknowledgment

We are grateful to H. Krath for excellent technical assistance and to Prof. P. Wyder for his interest in this work. Calculations have been supported by the CCVR, Palaiseau (France) and by the SISSA-CINECA joint project sponsored by the Italian Ministry of Education.

REFERENCES

1.   R.J. Elliot, **Phys. Rev. B**, 96:266 (1954); Y. Yafet, **in**: "Solid State Physics", F. Seitz and D.Turnbull, eds., Academic, New York, 14:1 (1963)

2. G.L. Bir, A.G. Aronov, G.E. Pikus, **Zh. Eksp. Teor. Fiz.**, 69:1382 (1975) [**Sov. Phys. JETP**, 42:705 (1976)]

3. M.I. D'yakonov and V.I. Perel, **Zh. Eksp. Teor. Fiz.**, 60:1954 (1971), [**Sov. Phys. Solid State**, 13:3023 (1972)]

4. "Optical Orientation, Modern Problems in Condensed Matter Sciences", Vol. 8, F. Meier and B.P. Zakharchenya, eds., North-Holland, Amsterdam, (1984)

5. K. Zerrouatti, F. Fabre, G. Bacquet, J. Frandon, G. Lampel, D. Paget, **Phys. Rev. B**, 37:1334 (1988)

6. G. Fishman, G. Lampel, **Phys. Rev. B**, 16:820 (1977)

7. R.C. Miller, D.A. Kleinman, W.A. Nordland, Jr., and A.C. Gossard, **Phys. Rev. B**, 22:863 (1980)

8. C. Weisbuch, R.C. Miller, R. Dingle, A.C. Gossard and W. Wiegmann, **Solid State Commun.**, 37:219 (1981)

9. W.A.J.A. van der Poel, A.L.G.J. Severens, H.W. van Kesteren and C.T. Foxon, **Superlatt. Microstr.**, 5:115 (1989)

10. See e.g.: Proceedings of Symposium "Alfred Kaster", **Ann. Phys. Fr.**,Vol. 10 (1985)

11. M. Potemski, J.C. Maan. A. Fasolino, F. Ancilotto, K. Ploog, G. Weimann, p.235 **in**: "Proc. 19th Intl. Conf. Phys. Semiconductors", Warsaw 1988, W. Zawadzki, ed., IPPAS, (1989); **Phys. Rev. Lett.**, 63:2409 (1989)

12. R.C. Miller, C.W. Tu, S.K. Sputz and R.F. Kopf, **Appl. Phys. Lett.**, 49:1245 (1986)

13. X. Liu, A. Petrou, B.D. McCombe, J. Ralston and G. Wicks, **Phys. Rev. B**, 38:8522 (1988)

14. R.C. Miller, D.A. Kleinman, A.C. Gossard and O. Munteanu, **Phys. Rev. B**, 25:6545 (1982); S. Charonneau, T. Steiner, M.L.W. Thewalt, E.S. Koteles, J.Y. Chi and B. Elman, **Phys. Rev. B**, 38:3583 (1988)

15. R. Bauer, D. Bimberg, J. Christen, D. Oertel, D. Mars, J.N. Miller, T. Fukunaga, H. Nakasima, p.525, **in**: "Proc. ICPS18, Stockholm", O. Engström,ed., World Scientific, Singapore, (1987)

16. W. Ossau, B. Jäkel, E. Bangert, G. Landwehr and G. Weimann, **Surf. Sci.**, 174:188 (1986)

17. G.E.W. Bauer and T. Ando, **Phys. Rev. B**, 37:3130 (1988)

18. S-R.E. Yang and L.J. Sham, **Phys. Rev. Lett.**, 58:2598 (1987)

19. J.C. Maan, G. Belle, A. Fasolino, M. Altarelli and K. Ploog, **Phys. Rev. B**, 30:2253 (1984); L. Vina, M. Potemski, J.C. Maan, G.E.W. Bauer, E.E. Mendez and W.I. Wang, **Superl. Microstr.**, 5:371 (1989)

20. F. Ancilotto, A. Fasolino and J.C. Maan, **Phys. Rev. B**, 38:1788 (1988)

21. Y. Chen, B. Gil, P. Lefebre and H. Mathieu, **Phys. Rev. B**, 37:6429 (1988)

22. M. Potemski, J.C. Maan, A. Fasolino, K. Ploog and G. Weimann, **Surface Sci.**, 229:151 (1990)

23. B.R. Salmassi and G.E.W. Bauer, **Phys. Rev. B**, 39:1970 (1988)

24. B.C. Cavenett and E.J. Pakulis, **Phys. Rev. B**, 32:8449 (1985)

25. M. Dobers, K. v.Klitzing and G. Weimann, **Phys. Rev. B**, 38:55453 (1988)

26. H.W. v.Kesteren, E.C. Cosmas, F.J.A.M. Greidanus, P. Dawson, K.J. Moore and C.T. Foxon, **Phys. Rev. Lett.**, 61:129 (1988)

27. M. Dobers, K. v.Klitzing, J. Schneider, G. Weimann and K. Ploog, **Phys. Rev. Lett.**, 61:1650 (1988)

SEMIMETAL-SEMICONDUCTOR TRANSITION IN II-VI COMPOUND TYPE III

SUPERLATTICES

Y. Guldner, J. Manassès, J.P. Vieren, M. Voos and
J.P. Faurie*

Laboratoire de Physique de la Matière Condensée de
l'Ecole Normale Supérieure, 24 rue Lhomond, 75231
Paris Ced.05 France
*University of Illinois, Chicago, Ill., 60680 USA

ABSTRACT

Magneto-optical experiments on n-type $Hg_{1-x}Zn_xTe$-CdTe
superlattices yield the determination of the conduction band
dispersion. Comparison with theoretical calculations
demonstrates the semimetallic character at low temperature of
the superlattices and supports a rather large value of the
valence band offset. Experiments performed in the temperature
range (2 - 200 K) reveal a semimetal-semiconductor transition
induced by the temperature in agreement with theoretical
calculations. Similar experiments performed on a $Hg_{1-x}Mn_xTe$-
CdTe superlattice show evidence of the exchange interaction in
this semimagnetic material.

1.   INTRODUCTION

II-VI superlattices (SL) formed with CdTe and a zero-gap
mercury based compound have been termed "Type III"
heterostructures because of the unique inverted band structure
of the zero-gap material. They can be either semiconducting or
semimetallic[1], depending on the layer thicknesses and we show
here that their band structure can be determined by far-infrared
magneto-absorption experiments.

We report cyclotron resonance measurements on high electron
mobility HgZnTe-CdTe SLs grown by molecular beam epitaxy in the
(100) direction. The SL electron cyclotron resonance is
measured as a function of θ, which is the angle between the
direction of the magnetic field and the SL axis. Comparison of
the results with band structure calculations clearly demonstrates the semimetallic nature of the SLs at low temperature
and support a large valence band offset value ($\geq$ 300 meV)
between $Hg_{1-x}Zn_xTe(x < 0.1)$ and CdTe. Experiments performed in
the temperature range (2 - 200 K) reveal a semimetal-semiconductor transition induced by the temperature in agreement
with theoretical band structure calculations.

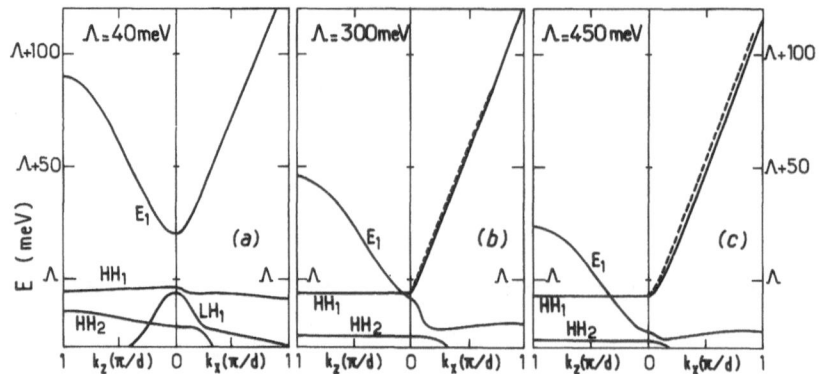

Fig. 1. Calculated band structure of S_1 for T = 2 K in
the plane of the layers (k_x) and along the
growth axis (k_z) for Λ = 40 meV (a), Λ = 300 meV
(b) and Λ = 450 meV (c). The dashed lines in
Figures 1b and 1c represent the in-plane
dispersion for $k_z = k_c$.

We also discuss far-infrared magneto-absorption experiments
performed on a $Hg_{0.96}Mn_{0.04}Te$-CdTe SL over a temperature range
from 1.5 to 10 K. The temperature dependence of the electron
cyclotron resonance shows evidence of the exchange interaction
between the localized Mn d-electrons and the conduction band
electrons in the semimagnetic SL. The results are interpreted
from superlattice band structure calculations which include the
magnetization effects.

2.   SEMIMETAL-SEMICONDUCTOR TRANSITION IN HgZnTe-CdTe
SUPERLATTICES

We present here results obtained on two SLs (S_1 and S_2)
grown on a (100) GaAs substrate with a 2 μm CdTe buffer[2] and
consisting of 100 periods of $Hg_{1-x}Zn_xTe$-CdTe, the CdTe layers
containing 15% HgTe[3]. For each sample, the layer thicknesses
and the alloy Zn composition x, as well as the electron mobility
μ and concentration at 25 K are listed in Table I. The samples
were prepared intentionally with thin CdTe barriers to study the
carrier transport along the SL axis and were n type in the
temperature range investigated (2 - 300 K), with Hall mobilities
in excess of 2 x 10^5 cm^2/Vs at low temperature.

We first discuss the SL band structure and Landau levels.
All the calculations are done by using a 6 x 6 envelope function
Hamiltonian[4], taking for CdTe the band parameters give in Ref.
4. For HgZnTe, we obtained the Γ_6-Γ_8 energy separation by using
x = 0.12 for the semimetal-semiconductor transition at 2 K[5],
the other parameters being identical to those of HgTe. Figure 1
shows the calculated band structure of S_1 in the plane of the
layers (k_x) and along the SL growth axis (k_z) using T = 2 K.
Reported experimental and theoretical values of the valence band
offset Λ between CdTe and HgTe range between 0 and 500
meV[4,6,7] and, in the case of $Hg_{1-x}Zn_xTe$-CdTe SL with low Zn
content, we show the calculations for Λ = 40 meV(a), Λ = 300
meV(b) and Λ = 450 meV(c). The zero of energy is taken at the
CdTe valence band edge. E_1, LH_1 (HH_1, HH_2), respectively,

Table 1. Characteristics of the HgZnTe-CdTe superlattices used in this work. d_1 and d_2 are the HgZnTe and CdTe layer thicknesses, respectively, x is the Zn composition; n and μ are measured at 25 K.

Sample	d_1 (Å)	d_2 (Å)	x	μ(cm²/Vs)	n(cm⁻³)
S_1	105	20	0.053	2.1×10^5	3.7×10^{15}
S_2	80	12	0.053	2.1×10^5	2.6×10^{15}

denote the first light particles (heavy holes) bands for the motion along the growth axis. For Λ = 40 meV (Fig. 1a), the SL is a semiconductor, E_1 being the conduction band and HH_1 the highest valence band. For large valence band offset, the E_1-HH_1 separation decreases until finally E_1 and HH_1 meet and then cross in the k_z direction for $k_z = k_c$. For Λ = 300 meV (Fig. 1b), HH_1 is the conduction band and the superlattice is semimetallic with a nearly zero gap and $k_c = 0.1\pi/d$. For Λ = 450 meV (Fig. 1c), E_1 lies about 16 meV below HH_1 and $k_c = 0.3\pi/d$. The most significant change in the band structure when increasing Λ is the k_z dispersion of the SL conduction band. For Λ = 40 meV, the band is nearly isotropic around k = 0. On the contrary, for Λ = 300 meV and 450 meV, it is dispersionless for $0 < k_z < k_c$ corresponding to a strong anisotropy. In any case the in-plane mass m_x is extremely small near $k_x = 0$ ($m_x \approx 2.10^{-3} m_0$), as a result of the small HH_1-E_1 separation. The dashed line in Figs. 1b and 1c shows the in-plane dispersion relation calculated for $k_z = k_c$. One can see that m_x depends on the growth-direction wavevector because $m_x \propto |HH_1-E_1|$ which is a function of k_z. From Fig. 1, it is clear that the measurement of the conduction band anisotropy will provide the determination of the nature (semiconductor or semimetallic) of the SL. Similar calculations done for S_2, show that S_2 undergoes a semimetal-semiconductor transition for $\Lambda \sim 250$ meV at 2 K.

We have performed[9] magnetotransmission experiments at 2 K, using a molecular gas laser (λ = 41 - 255 μm). The transmission signal was measured at fixed infrared photon energies E, while B could be varied up to 12 T. Figure 2a shows transmission spectra of S_1 at λ = 118 μm for θ = 0 (Faraday geometry), 45° and 90° (Voigt geometry). For θ = 0, two well-developed resonance lines occur at low magnetic field (< 1 T). These two lines are shifted towards higher magnetic fields when θ is increased. The energy positions of the transmission minima as a function of B are shown in the Fig. 2c. AT θ = 0, the two observed transitions extrapolate to E ≈ 0 at B = 0 and are therefore attributed to intra-conduction band transitions, considering the n-type nature of the sample. To interpret these results, we have calculated[4] the SL Landau levels energy at θ = 0 which are shown in Fig. 3a for Λ = 360 meV. The solid line in Fig. 2c is the calcuated energy of the -1 → 0 transition. The agreement with the experimental points is quite satisfying, so that the low magnetic field transition can be reasonably attributed to the SL electronic cyclotron resonance. The dashed line in Fig. 2c corresponds to the calculated energy of the 0 → 1' transition, which is also allowed in Faraday geometry. The fact that both -1 → 0 and 0 → 1' transitions are

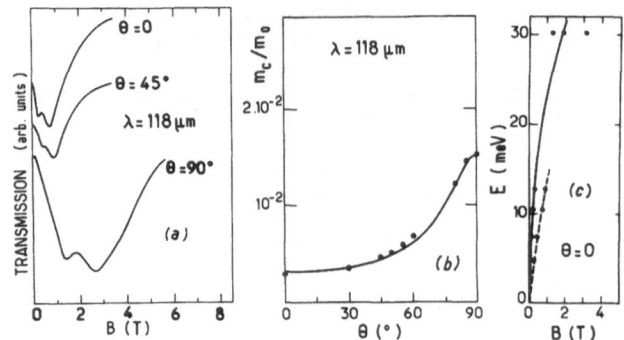

Fig. 2. (a) Typical transmission spectra of S_1 for λ = 118 μm and T = 2 K. (b) Cyclotron mass versus θ for λ = 118 μm and T = 2 K. The dots represent the experimental points and the solid line the calculated values in the ellipsoidal approximation. (c) Energy of the observed transmission minima versus B (dots). The solid and dashed lines are the calculated SL cyclotron resonances.

observed at the same magnetic field arises from the existence of an electron accumulation layer near the interface between the SL and the CdTe buffer layer. Such an electron accumulation results from charge transfer from CdTe towards the SL and is usually observed in type III heterostructures grown in the (100) direction in similar conditions[9], for instance in the HgCdTe-CdTe heterojunctions[10]. As a consequence, the electron concentration in the SL is not homogeneous, being higher near the interface where the $0 \rightarrow 1'$ transition can occur.

In the Voigt geometry (θ = 90°), the cyclotron resonance line takes place around 1 T for λ = 118 μm corresponding to a cyclotron orbit radius of 250 Å, so that electrons are forced to tunnel through several interfaces. In Fig. 2b, we have plotted the measured cyclotron mass m_c versus θ for λ = 118 μm. We assume an ellipsoidal dispersion relation for the conduction band near k = 0 so that the cyclotron mass is given[8] at low

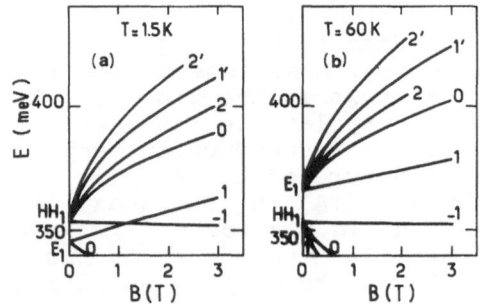

Fig. 3. Calculated Landau level energies in S_1 using Λ = 360 meV and θ = 0 in the (a) semimetallic and (b) semiconducting configurations.

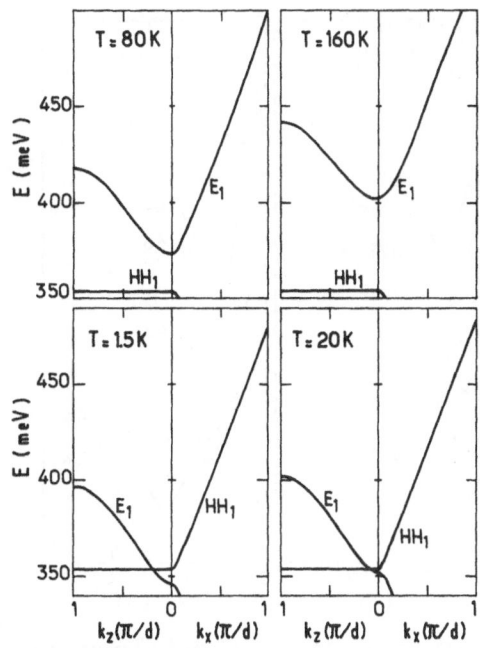

Fig. 4. Calculated band structure of S_1 for
various temperatures and $\Lambda = 360$ meV.

magnetic field by $m_c^2 = (m_x^2 m_z)/(m_z \cos^2\theta + m_x \sin^2\theta)$ where m_x is
the in-plane mass at the photon energy and m_z the mass along the
growth axis. The solid line in Fig. 2b, which is the calculated
cyclotron mass using $m_z = 0.065 m_0$, is in very good agreement
with experimental points. The measured mass anisotropy ratio
m_z/m_x is ~ 25 for $\lambda = 118$ μm. Comparison with band structure
calculations shown in Fig. 1 clearly supports the semimetallic
nature of the superlattice. Indeed, calculations done in the
semiconductor configuration for $\Lambda = 40$ meV (Fig. 1a) lead to an
anisotropy ratio of ≈ 1 at $\lambda = 118$ μm, while the anisotropy is
found to be ≥ 10 for $\Lambda \geq 300$ meV. Similar experiments performed
on sample S_2 lead to a mass anisotropy ratio ~ 6 at $\lambda = 118$ μm,
again consistent with the calculated band structure obtained for
$\Lambda \geq 300$ meV. We have not observed any plasma shift of the
cyclotron resonance for $\theta \neq 0$ in the photon energy range used in
our experiments ($\hbar\omega \geq 5$ meV)[9].

Finally, these results demonstrate that S_1 and S_2 are
semimetallic at 2 K and that the valence band offset is large,
in agreement with XPS measurements[7].

We now discuss the temperature dependence of the SL band
structure and cyclotron resonance. Figure 4 presents the calculated band structure of S_1 for various temperatures in the range
(1.5 -160 K) using $\Lambda = 360$ meV and a linear temperature varia-
tion of 0.57 meV K^{-1} for the Γ_8 - Γ_6 energy separation in $Hg_{1-x}Zn_xTe$ with low Zn content[11]. S_1 is semimetallic up to
$T = 25$ K and becomes semiconductor for $T \geq 25$ K, the SL band gap
(E_1 - HH_1) increasing with the temperature. From Fig. 4 it is
clear that when T is raised, a lightening of m_z, the mass along
the SL axis, occurs around 25 K and that the in-plane mass m_x
increases for $T > 25$ K, resulting from the increasing E_1 - HH_1
separation.

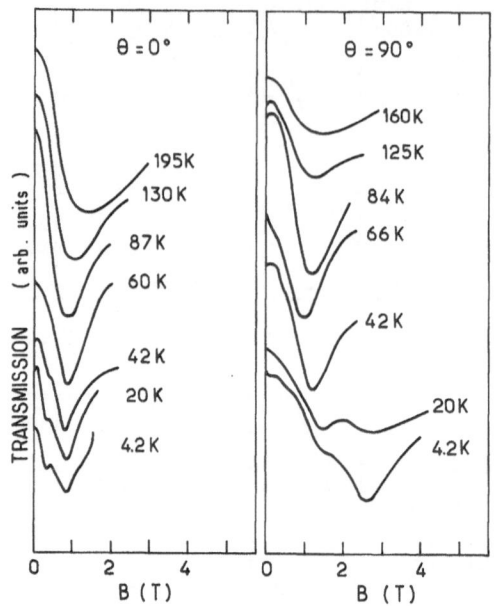

Fig. 5. Transmission spectra obtained on S_1 at $\lambda =$ 118 μm for several temperatures.

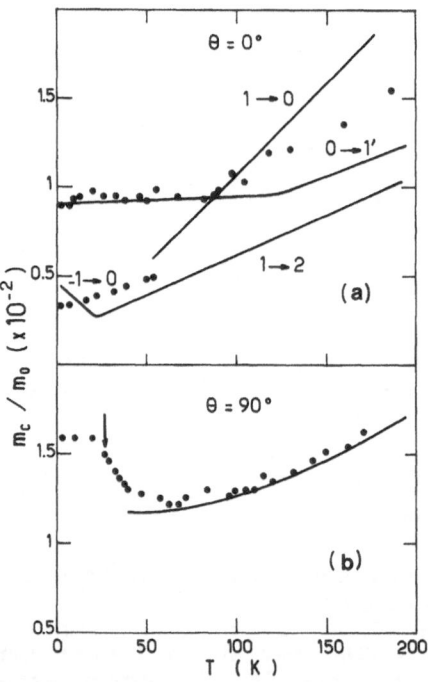

Fig. 6. Cyclotron masses corresponding to the different transmission minima observed at $\lambda = 118$ μm in the (a) Faraday and (b) Voigt geometry (dots). The solid lines are the calculated masses from the Landau level energies at $\theta = 0$ and from the band structure at $\theta = 90°$, as described in the text.

We have performed FIR magneto-optical experiments between
1.5 K and 200 K for $\theta = 0$ and $\theta = 90°$ and typical transmission
spectra obtained for $\lambda = 118$ μm on S_1 are shown in Fig. 5. For
$\theta = 0$, two resonance lines appear at low temperature and shift
towards higher magnetic fields with increasing temperature, the
lower field minimum vanishing progressively. At $\theta = 90°$, only
the lower field transition observed at low temperature has an
energy which extrapolates to E = 0 at B = 0 and is therefore
attributed to the SL electron cyclotron resonance. The transmission minimum first shifts to low fields, when T is raised up
to 60 K, and then moves to higher fields for further temperature
increase. The cyclotron masses corresponding to the different
transmission minima observed for $\theta = 0$ and $\theta = 90°$ at $\lambda = 118$ μm
are plotted on Fig. 6 as a function of the temperature. Note
that a drop of ~ 25% is observed around 25 K in the curve m_c(T)
at $\theta = 90°$ shown on Fig. 6b, in agreement with the calculated
semimetal-semiconductor transition for Λ ~ 350 meV which is
indicated by the arrow. The solid lines in Fig. 6a are the
calculated cyclotron masses corresponding to the main intraband
transitions using the Landau level calculations shown in Fig. 3.
At low temperature, the two resonances observed for $\theta = 0$ are
attributed to the $-1 \rightarrow 0$ and $0 \rightarrow 1'$ transitions which are the
two ground interband transitions in the semimetallic configur-
ation (Fig. 3a). For higher temperatures, the two minima
correspond to the $1 \rightarrow 2$ and $0 \rightarrow 1'$ transitions, as expected in
the semiconducting configuration (Fig. 3b). For T $\geq$ 100 K, the
thermal energy kT becomes comparable to the photon energy and
one observes only a single broad minimum which corresponds to
the mixing of the different intraband transitions. The
agreement between the calculated masses and the experiments
obtained for $\Lambda = 360$ meV is quite satisfying. To interpret the
data at $\theta = 90°$, we assume an ellipsoidal dispersion relation
for the conduction band near k = 0, so that the cyclotron mass
is given by $m_c = (m_x m_z)^{1/2}$. The solid line on Fig. 6b presents
the theoretical variation m_c(T) in the semiconducting situation.
In these calculations, the mass m_z at the photon energy in the
conduction band is obtained for each temperature from band
structure calculations shown in Fig. 4, while the mass m_x is
simply deduced from the cyclotron mass at $\theta = 0$, i.e. from the
energy of the $0 \rightarrow 1'$ transition which corresponds to the
dominant minimum observed in the spectra. Finally the drop
around 25 K as well as the monotonic increase of m_c observed for
T > 50 K, are well interpreted by the band structure calculations using $\Lambda = 360$ meV. Similar experiments were done on S_2
and a drop in the curve m_c(T) at = 90° is observed around 40 K,
also in agreement with band structure calculations using Λ ~ 350
meV.

In conclusion, our data provide an accurate determination
of the SL conduction band dispersion and the observed
temperature dependences of the effective masses m_x and m_z are
explained by the temperature-induced semimetal-semiconductor
transition occuring in these type III heterostructures.

3.   EXCHANGE INTERACTION IN HgMnTe-CdTe SUPERLATTICES

Much work has been directed toward semimagnetic SLs in
search of low dimensionality magnetic effects[12-15]. The type
III system $Hg_{1-x}Mn_xTe$-CdTe is unique because the two-dimensional
electron confinement is in the semimagnetic layers, and this

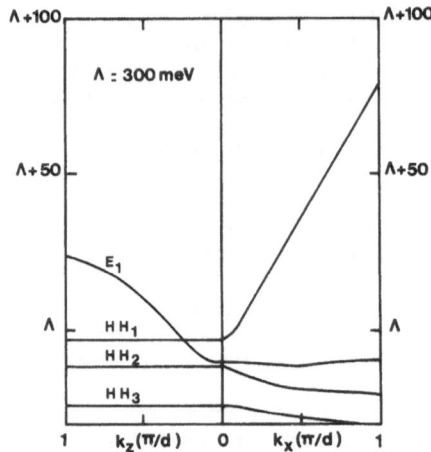

Fig. 7. Calculated band structure along the k_x and
k_z of the $Hg_{0.96}Mn_{0.04}Te$-CdTe superlattice
for T = 2 K.

system should be particularly interesting to probe the exchange
interaction between the localized Mn magnetic moments and the SL
conduction band electrons. In the HgMnTe-CdTe SLs, a strong
effect of the exchange interaction on the Landau level energy is
expected which can be evidenced by the temperature dependence of
the electron cyclotron resonance.

    We report here results obtained on a $Hg_{0.96}Mn_{0.04}Te$-CdTe SL
grown by molecular beam epitaxy on a (100) GaAs substrate with a
2 μm CdTe (111) buffer layer[16]. The SL consists of 100
periods of 168 Å thick $Hg_{0.96}Mn_{0.04}Te$ layers interspaced by 22 Å
thick CdTe barriers and is n type with electron density n = 6 x
$10^{16} cm^{-3}$ and mobility μ = 2.7 x 10^4 cm^2/Vs at 25 K.

    The SL band structure and Landau level calculations are
done by using a 6 x 6 envelope function Hamiltonian[4,17]. The
$\Gamma_6 - \Gamma_8$ energy separation is taken to be - 130 meV in
$Hg_{0.96}Mn_{0.04}Te$ at low temperature. For the $Hg_{0.96}Mn_{0.04}Te$ layers,
the model is modified to take into account the exchange

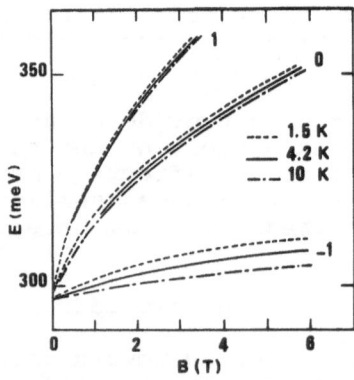

Fig. 8. Calculated conduction Landau levels for
different temperatures using Λ = 300 meV.

Table 2. Magnetization A given in meV/T at various magnetic fields and temperatures for bulk $Hg_{0.96}Mn_{0.04}Te$.

T(K) \ B(T)	0.5	3	5	7	10
1.5	-2.3	-1.3	-1.0	-0.8	-0.6
4.2	-1.4	-1.1	-0.9	-0.7	-0.6
10.0	-0.5	-0.5	-0.45	-0.45	-0.4

interactions between localized d electrons bound to the Mn^{2+} ions and the Γ_6 and Γ_8 s- and p-band electrons. The s-d and p-d interactions are introduced in the molecular field approximation through two additional parameters, r and A, where $r = \alpha/\beta$ is the ratio between the Γ_6 and Γ_8 exchange integrals, α and β, respectively[18,19]. Here A is the normalized magnetization defined by $A = (1/6)\beta x N_0 <S_z>$ where N_0 is the number of unit cells per unit volume of the crystal, x is the Mn composition (x = 0.04), and $<S_z>$ is the thermal average of the spin operator along the direction of the applied magnetic field B. Extensive magneto-optical data[18,19] and direct magnetization measurements [20,21], obtained on bulk alloys with dilute Mn concentration, have shown that $<S_z>$ is well described by a modified spin - 5/2 Brillouin function, where the temperature T is replaced with an empirical effective temperature $(T + T_0)$. T_0 is found[21] to be ~ 5 K for x = 0.04. The exchange integral[19] $N_0\beta$ ~ 1 eV and r ~ -1. The magnitude of A which satisfies the modified Brillouin function for x = 0.04 is given in Table II for various magnetic fields and temperatures between 1.5 and 10 K. The resulting band structure along k_x and k_z, calculated for zero magnetic field using a valence band offset $\Lambda = 300$ meV, is shown in Fig. 7 and one can note that the SL is semimetallic. Figure 8 shows the SL conduction Landau levels calculated for T = 1.5, 4.2 and 10 K using the bulk $Hg_{0.96}Mn_{0.04}Te$ magnetization. Note that as T decreases, the magnetization |A| is increased (see Table II) and all the conduction levels increase in energy.

FIR magneto-absorption experiments were done with temperature range 1.5 - 10 K. Figure 9 shows transmission spectra obtained at $\lambda = 41$ μm for several temperatures. The main absorption minimum corresponds to the $-1 \rightarrow 0$ transition (Fig. 8) which is the ground electron cyclotron resonance. For each temperature, the calculated field position of the line is indicated by the arrows on Fig. 9 and one can see that the agreement between experiments and Landau level calculations is quite good. The most striking point is that the $-1 \rightarrow 0$ transition shifts to lower B with increasing temperature, the variation being ΔB ~ -0.6T between 1.5 and 10 K. There are two contributions to this temperature dependent shift: the temperature dependence of the $Hg_{0.96}Mn_{0.04}Te$ energy gap and the temperature dependent Mn magnetization. In bulk HgTe, as T increases from 1.5 K to 10 K, the magnitude of the energy gap diminishes by less than 3 meV. The effect in $Hg_{0.96}Mn_{0.04}Te$ is expected to be similar. Such a small shift in energy gap is calculated to have only a slight effect on the cyclotron resonance transition, at most - 0.05 T. The observed shift of

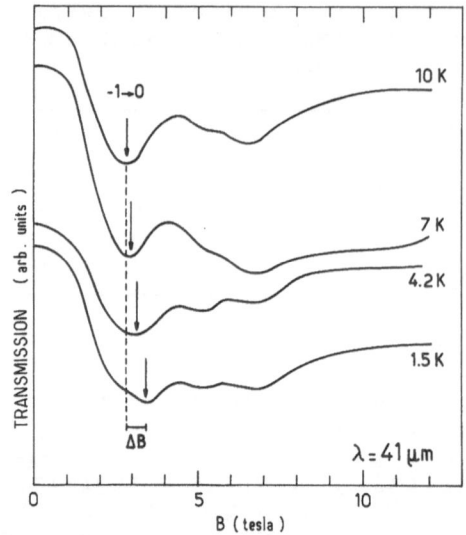

Fig. 9. Far-infrared magnetotransmission spectra
at $\lambda = 41$ μm for temperatures from 1.5 to
10 K. The arrows correspond to the calculated cyclotron resonance position. The
marker above ΔB indicates the magnitude of
the calculated shift in the -1 → 0 cyclotron resonance transition.

the -1 → 0 transition, is an order of magnitude larger than
could be due to the temperature dependence of the energy gap
alone and the Mn magnetization can account for the enhanced
temperature dependence of the cyclotron resonance line. As T
increases, the Mn magnetization decreases dramatically at lower
magnetic fields (see Table II). This has a direct effect on the
Landau levels in the superlattice as shown in Fig. 8. In
particular, the N = -1 level is depressed more than the N = 0
level, increasing the cyclotron resonance energy as temperature
is increased. In bulk semimagnetic semiconductors, the exchange
interaction has been observed by magneto-absorption primarily in
the spin and combined (cyclotron plus spin) resonances. The
observation of the exchange interaction in cyclotron resonance
is a result of the increased band mixing in a superlattice[17].

To summarize the results, we observe cyclotron resonance in
a $Hg_{0.96}Mn_{0.04}Te$-CdTe superlattice which is in good agreement
with band structure calculations in the envelope function
approximation using a valence band offset $\Lambda \sim 300$ meV. The
temperature dependence of cyclotron resonance transitions
evidences the effect of the exchange interaction in the
semimagnetic semiconductor superlattice. The magnetization
observed in the superlattice is consistent with the bulk
$Hg_{0.96}Mn_{0.04}Te$ magnetization, due to the thick $Hg_{0.96}Mn_{0.04}Te$
layers and thin CdTe barriers in our superlattice.

Acknowledgment

This work was partly supported by the NATO research grant
No. 9.13/890519.

REFERENCES

1. N.F. Johnson, P.M. Hui and H. Ehrenreich, **Phys. Rev. Lett.**, 61:1993 (1988)
2. X. Chu, S. Sivananthan and J.P. Faurie, **Superlattices and Microstructures**, 4:175 (1988)
3. J. Reno, R. Sporken, Y.J. Kim, C.Hsu and J.P. Faurie, **Appl. Phys. Lett.**, 51:1545 (1987)
4. J.M. Berroir, Y. Guldner, J.P. Vieren, M. Voos and J.P. Faurie, **Phys. Rev. B**, 34:891 (1986)
5. B. Toulouse, R. Granger, S. Rolland and R. Triboulet, **J. Phys.**, 48:247 (1987)
6. Y. Guldner, G. Bastard, J.P. Vieren, M. Voos, J.P. Faurie and A. Million, **Phys. Rev. Lett.**, 51:907 (1983)
7. S.P. Kowalczyk, J.T. Cheung, E.A. Kraut and R.W. Grant, **Phys. Rev. Lett.**, 56:1605 (1986); Tran Minh Duc, C. Hsu and J.P. Faurie, ibid. 58:1127 (1987)
8. N.W. Ashcroft and N.D. Mermin, "Solid State Physics", Holt, Rinehart and Winston, New York (1976) p. 571
9. J.M. Berroir, Y. Guldner, J.P. Vieren, M. Voos, X. Chu and J.P. Faurie, **Phys. Rev. Lett.**, 62:2024 (1989)
10. Y. Guldner, G.S. Boebinger, J.P. Vieren, M. Voos and J.P. Faurie, **Phys. Rev. B**, 36:2958 (1987)
11. M.H. Weiler, **in**: Semiconductors and Semimetals, 16:119, R.K. Willardson and A.C. Beer, eds., Academic, New York (1981)
12. R.N. Bicknell, R.W. Yanka, N.C. Giles-Taylor, E.L. Buckland and J.F. Schetzina, **Appl. Phys. Lett.**, 45:92 (1984)
13. L.A. Kolodziejski, T.C. Bonsett, R.L. Gunshor, S. Datta, R.B. Bylsma, W.M. Becker and N. Otsuka, **Appl. Phys. Lett.**, 45:440 (1984)
14. K.A. Harris, S. Hwang, Y. Lansari, J.W. Cook Jr., and J.F. Schetzina, **Appl. Phys. Lett.**, 49:713 (1986)
15. D.D. Awschalom, J.M. Hong, L.L. Chang and G. Grinstein, **Phys. Rev. Lett.**, 59:1733 (1987)
16. X. Chu, S. Sivananthan and J.P. Faurie, **Appl. Phys. Lett.**, 50:597 (1987)
17. G.S. Boebinger, Y. Guldner, J.M. Berroir, M. Voos, J.P. Vieren and J.P. Faurie, **Phys. Rev. B**, 36:7930 (1987)
18. G. Bastard, C. Rigaux, Y. Guldner, J. Mycielski and A. Mycielski, **J. Phys. (Paris)**, 39:87 (1978)
19. G. Bastard, C. Rigaux, Y. Guldner, A. Mycielski, J.K. Furdyna and D.P. Mullin, **Phys. Rev. B**, 24:1961 (1981)
20. W. Dobrowolski, M. von Ortenberg, A.M. Sandauer, R.R. Galazka, A. Mycielski and R. Pauthenet, **in**: "Physics of Narrow Gap Semiconductors", 152:302 of Lecture Notes in Physics, E. Gornik, ed., Springer, Heidelberg, (1982)
21. J.R. Anderson, M. Gorska, L.J. Azevedo and E.L. Venturini, **Phys. Rev. B**, 33:4706 (1986)

OPTICAL PROPERTIES OF SEMICONDUCTOR SYSTEMS OF LOW

DIMENSIONALITY

J.M. Worlock

Bellcore, Red Bank
New Jersey, USA

ABSTRACT

This is a review of the techniques employed to produce
interesting semiconductor structures of low dimensionality. In
this context, interesting means having demonstrable *quantum
confinement* of free excitons. We begin with the epitaxial
growth of simple two-dimensional *quantum wells*. The approach to
even lower dimensionality, seeking optical *quantum wires and
quantum dots*, involves a variety of techniques, including post-
growth modification of quantum wells and direct growth on
structured or textured surfaces. The study of excitons of low
dimensionality feeds back into the effort to understand the
fundamental processes of epitaxial growth and post-growth
patterning. Much of this work is driven by the desire to insert
quantum wells, quantum wires and quantum dots into useful
semiconductor technology. Illustrations are chosen where
possible from ongoing research at Bellcore.

1.   INTRODUCTION

Ever since discovering a little book called "Flatland"[1]
early on in my education, I have been fascinated with the
possibilities inherent in universes of both lower and higher
dimensionality than our own apparently 3D world. In recent
years the world of semiconductor physics recognized the
fundamental differences in phenomena taking place in restricted
geometries: two salient examples of unexpected behavior are the
fractional quantum Hall effect[2], and fractional statistics[3].
I am extremely pleased to be involved professionally in the
current communal attempt to explore the new possibilities
afforded by semiconductor microstructures of low dimensionality.

Our test probe is the Wannier exciton, the semiconductor
physicist's hydrogen atom. Creation and annihilation of the
exciton are accomplished by the absorption and emission of
photons in the energy range near the semiconductor's fundamental
energy gap, so that spectroscopy in this region can be used to
study the exciton's circumstances. Furthermore, we know already

a lot about the effects of dimensionality on the hydrogenic spectrum: if in 3D the binding energy is E_b (one Rydberg), then in 2D it is $4E_b$, and in 1D it becomes indefinitely large. Magnetic and electric response of hydrogen is also extremely well documented.

In order to limit the scope of this review to areas in which I have some personal knowledge, I shall not deal with the very interesting structures made by precipitating colloids of semiconductors in transparent matrices[4], though they have been studied widely, and their nonlinear optical properties demonstrated. Likewise, I shall avoid mention of those interesting quantum wires and dots designed to confine electrons or holes alone, leaving that task to others at this Workshop. Detlev Heitmann has further simplified my task by giving an excellent account of the accomplishments of his Stuttgart group in making quantum wires by patterned etching of high quality quantum well heterostructures, and their subsequent study by optical and magneto-optical techniques[5]. For a careful, timely review of "Optical Properties of III-V Semiconductor Quantum Wires and Dots", I recommend a review with that title, by Kathleen Kash, in Journal of Luminescence[6]. After a quick review of some properties of 3D and 2D excitons, I shall devote my remarks to several alternative techniques for making wires and dots which are currently being strongly pursued at Bellcore.

2. THREE-DIMENSIONAL EXCITONS

An excellent picture of the classical 3D Wannier exciton in GaAs was given to us in 1962 by Sturge[7]. Figure 1 displays the absorption edge of GaAs at several temperatures, showing a well developed exciton absorption peak at low temperatures broadened at room temperature, almost beyond recognition, by ionization lifetime. Absorption at the peak corresponds to creation of an exciton in the 1s state: higher s-states also participate, but are blended into the continuum and are not resolved. Because of the lightness of the effective masses and because the dielectric constant of GaAs weakens the electron-hole attraction, the binding energy of the pair is only ~ 5 meV. The weak binding is accompanied by a large Bohr radius on the order of 10 nm, so the exciton averages nicely over many unit cells and can be comprehended in an effective mass approach. All the III-V semiconductors have similar Wannier excitons.

By 1974, molecular beam epitaxy was sufficiently under control[8] to give us semiconductor heterostructures with "quasi" 2D excitons, as depicted in Fig. 2. This was an historically important figure, since it illustrated the power of epitaxial growth and brought the square-well potential (heretofore a textbook exercise) into the laboratory. For this experiment planar layers (thickness L_z) of GaAs were grown epitaxially embedded in the alloy AlGaAs which has a somewhat larger band gap. As a result both the hole and the electron are confined in a thin GaAs layer, and can orbit around each other only within that thin layer. The confinement is demonstrated by the offset of the absorption peak to higher energies in the thinner layers. Notice, too, that confinement has split the degeneracy of the heavy and light holes so that for each particle-in-a-box state (n = 1,2,3) there are two 2D excitons ("quasi" has by now passed from common usage). Spectra similar

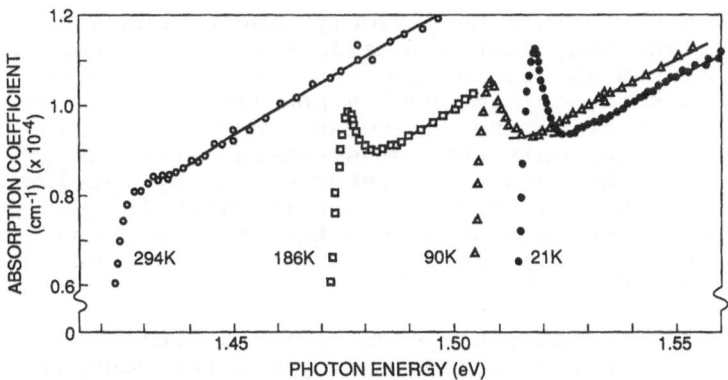

Fig. 1. Three-dimensional Wannier excitons in GaAs as
evidenced in absorption spectra near the fundamental band edge. As the temperature is reduced
the band edge shifts to higher energy, and the
exciton absorption peak becomes more pronounced.
After M. Sturge, Ref. 7.

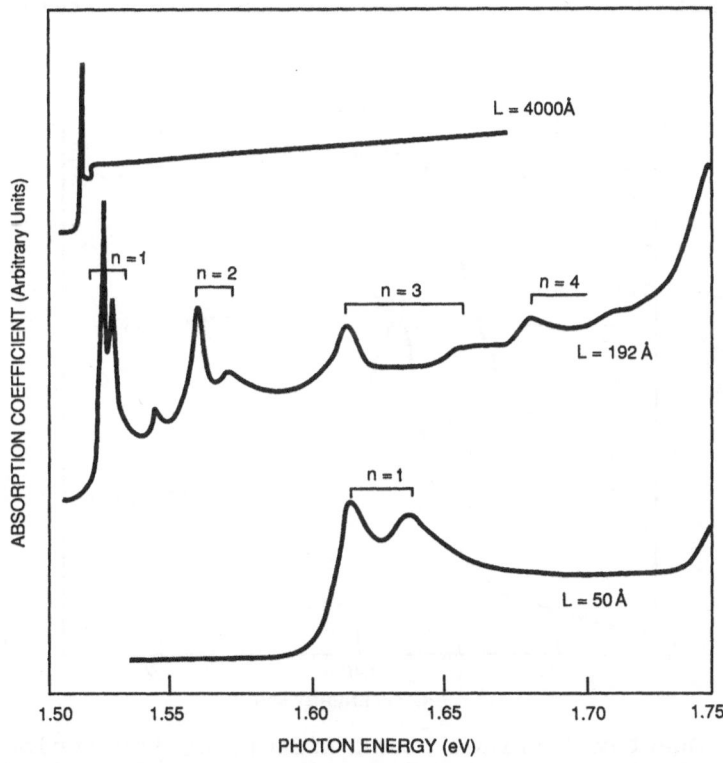

Fig. 2. Two-dimensional Wannier excitons in quantum
wells of GaAs embedded in AlGaAs, at low temperature. Easily observable are the splittings
into pairs (heavy and light hole excitons) and
the systematic shift to higher energies in the
thinner wells. The number "n" denotes the
transverse quantum state or subband to which the
exciton belongs. After A.C. Gossard, Ref. 34.

to those shown have since been widely studied and analyzed in
great detail, and many subtle details are now understood[9].
The inverse of the absorption process, or recombination
radiation (luminescence) is equally popular, but has the
disadvantage that at low temperatures, where the luminescent
efficiency is large, only the lowest-energy excitons are
evident. One of the remaining mysteries is why intrinsic
exciton recombination emission is so efficient in quantum
wells[10]: the corresponding 3D excitons have a strong tendency
to become trapped at impurities and recombine either as bound
excitons or nonradiatively.

    Exciton spectroscopy has been useful as well as
fascinating. Since the exciton energy depends sensitively on
the thickness of the quantum well, fluctuations in thickness can
be estimated from the (inhomogeneous) linewidth, as shown
originally by Weisbuch et al.[11]. This simple spectroscopic
technique has been used by virtually every grower of quantum
wells as a diagnostic: narrow linewidth is good and broad
linewidth is bad, and the measurement is an easy one. Standard
growth techniques are now expected to produce thickness
fluctuations on the order of one monolayer or less. The samples
used in the experiments described by Heitmann[5] are of
especially high quality judged by the linewidth criterion.

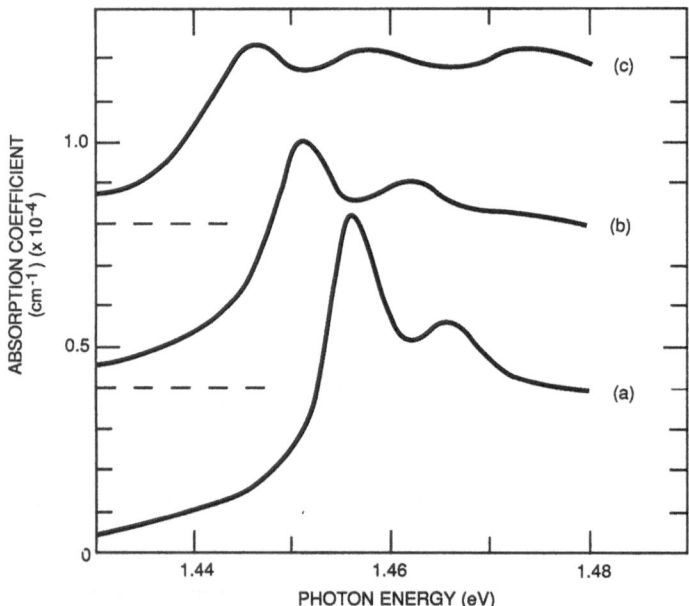

Fig. 3. Quantum Confined Stark Effect, or the tuning of
        the two-dimensional exciton by an electric field
        perpendicular to the quantum well. Shown are
        absorption spectra taken at three different
        field strangths: (a) 1×10^4 V/cm; (b) ~ 4.7 x
        10^4 V/cm; and (c) ~ 7.3 x 10^4 V/cm. Because of
        the confinement, the exciton does not become
        unstable in the electric field, but tunes
        through several linewidths. After Miller, et
        al. Ref. 13.

As suggested in the introduction, we can expect stronger binding as the dimensionality decreases. Stronger binding implies better electron-hole overlap and hence excitons with larger oscillator strength as well as more resistant to thermal ionization. As a result, many workers have come to regard quantum well excitons as candidates for room temperature device applications[12]. Both electro-optical and nonlinear optical-optical effects have been demonstrated. The quantum confined Stark effect[13], illustrated in Fig. 3, is an interesting example. Because of the one-dimensional confinement, an electric field in that direction will not ionize the exciton, but rather cause its absorption to shift to lower energy, as depicted in the sequence from zero field (a) to nearly 10^7 V/m(c). The modulation of transmission at 1.445 eV is evident, and probably useful. Finally, although it is not directly related to excitonic properties, it is important to note that the use of quantum wells in semiconductor lasers is widespread. The confinement-enforced overlap of injected electrons and holes and the 2D densities of states both aid in establishing the requisite gain.

3.   THE YEARNING FOR LOWER DIMENSIONALITY

Despite the excitement, or maybe because of it, one could sense a restlessness in the early 80's, a yearning for something beyond 2D excitons. Epitaxial growth and fine-scale hetero-structures had revitalized semiconductor physics, as suggested by Benoit a la Guillaume in his concluding remarks at the 16th International Conference on Physics of Semiconductors[14]. But as eating never satisfies an appetite for long, soon the hunger for novelty returned. One consequence was that when Bellcore was founded in 1984 we were determined to be in the vanguard of those pushing toward lower dimensionality. One hardly needed to be explicit about the technological benefits: it was necessary to *go there* - to make structures of lower dimensionality - to see what could be done. Little of what happens in 2D had been predicted in advance. Could 1D be less rich in rewards?

4.   ONE- AND ZERO-DIMENSIONAL EXCITONS

4.1 Patterned Etching

The straightfoward way of reducing dimensionality, as we percieved it at that time, was to start with a 2D quantum well and etch away material in fine scale patterns so as to leave narrow *quantum wires* and/or small *quantum dots*. The extra confinement would be provided by free space. Our first efforts at Bellcore were met with brilliant, if brief, success[15]. Dots and wires of diameter ~ 45 nm in free-standing ridges and posts left after reactive ion etching of GaAS-AlGaAs quantum well heterostructures gave strong photoluminescence signals, but showed no extra confinement energy. We did not expect this result, and never have we been able to duplicate this early feat. We are quite unable to account for why it happened.

Until the experiments of Heitmann et al.[5] we supposed that the GaAs-AlGaAs system was doomed by surface recombination, if not by damage created during the reactive ion etching. Our general conclusion, from Clausen's measurements of the

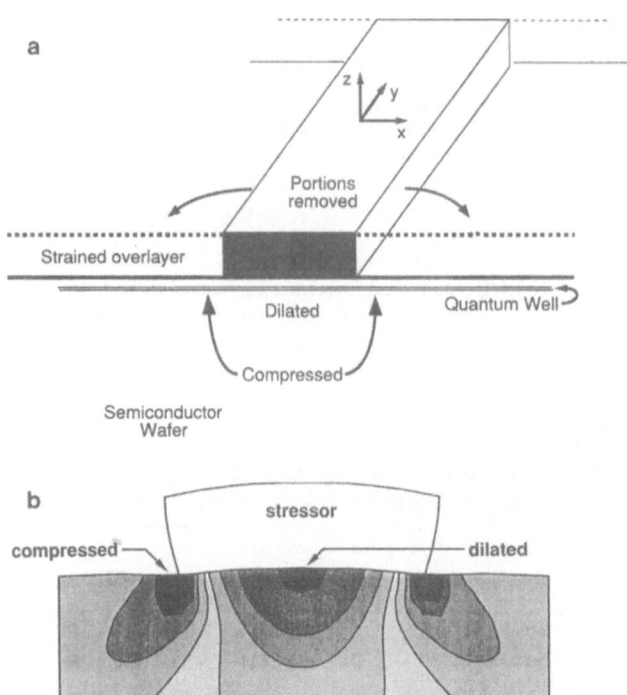

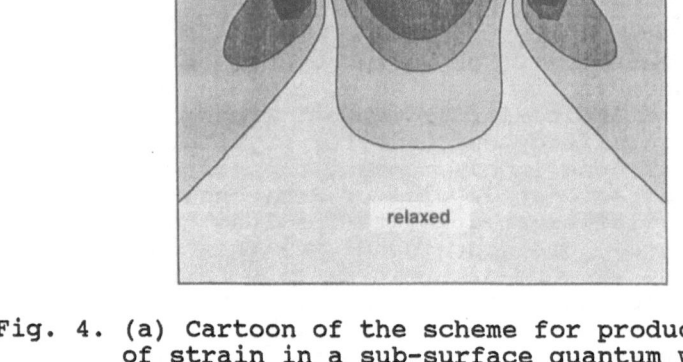

Fig. 4. (a) Cartoon of the scheme for producing patterns
of strain in a sub-surface quantum well - to
produce a quantum wire. A "Stressor" is formed
by removing portions of a uniform strained
layer, so that some of the strain can be transferred to the substrate. (b) Contour plots of
the hydrostatic strain patterns underneath a
stressor wire, showing the dilated and compressed regions. Strain patterns must be
computed numerically. The shear components of
the strain can lead to valence band splittings
and to piezoelectric effects.

efficiency of cathodoluminescence as a function of disk
size[16], was that normal etching produced a circumferential
region of damage in which nonradiative recombination was
essentially instantaneous, and that this damage was sufficiently
deep to preclude the observation of quantum confinement.
Perhaps gentler etching would not produce damage, but surface
recombination (nonradiative) at even the best GaAs surfaces is
known to be extremely efficient. I will not attempt to explain
the unexpected success of Heitmann et al.

Since the epitaxial interface of GaAs with AlGaAs can be
grown free of recombination centers, one of the obvious
solutions is to cover the offending surfaces with epitaxial

layers.   Until now, however, we have not learned how to prepare
those lateral surfaces free of chemical contamination so as to
achieve good epitaxy at the interface[17].   An alternative is to
work in another semiconductor system, such as InP-InGaAs, which
is known to have slower surface recombination.   This is the
approach taken by Gershoni et al.[18], who have had some good
success in making optical quantum wires.

4.2   Inhomogeneous Strain Patterns

Another approach to lateral quantum confinement of excitons
was discovered[19] and developed at Bellcore by K. Kash and her
co-workers.   Thin overlayers of strained material can transfer
some of their strain into the underlying semiconductor if the
overlayers are appropriately patterned, as portrayed
suggestively in Fig. 4a.   Shown in cross-section is a
semiconductor wafer, with a quantum well grown in just beneath
the surface.   Portions of a (compressed) strained layer
deposited on top have been removed, allowing the remaining
portion of the layer (called the stressor) to expand, dilating
the semiconductor in the central region.   Kash and her co-
workers have suggested and tried a variety of strained layers,

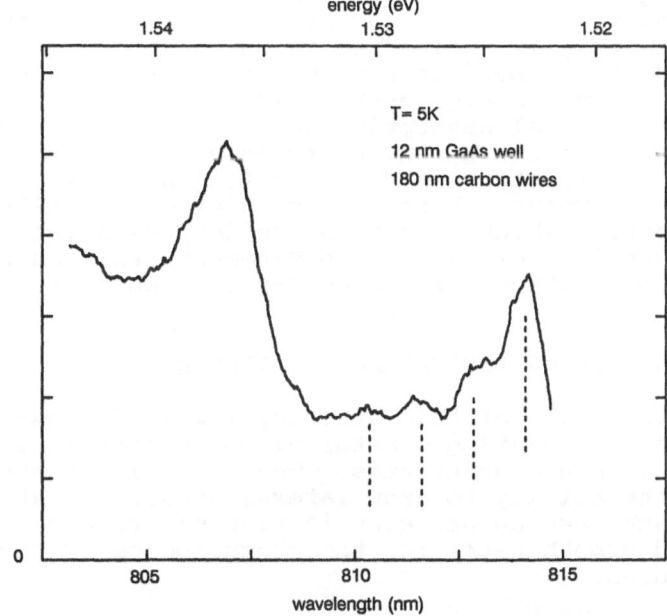

Fig. 5. A small portion of the photoluminescence
        excitation spectrum of a quantum wire formed in
        a 12 nm GaAs quantum well, using as a stressor a
        wire of amorphous carbon film 180 nm across.
        The heavy hole exciton appears at 1.523 eV -
        squeezed upward in energy by the 12 nm quantum
        well, red shifted by the dilation, and finally
        slightly blue shifted by lateral quantum con-
        finement.   Peaks marked with lines correspond to
        higher lateral subbands.   After Kash, et al.
        Ref. 21.

and have calculated the strain patterns, solving the equations of elasticity by numerical means. Figure 4b gives the results of one such calculation, in a two-dimensional plot of the hydrostatic strain pattern. The calculations show, as expected, dilation extending well into the semiconductor, with regions of compression near the edges of the stressor.

In the simplest approximation, the dilatation reduces the band gap. 2D excitons are attracted to the region of low energy and they become trapped or confined under the stressor. Luminescence emanating from that region should reflect the reduction in band gap. Red shifts of as much as 60 meV, corresponding to dilatation on the order of 1/2 to 3/4 percent, have been observed.

At the next level of approximation, the valence band states are mixed and shifted by the components of shear strain which are unavoidable when dilatation is inhomogeneous. Some of the consequences of this mixing have been observed by Kash et al., in the optical anisotropy of quantum wires[20]. Another serious result of shear strain is piezoelectricity, with polarization fields leading to sizeable electric charge densities. These effects have not yet been systematically studied.

In their most recent published work, Kash et al. have succeeded in generating true quantum wires, which is to say that they have observed and measured the lateral confinement energy of the 1D excitons[21], using the technique of photoluminescence excitation spectroscopy. In Fig. 5 are spectra taken in that mode that show, on the high energy side of the redshifted ground state exciton, several absorption peaks which can be understood to be those of excitons of the upper *lateral* subbands of the quantum wire. This story is far from finished, since there are complications of valence band mixing, piezoelectricity, and elastic anisotropy which cannot be avoided, especially in cases where two dimensions of lateral confinement are required, such as for quantum dots and structures for 1D transport.

5.   DIRECT GROWTH ON VICINAL AND PATTERNED SUBSTRATES

In the remainder of this article, I will discuss striking progress toward generating quantum wires by epitaxial growth on vicinal and patterned substrates. The epitaxial techniques are varied, but the ability to grow lateral structures at will depends in each case on new quantitative understanding of the diffusion and growth rates for the constituents of the heterostructures.

5.1. Vapor Levitation Epitaxy - Spontaneous Quantum Wire Arrays

Exciton spectroscopy was instrumental in discovering a new mechanism for forming wire-like arrays of InGaAs embedded in InP during growth by Vapor Levitation Epitaxy (VLE)[22]. Morais and Cox first found optical evidence for ultra thin quantum wells with thicknesses corresponding to 2-6 monolayers of InGaAs[23]. Then using extremely short growth cycles, under conditions where growth proceeds by lateral advancement from atomic steps, they were able to reduce the widths of the quantum wells in a lateral direction[24]. Subsequent analysis suggested that the first monolayer (heteroepitaxy) of InGaAs was growing significantly

Fig. 6. Transmission electron micrograph of arrays of
quantum wires (dark regions) in InGaAs embedded
in InP, grown using a morphological instability
starting from a nominally flat vicinal
substrate. The long axes of the wires are
perpendicular to the figure. After Cox, et al.
Ref. 25.

faster than subsequent ones (homoepitaxy). The differential
growth velocities have now been used to explain and control a
morphological instability that produces arrays of wires of
InGaAs after a few iterations of periodic exposure to InP
nutrients/InGaAs nutrients, starting from a smooth vicinal
surface 3° toward (010) from (001)[25,26]. Figure 6 is a
transmission electron micrograph of such a structure. The dark
regions are InGaAs wires with long dimensions perpendicular to
the plane of the figure. Atomic steps on the surface,
approximately evenly spaced to start at an average distance of
84 Å, congregate more and more with each growth iteration into
regularly spaced macrosteps. As the period of the macrostep
spacing is related to the relative lateral growth velocities and
times of the periodic exposures, a regular coherent array of
wires is produced. I suggest, somewhat wishfully, that a
variant of this technique will shortly be found which yields
arrays of quantum dots. Finally, although one might have
supposed that perfection of epitaxial growth was synonymous with
atomically smooth interfaces, it is clear from this striking
example that a controlled instability can lead to equally
important, though unexpected, structures.

5.2 Growth on Grooved Substrates

Another approach to the formation of arrays of quantum
wire-like heterostructures is under development by Colas and his
coworkers[27], using organometallic chemical vapor deposition.
Although this process begins with an etched groove in a vicinal
substrate, the formation of the wires takes place at some
distance and therefore does not depend on damage-free etching.
As depicted in Fig. 7, the anisotropic etch exposes a variety of
crystallographic surfaces at a groove. Only one of the edges

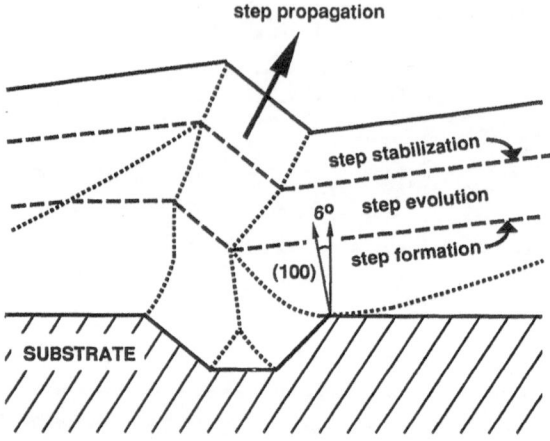

Fig. 7. Natural facets at a groove in a vicinal
substrate, illustrating the growth and
stabilization of a bend or a step at the surface
which is more or less independent of the details
of the groove. After Colas, et al. Ref. 26.

contains the (100) surface, giving the groove an inherent
asymmetry. Because of its higher reactivity, the (111)A plane
forms a stable step between bounding (100) surfaces. The shape
and size of the step depend on the distance to the next groove
and the vicinal angle, i.e. it does not depend on the details of
the groove formation. The step formation takes place with
simple homoepitaxy, depending only on sufficient diffusion to

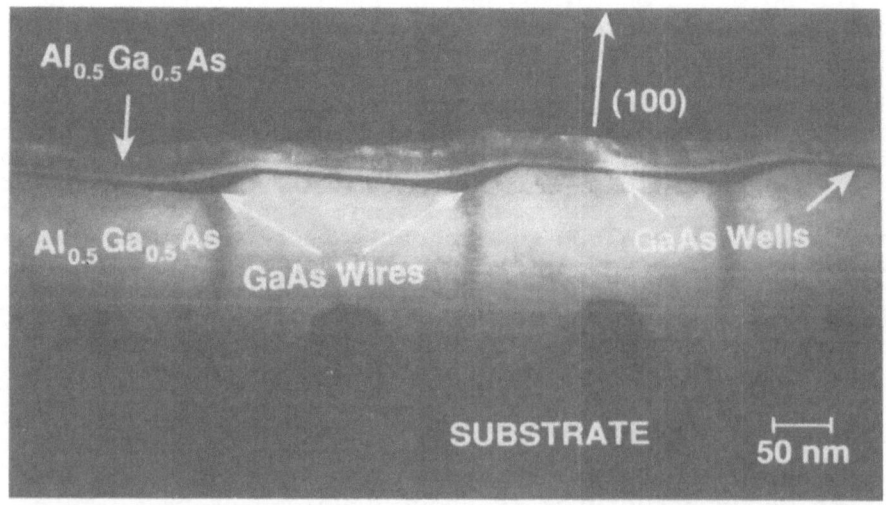

Fig. 8. Transmission electron micrograph of surface
arrays of quantum wires, connected with thinner
quantum well regions all seen in cross-section:
GaAs in AlGaAs. The submicron scale is shown in
the figure. After Colas, et al. to be
published.

develop the favored facets. Quantum wires can then be grown at
the steps by taking advantage of the same differences in
diffusion and reactivity. Figure 8 is a transmission electron
micrograph of the cross section of an array of wires grown using
this technique, using a pattern of grooves with period ~ 300 nm.
In this figure, the quantum wires are the somewhat thicker
regions connected by thinner parts of the continuous quantum
well. Changes in the growth conditions should result in better
separation of the wires, i.e. thinner quantum well material
connecting them.

This technique is rather closely related to that by which
Kapon and his co-workers have grown patterned quantum wells and
quantum wires in etched grooves on GaAs substrates, using both
molecular beam epitaxy[28] and organometallic chemical vapor
epitaxy[29]. Their motivation is to make use of the predicted
reduction in threshold for quantum wire lasers[30], and in this

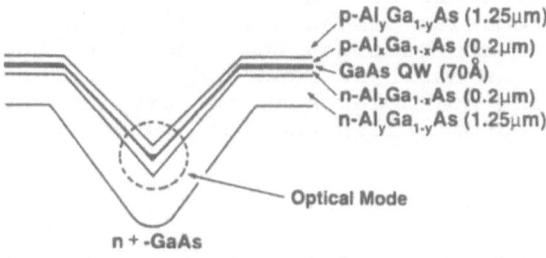

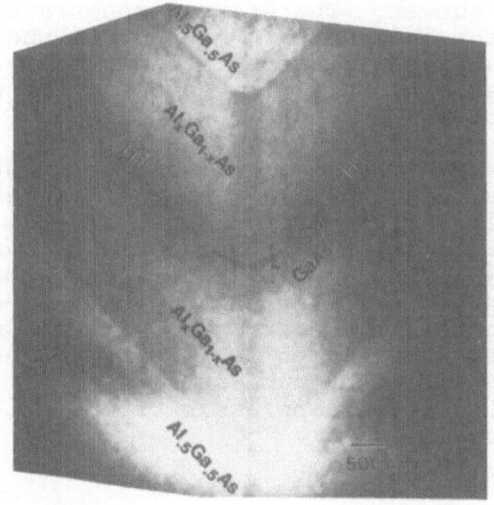

Fig. 9. Transmission electron micrograph of a quantum
         wire laser structure in cross-section: a
         crescent shaped wire of GaAs forms at the bottom
         of the groove because the AlGaAs grows faster on
         the (111)-sidewalls, sharpening the groove, than
         does the GaAs, which grows more thickly in the
         bottom. After E. Kapon, et al. Ref. 35.

they have succeeded very well, achieving record low threshold currents. Figure 9 shows a transmission electron micrograph of one of their crescent-shaped quantum wires. The facets in the groove are {111} crystal planes on which the AlGaAs grows especially rapidly, keeping the groove sharp as the growth front advances. GaAs, on the other hand, grows faster along the [100] direction, rounding the corner of the groove and fattening the center of the crescent. Since the groove recovers with AlGaAs growth in the next layer, we suppose that it will be possible to grow reasonably uniform vertical arrays of such quantum wires.

5.3 Vertical Superlattices

Petroff et al.[31] proposed an interesting scheme for growing "vertical superlattices" directly on periodically terraced substrates. Fractional monolayers of the heteroconstituents (GaAs and AlAs, for example, and with the fractions summing to ~ unity) should be deposited alternately and with enough mobility that they grow coherently at the terrace steps. Then the fractional monolayers of GaAs pile exactly on top of each other, and likewise with the AlAs, leading to vertical slabs of GaAs separated by those of AlAs. In spite of the precision demanded in this method in both the periodicity of the substrate and the control of the coverage, some striking success has been reported, both by Petroff and his co-workers[32] using molecular beam epitaxy, and by Fukui and Saito[33] using organometallic chemical vapor deposition.

CONCLUSIONS

The progress Heitmann and I have described here toward demonstration of reliable and reproducible quantum wires and quantum wire arrays has taken many paths, not all of them easily foreseeable at the outset. The story demonstrates that in this kind of research, the obvious path may not be the best. We have sometimes succeeded by being alert to the creative possibilities of apparent mistakes, and by taking advantage of what appeared at first to be problems.

Technology has rapidly made good use of quantum wells (2D) in a variety of devices, including optical modulators and lasers. Quantum wires (1D) have appeared already in lasers, but have yet to find application in other optical device structures. Quantum dots (zero-D), though they have been probably more thoroughly treated theoretically than quantum wires, have not been so widely sought in heterostructures. This may be in part because we don't yet know how to look for them, especially if they are hidden and buried by subsequent layers. Development of diagnostic techniques is perhaps as important as inventing those of growth and production.

Acknowledgment

I wish to acknowledge the benefit of a long term collaboration with Kathy Kash and an exciting episode with Herb Cox and Paulo Morais. I have been aided in preparing this reviewed by Herb Cox, Etienne Colas, Dah-Min Hwang and Eli Kapon.

REFERENCES

1. Edwin A. Abbot, "Flatland, a Romance of Many Dimensions", first published in 1884, but still in print in several editions
2. D.C. Tsui, H.L. Stormer and A.C. Gossard, **Phys. Rev. Letts.**, 48:1559 (1982)
3. See, for example, the semipopular article by G.S. Canright and S.M. Girvin, **Science**, 247:1197 (1990)
4. For a recent review, see M. Bawendi, M.L. Steigerwald and L.E. Brus, **Ann. Rev. of Physical Chemistry**, 41 (1990)
5. D. Heitmann, this workshop
6. K. Kash, **J. Lumin.**, 46:69 (1990)
7. M.D. Sturge, **Phys. Rev.**, 127:768 (1962)
8. R. Dingle, W. Wiegmann and C.H. Henry, **Phys. Rev. Letts.**, 33:827(1974)
9. See, for example, the contributions of G.E.W. Bauer, Y.-C. Chang, G. Duggan, J.C. Maan, R.J. Nicholas and L. Vina at this workshop
10. J. Feldmann, G. Peter, E.O. Gobel, P. Dawson, K. Moore, C. Foxon and R.J. Elliott, **Phys. Rev. Letts.**, 59:2337 (1987)
11. C. Weisbuch, R. Dingle, A.C. Gossard and W. Wiegmann, **Solid State Commun.**, 38:709 (1981)
12. D.S. Chemla, p.423 **in**: "Physics and Applications of Quantum Wells and Superlattices", Proc. of NATO Adv. Res. Inst., Erice, Italy, 1987, Plenum, New York, (1987)
13. D.A.B. Miller, D.S. Chemla, T.C. Damen, A.C. Gossard, W.Wiegmann, T.H. Wood and C.A. Burrus, **Phys. Rev. B**, 32:1043 (1985)
14. C. Benoit a la Guillaume, **Physica**, 117B-118B:1035 (1983)
15. K. Kash, A. Scherer, J.M. Worlock, H.G. Craighead and M.C. Tamargo, **Appl. Phys. Letts.**, 49:1043 (1986)
16. E.M. Clausen, Jr., H.G. Craighead, J.M. Worlock, J.P. Harbison, L.M. Schiavone, L. Florez and B. Van der Gaag, **Appl. Phys. Letts.**, 55:1427 (1989)
17. An important *tour de force* in regrowth on a surface formed by cleaving *in situ* has very recently been reported: L. Pfeiffer, K.W. West, H.L. Stormer, J.P. Eisenstein, K.W. Baldwin, D. Gershoni and J. Spector, **Appl. Phys. Letts.**, 56:1697 (1990)
18. D. Gershoni, H. Temkin, G.J. Dolan, J. Dunsmuir, S.N.G. Chu and M.B. Panish, **Appl. Phys. Letts.**, 53:995 (1988)
19. K. Kash, J.M. Worlock, P. Grabbe, A. Scherer, J.P. Harbison, H.G. Craighead, P.S.D. Lin and M.D. Sturge, **Appl. Phys. Letts.**, 53:782 (1988); K. Kash, R. Bhat, D.D. Mahoney, P.S.D. Lin, A. Scherer, J.M. Worlock, B.P. Van der Gaag, M. Koza and P. Grabbe, **Appl. Phys. Letts.**, 55:681 (1989)
20. K. Kash. J.M. Worlock, A.S. Gozdz, B.P. Van der Gaag, J.P. Harbison, P.S.D. Lin and L.T. Florez, **Surface Science**, 229:245 (1990)
21. K. Kash, B.P. Van der Gaag, J.M. Worlock, A.S. Gozdz, D.D. Mahoney, J.P. Harbison and L.T. Florez, **to be published in**: "Localization and Confinement of Electrons in Semiconductors", Springer-Verlag Series in Solid State Sciences (1990)
22. H.M. Cox, **J. Cryst. Growth**, 69:641 (1984); H.M. Cox, S.G. Hummel and V.G. Keramidas, **J. Cryst. Growth**, 79:900 (1986)
23. P.C. Morais, H.M. Cox, P.L. Bastos, D.M. Hwang, J.M. Worlock, E. Yablonovitch and R.E. Nahory, **App. Phys. Letts.**, 54:442 (1989)

24. H.M. Cox, P.C. Morais, D.M. Hwang, P. Bastos, T.J. Gmitter, L. Nazar, J.M. Worlock, E. Yablonovitch and S.G. Hummel, **Inst. Phys. Conf. Ser.**, 96:119 (1989)

25. S.J. Allen, Jr., P. Bastos, H.M. Cox, F. DeRosa, D.M. Hwang and L. Nazar, **J. Appl. Phys.**, 66:1222 (1989)

26. H.M. Cox, D.E. Aspnes, S.J. Allen, P. Bastos, D.M. Hwang, S. Mahajan, M.A. Shahid and P.C. Morais, to be published

27. E. Colas, E. Kapon, S. Simhony, H.M. Cox, R. Bhat, K. Kash and P.S.D. Lin, **Appl. Phys. Letts.**, 55:867 (1989)

28. E. Kapon, M.C. Tamargo and D.M. Hwang, **Appl. Phys. Letts.**, 50:347 (1987)

29. R. Bhat, E. Kapon, D.M. Hwang, M.A. Koza and C.P. Yun, **J. Cryst. Growth**, 93:850 (1988)

30. Y. Arakawa and A. Yariv, **IEEE J. Quantum Electron.**, QE-21:1666 (1985)

31. P. Petroff, A.C. Gossard and W. Wiegmann, **Appl. Phys. Letts.**, 45:635 (1984)

32. J.M. Gaines, P.M. Petroff, H. Kroemer, R.J. Simes, R.S. Geels and J.H. English, **J. Vac. Sci. Technol.**, B6:1378 (1988)

33. Takashi Fukui and Hisao Saito, **J. Vac. Sci. Technol.**, B6:1373 (1988)

34. A.C. Gossard, Molecular Beam Epitaxy of Superlattices and Thin Films, **in**: "Preparation and Properties of Thin Films", K.N. Tu and R. Rosenberg, eds., Academic Press, New York (1982)

35. E. Kapon, D.M. Hwang and R. Bhat, **Phys. Rev. Lett.**, 63:430 (1989)

FROM QUANTUM WELL EXCITON POLARITONS TO ONE-DIMENSIONAL EXCITONS

M.Kohl, D. Heitmann, P. Grambow and K. Ploog

Max-Planck-Institut für Festköperforschung
Heisenbergstrasse 1, 7000 Stuttgart 80
West Germany

ABSTRACT

We review quantum well exciton polaritons and one-dimensional quantum confined excitons which were studied by photoluminescence spectroscopy of laterally microstructured $Al_xGa_{1-x}As$ - GaAs quantum well systems and quantum wires as narrow as 60 nm. Freely propagating quantum well exciton polaritons were investigated in quantum well samples with microstructured cap layers. Their radiative decay was mediated by the grating coupler effect of the periodic corrugation. It is observed that also the photoluminescence spectra of wide quantum well wires are dominated by quantum well exciton polariton emission. When decreasing the wire width below $L_x \approx 150$ nm the polariton-specific features disappear in the spectra. Below $L_x = 100$ nm, the concept of quasi-one-dimensional excitons is an appropriate model to describe the behavior of the observed transitions. The 1D character of these excitons is manifested in the observation of several exciton resonances which are related to different energetically separated 1D quantum confined subbands, by a reduced diamagnetic shift, and by a 15% increased binding energy with respect to 2D reference samples which is particularly reflected in magneto-optical experiments.

1. INTRODUCTION

A broad range of fundamental research and novel applications in many different fields was initiated by the recent progress in crystal growth techniques which made it possible to fabricate layered semiconductor heterostructures, quantum wells and superlattices with atomically flat interfaces. These novel systems have unique physical properties which arise from the quasi-two-dimensional (2D) behavior and, in superlattices, from coupling between these 2D layers. Optical spectroscopy, in particular photoluminescence (PL), is a powerful tool to study these systems. The PL spectra of quantum well (QW) systems are governed by efficient intrinisic radiation of free excitons and exhibit a superior optical performance[1] (for a recent review see e.g. Ref. 2). One of the challenging topics of current interest involves systems of further reduced dimensionality, namely 1D quantum wires and 0D quantum dots. Besides the expectation of novel physical phenomena it is hoped that the optical properties, i.e. oscillator strengths, nonlinear coefficients and others, can be further improved with respect to the 2D systems to realize high performance optical devices. 1D and 0D systems have been prepared by starting from a 2D-layered system and employing modern micro and nanometer lithographic techniques[3-16] or by using novel growth techniques[17-19]. An excellent review has recently been given by Kash[20] where these different preparation techniques are described extensively, and I would also like to draw attention to John Worlock's fascinating review on optical spectroscopy in this volume.

In this paper we will discuss quantum well exciton polaritons (QWEP) and 1D quantum confined excitons in quantum wire structures which are inherent excitations of microstructured QW systems. We have studied the optical excitations in microstructured AlGaAs-GaAs quantum well systems which have been prepared by deep mesa etching techniques[10,11,13,14,21]. We have utilized the grating coupler effect of microstructured cap layers to study the resonantly enhanced emission of freely propagating QWEP. We will demonstrate that these excitations also dominate the optical spectra of wide QW wire systems. For very narrow wire widths, $L_x < 100$ nm, the polariton features change drastically and the observed resonances are better described as 1D excitons, i.e. excitons related to energetically separated 1D quantum confined subbands.

2.   SAMPLE PREPARATION AND EXPERIMENTAL TECHNIQUES

The samples were grown by molecular beam epitaxy (MBE) at 610°C on semi-insulating (001) GaAs substrates. They consisted typically of five (three) identical GaAs QWs with well widths $L_z = 12$ nm ($L_z = 14$ nm) separated by 10 nm thick $Al_xGa_{1-x}As$ barriers. The Al concentration of the different samples was kept in the range $0.26 < x < 0.37$. On top of the samples was a cap layer of 50 nm $Al_xGa_{1-x}As$ and 2 nm GaAs. For the investigation of the QWEP in QW structures a thicker cap layer of 170 nm thickness was used. A mask consisting of periodic photoresist lines was prepared by holographic lithography. The photoresist was exposed to the superposition of two expanded coherent laser beams (wavelength $\lambda = 458$ nm), as sketched in Fig. 1, which gives rise to a sinusoidal intensity modulation and, after development, to a spatially modulated photoresist profile. The width of the photoresist lines was adjusted by the development time of the resist. With subsequent reactive ion etching in a $SiCl_4$ plasma nearly rectangular grooves of about 170 nm (130 nm) depths were etched into the samples. The resulting wire widths, L_x, ranged from 300 nm down to 60 nm, which was determined by a scanning electron microscope. On each QW wire sample a small part was left unpatterned to take reference spectra.

We have prepared different types of samples. For the investigation of QWEP we have started from samples with thick cap layers and have only etched this cap layer as shown in Fig. 2a. The periodic corrugation of this cap layer then acts as a

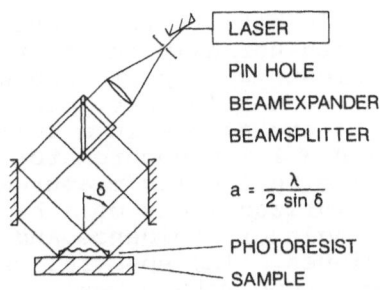

LASER

PIN HOLE

BEAMEXPANDER

BEAMSPLITTER

$$a = \frac{\lambda}{2 \sin \delta}$$

PHOTORESIST

SAMPLE

Fig. 1. Sketch of a set up for holographic lithography
to prepare arrays of linear submicron wires on
GaAs QW structures. The superposition of two
coherent laser beams with angle of incidence δ
leads to a sinusoidal intensity modulation with
period a. The exposed and developed photoresist
acts as a mask to transfer the wire pattern with
etching techniques into the sample.

grating coupler (see below) which couples QWEP with light,
without directly influencing the electronic properties of the
QW. We have also prepared QW wire structures by etching through
the active GaAs QW as shown schematically in Fig. 2b. The
important point of all these preparation techniques is to
optimise all process parameters in such a way that even at very
small wire widths there is still strong enough PL and that not
all radiative transitions are quenched due to nonradiative
processes. This is a particularly hard problem for the GaAs
systems and it took us an enormous effort to overcome many
preparative problems. More details of our preparation process
are given in Ref.22.

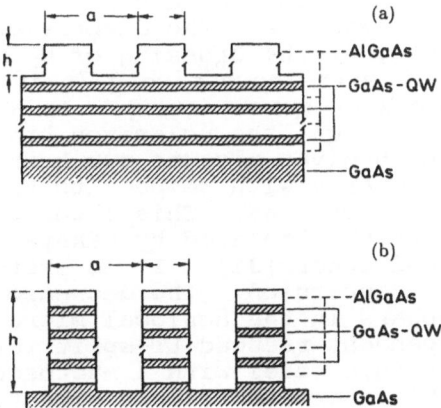

Fig. 2. (a) Schematic cross-section of the QW
structures. On top of the GaAs QW layers
(dashed regions) is a laterally $Al_xGa_{1-x}As$ cap
layer of thickness h = 170 nm. a is the grating
period and L_x the wire width. (b) Schematic
cross-section of the QW wire structures. The
heights h of the wires are chosen between 130 nm
and 170 nm, the wire-widths L_x between 300 nm
and 60 nm.

The PL was excited, if not otherwise stated, at the 647.1 nm line of a Kr^+-laser with normally incident light. PL excitation (PLE) measurements were performed with a Styryl 9 dye laser, which was pumped by an Ar^+-laser. The luminescence light was collected normally to the sample in a solid angle of 30° opening and analyzed with a 1 m monochromator and photon counting techniques. Also reflection measurements were performed with the monochromatized light of a halogen lamp exciting the samples with a 45° angle of incidence and detecting the signal by lock-in techniques. The spectral resolution was set to 0.05 nm and the sample temperature was 4.2 K if not otherwise indicated.

3.   SURFACE POLARITONS AND QUANTUM WELL EXCITON POLARITONS

In the following we wish to discuss the concept of quantum well exciton polaritons (QWEP). QWEP are a special type of 'surface polaritons'[23], i.e. the quanta of electrodynamic excitations which exist at the boundary or in thin films of materials when the real part of the dielectric function, $\epsilon_r(\omega)$, is negative; e.g., in metal-like systems 'surface plasmon polaritons' (e.g. Refs. 23,24) can be excited below the plasma frequency of the free electron gas. In polar materials 'surface phonon polaritons' exist in the frequency regime between the transverse and longitudinal eigenfrequencies, ω_T and ω_L, respectively[23,25]. Equivalent surface excitations in a material with excitonic behavior are surface exciton polaritons[26,27]. The dielectric response of such a material can be described by a Lorentzian resonance expression

$$\epsilon(\omega) = \epsilon_\infty [1 + (\omega_L^2 - \omega_T^2)/(\omega_T^2 + \beta Q^2 - \omega^2 - i\omega\Gamma)] \qquad (1)$$

where β governs the spatial dispersion, $\vec{Q}$ is the wave vector, Γ the damping and ϵ_∞ the high frequency dielectric function. The dispersion of these excitations for a semi-infinite excitonic material is sketched in Fig. 3a. The dispersion starts, if retardation is included, at the crossing of ω_T and the light line and approaches ω_L with increasing q (q is a wave vector in the plane, q_p the polariton wave vector). In a thin slab, as in the case of the thin GaAs QW, the polariton branches of the two interfaces couple, which gives rise to two branches, one of which increases in frequency with respect to the semi-infinite medium while the other decreases. This lower branch becomes the X-mode QWEP that has been calculated by Nakayama et al.[28,29], Andreani et al.[30] and others[31]. It is sketched in Fig. 3b including the spatial dispersion. The decrease of the frequency of the lower mode appears in the nonlocal microscopic approach of Ref. 30 as a q dependent reduced LT-split for polaritons in a thin slab. Note that in a model with a macroscopic dielectric function (Eqn. 1), this decrease follows directly from the electrodynamics of the thin slab without changing ω_L and ω_T. It should be noted that in the case of QW systems the dielectric function, Eqn. 1, is more complicated, in particular strongly anisotropic with respect to excitations parallel or perpendicular to the layered planes. In addition, due to the different symmetries of the light hole wave functions and the heavy hole wave functions, pronounced differences occur in the optical response of electron light hole (e-lh) and electron heavy hole (e-hh) transitions[32]. This is also reflected in certain properties of the QWEP[31].

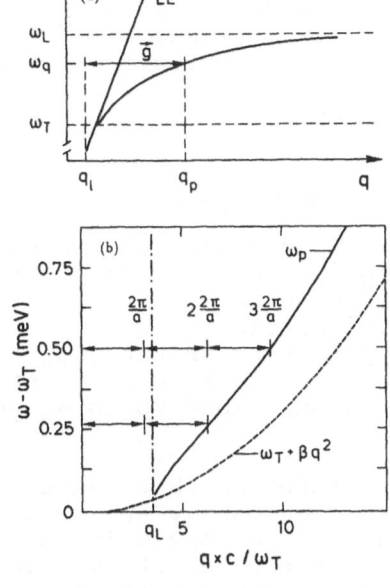

Fig. 3. Schematic dispersion of surface polariton at the
interface of an excitonic medium which is
characterized by a longitudinal (ω_L) and
transversal frquency (ω_T) (without spatial
dispersion). LL is the light line $\omega = qc/n$ where
n is the index of refraction of the adjacent
medium. Due to retardation the dispersion
starts at the crossing of ω_T and the light line
and approaches ω_L with increasing q.   (b)
Schematic diagram of the dispersion of X-mode-
QWEP (solid line) and photons in GaAs (dashed-
dotted line) according to Ref. 28.   The
dispersion of the QWEP and free excitons (dotted
line) bends to higher frequencies for increasing
q due to the spatial dispersion. By momentum
transfer $g = \pm n 2\pi/a$ mediated by the grating
structure polaritons with momentum q_p can be
scattered into the radiative regions on the left
side of the light line. The parameters of the
dispersion are chosen for GaAs; ω_T is 1.54 meV
and the grating period a is 250 nm.

There are three important aspects for surface polaritons
and in particular also for QWEP:

(a)   In laterally homogeneous systems these excitations are
nonradiative due to momentum conservation since the wave vector
$|\vec{q}_p| > \omega/c$. The electromagnetic fields decay exponentially from
the interface. They do not couple with freely propagating waves
($|\vec{q}_x| < \omega/c$) and thus do not contribute to the response in PL or
reflection measurements. However, any spatial modulation of the
optical properties in the near field of the interface, in
particular a periodic corrugation of the surface with
periodicity a, will cause a momentum transfer

$$q_{xn} = q_p + n \cdot 2\pi/a, \quad n = \pm 1,2,3\ldots \;. \tag{2}$$

We consider here electrodynamic excitations propagating in the x direction and grating rules parallel to the y direction. This is also the experimental arrangement. If $|\vec{q}_{xn}| < \omega/c$, the grating coupler effect of the periodic corrugation gives rise to radiative decay of the surface polaritons. The radiation is emitted for an angle θ_n with respect to the surface normal and is governed by

$$(\omega/c) \sin \theta_n = q_{xn} = q_p - n \cdot 2\pi/a \qquad (3)$$

in the case of polaritons propagating perpendicular to the grating rules. Grating couplers, or very similar prism arrangements, have been used to investigate different kinds of surface polaritons, e.g. surface plasmon polaritons[23,24], 2-D plasmons[33], surface phonon polaritons[25] and surface exciton polaritons[27]. In particular, Agranovich et al. have pointed out that a corrugation should lead to polariton emission in the photoluminescence[34].

(b)   A second important point is that these excitations are associated with a strong resonantly enhanced field strength at the interface which has been observed in several experiments [23,24]. In particular, the interesting 'Giant Raman'[23] effect has been associated with a surface polariton-type of field strength enhancement.

(c)   A third point is that the resonantly enhanced polariton excitation occurs only for p-polarization, i.e. for excitations where the electric field is polarized in the plane of the polariton wave vector and surface normal. If we observe the radiative decay in the x-z plane, i.e., perpendicular to the grating rules, the grating does not change the polarization and the emitted radiation is also p-polarized. This polarization selection rule is an important feature to identify QWEP resonances.

4.   EXCITON POLARITONS IN QUANTUM WELLS

We first want to present polarization-dependent PL measurements on QW systems with a microstructured cap layer[11]. The microstructure acts as a grating coupler but does not directly influence the electronic properties of the QW. The structure of this type of sample is shown schematically in Fig. 2a. As reference samples we have used (i) unprocessed QW samples and (ii) QWs where the cap layers has been homogeneously etched away almost down to the QW layers. A comparison of the PL results for the grating structure with the spectra of the two reference samples thus allows us to distinguish between effects which are induced by the geometry of the cap layer and those that are directly related to etching induced defect states.

The upper part of Fig. 4 shows p- (solid curve) and s-polarized (dashed curve) PL spectra of a QW sample with a grating structure on top. The s-polarized PL exhibits two maxima at 805.9 nm and 807 nm. This spectrum is very similar to that of the reference sample (ii) not shown here, while the spectrum of sample (i) exhibits only the maximum at 805.9 nm. The different origin of the two transitions is particularly reflected in intensity and temperature dependent measurements[11]. The first transition at 805.9 nm shows an almost

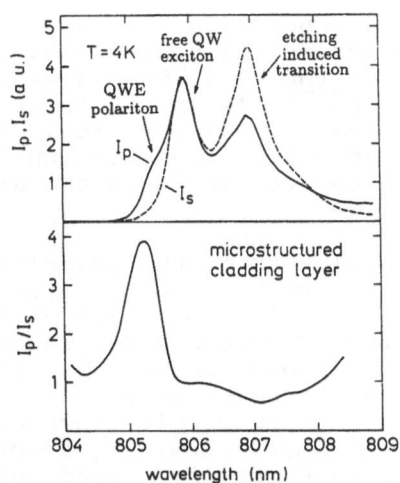

Fig. 4. Experimental luminescence spectra of a
microstructured GaAs-QW sample as shown in Fig.
2a with a = 500 nm and L_x = 200 nm. I_p and I_s
denote the p- and s-polarized luminescence
intensity, respectively. Additional emission
occurs in p-polarization as a shoulder at 805.4
nm on the low wavelength side of the free
exciton emission which arises from the
resonantly enhanced radiative decay of QWEP.
The resonant character is in particular
reflected in the I_p/I_s ratio in the lower part
of the figure. The excitation intensity was 10
mW/cm^2. (From Ref. 11)

linear intensity dependence and can be observed at temperatures
higher than 60 K. This behavior is typical for the recombination of free excitons. The transition at 807 nm, however,
saturates at high excitation intensities and disappears above
40 K. A comparison with the two reference samples indicates
that the second transition arises from defect states which are
induced by etching. Its temperature dependence reflects the
thermo-ionization of a carrier-defect complex. A calculation of
the transition energies with a finite potential well model
allows the assignment of the first transition to e_1-hh_1 (hh =
heavy hole) free excitons. The PL efficiency of this transition
is comparably high for all three types of samples . This
demonstrates that the QW layers are not affected by the etching
process.

A very interesting result is obtained, when we record the
p-polarized PL. As can be seen for the solid curve in the upper
part of Fig. 4 an additional PL transition contributes as a
shoulder on the high-energy side of the free QW exciton transition. This transition is obviously caused by the radiative
decay of QWEP, since all the characteristic features as
discussed above are observable:

(a) The additional transition occurs only in samples with a
grating shaped cap layer. Therefore, as expected, the grating
is an essential condition for the observation of QWEP.

(b) The polarization dependence of the additional transition corresponds to the selection rules discussed above. The strong p-polarized character of the PL in the wavelength regime of the QWEP emission is best observable in the lower part of Fig. 4, where we have plotted the intensity ratio I_p/I_s of the two differently polarized spectra. The emission efficiency of the polariton signal turns out to depend on the amplitude and shape of the grating profile[10].

(c) The transition wavelength of the polariton signal λ_p is smaller than the e_1-hh_1 transition wavelength λ_T. The corresponding energy separation $\hbar(\omega_p - \omega_T)$ is approximately 0.8 meV. This is within the expected energy range for the QWEP as can be seen in Fig. 3b, if one assumes that the free exciton transition occurs at the transversal eigenenergy $\hbar\omega_T$[26]. As sketched in Fig. 3b, a grating coupler induces coupling via several harmonics $|\vec{g}_n| = n2\pi/a$. However, in general the higher Fourier components decrease rapidly. Concerning the spatial dispersion, Fig. 3b is roughly drawn to the right scale for a period $a = 250$ nm. We see from this figure that several grating vectors $\vec{g}$ are needed to couple to the QWEP dispersion and the spatial dispersion can well explain the observed frequency shift of 0.8 meV. Theoretically, an enhancement of the LT-split compared to the bulk value of 0.1 meV is expected because of the enhanced oscillator strengths of 2D confined systems[2]. However, we are not sensitive to a change of the LT-split and cannot give any statement on this interesting property. This is, first, due to the strong influence of the spatial dispersion, secondly, due to the decreased frequency of the X-polariton mode in a thin slab and thirdly, due to the limited experimental resolution, since the linewidth of about 2 meV is larger than the energy shift.

The intrinisic nature of the observed QWEP transition is in particular reflected in temperature dependent measurements. The etching induced transition at 807 nm on the one hand disappears above 40 K, the free hh excitons and the QWEP transition on the other hand can easily be observed up to 60 K, in particular in the I_p/I_s plot. In the regime between 808 nm and 814 nm a broad polarization dependent PL-band is observable, which is also related to etching induced defect states[11].

Very recently QWEP have also been observed in time-of-flight measurements[35,36]. In these experiments the QWEP were excited by light incident directly onto the edge of a specially designed multi-QW sample.

5.   PHOTOLUMINESCENCE OF WIDE QUANTUM WELL WIRES

We would now like to discuss systems as described in Fig. 2b, where we have etched all the way through the five QW[10]. The lateral width of these QW wires was $L_x = 150$ nm or larger. As we will discuss below, this confinement leads in principle to a 1D quantization of the electronic wavefunctions. However, 150 nm is still too large to resolve 1D quantization phenomena. Here we are only sensitive to electrodynamic effects. When detecting the PL radiation with the electric field vector parallel to the wires (s-polarization) the spectrum shows two maxima, see Fig. 5. The maximum at 802.5 nm is due to the e_1-hh_1 free QWE transition, as can be deduced from the temperature

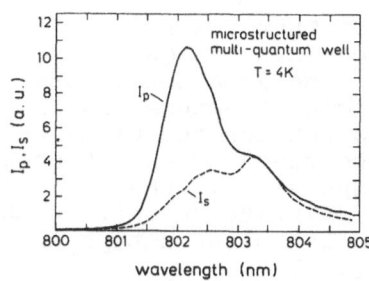

Fig. 5. Experimental luminescence spectra of a
microstructured GaAs-QW wire sample as shown in
Fig. 2b with a = 350 nm and L_x = 150 nm. I_p and
I_s denote the p- and s-polarized luminescence
intensity, respectively. The strong transition
at 802.1 nm, which only occurs in p-
polarization, arises from the radiative decay of
QWEP. Thus the luminescence spectra of wide QW
wire structures are also dominated by QWEP
emission. The excitation intensity was 10
mW/cm^2. (From Ref. 10)

and intensity dependence. The corresponding transition wavelength of about 802.2 nm is shifted by 0.1 nm with respect to
the free QWE on the unstructured reference sample. The
smallness of this shift indicates that stress effects[37], which
could have been induced by the etching, are negligible in our
samples and will not cause any polarization effects. The peak
at 803.4 nm arises from etching-induced states, as discussed
above. The luminescence efficiency for the free QWE is reduced
by more than one order of magnitude as compared to the corresponding transition in the unstructured reference sample. In the
p-polarized spectrum an additional strong transition shows up on
the high energy side of the free QWE transition as in Fig. 4 and
strongly dominates the PL spectrum. We conclude that this p-
polarized emission again arises from the radiative decay of
QWEP-type excitations, which thus are also present in QW wire
structures. It is important to note that this enhanced p-
polarized emission does *not* depend on the polarization of the
incident exciting radiation with respect to the wire structure.
Temperature dependent measurements reveal that the QWEP emission
can be observed up to more than 60 K. This clearly demonstrates
the intrinsic character of the emission. Thus we can conclude
that also for the PL of wide quantum well wires QWEP emission
can give a significant contribution.

From the grating coupler condition (Eqn. 3) we can, in
principle, determine the dispersion $\omega_p(q_p)$ of QWEP by scanning ω
and the angle of emission θ. It is interesting to consider the
dispersion of QWEP in more detail. The dispersion in laterally
homogeneous QW systems is shown in Fig. 3b. In laterally
microstructured QW systems, as discussed here, additional
effects should arise. If the influence of the lateral
microstructure is small we expect gaps in the dispersion at
$a_p = m \cdot \pi/a$, m = ±1,2,3... due to the lateral superlattice effect
which produces new Brillouin zones. Such lateral superlattice
effects have e.g. been observed in the 2D plasmon dispersion of
laterally density modulated 2D electronic systems[38]. In a

wire structure we expect that the boundaries of the wire cause a confinement of the electrodynamic QWEP resonance. In a certain sense we expect 'polaritons in a box'. However, if the wires are arranged in a periodic structure, as in our experiments, electrodynamic coupling between the wires occurs, which influences the resonance frequencies. Such a coupled wire system even shows a dispersion if the different wires are excited with different relative phases. Such an effect has recently been discussed for plasmon-type excitations in arrays of 1D electron systems[39].

Unfortunately, all these detailed effects on the dispersion that we have discussed above cannot be resolved in the experiments so far, since the expected splitting is smaller than the linewidth of the observed transitions. However, if the lateral dimensions become very small then we cannot neglect that we also change the electronic properties of our system, i.e. we expect that quantum confinement effects on the electrons and holes become increasingly important and a continuous transition to 1D excitons occurs which is in this limit a better description of the excitation. The analogous transition from '3D polaritons in a box' to confined 2D excitons has recently been very nicely demonstrated for a transition from 3D exciton polaritons to 2D excitons in CdTe-CdZnTe superlattices with different well widths[40]. In the next paragraph we will demonstrate that we can achieve indeed the same transition from a 2D QWEP to a 1D exciton in very narrow quantum wire structures.

Here we like to note that confinement energies, energy levels, oscillator strengths and polarization dependences for 1D excitons have been calculated by several authors (e.g. Refs.41-45 and references therein). To our knowledge, the inter-action between neighboring quantum wires, which is present in all experiments that have been reported on 1D excitons so far, and which in a certain sense is a remainder of the polariton model, has not been calculated. In particular, the influence on the polarization dependence would be of great interest, since the experimental polarization dependence might be used as a proof to confirm the 1D character of the excitons (e.g. Ref. 17).

6.   QUASI-1D EXCITONS IN NARROW QUANTUM WIRES

So far we have restricted our discussion to QW wire systems with rather large wire widths ($L_x > 150$ nm), where the transition wavelengths and emission efficiencies are comparable to the results found for the exciton polaritons in QW systems. In the following we want to discuss QW wire systems with wire widths below $L_x = 100$ nm[13,14]. For these systems lateral quantum confinement effects become important.

Polarization-dependent PLE spectra obtained from a QW wire sample with $L_x = 70$ nm are shown in Fig. 6, together with the corresponding reference spectrum. The transitions in the reference spectrum are the e_1-hh_1 transitions (808.525 nm, 808.755 nm) and the e_1-lh_1 transitions (804.975 nm, 805.225 nm). From calculations we can ascribe the small splitting of these transitions to QW-thickness fluctuations of one monolayer[46]. The PLE spectrum of the QW wire sample exhibits a different

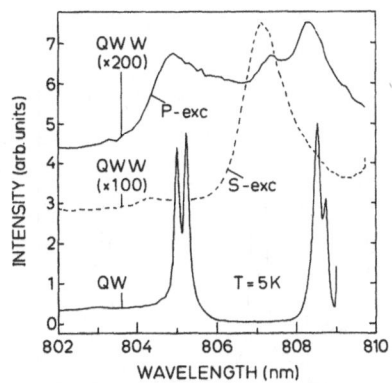

Fig. 6. PLE spectra of a QW wire (QWW) with $L_x \approx 70$ nm
and of the corresponding reference QW[13]. The
spectra are shifted vertically with respect to
each other for clarity and show the dependence
on the polarization of the exciting laser light.
For s-polarized (p-polarized) light $\vec{E}$ is
parallel (perpendicular) to the wires. The
short wavelength resonance of the reference QW
represents a light hole exciton excitation. The
long wavelength resonance a heavy hole exciton.
(The small splitting of the peaks in the
reference sample arises from monolayer
fluctuations.) The heavy hole excitons of the
QW wire samples is split into two resonances
which represent 1D excitons, i.e. excitons
related to energetically separated quantum
confined 1D subbands. The strong polarization
dependence reflects the different symmetry of
the wavefunctions. The excitation intensity was
about 1 mW/cm^2 for the QW wires.

behavior, which is in particular strongly dependent on the
polarization of the exciting laser light. Let us first
concentrate on the p-polarized spectrum. Three transitions can
be resolved. Two transitions occur in the wavelength regime of
the excitonic hh- and lh-groundstates of the reference sample.
Note the shift of these transitions of about 0.2 nm (0.4 meV) to
smaller wavelengths with respect to the corresponding reference
transitions. A third additional transition can be observed,
which is 1.3 nm away on the smaller wavelength side of the hh-
groundstate at 807 nm. These transitions represent quasi-1D
exciton transitions.

To demonstrate this we have performed calculations of the
confinement energies using a 2D finite potential well model.
Taking the values for the exciton binding energies into account,
which we deduced from magneto-optical experiments (see below),
we can indeed ascribe the observed transitions to excitonic
transitions related to quantum confined 1D subbands. In
particular, the third new transition, which will be labeled hh_{12}
in the following, is due to the second 1D subbands in the
conduction and valence band. The indices of hh_{ij} denote the
quantum numbers for the z and x direction, respectively. From
our modeling we obtain a value for the blue shift of 0.2 meV,

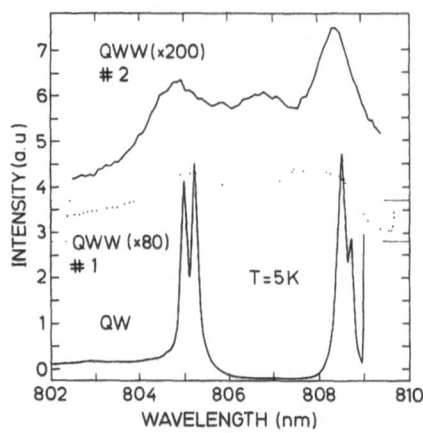

Fig. 7. Wire widths dependence of 1D excitons[14]. PLE
spectra of two QW wires (QWW) with $L_x \approx 60$ nm
(#2) and $L_x \approx 100$ nm (#1) and of the
corresponding reference QW. The spectra are
shifted vertically with respect to each other
for clarity. The energy separation of the L_x =
60 nm wire is 3.3 meV (1.65 nm), the energy
separation of the L_x = 100 nm wire is 1.5 meV
(0.75 nm).

which is in reasonable agreement with the experimental finding.
We therefore conclude that stress effects[37] play a minor role
even in these very narrow QW wire structures.

As another direct manifestation of the quantum confinement
one expects that the energy separation between the hh_{11} and hh_{12}
transitions should increase with decreasing wire width L_x. We
indeed find this wire width dependence if we compare the PLE
spectra of QW wire samples with different values of L_x in Fig.
7. The lower curve is the corresponding reference spectrum of a
QW sample from the same wafer. The two peaks in the reference
spectrum at 808.93 nm and 805.48 nm are the e_1-hh_1 and e_1-lh_1
transitions, respectively. Both QW wire spectra exhibit three
transitions. The two groundstate transitions, hh_{11} and lh_{11},
are blue shifted with respect to the corresponding reference
transitions. For the case of L_x = 100 nm, the third transition,
hh_{12}, is only separated by 0.75 nm (1.5 meV) from the hh_{11}
transition. With decreasing wire width this separation
increases to 3.3 meV for L_x = 60 nm. In addition, the energy
separation between the hh_{11} and lh_{11} excitons increases. The
latter effect is rather small; nevertheless it seems to indicate
a systematic wire width dependence of the confinement for the
hh_{11} and lh_{11} states and of the binding energies of the
corresponding excitons.

The intrinsic nature of the hh_{12} transition has further
been confirmed by temperature dependent PLE spectroscopy and a
series of PLE measurements, where we set the detection wavelength to the different resonance positions. However, in all
these measurements we were not able to resolve the lh_{12}
transition, which should be located for L_x = 70 nm at about 803
nm. The weakness of transitions between higher 1D electron and

134

lh subbands has also been observed for InP/InGaAs QW wire samples[8].

A very interesting observation for the QW wire samples is a characteristic polarization dependence. As can be seen in Fig. 6, the lh_{11} and hh_{11} transitions are dominant when we excite the QW wire with p-polarized laser light. However, the shape of the spectra drastically changes when we switch to s-polarized excitation. In this case the spectrum is dominated by the hh_{12} transition such that the hh_{11} transition can only be observed as a shoulder. The lh_{11} transition is nearly absent in the spectrum. The strong polarization dependence, in particular the different behavior of the hh_{11} and hh_{12} transitions shows that this is an intrinsic property of the 1D system, which must be related to the symmetry of the 1D wavefunctions. The anisotropy of the groundstate transitions in a QW wire has been discussed in Ref. 17. The exact shape of the exciton wave functions depends strongly on the boundary conditions at the etched sidewalls, which are so far not exactly known. It is very possible that the symmetry of the wave functions is influenced by surface states and the band bending near the etched surfaces. For a quantitative explanation of the polarization dependence observed here, in particular of the hh_{12} transition, one has in addition to keep in mind that it depends not only on the exciton wave function symmetry but also on the electrodynamic effects, as discussed above. The adjacent wires are coupled via electromagnetic fields and the polarization of the excited and emitted radiation is dependent on the grating coupler effect of the strong corrugation of the samples. We expect that these effects are particularly strongly pronounced in our deeply etched grating structures. Thus, without a complete grating coupler theory, which is a very complex problem, we cannot predict the polarization dependence quantitatively.

For a further confirmation of the 1D character and for additional quantitative information we have performed magneto-optical PLE experiments in magnetic fields oriented perpendicularly to the layers[13,14]. This method is well established for 2D QW excitons[47-49]. The exciton energies in a magnetic field are determined by an interplay between the excitonic Coulomb interaction, governed by the effective Rydberg energy and by the magnetic energy which is determined by the cyclotron resonance energy[50]. For low exciton states, in particular for the ground state and small magnetic fields, the Coulomb interaction dominates and the exciton energy is only affected by the diamagnetic shift. This diamagnetic shift is proportional to the square of the transverse extent of the exciton wavefunction and as such can be used to study the expected shrinkage of the 1D exciton wavefunction. For higher exciton states and larger B, the magnetic energy dominates and the transitions take the character of Inter-Landau-Level transitions with a linear B dependence. These Inter-Landau-Level transitions can be easily observed with increasing magnetic field. Extrapolation to B = 0 then allows a determination of the exciton binding energy.

The energy positions in the PLE spectra of a QW wire with L_x = 70 nm and of the corresponding reference QW in a magnetic field are summarized in Fig. 8. For clarity we did not plot the magnetic field dispersion of the QW in discrete data points but as solid lines. The magneto-dispersion of the reference sample

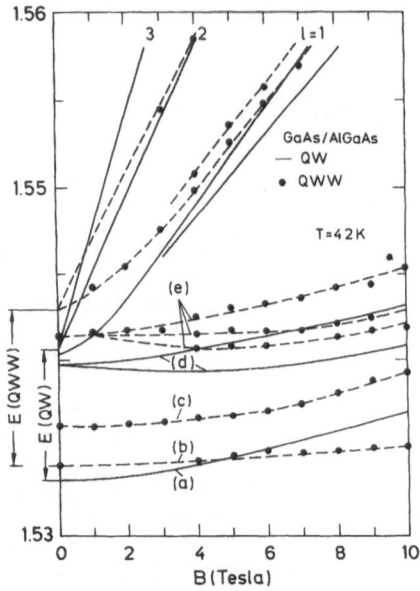

Fig. 8. Energy positions of the maxima in the magnetic-
field dependent PLE-spectra of a $L_x \approx 70$ nm QW
wire (QWW)[13]. The solid lines indicate the
magnetic field dispersion for the transitions of
the corresponding reference QW. (a), (b), (c),
(d) and (e) denote the hh_1, hh_{11}, hh_{12}, lh_1 and
lh_{11} transitions, respectively; the index l
labels the order of the Inter-Landau-Level
transitions. In addition, the exciton binding
energies for the QW and the QW wire, E(QW) and
E(QWW), respectively, which were deduced from
zero-field extrapolations, are indicated. They
show directly the increased binding energies of
the 1D excitons. The reduced dimensionality is
also reflected in the smaller diamagnetic shift
of the QW wire sample.

is very similar to earlier measurements on QW systems, e.g.
Refs. 47-49. The hh_1 and lh_1 transitions show a diamagnetic
shift which is stronger for the hh_1 transition reflecting the
weaker Coulomb interaction of the hh_1 exciton. For the lh_1
transition the spin-splitting is resolved. The magnetic field
dependence of the QW wires is strikingly different. The hh_{11}
transition of a QW wire shows a smaller diamagnetic shift as
compared to the reference sample, indicating a stronger exciton
binding energy. The diamagnetic shift of the hh_{12} transition
demonstrates its excitonic character, which is less pronounced
than that of the groundstate exciton. These features are
qualitatively expected for 1D magneto-excitons and confirm our
interpretation of the hh_{11} and hh_{12} transitions.

We analyzed the magnetic field dependence following similar
evaluations for 2D systems, e.g. Refs. 48,49. The analysis was
performed at sufficiently low magnetic fields (below 3 T), where
we observed no significant deviation from a quadratic energy
shift. The diamagnetic shift ΔE of the hh-groundstate exciton

of the reference sample was 0.51 meV at 3 T, which is consistent with published values[48,49]. The corresponding exciton binding energy is E_B of 7.5 meV, which was determined by extrapolating the Inter-Landau-Level transitions to B = 0. The value of 7.5 meV agrees very well with reported binding energies of comparable QW samples[51,52]. For the hh_{11} transition of the QW wire we could determine E_B = 8.7 meV, which indicates an enhancement of the groundstate excitonic interaction in a QW wire by about 15% with respect to the 2D QW reference sample. From the diamagnetic shift of the excitons we evaluate that the transverse extension of the hh_1 exciton of the reference QW was 1.09 a_B, if we follow the analysis of Ossau et al.[49]. This value is in excellent agreement with the calculated value of Bastard[53] for a 14 nm QW. The corresponding value of the QW wire hh_{11} exciton (0.97 a_B) is by about 11% reduced, which directly indicates the shrinkage of the excitonic wave function due to the additional lateral confinement. It was also possible to determine the exciton binding energy of the hh_{12} transitions for the QW wire. It was found to be E_B = 8.1 meV.

CONCLUSIONS

Quantum well exciton polaritons and one-dimensional quantum confined excitons have been studied by photoluminescence spectroscopy of laterally microstructured $Al_xGa_{1-x}As$ - GaAs quantum well systems and quantum wires as narrow as 60 nm. The resonantly enhanced radiative decay of freely propagating quantum well exciton polaritons was investigated in quantum well samples via the grating coupler effect of the periodically corrugated cap layer. It was observed also that the photoluminescence spectra of wide quantum well wires is dominated by quantum well exciton polariton emission. When decreasing the wire width below $L_x \approx$ 150 nm the polariton-specific features disappear from the spectra. Below L_x = 100 nm, the concept of quasi-one-dimensional excitons is an appropriate model to describe the behavior of the observed transitions. The 1D character of these excitons is, for example, manifested in the observation of several exciton resonances which are related to different, energetically separated quantum confined 1D subbands, by a reduced diamagnetic shift in a magnetic field, and by a 15% increased binding energy with respect to 2D reference samples.

Acknowledgment

This work was supported by the Bundesministerium für Forschung und Technologie.

REFERENCES

1.   R. Dingle, W. Wiegmann and C.H. Henry, **Phys. Rev. Lett.**, 33:827 (1974)
2.   C. Weisbuch, p261 **in**: "Physics and Application of Quantum Wells and Superlattices", E.E. Mendez and K. von Klitzing, eds., Plenum Press, New York (1987)
3.   K. Kash, A. Scherer, J.M. Worlock, H.G. Craighead and M.C. Tamargo, **Appl. Phys. Lett.**, 49:1043 (1986)
4.   J. Cibert, P.M. Petroff, G.J. Dolan, S.J. Pearton, A.C. Gossard and J.H. English, **Appl. Phys. Lett.**, 49:1275 (1987)

5. M.A. Reed, R.T. Bate, K. Bradshaw, W.M. Duncan, W.R. Frensley, J.W. Lee and M.D. Shih, **J. Vac. Sci. Technol.**, B4:358 (1986)

6. H. Temkin, G.J. Dolan, M.B. Panish and S.N.G. Chu, **Appl. Phys. Lett.**, 50:413 (1987)

7. Y. Hirayama, S. Tarucha, Y. Suzuki and H. Okamoto, **Phys. Rev. B**, 37:2774 (1988)

8. D. Gershoni, H. Temkin, G.J. Dolan, J. Dunsmuir, S.N.G. Chu and M.B. Panish, **Appl. Phys. Lett.**, 53:995 (1988)

9. H.E.G. Arnot, M. Watt, C.M. Sotomayor-Torres, R. Glew, R. Cusco, J. Bates and S.P. Beaumont, **Superlattices and Microstr.**, 5:459 (1989)

10. M. Kohl, D. Heitmann, P. Grambow and K. Ploog, **Phys. Rev. B**, 37:10927 (1988)

11. M. Kohl, D. Heitmann, P. Grambow and K. Ploog, **Superlattices and Microstr.**, 5:235 (1989)

12. A. Forchel, H. Leier, B.E. Maile and R. German, "Advances in Solid State Physics", U. Rössler, ed., Vieweg, Braunschweig (1988)

13. M. Kohl, D. Heitmann, P. Grambow and K. Ploog, **Phys. Rev. Lett.**, 63:2124 (1989)

14. M. Kohl, D.Heitmann, P. Grambow and K. Ploog, **Surf. Sci.**, 229:248 (1990)

15. K. Kash, J.M. Worlock, M.D. Sturge, P. Grabbe, J.P. Harbison, A. Scherer and P.S.D. Lin, **Appl. Phys. Lett.**, 53:782 (1988)

16. K. Kash, J.M. Worlock, A.S. Gozdz, B.P. Van der Gaag, J.P. Harbison, P.S.D. Lin and L.T. Florez, **Surf. Sci.**, 229:245 (1990)

17. M. Tsuchiya, J.M. Gaines, R.H. Yan, R.J. Simes, P.O. Holtz, L.A. Coldren and P.M. Petroff, **Phys. Rev. Lett.**, 62:466 (1989)

18. M. Tanaka and H. Sakaki, **Appl. Phys. Lett.**, 54:1326 (1989)

19. E. Kapon, D.M. Hwang and R. Bhat, **Phys. Rev. Letts.**, 63:430 (1989)

20. K. Kash, **J. Luminescence**, 46:69 (1990)

21. D. Heitmann, M. Kohl, P. Grambow and K. Ploog, p.255 **in:** "Science and Engineering of One- and Zero-Dimensional Semiconductor Systems", S. Beaumont and C.M. Sotomajor Torres, eds., Plenum Press (1990)

22. P. Grambow, T. Demel, D. Heitmann, M. Kohl, R. Schule and K. Ploog, **Microelectronic Engineering**, 9:357 (1989)

23. For monographs on surface excitations and coupling processess, see e.g. "Electromagnetic Surface Moes", A.D. Broadman, ed., Wiley, New York (1982); "Surface Polaritons", V.M. Agranovich and D.L. Mills, eds., North-Holland, Amsterdam (1982)

24. D. Heitmann and H. Raether, **Surf. Sci.**, 59:17 (1976)

25. N. Marshall and B. Fischer, **Phys. Rev. Lett.**, 28:811 (1972)

26. D.D. Sell, S.E. Stokowski, R. Dingle and J.V. DiLorenzo, **Phys. Rev. B**, 7:4568 (1973)

27. J. Lagois and B. Fischer, **Phys. Rev. Lett.**,36:680 (1976)

28. M. Nakayama, **Solid State Commun.**, 55:1053 (1985)

29. M. Nakayama and M. Matsuura, **Surf. Sci.**, 170:641 (1986)

30. R. Del Sole and A. D'Andrea, p289 **in:** "Optical Switching in Low Dimensional Systems", H. Haug and L. Bányai, eds., Plenum Press, New York (1989); L.C. Andreani and F. Bassani, **Phys. Rev.B**, in press.

31. C. Zhang, M. Kohl and D. Heitmann, **Superlattices and Microstr.**, 5:65 (1989)

32. C. Weisbuch, R.C. Miller, R. Dingle, A.C. Gossard and W.

Wiegmann, **Solid State Commun.**, 37:219 (1981)

33. D. Heitmann, **Surf. Sci.**, 170:332 (1986)
34. V.M. Agranovich and T.A. Leskova, **Pis'ma Zh. Eksp. Teor. Fiz.**, 29:151 (1979); **JETP Lett.**, 30:538 (1979)
35. K. Ogawa, T. Katsumura and H. Nakamura, **Appl. Phys. Lett.**, 53:1077 (1988)
36. K. Ogawa, T. Katsumura and H. Nakamura, **Phys. Rev. Lett.**, 64:796 (1990)
37. C. Jagannath, E.S. Koteles, J. Lee, Y.J. Chen, B.S. Elman and J.Y. Chi, **Phys. Rev. B**, 34:7027 (1986)
38. U. Mackens, D. Heitmann, L. Prager, J.P. Kotthaus and W. Beinvogl, **Phys. Rev. Lett.**, 53:1485 (1984)
39. W. Que and G. Kirczenow, **Phys. Rev. B**, 37:1589 (1988)
40. H. Tuffigo, R.T. Cox, F. Dal'bo, G. Lentz, N. Magnea, H. Mariette and C. Grattepain, **Superlattices and Microstr.**, 5:83 (1989); H. Tuffigo, B. Lavigne, R.T. Cox, G. Lentz and N. Magnea, **Surf. Sci.**, 229:480 (1990)
41. M.H. Degani and O. Hipolito, **Phys. Rev. B**, 35:9345 (1987)
42. I. Suemune and L.A. Coldren, **IEEE QE**, 24:1778 (1988); Suemune, L.A. Coldren, and S.W. Corzine, **Superlattices and Microstr.**, 4:19 (1988)
43. G. Bastard, p21 **in:** "Physics and Application of Quantum Wells and Superlattices", E.E. Mendez and K. von Klitzing, eds., Plenum Press, New York (1987)
44. J.W. Brown and H.N. Spector, **Phys. Rev. B**, 35:3009 (1987)
45. G.W. Bryant, **Phys. Rev. B**, 37:8763 (1988)
46. M. Kohl, D. Heitmann, S. Tarucha, K. Leo and K. Ploog, **Phys. Rev. B**, 39:7736 (1989)
47. J.C. Maan, G. Belle, A. Fasolino, M. Altareli and K. Ploog, **Phys. Rev. B**, 30:2253 (1984)
48. D.C. Rogers, J. Singleton, R.J. Nicholas, C.T. Foxon and K. Woodbridge, **Phys. Rev. B**, 34:4002 (1986)
49. W. Ossau, B. Jäkel, E. Bangert, G. Landwehr and G. Weimann, **Surf. Sci.**, 174:188 (1986)
50. O. Akimoto and H. Hasegawa, **J. Phys. Soc. Japan**, 22:181 (1967)
51. R.C. Miller, D.A. Kleinman, W.T. Tsang and A.C. Gossard, **Phys. Rev. B**, 24:1134 (1981)
52. U. Ekenberg and M. Altarelli, **Phys. Rev. B**, 35:7585 (1987)
53. G. Bastard, E.E. Mendez, L.L. Chang and L. Esaki, **Phys. Rev. B**, 26:1974 (1982)

THE OPTICAL PROPERTIES OF NARROW-GAP LOW DIMENSIONAL STRUCTURES

R.A. Stradling

London University Interdisciplinary Research Centre in
Semiconductor Materials and the Blackett Laboratory
Imperial College, London SW7 2BZ, UK

ABSTRACT

The paper reviews recent developments in the optical
properties of narrow-gap semiconductors. The extension of
optoelectronics to wavelengths beyond 1.5 μm has required the
development of novel device concepts and, unlike the situation
with GaAs based systems, new materials technologies. A case in
question is the interest in replacing mercury cadmium telluride
alloys for devices operating in the region of 10 μm wavelength,
because of the severe structural and stability problems. The
following examples are given of differing approaches currently
being investigated.

Mismatch epitaxy is now widely employed either to modify
the band structure by fabricating strained layer superlattices
or to integrate grossly mismatched materials such as InSb with
GaAs or GaAs with silicon. In the latter case the strain is
relaxed by generation of large concentrations of misfit
dislocations. Nevertheless it is shown that the carrier
mobility need not be strongly degraded even close to the
interface between the mismatched materials.

Strained layer superlattices based on the alloy system
In(As,Sb) have been used to generate band edge photoluminescence
at wavelengths beyond 10 μm. Examples are given of structural
problems encountered with these III-V alloys. Remarkable
photoluminescence results have also been obtained with single
monolayers of InAs inserted into GaAs where intense sub band gap
emission is observed with samples grown by flow-modulation
MOCVD.

Other novel narrow-gap systems which are currently being
investigated include alpha-tin and its alloys with germanium
where the alpha phase is stabilized by built-in strain, and
alloys of InSb and InAs with bismuth.

Alternative approaches to midinfrared photonics include i)
the use of doping superlattices (n-i-p-i structures) to modify
the band gap and to produce highly nonlinear optical effects,

ii) the exploitation of intersubband absorption in quantum well structures.

1.   INTRODUCTION

    The most developed low dimensional semiconductor structures, both for fundamental studies and in device applications, are those based on abrupt heterostructures between GaAs and AlGaAs. This system depends for its success on the near-perfect match between the binary components, GaAs and AlAs. However, neither silicon nor GaAs has a band gap suitable for long wavelength optical sources, signal processing devices or detectors. Consequently developments in optoelectronics have produced demands for new semiconductor materials and device concepts. This article reviews advances in the optical properties of low dimensional narrow-gap semiconductors which have taken place since the NATO Advanced Research Workshop of that name was held in St. Andrews, Scotland in 1986[1].

    A variety of new materials systems are now emerging where the components are deliberately chosen to be grossly mismatched. In many cases mismatch epitaxy is used deliberately to change the symmetry of the bands and to modify the band gaps (e.g. $Ga_{1-x}Al_xAs/GaAs$ and $Si/Si_{1-x}Ge_x$ strained layer superlattices). In the case of $InAs_{1-x}Sb_x$ strained layer quantum wells the long wavelength barrier of 10 μm can be breached although both components of the heterostructures have band gaps corresponding to shorter wavelengths. However, as will be discussed in this paper, there are severe metallurgical problems in the mid-alloy range with this system.

    A novel development in strained layer systems concerns the incorporation of single monolayers of grossly mismatched materials in a matrix of another material (e.g. single layers of InAs or GaSb in GaAs - good quality samples of both systems have been demonstrated). It becomes a semantic question as to whether this should be considered as delta isoelectronic doping or an ultrathin strained layer system.

    The strain can also be used to increase the phase stability of one component as for the narrow-gap system consisting of alpha-tin grown on InSb or CdTe. This system shows a very pronounced two-dimensional electron gas at the interface of a single heterostructures.

    There is a need with the new materials systems to integrate structures onto well established materials such as silicon or GaAs. With some systems (e.g. InSb or InAs on GaAs or GaAs on Si) the mismatch is too great to permit the growth of thick epitaxial layers although it is desirable for other reasons to create structures of these combinations. In these cases the layers are allowed to relax and a high density of misfit and threading dislocations form at the interface. A surprising result is that strong Shubnikov-de Haas oscillations and mobilities close to the bulk value can be observed in lightly doped ultrathin sample of InSb grown on GaAs substrates. This result suggests that the dislocations are only weakly charged and do not therefore degrade the majority carrier properties significantly.

Spike or delta doping where the dopants are deposited on a single atomic plane gives rise to the possibility of building in electric fields as high as 10^6V/cm into semiconductor structures. The fabrication of n-i-p-i structures can give rise to large reductions in band gap, long lifetimes and big optical nonlinearities.

Another alternative to narrow-gap semiconductors for devices operating in the long wavelength regime is the use of intersubband transitions in quantum wells or doping structures.

2. NOVEL MATERIALS FOR INFRARED OPTOELECTRONICS

The alloy system $InSb_{1-x}As_x$ is of considerable interest as strained layer superlattices with band gaps corresponding to wavelengths longer than 10 μm have been grown with this sytem[2]. However, the metallurgical problems encountered with this system provide a good example of the difficulties associated with developing a new materials technology.

As a result of accurate calibration of the Group V fluxes by in-site RHEED we are now able to reproducibly grow superlattices of $InAs_xSb_{1-x}/InAs_{1-y}Sb_y$ over a wide range of thickness or composition. Depending upon the growth temperature, the 77K Hall mobility of these alloys can drop by as much as three orders of magnitude in mid-range compared with the mobilities measured for the binary compounds (Fig. 1). This is a surprising result because the level of contamination in the growth system is very low as demonstrated by the high quality of both the InAs and InSb samples produced. It is believed that this drop of mobility in the MBE alloys arises from structural problems. High on the list of possible mechanisms for the drop in the mobility in the alloys are the ordering effects which have recently been observed[3] in transmission electron diffraction which include localized order, the formation of clusters and even phase separation in the mid-alloy range. This type of electrical inhomogeneity on the scale of the electron mean free path will severely distort the current flow in the sample and give rise to highly misleading Hall measurements. We are now carrying out a careful comparison of the electrical, optical and structural properties of the alloys in order to establish the correlation with alloy ordering clustering and phase separation.

It should be noted that alloy ordering was not reported in Ref. 3 for the low arsenic concentrations (x < 0.2) employed to fabricate the strained layer superlattices which showed band gap photoluminescence at wavelengths longer than 10 μm as discussed in Section 3.3. Nevertheless, a pronounced reduction in mobility is found on replacing only a small proportion of the Sb by As as is apparent in Fig. 1. This result suggests that alloy scattering is strong even for low values of x. It is interesting to note that polycrystalline alloy samples prepared on glass substrates by the relatively crude technique of flash evaporation followed by recrystallization showed very similar mobilities as reported at the NATO Workshop held in St. Andrews[4] (e.g. 35,000 cm^{-2}/Vs for x = 0.15). In this case the Hall mobility could be shown to be the true bulk mobility by the observation of a well-resolved cyclotron resonance line with the expected line width and effective mass (see Fig. 2).

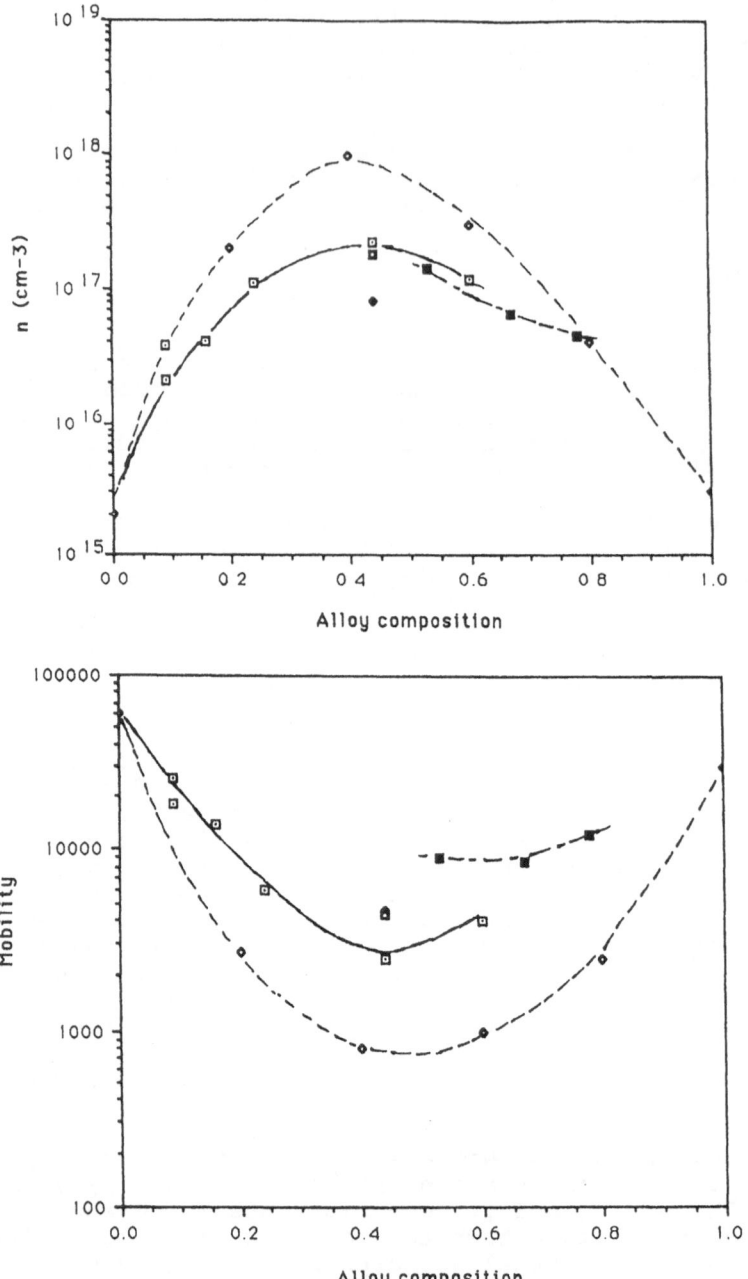

Fig. 1. Shows the variation of Hall mobility and carrier
concentration found at 77 K for a series of
$InAs_{1-x}Sb_x$ alloy samples prepared by MBE. The
upper set of points are for films grown at
410°C, the lower for samples grown at 370°C.
The lower set of samples exhibit the phase
separation effects shown in Fig. 5 for
compositions $(0.2 \leq x \leq 0.8)$ (I. Ferguson, A.
d'Oliveira & R.A. Stradling to be published).

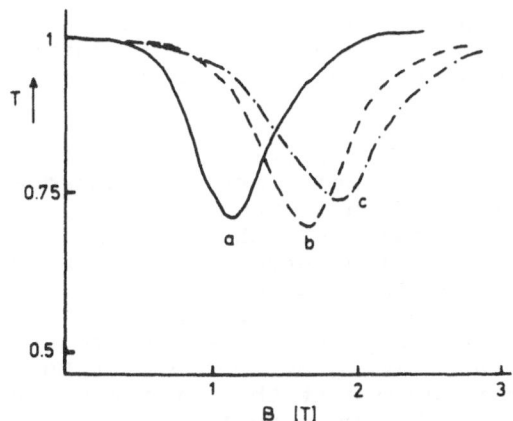

Fig. 2. Shows the results of cyclotron resonance
experiments with a polycrystalline film of
$InAs_{0.15}Sb_{0.85}$ deposited on a glass
substrate (taken from Ref. 4). The
recordings are taken with a laser
wavelength of 118 μm and at three
different values of hydrostatic pressure
(0, 6.7 and 10.3 kbar).

The effective mass for zero hydrostatic pressure derived
from the data shown in Fig. 2 is 0.0105 m_e. The extrapolation
of the masses obtained over a range of laser wavelengths (32 to
350 μm) to zero energy give a band edge mass of 0.0088m_e. This
result is that expected from a four-band k.p model with momentum
matrix elements (P^2) independent of alloy composition, i.e. that
the x-dependence of the band edge effective mass shows very
similar 'bowing' to that found with the band gap. Earlier
cyclotron resonance experiments with wider band gap III-V alloys
(e.g. $InAs_{1-x}P_x$) showed a linear dependence of band edge mass on
x despite pronounced bowing being apparent in the variation of
band gap with alloy composition[5]. The latter result is
thought to arise from an increasing admixture of higher-order
bands by alloy disorder. With the narrow-gap system $InAs_xSb_{1-x}$
the effective mass is dominated by the Γ_8-Γ_6 gap and relatively
weakly affected by admixture of higher order bands by alloy
disorder.

Alternative III-V systems for the mid-infrared wavelength
range are $InSb_{1-x}Bi_x$[6] and $InAs_{1-x}Bi_x$. With the latter system
the band gap of InAs was reduced by 25% on introducing 2% Bi[7].
An even more exotic possibility for this wavelength region is
the alpha-tin/germanium alloy system which can be combined with
InSb or CdTe as discussed in Section 4. With both the SnGe
alloy system and the narrow-gap III-Vs alloyed with Bi, the
alloys are metastable but can be grown by thin-layer epitaxial
techniques.

Another alloy system where ordering and clustering can be a
problem is $Al_{1-x}In_xAs$ as there are large differences in the In
and Al related bond energies. The interest in this system is
that it is lattice matched to InP at a composition x = 0.5 and
(GaIn)As/(AlInAs) HEMTs show superior microwave performance to
(GaInAs/InP) structures, probably because of the larger

conduction band offsets. This large offset for (GaIn)As/
(AlIn)As has enabled intersubband multiple-quantum-well
detectors operating at wavelengths as short as 5 μm to be
fabricated[8].

A further interesting narrow-gap combination is InAs/AlSb.
These two compounds are closely lattice matched and have a
conduction band offset which is thought to be in the range 1.35
to 1.9 eV, i.e. even larger than that for the (GaIn)As/(AlIn)As
system. Excellent electrical properties including quantum Hall
measurements have been reported for this system[9]. However,
both the anion and cation change across the interface and the
electrical quality depends strongly on the sequence in which the
quantum well interfaces are grown because of the possibility of
formation of antisite donors at the interface. High mobilities
are only seen when the bottom interface is InSb-like.

In the search for materials having optimal properties,
GaSb/AlSb, GaSb/InAs and InSb/In$_{1-x}$Al$_x$Sb are being investigated
for long wavelength applications and the quaternary systems
(GaIn)(AsP) and (GaIn)(AsSb) for applications in the 1.3 to 1.6
μm wavelength band.

3.   MISMATCH EPITAXY

Mismatch epitaxy is now widely employed to fabricate
heterostructures where the lattice constants of the two
component materials differ by $\geq$ 1/2%. In strained layer
superlattices (SLS) the epitaxial layers of the two components
are thinner than the critical thickness for the formation of
misfit dislocations with the result that the layers remain in
registry. In this case the resulting strain is deliberately
used to modify the band structure of the system. With SLS the
materials are generally chosen so that the mismatch is
relatively small so that quite thick layers can be grown without
dislocations forming. In many other cases, it may be desirable
to combine materials with very different electrical or optical
properties despite the large mismatch precluding the pseudomorphic growth of more than a few monolayers (i.e. when the
mismatch exceeds a few percent). Examples of such materials
combinations are InSb or InAs on GaAs where GaAs substrates are
chosen because of their cheapness and their electrical
insulating properties; and GaAs on Si where the aim is to
integrate optoelectronic and microelectronic devices on the same
chip. With these systems the strain is relieved in the
interface region by the formation of a high density of misfit
dislocations and relatively thick layers may then be grown with
reasonable electrical and optical quality with the epitaxial
material in the almost completely relaxed state.

3.1 Electrical Measurements in Relaxed Films of InSb and InAs

If the components are grossly mismatched and epitaxial
films are grown at thicknesses much greater than the critical
thickness above which the lowest energy state is the layer
having its natural lattice spacing, a very high density of
misfit dislocations (10^{11} cm^{-2}) will be found close to the
interface and the film is almost completely relaxed. The
dislocation density falls below 10^7 cm^{-2} after growth has
continued for a few microns. There is an almost universal

belief that a high density of dislocations will dramatically
degrade the electrical properties of semiconductors and there
are many reported cases of materials systems where the Hall
mobility falls rapidly as the thickness of the epitaxial layers
is decreased. The dependence of the Hall effect, conductivity
and magnetoresistance on temperature and film thickness has been
investigated for a series of thin layers of InSb grown by
Molecular Beam Epitaxy (MBE) on GaAs substrates where the
dislocation density is high throughout the films (the lattice
mismatch is 14% for this system). GaAs is a popular substrate
for narrow-gap semiconductors such as InSb and InAs as it
provides good isolation for electrical measurements.

The average carrier concentration (n) determined by the
Hall effect increased rapidly with decreasing thickness and the
mobility (μ) fell at a slightly greater rate (see Fig. 3). The
magnetic field employed for the Hall measurements was below
0.05 T to ensure that a low field analysis could be applied. No
conductivity could be measured at thicknesses below 0.08 μm.
Also included in the figure are 77 K mobility results for
magnetron sputtered samples of lower purity than the MBE
samples[10].

However, measurements with samples backdoped with silicon
show that the bulk electrical parameters can be recovered even
with very thin samples at modest doping levels. The mobility at
77 K is within a factor of two of the normal ionised impurity
mobility for bulk samples at the doping concentration concerned.
The doping concentration chosen and measured in the Hall
experiment was close to the value measured with thin undoped
samples of thickness 0.1 μm. Yet the mobility measured with the
thin undoped samples (100 cm^{-2}/Vs at 1500 Å and 20 cm^{-2}/Vs at
800 Å) was more than three orders of magnitude lower.

It is very clear that the Hall measurements obtained with
the doped samples are in conflict with those for the undoped
samples. Shubnikov-de Haas measurements provide an independent
check on both the microscopic carrier concentrations and
mobilities in the doped samples. With all the samples studied
good quality Shubnikov-de Haas peaks were observed at 4 K. The
periods observed at high fields were consistent with the silicon
doping levels.

These results show clearly and unambiguously that the Hall
measurements on the thin undoped samples are misleading when
interpreted in terms of a single carrier model for InSb grown on
GaAs substrates and that the local mobility can be close to the
bulk value even when the conducting region is very close to the
interface demonstrating that a high density of dislocations has
little effect on the electrical properties of the majority
carriers.

The contradictory results for the doped and undoped samples
can be reconciled if it is assumed that the apparent cata-
strophic fall in the mobility in the undoped samples as the film
thickness decreases is an artefact arising from the presence of
a p-type region close to the interface. The resultant two
carrier conduction resulting from such a layer would cause the
Hall voltage to decrease rapidly with decreasing thickness while
maintaining the local conductivity approximately constant as is
observed experimentally with the undoped structures.

The conclusion therefore is that, even when conduction is confined to within 1000 Å of the mismatched interface, the high density of dislocations have only a small effect on the carrier mobility. It was shown by Kimmerling and Patel[11] that in silicon only a small proportion of the sites at the dislocation core are electrically active. It therefore implies that virtually all the dislocation core is reconstructed and neutral. Recombination site spacings of the order of 200 Å along the dislocation cores were estimated. Consequently Kimmerling and Patel proposed that the electrically active sites are located at the dislocation kinks. The situation should be similar in III-V materials[12]. In a sufficiently narrow quantum well super-lattice the majority of the dislocation kinks can be expelled from the wells into the barriers with the result that good luminescence efficiency can be obtained from GaAs/AlGaAs quantum well lasers grown on Si substrates[13].

With our thin InSb on GaAs films, a doubly-charged dangling bond at each kink, a kink separation of 200 Å and a dislocation density of 10^{10} cm^{-2} would give rise to 10^{11} charges cm^{-2} in a 1000 Å film or a volume density of about 3×10^{16} cm^{-3}. This is less than the silicon levels used for the doped InSb samples and therefore in agreement with experiment would not degrade the mobility significantly. Electrical measurements with epitaxial InAs films provide another good example of how Hall measurements can be misinterpreted if simple one-carrier theory is applied on the assumption that the conducting medium is homogeneous. The InAs surface contains a high density of surface states which normally lead to a conducting skin which is formed by an electron accumulation layer.

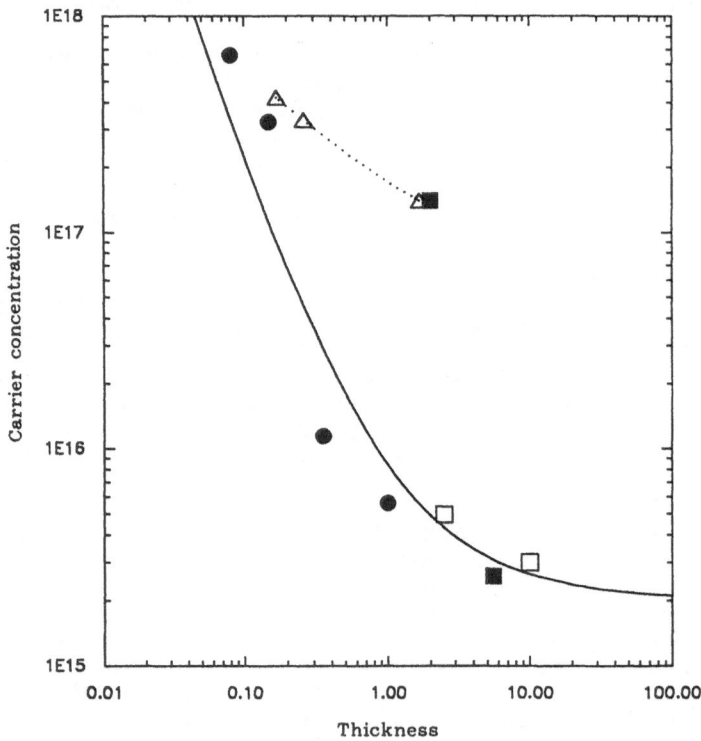

Fig. 3. (continued on facing page)

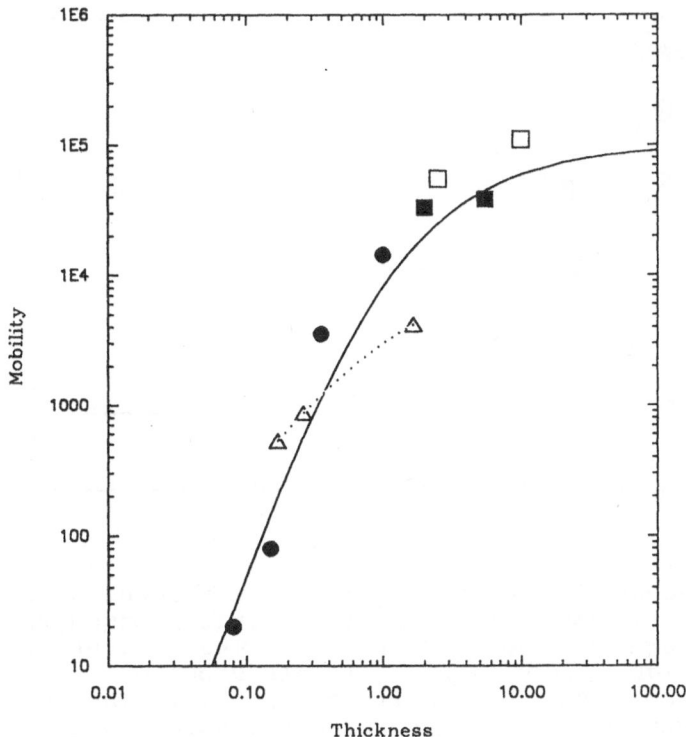

Fig. 3. (Continued) Shows the Hall mobility and carrier
         concentration measured for a range of InSb on
         GaAs heterostructures of differing thicknesses
         (S.J. Patel - MSc thesis, Imperial College).
         The majority of the samples were grown by MBE
         but the dotted curves are the results for
         magnetron sputtered samples (Ref. 10). The
         latter samples have two orders of magnitude
         higher carrier concentration (n) and an order of
         magnitude lower mobility (μ) at thicknesses
         greater than about 1 μm indicating a lower base
         purity than the MBE samples. However, the
         values of n and μ for the samples prepared by
         the different techniques are virtually identical
         at thicknesses of the order of 0.1 μm because of
         the influence of the interface.

     Recently some very high mobility films of n-InAs were grown
on GaAs substrates in the Imperial College MBE Facility[14]. As
can be seen in Fig. 4, the mobility derived from the product
(Rσ) decreases rapidly with decreasing film thickness. A
similar variation had been noted by another group and attributed
to the onset of scattering by misfit dislocations located close
to the interface. The farinfrared magneto-optical spectrum was
very informative, showing a very broad but strong line at high
field together with a number of weak but sharp lines at lower
field (Fig. 5). The broad intense line was found to move
upwards in field on tilting the magnetic field with respect to
the normal to the epitaxial film with the characteristic 1/cosθ
dependence of a two-dimensional electron gas. This line can
therefore be interpreted as arising from electron accumulation

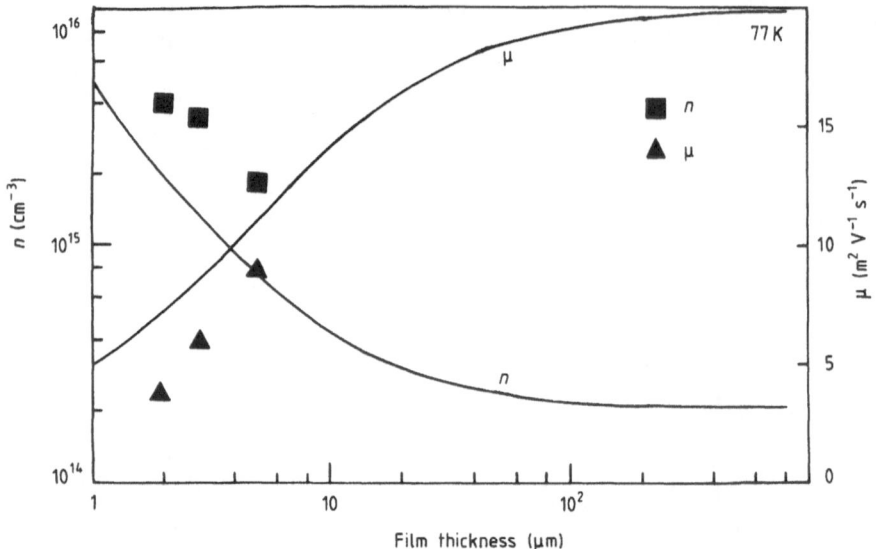

Fig. 4. Shows the variation with film thickness of the
mobility (triangles) and carrier concentration
(squares) deduced from Hall measurements with
high purity samples of InAs grown on GaAs
substrates by MBE (from Ref. 14). The full
curves are the predicted variations of Hall
mobility and carrier concentration when the
parallel conductance from the surface accumulation layer is taken into account. The
concentration and mobility of this two-
dimensional electron gas (2DEG) at the surface
and of the high-mobility bulk region are deduced
from farinfrared magneto-optical measurements
such as those shown in Fig. 5 and from
Shubnikov-de Haas studies of the 2DEG.

caused by the presence of surface states. The sharp lines did
not change in position on tilting the magnetic field and arise
from a high mobility region in the center of the film. The
highest field line in this group is due to the bulk cyclotron
resonance (m*/m = 0.0236) and the lines to lower field are the
1s-2p$_+$ and higher-order shallow donor lines. These lines are
almost as sharp as in the highest mobility GaAs. Central cell
structure can be observed on the 1s-2p$_-$ line and two different
impurity species can be detected. MBE GaAs grown with similar
As source material also showed the presence of two residual
contaminating impurities which were identified as sulfur and
selenium[15]. Consequently it is believed that the same two
contaminants are also present in MBE InAs.

No evidence for dislocation scattering was found as the
cyclotron resonance line width was independent of the film
thickness (see Fig. 5). From the integrated intensity of the
2DEG cyclotron resonance line shown in Fig. 5 it is clear that
the low-mobility surface will dominate the bulk in electrical
measurements with thin samples. The mobilities in the two
regions can be estimated from the magneto-optical line widths
and the number of carriers in each of the two regions could be

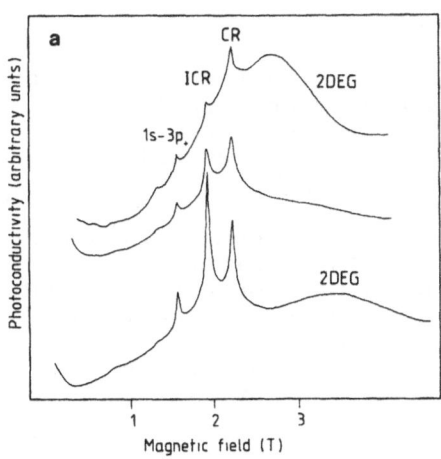
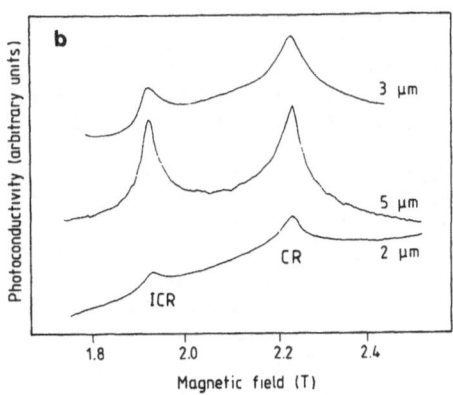

Fig. 5. (a) Shows the results of farinfrared magneto-
optical measurements for high purity samples of
InAs grown on a GaAs substrate by MBE. The
laser wavelength is 118 μm. The upper recording
is for a 2 μm thick sample and the spectrum is
dominated by the broad intense line to high
field from the surface accumulation layer
(2DEG). The middle recording is for a 3 μm
thick sample and the lower recording is for the
same sample tilted at 45° with respect to the
magnetic field in order to move the line due to
the 2DEG to higher magnetic field. (b) Is an
expanded field recording showing the bulk free
carrier cyclotron resonance (CR) and the 1s-2p$_+$
donor impurity line which is also known as the
'impurity-shifted' cyclotron resonance (ICR).
Although the intensities of these two lines with
respect to the underlying broad line due to the
2DEG decreases with decreasing film thickness,
the widths of the lines are independent of
thickness demonstrating that dislocation
scattering is unimportant down to thicknesses of
at least 1 μm.

determined; in the one case from the Shubnikov-de Haas effect
observed by the 2DEG formed by the surface accumulation layer
and, for the bulk electrons, from the absolute value of the
absorption coefficient found from the intensity of the cyclotron
resonance line. The full lines in Fig. 4 for the variation of
measured Hall mobility with sample thickness are generated from
the values of surface and bulk carrier concentration and
mobility (assumed independent of thickness) required to fit the
magneto-optical and Shubnikov-de Haas measurements without
adjustable fitting parameters. Within the experimental errors
associated with these measurements, agreement between experiment
and theory assuming two carrier conduction is good.

Qualitatively the dislocation structures found in relaxed
films of III-V materials on other grossly mismatched substrates
appear very similar. Consequently other materials systems such
as GaAs or InP on silicon should also show little degradation in
the microscopic mobilities even at high dislocation densities.

3.2 Strained Layer Superlattices Based on GaAs/InAs

The most extensive studies of strained layer superlattices incorporating strains greater than 1% have been with InAs/GaAs binary and alloy combinations on InP or GaAs substrates (e.g. see review by Marzin[16]). Good electrical and optical properties can be obtained at moderate strains.

Early attempts to grow more strongly strained combinations of GaAs/InAs by MBE were disappointing. Three-dimensional growth was found and indium rich clusters were observed in TEM studies. There was a shift to lower energy and degradation of the absorption spectrum with such structures (e.g. a super-lattice structure consisting of two monolayer InAs wells separated by 200 Å GaAs barriers showed a 100 meV wide emission line at 1.15 eV).

Recently, however, remarkable results have been obtained both by MBE[17] and by flow-modulation MOCVD growth[18] where the enhanced migration of the In atoms stimulated by the interrupted flow is thought to be responsible for the improved heterostructure quality. Extremely sharp photoluminescence lines (0.4 meV full width at half maximum) are seen with InGaAs/GaAs quantum wells grown by this technique.

Even the replacement of a single plane of Ga atoms by In (i.e. a single monomolecular well of InAs inserted into a 0.5 μm thick GaAs layer) reduces the GaAs band edge emission in the region of 1.510 eV in photoluminescence experiments. The free carriers created in the GaAs diffuse into the InAs monomolecular well and recombine to give a sharp line at 1.483 eV of 3 meV total width. The intensity of this line is some five hundred times greater than the GaAs band edge emission from the same sample and two hundred times stronger than that from GaAs layers grown without the single In plane.

Samples were also fabricated consisting of two InAs monomolecular wells separated by up to 80 GaAs monolayers. With decreasing separation between the wells the emission shifted from 1.483 eV to 1.34 eV. The observed variation was fitted rather by a simple model which assumed that the electronic wavefunctions were located only in the GaAs.

3.3 Strained Layer Superlattices based on InSb$_{1-x}$As$_x$

The InSb$_{1-x}$As alloy system exhibits very pronounced band bowing which produces a minimum band gap at a composition of x = 0.4 which is much less than the band gap of either InSb or InAs and which corresponds at low temperatures to a wavelength of about 7 μm. It has been estimated that by using strained layer techniques it may be possible to reduce the band gap by about a factor of two so that the associated wavelength response would be extended out to 14 μm wavelength[19]. However, as discussed in section 2, this alloy system is prone to severe structural problems in the mid alloy range. Nevertheless it has proved possible to obtain strained layer superlattice structures with good quantum well photoemission in the 10 μm wavelength region in structures grown by both MOCVD and MBE. A type II superlattice is obtained. By varying the period of the superlattice it was possible to use the confinement energy to tune the emission wavelength. As can be seen from Fig. 6[2,20],

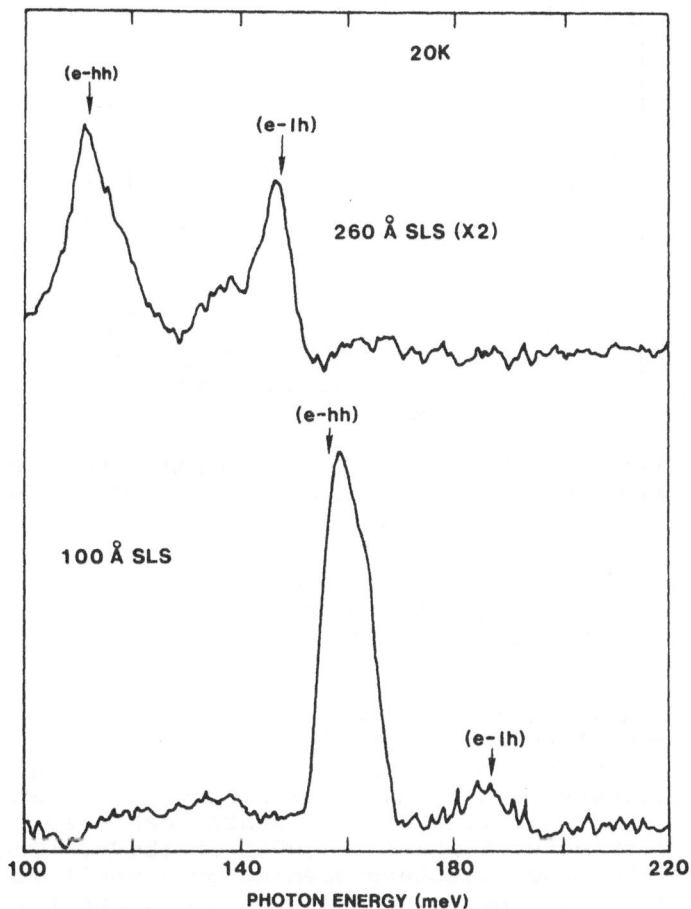

Fig. 6. Shows the photoemission from two strained layer
superlattices consisting of $InAs_{0.15}Sb_{0.85}$/InSb
multiple quantum well.s The emission between
the lowest subbands is at a wavelength of 11.3
μm for the structure with the wider wells (260 Å
repeat distance) (from Ref. 2).

a sharp photoemission peak corresponding to transitions between
the lowest subbands was observed at an energy of 110 meV which
corresponds to a wavelength of 11.3 μm. This system is
therefore promising for long wavelength detection.

4.   THE GROWTH OF α-Sn AND ALLOYS

    Tin can be found in nature in the alpha or 'grey' form
having diamond structure in addition to its more familiar
metallic phase. In a series of pioneering experiments some
twenty years ago Paul and other workers[21-23] demonstrated that
low impurity concentrations ($< 5 \times 10^{14}$ cm^{-3}) and high
mobilities ($> 10^{5}$ cm^{2}/Vs) were achievable with this material.
The band structure was of semimetallic form giving the
possibility that a very small energy gap might be created either
by the application of uniaxial stress or by alloying with

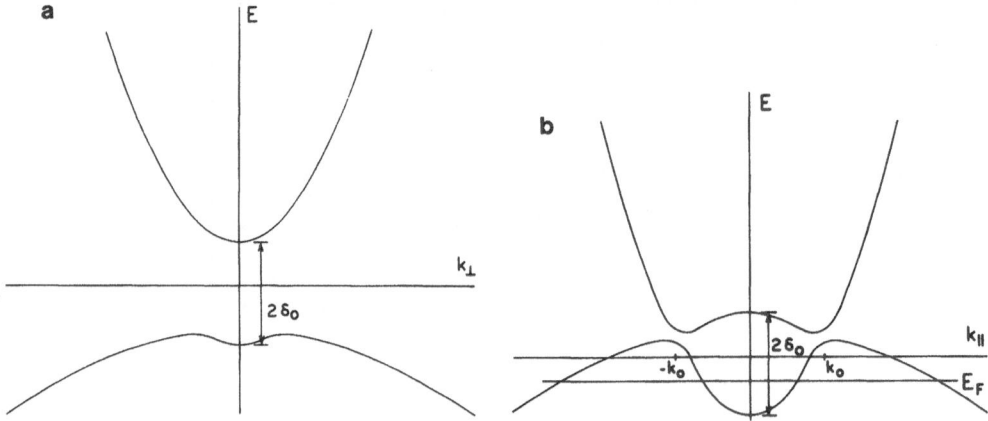

Fig. 7. Shows the band structure of alpha tin under
uniaxial compressive and tensile stress[30].

germanium. However, very little development work was
subsequently performed on this material following on the initial
demonstration of its electronic properties despite its potential
device applicability.

Farrow at RSRE Malvern[24] using MBE methods demonstrated
two key features of thin-film growth of this material and
stimulated a resurgence of interest in its properties. Strain
introduced by growth on a slightly mismatched substrate (such as
InSb or CdTe) or by alloying with germanium can stabilize the
alpha (diamond structure) phase to 200°C or above. The strain
also modifies the band structure opening up a small direct
energy gap (Fig. 7). The most direct evidence for the existence
of such an energy gap has been the identification of a loss
feature in High Resolution Energy Loss Spectroscopy studies[25]
for α-Sn samples grown on CdTe substrates. These results and
theoretical estimates of the variation of direct energy gap with
film thickness are shown in Fig. 8[26]. At thicknesses greater
than fifty monolayers the band gap is determined by the strain
and levels off at about 100 meV. Quantum confinement increases
the band gap to a maximum of about 500 meV at 20 monolayers. At
smaller thicknesses leakage into the surface regions start to
decrease the band gap. A very similar dependence of energy gap
on thickness was found from fitting the intrinsic carrier
concentration in a series of tin films on CdTe of differing
thicknesses[27].

Using a refined substrate cleaning procedures for InSb
substrates, we have grown high quality alpha tin layers on InSb
which have shown excellent interface properties. A good
understanding of the surface reconstruction and growth processes
has been obtained by means of RHEED studies[28]. We also are
the first group to observe a high mobility two-dimensional
electron gas (2DEG) at the interface between the α-Sn and the
substrate[29]. Shubnikov-de Haas measurements showed that at
least six subbands were occupied (see Fig. 9). Rotation of the
magnetic field away from the perpendicular to the surface caused
the peaks to move upwards in field with the $1/\cos\theta$ dependence
characteristic of a 2DEG.

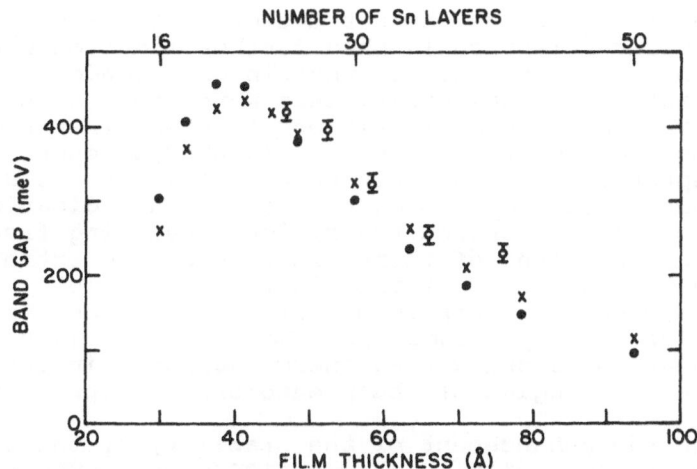

Fig. 8. Shows the theoretical and measured variation in band gap as a function of film thickness for α-Sn samples grown on CdTe substrates[25,26].

The carrier concentration measured in the Shubnikov-de Haas effect is extremely high. It would therefore seem from the high electron concentration found in the 2DEG that the offsets involved are playing only a minor role in creating the 2DEG. A perfect polar-nonpolar interface by itself should produce an even higher density of interface charge and the lower carrier density observed may result from deviations from ideality such as the diffusion of one component across the interface. For example, the diffusion of Sn by only a single lattice spacing into the InSb would result in an asymmetric δ-doped well in the substrate. Similarly, diffusion of either the In or Sb into the Sn could produce high local doping of the epilayer. The occupancy of the subbands was dependent on the thickness of the tin film indicating that the 2DEG was located at least in part in the tin.

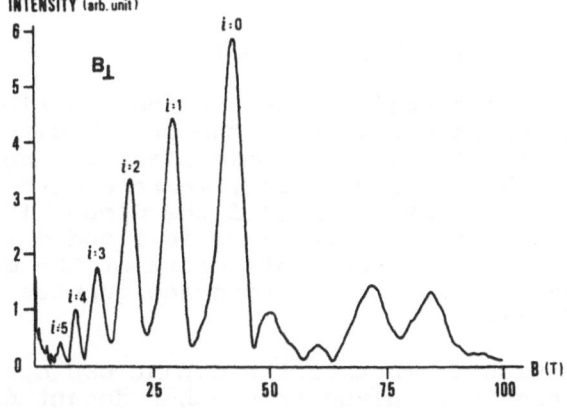

Fig. 9. Shows the Fourier analysis of the perpendicular field Shubnikov-de Haas effect for an epitaxial α-Sn sample grown on an InSb substrate[29].

In order to determine whether the effective mass of the
two-dimensional carriers could shed further light on whether
2DEG was completely localized in the tin or extended across the
interface, magneto-optical experiments were undertaken. In
addition to the bound hole transitions from the InSb substrate
which extrapolate back to the zero field binding energy of the
cadmium acceptors, a considerable number of lines could be
observed which extrapolated back to the origin indicating that
they were cyclotron resonance transitions involving free
carriers. The strengths of these lines were also extremely
dependent on the thickness of the tin film (Fig. 10). With a
thin sample a group of lines is observed with effective masses
of the order of 0.03 m_e. These are thought to arise from the
strongly nonparabolic and warped energy surfaces in the highly-
strained unrelaxed region of the pseudomorphic alpha tin[30].

Although the enhancement of the stability of the alpha
phase to temperatures of the order of 100°C by growth of thin
films on InSb or CdTe has been extremely important, the key to
future developments is to increase the stability range even
further by alloying the tin with germanium. At first sight this
does not seem to be a very promising approach as the alloys are
thermodynamically immiscible in the bulk at concentrations of
greater than 1% Ge in tin or 2% tin in germanium. Farrow et
al.[24] were able to observe the stabilizing effect of
introducing Ge into MBE growth of tin and noted that the bulk
immiscibility limit could be overcome. Fitzgerald, Kimerling et
al.[31] were able to grow good quality metastable films with up
to 8% germanium and were subsequently able to extend this to 13%
germanium[32,33]. 1200 Å thick films of this composition were
stable up to 120-130°C and 50 Å films were stable to 220°C. By
choosing the correct combination of alloy composition and
substrate it has even been possible to grow InSb on top of a
diamond structure Sn/Ge alloy[31] leading to the prospects of
superlattices of the alloy combined with InSb. With germanium
rich alloys grown on Si substrates, Pukite et al.[33] report
mirror smooth films stable up to 140°C with a film thickness of
1600 Å with 30% of tin and 70% germanium. Finally, Abstreiter
has reported at this workshop the first growth of superlattice
structures based on tin-germanium alloys.

5.   ATOMIC PLANE, DELTA OR SPIKE DOPING

Modern epitaxial techniques permit the creation of
extremely abrupt doping profiles in the growth direction. The
most extreme example of this is atomic plane doping (sometimes
referred to as delta or spike doping) where growth is
interrupted (usually by switching off the Group III source in
the case of III-Vs) and a single plane is flooded with the
required dopant. The electronic structure of the doping layer
so produced shows a number of striking new quantum features as
discussed in Ref. 34.

A major question is whether the dopant can stay localized
on the initial deposition plane or whether dopant diffusion or
surface segregation occurs. The Shubnikov-de Haas effect is
extremely powerful in providing very accurate values for the
individual subband occupancies. The fractional occupancy of the
$i = 0$ subband is very sensitive to the diffusion of the dopant
as the local potential at the initial doping plane ($z = 0$) loses

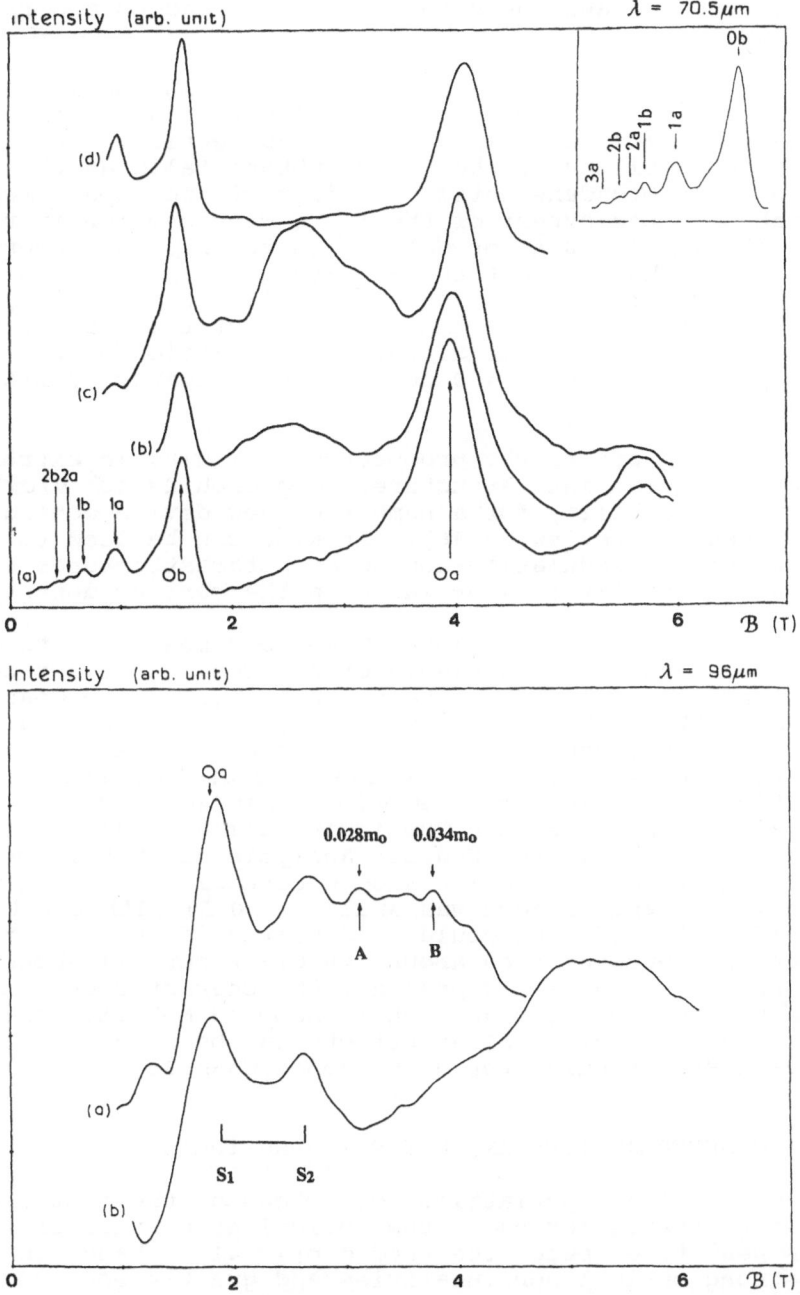

Fig. 10. Farinfrared magneto-optical transitions observed with
an α-Sn on InSb heterostructure taken at wavelengths of
70 and 96 μm. The lower recording in the top set is
for bulk p-InSb and the strong sharp lines are acceptor
transitions. Curves b) and c) are for two α-Sn films
of differing thickness and d) for a similar specimen
except that the α-Sn has been etched away. The broad
lines at about 2.5 T clearly arise from the tin film.
In the lower set of recordings taken at the longer
wavelength the acceptor transitions have moved to lower
field revealing a complicated series of lines for the
two samples with differing thicknesses of tin.

its cusp-like shape and the effective well broadens and becomes more shallow.

In addition the mobility of the carriers in the i = 0 subband is rather sensitive to the local potential as these carriers are the most localized in the z-direction. In particular the mobility of the i = 0 subband falls as the impurity profile broadens until the width of the dopant region is comparable to the extent of the i = 0 wavefunction in the z-direction (typically 30 Å for GaAs). In the Fourier spectrum of the Shubnikov-de Haas effect the amplitude of the i = 0 peak is rather weak because of diffusion of the silicon. However, the mobility of this subband and the amplitude of this peak recover on applying pressure because of the neutralization of the silicon dopants close to z = 0 due to the occupancy of D(X) centers[35].

Delta doping offers the prospect of building-in extremely high electric fields into structures (approaching 10^6 V/cm) and provides the possibility for a number of new device config-urations - see the review in[36]. It also can be used to optimize remote or modulation-doping of heterostructures either to improve the mobility[37] or increase the carrier density[38]

In our own laboratory Quantum Transport measurements have been made for single doping planes of silicon donors introduced into InSb and InAs films grown heteroepitaxially onto GaAs substrates by MBE. Up to five subbands are occupied (Fig. 11). The free electron concentration saturates at carrier concentrations of 4×10^{12} cm^{-2} in the case of InSb but a concentration of 2×10^{13} cm^{-2} is achieved with InAs[39]. Good agreement is achieved between the individual subband carrier concentrations deduced from Fourier analysis of Shubnikov-de Haas measurements and the calculated occupancies indicating that the diffusion of the silicon was small ($\leq$ 50 Å) although with InSb significant diffusion could be detected if the growth temperature was greater than about 340°C. Figure 11 shows the shift downwards of the i = 0 peak and the characteristic drop in amplitude t = due to the reduction in subband mobility arising from dopant diffusion. Similar effects can be seen in the Shubnikov-de Haas results for delta-doped GaAs[35].

6.   DOPING SUPERLATTICES AND N-I-P-I STRUCTURES

N-i-p-i doping superlattices offer one of the most flexible approaches to the tailoring of the optical properties of materials near to or below the band gap[40,41]. Band gap tuning, strong optical nonlinearities and greatly enhanced carrier lifetimes have all been reported (although the latter is accompanied by a decreased absorption coefficient). Figure 12 shows the change in the photoluminescence as a function of pump power which is characteristic of a n-i-p-i structure[42]. The structure is a GaAs short-period ('type A') superlattice where the recombination time is relatively short and the band gap is reduced by about 40% due to the saw-tooth doping potential. At low pump powers there is virtually no emission at the normal GaAs band edge wavelength of 850 nm but instead a strong series of lines are observed which arise from transitions between different subbands in the n-i-p-i structure[43]. As the pump power is raised to an intermediate level these lines all shift

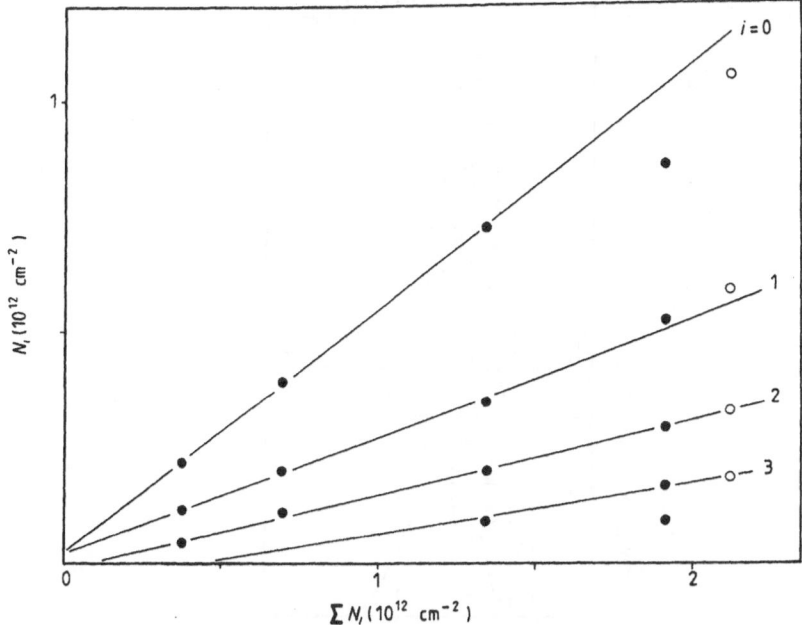

Fig. 11. Shows the results of Fourier analysis of the
Shubnikov-de Haas effect in atomic plane doped
InSb. The upper recording is for a sample
grown at 304°C and the i = 0 subband is shifted
to lower field (occupancy) and is weaker
compared with the results for the sample grown
at 240°C (lower recording). The features are
characteristic of the silicon dopant diffusing
at the higher temperature (from Ref. 39).

to shorter wavelength because the saw-tooth potential is
partially screened by the photoexcited carriers. At the highest
power levels the band edges have virtually straightened out
because of the large number of photoexcited carriers present and
virtually all the emission is at the GaAs band edge.

A GaAs laser structure incorporating a n-i-p-i region has
been reported to be tunable over a wavelength range of 35 Å[44]
and an all GaAs technology for the minimum dispersion region of
silicon fibres (1.3 μm) based on n-i-p-i devices has been
proposed. Perhaps the spectral region which has the most
potential for n-i-p-i structures is the mid-infrared [45-47].
In our laboratory we have grown n-i-p-i structures in both InSb
and InAs. Our InSb n-i-p-i structures have shown all of the
characteristic optical properties expected of such structures
(i.e. long recombination times, strong subband-gap absorption
and strong optical nonlinearities in the subband-gap region)[48]
- see also Fig. 13. Detectivities (D*) values of 8 x 10^{10} cm/Hz
are anticipated[49].

7.   INTERSUBBAND QUANTUM WELL DETECTORS

Electromagnetic transitions between the size-quantized
subbands in quantum wells have been proposed as the basis for

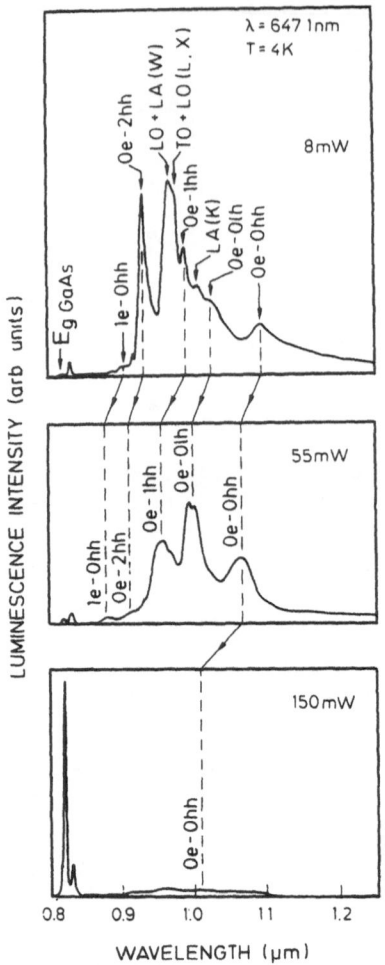

Fig. 12. Shows the photoluminescence from a short-period
        GaAs n-i-p-i structure as a function of the
        laser pump power demonstrating strong subband
        gap luminescence at low power levels. These
        lines progressively shift and bleach with
        increasing power until at high power levels the
        GaAs band edge emission is seen (from Ref. 42).

infrared detection. In order for there to be a matrix element,
the radiation must be polarized with the electric field vector
perpendicular to the plane of the well which is experimentally
inconvenient but can be achieved by grating coupling or by
introducing the radiation from the side in a waveguide geometry.
Oblique incidence can also be employed but this has low
efficiency because of the large refractive indices of common
semiconductors. With GaAs quantum wells the 10 μm region can be
covered and detectivities of about 10^{10} cm/Hz/W have been
reported[45]. Advantage could also be taken of the mature GaAs
growth and processing technologies giving the prospect of
integration with high-speed FETs and enabling large area focal
plane arrays of detectors to be considered.

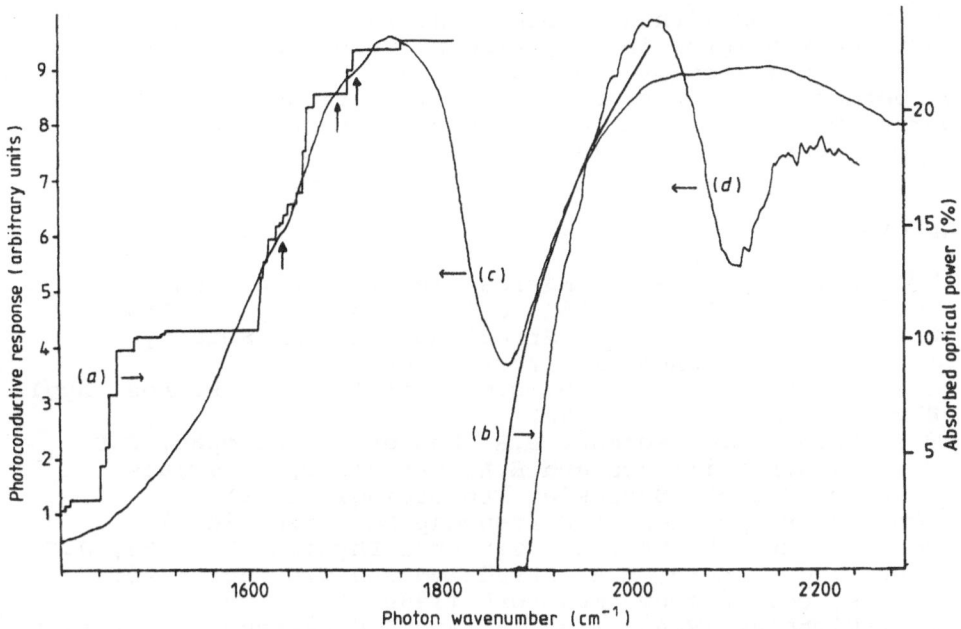

Fig. 13. Shows the characteristic photoresponse below
the band edge expected for a n-i-p-i structure.
The sample was grown by MBE at Imperial College
(from Ref. 45).

However, the detectivities achieved are still far less than
the industry standard system which is $Hg_{1-x}Cd_xTe$ where
detectivities of 3×10^{12} cm./Hz/W are possible at 77K[49].
Furthermore, the technology for HgCdTe is now extremely well
advanced and, by using nonequilibrium depletion techniques to
suppress Auger noise, it is now possible to achieve comparable
detectivities close to room temperature[50].

It should be possible by going to more refined structures
such as those using grating enhancement[51] or exploiting Fabry-
Perot structures[52] to improve the performance of intersubband
detectors.

Generally a sharply-peaked response is observed (typically
6 meV or about 50 cm^{-1} full width for the i = 0 to 1 transition
in GaAs[52,53] with more than double this width being found with
InGaAs wells[54,55]). If interpreted as a natural width 6 meV
would correspond to a lifetime of 0.1 ps. However, saturation
experiments give decay times of up to 15 ps[56]. Thus the
observed linewidth probably arises from variation in well width.

Acknowledgment

The work and assistance of the following past and present
members of Imperial College is gratefully acknowledged:
R. Droopad, I. Ferguson, C.C. Hodge, E.J. Johnson, B.A. Joyce,
W. Liu, A. MacKinnon, R. Newman, A. d'Oliveira, S.D. Parker,
S. Patel, C.C. Phillips, P.D. Wang, R.L. Williams, W.T. Yuen and
Z. Wasilewski

Conversations with H. Kroemer, J. Merz and W. Walukiewicz were extremely helpful in assisting my understanding of the electrical properties of dislocations and the properties of native defects and interfaces and conversations with R. Booker, A. Norman and T. Seong on the structural properties of alloys are also gratefully acknowledged.

REFERENCES

1. "Optical Properties of Narrow-gap Low Dimensional Semiconductors", NATO ASI Series B Physics, Vol 152, J.C. Portal, J.C. Maan, R.A. Stradling and C.M. Sotomayor Torres, ed., Plenum, New York (1986)

2. S.R. Kurtz, G.C. Osbourn, R.M. Biefeld and S.R. Lee, **Appl. Phys. Lett.**, 53:216 (1988)

3. T.Y. Seong, A.G. Norman, G.R. Booker, R. Droopad, S.D. Parker, R.L. Williams and R.A. Stradling, Materials Research Society Symposium Proceedings (1989)

4. "Optical Properties of Narrow-gap Low Dimensional Semiconductors", NATO ASI Series B Physics, Vol 152, J.C. Portal, J.C. Maan, R.A. Stradling and C.M. Sotomayor Torres, ed., Plenum, New York (1986)

5. R.J. Nicholas, R.A. Stradling and J.C. Ramage, **J. Phys. C**, 12:1641 (1979)

6. A.J. Noreika, W.J. Takei, M.H. Francombe and C.E.C. Wood, **J. Appl. Phys.**, 53:4932 (1982)

7. K.Y. Ma, Z.M. Fang, D.J. Jaw, R.M. Cohen, G.B. Stringfellow, W.P. Kosar and D.W. Brown, **Appl. Phys. Lett.**, 55:2420 (1989)

8. G. Hasnain, B.F. Levine, D.L. Sivco & A.Y. Cho, **Appl. Phys. Lett.**, 56:770 (1990)

9. G. Tuttle, H. Kroemer and J.H. English, **J. Appl. Phys.**, 65:5239 (1989); **J. Appl. Phys.**, (1990)

10. T.S. Rao, C. Halpin, J.B. Webb, J.P. Noad and J. McCaffrey, **J. Appl. Phys.**, 65:585 (1989); J.B. Webb, M. Paiment and T.S. Rao, **Solid State Commun.**, 71:871 (1989)

11. L.C. Kimmerling and J.R. Patel, **VLSI Electronics**, 12:223 (1985)

12. H. Kroemer, T.Y. Liu and P.M. Petroff, **J. Cryst. Growth**, 95:96 (1989)

13. G. Griffiths, K. Mohammed, S. Subbana, H. Kroemer and J.L. Merz, **Appl. Phys. Lett.**, 43:1059 (1983)

14. S.N. Holmes, R.A. Stradling, P.D. Wang, R. Droopad, S.D. Parker and R.L. Williams, **Semicond. Sci. & Tech.**, 4:303 (1989)

15. S.N. Holmes, C.C. Phillips, R.A. Stradling, Z. Wasilewski, R. Droopad, S.D. Parker, W.T. Yuen, P. Balk, A. Brauers, H. Heinecke, C. Plass, M. Weyers, C.T. Foxon, B.A. Joyce, G.W. Smith and C.R. Whitehouse, **Semicon. Sci. & Tech.**, 4:782 (1989)

16. "Optical Properties of Narrow-gap Low Dimensional Semiconductors", NATO ASI Series B Physics, Vol 152, J.C. Portal, J.C. Maan, R.A. Stradling and C.M. Sotomayor Torres, eds., Plenum, New York (1986)

17. J.M. Gerard and J.Y. Marzin, **Appl. Phys. Lett.**, 53:568 (1988)

18. M. Sato & Y. Horikoshi, **J. Appl. Phys.**, 66:851 (1989)

19. G.C. Osbourn, **J. Vac. Sci. & Tech. B**, 2:176 (1984)

20. J.R. Dawson, **J. Cryst. Growth**, 98:220 (1989)

21. S. Groves and W. Paul, **Phys. Rev. Lett.**, 11:194 (1963)

22. B.L. Booth and A.W. Ewald, **Phys. Rev.**, 186:770 (1969)
23. S. Groves, Physics of Semimetals & Narrow-gap Semiconductors, D.L. Carter and R.T. Bate, eds., Pergamon (1971) p.447
24. R.F.C. Farrow, D.S. Robertson, G.M. Williams, A.G. Cullis, G.R. Jones, I.M. Young and P.N.J. Dennis, **J. Cryst. Growth**, 54:507 (1981)
25. S. Takatani and Y.W. Chung, **Phys. Rev.B**, 31:2290 (1985)
26. B.I. Craig & B.J. Carrison, **Phys. Rev.**, 33:8130 (1986)
27. Li-Wei Tu, G.K. Wong and J.B. Ketterson, **Appl. Phys. Lett.**, 55:1327 (1989)
28. W.T. Yuen, W.K. Liu, B.A. Joyce and R.A. Stradling, **Semicond. Sci. & Tech.**, 5:373 (1990)
29. W.T. Yuen, W.K. Liu, S.N. Holmes and R.A. Stradling, **Semicond. Sci. & Tech.**, 4:819 (1989)
30. L. Liu, **Physics Lett. A**, 45:285 (1973); **Solid State Comm.**, 16:285 (1975)
31. E.A. Fitzgerald, P.E. Freeland, M.T. Asom, W. Lowe, R. MacHarrie, A.R. Kortan, Y.H. Xie, F.A. Thiel, A.M. Sargent, L. Cooper, G.A. Thomas, K.A. Jackson, B.E. Weir, G.P. Schwartz, G.J. Gualtieri and L.C. Kimmerling, Abstracts 119th TMS Annual Meeting, p31 (1990)
32. M.T. Asom, E.A. Fitzgerald, A.R. Kortan, B. Spear and L.C. Kimmerling, **Appl. Phys. Lett.**, 55:578 (1989)
33. P.R. Pukite, A. Harwit and S.S. Iyer, **Appl. Phys. Lett.**, 54:2142 (1989)
34. A. Zrenner, F. Koch and K. Ploog, **Surface Science**, 196:671 (1988)
35. A. Zrenner, F. Koch, R.L. Williams, R.A. Stradling, K. Ploog and G. Weimann, **Semicond. Sci. & Tech.**, 3:1203 (1988)
36. K. Ploog, M. Hauser and A. Fischer, **Appl. Phys. A**, 45:233 (1988)
37. T. Ishikawa, K. Ogasawara, T. Nakamura & K. Kondo, **J. Appl. Phys.**, 61:1937 (1987)
38. E.F. Schubert, J.E. Cunningham, W.T. Tsang and G.L. Timp, **Appl. Phys. Lett.**, 51:1170 (1987)
39. R.L. Williams, E. Skuras, R.A. Stradling, R. Droopad, S.N. Holmes and S.D. Parker, **Semicond. Sci. & Tech.**, 5:S338 (1990)
40. G.H. Dohler, **Phys. Stat. Solidi B**, 52:79 (1972)
41. K. Ploog and G.H. Dohler, **Adv. Phys.**, 32:285 (1983)
42. B. Ullrich, C. Zhang and K. von Klitzing, **Appl. Phys. Lett.**, 54:1133 (1989)
43. E.F. Schubert, T.D. Harris, J.E. Cunningham and W. Jan, **Phys. Rev. B**, 39:11011 (1989)
44. E.F. Schubert, J.E. Cunningham and W.T. Tsang, **Appl. Phys. Lett.**, 51:817 (1987); E.F. Schubert to be published.
45. C.C. Phillips, C. Hodge, R. Thomas, S.D. Parker, R.L. Williams and R. Droopad, **Semicond. Sci. & Tech.**, 5:S319 (1990)
46. C.C. Phillips, **Appl. Phys. Lett.**, 56:151 (1990)
47. J. Maserjian, F.J. Grunthaner and C.T. Elliott, **Infrared Physics**, 30:27 (1990)
48. B.F. Levine, C.G. Bethea, G. Hasnain, J. Walker and R.J. Malik, **Appl. Phys. Lett.**, 53:296 (1988)
49. M.A. Kinch and A. Yariv, **Appl. Phys. Lett.**, 55:2093 (1989)
50. C.T. Elliott, A. Davis and A.M. White, **Semicond. Sci. & Tech.**, 5 (1990)
51. K.W. Goossen, S.A. Lyon and K. Alavi, **Appl. Phys. Lett.**, 53:1027 (1988)
52. D.R.P. Guy, N. Apsley, L.L. Taylor, S.J. Bass and P.C.

Klipstein, in: "Quantum Well and Superlattice Physics",
G.H. Dohler and J.N. Schulman, eds., SPIE (1987)

53. J.Y. Andersson and G. Landgren, **J. Appl. Phys.**, 64:4123
(1988)

54. F. Muller, V. Petrova-Koch, M. Zachau, F. Koch, D.
Grutzmacher, R. Meyer, H. Jurgensen and P. Balk, **Semicond.
Sci. & Tech.**, 3:797 (1988)

55. M.J. Kane, L.L. Taylor, N. Apsley and S.J. Bass, **Semicond.
Sci. & Tech.**, 3:586 (1988)

56. F.H. Julien, J.M. Lourtioz, N. Herschkorn, D. Delacourt,
J.P. Pocholle, M. Papuchon, R. Planel and G. Le Roux, **Appl.
Phys. Lett.**, 53:116 (1988)

OPTICAL AND MAGNETIC PROPERTIES OF DILUTED MAGNETIC

SEMICONDUCTOR HETEROSTRUCTURES*

L.L. Chang, D.D. Awschalom, M.R. Freeman and L. Vina†

IBM T.J. Watson Research Center
Yorktown Heights, New York, USA
† Instituto de Ciencia de Materiales - CSIC and
Departamento de Fisica Aplicada, Universidad Autonoma
Madrid, Spain

ABSTRACT

Diluted magnetic semiconductors offer the unique medium for the study of semiconductor physics and magnetism through the electron-ion exchange interaction. Precisely controlled heterostructures alter the band structure, the interaction scheme and the dimensionality, affecting both the underlying electronics and magnetics. Focusing on the typical system of CdTe-CdMnTe, we have used static and dynamic techniques to pursue optical and magnetic properties from luminescence, spin-polarization and magnetization experiments. We present results of the dimensional effect on electric confinement and magnetic interaction, as well as spin exchange and relaxation of the electronic and ionic systems.

1.   INTRODUCTION

With recent advances in epitaxial techniques, there has been a renewed effort on II-VI semiconductors and, in particular, their heterostructures. This group of materials covers a wide range of energy gaps and lattice constants[1], leading to interesting physical and electronic properties in many areas. The wide-gap semiconductors, such as the Zn-compounds, represent one end in the visible spectrum, which are attractive for light-emitting and display applications. On the other end are the Hg-compounds with energies in the infrared range, which forever draw attention as detectors and imaging devices. The third area consists of materials which form solid solutions with magnetic elements, known as the diluted magnetic semiconductors or DMS[2]. Notable examples are CdMnTe, ZnMnSe and HgMnTe.

Like other alloys, the DMS materials provide a variety of properties with a change in the alloy composition. More

---

*Work supported in part by the U.S. Army Research Office.

importantly, they offer the unique opportunity for combined studies of semiconductor physics and magnetism. The strong exchange interaction between the local moments of the magnetic ions and the spins of the band electrons gives rise to a dramatic enhancement of the Zeeman splitting. The exchange interaction among the ions themselves leads to a wealth of phenomena involving magnetic phase transitions and excitations[2,3]. In a heterostructure environment, the structural configuration can be precisely controlled to vary the band schemes and dimensionality[4]. This will in turn affect both the underlying electronic and magnetic systems as well as their interactions.

In this work, we will focus on the system made of CdTe-CdMnTe, which represents the first heterostructure achieved in this group of materials[5,6] and has since been the most extensively studied[7,8]. With a brief description of the fabrication of the structure, we will start with optical properties commonly pursued in semiconductor quantum wells and superlattices. This will be followed by dynamic properties associated with carrier spin polarization, and both static and time-resolved experiments in magnetic spectroscopy.

2. STRUCTURE FORMATION AND MEASUREMENT

Although both metalorganic chemical vapor deposition[9] and atomic layer epitaxy[10] have been employed for the growth of II-VI heterostructures, molecular beam epitaxy remains the dominant technique[1,11]. We used a single compound source of CdTe, which ideally exhibits congruent evaporation. Because the component vapor pressures over the compound are much lower than those over the elements, the impinging fluxes on the substrate are expected to re-evaporate unless reaction occurs to form the compound. In other words, stoichiometry is generally assured, and additional sources are added only for the purpose of fine adjustment. In our case, we used a separate Te source in addition with the introduction of Mn, a necessary provision for the growth of CdMnTe with high Mn compositions. Typical values of the various source temperatures are 500°C for CdTe, 800°C for Mn and 300°C for Te. These result in a growth rate of 1.5 Å/sec and an Mn composition of 0.2.

The substrates cover a variety of materials, including GaAs and CdTe itself. Homoepitaxy in principle is the most desirable, but GaAs remains the most commonly used. Considerations of perfection in crystalline quality and reproducibility in surface preparation override issues related to lattice mismatch. The GaAs (100) substrates were etched with the standard peroxide-sulfuric acid, and the oxide was subsequently desorbed at 580°C. This procedure invariably led to the growth of CdTe (111) with [01$\bar{1}$] ∥ GaAs [01$\bar{1}$] and [211] ∥ GaAs [100]. The growth temperature was typically 300°C. A buffer layer of 2000 Å of CdTe was initially deposited to relieve the strain and to smooth out the surface, prior to the growth of the heterostructure. The structures include both CdTe and CdMnTe wells and CdMnTe barriers with a layer thickness from 10 to 200 Å and an Mn composition up to 0.4. Evaluations were made with electron microscopy and X-ray diffraction which showed well-defined satellite peaks in the spectrum in the case of super-lattices[8].

Standard optical experiments were used to characterize the properties of the heterostructures, including luminescence excitation and Raman scattering. In addition, dynamic measurements were performed both in luminescence spectroscopy to monitor the carrier spin polarization and in direct magnetic spectroscopy to probe the ion magnetization. The pump-probe method was used for the time-resolved measurements in magnetization with a SQUID susceptometer serving as an integrating detector in a boxcar averaging scheme[12,13]. In luminescence, the time resolution was achieved through the sum-frequency generation technique. The dynamic polarization behaviors of the carriers were studied either by their creation with linearly polarized excitation in a magnetic field or by their decay after excitation with circularly polarized radiation[14]. A tunable dye laser, synchronously pumped by a mode-locked Nd:YAG laser, was used as the source. Most of the measurements were carried out at low temperatures at or below 4 K.

3.   OPTICAL SPECTROSCOPY

Because of their spectroscopic nature which can usually be related to band structure and lattice energies, optical measurements are the most commonly used to probe electronic properties in quantum wells and superlattices. This is particularly true for wide-gap II-VI compounds, including CdTe-CdMnTe, for which transport experiments cannot be carried out systematically for lack of doping control. The CdTe-CdMnTe is a type-I heterostructure, similar to GaAs-GaAlAs, in which the

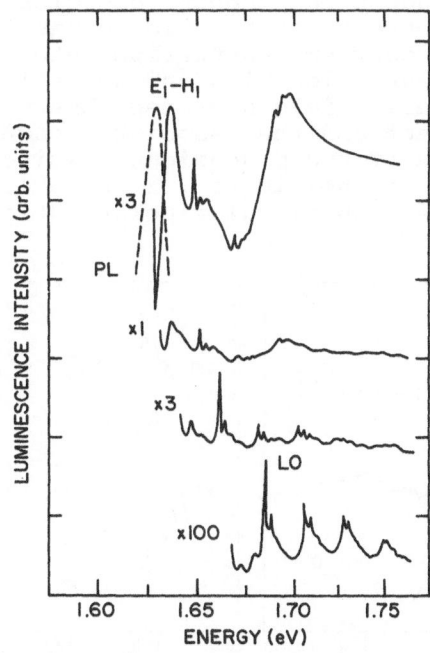

Fig. 1. Excitation spectra of CdTe (100 Å) - $Cd_{0.87}Mn_{0.13}Te$ (100 Å) taken at different energies. The luminescence is shown in dashed lines. Excitonic transition of the ground state (E_1-H_1) and CdTe phonon (LO) are indicated.

CdTe layers serve as potential wells for both electrons and holes. Unlike the situation in GaAs-GaAlAs, however, the valence band offset is rather small, although its value is not precisely known. Also, CdMnTe remains a direct-gap semi-conductor even for large concentrations of Mn.

The simplest optical techniques for evaluation are photo-luminescence and its related excitation spectroscopy. For pure CdTe, the luminescence spectrum consists of fine impurity-exciton features with a linewidth below 1 meV. The spectrum of CdTe-CdMnTe is usually dominated by a single peak which is Stokes-shifted from the ground heavy hole exciton[15], as indicated in Fig. 1. The linewidth is typically of the order of 10 meV, widening further with narrowing well thickness and higher Mn composition as a result of inhomogeneous broadening. Other peaks at high energies are usually ascribed to excitons associated with light holes and excited subbands by comparing their energy positions with those calculated. Such identi-fications are less certain, and must be corroborated by other evidence. The peak at 1.65 eV, for example, assigned to ground light hole exciton, may also involve the second heavy hole state or the continuum of the ground heavy hole state[16].

With the application of a magnetic field, the excitonic transition energies can in principle be affected in a number of ways. However, both the usual Landau splitting and the classical Zeeman splitting give rise to relatively small effects. The exchange interaction in CdMnTe is significant in affecting the barrier energies and thus the absorption characteristics[17]. The changes in barriers for different spin components also influence the subband energies in CdTe in most cases. The most important contribution comes from exciton-ion interaction when the carrier penetration into the barrier region is substantial. Sizeable lowering and splitting of the exciton energies by the magnetic field has been observed[18,19]. The magnitude of the effect and the associated magnetic anisotropy led to the conclusion of two-dimensional excitons localized near the interfaces[18]. We show in Fig. 2 the excitation spectra under magnetic fields with circular polarization[16]. The

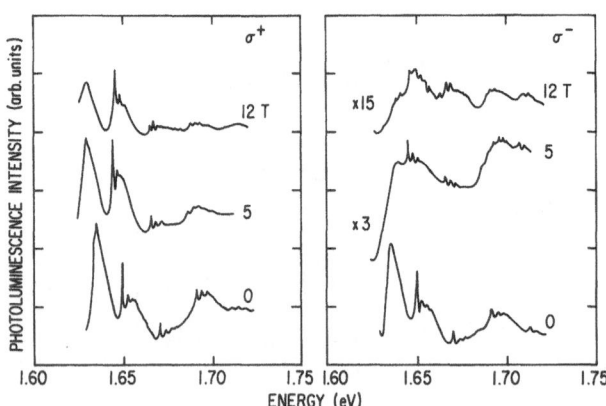

Fig. 2. Excitation spectra of the sample in Fig. 1 for
both left and right circularly polarized
components under a magnetic field.

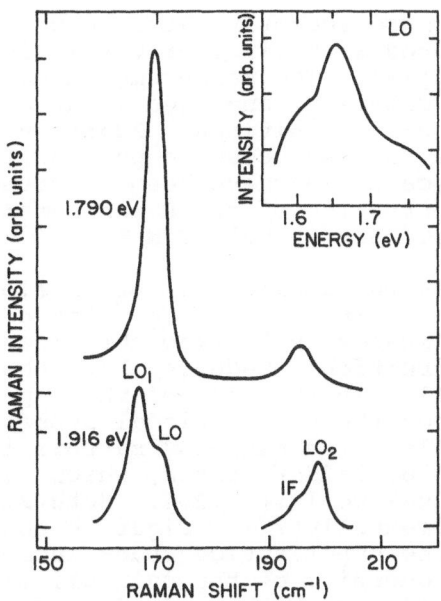

Fig. 3. Raman spectra of CdTe (86 Å) – $Cd_{0.80}Mn_{0.20}Te$ (86 Å) at different excitation energies. The phonon features are indicated. The inset shows the resonant behavior of LO(CdTe).

lowest state shifts in opposite directions for the two components, leading to an effective g-factor of 25, as a result of the strong interaction. The shifts to lower energies, similar in the two cases for the second lowest state which was assigned earlier as the light hole ground state, are more complicated[16,20].

The sharp structures in the excitation spectra in Fig. 1 correspond to phonons, which become progressively pronounced as the detection energy is moved to higher values. They are identified as LO phonons of CdTe, MnTe and interface modes as well as their higher orders and combinations. Raman experiments have revealed the resonant behaviors[15], as shown in Fig. 3, where the various phonons are indicated. As the excitation approaches the energy of the ground-state of the superlattice, both LO (CdTe) and IF (interface phonons) are greatly enhanced, while LO_2 (MnTe-like) is suppressed because of its confinement in CdMnTe. The LO_1 (CdTe-like), which propagates in both layers is merged with LO on its low energy side. The resonance is illustrated in the inset of Fig. 3, where the main peak and its low-energy shoulder are due to the outgoing and incoming channels, while the high energy features, again, are presumably due to light hole and excited heavy hole states. The results of the various phonons observed here for the (111) orientation are similar to those reported for the (100) orientation[21].

The Raman profiles, measured for samples with the same CdTe layer but decreasing CdMnTe layers, show broadened resonance with peak energies shifting to lower values[22]. These are consistent with the formation of superlattice miniband with an increase in band width, as the coupling between neighboring wells is enhanced. More recently, Raman experiments have been performed and monitored with circularly polarized radiation under magnetic field[16]. The measurements complement those by photoluminescence excitation, and shed light on the detailed nature of the excitonic transitions. Additional Raman measurements involving zone-folding acoustic phonons[21] and exchange enhanced spin-flip scattering have been reported[23]. Also reported was the observation of the Raman paramagnetic resonance, similar to that in bulk CdMnTe.

Experiments with the application of an electric field have recently been performed in CdTe-CdMnTe. The effect on quantum wells with isolated states is to skew the electron and hole wells in opposite directions, reducing both the emission energy and the intensity[24]. In the case of the superlattice, the field tends to destroy the translational symmetry and reduce the coupling between wells. The miniband narrows in energy and breaks into a series of ladder states, which eventually become localized in individual wells.[25,26]. Both situations have been realized experimentally[27]. Figure 4 shows the luminescence spectra of the superlattice under increasing fields. The shift to higher energies of the main emission peak is obvious, and it persists to higher fields and saturates around 50 kV/cm when adjacent wells are decoupled. Also seen in Fig. 4 is the broad feature at lower energies, which shifts with fields in the opposite direction. This suggests recombination between two neighboring wells, or transitions from the first ladder. That no higher orders of the ladders are observed, as in the case of GaAs-GaAlAs, is a reflection of the much shorter coherent length in the present case.

4. SPIN POLARIZATION SPECTROSCOPY

While static optical measurements reveal electronic properties associated with band and phonon structures, experiments under pulsed conditions probe carrier dynamics.

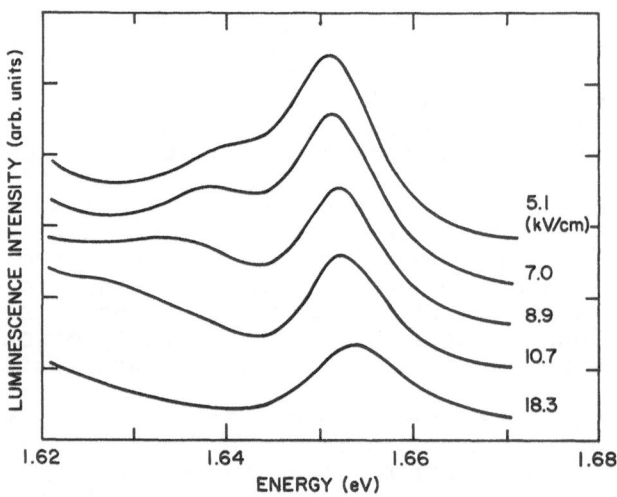

Fig. 4. Luminescence spectra of CdTe (42 Å) – $Cd_{0.87}Mn_{0.13}Te$ (24 Å) under an electric field.

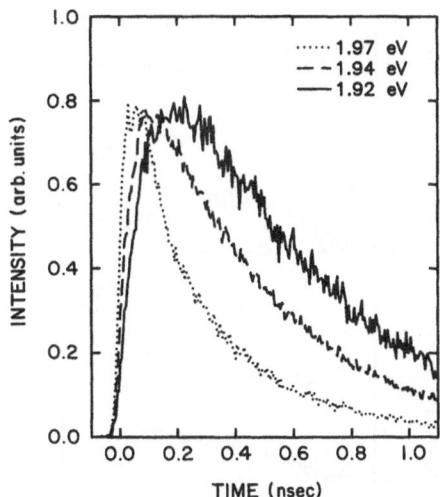

Fig. 5. Time-resolved photoluminescence measured at different energies within the luminescence linewidth of $Cd_{0.935}Mn_{0.065}Te$ (21 Å) – $Cd_{0.62}Mn_{0.38}Te$ (86 Å).

Here carriers are generated optically with a finite polarization. As they undergo the fast energy-loss cascade process and eventually recombine, they experience spin relaxation and may transfer some of their spin alignment to the Mn ions, which also relax subsequently. Each of the processes has a characteristic time constant which may be quite different. Figure 5 shows the time resolved photoluminescence with different detection energies which are within the luminescence width[28]. The time scale is a fraction of a nanosecond, which is close to the carrier lifetime in this or other types of quantum wells such as GaAs-GaAlAs[29]. The shifts with decreasing energy are due to the lowering of the exciton energy during the formation of the bound magnetic polaron[30]. This effect will be discussed further in the next section.

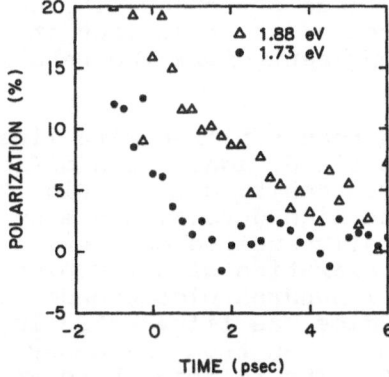

Fig. 6. Luminescence polarization decay of $Cd_{0.935}Mn_{0.065}Te$ (86 Å) – $Cd_{0.62}Mn_{0.38}Te$ (86 Å) with excitation energies at 2.0 eV and detection energies as indicated.

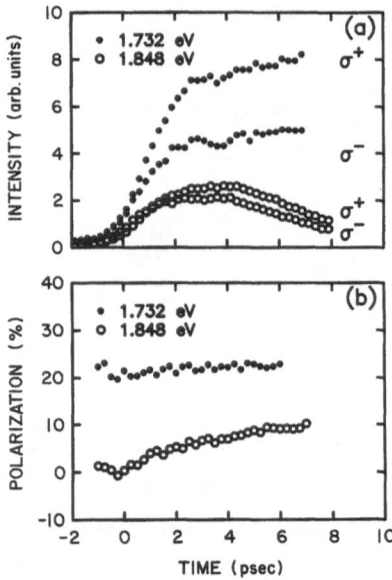

Fig. 7. Time-dependence of luminescence intensity (a)
and polarization (b) of the sample in Fig. 6
at a magnetic field of 0.5 T. The detection
energies are indicated.

Of more interest is the carrier spin relaxation time, shown
in Fig. 6 for a magnetic quantum well[28]. The sample is pumped
with circularly polarized radiation at 2.0 eV, about 0.25 eV
above the ground state. The luminescence is measured for both
circular components, and the difference divided by the sum of
the two is plotted as the polarization. That the initial
polarization is close to 0.25 is consistent with the notion that
the strong spin-orbit coupling in the valence band results in
extremely fast scattering of holes[31,32]. From Fig. 6, it is
clear that the electron spin relaxation time is very short, of
the order of a few picoseconds, and that the times are different
for the two detection energies with one on the high-energy side
above the ground state. What is surprising is that the spin
relaxes slower for the case of nonequilibrium population of
electrons and holes created at an earlier stage than at the
ground state, which represents more the behavior of bound
excitons.

The situation has been explored with linearly polarized
light under a magnetic field, again with different energies of
measurement[14]. The intensity data in Fig. 7a clearly show the
short decay time at the high detection energy as a result of the
steep cascade of optically excited carriers on these time
scales. Their rapid relaxation of a few picoseconds is in sharp
contrast to the several hundred picoseconds shown earlier in
Fig. 5. The spin dynamics, as illustrated in Fig. 7b, is
consistent with that obtained from the decay of the spin
polarization in Fig. 6. The response here at the low energy is
virtually instantaneous, and it becomes slower at the high
energy where we see the spin-lattice relaxation of the electrons
as they settle into Zeeman-split levels. The observation cannot
be reconciled with the qualitative picture in which electron

spin orientation is erased by spin scattering processes that become slower with decreasing carrier energies, while the dominant energy-loss scattering by optical phonons is spin-conserving.

The spin relaxation time of a few picoseconds agrees well with that calculated theoretically based on spin-flip exchange with the Mn-ions[33]. This agreement, however, is likely to be accidental. The theory predicts a much longer time for non-magnetic wells since the process depends solely on the exchange mechanism which involves here only the exponential tail of the electron penetration into the barrier. The relaxation time, surprisingly, is found to be independent of the Mn composition over the range of 0 - 0.13. To shed light on this puzzling discrepancy, comparative experiments have been performed under steady-state conditions between a quantum-well and a super-lattice sample. Figure 8 shows the polarization decay as a function of excess excitation energy above the detection energy near the luminescence peak. The quantum well exhibits instan-taneous depolarization, while the superlattice shows long spin relaxation at low excess energies, similar to the situation in bulk materials[34]. This leads us to believe that the origin of the fast relaxations lies in the confinement of the carriers in the two-dimensional system. The large binding and exchange energies of the excitons in this case[35], coupled with the extremely fast relaxation of holes, are believed to be the origin that causes rapid electron relaxation.

5.   MAGNETIC SPECTROSCOPY

In DMS structures, one can make direct magnetic measure-ments which have no counterpart in ordinary semiconductors. While spin polarization probe optically the magnetic state at the level of photo-generated carriers, here we investigate magnetization of the ions as a result of their interaction with the carriers. Figure 9 shows the magnetizations induced through

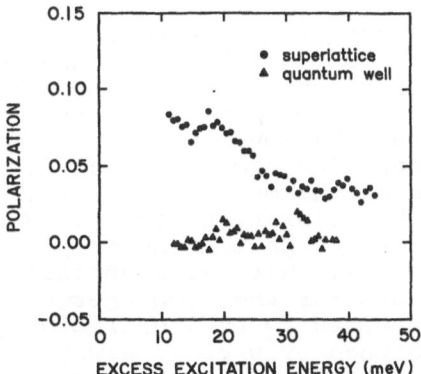

Fig. 8. Time-integrated luminescence polarization with excitation at excess energies above the detection point. The samples are: CdTe (90 Å) - $Cd_{0.85}Mn_{0.15}Te$ (18 Å) superlattice and $Cd_{0.935}Mn_{0.065}Te$ (86 Å) - $Cd_{0.62}Mn_{0.38}Te$ (86 Å) quantum well.

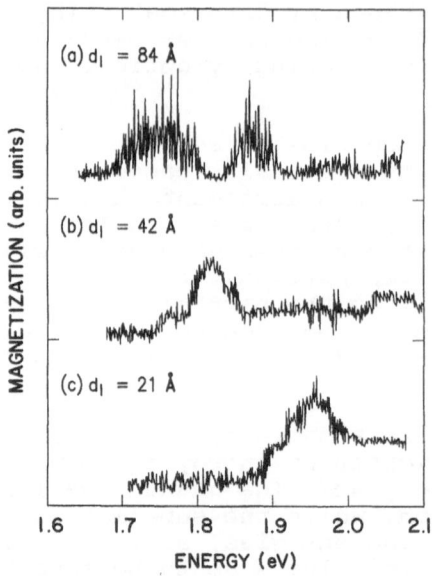

Fig. 9. Magnetic spectroscopy of $Cd_{0.935}Mn_{0.065}Te$
(84 Å, 42 Å, 21 Å) - $Cd_{0.62}Mn_{0.38}Te$ (84 Å).

the strong spin-exchange interaction in several quantum wells as
a function of the exciting photon energy[13]. These spectra
display subband transitions in the same manner as photoluminescence excitation commonly used, amounting to magnetic
manifestation or detection of electric quantization. The
strength of the signal depends on a combination of parameters,
including the density of states, the spatial extent of the
carrier wavefunctions and the spin relaxation rate. It is seen
clearly that, as the well width is widened, the ground-state
transition shifts to lower energies, and more transitions from
excited subbands at higher energies are observed. The small
features prior to the main peaks are likely to be associated
with impurities.

The dynamic response of the magnetizations of the ions are
illustrated in Fig. 10, using again polarized carriers with
time-resolved magnetic spectroscopy. As the radiation energy is
successively tuned to the subband energy from the ground to the
first and second state, the ionic relaxation time shortens
considerably. This dependence can be understood directly from
the strong dependence of the magnetic relaxation itself on the
magnetic dilution. The relaxation time drops from tens of nano-
seconds at a few percent of Mn to tens of picoseconds at a few
tens of a percent.[12,36]. With substantially more penetration
into the high Mn regions for the high energy states, the average
relaxation time is expected to be significantly reduced. The
same conclusion is reached by probing always the ground state
which, however, is varied in energy by varying the well width in
different heterostructures[13]. It is interesting to note that
the magnetic relaxation time can be much longer than the carrier
spin relaxation time and also the carrier lifetime. The
carriers, in other words, impart a magnetic imprint on the
system which evolves long after the carriers themselves have
relaxed and recombined.

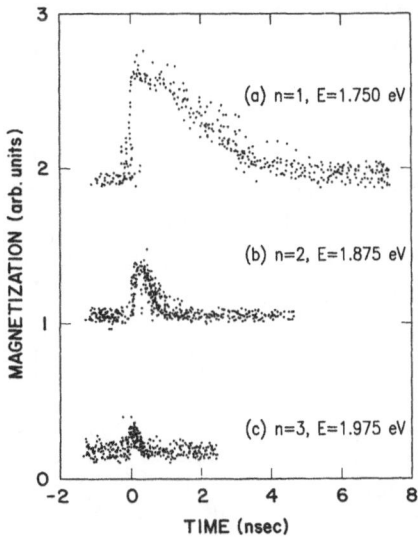

Fig. 10. Time-resolved magnetic spectroscopy of the
sample in Fig. 9(a) at energies corresponding
to the magnetization peaks, as indicated.

In addition to transferring their spins to the ions through
spin-flip interaction, the carriers can also create a local
field for spin orientation. The phenomenon of magnetic polaron
has been of considerable interest, and its presence has been
demonstrated with photo-excitations at energies near the
fundamental gap[6,12]. Using a superlattice with a thin CdMnTe
layer for electron penetration, we show in Fig. 11 both the
static and dynamic data. The time-integrated spectrum in (a) is
similar to those in Fig. 9, showing the subband transition peak.
The dynamic spectra in (b) and (c), taken at energies as
indicated in (a), show drastically different behavior. While
the spins are transferred instantaneously to the ions at the
high energy, the initial, gradual rise at the low energy near
the ground state demonstrates the dynamics of the local spin
organization as the polaron is formed. Both the polaron
formation and the ion relaxation are governed by the spin-
lattice process. The time scale is similar to that observed in
the bulk[12], which is consistent with other measurements by
monitoring the evolution of the polaron binding energy[30].

We discuss in passing a set of experiments which deal
exclusively with interactions among the ions themselves. The
short-range, antiferromagnetic interaction in DMS leads to a
magnetic phase transition from paramagnetism to spin-glass
ordering as the ion concentration is increased and the
temperature is lowered[2,3]. In a heterostructure
configuration, what is of interest in this context is not
electric quantization but the dimensionality effect, in terms of
the width of the magnetic layer, on the phase transition.
Figure 12 shows the susceptibility for structures with the same
CdTe but decreasing layer thickness of CdMnTe[37]. The well-
defined cusp in (a), reflecting the spin-glass transition as in
the bulk, becomes broadened in (b), and eventually vanishes in
(c) with the simultaneous appearance of hysterisis. Since the

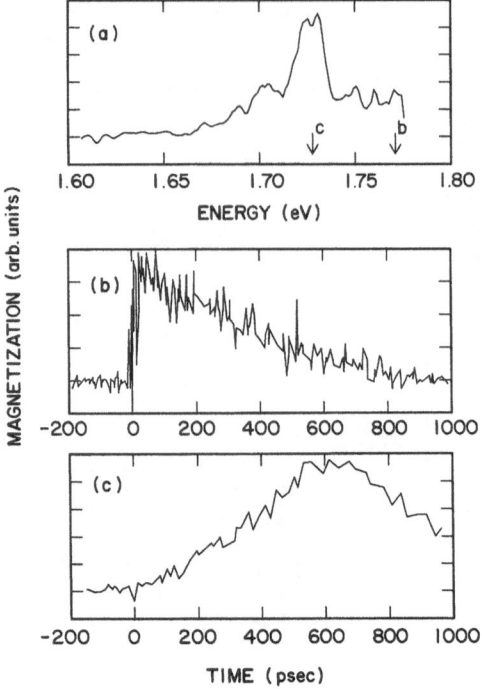

Fig. 11. Magnetic spectroscopy of CdTe (40 Å) –
Cd$_{0.77}$Mn$_{0.23}$Te (20 Å): (a) time-integrated
spectrum; (b) and (c) time-resolved
spectra at excitation energies indicated
in (a).

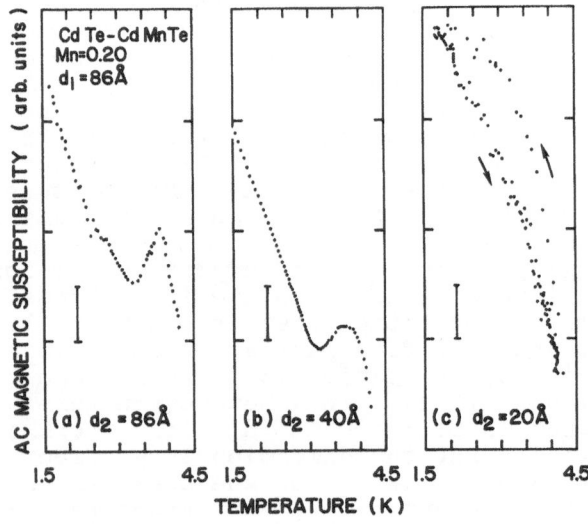

Fig. 12. Magnetic susceptibility of CdTe (86 Å) –
Cd$_{0.80}$Mn$_{0.20}$Te (86 Å, 40 Å, 20 Å). The
vertical scale corresponds to
2×10^{-6} emu/g.

percolation threshold is smaller than or comparable to the sample scale in the structures under consideration[38], the results in Fig. 12 indicate that spin-glass order cannot be maintained in a two-dimensional magnetic system. This conclusion, further supported by a systematic study with different layer thicknesses and Mn compositions, provides an answer to the question of long theoretical interest[39]. The hysteretic behavior can be understood on the basis of trapped short-range spin-glass order not responding to external fields.

CONCLUDING REMARKS

Since their first successful formation, quantum wells and superlattices made of diluted magnetic semiconductors have experienced substantial progress, both in quality of materials and structures and in observations of electronic and magnetic properties. Yet in view of the vast opportunities offered by this class of semiconductors, what has been achieved and understood must be considered rather limited. With growing interest and effort, activities in this area are expected to accelerate in scope as well as in depth. Already, investigations have been expanded to systems other than CdTe-CdMnTe. The success in ZnSe-ZnMnSe and HgTe-HgMnTe covers the range of both high- and low-gap regimes[40,41]. To be added to this list are the interesting IV-VI compounds such as PbTe-PbMnTe[44] and even III-V compounds, for example, InAs-InMnAs which has recently been successfully synthesized[45]. While the expanding front in the materials arena is no doubt exciting, the long-standing problem of conductivity control largely remains in wide-gap II-VI compounds with or without the magnetic element. Recent observations in connection with photo-assisted deposition[46] and Li-N doping[47] are encouraging. But significant progress in this area is critically needed, before this group of materials in general will become truly viable.

Acknowledgment

We would like to thank J. Yoshino, H. Munekata, J.M. Hong and C. Hsu for their collaboration in providing heterostructures used in this work.

REFERENCES

1. L.L. Chang, **Proc. Mat. Res. Soc. Symp.** 56:269 (1986)
2. N.B. Brandt and V.V. Moshchalkov, **Adv. Phys.**, 33:193 (1984)
3. J.K. Furdyna, **J. Appl. Phys.**, 61:3526 (1987)
4. L.L. Chang and E.E. Mendez, ch. 4 **in**: "Synthetic Modulated Structures", L.L. Chang and B.C. Giessen, eds., Academic Press, Orlando (1985)
5. R.N. Bicknesll, R.W. Yanka, N.C. Giles-Taylor, E.L. Buckland and J.F. Schetzina, **Appl. Phys. Lett.**, 45:92 (1984)
6. L.A. Kolodziejski, T.C. Bonsett, R.L. Gunshor, S. Datta, R.B. Bylsma, W.M. Becker and N. Otsuka, **Appl. Phys. Lett.**, 45:440 (1984)
7. R.L. Gunshor, L.A. Kolodziejski, N. Otsuke and S. Datta, **Surf. Sci.**, 174:522 (1986)
8. L.L. Chang, **Superlattices and Microstructures**, 6:39 (1989)
9. D.W. Kisker, P.H. Fuoss, J.J. Krajewski, P.M. Amirtharaj,

S. Nakahara and J. Menendez, **J. Crystal Growth**, 86:210 (1988)

10. M. Pesse, J. Lilja, O. Jylha, M. Ishiko and H. Asonen, **Proc. Mat. Res. Soc. Symp.**, 89:303 (1987)

11. R.F.C. Farrow, ch. 7, **in**: "Molecular Beam Epitaxy and Heterstructures", L.L. Chang and K. Ploog, eds., Martinus Nijhoff, Dordrecht, (1985)

12. D.D. Awschalom, J. Warnock and S. von Molnar, **Phys. Rev. Lett.**, 58:812 (1987)

13. D.D. Awschalom, J. Warnock, J.M. Hong, L.L. Chang, M.B. Ketchen and W.J. Gallagher, **Phys. Rev. Lett.**, 62:199 (1989)

14. M.R. Freeman, D.D. Awschalom, J.M. Hong and L.L. Chang, **Phys. Rev. Lett.**, 64:2430 (1990)

15. L. Vina, L.L. Chang and J. Yoshino, **J. Physique**, C5:317 (1987)

16. L. Vina, F. Calle, J.M. Calleja, F. Mesequer, L.L. Chang, J. Yoshino and M.J. Hong, presented at the NATO ARW on Light Scattering in Semiconductor Structures and Superlattices, Mt. Tremblant, March 1990 (to appear in proceedings).

17. G. Couturier, J. Yoshino, L.L. Chang and S. von Molnar, **J. Physique**, C5:341 (1987)

18. X.C. Zhang, S.K. Chang, A.V. Nurmikko, L.A. Kolodziejski, R.L. Gunshor and S. Datta, **Phys. Rev. B**, 31:4056 (1985)

19. J. Warnock, A. Petrou, R.N. Bicknell, N.C. Giles-Taylor, D.K. Blanks and J.F. Schetzina, **Phys. Rev. B**, 32:8116 (1985)

20. S.K. Chang, A.V. Nurmikko, J.W. Wu, L.A. Kolodziejski and R.L. Gunshor, **Phys. Rev. B**, 37:1191 (1988)

21. E.K. Suh, D.V. Bartholomew, A.K. Ramdas, S. Rodriguez, S. Venogupalan, L.A. Kolodziejski and R.L. Gunshor, **Phys. Rev. B**, 36: 4316 (1987)

22. L. Vina, F. Calle, J.M. Calleja, C. Tejedor, J.M. Hong and L.L. Chang, p.819 **in**: Proc. 19th Int. Conf. Phys. Semicond., Inst. Phys. Polish Academy of Sciences, Warsaw, (1988)

23. E.K. Suh, D.V. Bartholomew, A.K. Ramdas, R.N. Bicknell, R.L. Harper, N.C. Giles and J.F. Schetzina, **Phys. Rev.B**, 36:9358 (1985)

24. E.E. Mendez, G. Bastard, L.L. Chang, L. Esaki, H. Morkoç and R. Fischer, **Phys. Rev.B**, 26:7101 (1982)

25. E.E. Mendez, F. Agullo-Rueda and J.M. Hong, **Phys. Rev. Lett.**, 60:2426 (1988)

26. P. Voisin, J. Bleuse, C. Bouche, S. Gaillard, C. Alibert and A. Regreny, **Phys. Rev. Lett.**, 61:1639 (1988)

27. A. Harwit, C. Hsu, F. Agullo-Rueda and L.L. Chang, **Appl. Phys. Lett.**, in press.

28. D.D. Awschalom and M.R. Freeman, **J. Luminescence**, 44:399 (1989)

29. F.O. Gobel, J. Kuhland, R. Hoger, **J. Luminescence**, 30:541 (1985)

30. J.H. Harris and A.V. Nurmikko, **Phys. Rev. Lett.**, 51:1472 (1983)

31. See, for example, M.I. Dyakonov and V.I. Perel, **in**: "Optical Orientation", F. Meier and B.P. Zakharchenya, eds., Elsevier, Amsterdam, (1984)

32. H. Krenn, W. Zawadzki and G. Bauer, **Phys. Rev. Lett.**, 55:1510 (1985)

33. G. Bastard and L.L. Chang, **Phys. Rev.B**, 41:7899 (1990)

34. J. Warnock, R.N. Kershaw, D. Ridgley, K. Dwight, A. Wold and R.R. Galazka, **Solid State Commun.**, 54:215 (1985)

35. Y. Chen, B. Gil, P. Lefebvre, H. Mathieu, T. Jukenuaga and H. Nakashima, in: "Excitons in Confined Systems", R.Del Sole, A. D'Andrea and A. Lapiccirella, eds., Spring-Verlag, Berlin (1988)

36. E.A. Harris and K.S. Yngvesson, **J. Phys. C**, 1:990 (1968)

37. D.D. Awschalom, J.M. Hong, L.L. Chang and G. Grinstein, **Phys. Rev. Lett.**, 59:1733 (1987)

38. K. Binder and A.P. Young, **Rev. Mod. Phys.**, 58:801 (1986)

39. J.J. Zayhowski, C. Jagannath, R.N. Kershaw, D. Ridgely, K. Dwight and A. Wold, **Solid State Commun.**, 55:941 (1985)

40. L.A. Kolodziejski, R.L. Gunshor, T.C. Bonsett, R. Venkatasubramanian, S. Datta, R.B. Bylsma, W.M. Becker and N. Otsuka, **Appl. Phys. Lett.**, 47:169 (1985)

41. K.A. Harris, S. Hwang, R.P. Burns, J.W. Cook and J.F. Schetzina, **Proc. Mat. Res. Soc. Symp.**, 89:255 (1987)

42. X. Chu, S. Sivananthan and J.P. Faurie, **Appl. Phys. Lett.**, 56:597 (1987)

43. R.G. Alonso, E.K. Suh, A.K. Ramdas, N. Samarth, H. Luo and J.F. Furdyna, **Phys. Rev. B**, 40:3720 (1989)

44. H. Clemens, H. Krenn, P.C. Weilgani, U. Stromberger and G. Bauer, **Surf. Sci.**, 228:236 (1990)

45. H. Munekata, H. Ohno, S. von Molnar, A. Segmuller, L.L. Chang and L. Esaki, **Phys. Rev. Lett.**, 63:1849 (1989)

46. R.N. Bicknell, N.C. Giles, J.F. Schetzina and L.C. Hitzman, **J. Vac. Sci. Technol. A**, 5:3059 (1987)

47. T. Yasuda, I. Mitsuishi and H. Kukimoto, **Appl. Phys. Lett.**, 52:57 (1988)

TIME-RESOLVED PHOTOLUMINESCENCE DETERMINATION OF X_z-$X_{x,y}$ ENERGY SEPARATION AND X_z STATE LIFETIMES IN AlGaAs/AlAs TYPE II MULTIPLE QUANTUM WELLS

Jeff F. Young, S. Charbonneau and P. Coleridge

Division of Physics
National Research Council
Canada, K1A OR6

ABSTRACT

Results of photoluminescence (PL) (steady-state and time-resolved with 150 ps resolution) experiments on a series of type II $Al_{0.15}Ga_{0.85}As$/AlAs multiple quantum wells with a (nominally) fixed alloy thickness of 4.0 nm and AlAs thicknesses ranging from 4.8 to 7.8 nm are reported. Samples with AlAs thickness less than 6.2 nm have the X_z state as the lowest lying conduction band level while for thicker AlAs layer the $X_{x,y}$ states are lowest in energy. For samples with AlAs layers thicker than the critical value, a time-windowing technique is used to simultaneously observe PL from the X_z and lower-lying $X_{x,y}$ states immediately following excitation by picosecond laser pulses. The energy separation of the two no-phonon lines directly yields the X_z- $X_{x,y}$ separation, and the decay rate from X_z to $X_{x,y}$ can be obtained by time-resolving the hot X_z PL. The lifetime of the nonequilibrium X_z states in samples above the critical thickness is measured as a function of barrier thickness and uniaxial stress: it was found to depend weakly on the X_z- $X_{x,y}$ separation from 3 - 9 meV, with values between 1 and 1.5 ns.

1.   INTRODUCTION

Semiconductor heterostructures have now found wide application in fields ranging from optical waveguides to high-mobility electronic devices. Many different constituent materials are available, and it is often possible to optimize a particular response characteristic by appropriate choice of the semiconductor material. In a number of cases, such as short-period superlattices[1,2] and double-barrier resonant tunneling structures[3], the large direct band gap of AlAs makes it an attractive companion material for GaAs and low Al concentration AlGaAs alloys.

Unfortunately, relatively little is known about the basic properties of AlAs due to the difficulties involved in obtaining good quality samples in bulk form. However, it is well established that the conduction band minimum is at the X point of the Brillouin zone, and there are indications that these X-valley states influence the tunneling behavior of Γ electrons from GaAs through AlAs[3].

The study of electronic properties of X states in AlAs can be facilitated by growing GaAs/AlAs or AlGaAs/AlAs superlattices (or multi-quantum-wells) with sufficiently thin GaAs or AlGaAs layers such that the confinement energy of the Γ electrons in these layers raises their energy above that of the relatively weakly confined X states in the AlAs. These so-called type II superlattices have the X states in the AlAs as their lowest conduction band minimum, while the lowest energy hole band remains in the GaAs or AlGaAs layer.

The properties of these AlAs-based type II superlattices have been studied extensively over the past four years, primarily through optical probing[1,2,4-13] but also using magneto-transport techniques[14]. Much of the effort in the optical domain has focused on identifying and quantifying the transition from type I to type II behavior as the layer thicknesses are varied[13,15,16], based primarily on the photo-luminescence (PL) kinetics. In type I material the exciton PL is very short-lived ($\leq$ 1 ns) due to the large oscillator strength associated with the direct gap. In type II materials the PL lifetime can be in the millisecond regime due both to the fact that the electrons and holes become spatially separated in different layers, and that the transition becomes indirect in momentum space.

Recent experimental progress has been made in studying the ultra-fast transfer of electrons, initially excited in the Γ states of the GaAs layer, to the AlAs layers[17,18]. This time, typically less than 1 ps, provides a measure of the Γ-X mixing of subband states in the two materials.

Another issue which has recently received attention concerns the lifting of the bulk degeneracy of the six equivalent X minima in the first Brillouin zone[6,9,10]. At least three independent studies have now demonstrated that this degeneracy is lifted due both to confinement effects, and to residual strain in the AlAs resulting from the (small) lattice mismatch with the GaAs substrate material. The two effects work in opposition to each other: the confinement preferentially raises the energy of the in-plane X_x and X_y minima with respect to the X_z minima aligned along the z growth axis, since the effective electron mass along z is much higher, ~ $1.1m_0$, for X_z minima as opposed to ~ $0.2m_0$, for $X_{x,y}$ minima. However, the biaxial compressive stress tends to raise the X_z minima with respect to the $X_{x,y}$ minima by an amount which has been estimated by various groups to be between 19 and 23 meV[9,10]. One therefore expects that for a fixed GaAs or AlGaAs layer thickness, there should be some critical AlAs thickness, t_c, below which the minimum energy X states will be X_z-like, and above which the $X_{x,y}$-like states will be lowest in energy.

There are many interesting issues associated with the nature of $X_{x,y}$ and X_z states in AlAs which are just beginning to receive attention[6,9,10]. It is well established that there exist quite significant localization energies for photo-excited exciton states in these type II structures, most likely associated with potential fluctuations near the interfaces[4].

Such localizing potentials can in principle mix the various X minima, in analogy with the valley-orbit mixing of impurity states in Ge and Si[9,19]. Just as the mixing of Γ and X states is crucial to many of the unique type II properties, any complete understanding of the electronic band structure of these materials will have to include the mixing of various X minima.

The purpose of the present paper is to report on a technique which is well suited to study the static and dynamic properties of **both** $X_{x,y}$-like **and** X_z-like exciton states in type II structures with AlAs thicknesses greater than or equal to t_c. A series of $Al_{0.15}Ga_{0.85}As$/AlAs multi-quantum-well structures with a (nominally) fixed alloy thickness of 4.0 nm, and AlAs thicknesses ranging from 4.8 to 7.8 nm, have been studied using a time-windowing PL technique with 150 ps time-resolution. In three of the samples with AlAs thicknesses greater than 6.2 nm, two distinct sets of PL transitions are observed at short ($\leq$ 2ns) times directly following excitation with intense pico-second laser pulses at temperatures less than 20 K. Detailed study of these PL features, both as a function of temperature and uniaxial stress, confirms that one set behaves as though derived from an X_z-like state, and the other as though derived from an $X_{x,y}$-like state, so long as the X_z-like state is more than 3 meV above the $X_{x,y}$-like state. In addition to serving as a tool for measuring the X_z-$X_{x,y}$ energy separation (ΔE) as a function of AlAs thickness and applied stress, this technique also provides a direct measure of the relaxation rate out of the high-lying X_z-like state, and the dependence of this relaxation rate on ΔE.

2. EXPERIMENT

A description of the samples used in the present study is given in Table 1. All experiments were done with the samples mounted strain-free in a variable temperature cryostat with temperature control to ± 0.2 K. Steady-state PL spectra were obtained using 2 mW HeNe laser excitation, and time-resolved results were obtained using a synchronously pumped R640 dye laser which produced 4 ps pulses at 629 nm. An electro-optic modulator was used to obtain a 2 MHz repetition rate.

The PL detection system consisted of a single 0.64 m spectrometer with an imaging photomultiplier tube (PMT) used in a photon-counting mode. The 1" diameter PMT (ITT "Mepsicron") enabled parallel acquisition of the PL spectra, which when combined with a correlated photon-counting temporal

Table 1.   Sample Identification

Sample No.	AlAs Thickness (nm) ± 0.2 nm (from X-ray)
1	4.8
2	5.5
3	6.0
4	6.2
5	7.2
6	7.8

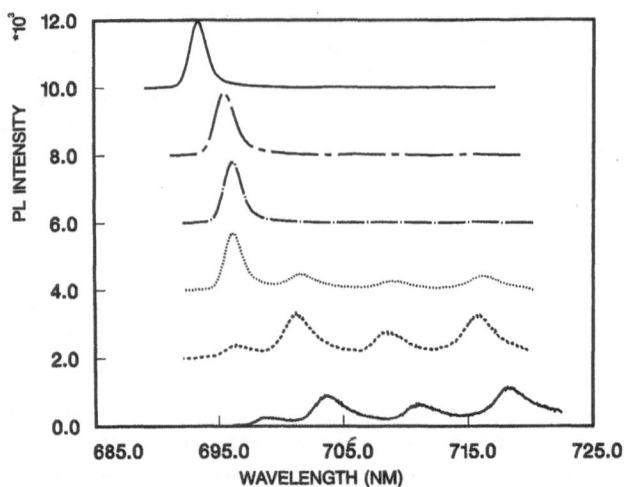

Fig. 1. Plots of the PL spectra from the samples listed
in Table 1, in the corresponding order (#1 at
top). The samples were mounted strain-free at
4.5 K.

discriminator, allowed entire spectra to be collected in eight
independent time-windows. For all of the time-windowed spectra
discussed here, the time-bins were 500 ps wide, directly
adjacent to one another, with the first corresponding to the
arrival of the laser pulse on the sample.

Uniaxial stress was applied by loading calibrated weights
on a piston that pushed vertically on the (110) cleaved surfaces
of the samples, each of which had an area 5 mm x 0.46 mm in
contact with the piston.

4. RESULTS

Figure 1 illustrates the cw PL spectra obtained from all of
the samples, mounted strain-free, at 4.5 K. Care must be
exercised in interpreting the absolute energies since small,
sample-dependent variations in the alloy composition and
thickness can modify the energy of the hole-state contribution
to the PL. The main point to be deduced from Fig. 1 is the
obvious qualitative difference between the spectra from samples
#1 - #3 versus those from #5 and #6. The dominant features in
samples #5 and #6 are the three phonon replicas of the weak, no-
phonon line at higher energy[2]. Similar phonon replicas can
also be observed in the spectra from samples #1 - #4 if an
expanded scale is used. These phonons, at energies of 14, 27
and 50 meV have been attributed by others[2] to TA, LA and LO
phonons at the X point of the AlAs Brillouin zone.

The relative strength of no-phonon to phonon-replica PL
intensities in the steady state spectra has been used by
Scalbert et al.[9] to categorize a given sample as to the lowest
lying X minima. Their basic argument, also discussed by
others[6,10], is that for X_z-like states, the superlattice
potential can mix the X_z with Γ-like conduction band states,

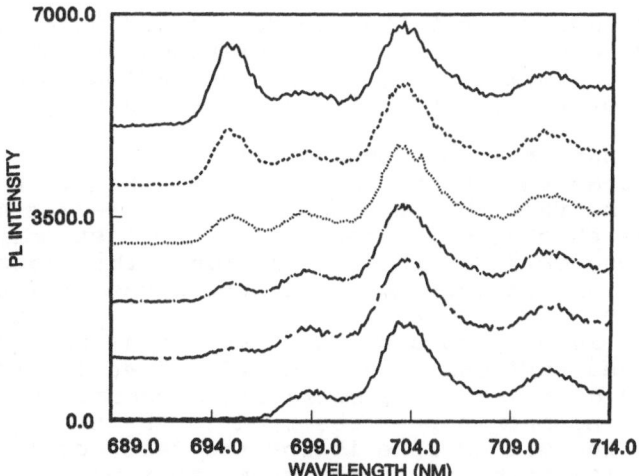

Fig. 2. A series of spectra from sample #6 under maximum
uniaxial stress (see Fig. 3). Each spectrum is
time-integrated over adjacent 500 ps windows,
with the top spectrum overlapping the arrival of
the laser pulse on the sample.

even in the absence of fluctuating potentials. Thus there will
be some "direct" oscillator strength associated with the X_z-like
states that enables radiative recombination of the X_z-like
exciton without the need for phonon participation. For the
$X_{x,y}$-like excitons, only perturbations which destroy the translational invariance in the layer-plane can couple the $X_{x,y}$ and Γ
states, and these are expected to be weak compared to the
superlattice potential, hence the dominance of phonon-assisted
transitions when the lowest lying X state is $X_{x,y}$-like. On this
basis then, samples #5 and #6 are expected to have $X_{x,y}$-like
minima, and sample #4 should be close to the critical thickness,
with almost equal-energy X_z and $X_{x,y}$ states.

At 4.5 K the lifetime of the PL from all of the samples is
longer than 10 μs, and cannot be efficiently resolved using the
correlated photon-counting technique. For samples #1 - #3, the
spectra remain essentially invariant, the lifetime remains
longer than 10 μs, and there is no sign of a short-lived
component for laser-pulse fluences up to ~ 40 μJ/cm^2. For
samples #4 - #6 the overall lifetime also remains longer than
10 μs up to the same peak fluence; however, in these samples
with thicker AlAs layers, a short-lived feature appears in the
spectra for incident fluences above a threshold level which is
sample-dependent, but always less than 40 μJ/cm^2. Figure 2
shows a series of six spectra taken from sample #6 in adjacent
time-windows, 500 ps apart. The time-window corresponding to
the top spectrum coincides with the arrival of the laser pulse
on the sample. The feature near 695 nm appears immediately
following the laser pulse. It decays with an exponential time-
constant of 1.3 ± 0.2 ns with a spectral shape that does not
change during the decay process. All of the other features in
the spectra are time-invariant, and correspond to the no-phonon
and phonon-replica peaks observed in the cw spectrum of Fig. 1.
Similar sets of time-resolved spectra are obtained from samples

#5 and #6, with and without the application of uniaxial stress along the [110] direction; the only influence of sample or stress is on the absolute energies of the various peaks, and the absolute value of the decay time, which can vary by 50%.

The fact that the fast component in the PL is at slightly higher energy than the no-phonon feature in the steady-state PL, together with the fact that the separation increases with increasing AlAs thickness, suggests that the fast component is due to the no-phonon X_z exciton transition. This is confirmed by studying the effect that uniaxial stress has on the peak positions. Figure 3 illustrates the change in peak positions upon applying a 14.5 Kg mass on the 5 mm x 0.46 mm, (110)-oriented, cleaved surfaces of samples #5 and #6 (a nominal pressure of ~ 0.62 kbar). The solid lines correspond to the time-invariant, no-phonon peak observed in the steady-state spectra, while the dashed line is for the fast component[20]. Lefebvre et al.[6] have previously shown that uniaxial, compressive stress along the [110] direction shifts the X_z-like PL energies very little, but shifts the $X_{x,y}$-like PL peaks down in energy by ~ 4.5 meV/kbar, in a number of GaAs/AlAs type II superlattices. The shift to lower energy of the steady-state no-phonon line (the phonon-replicas shift the same) therefore confirms its identification as due to an $X_{x,y}$-like exciton. Similarly, the relatively weak variation of the fast component's energy with stress supports its association with X_z-like excitons.

Figure 4 shows the dependence of the PL intensity of the fast component (solid) and the steady-state component (dashed) from sample #6 as a function of the incident laser intensity. The data were obtained by placing neutral density filters in the beam's path, before it irradiated the sample. The strong saturation of the steady-state PL, while the fast component

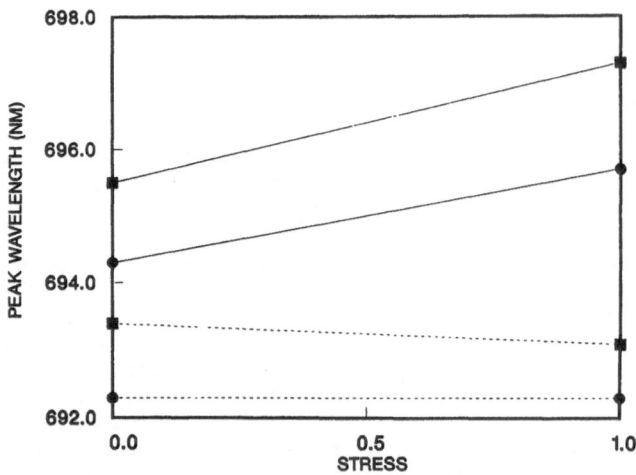

Fig. 3. Variation of the X_z-like (dashed) and $X_{x,y}$ (solid) PL energies from samples #5 (circles) and #6 (squares) as a function of applied stress along the [110] direction. Unity stress corresponds to 14.5 Kg applied on 5 mm and 0.46 mm cross sections of the cleaved samples.

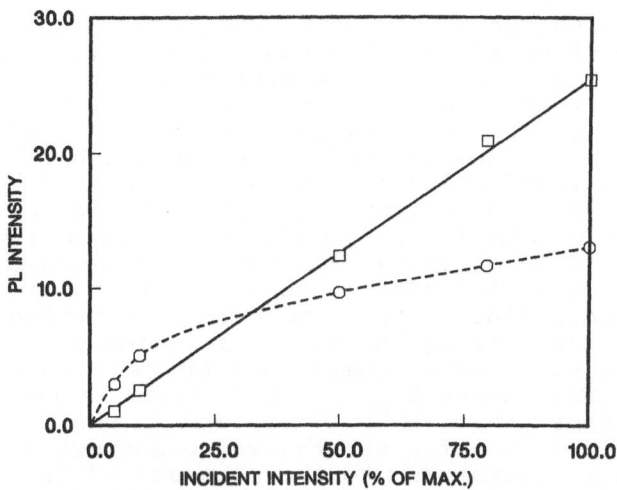

Fig. 4. Dependence of the peak PL intensity from the
fast X_z-like signal (solid) and from the first
phonon-replica of the slow $X_{x,y}$-like signal
(dashed) on incident laser power. The data were
taken under the conditions corresponding to Fig.
2 and the maximum laser pulse fluence
corresponds to ~ 40 $\mu J/cm^2$.

increases linearly, explains why the fast component is observed
only for incident fluences above a certain threshold value.

This data also aids in identifying the transitions that
give rise to the PL peaks in the time-windowed spectra, which
were all taken using laser intensities such that the fast
component was stronger than the slow component. In particular,
the strong saturation of the steady-state PL that occurs near
13% in Fig. 4 is accompanied by an increase in its peak energy.
These factors indicate that at low intensities, relatively low-
energy, localized $X_{x,y}$ exciton states are being filled. These
are relatively long-lived, isolated states at 4.5 K, and they
are rapidly filled at low incident fluences, so that at higher
intensities, $X_{x,y}$ excitons must occupy shallower localized
states, which have a much higher density of states, and are
therefore more susceptible to nonradiative recombination. The
linear, albeit reduced slope of the dashed curve in Fig. 4 above
~ 20%, together with a fixed peak PL energy over this intensity
range, therefore implies that the corresponding steady-state PL
is from $X_{x,y}$ excitons in a region of relatively high density of
states.

As for the fast X_z component, its linear intensity
dependence, together with the invariance of its peak position
with respect to both laser intensity and temperature, indicate
that it is associated with free or nearly free X_z excitons, in a
region having a high density-of-states.

4.   DISCUSSION

The above data implies that time-windowed PL spectra,
obtained at 4.5 K from samples with AlAs thicknesses greater

than that at which the X_z-$X_{x,y}$ crossover occurs, t_c, provide a direct measure of weakly localized $X_{x,y}$ **and** X_z exciton energies simultaneously. The energy difference between X_z and X_x-like excitonic PL features, ΔE, is in general due to differences in the strain- and confinement-related subband energies of the respective X minima, and to differences in the binding energies of the (nearly) free excitons. The strain-related shift should be independent of layer thickness, hence once the critical thickness t_c is obtained, width-dependent changes in the X_z-$X_{x,y}$ separation above t_c can be compared with theoretical estimates for the differences in confinement and exciton binding energies. To determine t_c, we note that application of less than 0.1 kbar stress along the [110] direction converts sample #4 from exhibiting behavior characteristic of minimum energy X_z states, to minimum energy $X_{x,y}$ states. (This is represented both in the steady state PL behavior, and in that the stress allows the observation of a short-lived component in the time-window spectra). Therefore, to the accuracy of the layer thicknesses, these results support those of Kesteren et al.[10] who estimated t_c = 6.0 nm. Using the AlAs-width dependent confinement energies reported by Kesteren et al.[10] and a t_c of 6.2 nm, the X_z-$X_{x,y}$ energy separation of 5.2 and 5.4 meV measured with no stress applied to samples #5 and #6 would suggest AlAs thicknesses of ~ 7.3 and 7.5 nm, in quite good agreement with the X-ray data in Table 1.

The other significant result from these studies is that the decay-time of the hot X_z PL is fixed at 1.3 ± 0.3 ns, independent of ΔE in the range 3 – 9 meV. Within this range the lifetime is not sensitive to factor-of-five changes in laser intensity, or to temperature changes of 5 K. For ΔE less than ~ 3 meV (obtained by applying uniaxial compressional stress along the [110] axis of sample #4), the decay-time of the fast PL tends to increase, and becomes very sensitive to stress, laser intensity and temperature.

When the energy of the X_z-like state is above that of the $X_{x,y}$-like states, there exist numerous possible channels through which it may decay. One obvious possibility is that a new, fast channel opens up as soon as the $X_{x,y}$-like states become available at lower energy. This scenario will be considered below. However, even when the $X_{x,y}$-like states are higher in energy, the essentially-free X_z-like state can nonradiatively decay through channels having nothing to do with $X_{x,y}$-like states. It is therefore possible that the decay of the fast component in the PL spectra is a measure of the exponential, nonradiative decay rate of essentially-free X_z-like excitons in these samples. In this scenario, the fast component is observable as hot PL when X_z is above $X_{x,y}$ because there is a relatively large density of states for the free X_z exciton, which has a relatively large oscillator strength, and there is no competing PL mechanism at the same energy. One then has to explain the absence of this fast decay component in samples where X_z is the lowest energy state. In such samples there is strong, long-lived PL from localized X_z exciton states at energies shifted below the free exciton energy by the local-ization potential which is comparable to the PL linewidth. This could simply overwhelm the hot PL from the nearby free exciton, which would show up as a very weak, high energy shoulder. The extreme sensitivity of the fast component's decay rate when ΔE is less than 3 meV would then be due to a situation where the PL contributions from essentially-free and weakly localized X_z excitons are comparable, resulting in a net average of fast and slow lifetimes.

Since the above scenario does not involve details of $X_{x,y}$-like states, it is consistent with an exponential lifetime which is essentially independent of ΔE as long as the hot PL spectrum is isolated from the long-lived portion. A possible difficulty with this interpretation is that in samples #1 - #3, the lifetime of the X_z-like state decreases with increasing temperature, tending to saturate at temperatures above 30 K, but at sample-dependent decay times that are always longer than 30 ns. This saturation in lifetime is accompanied by the saturation of a blue-shift in the PL peak position. One might expect the saturated lifetime to be closer to 1 ns, and the same for all samples, if the decay of the hot X_z Pl in samples #5 and #6 is indeed due to nonradiative recombination.

If the hot PL decay is not due to general nonradiative recombination, and exists only for AlAs thicknesses greater than t_c, it must be associated with scattering from the X_z-like states into the lower-lying $X_{x,y}$-like states. Since the hole state involved in the PL is common to X_z and $X_{x,y}$ transitions, this decay process is essentially due to X_z-$X_{x,y}$ intervalley scattering. As the maximum ΔE is less than any X point AlAs phonon energy, such an intervalley scattering mechanism must be mediated by short-range, in-plane potential fluctuations. In this scenario, the weak energy dependence of the decay time for $\Delta E \geq 3$ meV would be consistent with roughness-induced intervalley scattering of free X_z electrons to $X_{x,y}$ valleys which offer a constant, two-dimensional density of final states. The extreme sensitivity of the decay time in the sample with AlAs thickness close to t_c could then be due to strong mixing of X_z and $X_{x,y}$-like states when they are close in energy, in analogy with the valley-orbit mixing for impurity states in Si and Ge. If this is the cause, an upper limit of 3 meV could be put on the characteristic mixing energy, which is not inconsistent with the mixing energy of 2 meV deduced by Lefebvre et al.[6] from their steady-state PL stress measurements.

CONCLUSIONS

A series of $Al_{0.15}Ga_{0.85}As$/AlAs multi-quantum wells have been studied using steady-state and time-resolved PL. The alloy layer thickness was nominally fixed at 4.0 nm for all samples, while the AlAs layer thickness varied from 4.8 to 7.8 nm. Both steady-state and time-resolved data suggest that the critical AlAs thickness at which the minimum X state changes from X_z to $X_{x,y}$ is ~ 6.2 nm. By using a time-windowing PL technique with 150 ps time-resolution, it is possible to simultaneously observe $X_{x,y}$ and hot X_z PL from samples with AlAs thickness greater than the critical thickness. This allows the determination of $X_{x,y}$ and X_z energies even in samples in which these states are separated in energy by more than 10 meV. The decay time of the hot X_z PL is found to be weakly dependent on the sample and energy separation from 3 to 9 meV. The rate-limiting mechanism responsible for this decay time has not been unambiguously identified. It may represent the nonradiative lifetime of free X_z excitons, or it may be related to intervalley X_z-$X_{x,y}$ scattering of electrons or excitons.

REFERENCES

1. E. Finkman, M.D. Sturge, M.H. Meynadier, R.E. Nahory, M.C.
    Tamargo, D.M. Hwang and C.C. Chang, **J. of Lumin.**, 39:57
    (1987)
2. B.A. Wilson, **IEEE J. Quan. Elect.**, 24:1763 (1988)
3. D.Z.Y. Ting, M.K. Jackson, D.H. Chow, J.R. Soderstrom, D.A.
    Collins and T.C. McGill, **Sol. Stat Elect.**, 32:1513 (1989)
4. F. Minami, K. Hirata, K. Era, T. Yao and Y. Masumoto, **Phys.
    Rev. B**, 36:2875 (1987)
5. K.J. Moore, G. Duggan, P. Dawson and C.T. Foxon, **Phys. Rev.
    B**, 38:5535 (1988)
6. P. Lefebvre, B. Gil, H. Mathieu and R. Planel, **Phys. Rev.
    B**, 39:5550 (1989)
7. M.S. Skolnick, G.W. Smith, I.L. Spain, C.R. Whitehouse,
    D.C. Herbert, D.M. Whittaker and L.J. Reed, **Phys. Rev. B**,
    39:11191 (1989)
8. P. Dawson, K.J. Moore, C.T. Foxon, G.W. 't Hooft and R.P.M.
    van Hal, **J. Appl. Phys**, 65:3606 (1989)
9. D. Scalbert, J. Cernogora, C. Benoit a la Gullaum,
    M. Maaref, F.F. Charfi and R. Panel, **Sol. State Comm.**,
    70:945 (1989)
10. H.W. van Kesteren, E.C. Cosman, P. Dawson, K.J. Moore and
    C.T. Foxon, **Phys. Rev.B**, 39:13426 (1989)
11. B.A. Wilson, C.E. Bonner, R.C. Spitzer, R. Fischer, P.
    Dawson, K.J. Moore, C.T. Foxon and G.W. t'Hooft, **Phys. Rev.
    B**, 40:1825 (1989)
12. R. Cingolani, L. Tapfer, Y.H. Zhang, R. Muralidharan, K.
    Ploog and C. Tejedor, **Phys. Rev. B**, 40:8319 (1989)
13. G. Li, D. Jiang, H. Han, Z. Wang and K. Ploog, **Phys. Rev.
    B**, 40:10430 (1989)
14. T.P. Smith III, W.I. Wang, F.F. Fang, L.L. Chang, L.S. Kim,
    T. Pham and H.D. Drew, **Surf. Science**, 196:287 (1988)
15. G. Danan, B. Etienne, F. Mollot, R. Planel, A.M. Jean-
    Louis, F. Alexandre, B. Jussurand, G. LeRoux, J.Y. Marzin,
    H. Savary and B. Sermage, **Phys. Rev. B**, 35:6207 (1987)
16. R. Cingolani, L. Baldassarre, M. Ferrara, M. Lugara and K.
    Ploog, **Phys. Rev.B**, 40:6101 (1989)
17. J. Feldmann, R. Sattmann, E.O. Gobel, J. Kuhl, J. Hebling,
    K. Ploog, R. Muralidharan, P. Dawson and C.T. Foxon, **Phys.
    Rev. Lett.**, 62:1892 (1989)
18. P. Saeta, J.F. Federici, R.J. Fischer, B.I. Greene, L.
    Pfeiffer, R.C. Spitzer and B.A. Wilson, **Appl. Phys. Lett.**,
    54:1681 (1989)
19. H. Fritzsche, **Phys. Rev.**, 125:1560 (1962)
20. This data was taken on different pieces of wafers #5 and #6
    from those used to obtain the spectra in Fig. 1.

MECHANISM OF Γ-X ELECTRON TRANSFER IN TYPE II (Al)GaAs/AlAs

SUPERLATTICES AND MULTIPLE QUANTUM WELL STRUCTURES

J. Feldmann, E. Göbel, J. Nunnenkamp,† J. Kuhl,†
K. Ploog,† P. Dawson‡ and C.T. Foxon‡

Fachbereich Physik, Philipps Universität Marburg
3550 Marburg, Federal Republic of Germany
† Max-Planck-Institut für Festkörperforschung
7000 Stuttgart 80, Federal Republic of Germany
‡ Philips Research Laboratories, Redhill
Surrey RH15HA, UK

1. INTRODUCTION

In the conduction band of III-V semiconductors having cubic
zinc-blende structure there are local energetic minima at the
Γ, X and L points of the Brillouin zone. The idea that
electrons in the conduction band are scattered by phonons
between these valleys first became important in theories of the
Gunn effect[1,2]. In recent years the intervalley scattering
process has been extensively investigated in III-V semiconductors[3] mainly because of two reasons: first, ultrafast
optical spectroscopy became available for investigating
scattering processes which take place on a femtosecond
timescale. Secondly, the transport properties of high-speed
electronic semiconductor devices[4] are influenced by
intervalley scattering.

The continuous progress in the field of growing different
semiconductor materials on each other, which nowadays can be
done by molecular beam epitaxy (MBE) with almost atomic
precision, is another important development of the recent years.
In this article we deal with the special case, where the direct
band gap semiconductor materials GaAs or $Al_xGa_{1-x}As$ with x < 0.4
are grown on the indirect band gap semiconductor material AlAs.
Mendez et al.[5] performed tunneling experiments on GaAs-AlAs
heterostructures and found that electrons of the Γ valley of the
GaAs can transfer to the X states of the AlAs across the
heterointerface. This process is due to elastic intervalley
scattering caused by mixing of Γ and X states at the heterointerface[6]. In analogy, an electron transfer across the
heterointerface from Γ-related states confined in the GaAs layer
to X-related states confined in the AlAs layer was observed for
so-called type II GaAs/AlAs short period superlattices (SPS) in
cw luminescence experiments[7-9] and more clearly in time-
resolved luminescence experiments[10]. In spite of the spatial
charge transfer the Γ-X scattering rate in all binary GaAs/AlAs

type II SPS can be extremely high and comparable to the Γ-X scattering rate in GaAs bulk crystal[11,12]. Furthermore, we have demonstrated[13] that the Γ-X scattering rate, which is an intrinsic constant for bulk GaAs, can be varied systematically within a broad range (100 fs - 20 ps) by changing the geometrical dimensions of the sample structure.

In this article we present experimental studies of the scattering time constants with varying excitation intensity and sample temperature which provide information about the nature of the scattering mechanism. The next section deals with the linear optical properties of type II (Al)GaAs/AlAs superlattices. It is shown that luminescence experiments already give some hints for the underlying scattering mechanism. In Section 3 we shortly describe the time-resolved pump and probe experiments which allow the direct determination of the Γ-X transfer times. Our experimental results presented in Section 4 lead to the conclusion that electron-phonon scattering is the dominant scattering mechanism for samples with (Al)GaAs layer thickness > 100 Å. However, for samples with (Al)GaAs layer thickness < 35 Å interface scattering due to Γ-X mixing probably becomes dominant.

2. OPTICAL PROPERTIES

For type II (Al)GaAs/AlAs superlattices the lowest lying electronic states of the conduction band are confined in the AlAs layers, whereas the highest lying states of the valence band are confined in the (Al)GaAs layers as shown in Fig. 1. This staggered band alignment is achieved if the energy of the lowest confined Γ level in the (Al)GaAs becomes sufficiently high, which in an all binary $(GaAs)_m(AlAs)_n$ SPS (m,n give the number of monolayers with thickness $a_0 = 2.83$ Å) is fulfilled[8] for m smaller than 12. Type II behavior can be maintained in the ternary/binary $(Al_xGa_{1-x}As)_m(AlAs)_n$ superlattice for any thickness provided the composition x is chosen properly[7].

The lowest energy band gap of these type II superlattices is determined by transitions involving the spatially separated Γ

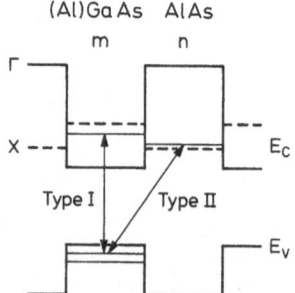

Fig. 1. Bandscheme of a type II (Al)GaAs/AlAs
        superlattice in growth direction z. The direct
        type I and indirect type II optical transitions
        between the quantized valence band and
        conduction band states are indicated by the
        arrows.

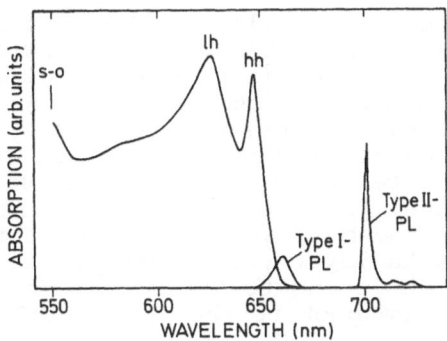

Fig. 2. Low temperature absorption spectrum of the (m,n)
= (10.3,17.1) GaAs/AlAs SPS together with the
corresponding PL spectrum (for details see
text).

valence band states of the GaAs and the confined X conduction
band states of the AlAs as schematically depicted in Fig. 1.
The respective oscillator strengths of these Γ-X or type II
transitions, however, are very small compared to the direct Γ-Γ
or type I transitions at higher energies because they are
indirect in real and in k-space.

The absorption spectrum of a type II superlattice is thus
determined by the direct type I transitions involving the Γ
conduction and valence band states in the (Al)GaAs. The low
temperature spectrum of a $(GaAs)_m(AlAs)_n$ SPS with (m,n) =
(10.3,17.1) is depicted in Fig. 2[14]. It reveals the heavy
and light hole as well as the onset of the split-off excitonic
transitions. The absorption coefficient at lower energies
corresponding to transitions between the spatially separated Γ
valence band states confined in the GaAs layer and the X
conduction band states confined in the AlAs is smaller by orders
of magnitude.

The photoluminescence (PL) spectrum (also depicted in
Fig. 2) shows a spectrally broad emission peak in the low-energy
tail of the direct absorption and a main emission feature at
lower energy. The high-energy type I PL is attributed to
recombination of direct heavy hole excitons. This type I
luminescence decays with a time constant less than 20 ps and
reveals directly the fast spatial transfer of electrons from the
Γ-states of the GaAs to the X states of the AlAs[10].

The main low-energy PL corresponds to recombination of
indirect type II excitons involving spatially separated
electrons in the X conduction band states of the AlAs and holes
in the Γ valence band states of the GaAs. In order to explain
the spectral structure of the type II luminescence we have to
consider the characteristic electronic properties of the X-
related states of the AlAs layers in more detail. This will
also provide some information about the scattering mechanism
involved in the Γ-X transfer of electrons.

There are two different subbands for the lowest AlAs
conduction band states due to the anisotropy of the X minimum in
AlAs. Since the structures are grown in [001] direction, the

heavy longitudinal mass m_l = 1.1m_0 is the quantization mass for the X_z minimum aligned along this growth direction, whereas the light transverse mass m_t = 0.2m_0 is the quantization mass for the X_x and X_y minima aligned along the [100] and [010] directions, respectively. Consequently, the X_z state is expected to be the lowest confined X sublevel located in the AlAs layer. However, recent ODMR studies by van Kesteren et al.[15] have shown that the $X_{x,y}$ states with the light quantization mass m_t become lowest in energy for type II GaAs/AlAs SPS with AlAs layer thicknesses exceeding 20 monolayers. This fact is due to the strain-induced lowering of the $X_{x,y}$ valleys with respect to the X_z valley.

The indirect type II recombination can be described as a two-step process via an intermediate Γ conduction band state. Electrons of the lowest lying X conduction band states ψ_c^X are scattered by a perturbation potential V into virtual conduction band states at the Γ point ψ_c^Γ and then recombine with holes of the Γ valence band states ψ_v^Γ. The corresponding matrix element can be written as[16]

$$M_{cv} = \frac{\langle\psi_c^X|V|\psi_c^\Gamma\rangle \cdot \langle\psi_c^\Gamma|p|\psi_v^\Gamma\rangle}{\Delta E} \tag{1}$$

where p is the dipole operator and ΔE is the energy difference between the Γ point and the virtually occupied Γ state. The scattering potential V can be the electron-phonon interaction as for indirect optical transitions in bulk semiconductors. In addition, the so-called interface potential induces strong mixing between Γ and X_z states[6] and the potential fluctuations due to interface roughness also lead to mixing between Γ and $X_{x,y}$ states. The scattering of electrons due to the interface potential (Γ-X_z mixing) and due to the interface roughness (Γ-$X_{x,y}$ mixing) thus gives rise to a zero-phonon line in the type II luminescence spectrum[17]. Since the X_z state is lowest for the (10.3,17.1) SPS, the main type II PL-peak at about 700 nm in Fig. 2 is the zero-phonon line due to Γ-X_z mixing, whereas the weaker type II PL-peaks at longer wavelength are ascribed to indirect Γ-X_z transitions with participation of Brillouin zone boundary phonons of the AlAs. The type II luminescence spectrum thus suggests that interface scattering may be the dominant mechanism for the spatial transfer of electrons from the Γ-related states of the GaAs layer, where the electrons are photocreated, to the X-related states of the AlAs layer. We have measured the spatial Γ-X transfer by time-resolved optical pump and probe experiments in order to investigate directly the underlying scattering mechanism.

3.   EXPERIMENTAL METHODS

We have studied 5 type II $(GaAs_m)(AlAs)_n$ SPS samples with $m \leq 12$ and 2 type II $(Al_xGa_{1-x}As)_m(AlAs)_n$ multiple quantum well (MQW) samples with $m \gg 12$, which have been grown by MBE.

In the case of the all binary GaAs/AlAs SPS samples we have performed nonresonant optical pump and probe experiments with 100 fs time resolution in order to determine the subpicosecond Γ-X time constants[11,13]. For each sample the excitation pulse at 620 nm nonresonantly excites electrons as well as heavy and light holes in the Γ conduction and valence band states of the GaAs, respectively. Bleaching of the absorption, which is probed by a white light continuum pulse, arises instantaneously with the pump pulse at the spectral positions of the heavy hole, light hole and split-off excitons and is due to phase space filling, Coulomb screening and band gap renormalization. The bleached absorption at each excitonic transition then partly recovers to a quasi-constant value, which is mainly due to a loss in phase space filling when electrons scatter from the Γ-related states of the GaAs to the X-related states of the AlAs. It turns out that the partial decay of the differential transmission $\Delta T/T_o$, which is the relative difference in transmission with and without the pump pulse, reflects the Γ-X transfer of electrons and is best observed at the spectral position of the split-off exciton[11,13].

In order to determine the time constants of the Γ-X transfer for the ternary/binary AlGaAs/AlAs MQW samples we have performed resonant pump and probe experiments using a cavity-dumped hybridly mode-locked dye laser with the ability to tune the laser emission wavelength. This experimental setup provides a time resolution of 1 ps. The laser wavelength was tuned to excite and probe resonantly direct type I heavy hole excitons for each sample. Again we observe an initial partial recovery of the bleached absorption due to the Γ-X transfer of electrons and thus are able to determine the respective time constants[13].

4.   RESULTS AND DISCUSSION

In a recent publication[13] we have demonstrated that the time constants for the spatial Γ-X transfer drastically increase with increasing thickness of the (Al)GaAs layer, but that the scattering times do not show a significant dependence on the thickness of the AlAs layer. The time constants increase from about 100 fs for a $(GaAs)_m(AlAs)_n$ SPS with m = 7.6 up to 22 ps for an $(Al_xGa_{1-x}As)_m(AlAs)_n$ MQW with m = 68. It turns out that the Γ-X transfer rate is determined by the spatial overlap of the Γ and X wavefunctions confined in the different layers. In the following, we present experimental results of intensity and

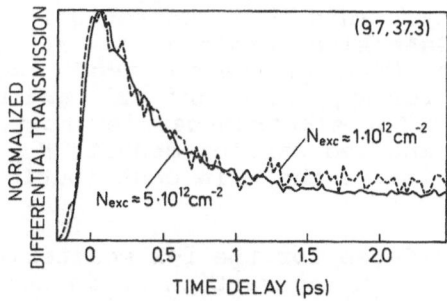

Fig. 3. Time-resolved differential transmission at the split-off exciton at T = 10 K for an (m,n) = (9.7,37.3) GaAs/AlAs SPS for different initial carrier densities N_{exc}.

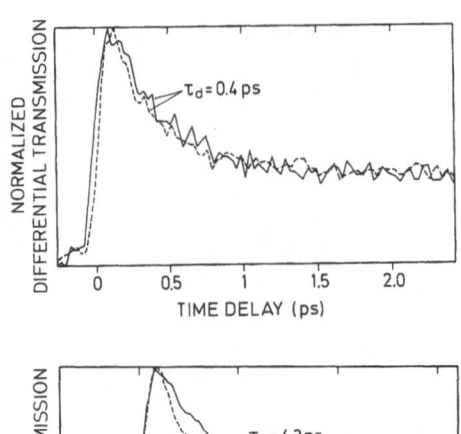

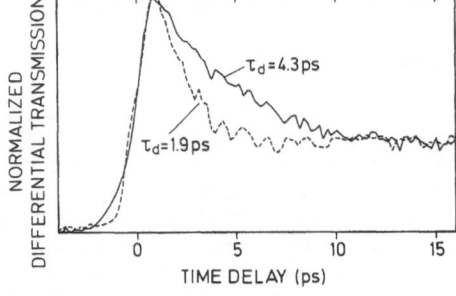

Fig. 4. Time-resolved differential transmission a) at
the split-off excitonic transition of the (m,n)
= (10.3,17.1) GaAs/AlAs SPS and b) at the heavy
hole exciton for an (m,n) = (39,33.5)
$Al_{0.36}Ga_{0.64}As$/AlAs MQW for T = 295 K (dashed
line) and for T = 5 K (solid line).

temperature dependent measurements of the Γ-X time constants,
which provide insight into the scattering mechanism causing the
spatial Γ-X transfer in type II (Al)GaAs/AlAs superlattices.

The initially induced carrier density N_{exc} was varied
between 1 x 10^{12} cm^{-2} and 1 x 10^{13} cm^{-2} in the case of the nonresonant pump and probe experiments for type II GaAs/AlAs SPS.
In the resonant pump and probe experiments of the type II
$Al_xGa_{1-x}As$/AlAs MQW the carrier density N_{exc} was varied between
1 x 10^{10} cm^{-2} and 5 x 10^{12} cm^{-2}. In both cases, however, the Γ-X
transfer times derived from the initial partial recovery of the
differential transmission do not change within the limits of the
experimental accuracy. This is illustrated in Fig. 3 for an
(m,n) = (9.7,37.3) GaAs/AlAs sample at T = 10 K, where the
temporal evolution of $\Delta T/T_o$ at the split-off excitonic
transition is shown for N_{exc}= 1 x 10^{12} cm^{-2} and N_{exc}= 5 x 10^{12}
cm^{-2}. Since the rate for electron-carrier scattering is
proportional to the induced carrier density N_{exc}, we can rule
out that this mechanism is the main contribution to the Γ-X
transfer.

As further candidates for the Γ-X scattering mechanism
there are electron-phonon interaction as in bulk III-V
semiconductors[18] and interface scattering. In order to
distinguish between these two interaction potentials we have
carried out temperature dependent experiments. The time-
resolved bleaching curves of nonresonant pump and probe
experiments at the split-off excitonic transition are

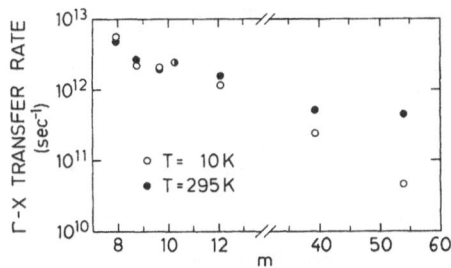

Fig. 5. Experimental Γ-X transfer times at 10 K (open
circles) and 295 K (full circles) versus the
number m of monolayers (Al)GaAs.

illustrated in Fig. 4a for an $(m,n) = (10.3,17.1)$ GaAs/AlAs
sample for T = 10 K (solid curve) and for T = 295 K (dashed
curve). Obviously, the initial partial recovery of the bleached
absorption is independent of temperature. For both temperatures
the Γ-X time-constants amount to 0.4 ps. However, for an (m,n)
$= (39,33.5)$ $Al_{0.36}Ga_{0.64}As$/AlAs sample the time constants for the
partial decay of $\Delta T/T_o$ at the heavy hole excitonic transition
are different for T = 10 K and T = 295 K. As illustrated in
Fig. 4b the Γ-X transfer times amount to 1.9 ps and 4.3 ps for T
= 295 K and T = 10 K, respectively. This increase of the Γ-X
scattering rate with increasing temperature is even more
pronounced for an $(m,n) = (68.3,36.3)$ $Al_{0.37}Ga_{0.63}As$/AlAs sample,
where the Γ-X time constants are equal to 2.3 ps and 22 ps for T
= 295 K and T = 10 K, respectively.

We have determined the Γ-X time constants at room
temperature and low temperature for 7 different type II
(Al)GaAs/AlAs samples. The results are illustrated in Fig. 5,
where the experimentally determined Γ-X time constants for
T = 10 K (open circles) and T = 295 K (full circles) are plotted
versus the number m of monolayers (Al)GaAs. As an experimental
result we find that the Γ-X scattering rate depends on
temperature only for type II samples with thick (Al)GaAs layers
(m >> 12), whereas for samples with thin (Al)GaAs layers
(m < 12) there is no temperature dependence within the accuracy
of the experimental data.

On the basis of the experimental findings we assume
interface scattering to be the temperature independent
scattering mechanism in case of the type II GaAs/AlAs SPS with
thin GaAs layers (m < 12). Interface scattering is due to the
potential fluctuations caused by the interface roughness ($Γ-X_{x,y}$
mixing) and/or due to the interface potential which effectively
mixes Γ states of the GaAs with X_z states of the AlAs as already
discussed in Section 2.

A different temperature dependent scattering mechanism
becomes dominant for type II AlGaAs/AlAs MQW with thick AlGaAs
layers (m > 12). The decrease of the Γ-X scattering rate with
decreasing crystal temperature can be explained if electron-
phonon interaction is assumed to be the dominant scattering
mechanism for these type II samples. At room temperature
electrons can be scattered both by absorption and by emission of
phonons, whereas at low temperature the phonon population is
small and as a consequence, the scattering by absorption of

197

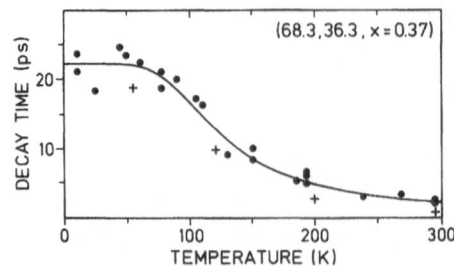

Fig. 6. Experimentally determined Γ-X transfer times
versus crystal temperature for the (68.3,36.3;
x = 0.37) MQW obtained by resonant (circles)
and nonresonant (crosses) pump and probe
experiments. The solid curve represents the
theoretically expected temperature dependence
according to Eqn. 2.

phonons is no longer important. Thus the Γ-X scattering rate is
reduced when the crystal temperature decreases.

In order to find the functional dependence of the Γ-X
scattering rate on the crystal temperature we have measured the
Γ-X time-constants for the type II samples with thick AlGaAs
layers over the whole temperature range from T = 10K up to 295K.
In Fig. 6 the experimental results for the type II
$Al_{0.37}Ga_{0.63}As$/AlAs sample with (m,n) = (68.3,36.3) are depicted.
The results of resonant pump and probe experiments (full points)
as well as the results of nonresonant pump and probe experiments
(crosses) both show the drastic increase of the Γ-X time-
constants with decreasing temperature. We assume that electron
LO-phonon deformation potential interaction is the dominant
interaction potential. The Γ-X scattering rate τ^{-1} is then
given by

$$\tau^{-1} \ \alpha \ \ R \ N_{ph} + (N_{ph} + 1) \tag{2}$$

with the phonon occupation probability

$$N_{ph} = [(\exp(h\nu_{ph}/kT)-1]^{-1}. \tag{3}$$

The first and second term of Eqn. 2 accounts for absorption and
emission of a phonon with energy $h\nu_{ph}$, respectively. The factor
R represents the relative strength for the absorption and
emission of an LO-phonon.

In Fig. 6 we have normalized τ of Eqn. 2 to the measured
Γ-X time-constant at 295 K. The parameter R determines the
difference of the scattering times at 10 K and 295 K, whereas
the LO-phonon energy $h\nu_{ph}$ determines the form of the curve $\tau(T)$.
If we choose R = 31 and $h\nu_{ph}$ = 39 meV we end up with the solid
curve shown in Fig. 6. This value of 39 meV for $h\nu_{ph}$ is exactly
the effective LO-phonon energy for $Al_{0.37}Ga_{0.63}As$[19]. Thus the
freezing-out of LO-phonons according to the Bose-Einstein
distribution N_{ph} describes the experimental data very well. We
therefore conclude that the Γ-X transfer of electrons in these
type II AlGaAs/AlAs MQW with relatively thick layers of the
AlGaAs is due to electron LO phonon interaction.

In conclusion, we have identified interface scattering to cause the spatial Γ-X transfer of electrons in type II (Al)GaAs/AlAs SPS with thin (Al)GaAs layers (< 35 Å). For type II (Al)GaAs/AlAs MQW with thick AlGaAs layers (> 100 Å) electron LO-phonon scattering is the dominant scattering process.

Acknowledgment

We would like to thank M. Preis and A. Schulz for excellent technical assistance. We gratefully acknowledge the work of A. Fischer and R. Muralidharan with the sample growth. The valuable and helpful discussions with C. Anthony, P. Thomas and W.W. Rühle are also gratefully acknowledged. The work in Marburg is supported by the "Deutsche Forschungsgemeinschaft". Parts of the work in Stuttgart are sponsored by the "Bundesminsterium f. Forschung und Technologie".

REFERENCES

1. P.N. Butcher and W. Fawcett, **Phys. Lett.**, 21:489 (1966)
2. E.M. Conwell and M.O. Vassell, **IEEE Trans. Electron Devices E-D**, 13:22 (1966)
3. J. Shah, **Sol. State Elect.**, 32:1051 (1989)
4. M. Heiblum. M.I. Nathan, D.C. Thomas and C.M. Knoedler, **Phys. Rev. Lett.**, 55:2200 (1985)
5. E.E. Mendez, W.I. Wang, E. Calleja and C.E.T. Goncalves, **App. Phys. Lett.**, 50:1263 (1987)
6. H.C. Liu, **Appl. Phys. Lett.**, 51:1019 (1987)
7. E. Finkman, M.D. Sturge, M.H. Meynadier, R.E. Nahory, M.C. Tamargo, D.M. Hwang and C.C. Chang, **J. Lum.**, 39:57 (1987)
8. B.A. Wilson, **IEEE J. Quantum Elect.**, QE-24:1763 (1988)
9. K.J. Moore, P. Dawson and C.T. Foxon, **Phys. Rev.B**, 38:3368 (1988)
10. G. Peter, E. Göbel, W.W. Rühle, J. Nagle and K. Ploog, **Superl. and Microstr.**, 5:197 (1989)
11. J. Feldmann, R. Sattmann, E. Göbel, J. Kuhl, J. Hebling, K. Ploog, R. Muralidharan, P. Dawson and C.T. Foxon, **Phys. Rev. Lett.**, 62:1892 (1989)
12. P. Saeta, J.F. Federici, R.J. Fischer, B.I. Greene, L. Pfeiffer, R.C. Spitzer and B.A. Wilson, **Appl. Phys. Lett.**, 54:1681 (1989)
13. J. Feldmann, R. Sattmann, E. Göbel, J. Nunnenkamp, J. Kuhl, J. Hebling, K. Ploog, R. Muralidharan, P. Dawson and C.T. Foxon, **Sol. Stat Electr.**, 32:1713 (1989)
14. The values for m and n have been determined by double crystal X-ray diffraction
15. H.W. van Kesteren, E.C. Cosman, P. Dawson, K.J. Moore and C.T. Foxon, **Phys. Rev. B**, 39:13426 (1989)
16. M.S. Skolnick, G.W. Smith, I.L. Spain, C.R. Whitehouse, D.C. Herbert, D.M. Whittaker and L.J. Reed, **Phys. Rev. B**, 39:11191 (1989)
17. P. Dawson, C.T. Foxon and H.W. van Kesteren, **Semic. Sci. Techn.**, 5:54 (1990)
18. J.L. Birman, M. Lax and R. Loudon, **Phys. Rev.**, 145:620 (1966)
19. S. Adachi, **J. Appl. Phys.**, 58:R1 (1985)

SPECTROSCOPIC STUDIES OF MINIBAND STRUCTURE AND BAND MIXING IN SUPERLATTICES

R.J. Nicholas, N.J. Pulsford, G. Duggan,†
C.T. Foxon,† K.J. Moore,† C. Roberts† and
K. Woodbridge†

、 Clarendon Laboratories
  Parks Rd., Oxford, OX13PU, UK
† Philips Research Laboratories
  Redhill, Surrey, RH15HA, UK

1.   INTRODUCTION

True electronic superlattice structures have been relatively little studied to date, in contrast to the considerable body of work on isolated quantum wells and multiple quantum well (MQW) structures[1]. A superlattice in which the inter-well coupling is significant leads to a new band structure in the superlattice growth direction, and the structure can then be viewed as an anisotropic material whose properties are adjusted by the layer thicknesses and compositions. The most widely studied such system is GaAs/(Ga)AlAs, although for the pure binary superlattice, inter-well coupling only gives significant band widths (of order 10 meV) for barriers of 20 Å and below. For such short period structures it has been argued (see e.g. the review by Sham and Lu[2]) that the widely used effective mass approach (as summarized by the elegant book by Bastard[3]) breaks down and cannot cope with the major perturbations to the electronic structure introduced by the interfaces. The most significant examples of this are the mixing of the Γ and X Bloch functions, which is thought to occur at the interfaces of GaAs/AlAs structures, and the zone folding thought to occur in Si/Ge superlattices. Once states from different bands become involved, it is also highly important to consider the symmetry of both the new bands, and of the new superlattice structures, which can depend crucially upon the exact layer sequences and thicknesses[2]. The methods employed to calculate the band structure in short period superlattices such as tight-binding, pseudopotentials, or the local density approximation (see the review by Sham and Lu[2]), involve a greater or lesser amount of empirical input, and often suffer from numerical complexity and a lack of the absolute accuracy necessary for a comparison with experiment.

An elegant and useful generalization of the effective mass approach has been described by Ando and Akera[4]. They examine the connection of an envelope function between different materials when written for a manifold covering several bands. This allows the conceptual and computational simplicity of the effective mass approximation to be retained, with the advantage that exact comparisons can easily be made with experiment. Pulsford et al.[5] have recently shown that this technique can be very simply applied to the case of Γ-X mixing in GaAs/AlAs structures. All that is needed is the definition of an interface matrix which determines the amount of Bloch function mixing.

In this paper we review the use of a matrix approach to the calculation of superlattice properties, including the effects of band mixing. As examples of the use of this we describe experiments on both GaAs/AlAs systems, where the well and barrier heights are large and the interfaces play a large role, and strained GaAs/In$_x$Ga$_{1-x}$As superlattices where the barriers are lower and superlattice periods larger so providing a good testbed for the simple effective mass approach.

2.   CALCULATIONS

A matrix approach to effective mass calculations is a straightforward way to calculate band structure. For a single band we use a basis of the envelope function and its derivative (ψ, am$_e$/m$^*\nabla\psi$) where a is the lattice constant and m* the effective mass, giving spatial transfer, $T(1)$, and interface, T_{AB}, matrices of

$$T(1) = \begin{bmatrix} \cos kl & m^*/kam_e \sin kl \\ -kam_e/m^* \sin kl & \cos kl \end{bmatrix} \text{ and } T_{AB} = \begin{bmatrix} 1 & 0 \\ 0 & 1 \end{bmatrix}. \tag{1}$$

The unit interface matrix results from the choice of basis functions, and corresponds to the conditions for conservation of wavefunction and probability current across the interface. The band structure of the constituent materials is contained in the Energy-momentum (E-k) relation (which can often be approximated as $E = E_0^{A,B} + \hbar^2 k_{A,B}^2/2m_{A,B}^*$, or can include the fuller results from k.p theory[3]) and allowing k to become imaginary below the band edge, thus transforming the sin terms into sinh etc. The superlattice band structure can now be generated by solving the Bloch condition

$$|T(1_A) \, T_{AB}T(1_B)T_{BA}| = \exp\{ik(1_A + 1_B)\} \tag{2}$$

to give the superlattice E-k dispersion.

We now consider ways to introduce modifications to the envelope function approach caused by the presence of the interfaces. A delta function of magnitude $V_0\delta(x)$ at the interface will give[6]

$$T_{BA} = \begin{bmatrix} 1 & 0 \\ 2m^* aV_0/\hbar^2 & 1 \end{bmatrix} \tag{3}$$

which will mix the wavefunction with its gradient.

The next step is to include mixing with further bands. For GaAs/AlAs we choose to include the X-minima, so the new basis used includes envelope functions from both Γ and X valleys $(\psi^\Gamma, \psi^X, am_e/m_\Gamma^* \nabla\psi^\Gamma, am_e/m_X^* \nabla\psi^X)$. By analogy with the single band case (Eqn. 3) an intervalley transfer potential of $\alpha\delta(x)$ at the interface[7] will then give

$$
T_{BA} = \begin{bmatrix} 1 & 0 & 0 & 0 \\ 0 & 1 & 0 & 0 \\ 0 & -\alpha & 1 & 0 \\ -\alpha & 0 & 0 & 1 \end{bmatrix}
\tag{4}
$$

which was used by Liu[7] to interpret Γ-X transfer in tunneling measurements[8].

A more fundamental approach is to examine the connectivity of the Bloch functions on going from one material to another, and look at how this influences the envelope function approximation. A number of authors have examined this problem (see [4] for more discussion), and in particular Ando and Akera[4,9] have studied the case of GaAs/AlAs for both the even (X_1) and odd (X_3) symmetry X-valleys. It should be noted however that the symmetry of the X minima can be reversed[2] when the number of layers of the corresponding material is odd. For even (g) symmetry the two Bloch functions mix, while for odd (u) symmetry the gradients are mixed, giving:

$$
\begin{array}{cc}
\text{even} & \text{odd} \\
T_{BA}^g = \begin{bmatrix} t_{11}^\Gamma & t_{11}^{\Gamma g} & 0 & 0 \\ t_{11}^{g\Gamma} & t_{11}^g & 0 & 0 \\ 0 & 0 & t_{22}^\Gamma & t_{22}^{\Gamma g} \\ 0 & 0 & t_{22}^{g\Gamma} & t_{22}^g \end{bmatrix} &
T_{BA}^u = \begin{bmatrix} t_{11}^\Gamma & 0 & 0 & t_{12}^{\Gamma u} \\ 0 & t_{11}^u & t_{12}^{u\Gamma} & 0 \\ 0 & t_{21}^{\Gamma u} & t_{22}^\Gamma & 0 \\ t_{21}^{u\Gamma} & 0 & 0 & t_{22}^u \end{bmatrix}
\end{array}
\tag{5}
$$

while wavefunction and flux matching impose constraints on the values of the elements t_{ij}[4]. In practice experiments to date can usually only measure a single property, so the matrices can be simplified for most purposes into either an odd acting form such as given by T_{BA}^α (Eqn. 4) or a direct Bloch function mixing for even states T_{BA}^γ, as used by Pulsford et al.[5]:

$$
T_{BA}^\gamma = \begin{bmatrix} \sqrt{(1-\gamma^2)} & -\gamma & 0 & 0 \\ +\gamma & \sqrt{(1-\gamma^2)} & 0 & 0 \\ 0 & 0 & \sqrt{(1-\gamma^2)} & -\gamma \\ 0 & 0 & +\gamma & \sqrt{(1-\gamma^2)} \end{bmatrix}
\tag{6}
$$

These give a single factor $\gamma(\%)$ or $\alpha(meV\text{Å})$ to be fitted to experiment. The translation matrix (Eqn. 1) is easily expanded to cover both bands, using the appropriate masses and dispersion relations, and the superlattice properties can be calculated from Eqn. 2.

It should be emphasised in all of the above discussion that the mixing of different bands is a result of the presence of **perfect** interfaces, which give rise to a coherent mixing of the character of the electrons, because the Bloch functions are not identical in the different materials. The superlattice structures have broken the translational symmetry of the underlying materials. It is not yet clear however that the structures studied are always sufficiently well controlled for the exact superlattice symmetry (e.g. an odd or an even number of monolayers) to be perfectly defined.

3.   GaAs/AlAs SUPERLATTICES

Two superlattices were studied with 60 periods of 25 Å GaAs wells and either 5 Å or 8 Å AlAs barriers. Interband magneto-reflectivity was used to measure the dispersion relation for the superlattices in both the in-plane and superlattice directions, by aligning the magnetic field parallel or perpendicular to the superlattice direction. Excitonic transitions are observed, however the dominant contribution to the energies comes from the much lighter electrons. The cyclotron radius is typically of order 100 Å, so the coherent quantized electron-hole motion can sample a large number of interfaces when the superlattice period is very short as in our case. This technique has been used by Belle et al.[10] to study conduction band anisotropy at lower energies in alloy barrier structures.

Some typical magnetoreflectivity spectra are shown in Fig. 1, for magnetic fields parallel ($B_\parallel$) to the layers. A series of excitonic Landau level transitions are observed to evolve out of the HH1 (heavy hole) - E1 (electron) transition at 1.805 eV. The exact lineshapes of the transitions are modeled by taking a Lorentzian response for the superlattice excitons[11], with arrows indicating the actual positions of the transitions. A similar series of transitions is observed for magnetic field perpendicular to the superlattice ($B_\perp$). Figure 2 shows plots of the measured transition energies against magnetic field for the two orientations for the 8 Å barrier sample. The Landau splitting is smaller in $B_\parallel$ as the excitonic cyclotron motion samples both the in-plane reduced mass ($m_\parallel$) and the heavier perpendicular miniband mass ($m_\perp$) whereas in $B_\perp$ only the lighter in-plane mass ($m_\parallel$) is seen. The excitonic Landau levels are fitted by adapting a 3D exciton model[12], with the effective masses and binding energy as variable parameters. The modeled reduced masses give a 50% mass anisotropy and the fitted excitonic binding energies (≈ 8.5 meV) are closer to the corresponding AlGaAs alloy than an isolated 25 Å quantum well, thus indicating an anisotropic 3D band structure, as expected in a strongly coupled superlattice.

In the $B_\parallel$ spectra, no Landau levels are observed in the range 1.867 to 1.878 eV: as a transition approaches this range, its optical intensity decreases and its cyclotron energy increase slows down, showing typical level crossing behavior.

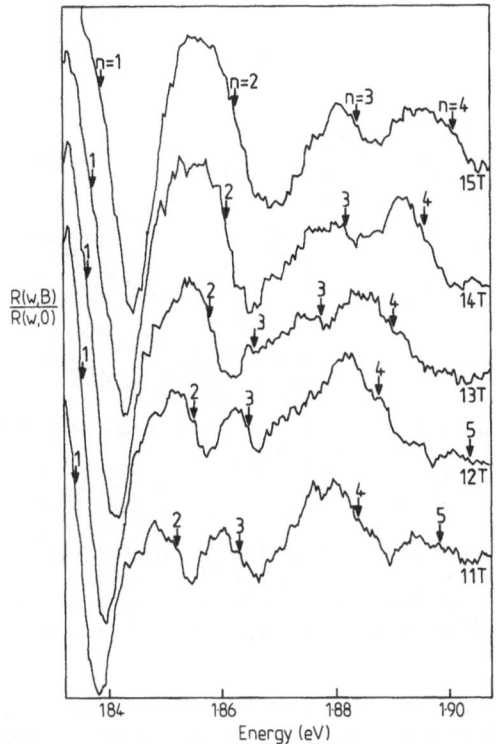

Fig. 1. Ratioed experimental reflectivity spectra for
magnetic fields parallel to the superlattice
layers. Transition energies, indicated by
arrows, are obtained by modeling. The n = 3
Landau level displays a typical level crossing
behavior around 1.87 eV.

This is only observed for $B_\parallel$ indicating that it is due to the
superlattice dispersion, as the cyclotron motion only samples
the in-plane dispersion for $B_\perp$. This is the energy range where
the confined states due to the X minima in the conduction bands
are expected to lie and it is expected that the Γ- and X-related
states will interact[2-5,7-9,13,14], producing anti-crossing
effects in the superlattice dispersion. No anti-crossing
effects were observed in the 5 Å barrier samples as the X-states
lie at a higher energy, beyond the range of the excitonic Landau
levels.

The superlattice band structure was modeled using the
transfer matrix approach to the envelope function approximation
described above, including the Γ-X mixing in the even Bloch
function approximation of Pulsford et al.[5] with T_{BA}^7. A
calculated band structure is shown in Fig. 3 for both the in-
plane and superlattice directions. The effect of the Γ-X mixing
is to produce an anti-crossing behavior where the two states
meet in the dispersion curve. In the superlattice direction the
X-state is flat reflecting the heavy longitudinal X_z mass
$(1.1m_e)$ tunneling through 25 Å of GaAs, whereas the Γ state has
a strong dispersion with a miniband width of 175 meV due to the
thin 8 Å AlAs barriers. In contrast, in the in-plane direction,

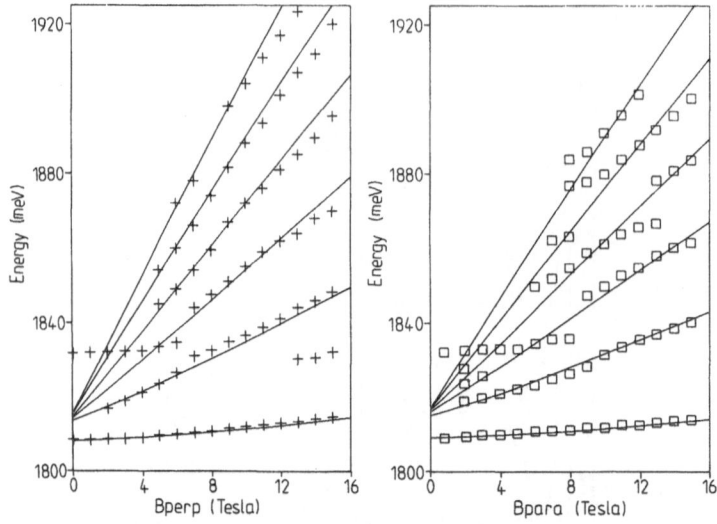

Fig. 2. Fan diagrams showing the transition energies for
magnetic fields $B_\perp$ (crosses) and $B_\parallel$ (squares).
The Landau levels are closer for $B_\parallel$ indicating
the heavier mass in the superlattice direction.

the X-state rises with the lighter transverse mass ($0.19m_e$) and
does not meet the Γ-state until a much higher energy. The hole
energies were also modeled using a superlattice calculation,
however they only contribute about 15% to the combined subband
energies and are a relatively unimportant part of the overall
picture.

The precise fitting of the data to the calculated bandstructure was done from a plot of energy versus $\sqrt{\{(n+\frac{1}{2})B\}}$.
Semiclassical quantization of the orbit in k-space gives

$$\frac{\hbar^2 k_i k_j}{2m} = \frac{\hbar eB(n+\frac{1}{2})}{m} \qquad \sqrt{k_i k_j} = \sqrt{\{2e(n+\frac{1}{2})B/\hbar\}} \ . \tag{7}$$

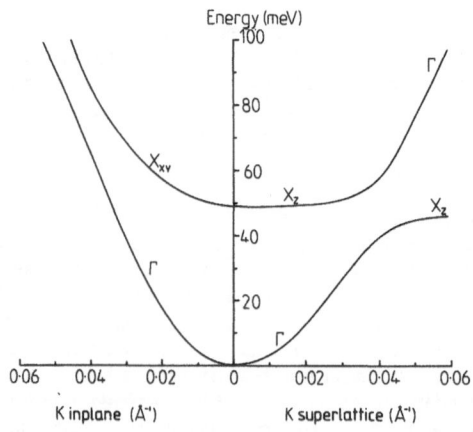

Fig. 3. Calculated miniband structure for 4% Γ-X mixing
for in-plane and superlattice directions. The
dominant character of each branch is shown.

This is shown in Fig. 4 for both the orientations together with a modeled fit. The magnitude of the anti-crossing in $B_\parallel$ is determined by the strength of the Γ-X mixing at the GaAs-AlAs interfaces. The fit shown in Fig. 4 uses a 4% mixing (γ = 0.04). In comparison the mean value of the $t^{\Gamma u}$ matrix elements calculated by Ando and Akera[4] gives 3.8%.

The alternative approach to the interface matrix is the elastic intervalley transfer potential, T^u_{BA}, as proposed by Liu[7]. Using this method, our experimental data is fitted with a transfer potential of α = 180 meVÅ, which compares with a value of 155 meVÅ deduced by Liu from tunneling simulations, but with a much higher theoretical estimate of 700 meVÅ from Ando and Akera[4].

Other measurements of Γ-X mixing come from the work of Meynardier et al.[15], who studied a zone center anti-crossing effect induced by an external electric field in a weakly coupled 35 Å GaAs-80 Å AlAs p-i-n superlattice and also type II (X-minimum) luminescence decay times. The luminescence decay times were studied in more detail by Dawson et al.[16], and both sets of authors interpreted the Γ-X mixing as a perturbation on the pure envelope function model determined by a matrix element $V_m = \langle \psi \ \sigma^\Gamma |V| \psi^X \sigma^X \rangle$, where the σ represent the Bloch functions. Inclusion of the envelope function overlap[16] led to a reasonably consistent set of values, however there still appears to be a significant structure dependence in the final values. This is eliminated when the interface matrix is used to simulate these results. Table 1 shows a summary of the values of γ and α (used as alternatives) deduced from both experiment and theory. It can be seen that the data is remarkably consistent, both between experiments and when compared with theory, and gives a value for γ = 4 ± 0.4%. The values for a α show a rather larger spread, but are still consistent with a single value of order 200 meVÅ. Overall this suggests that the best

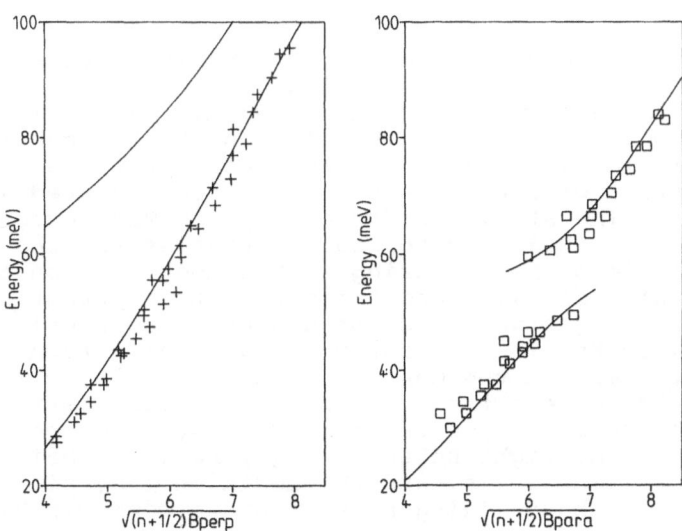

Fig. 4. Plots of energy against $\sqrt{\{(n + \frac{1}{2})B\}}$. The solid lines are calculated using the superlattice band structure of Fig. 3.

Table 1. Experimental and theoretical values of the Γ-X mixing parameters.

Measurement	Ref.	$d_{GaAs}(\text{Å})$	$d_{AlAs}(\text{Å})$	$\gamma(\%)$	$\alpha(\text{mevÅ})$
Pulsford et al	5	24	10	4	180
Liu et al	7,8	-	31,41,52	-	155
Meynadier	15	35	80	4.2	290
Ando, Akera	4	-	-	3.8	700
Dawson et al	16	22	41	3.7	245
		23	27	3.6	235
		24	18	3.3	195
Jian-Bia Xia	29	34	34	-	330

picture for Γ-X mixing is with a symmetric X_1 minimum as the lowest state, however the uncertainties in the exact band structure of the basic materials (in particular AlAs), and the exact number of monolayers and thus the layer symmetry, make this conclusion only tentative. We see that we have a numerically straightforward calculation technique which can be easily implemented on a small desktop computer, and which gives a comparable level of agreement to other semi-empirical techniques, such as the use of pseudopotentials[17].

4.   InGaAs-GaAs STRAINED LAYER SUPERLATTICES

The InGaAs-GaAs system is particularly interesting to study both because of the relatively shallow potentials produced by small indium content, and due to the opposite behavior of the light and heavy holes, which form wells in alternate layers and hence show both type I and type II superlattice configurations simultaneously. This has been elegantly demonstrated by Gerard and Marzin[18] for structures grown at the GaAs lattice constant.

Results from four strained layer superlattices are described, with structures consisting of 20 periods of 100 Å GaAs, and $In_xGa_{1-x}As$ wells of thicknesses ranging from 25 Å to 200 Å. The indium content x = 9%[19], and as the superlattice is epitaxially lattice matched to the GaAs substrate the tetragonal strain of approximately 1% is accommodated in the InGaAs layers. The strain shifts and splits the degenerate valence band edge [20], and for compressive strain the heavy hole band edge is highest. Using a strained electron to hole band offset ratio of 67:33 and the deformation potentials of Gershoni et al.[21] gives electron and heavy hole well depths of 71 and 36 meV in the InGaAs, but the light holes form a shallow well of about 4 meV in the GaAs and are hence a type II system with respect to the electrons. This factor considerably simplifies the optical spectra for this system, since the type II character of the light holes strongly reduces their absorption for wider wells, and the additional strain splitting moves any valence band mixing effects up to considerably higher energies.

Figure 5 shows the results of an effective mass calculation of the band structure for these superlattices, neglecting any

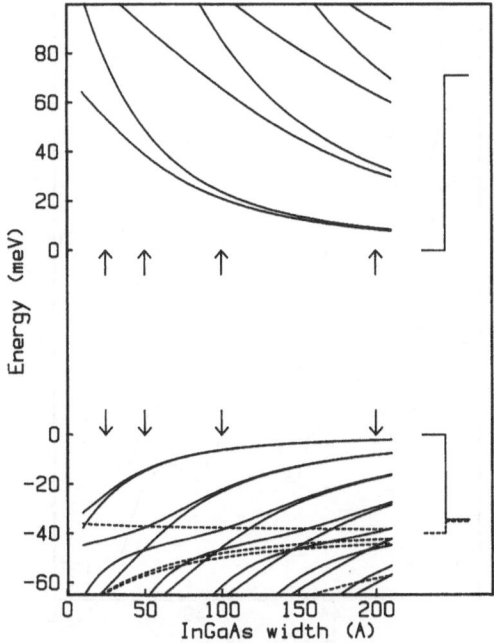

Fig. 5. The calculated superlattice band structure for
100 Å GaAs layers as a function of $In_xGa_{1-x}As$
thickness for $x = 9\%$. The solid lines show the
band edges of the electron and $m_J = \frac{3}{2}$ heavy
holes which form a type I system. The dashed
lines show the $m_J = \frac{1}{2}$ light holes which have
type II character as indicated on the right.
The arrows show the barrier thicknesses studied
in this work.

band mixing. The miniband edges are plotted as a function of
InGaAs width for 100 Å GaAs layers. The band masses have been
adjusted to include the effects of strain[21]. For the thinner
InGaAs layers the electron and heavy hole states are near to the
top of the barriers and are thus strongly coupled into wide
minibands, but for the 200 Å layers the lowest levels are
virtually uncoupled.

The magneto-optical properties were mainly studied in
transmission, as shown in Fig. 6 for the 200 Å sample, as the
majority of structure was below the GaAs substrate band edge,
and reflectivity was used for higher energies.

4.1 In-plane Studies with B⊥

A typical fan diagram is shown in Fig. 7 for B⊥ and the
25 Å well. The exciton binding energies were determined by
extrapolation of the low field magnetoexciton states[22], using
a 3D magnetoexciton model[12] shown by the dashed and solid
lines(2D excitons also give very similar results, but are
thought to be a less exact description due to the finite
superlattice band widths). The dashed lines correspond to a
'light' hole mass of $0.19m_e$, and give a good fit at low
energies, while the solid lines fit better at higher energy

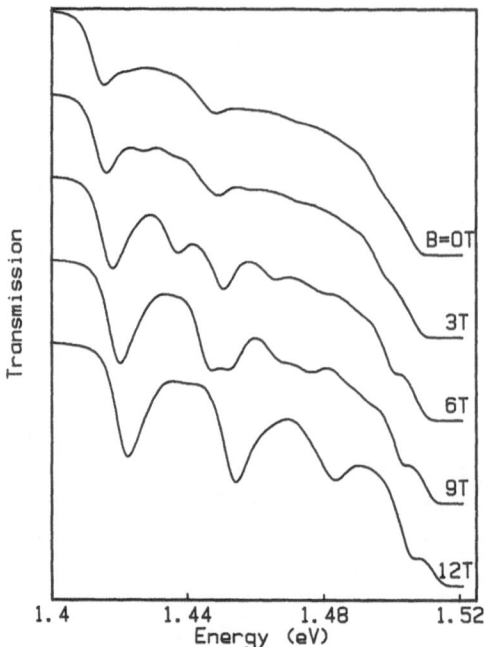

Fig. 6. Experimental traces of the magnetotransmission
of the 200 Å InGaAs sample for $B_\perp$. The cut off
above 1.52 eV is duie to the substrate.

using a mass of $0.38m_e$. This is due to the strain induced
decoupling of the valence bands which gives a 'light' band edge
mass for in-plane motion of the $m_J = \frac{3}{2}$ heavy hole [23,24], which
breaks down away from $k_\perp = 0$ due to valence band mixing.

Figure 8 shows the exciton binding energies deduced from
the low field extrapolations as a function of InGaAs width. For
25 Å the energy of 4.5 meV is little more than that of bulk GaAs
(4.2 meV) indicating that the structure is acting as a weakly
anisotropic 3D solid. By 100 Å the binding energy has doubled
as the wells decouple and excitons become more two-dimensional,
and finally as the wells become wider the binding energy again
falls towards the 3D value as the exciton radius begins to fit
inside the well.

4.2 Superlattice Dispersion with $B_\parallel$

The superlattice miniband dispersion has been studied in
more detail using the $B_\parallel$ orientation. Figure 9 shows the Landau
levels for the 25 Å sample, for which the envelope function
calculations predict miniband widths of 20.3 and 1.6 meV for the
electrons and heavy holes. It was shown by Belle et al.[10] in
GaAs/GaAlAs structures that Landau states were not observed
above the top of the miniband edges for $B_\parallel$, and for this sample
no transitons were seen beyond about 20 meV from the subband
edge, in contrast to the results shown in Fig. 7 for $B_\perp$ (but the
presence of the bulk GaAs exciton prevents levels being detected
from 1.51 to 1.52 eV). Below this the levels were again fitted
with a 3D exciton model with the cyclotron mass as a geometrical
average of the in-plane and superlattice masses for each carrier

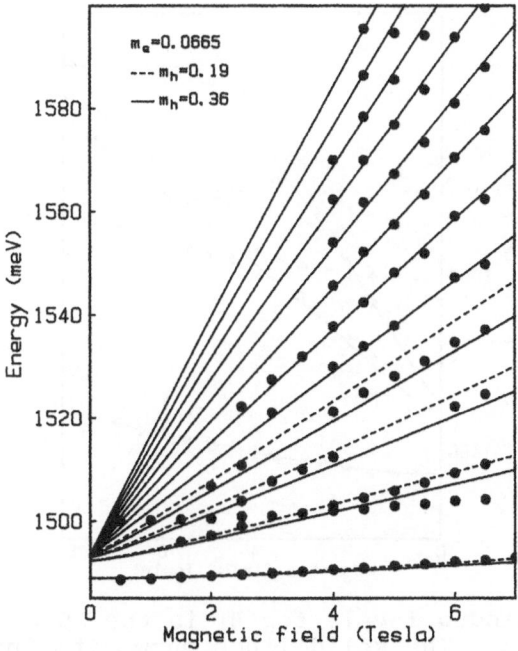

Fig. 7. The magnetoexciton positions for the 25 Å
barrier sample with B⊥. The solid lines show
fits using a 'heavy' heavy hole mass of 0.38 m_e
which is appropriate at high energies once the
valence bands are strongly mixed. This happens
quite quickly in this sample due to the large
'heavy' hole confinement energy. At low
energies a lighter 'decoupled' mass of 0.19 m_e
gives a better fit (dashed lines).

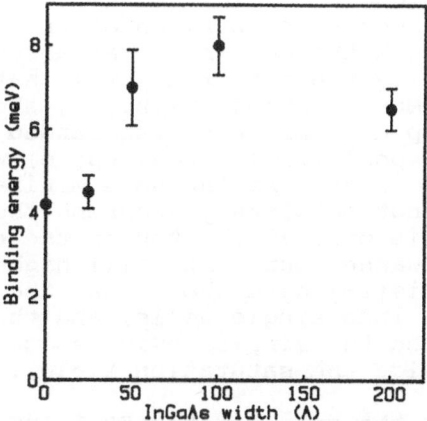

Fig. 8. The N = 1 heavy hole exciton binding
energies as a function of barrier
thickness.

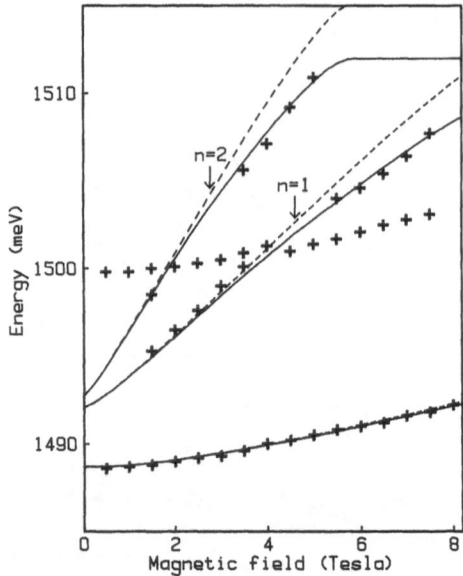

Fig. 9. The Landau levels for $B_\parallel$ in the 25 Å barrier
sample. The solid lines show fits including the
full miniband structure, while the dashes assume
a constant hole mass. The arrows show where the
hole miniband saturates.

type, as taken from the envelope function calculations. The
magnetic field quantization is thus semi-classical. The fit is
conventional at low energies, as shown by the solid line, and
confirms the calculated reduced mass anisotropy of $\mu_\perp/\mu_\parallel = 1.29$.
Above 1.50 eV, however, the gradient of the levels decreases,
due to the hole levels reaching the top of the miniband. This
occurs first for the holes due to their much greater anisotropy
$(\eta_h = m_\perp/m_\parallel \approx 3.4)$, at fields given by $(n + \frac{1}{2})B_\perp = 6.9\ T =$
$\hbar\pi^2/\ 2e\eta L^2$, where L is the superlattice period. Above this the
holes make no further contribution to the transition energies,
so the gradient at higher energies is due to the electrons
alone. This can be seen by comparison of the full fits (full
line) with one in which the hole mass was assumed to be constant
(dashed line). Finally the electron levels saturate at the top
of their miniband. Maan's calculations[25] showed that for
levels around the top of a miniband the Landau level energy
depends on the orbit position (in real space) which makes the
transitions broad and weak. As the hole miniband is narrow
(1.6 meV) this does not completely suppress the excitonic
transitions, and it is only at the top of the electron miniband
that the levels are washed out. At still higher magnetic fields
Berezhovskii and Suris[26] have shown that the extended miniband
states are localized into single wells, and the band structure
is thus changed beyond the simple semiclassical quantization
picture (at about twice the saturation field).

Figure 10 shows the data for the 50 Å InGaAs sample for $B_\parallel$.
The heavy holes in this sample are essentially two-dimensional,
as the miniband width is only ≈ 0.5 meV, and do not contribute
to the energy for $B_\parallel$. Two HH1-E1 levels are observed for $B_\parallel$,
and the fit confirms the reduced mass anisotropy $(\mu_\perp/\mu_\parallel)$ which

212

is calculated to be 1.56 at the miniband edge. The upper level again starts to saturate around the top of the miniband, which is approximately 10 meV wide. In this case, however, the upper level (at round 1.47 eV) moves through the top of the band at high fields. The position is intermediate between the miniband calculation (solid line) and that for a constant electron effective mass (dashed line). We attribute this to some form of breakdown of the superlattice miniband, in the region where the cyclotron radius is becoming smaller than the superlattice period, which may be caused by imperfections in the superlattice periodicity[27].

No higher excitonic Landau levels are observed for $B_\parallel$ in the two weakly coupled samples with 100 Å and 200 Å layers. However, we can observe the change-over from superlattice to quantum well behavior by comparing the diamagnetic shifts for the four samples in this orientation, as shown in Fig. 11. For the strongly coupled 25 and 50 Å samples the shift is less than the miniband width, and both show a sub-parabolic, almost linear field dependence which is characteristic of a high field 3D exciton. As the exciton becomes strongly confined to a single well, as in the 100 Å sample, the shift becomes much smaller and is parabolic. This can be modeled by perturbation theory[28], which then predicts an increase again as the well becomes wider due to the spreading of the exciton wavefunction as occurs for the 200 Å sample.

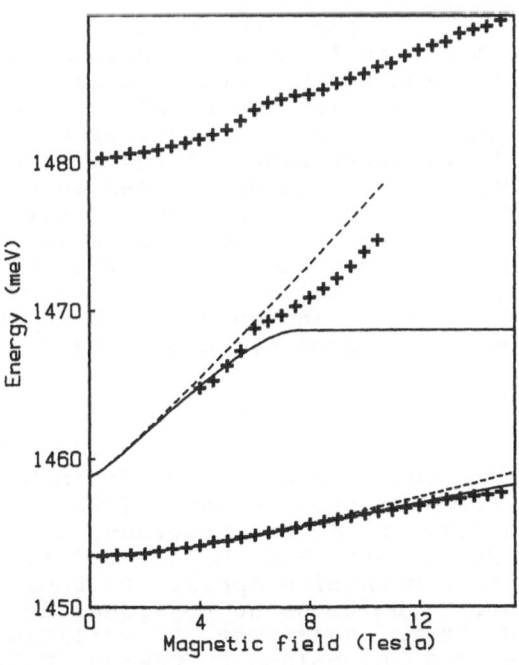

Fig. 10. The Landau levels for $B_\parallel$ in the 50 Å barrier sample. The solid line is calculated with the full miniband dispersion, while the dashed line assumes a constant electron mass.

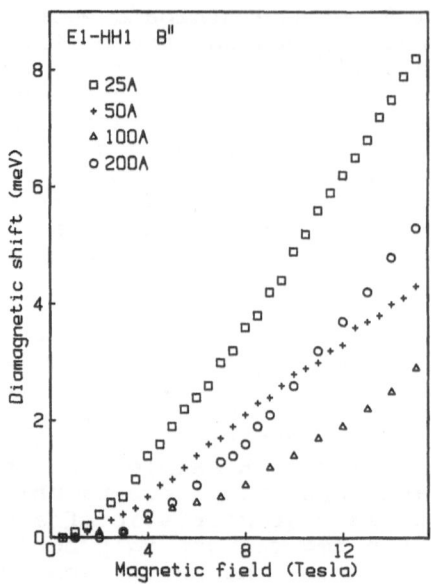

Fig. 11. The diamagnetic shifts for B∥ for four
different samples. The two thinnest barrier
samples show an almost linear shift which is
less than the miniband width.

SUMMARY

We have shown that it is now possible to use magneto-
optical techniques to test the theoretical descriptions of the
band structure which occurs in true electronic superlattice
structures. Introduction of the interface matrix into the
envelope function approximation allows us to extend the simple
calculation techniques based on the transfer matrix to very
short period structures where zone folding and band mixing
effects occur. Work on the strained system GaAs/InGaAs is
proving particularly interesting, as the relatively shallow
potential profiles allow one to observe superlattice phenomena
in relatively large scale structures and at lower magnetic
fields. The interpretation is also simplified by the relatively
large splitting of the light and heavy holes which removes some
of the complications introduced by valence band mixing.

REFERENCES

1.   C. Weisbuch **in**: "Semiconductors and Semimetals", 24:1, R.
      Dingle, ed., Academic Press, London (1987)
2.   L.J. Sham and Y.T. Lu, **J. Luminescence**, 44:207 (1989); Y.T.
      Liu and L.J. Sham, **Phys. Rev. B**, 40:5567 (1989)
3.   G. Bastard, "Wave Mechanics Applied to Semiconductor
      Heterostructures", Editions de Physique, Les Ulis. (1988)
4.   T. Ando and H. Akera, **Phys. Rev. B**, 40:11619, 11604 (1989)
5.   J.J. Pulsford, R.J. Nicholas, P. Dawson, K.J. Moore, G.
      Duggan and C.T. Foxon, **Phys. Rev. Lett.**, 63:2284 (1989)
6.   Q.G. Zhu and H. Kroemer, **Phys. Rev. B**, 27:3519 (1983)
7.   H.C. Liu, **Appl. Phys. Lett.**, 51:1091 (1987)
8.   D. Landheer, H.C. Liu, M. Buchanan and R. Stoner, **Appl.
      Phys. Lett.**, 54:1784 (1989)

9.   H. Akera, S. Wakakhara and T. Ando. **Surf. Sci.**, 196:694 (1988)

10. G. Belle, J.C. Maan and G. Weimann, **Solid State Commun.**, 56:65 (1985)

11. N.J. Pulsford, R.J. Nicholas, P. Dawson, K.J. Moore and C.T. Foxon, **Superlattices and Microstructures**, 6:51 (1989)

12. P.C. Makado, **Physica B & C**, 132:7 (1985)

13. E.E. Mendez, E. Calleja, C.E.T. Gonçalves da Silva, L.L. Chang and W.I. Wang, **Phys. Rev. B**, 33:7368 (1986)

14. A.R. Bonnefoi, D.H. Chow, T.C. McGill, R.D. Burnham and F.A. Ponce, **J. Vac. Sci. Technol. B**, 4:988 (1986)

15. M.-H. Meynadier, R.E. Nahory, J.M. Worlcok, M.C. Tamargo, J.L de Miguel and M.D. Sturge, **Phys. Rev. Lett.**, 60:1338 (1988)

16. P. Dawson, K.J. Moore, C.T. Foxon, G. W. t'Hooft and R.P.M. van Hal, **J. Appl. Phys.**, 65:3606 (1989)

17. M.A. Gell, D. Ninno, M. Jaros, D.J. Wolford, T.F. Keuch and J.A. Bradley, **Phys. Rev. B**, 35:1196 (1987)

18. J.M. Gerard and J.Y. Marzin, **Phys. Rev. B**, 40:6450 (1989)

19. K.J. Moore, G. Duggan, K. Woodbridge and C. Roberts, **Phys. Rev. B**, 41:1095 (1990)

20. C.G. Van der Walle, **Phys. Rev. B**, 39:1871 (1989)

21. D. Gershoni, H. Temkin, M.B. Panish and R.A. Hamm, **Phys. Rev. B**, 39:5531 (1989)

22. D.C. Rogers, J. Singleton, R.J. Nicholas, C.T. Foxon and K. Woodbridge, **Phys. Rev. B**, 34:4002 (1986)

23. D. Lancefield, W. Batty, C.G. Crookes, E.P. O'Reilly, A.R. Adams, K.P. Homewood, G. Sundaram, R.J. Nicholas, M. Emeny and C.R. Whitehouse, **Surf. Sci.**, 229:122 (1990)

24. E.D. Jones, H. Ackerman, J.E. Schirber, T.J. Drummond, L.R. Dawson and I.J. Fritz, **Solid State Commun.**, 55:525 (1985)

25. J.C. Maan, **Festkörperprobleme**, 27:137 (1987)

26. A.M. Berezhkovskii and R.A. Suris, **Sov. Phys. JETP**, 59:109 (1984)

27. N.J. Pulsford, R.J. Nicholas, R.J. Warburton, G. Duggan, K.J. Moore, K. Woodbridge and C. Roberts, **Phys. Rev. B**, in press (1990)

28. N.J. Pulsford, J. Singleton, R.J. Nicholas and C.T. Foxon, **J. Phys. (Paris) Colloq.**, 48:C5-231 (1987)

29. J.B. Xia, **Surf. Sci.**, 228:476 (1990)

SADDLE-POINT EXCITONS AND FANO RESONANCES IN SEMICONDUCTOR

SUPERLATTICES

Yia-Chung Chang and Hanyou Chu*

Department of Physics and Materials Research
Laboratories, University of Illinois at Urbana-
Champaign, Urbana, Illinois 61801, USA

ABSTRACT

Theoretical investigations of optical properties in
semiconductor quantum wells and superlattices are presented. In
particular, we introduce a line-shape theory for treating
excitonic absorption in superlattices. This theoretical method
allows one to accurately and efficiently calculate the Green's
function associated with the excitonic Hamiltonian of a
realistic superlattice, including the mixing of heavy and light
hole states (the valence band mixing) and inter-well tunneling
of excitons. Realistic absorption spectra for a number of GaAs-
$Ga_{1-x}Al_xAs$ superlattices were obtained with this method.
Interesting features such as the saddle-point excitons and Fano
resonances due to the interaction of discrete exciton states
with continuum states are discussed. The theoretical
predictions are in excellent agreement with available data.

1.   INTRODUCTION

Photo-absorption and photoluminescence excitation spectra
are widely used for probing the electronic states of
semiconductor superlattices. Comparisons between these
experimental results with the absorption coefficient predicted
by realistic model calculations can provide better understanding
of the electronic structures of these materials. In the
superlattices of interest here, each superlattice period
consists of a well material (e.g. GaAs) and a barrier material
(e.g. $Ga_{1-x}Al_xAs$). A superlattice with wide barrier layers is
considered as a collection of independent quantum wells. The
general features in the absorption spectra for a quantum well
can be interpreted by using a simple effective-mass (particle-
in-a-box) model[1], which assumes parabolic band structures for
both the electrons and holes. In this model, the oscillator
strengths for the band-to-band transitions are proportional to
the squared overlap between the envelope functions of the

---

*Present address: Division of Physics, National Research
Council, Ottawa, Canada K1A0R6.

Condensed Systems of Low Dimensionality
Edited by J. L. Beeby *et al.*, Plenum Press, New York, 1991

electron and hole subbands respectively. The electron and hole envelope functions are approximately described by cosine or sine functions with the number of nodes equal to the principal quantum number n minus 1.   Thus the interband transitions are nearly forbidden unless the conduction and valence subband states have the same principal quantum number. This is called the Δn = 0 selection rule[2]. With this selection rule, the absorption spectrum due to the band-to-band transitions is given by the sum of a series of step functions, one for each pair of conduction and valence (heavy or light hole) subbands with a given n.   Each step function results from the constant joint density of states for two-dimensional subbands.

Associated with each pair of conduction and valence subbands there exists an exciton state which appears in the absorption spectrum as a prominent peak at an energy lower than the onset of the band-to-band transition by a binding energy. Ignoring the coupling among excitonic states derived from different pairs of conduction and valence subbands, one expects the Δn = 0 selection rule to hold for the excitonic transitions as well.   Indeed, most experimental data indicate than Δn = 0 excitonic transitions are at least an order of magnitude stronger than the other excitonic transitions which violate this selection rule.

Recent studies[3,4] of the electronic and optical properties of semiconductor quantum wells have revealed that the mixing of heavy and light hole components (valence band mixing) in the quantum well (or superlattice) states can lead to Δn ≠ 0 (forbidden) inter-band transitions with strengths much larger than those expected from the simple particle-in-the-box model. Two most pronounced Δn ≠ 0 transitions which are observed experimentally are the HH3-CB1[5] and LH1-CB2 exciton states[6]. Here HHn, LHn and CBn denote the n-th subbands associated with the heavy hole, light hole and conduction bands, respectively. The labels are based on the symmetry properties of the subbands at $\vec{k}_\parallel$ = 0.   For $\vec{k}_\parallel$ ≠ 0 the superlattice valence band states in general contain admixture of heavy hole and light hole characters.   The HH3-CB1 exciton ground state rides on the HH1-CB1 continuum, and the line shape of the absorption peak near the HH3-CB1 exciton is asymmetric due to the interference effect known as the Fano-resonance[7].

In the calculations of electronic structures of excitons one generally ignores the coupling of exciton states derived from one pair of subbands with those derived from other pairs of subbands.   This approach usually works well for the Δn = 0 excitons except the HH2-CB2 exciton.   The HH2-CB2 exciton is a special case, because the first light hole subband (LH1) and the second heavy hole subband (HH2) have nearly the same energy and they are strongly coupled by the off-diagonal term in the Kohn-Luttinger Hamiltonian[8].   To obtain accurate line shapes for all excitons which ride on the continuum of transitions associated with other pairs of subbands, one must take into account their interactions, i.e. the Fano-resonance effect. This can be accomplished by directly calculating the exciton Green's function with a numerical method.   Such a method is the central issue of this paper.

In superlattices with a thin barrier layer, the subband dispersion in the growth direction cannot be ignored, and the

218

optical properties are somewhat different. Along the growth
direction (taken as the z axis), each subband has a minimum and
a maximum at zone edges. At the zone edge where a conduction
subband is a maximum, the three-dimensional band structure must
have a saddle shape, since the energy band along the
perpendicular (x and y) directions always curve upward (within
the effective-mass approximation). The density of states at the
saddle point is characterized by an M_1 singularity[9].
Similarly, the joint density of states of a superlattice will
exhibit a series of M_1 singularities. Coulomb interaction
between electrons and holes associated with these saddle-point
states gives rise to excitonic states below the energy of the
saddle point.

In the past, excitonic effects associated with the M_1
singularity in the joint density of states of solids have
attracted a great deal of interest both theoretically and
experimentally[10-15]. Kane[13] studied the saddle-point
excitons with Coulomb interactions by an adiabatic
approximation. Such an approximation is valid only if the mass
anisotropy is sufficiently large (i.e. $m_3/m_1 \gg 1$). Velicky and
Sak[11] and Kamimura et al.[12] have studied the absorption
spectra associated with the M_1 singularity by approximating the
electron-hole Coulomb interaction with a contact potential.
Although qualitative understanding of this phenomenon can be
obtained via a contact-potential model[11], quantitative
calculations for the Coulombic potential are desired. Chu and
Chang[16] have recently performed quantitative calculations for
the line shapes of photo-absorption associated with saddle-point
excitons in semiconductors.

Semiconductor superlattices are ideal materials for
studying the structure of saddle-point excitons, because the
band parameters (or mass anisotropy) near the saddle point can
be tailored by varying the layer thicknesses and band gaps of
the constituent materials. Detailed variation of the line
shapes of the absorption peaks derived from subband states near
the saddle-point can be calculated by a numerical method ($\vec{k}$-
space sampling method) discussed in the following sections[17].
Many different line shapes associated with saddle-point states
for a number of superlattices are predicted by the numerical
calculation and they all agree well with available data.

2.   ELECTRONIC STRUCTURES OF EXCITONS

In a simple model system, where the electron and hole are
described by spherical effective masses, the exciton in its
center-of-mass frame is analogous to a hydrogen atom.
Bastard[18] has studied the variational solution to a hydrogen
atom placed at the center of an infinitely-deep quantum well
with variable width. It is found that the binding energy of the
system increases monotonically from one rydberg in the wide-well
(3D) limit to four rydbergs in the narrow-well (2D) limit. This
indicates the importance of excitons in quantum wells with well
width comparable to the Bohr radius of the exciton (around 100 Å
for GaAs).

Binding energies of excitons in quantum wells with finite
barrier height were first studied by Greene and Bajaj[19] in an
effective-mass model which ignores the coupling between the

heavy hole and light hole bands. They found that the binding
energies of both the heavy hole and light hole excitons as
functions of the well width have a maximum somewhere between
30 Å and 70 Å. The maximum exciton binding energy in a quantum
well is about twice as large as the bulk value. The light hole
exciton tends to have larger binding energy than the heavy hole
exciton. This is because the in-plane effective mass for the
light hole band (with lighter effective mass in the growth
direction) is actually heavier than that for the heavy hole
band.

Sanders and Chang[20] and Broido and Sham[21] have
considered the effect of band hybridization on the excitonic
states. An approximation was adopted, in which the exciton is
assumed to be derived from a definite pair of electron and hole
subbands. Coupling of excitons derived from different pairs of
subbands are then included by a perturbation method. The band
hybridization comes from the mixing of the heavy and light hole
states due to the off-diagonal terms of the Luttinger-Kohn
Hamiltonian[8] defined as

$$
H^{(off)} = \frac{\hbar^2}{2m}
\begin{bmatrix}
A_+ & L & M & 0 \\
L* & A_- & 0 & M \\
M* & 0 & A_- & -L \\
0 & M* & -L* & A_+
\end{bmatrix},
$$
(1)

where

$$
A_\pm \equiv (\gamma_1 \pm \gamma_2)(k_x^2 + k_y^2) + (\gamma_1 \mp 2\gamma_2)k_z^2,
$$

$$
L = -2i\sqrt{3}\bar{\gamma}_3 K^- k_z,
$$

$$
M = -\sqrt{3}\bar{\gamma}K^{-2} + (\gamma_2 - \gamma_3)k_x k_y,
$$

$$
K^\pm \equiv (k_x \pm ik_y),
$$

and

$$
\bar{\gamma} = (\gamma_2 + \gamma_3)/2 .
$$

The mismatch between the valence band parameters for the well
and barrier material is ignored. Because the wave functions of
interest are mostly confined in the well materials, the error
introduced by this approximation is small. The hole effective-
mass equation including the valence band mixing can be solved by
a variational method in which the real-space envelope functions
are written in terms of linear combinations of Gaussian-type
orbitals. Detailed results for the valence subband structures
of GaAs-Ga$_{1-x}$Al$_x$As quantum wells can be found in Refs. 20-22.

The excitonic states associated with the n-th conduction
subband and the m-th valence subband in a quantum well can be
written as

$$\Psi_X^{nm} = \sum_{n,m} \sum_{\vec{k}_{\parallel}} G_{nm}(\vec{k}_{\parallel}) \Psi_{n,\vec{k}_{\parallel}}^{e} \Psi_{m,-\vec{k}_{\parallel}}^{h}, \qquad (2)$$

where $\Psi_{n,\vec{k}_{\parallel}}^{e}$ denotes the n-th conduction subband state and $\Psi_{m,-\vec{k}_{\parallel}}^{h}$ denotes the m-th hole subband state. Substituting the expansion into the Schrödinger equation yields an effective-mass equation for the exciton envelope function, which can be solved by the variational method.

In the calculations of Ref. 20 and 21, the angular dependence in the subband energies and the expansion coefficients is ignored. The conduction subbands have circular symmetry within the effective-mass approximation. The valence subbands typically have about 20% angular deviation for GaAs-$Al_xGa_{1-x}As$ quantum wells[20]. Altarelli has introduced an axial approximation[23] [with $\gamma \equiv (\gamma_2 + \gamma_3)/2$] which gives rise to circular valence subbands. The circular approximation simplifies the calculation substantially and provides some insight into the problem.

Several recent calculations[24-27] have taken into account the band anisotropy effect. The differences among the exciton binding energies obtained by the calculations with and without axial approxixmation are typically less than 10%. However, the angular variation in the exciton expansion coefficient $G_{nm}(\vec{k}_{\parallel})$ has important implications on the oscillator strengths[25]. Zhu[25] has shown that the four components of the envelope function in a hole-subband state (say m) have different angular dependences given explicitly by

$$g_{m,\vec{k}_{\parallel}}^{\nu}(k_z) = g_{m,\vec{k}_{\parallel}}^{-\nu}(k_z)\exp\{i(\tfrac{1}{2} - \nu)\phi\},$$

where $\nu = \tfrac{3}{2}, \tfrac{1}{2}, -\tfrac{1}{2}, -\tfrac{3}{2}$, and ϕ is the angle between $\vec{k}_{\parallel}$ and the x-axis (the azimuthal angle for $\vec{k}$). Thus in an exciton state, different components of the valence band envelope function must be associated with different angular momenta. For example, an s-like $\nu = \tfrac{3}{2}$ (heavy hole-like) envelope function must be coupled to a p-like $\nu = \tfrac{1}{2}$ (light hole-like) envelope function. As a consequence, the parity-forbidden LH1-CB2 excitonic transition must have an orbital angular momentum l = 1, as it is coupled to the s-like HH2-CB2 excitonic transition.

For superlattices with thin barriers, the electronic subbands have substantial dispersion in the growth direction which will affect the exciton binding energies. A variational calculation on the exciton binding energies for thin-barrier superlattices was carried out by Chomettee et al.[28]. It is found that the exciton binding energy reduces as the barrier width decreases, since the quasi-two-dimensional system approaches the three-dimensional system in the thin-barrier limit (with large subband dispersion).

3. ABSORPTION SPECTRA

Two possible approaches can be used to calculate the absorption spectra including the excitonic effect. One is the direct diagonalization approach. The other is the Green's function approach. In the direct diagonalization approach, one finds the eigenvalues and eigenvectors of interest by diagonalizing the exciton Hamiltonian defined in a finite set of basis states. The exciton is a two-particle system which has six degrees of freedom. We can immediately eliminate three degrees of freedom by considering only the excitonic states with zero total momentum. Namely, we only worry about the electron-hole relative motion, not the center-of-mass motion of the exciton. This is because the incident photon which creates the electron-hole pair in semiconductors has negligible momentum. In principle, one can always construct basis states for the zero-momentum exciton states in a $\vec{k}$-space which describes the electron-hole relative motion. We call this a $\vec{k}$-space sampling method[16]. A finite mesh in the three-dimensional $\vec{k}$-space is chosen so that the corresponding energies cover the range of interest. The spacing between mesh points is dictated by the line widths of the distinct structures in the absorption spectrum we wish to predict. If many fine structures are expected in the energy range of interest, then a very small mesh spacing is needed. This usually translates to a large number of basis states (more than 1000). In order to preserve the density of states, each basis state at a mesh point $\vec{k}_i$ is defined to be the sum of all states $\vec{k}$ within a "cell" of the mesh lattice centered at $\vec{k}_i$. This also avoids possible singularities in the matrix elements of the Hamiltonian. Exploiting the symmetry of the system can minimize the number of basis states needed in the calculation. For systems with spherical symmetry, we only need to consider states with zero angular momentum (i.e. s-like). This is because the incident photon has angular momentum $l = 1$ and the Bloch state contribution to the electron-hole pair also has angular momentum $l = 1$.

For superlattices, excitons associated with different pairs of subbands are coupled via the Coulomb interaction. For many superlattices of interest (with well width less than 100 Å and barrier width greater than 30 Å) the energy separations between two consecutive conduction subbands are much larger than the exciton binding energy, we can ignore the coupling of excitons associated with different conduction subbands. However, we have to include the coupling of several valence subbands, because their energy spacings are much smaller. The off-diagonal term of the Kohn-Luttinger Hamiltonian further couples excitons associated with different valence subbands and with different angular momenta. Thus, even with the full use of symmetry, we still need a large number (a few thousand) of basis states for calculating the absorption spectra. This makes the direct diagonalization approach unattractive. We thus resort to the Green's function approach.

In the Green's function approach, we write the absorption coefficient in terms of the imaginary part of the exciton Green's function, viz.

$$\alpha(\hbar\omega) = \frac{C}{\omega} \sum_i |\langle \Psi_i | \hat{\epsilon} \cdot \vec{p} | 0 \rangle|^2 \delta(E_i^{ex} - \hbar\omega)$$

$$= \frac{C}{\omega} \text{Im} \sum_i \langle 0 | \hat{\epsilon} \cdot \vec{p} G(\hbar\omega) \hat{\epsilon} \cdot \vec{p} | 0 \rangle.$$

where $|\Psi_i>$ denotes the i-th excitonic states (either discrete or in the continuum), $|0>$ designates the ground state of the solid, $\hat{\epsilon}$ is the polarization vector for the incident phonon, C is a constant and G is the Green's function. To introduce broadening in the calculated spectra, we replace E with $E + i\Gamma$, where Γ is the half width for broadening. A recursion method[17] can be used to calculate $\alpha(\hbar\omega)$ directly.

The recursion method, or alternatively the Lanczos method[29], is not suitable for obtaining exact energy eigenvalues except for the lowest-lying and highest-lying states. But it gives a good description of a projected density of the states and this is just what we are after. In this method, we start from an initial state, $|0'> \equiv \hat{\epsilon}\cdot\vec{p}|0>$ on which the Green's function is to be projected. The state is then multiplied by the Hamiltonian operator of the system and the resulting state is orthogonalized with the previous ones. This is repeated for N steps, where N is the dimension of the system. The Hamiltonian is thus tridiagonalized and the Green's function can be calculated directly by a continued fraction scheme. The algorithm is stated in the following. Let $|n>$ be the basis state generated at the n-th step and $|u_n> \equiv H|n>$. We define

$$|1> = \frac{1}{b_0} [|u_0> - a_0 |0'>]$$

and

$$|n + 1> = \frac{1}{b_n} [|u_n> - a_n |n> - b_{n-1} |n - 1>] \text{ for } n > 1$$

with

$$a_n = <n|H|n> = <n|u_n> \text{ and } b_n = (<u_n|u_n> - a_n^2 - b_{n-1}^2)^{1/2} ,$$

where $|0'>$ is the initial state. The iteration is stopped when $b_n = 0$. The Hamiltonian matrix in the new set of basis states $\{|n>\}$ becomes

$$H = \begin{pmatrix} a_0 & b_0 & & & & \\ b_0 & a_1 & b_1 & & & \\ & b_1 & a_2 & b_2 & & \\ & & \cdots & \cdots & \cdots & \\ & & & b_{n-1} & a_n & b_n \\ & & & & \cdots & \cdots & \cdots \end{pmatrix} .$$

The Green's function projected in state $|0'>$ is then given by

$$\langle 0' | G(E) | 0' \rangle = \cfrac{1}{E-a_0 - \cfrac{b_0^2}{\cfrac{\vdots}{E-a_N-b_N^2}}} \quad .$$

Mathematically, this algorithm is exact, but numerically it is unstable due to the round off error so that only the smallest and the largest eigenvalues are reliable. Many spurious eigenvalues can be generated. However, this method gives the projected density of states with good accuracy. If one is to analyze this algorithm, it can be seen that the Lanczos method is very efficient both in speed and in storage space. With the use of the recursion method, our problem is substantially simplified.

Similar to the direct-diagonalization method, we select a nearly complete (but finite) basis in which the exciton Hamiltonian is evaluated. The basis contains discrete mesh points denoted $|\vec{k}_i\rangle$ in the $\vec{k}$-space for electron-hole product states. Linear combinations of these states $|\vec{k}_i\rangle$ with proper symmetry are used to reduce the size of the Hamiltonian matrix. The basis states are defined as

$$|nm\nu, \vec{k}_i\rangle_{HH} = \sum_{\vec{k}}^{(i)} \exp\{i(\tfrac{3}{2} - \nu)\phi\} \psi_{n\vec{k}}^e \psi_{m\nu, -\vec{k}}^h$$

for the heavy hole (HH) series, and

$$|nm\nu, \vec{k}_i\rangle_{HH} = \sum_{\vec{k}}^{(i)} \exp\{i(\tfrac{1}{2} - \nu)\phi\} \psi_{n\vec{k}}^e \psi_{m\nu, -\vec{k}}^h$$

for the light hole (LH) series, where the symbol $\sum^{(i)}$ denotes summation over states in the "cell" centered at $\vec{k}_i$. The HH series contains electron-HH ($\nu = \tfrac{3}{2}$) product states with zero angular momentum and other electron-hole product states ($\nu = \tfrac{1}{2}$, $-\tfrac{1}{2}, -\tfrac{3}{2}$) with nonzero angular momenta. The LH series contains electron-LH ($\nu = \tfrac{1}{2}$) product states with zero angular momentum and other electron-hole product states ($\nu = \tfrac{3}{2}, -\tfrac{1}{2}, -\tfrac{3}{2}$) with nonzero angular momenta. The two series are decoupled due to symmetry. Thus, in the calculation of absorption spectra, we can compute the contributions from the HH and LH series separately and add them together.

Below, we show theoretical predictions for the absorption spectra obtained by this method for some representative quantum

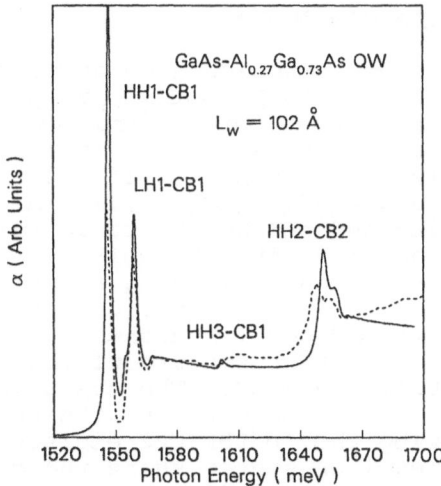

Fig. 1. Theoretical absorption spectrum (solid
curve) and experimental excitation
spectrum (dashed curve) (from Ref. 6) for
a 102 Å GaAs-Al$_{0.27}$Ga$_{0.73}$As quantum well.

wells and superlattices and compare them with the available
photoluminescence excitation (PLE) spectra. The PLE experiment
measures the product of the absorption coefficient and the rate
that a photo-excited electron-hole pair relaxes to the quasi-
equilibrium state in which the electron and hole are in their
lowest energies. Since the relaxation rate is expected to be a
smooth function of the incident photon frequency, the PLE
spectra should contain all the salient features in the
absorption spectra.

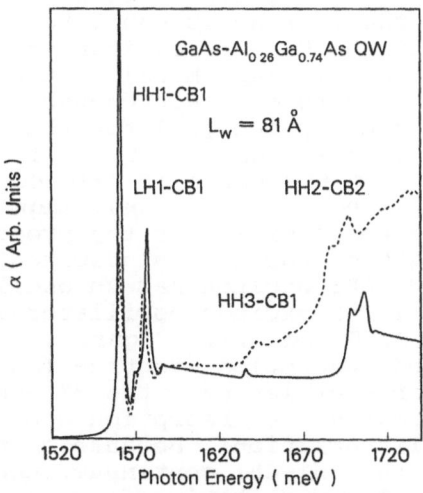

Fig. 2. Theoretical absorption spectrum (solid
curve) and experimental excitation
spectrum (dashed curve) (from Ref. 6) for
an 81 Å GaAs-Al$_{0.26}$Ga$_{0.27}$As quantum well.

3.1. GaAs-Al$_x$Ga$_{1-x}$As Quantum Wells

Figures 1 and 2 show the calculated absorption spectra (solid curves) of a 102 Å GaAs-Al$_{0.27}$Ga$_{0.73}$As quantum well and an 81 Å GaAs-Al$_{0.26}$Ga$_{0.74}$As quantum well, respectively. The Luttinger parameters used here are taken from Ref. 30 and the valence band offset is taken to be 31% of the band gap difference of GaAs and Ga$_{1-x}$Al$_x$As[31]. The calculations include couplings of all s-like and p-like excitonic (discrete plus continuum) states associated with the HH1-CB1, LH1-CB1, HH2-CB1, HH3-CB1, HH1-CB2, HH2-CB2, and LH1-CB2 transitions. The corresponding photoluminescence excitation spectra obtained by Miller et al.[6] are reproduced (in dashed curves) for comparison. The most important feature of these spectra is the doublet structure labeled HH2-CB2. The splitting is a result of strong mixing of the 1s HH2-CB2 exciton with the 2p LH1-CB2 exciton. Without mixing the 2p LH1-CB2 exciton is forbidden. With mixing, the two exciton states share the oscillator strength of the HH2-CB2 transition. The over-all absorption spectra predicted by the theory (with no adjustable parameters other than a uniform broadening of 1.2 meV for CB1 related transitions and 2.4 meV for CB2 related transitions) agree well with the experiment for both samples. In particular, the theory correctly predicts the variation of line shapes for the doublet structure associated with the HH2-CB2 transition when the well width changes from 102 Å to 81 Å. It should be noted that it is important to include the coupling of discrete exciton states (e.g. p-like LH1-CB2) with the continuum states of other excitons (e.g. s-like HH2-CB2). If one only includes the coupling of discrete excitonic states as was done in the variational calculations of Zhu[25], the oscillator strength of the 2p LH1-CB2 exciton would be too small to account for the experimental data.

Another important feature in these figures is the small hump due to the HH3-CB1 exciton. This exciton is allowed even in the simple effective-mass model without valence-band mixing, because the overlap of HH3 and CB1 envelope functions is non-zero. The square of the overlap is equal to the ratio of oscillator strength of the HH3-CB1 exciton to that of the HH1-CB1 exciton in the effective-mass model. It can be shown that this overlap is sensitive to the valence-band offset. For example, the squared overlap integral for the 102 Å case is 0.00018, 0.0062 and 0.011 for Q_v = 0.15, 0.31 and 0.4, respectively, where Q_v is the ratio of valence band offset to the difference of GaAs and Ga$_{1-x}$Al$_x$As band gaps. The valence-band mixing effect is responsible for the predominant portion of the oscillator strength of the HH3-CB1 exciton[3,20]. The net effect is that the HH3-CB1 exciton has an oscillator strength of roughly 10% of the HH1-CB1 exciton oscillator strength, for values of Q_v between 0.15 and 0.4. Figure 3 shows an enlarged plot of the calculated absorption spectrum (solid curve) of the 102 Å quantum well for energies near the HH3-CB1 transition. The dashed curve indicates the absorption spectrum without including the HH3-CB1 transition. Because of the mixing of the HH3-CB1 exciton with the HH1-CB1 continuum, the resulting line shape is asymmetric. The qualitative feature can be understood by Fano's theory[7]. This particular Fano resonance is more clearly observed in a 210 Å GaAs-Al$_{0.3}$Ga$_{0.7}$As quantum well[32]. The comparison between the experimental absorption spectrum with the prediction by the present theory is shown in Fig. 4. The

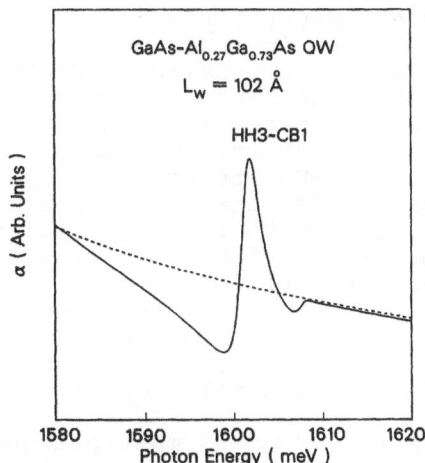

Fig. 3. Theoretical absorption spectra for a 102 Å GaAs-
$Al_{0.27}Ga_{0.73}As$ quantum well near the HH3-CB1
transition with (solid curve) and without
(dashed curve) the HH3-CB1 contribution.

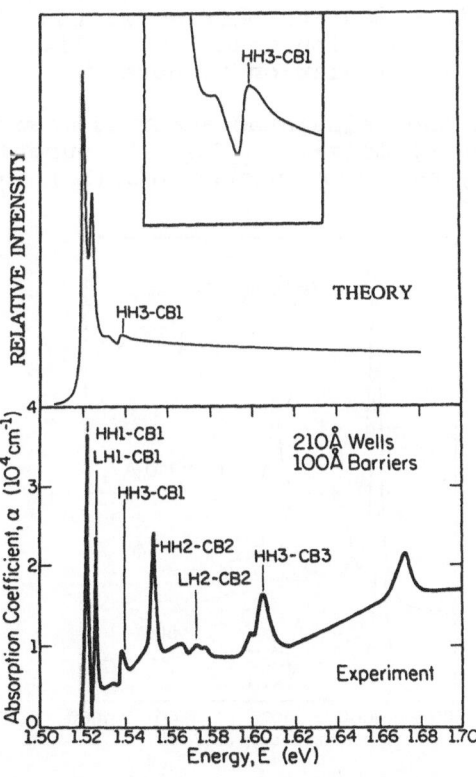

Fig. 4. Theoretical (upper curve) and measured (lower
curve) absorption spectra of a 210 Å GaAs-
$Al_{0.3}Ga_{0.7}As$ quantum well. Data taken from Ref.
32. Only a few transitions to the CB1 subband
are included in the theory.

inset shows the enlarged spectrum for energies near the HH3-CB1 transition. The calculation includes transitions associated with HH1-CB1(s), LH1-CB1(s), HH2-CB1(p), HH3-CB1(s), and LH2-CB1(p). s and p in the parentheses denote the exciton angular momenta. The dip structure preceding the HH3-CB1 peak agrees with the experimental observation. However, the calculated absorption peak for the HH3-CB1 transition appears to be a factor of two or three smaller than the experimental results. We attribute this discrepancy to the insufficient number of states included in the calculation. For example, the HH2-CB1(s), LH1-CB1(d), and HH3-CB1(d) excitonic states (including continuum states) may have effect on this resonance structure. Further research on this needs to be done in the future.

3.2. Superlattice Effect on Optical Properties

Superlattices can be viewed as multiple quantum wells if the barrier material in each superlattice unit cell is sufficiently wide that interactions between electronic states associated with one well and those with adjacent wells are negligible. For superlattices with narrow barrier width, the energy dispersion for wave vectors along the growth direction is large and its effect on the absorption coefficient is important. One expects the line shape of the absorption spectrum to change gradually from a three-dimensional character in the ultra-thin barrier case to a quasi-two-dimensional character in the wide barrier case. For intermediate barrier widths it is difficult to predict what the absorption spectrum is like, unless a realistic theoretical calculation is done.

Figure 5 shows the calculated absorption spectra (solid curves) for a series of GaAs-$Al_{0.23}Ga_{0.77}As$ superlattices with different $Al_{0.23}Ga_{0.77}As$ layer thicknesses (L_B) and the same

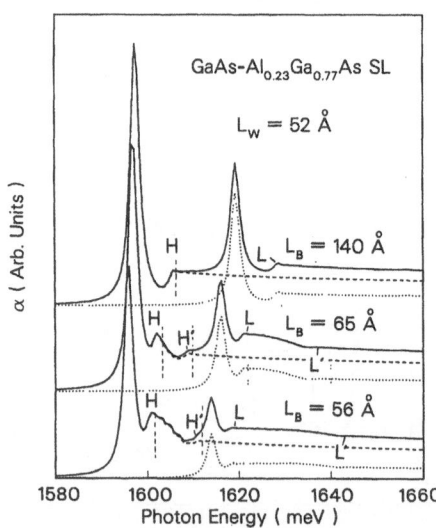

Fig. 5. Theoretical absorption spectra for GaAs-$Al_{0.23}Ga_{0.77}As$ superlattices with well width L_W = 52 Å and barrier widths L_B = 56 Å, 65 Å and 140 Å. Dashed curves: HH1-CB1 contribution. Dotted curves: LH1=CB1 contribution. Solid curves: total.

GaAs layer thickness (L_W = 52 Å). The dashed and dotted curves
are the contributions to the absorption spectra from the HH-
series states and LH-series states, respectively. The parallel
vertical lines illustrate the subband widths. For the 140 Å
case, the subband dispersion is negligible and we have
essentially uncoupled quantum wells. The shoulder structure
just below the LH1-CB1 exciton (marked H) is associated with the
2s excited state of the HH1-CB1 exciton. For the 65 Å case, the
subband dispersion is substantial (about 8 meV for HH1-CB1 and
18 meV for LH1-CB1), and we observe line shape change due to the
tunneling of excitons from one well to another. The most
noticeable is the asymmetric line shape of the secondary peak
structures marked H and L. These structures contain the closely

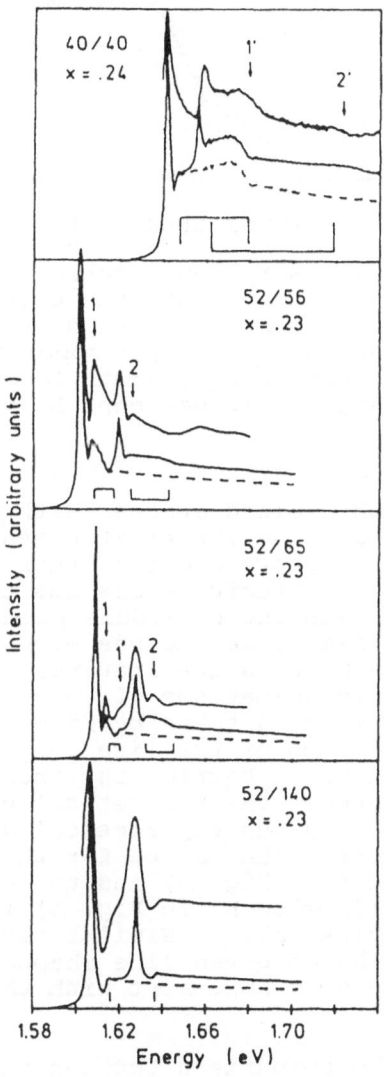

Fig. 6. Theoretical absorption spectra (lower curve) and
        PLE spectra (upper curve) (from Ref. 33) for
        GaAs-$Al_{0.23}Ga_{0.77}As$ superlattices with (L_W = L_B)
        = (40/40), (52/56), (52/65) and (52/140) (in
        units of Å).

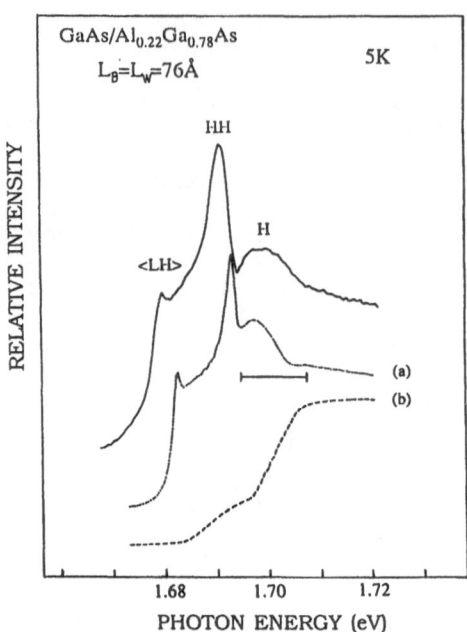

Fig. 7. Theoretical absorption spectra with (dotted curve) and without excitonic effects (dashed curve) and PLE spectra (solid curves) (from Ref. 35) for GaAs-Al$_{0.22}$Ga$_{0.78}$As superlattices with equal well and barrier width L$_W$ = L$_B$ = 76 Å. The theoretical fit was done by using L$_W$ = L$_B$ = 76 Å.

spaced discrete exciton excited states (with predominent contribution from the 2s) and the exciton resonances near the onsets of the continua. There are two other weak structures marked H′ and L′. These structures are due to exciton resonances associated with the M$_1$ saddle points. The structure L′ becomes more noticible if we include more $\vec{k}$ points in our calculation. The 56 Å case is qualitatively similar with larger subband dispersion. The comparison of theoretical spectra with the corresponding PLE spectra taken by Deveaud et al.[33] is shown in Fig. 6. Variation of the line shape of the absorption spectra due to the change of barrier thickness is apparent. Very good agreement between the theoretical predictions (with no adjustable parameters) and the experimental data are found. In particular, the asymmetric line shape for the structures labeled 1 and 2 (labeled H and L in Fig. 5) and the saddle-point derived structure labeled 1′ (labeled H′ in Fig. 5) for the 65 Å sample are clearly seen experimentally. Similar features were observed by Song et al.[34]. The observed line shapes for exciton resonances are also in good agreement with the theoretical predictions.

Figure 7 shows the comparison between the calculated absorption spectrum and measured PLE spectrum associated with transitions to the second conduction subband (CB2) in a GaAs-Al$_{0.22}$Ga$_{0.78}$As superlatice with L$_B$ = L$_W$ = 76 Å. The data were taken by Zhou et al.[35]. The calculation was done for L$_B$ = L$_W$ = 79 Å to compensate for the nonparabolicity effect, which is not included in the calculation and is important at high

transition energies as shown here. Again, the theoretical
results agree well with the observed exciton line shapes.

SUMMARY

     In summary, we have described recent theoretical develop-
ment for calculating the absorption spectra of realistic
semiconductor quantum wells and superlattices. The band-mixing
effect as well as the Fano-resonance effect give rise to the
interesting line shapes of the excitonic peaks in the absorption
spectra. The line shapes predicted by the theoretical calculations are found in excellent agreement with those observed in
the photo-absorption and photoluminescence excitation (PLE)
experiments.

Acknowledgment

     We would like to thank J.J. Song, H. Morkoç and B. Deveaud
for providing us with their data and for fruitful discussions.
We also thank G.Z. Wen for assistance in computation. This work
was supported by the Office of Naval Research (ONR) under
Contract N00014-89-J-1157 and the University of Illinois
Materials Research Laboratory under National Science Foundation
(NSF) Contract No. NSF-DMR-86-12860.

REFERENCES

1.   A.C. Gossard, P.M. Petroff, W. Wiegman, R. Dingle and A.
     Savage, **Appl. Phys. Lett.**, 29:323 (1976); E.E. Mendez,
     L.L. Chang, C.A. Chang, L.F. Alexander and L. Esaki, **Surf.
     Sci.**, 142:215 (1984)
2.   R. Dingle, **in**: "Festkorperprobleme", J. Trensch, ed.,
     Pergamon, New York 15:21 (1975)
3.   Y.C. Chang, J.N. Schulman, **Appl. Phys. Lett.**, 43:536
     (1983); **Phys. Rev.B**, 31:2069 (1985)
4.   G.D. Sanders, Y.C. Chang, **Phys. Rev.B**, 31:6892 (1985);
     Phys. Rev.B, 32:4282 (1985); **Phys. Rev.B**, 35:1300 (1987)
5.   A.C. Gossard, "Treatise on Materials Science and
     Technology", K.T. Tu, R. Rosenberg, eds., Academic, New
     York (1982), vol. 24
6.   R.C. Miller, A.C. Gossard, G.D. Sanders, Y.C. Chang, J.N.
     Schulman, **Phys. Rev. B**, 32:8452 (1985)
7.   U. Fano, **Phys. Rev.**, 124:1866 (1961)
8.   J.M. Luttinger and W. Kohn, **Phys. Rev.**, 97:869 (1956)
9.   F. Bassani and C.P. Parrasvacini, "Electronic States and
     Optical Properties in Solids", Pergammon, New York (1975)
10. J.C. Phillips, Advances in Physics 18:56 **in**: "Solid State
     Physics", Academic Press, New York (1966); **Phys. Rev.**,
     136A:1705 (1964)
11. B. Velicky, J. Sak, **Phys. Status Solidi**, 16:147 (1966)
12. H. Kamimura, K. Nakao, **J. Phys. Soc. of Japan**, 24:1313
     (1968)
13. E.O. Kane, **Phys. Rev.**, 180:852 (1969)
14. J.E. Rowe, F.H. Pollak, M. Cardona, **Phys. Rev. Lett.**,
     22:933 (1969)
15. S. Antoci, E. Reguzzoni, G. Samoggia, **Phys. Rev. Lett.**,
     24:1304 (1970)
16. H. Chu and Y.C. Chang, **Phys. Rev.B**, 36:2946 (1987)

17. H. Chu and Y.C. Chang, **Phys. Rev.B**, 39:10861 (1989)
18. G. Bastard, **Phys. Rev.**, 25:7584 (1982)
19. R.L. Greene, K.K. Bajaj, **Solid State Commun.**, 45:831 (1983)
20. G.D. Sanders and Y.C. Chang, **Phys. Rev. B**, 32:4282 (1985);
    Phys. Rev.B, 35:1300 (1987)
21. D.A. Broido and L.J. Sham, **Phys. Rev.B**, 34:3917 (1986)
22. A. Fasolino and M. Altarelli, **in**: "Two-dimensional Systems,
    Heterostructures, and Superlattices", G. Bauer, F. Kuchar
    and H. Heinrich, eds., Springer-Verlag, New York (1984)
23. M. Altarelli, **Phys. Rev. B**, 32:5138 (1985)
24. B. Zhu and K. Huang, **Phys. Rev. B**, 36:8102 (1987)
25. B. Zhu, **Phys. Rev.B**, 37:4689 (1988)
26. U. Ekenberg and M. Altarelli, **Phys. Rev. B**, 35:7585 (1987)
27. G. Bauer and T. Ando, **Phys. Rev. B**, 37:3130 (1988)
28. A. Chomette, B. Lambert, B. Deveaud, F. Clerot, A. Regreny
    and G. Bastard, **Europhys. Lett.**, 4:461 (1989)
29. G. Grosso and G. Pastori Parravicini, **in**: "Memory Function
    Approaches to Stochastic Problems in Condensed Matter",
    Advances in Chemical Physics, V.62; D.M. Woodruff, S.M.
    Anlage and D.L. Smith, **Phys. Rev.B**, 36:1725 (1987)
30. P. Lawaetz, **Phys. Rev. B**, 4:3460 (1971)
31. D.J. Wolford, T.F. Keuch, J.A. Bradley, M.A. Gell, D. Ninno
    and M. Jaros, **J. Vac. Sci. Technol.B**, 4:1043 (1986)
32. P.J. Pearah, W.T. Masselink, T. Henderson, C.K. Peng, H.
    Morkoç, G.D. Sanders and Y.C. Chang, **J. Vac. Sci. Technol.B**
    4:525 (1986)
33. B. Deveaud, A. Chomette, F. Clerot, A. Regreny, J.C. Maan,
    R. Romestain, G. Bastard, H. Chu and Y.C. Chang, **Phys. Rev.
    B**, 40:5802 (1989)
34. J.J. Song, P.S. Jung, Y.S. Yoon, H. Chu, Y.C. Chang and
    C.W. Tu, **Phys. Rev. B**, 39:5562 (1989)
35. J.F. Zhou, P.S. Jung, J.J. Song and C.W. Tu, **Appl. Phys.
    Lett.** (May issue, 1990) in press.

SEMICONDUCTOR TRANSPORT

PERSPECTIVES ON ELECTRON TRANSPORT IN SEMICONDUCTOR STRUCTURES

F. Koch

Technische Universität München
Physik-Department E16
8046 Garching, FRG

ABSTRACT

We examine current research on electronic transport in low dimensional semiconductor structures. It is argued that the transport phenomena have acquired a qualitatively new look in that quantum mechanical aspects have become very important. Better control of the defect scattering mechanisms has made possible significant increases in the length of coherent transport. For devices, both the small size of the structures and long mean free paths have led to great importance of carrier heating effects.

1.   INTRODUCTION

Transport of charge is the essence of active device structures. How electrons respond to applied voltages and fields determines the input-output characteristics and sets limits to the performance of devices. Transport is a multifaceted subject which builds on the knowledge of the electronic states and their interactions, among themselves, with the lattice and with all the many possible defects. For modern semiconductor structures it involves quantum modes of propagation. A study of transport requires proper account of wavefunctions and their coherent superposition after scattering from potential barriers and randomly positioned defects. Transport involves interference effects and the quantization of conductivity in units of e^2/h.

In today's world of lithographically tailored semiconductor structures and epitaxially grown heterolayer sequences, electronic transport is not what it used to be. Ohm's law $\sigma = ne\mu$, in which mobility is derived from the incoherent summation of scattering processes, is an anachronism. Devices today are so small that "fingerprint" patterns of the current-voltage relation from a distinct and unique arrangement of scatterers result. Noise is discussed in terms of individually identifiable switching events involving a single elementary unit of charge.

Condensed Systems of Low Dimensionality
Edited by J. L. Beeby *et al.*, Plenum Press, New York, 1991

Not only has the qualitative picture, the conceptual
structure of electron transport, been transformed, but there
have also occurred radical changes in the numbers which describe
transport. Understanding of scattering processes and precise
epitaxial growth procedures have combined to give mobilities in
the millions of cm^2/Vsec. Traditional mobility values for
semiconductor electrons were in the 10^3 cm^2/Vsec range. Along
with the progressive miniaturization of devices hot electron
effects have become a central issue. The small and beautiful
device structures, built by the millions into integrated
circuits, function with current densities and power dissipation
per unit of active volume which reach "astronomical" values in
the terawatt/cm^3 range. Understanding and controlling hot
electron effects is important business for the semiconductor
scientist working on devices.

In this lecture the intent is to illustrate the new
qualities of electronic transport in semiconductors. We want to
show the trends and set the stage for the scientific contributions of a number of colleagues working in the field. Their
subjects have been selected to demonstrate aspects of the "new
transport" and the physics that goes along with it. The choices
are exemplary of activity in the field.

The paper begins with a discussion of how quantum
mechanical propagation leads to a distinctly different way in
which defect scattering enters the conductivity. In Section 2,
under the heading "Nonohmic Scattering", we discuss how current
theoretical understanding and the availability of small device
structures have led to experiments which highlight the quantum
nature of scattering events. There follows in Section 3 an
introduction to the phenomenon of resistance quantization.
Section 4 discusses experimental strategies for mobility
improvement. Finally, Section 5 deals with some aspects of the
electron heating problem in devices.

2.   NONOHMIC SCATTERING

Electron transport in a semiconductor occurs by several
distinct mechanisms. The most elegant and direct way from here
to there is a straight line. This is the ballistic transport
mode where an electron moves in accord with the forces acting on
it by the external fields without being deflected or scattered
along its path. With the materials and structures becoming both
more perfect and smaller, ballistics plays an ever more
important role. It is not a subject with much physics in it.
Ballistics by itself is in the narrow sense of the word too
"straightforward" to show up the subtleties of quantum transport
physics. When ballistic motion is combined with reflection at a
distinct boundary it takes on a special significance. The
periodic movement of the electron between the boundaries of a
quantum well or in the self-consistent potential of a layer of
dopant atoms, leads by coherent superposition of incident and
reflected waves to a quantum state.

We are by now thoroughly familiar with quantized electron
states in planar structures such as those in Fig. 1. The
pioneering MOSFET experiments on the two-dimensional electron
gas (2DEG) by Fowler, Fang and Stiles[1] are a quarter of a
century old. The essential ingredient in forming a quantum

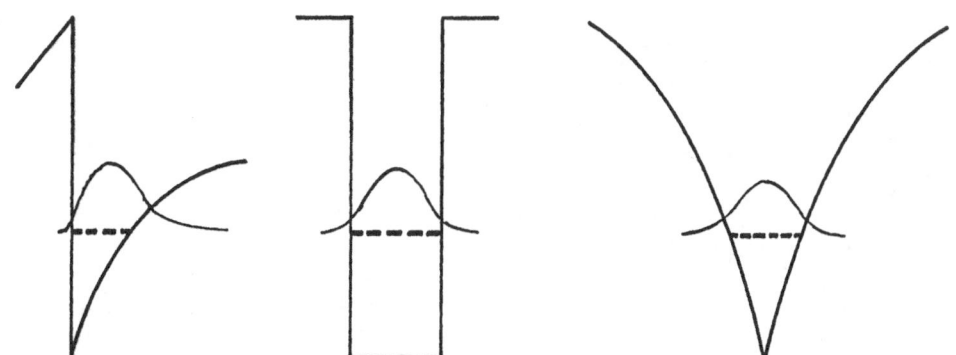

Fig. 1. Potential wells and ground states for three
examples of quantized, one-dimensionally
confined electron gas systems - the MOSFET,
quantum well and δ-layer. In-plane transport
may also show quantum mechanical aspects.

state is confinement by boundaries on a scale of length where
coherent ballistic motion takes place. We have learned to treat
the 2DEG in wave mechanical terms with distinct quantum levels.
The electronic transport within the plane, however, is
traditionally discussed in classical terms. Particles carrying
current are considered to be scattered incoherently between
randomly positioned defects.

In more recent work quantum aspects of the in-plane
transport have become apparent. There are two reasons for this,
both founded in the great advances in semiconductor technology.
For one, lithographic techniques have been perfected to the
stage where the making of μm-sized laterally confined 2DEG
structures is considered routine and 0.1 μm dimensions are
possible. On the other hand, the number of defects and
scattering centers in the plane has been greatly reduced. Thus
ballistic free paths are approaching the physical size of the
lateral structures. This necessarily brings on quantum physics.
In addition, the number of defects per device has become
enumerable and small. This makes apparent the scattering
physics of individual centers.

Wave aspects in planar transport actually became evident in
a way that did not even require matching up the physical size of
the device structure with the length of coherent motion.
Altshuler's theory of weak localization[2] and the negative
magnetoresistance phenomenon that follows from it, provided the
first evidence for nonohmic interaction of the electrons with
the scattering defects in 2D layer systems. The theory
considers a class of electronic paths that after a number of
elastic scattering events intersect themselves, thus forming
closed loops as in Fig. 2a. These loops have equal probability
for clockwise or counterclockwise traversal. The corresponding
electron waves interfere coherently in a given loop to form a
standing wave. The result is an enhanced backscattering and
therefore an additional resistance. In a magnetic field this
additional resistance is removed. The field breaks the
symmetry. Particles circling the loop in different directions
need no longer interfere constructively. Temperature T has a

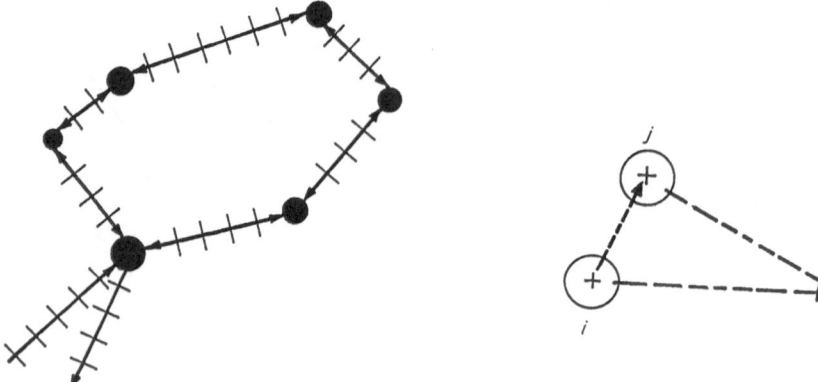

Fig. 2. Configurations of in-plane electron motion which
demonstrate wave aspects in transport.
a) Electron waves scattering elastically from
various defects and finally crossing their own
paths. b) Triangular configuration of impurity
sites arranged such that a particle starting at
i has two alternative tunneling paths to reach
k.

similar effect. With rising T, inelastic scattering is
increased. The number of loops that will be traversed
coherently is accordingly reduced. Both of these aspects of
weak localization have been observed and are now classic
evidence for quantum mechanical features of lateral transport in
the 2DEG.

     Weak localization is a concept that applies to particles
which propagate freely between scattering events. In wave
mechanics a tunneling mode of propagation, where the wave decays
exponentially into a potential barrier, is also possible. Such
transport applies to semiconductors at low T and is known as
hopping. Hopping is also a process for which interference may
occur. In an n-type material it involves electrons tunneling
between allowed impurity-ion sites. Phonons are emitted and
absorbed as required in order to match initial and final state
energies. Interference occurs when waves which have tunneled
along different paths are superposed[3,4]. In particular, Ref.
4 gives a simple discussion in terms of an elementary triangular
configuration of impurity sites such as in Fig. 2b. Here the
electron tunneling from an inititial site i to a final state k
directly is considered to interfere with itself as it tunnels on
an alternate path through the intermediate site j. The wave
scattered at j experiences a phase shift by the potential at
that site. It adds out of phase to the directly tunneled wave
and thus reduces the overall transmission rate. The application
of a magnetic field in this case also will produce a lowering of
the resistance. The experimentally observed negative magnetoresistance for hopping transport in a dilutely doped plane of
donor impurity atoms is in part caused by this effect[5].

     As the structures in which transport is studied have become
smaller, there is yet another way in which the nonohmic interaction of the electrons with the scatterers becomes apparent.
The usual description of conductivity not only involves

238

incoherent superposition of scattered intensities but also deals with macroscopically averaged quantities. Scatterers have a density. Their positioning is random and of no particular concern. As experimental samples approach linear dimensions of 0.1 μm and areas in the 10^{-9} to 10^{-10} cm^2 range, the number of scatterers becomes small and their relative placement begins to have special significance. Resistance in a small structure is the result of interfering scattered waves from a unique spatial arrangement of defect potentials. In such systems the action of a magnetic field cannot be described in terms of an average magnetoresistance. One observes in conducting semi- conductor layers fine-grained magnetoconductivity fluctuations as Altshuler-type loops are tuned by the magnetic flux going through them. The patterns of fluctuating conductance have been termed "magnetic fingerprints". A fingerprint is also obtained when the Fermi wavelength is swept by means of a gate voltage in FET-like structures[6]. As long as the transport is metallic and the cause of the fingerprint fine structure can be traced back to the loops, the average fluctuation amplitude is a universal e^2/h.

Along with the fingerprints must be considered the switching phenomenon which describes the time evolution of a system of a small number of scatterers which each have a statistical probability of being turned on and off. A trap, capable of holding an electron, may be occupied or empty. It will, in general, switch back and forth between the filled and empty state. The result is a switching between two specific resistance values. The noise is referred to as a random telegraph signal and has a Lorentzian spectrum. By adjusting the voltage settings of a FET the specific noise signal can be tuned in and out. In Fig. 3 we show an example of such a switching noise source. In the contribution of Timp and co-workers[7] in this volume will be found additional data. The switching noise amplitude depends on the exact placement of the scatterer and its interaction with the scattered electron waves. Thus even when there are several switching signals they are individually identifiable by their amplitudes.

3.   RESISTANCE QUANTIZATION

Of all the many ways that quantum physics enters in the transport of charge in a planar semiconductor structure, none is more striking than the quantization phenomenon associated with current flow through a point contact. Several of the contributions in this volume deal with this effect. We have previously mentioned G. Timp et al.[7] in connection with the switching signals. Molenkamp and co-authors[8] discuss a transverse voltage oscillation that also has its origin in the lateral quantization. Wharam[9] presents well resolved resistance steps with high quality plateau structure. There are two expert theoretical treatises of the effect in this volume[10,11]. Resistance quantization is a timely topic of current research on transport in low dimensional systems.

The experimental basis of all this work is founded in the principle of lateral confinement of the 2DEG by means of a structured gate. Samples are fabricated starting from the high quality, modulation doped GaAs heterostructures. The top layer is covered with a metallic gate and structured lithographically

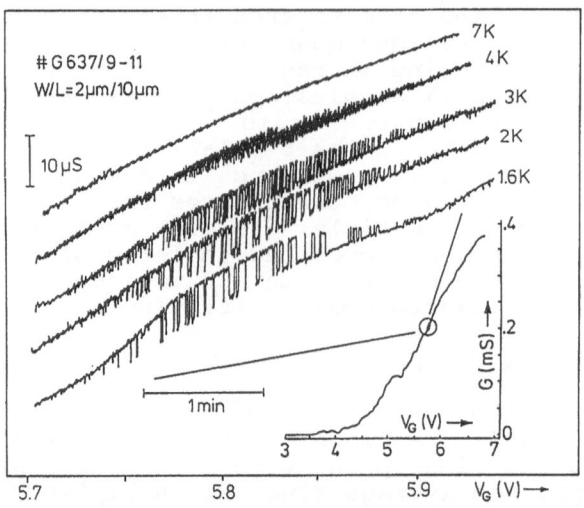

Fig. 3. A switching noise signal source in the
transconductance characteristic of a MOSFET.
The gate voltage is slowly swept to tune the
switching in and out at different temperatures.

into what is termed the split gate configuration. Electrons are
forced by the applied gate voltage into a narrow channel. The
"narrowness" criterion is that the Fermi surface electrons are
constrained transverse to the direction of current flow in a
channel whose width is a small number of wavelengths. They
occupy states in the quasi-parabolic potential minimum provided
by the superposed gate and doping layer potentials. Because of
the random statistical distribution of the positive ionized
donor charges, there are considerable potential fluctuations in
and about the minimum. The reader will find a theoretical
modeling of the potential in the contribution of Davies and co-
workers[12] in this volume.

With all the current excitement about resistance
quantization it is easy to lose the historical perspective on
the problem. Those who have been around for a while will
remember the work of Warren et al.[13]. Using a similar
approach to linear confinement, transport is studied in parallel
one-dimensional channels. It was the fashion to make long and
narrow electronic channels. One expected, and also clearly
found in Ref. 13, the quantization effect related to transverse
lateral confinement. The very long channels, however, did not
allow ballistic traversal and therefore resistance quantization
could not be observed.

The present work[14,15] relates to narrow constrictions and
the adiabatic spatial variation of the quantum state. Glazman
and co-workers[16] have shown that when the lateral confinement
varies on the scale of the Fermi wavelength then the transverse
energy modes are well defined. The number of subbands carrying
the current is that at the minimal width of the channel. The
quantization argument is straightforward. It depends on the
exact cancellation of the one-dimensional density of states at
the Fermi energy and the Fermi velocity. The current per
occupied, spin-degenerate subband is $\Delta n e v_F$, where Δn is the
number of electron states in the energy interval eV about the

240

Fermi energy. V is the applied voltage and v_F the Fermi
velocity. The number Δn is obtained from the density of states
as $\Delta n = (dn/dE)eV = (2/hv_F)eV$. The conductance G per subband is
the ratio of current to voltage, i.e. $2e^2/h$.

4.   MOBILITIES IN THE MILLIONS

     Along with the spectacular qualitative changes in our
understanding of transport there has been a long and steady
advance in the mobilities. The driving force behind this
development is the novel epitaxial growth technique available to
the experimenter today. Modulation doping, whereby the channel
electrons are physically separated from their ionized parent
donors, has become standard operating procedure for MBE growth.
A mobility value of 1×10^6 cm^2/Vsec is considered a good
standard for MBE GaAs. Because of the great perfection of the
interface morphology in the epitaxial growth process, there is
little left to do the scattering. This is what prompts Das
Sarma[11] to raise the question of the ultimate value of the
mobility of electrons in GaAs. At finite temperatures he argues
that only phonons and thus the strength of the deformation
potential coupling will give this ultimate value. To cite from
his work, for T = 10 K the predicted maximum is 4.0×10^6
cm^2/Vsec. It would occur for a density of 0.6×10^{11}
electrons/cm^2.

     The issue of how surface layer electrons interact with
phonons and convert the electrical energy into heat, is
addressed in the paper by Dietsche[17]. The reader will find
here an account of the details involved in the energy transfer
considerations. The article also serves as a good introduction
to the central theme of "hot electrons" in the next section.

     To balance the current excitement about epitaxially grown
materials it is only fair to look back and re-examine the status
quo for thermally grown Si-SiO$_2$ interfaces. Authors Kruithof
and Klapwijk have taken on this task[18] and summarized the
numbers. Charged impurity scattering in this system has always
been reduced by the "modulation doping" principle. Most
positive charge mostly is on the gate electrode and at least one
oxide thickness away from the interface. The problem of
achieving higher mobilities is recognized to lie with the
remaining oxide charge and its linkage with interface roughness.
Epitaxial growth produces a better interface structure than
thermal oxidation. There is always a remnant of structural
disorder at the thermally grown interface between crystalline Si
and amorphous SiO$_2$. Structural defects can be expected to trap
interface charge. There lies the dilemma which has so far
prevented even the most diligent and persistent Si-oxidizer to
reach a mobility of more than 50,000 cm^2/Vsec.

     Given all the many attempts at thermal oxidation of
(100)Si, it is unlikely that a significantly higher mobility for
the Si-SiO$_2$ interface will ever be achieved by that means. The
future challenge lies with the alternative means of growth, such
as epitaxial growth of Si which is then followed by chemical
vapor deposition (CVD) of the oxide. In view of the dramatic
recent improvements which have been made in the CVD of SiO$_2$ and
the known high quality of epitaxially grown Si layers, it is
likely that some day higher mobilities for the Si-SiO$_2$ interface
will be achieved.

5. HOT ELECTRONS

An aspect of electron transport which is becoming increasingly important is that of electron heating. In real device structures charges move in electric fields of megavolts/cm. Their average energies far exceed the thermal ambient. Especially in Si, where the gap energy is not particularly large and one lacks the efficient polar optical phonon mechanism of energy loss, the generation of electron-hole pairs is always something to be considered. A few electrons in the surface channel of a MOSFET even reach energies which enable them to surmount the 3.5 eV potential barrier and enter the SiO_2. Degradation of the devices and transport properties at high average energies of the electrons are a fact of life for device operation.

We illustrate these qualitative statements on carrier heating with some numbers. In typical field-effect devices such as the Si MOSFET , the GaAs MESFET or HEMT, typical channel lengths are in the sub-μm range. Values of ≈ 0.1 μm have been achieved. Since voltages applied are in the 1-10 Volts range, it follows that channel electric fields of order 10^6 V/cm are possible. Because of the inhomogeneous voltage drop along the channel for typical FET operation, the peak fields can even exceed the megavolt/cm value.

Typical current flow in a FET device with a channel width of 1 μm is 1 ma. For a drain voltage of +5 V the dissipation is 5 mW, but this seemingly small amount of electrical energy is deposited in a tiny volume. The current flow extends only over the channel depth of order 10^{-6} cm. Thus the current density is of the order mega-amps/cm^2. The power dissipation density reaches values in the terawatt/cm^3 range. Fortunately, the electrical dissipation is limited to very small volumes which are embedded in a single crystal substrate with very good thermal transport properties. Moreover, devices such as a dynamic memory are operated with a duty cycle of something like 10^{-6}. All of this means that the overall thermal budget can be managed. For steady state operation of a Si-MOSFET the warming effect is 10-100 K above room temperature.

It is important to know what the energy distribution of the carriers is in order to understand the consequences for the device operation. The transport is highly nonohmic. Carrier drift velocities are not linked with a mobility and actually become independent of the electric field. This velocity saturation phenomenon is well known. Values of order 1-2 x 10^7 cm/sec can be found in the literature. The kinetic energy associated with the drift will be of order 0.1 eV per carrier. Expressed as a thermal energy this is 10^3 K.

Precisely how carriers are accelerated, how they scatter and dissipate their energy, is a question of concern to the device scientist. Simulation calculations for the inner workings of devices are an important theme in current research. The basic input to such calculations is a correct microscopic understanding of scattering rates, a description of the interaction of carriers among themselves and with the lattice vibration. The carrier relaxation dynamics are discussed in the papers of Seilmeier[19] and Rühle[20]. Nevertheless, these studies provide only partial answers to the relevant questions

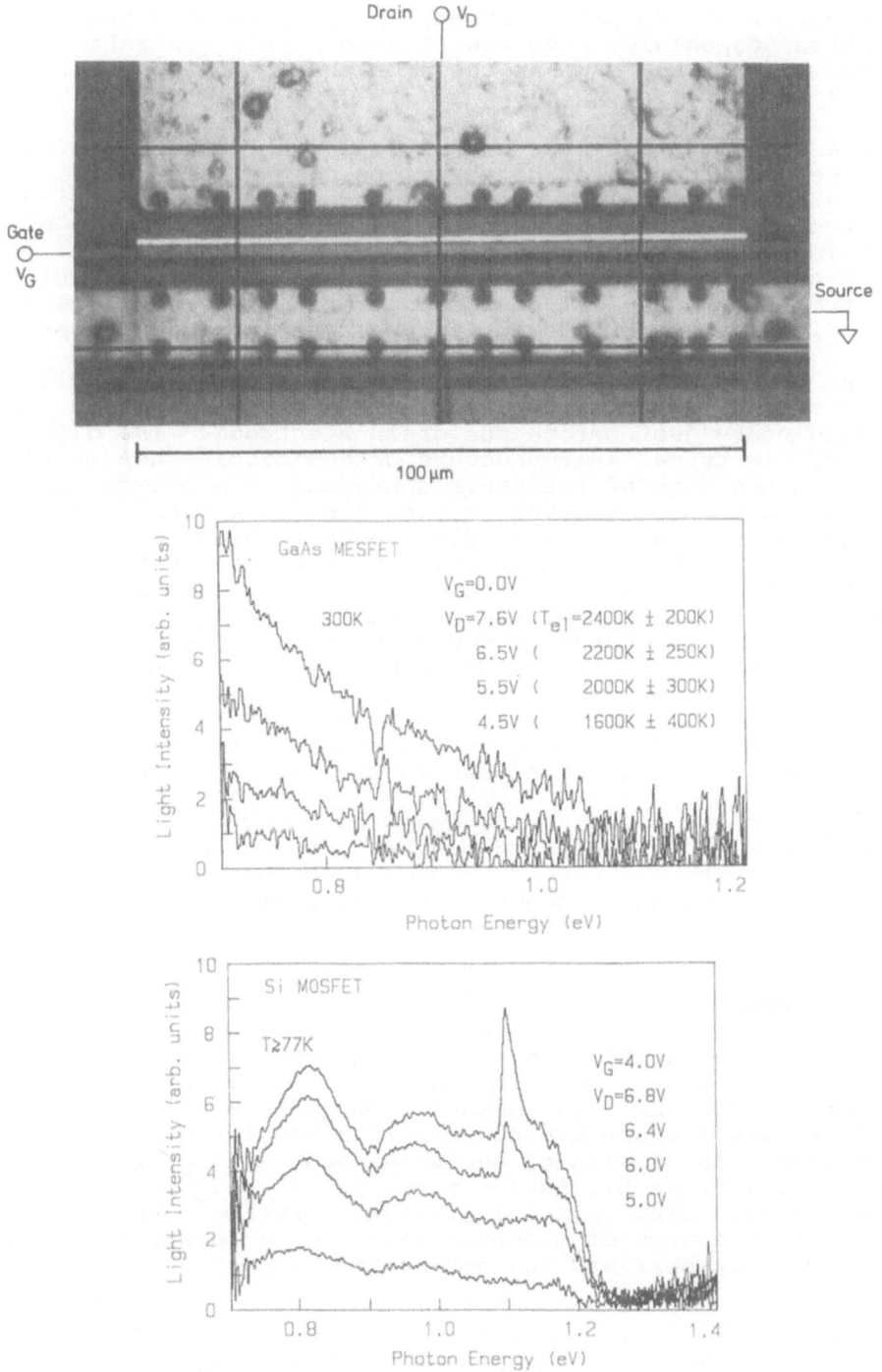

Abbildung 6.1: Photographie eines MOSFET der Serie G355 bei Zimmertemperatur. V_D=7.8 V, V_G=5.5 V

Fig. 4. Si-MOSFET test structure operated at room temperature with V_D = 7.8 V and V_G = 5.5 V. The infrared-sensitive film records a brightly lit stripe at the drain contact. Below are shown the emission spectra of a GaAs-MESFET and a Si-MOSFET.

of hot electron transport. It must be kept in mind that optical excitation, especially when electron-hole pair generation is involved, is not the same as the acceleration by an electric field in a device.

One experimental means whereby one can learn something about the distribution of hot carriers that exists inside a device structure is light emission spectroscopy. The principle is that of an optical pyrometer in which the temperature of an object is judged by the color of the light that it emits. It is not a great problem to observe the electroluminescence which comes from Si or GaAs FET devices operated at typical voltages. We show in Fig. 4 a MOS device as observed in a microscope using infrared sensitive film. Note the bright strip of light emerging on the drain side of the device. This gives evidence for the strongly asymmetric distribution of the electric field. The field peaks just before the drain electrode. The dissipation of electrical energy occurs predominantly when carriers scatter at the edge of the drain contacts. The energy spectrum of the light that is emitted in Fig. 4 shows a number of complicated spectral features[21]. Before relating such spectra to the energy distribution of the hot carriers one has to understand the microscopic mechanisms which cause the light emission. For the Si spectrum in the figure most of the radiation involves recombination with avalanche-generated minority holes. Bremsstrahlung caused by scattering of the hot carriers is seen much more clearly in the radiation of a GaAs MESFET. For this case it has been possible to assign an electron temperature which describes the average kinetic energy per carrier. Typical values are in the 2000 K range.

Light emission provides a means to study hot electrons in a device. Before such emission spectroscopy can be related quantitatively to the carrier energy distribution, it will be necessary to accurately identify the mechanism that causes the light.

6.   CONCLUDING REMARKS

In this perspective we have attempted to give an impression of what are current issues in transport for two-dimensional electron gas systems. The emphasis on quantum mechanical details of the transport processes is a revolutionary development. The limits of Ohm's law description are now evident. The other big issue in current work is hot carriers. The reader will find in the individual contributions of this volume a rich source of material relating to both the quantum physics of transport and the hot carrier aspect.

REFERENCES

1.   A.B. Fowler, F.F. Fang, W.E. Howard and P.J. STiles, **Phys. Rev. Lett.**, 16:901 (1966)
2.   B.L. Altshuler, **JETP Lett.**, 41:648 (1985)
3.   V.L. Nguyen, B.Z. Spivak and B.I. Shklovski, **JETP**, 62:1021 (1986)
4.   W. Schirmacher, **Phys. Rev. B**, 41:2451 (1990)
5.   Qiu-yi Ye, A. Zrenner, F. Koch and K. Ploog, **Phys. Rev. B**, 41:8477 (1990)

6. See for example: W.J. Skocpol, P.M. Mankiewich, R.E. Howard, L.D. Jackel and D.M. Tennant, p.1491 **in**: "Proc. 18th Int. Conf. Phys. of Semicond". (Stockholm 1986), O. Engström, ed., World Scientific, Singapore, p.1491

7. G. Timp, R.E. Behringer, J.E. Cunningham and E.H. Westerwick, p.347 of this volume

8. L.W. Molenkamp, H. van Houten, C.W.J. Beenakker, R. Eppenga and C.T. Foxon, p.335 of this volume

9. D. Wharam, p.359 of this volume

10. E. Tekman and S. Ciraci, p.369 of this volume

11. S. Das Sarma, p.261 of this volume

12. J.H. Davies, J.A. Nixon and H.U. Baranger, p.387 of this volume

13. A.C. Warren, D.A. Antoniadis and H.I. Smith, **Phys. Rev. Lett.**, 56:1858 (1986)

14. B.J. van Wees, H. van Houten, C.W.J. Beenakker, J.G. Williamson, L.P. Kouwenhove, D. van der Marel and C.T. Foxon, **Phys. Rev. Lett.**, 60:848 (1988)

15. D.A. Wharam, T.J. Thornton, R. Newbury, M. Pepper, H. Ahmed, J.E.F. Frost, D.G. Hasko, D.C. Peacock, D.A. Ritchie and G.A.C. Jones, **J. Phys. C**, 51:L209 (1988)

16. L.I. Glazman, G.B. Lesovik, D.E. Khmelnitskii and R.I. Shekter, **JETP Lett.**, 48:239 (1988)

17. W. Dietsche, p.327 of this volume

18. G.H. Kruithof and T.M. Klapwijk, p.247 of this volume

19. A. Seilmeier, p.317 of this volume

20. W.W. Rühle, J. Collet, M. Pugnet and K. Leo, p.61 of this volume

21. M. Herzog, M. Schels, F. Koch, C. Moglestue and J. Rosenzweig, **Sol. State Electronics**, 32:1065 (1989)

FOLKLORE AND SCIENCE IN HIGH MOBILITY MOSFETs

G.H. Kruithof and T.M. Klapwijk

Department of Applied Physics
Materials Science Center, University of Groningen
Nijenborgh 18, 9747 AG Groningen, The Netherlands

1.   INTRODUCTION

In the last decade a gradual increase in the mobility of
silicon MOSFETs has occurred.   Interest in higher mobilities is
primarily related to the discovery of the Quantum Hall effect in
1980, for which high mobilities are essential.   At the same time
improved control of GaAs molecular beam epitaxy started to
provide samples with much higher mobilities to study the
properties of the two-dimensional electron gas (2DEG).   Samples
with record values up to $12 \; 10^6 \; cm^2/Vs$ are now grown.
Understandably, a substantial part of the research on 2DEGs has
shifted towards GaAs.   Yet, although much less dramatic the
mobilities in Si MOSFETs have increased as well, from values
around 20,000 in 1980 to maximum values of $50,000 \; cm^2/Vs$ now.
Despite the impressive effort in the use of silicon in the IC
industry an obvious technological recipe to further increase the
mobility in MOSFETs has not emerged.   Higher mobilities would be
of interest to study material dependent properties of low
dimensional electrons.   They would also be of interest for
future developments of submicron devices based on, for example,
quantum ballistic transport.

In this paper we present an overview of experimental
results about high mobility MOSFETs obtained by various
researchers.   We will first summarize reported results, followed
by a comparison of our own data with theory.   Then we summarize
aspects of the technology and handling which are reported to be
important.   Finally, we will discuss options which have the
potential of leading to higher mobilities for silicon.

2.   EXPERIMENTAL RESULTS ON HIGH MOBILITY MOSFETs

A typical example of a mobility versus electron density
curve for high mobility MOSFETs is shown in Fig. 1 for three
different temperatures, 1.5, 2.5 and 4.2 K.   At low density a
peak in mobility is observed and around this peak the mobility
is a function of temperature.   In most experiments where the
highest mobilities are needed the carrier density is adjusted

Condensed Systems of Low Dimensionality
Edited by J. L. Beeby *et al.*, Plenum Press, New York, 1991

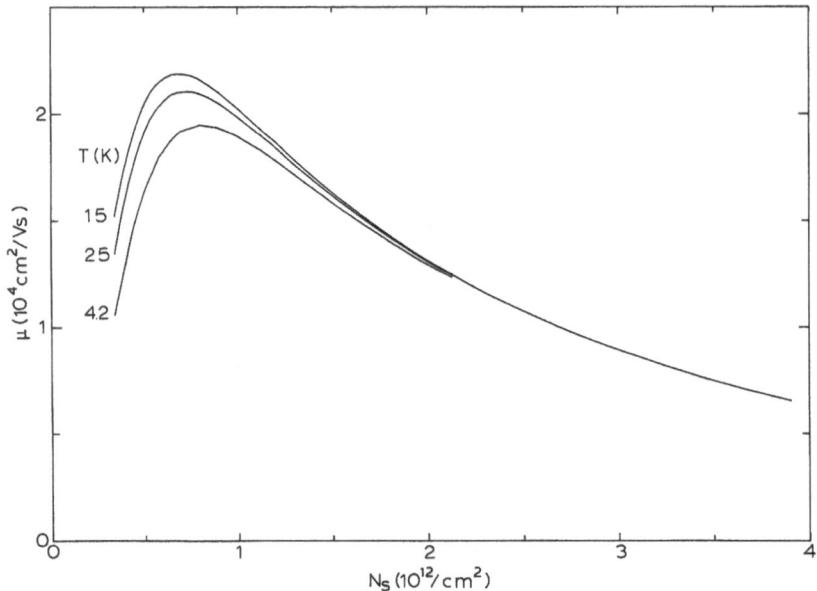

Fig. 1. Typical mobility vs carrier density curve for high mobility MOSFETs. Note the temperature dependence of the peak mobility.

around the peak and usually only information about the height μ_{max} and location N_{max} of the peak is given. In Table 1 we present a summary of data on high mobility MOSFETs. Only reports about mobilities exceeding 20,000 cm^2/Vs have been included. Except for our own work[1], we believe that only four laboratories have produced high mobility MOSFETs over a reasonably long period. Occasionally, there are reports from other groups which at a particular time have reported some high mobility results. In the absence of detailed information these reports have been ignored. Our own data will be discussed more extensively in Section 4.

The first entry is the result of Cham and Wheeler (Yale, New Haven) who, at an early stage[2], reported mobilities exceeding 20,000 cm^2/Vs. Their record value is reported[3] to be 28,000 cm^2/Vs at a temperature of 4.2 K (for a carrier density of 7 10^{11} cm^{-2}) and 39,000 cm^2/Vs at a temperature of 0.5 K (for a carrier density of 1.8 10^{11} cm^{-2}).

Dorda (Siemens) has supplied several high mobility MOSFETs to different research groups. The highest reported value, known to us, for a Siemens sample has been in the thesis of Smith[4] 31,000 cm^2/Vs at 1.3 K (carrier density 5 10^{11} cm^{-2}).

Researchers from the Institute of Solid State Physics (Academy of Sciences, Moscow) have had access to samples with the highest mobility reported to date. The most extensive data on these exceptionally high mobility MOSFETs is provided in a paper by Kukushkin, Timofeev and Cheremnykh[5]. They report that from a large number of MOSFETs the majority had mobilities from 1.3 to 2 10^4 cm^2/Vs with N_{max} of 6 to 8 10^{11} cm^{-2}. Of these samples μ_{max} and N_{max} were very weakly dependent on temperature.

Table 1. Compilation of reported data about high mobility MOSFETs. [μ_{max} is the peak mobility observed at N_{max}; T-temperature, t_{ox}-oxide thickness, N_{acc}-acceptor density, ρ_{substr}-substrate resistivity and V_{th}-threshold voltage].

AUTHORS	μ_{max} cm²/Vs	N_{max} 10^{11}/cm²	T K	t_{ox} μm	N_{acc} 10^{14} cm³	ρ_{substr} Ohm cm	V_{th} V
Wheeler[2]	22,000		4.2	0.25		30 - 40	
Kukushkin[23]	40,000	5.2	4.2	0.166			
Kukushkin[5]	40,000		4.2	0.16	7		-0.2
	52,000	2	1.5				
	30,000		4.2	0.18			-0.05
	36,000	3	1.5				
	23,000		4.2	0.13			+0.05
	28,000	4.5	1.5				
Kukushkin[24]	32,000	3.5	1.5		7		
	41,000	2.7	0.35				
Kukushkin[25]	31,000		1.6				
Furneaux, Yale[3]	22,000	6	4.2	0.1			
	24,000	4		0.2			
Furneaux, Seimens[26]	28,000	7	4.2				
	33,000	1.8		0.5			
Pudalov[27,28]	46,000		0.45				
Semenchinskii[29]	30-40,000	1					
Dolgopolov[19]	21,000	1	4.2	0.137			
	22,000	5	4.2	0.254			
	21,000	6	4.2				
	25,000	6	4.2				
Kruithof[30]	25,000	6	1.1	0.1	2	90	0.4
Kruithof[1]	31,000	5	1.1	0.1	2	90	0.4

Out of the large quantity of MOSFETs some selected specimens happened to have unusual transport properties. They showed the fractional Quantum Hall effect in agreement with the high mobility. At 4.2 K the highest mobility was 40,000 cm²/Vs, for samples of a size of 1.2 mm by 0.4 mm. At 1.5 K the mobility was as high as 52,000 cm²/Vs. This maximum was observed at a carrier density of $2 \cdot 10^{11}$ m^{-2}.

Kukushkin et al.[5] list a number of characteristic features which distinguish these high mobility samples from the others. First, a shift of μ_{max} to smaller values of N_{max}. Secondly, a much stronger dependence of mobility on bath temperature, i.e. an increase by 50% on a temperature change from 4.2 K to 1.5 K. Finally, a much stronger sensitivity to heating due to the electric field between source and drain. In the conventional samples a depression of mobility due to the electric field is observed for fields exceeding 10^{-1} V/cm. In the high mobility samples heating was already observable, from Shubnikov-de Haas measurements, for fields of 10^{-2} V/cm. They argue that this observation implies an order of magnitude smaller energy relaxation rate, namely about $3 \cdot 10^{-7}$ s.

An additional point to notice is that the MOSFETs happen to have a small to negative value of the threshold voltage (+0.05

to -0.2 V) and a relatively although not exceptionally thick oxide (0.13 to 0.18 μm). The 4.2 K record value is 40,000 cm^2/Vs. Samples of this kind are stored very carefully and many reported measurements are based on the same limited set of MOSFETs.

Researchers from the All-Union Scientific Research Institute for Metrological Service (V.M. Pudalov and S.G. Semenchinskii) obtain samples from the Institute of Microelectronics (Chernogolovka). The technology is described in a paper by Vernikov, Pazinich, Pudalov and Semenchinskii[6]. Very detailed information about the properties of these high mobility MOSFETs is not available. They report mobilities of 40,000 cm^2/Vs below 1 K, with a maximum of 46,000 cm^2/Vs at 0.45 K. The thickness of the oxide has at least in some cases been equal to 0.2 μm. The carrier density at which the peak mobility is observed is not given.

These high mobility MOSFETs represent the maximum values reported to date and are the end result of a long history of technological improvements in the fabrication of the Si/SiO$_2$ interface. Figure 2 summarizes data about μ_{max} at 4.2 K as a function of N_{max}. Obviously, the increase in peak mobility is accompanied by a decrease in N_{max}. In the upper left the 4.2 K data of Table 1 is included.

3.   SCATTERING POTENTIALS

The density of two-dimensional electrons N_s is varied by the potential on the gate V_g:

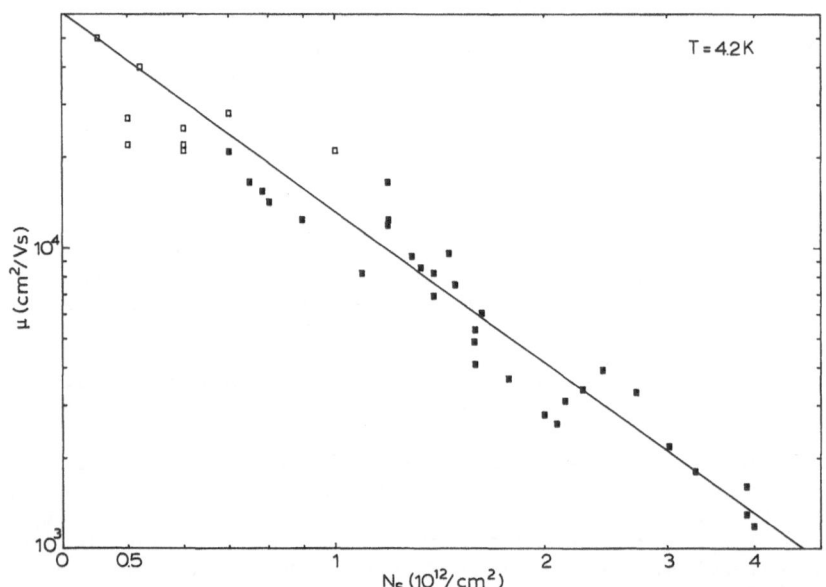

Fig. 2. A compilation of data about peak-mobilities observed at 4.2 K vs carrier density at the peak. The open squares are data included in Table 1. An increase in peak-mobility has been accompanied by a reduction in carrier density at the peak.

$$eN_s = \epsilon_0 \epsilon_{ox}/t_{ox}(V_g - V_t)$$ (1)

where t_{ox} is the oxide thickness, ϵ_{ox} is the relative dielectric constant, and V_t is the threshold voltage. Electrons in the lowest subband of the 2DEG move with a mobility determined by:

$$\mu = e\tau(\Gamma,T)/m^*$$ (2)

where m^* is the conduction effective mass and $1/\tau(\Gamma,T)$ the scattering rate of the electrons, with T the temperature and Γ the collisional broadening. Below 4.2 K the inelastic electron-phonon and electron-electron scattering rates are small compared to the elastic scattering and can be neglected[7]. The transport scattering rate is dependent on the scattering potential through:

$$1/\tau(\Gamma,T) = 1/(2\pi E_F)\int_0^{2k_F} dq \frac{q^2}{(4k_F{}^2 - q^2)^{\frac{1}{2}}} \frac{<|U(q)|^2>}{\epsilon(q,T,\Gamma)^2}$$ (3)

where E_F and k_F are Fermi energy and momentum respectively; $<|U(q)|^2>$ is the square of the matrix element for momentum transfer $\hbar q$. $\epsilon(q,T,\Gamma)$ is the dielectric response function which depends on temperature T and collisional broadening Γ. The temperature dependence dominates if the mobility is high. In general the total scattering probability is not identical to the probability for backward scattering, although for MOSFETs they will not differ very much.

The elastic scattering processes considered are charged Coulomb scattering[8] and interface or surface roughness (SR) scattering[9]. It is assumed that simultaneously present scattering rates can be added. The square of the matrix element for momentum transfer $\hbar q$ can then be written as:

$$<|U(q)|^2> = <|U_C(q)|^2> + <|U_{SR}(q)|^2>$$ (4)

where the subscripts refer to the scattering processes.

The scattering associated with Coulomb centers near the plane of the 2DEG can be separated into contributions from the depletion layer, the oxide charge and the interface charge. Although the first two will undoubtedly be present, it is common usage to consider only the latter, because only the charged centers near the plane of the 2DEG contribute effectively to the total scattering rate of electrons. Contributions to the scattering by depletion and oxide charge are assumed to give effectively an increased interface charge N_C. The substrate doping is chosen low to minimize scattering of electrons on depletion charge.

The stoichiometry of the oxide, especially of the first few layers adjacent to the Si-SiO$_2$ interface, and its dependence on oxide growth parameters has been extensively studied. The oxidation temperature and ambient as well as post oxidation anneal and removal from the furnace greatly influence the microscopic structure of the oxide and the interface. Direct comparison of the interface of *high* mobility samples, including ours, with High Resolution Transmission Electron Microscopy

(HRTEM) measurements is not available. However, extensive HRTEM
studies have been made of the properties of the Si/SiO_2
interface in *low* mobility samples. The interface is found to be
a fairly abrupt change from crystalline Si to amorphous SiO_2.
The location of this interface is not sharp and leads to
potential fluctuations associated with a roughness described
as[10]:

$$V(z + \delta(r)) = V(z) + \delta(r)\partial V(z)/\partial z . \qquad (5)$$

$\delta(r)$ is a measure of the roughness and is most conveniently
expressed in the autocovariance function of $\delta(r)$, which can be
measured using, for instance, HRTEM. The roughness is usually
assumed to be isotropic and Gaussian. The exact form of the
power spectrum and its anisotropy is determined by the oxidation
process and varies from sample to sample. Only precise
knowledge of the physics involved in oxide growth permits a more
detailed theoretical description of the roughness. The power
spectrum $S(q)$ is the two-dimensional Fourier transform of the
autocovariance function of $\delta(r)$. The average distance from the
electrons in the inversion layer to the oxide, and therefore the
scattering rate, depends on the electric field F_s perpendicular
to the interface. With this quantity the matrix elements for SR
scattering are given by:

$$<|U_{SR}(q)|^2> = S(q)e^2 F_s^2 . \qquad (6)$$

For Gaussian correlated roughness it is given by:

$$S_G(q) = \pi\Delta^2 L^2 \exp(-q^2 L^2/4) \qquad (7)$$

with L the correlation length, and Δ the rms value of the
roughness amplitude. Goodnick et al.[10] have made an extensive
analysis of HRTEM measurements in particular to test the
assumption of Gaussian correlation. They find that exponential
correlation describes the roughness much better than Gaussian
correlation irrespective of growth conditions. Roughly
speaking, it means that the interface may be regarded as
consisting of terraces of a new nm's size separated by atomic
steps of a few tenths of nm's. The power spectrum for
exponential correlation is given by:

$$S_E(q) = \pi\Delta^2 L^2/(1 + q^2 L^2/2)^{3/2} . \qquad (8)$$

Goodnick et al. report that for low mobility samples the
dependence of mobility on electron density can be fitted with
either power spectrum. Only the parameters characterizing the
roughness are slightly different if either the Gaussian or the
exponential correlation is used. As we will show, we find that
for high mobility samples exponential correlation is in much
better agreement with the transport data than Gaussian
correlation.

Recently a detailed study has been made of oxides grown
under MBE conditions on freshly prepared silicon surfaces. It
is found that on top of the silicon a thin layer of a
crystalline form of SiO_2 grows epitaxially. In subsequent work
on HRTEM studies of the Si-SiO_2 interface on extremely flat
silicon, made by MBE, it is found that the first few layers of
SiO_2 covering the Si have a crystalline form, tridymite, while
the bulk of SiO_2 is amorphous[20,21]. Although this work deals

with thin oxides grown on uniquely prepared wafers and under special oxidation conditions, it fits into the picture drawn by Goodnick et al. of an interface consisting of terraces separated by atomic steps. At monolayer steps, the crystalline phase of SiO_2 on the terraces is rotated over 90°, rendering the initial oxide into a polycrystalline layer. Further support of this picture has been obtained from grazing incidence X-ray scattering of the interface[22]. Clearly an unequivocal answer about the statistics of the interface roughness in high mobility samples is extremely important. With respect to Coulomb scattering the presence of an ordered crystalline oxide layer is of direct importance for the location of defects in the oxide, whether they are due to charged sodium or to unsaturated bonds.

The description of scattering summarized in the preceding paragraphs leads to a finite value of mobility with decreasing electron density. However, experimentally one finds that the mobility decreases much more rapidly, eventually entering the regime of thermally activated hopping. An approach to the problem of the behavior at low electron densities focuses on the density of states, $D(E)$. In two dimensions $D(E)$ is a stepfunction. In a real situation this step is likely to be rounded. Any arbitrary smearing of the band edge can be modeled in the following way[11]:

$$D(E) = \frac{2m^*}{\pi\hbar^2} \, p(E) \ . \tag{9}$$

$p(E)$ is the fraction of allowed space, which ideally is zero below the lowest subband edge and one above. For a random potential, which may be different from the one for surface roughness scattering because of the low electron density, $p(E)$ can be written as:

$$p(E) = 1/2 \ \mathrm{erfc}(-E/(\sqrt{2}\sigma)) \tag{10}$$

with σ a characteristic value for the amount of band tailing. When the Fermi level approaches the subband edge, the small value of the density of states leads to a smaller screening wavevector and therefore to a decrease of mobility.

The observed temperature dependence of the mobility is fully determined by the static dielectric response function appearing in Eqn. 3. Its value at $T = 0$ in the absence of collisional broadening $\epsilon(q,0,0)$ has been calculated by Stern[12]. At $q = 2k_F$ the polarizability shows a sharp edge. The dielectric response is related to the compressibility g_0 of the electron gas, which for $T = 0$ is equal to the density of states of the inversion layer, $2m^*/\pi\hbar^2$. Any broadening of the electron distribution changes the polarizability[13-15]. One of the important consequences is the rounding of the sharp corner in the polarizability at $q \approx 2k_F$ and hence a reduced screening of the scattering potentials.

Temperature is the dominant broadening mechanism. Maldague[16] has derived an expression for the temperature dependence of the compressibility:

$$g_0(q,T,E_F) = 1/(4kT) \int_0^\infty dE \, \frac{g_0(q,0,E)}{\cosh^2[(E_F - E)/2kT]} \quad .$$

(11)

Based on this expression Stern has calculated the explicit
temperature dependence of the mobility numerically leading to a
linear decrease of mobility with temperature. An analytic
expression of $\epsilon(q,T,0)$ has been derived by Gold and
Dolgopolov[17].

At low electron density in the presence of band tailing,
the equations describing the broadening dependence of the
dielectric response lose their validity. The simultaneous
application of band tailing as well as broadening by collisions
and temperature has not been fully established. We have assumed
that smearing of the subband edge introduces no additional
temperature dependence and we will use the model of band tailing
in combination with the broadening dependent dielectric response
for a comparison with the mobility measurements in this range of
electron densities.

4.   COMPARISON OF THEORY AND EXPERIMENT

To compare predictions of Eqn. 3 with experiment (i.e. Fig.
1), it is necessary to take into account the full broadening
dependence of the response function at a particular temperature.
For convenience we proceed by constructing a zero-broadening
mobility curve by extrapolating the data to T = 0, using the
linear temperature dependence of the mobility[1]. This allows
easier extraction of the scattering potentials, and gives the
possibility to compare the experiment to theories of which the
temperature dependence is not known explicitly. Once the
strength of the scattering potentials is determined, curves at
T > 0 will be evaluated.

At high electron densities surface roughness scattering
determines the mobility almost completely. In Fig. 3 the
measured mobilities extrapolated to T = 0 are given and compared
with calculated mobilities. Figure 3 shows the results of a
calculation assuming the usual Gaussian (Eqn. 7) as well as
exponential correlation (Eqn. 8) to model the roughness . In
this calculation only Coulomb scattering and surface roughness
scattering are used, ignoring deviations at the low electron
density edge. As is evident from the figure the description of
the mobility in these samples is in much better agreement if
surface roughness is modeled by exponential instead of Gaussian
correlation. The latter can only describe the data in a limited
part of the range of electron densities yielding a poor fit in
other ranges. Changing the parameters does not improve the fit,
it only moves the range of electron densities where the agreement is reasonable. For exponential correlation the agreement
is not perfect but the deviations from the measurement are far
less pronounced. As is evident from the parameters the correlation length and rms value of the roughness amplitude are
larger for exponential correlation than for Gaussian correlation. The best fit is obtained for L = 70 Å, which is much
larger than the numbers quoted for low mobility samples.
Similar large values for the correlation lengths have been found

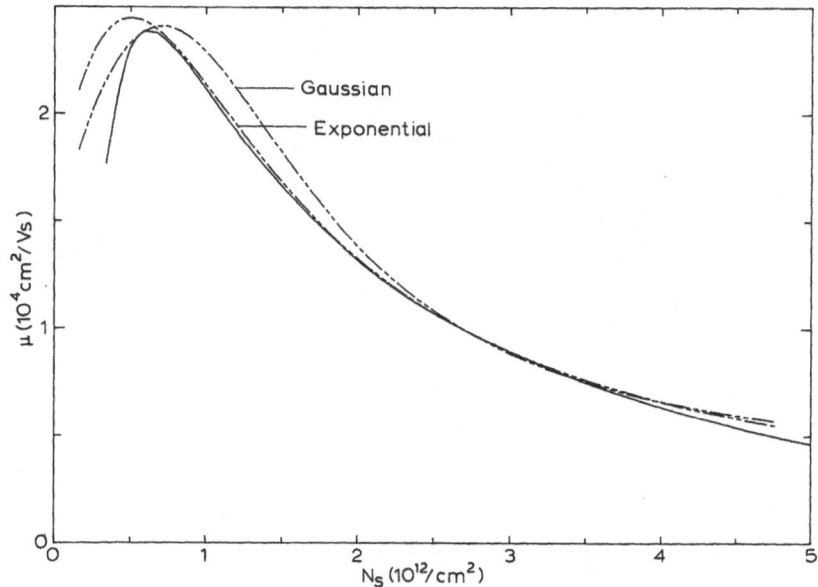

Fig. 3. A comparison of the experimental data shown in
        Fig. 1 with two different roughness models.
        Clearly, exponential roughness (with L = 70 Å, Δ
        = 3.8 Å and N_C = 2.4 10^{10} cm^{-2}) is in much
        better agreement with the data than Gaussian
        roughness (with L = 40 Å, Δ = 2.6 Å and N_C =
        2.8 10^{10} cm^{-2}).

for the high mobility MOSFETs studied by Vyrodov et al.[18],
although they have used Gaussian correlation to extract these
numbers. Since exponential correlation generally leads to
higher values of the correlation length we conclude that a large
correlation length appears to be characteristic for high
mobility samples in general.

The deviations at low electron density have been treated by
assuming a certain amount of band tailing. The result is shown
in Fig. 4 for T = 0. It is assumed that all Coulomb centers are
located at the interface. If we allow the centers to be located
a small distance from the interface, for example 4 Å in the
oxide, a good fit requires a density of Coulomb centers of
N_C = 3.0 10^{10} cm^{-2}, while the amount of band tailing is lowered
to σ = 1.55 meV. As one would expect the precise values are
fairly sensitive to assumptions made about the location of the
charged centers.

The temperature dependence observed in high mobility
samples is only contained in the dielectric response function
(Eqn. 3). We have used the analytic expressions derived by Gold
and Dolgopolov[17] to find from the T = 0 parameters the
mobility at T = 1.5, 2.5 and 4.2 K. For a given electron
density we generate the values at finite temperatures. The
resulting curves are also shown in Fig. 4 together with the
experimental curves. Over the entire range of electron density
the theory follows experiment rather closely. However, to
achieve agreement at the low electron densities, both for the
absolute value of the scattering rate as well as its temperature

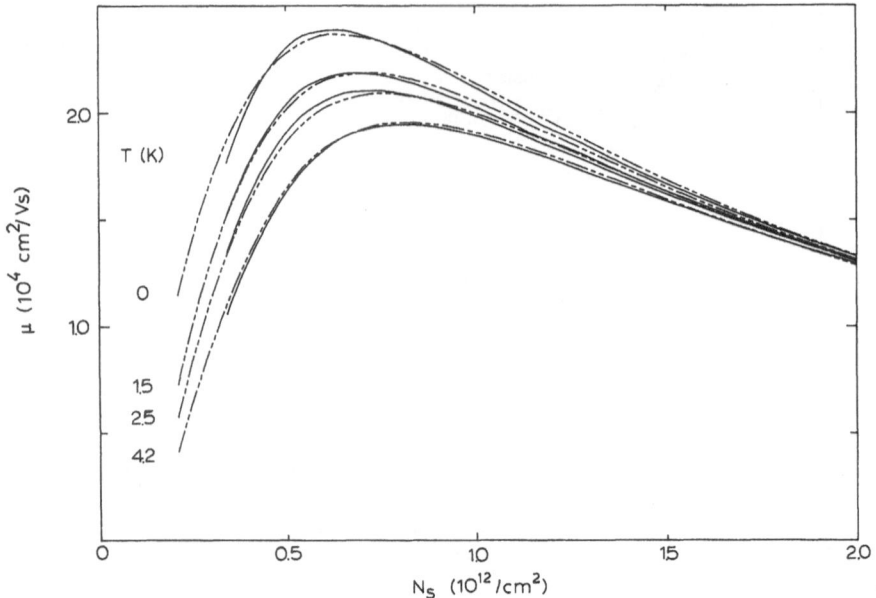

Fig. 4. Measured mobility data compared to theory. The
solid line, labeled T = 0, represents a linear
extrapolation of the finite temperature data.
The dashed-dotted lines are theoretical
dependences taking a surface roughness
correlation length of 70 Å and an rms value of
3.8 Å, whereas the Coulomb density equals 2.4
10^{10} cm^{-2}. The band tailing parameter has been
fixed at 1.8 meV.

dependence, the reduction in the density of states as contained
in the band tailing parameter is needed.

5.   CONTROL OF MOBILITY LIMITING FACTORS

Although the lower electron density side will require
further analysis, in general the dependence of mobility on
carrier density and temperature appears to be well understood by
assuming only elastic scattering due to surface roughness and to
charged interface states i.e. Coulomb centers. In particular
the reappearance of a temperature dependence in the mobility for
the high mobility samples was readily understood as due the
temperature broadening of the dielectric response function. In
our own work we find that specifically for high mobility samples
the surface roughness is best described by exponential
correlation, which is informative about the nature of the
interface. The interesting question is whether this understanding can be related to the technological control of both
surface roughness and the density of Coulomb centers used to
achieve high mobilities.

As far as we have been able to find out nobody has
performed a systematic study of the relationship between process
steps and low temperature mobilities exceeding 20,000 cm^2/Vs.
In general most researchers emphasize that mobilities exceeding

30,000 cm^2/Vs are not obtained routinely. A carefully executed program of fabrication leads to a large number of devices with mobilities close to 20,000 cm^2/Vs or just above with occasionally a few with higher mobilities. Many of the papers on high mobility MOSFETs are based on just a few lucky samples. Usually, one has varied certain parameters within the set of equipment available. Yet it appears that the basic ingredients are quite comparable. One starts with a sufficiently low doping level to guarantee that scattering from ionized donors is negligible. This is already the case for resistivities above 10 Ωcm. Having a good material to start with, the mobility will be determined by the technology and the handling. Control of the density of Coulomb centers as well as the interface roughness is needed, which may be related.

A rough surface, for example generated by wet oxidation, leads invariably to a reduced mobility. For high mobility samples the technological rule is that one should use a slowly growing, usually dry, oxidation process. In addition it is strongly recommended to try to remove the initial surface roughness by repeated oxidations and etching steps. The observed exponential correlation points towards an interface which consists of relatively large terraces separated by atomic steps. It is interesting to note that terraces of about 100 Å size are reported for well polished virgin wafers[19], although this situation might easily change during the growth of a 100 nm oxide. It would be of interest to check the interface roughness for high mobility samples by a HRTEM analysis. In this respect the observed variability over the wafer is of interest. It might be possible that certain parts, after repeated oxidation and etching steps, are much smoother than other parts.

The other side of the coin is the density of Coulomb centers. In a description of the scattering by Coulomb centers it is usually assumed that they are located at the interface. In principle it may be possible that the density of Coulomb centers varies with the distance from the interface. However, the contribution to the scattering decays exponentially with the distance from the interface and the original assumption that they are located at a fixed position with respect to the interface may be a fairly good approximation. However, there is no reason why they should be located at the interface itself. In view of a recent structural analysis[20,21] it might be that they will reside a few Ångstroms from the interface where the crystalline SiO$_2$ goes over into the amorphous SiO$_2$.

Furthermore it is unknown which feature of the interface, and of the oxide, contributes to the density of Coulomb centers and how to control it technologically. Part of the density may be caused by particular defects. However, from a variety of experiments it is clear that a certain fraction is due to the sodium impurities built in during the various steps of cleaning and etching the wafer and growing the oxide. It is found that application of a voltage at room temperature leads to a deterioration of the mobility because of the drift of the mobile ions in the oxide towards the interface. It is reported that the highest mobility devices (> 40,000 cm^2/Vs) are not stable if stored at room temperature. Kukushkin reports that devices of 40,000 cm^2/Vs had deteriorated, after one year of storage, to 30,000 cm^2/Vs. Storage at liquid nitrogen temperatures is recommended. For the Siemens samples it has been reported that the mobility varies from 30,000 to 20,000 cm²/Vs from one cooldown to another. The mobility can even change significantly due to relatively small voltage spikes applied when the sample is at liquid helium temperatures.

6. PROSPECTS

The final question is whether the experience built up over the past decade does provide a route to further enhance the mobilities in Si inversion layers. A change in interface roughness has only a minor effect on the peak mobility. An increase in the correlation length by a factor of 7 does not alter the peak mobility very much, although it has a drastic effect on the mobility at the higher electron densities (Fig. 5). A change in the average roughness has a much stronger effect. However, the latter quantity has presumably a lower limit of the size of one atomic step.

In practice a high mobility is achieved by a lowering of the density of Coulomb centers. This is illustrated in Fig. 2 where it is shown that a high mobility is generally associated with a shift of the peak mobility to a lower electron density. It is known that the lowest density of mobile sodium atoms, presumably the dominant contaminant due to clean room

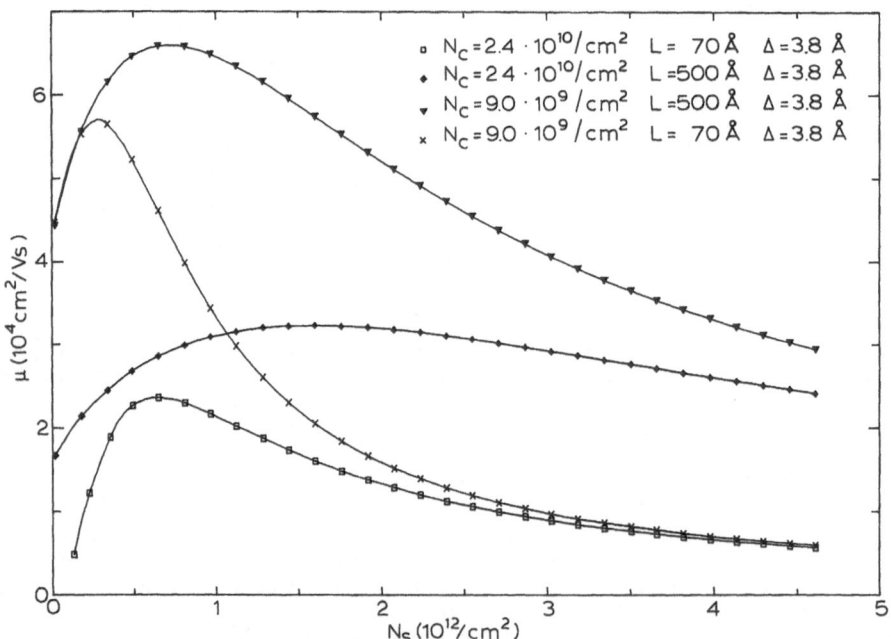

Fig. 5. Predicted mobility-curves taking the parameters for the experimental data (open squares) as a starting point. A change in roughness, which may be realized with MBE-flattened samples affects the mobility primarily at the high electron densities. A minor reduction in the density of Coulomb centers has a strong effect on the peak mobility.

processing, is about $3 \ 10^9$ cm^{-2}. If we compare this number with the values obtained for the samples shown in Figs. 1 and 4 we conclude that in the best high mobility samples the Coulomb centers are not due to mobile sodium but are built-in near the Si-SiO$_2$ interface itself. Given this observation it is understandable that improvements in clean processing have not provided a continued increase in mobilities. At present the mobility appears to be stagnant at 30,000 cm^2/Vs. In Fig. 5 we show that a reduction of the density of Coulomb centers by a factor of 3 would, for the same roughness parameters, lead to a substantial increase in mobility.

This hypothetical MOSFET has a mobility of 70,000 cm^2/Vs. As input parameters we have used the interesting results on oxides grown on MBE-flattened wafers. We have assumed that the interface roughness can be reduced to atomic terraces of 50 nm size with step heights of 0.38 nm. We have kept the Coulomb centers located at the interface. The number of Coulomb centers has been reduced to obtain the desired mobility. This means a value of $N_C = 9 \ 10^9$ cm^{-2}, which is a factor of 3 lower than the value found for the present high mobility MOSFETs. It is hard to believe that such changes in the numbers are impossible to reach.

Acknowledgments

We thank R.G. Wheeler, G. Dorda, V.I. Kukushkin, V.M. Pudalov and J.E. Furneaux for communicating their experience with high mobility MOSFETs.

REFERENCES

1. G.H. Kruithof, T.M. Klapwijk and S. Bakker, to be published.
2. K.M. Cham and R.G. Wheeler, **Phys. Rev. Lett.**, 44:1472 (1980)
3. J.E. Furneaux, D.A. Syphers, J.S. Brooks, G.M. Schmiedeshoff, R.G. Wheeler and P.J. Stiles, **Surf. Science**, 170:154 (1986)
4. R.P. Smith, Thesis Brown University (1986)
5. I.V. Kukushkin, V.B. Timofeev and P.A. Cheremnykh, **Zh. Eksp. Teor. Fiz.**, 87:2223 (1984) [**Sov. Phys. JETP**, 170:154 (1986)]
6. M.A. Vernikov, L.M. Pazinich, V.M. Pudalov and S.G. Semenchinskii, **Elektronnaya Tekhninka, Ser.2, Poluprovodinkovye pribory**, 6:27 (1985)
7. T. Ando, A.B. Fowler and F. Stern, **Rev. Mod. Phys.**, 54:437 (1982)
8. F. Stern and W.E. Howard, **Phys. Rev.**, 163:816 (1967)
9. T. Ando, **J. Phys. Soc. Jpn.**, 43:1616 (1977)
10. S.M. Goodnick, D.K. Ferry, C.W. Wilmsen, Z. Liliental, D. Fathy and O.L. Krivanek, **Phys. Rev.B**, 32:8171 (1985)
11. E. Arnold, **Surface Science**, 58:60 (1976)
12. F. Stern, **Phys. Rev. Lett.**, 18:546 (1967)
13. F. Stern, **Phys. Rev. Lett.**, 44:1469 (1980)
14. T. Ando, **J. Phys. Soc. Jpn.**, 51:3215 (1982)
15. S. Das Sarma, **Phys. Rev. Lett.**, 50:211 (1983)
16. P.F. Maldague, **Surface Science**, 73:296 (1978)
17. A. Gold and V.T. Dolgopolov, **Phys. Rev.B**, 33:1076 (1986)
18. E.A. Vyrodov, V.T. Dolgopolov, C.I. Dorozhkin and N.B.

Zhitenev, **Zh. Eksp. Teor. Fiz.**, 94:234 (1987) [**Sov. Phys. JETP**, 67:998 (1988)]

19. P.O. Hahn, M. Grundner, A. Schnegg and H. Jacob, p.401 **in:** "The Physics and Chemistry of SiO_2 and the $Si-SiO_2$ Interface", C.R. Helms and B.E. Deal, eds., Plenum, New York (1988)

20. A. Ourmazd, D.W. Taylor, J.A. Rentschler and J. Bevk, **Phys. Rev. Lett.**, 59:213 (1987)

21. A. Ourmazd and J. Bevk, p.189 **in:** "The Physics and Chemistry of SiO_2 and the $Si-SiO_2$ Interface", C.R. Helms and B.E. Deal, eds., Plenum, New York (1988)

22. P.H. Fuoss, L.J. Norton, S. Brennan and A. Fischer-Colbrie, **Phys. Rev. Lett.**, 60:600 (1988)

23. M.G. Gavrilov, Z.H. Kvon,, I.V. Kukushkin and V.B. Timofeev, **Pis'ma Zh. Eksp. Teor. Fiz.**, 39:420 (1984), [**JETP Lett.**, 39:420 (1984)]

24. I.V. Kukushkin and V.B. Timofeev, **Pis'ma Zh. Eksp. Teor. Fiz.**, 44:179 (1986) [**JETP Lett.**, 44:228 (1986)]

25. I.V. Kukushkin, **Pis'ma Zh. Eksp. Teor. Fiz.**, 45:222 (1987) [**JETP Lett.**, 45:276 (1987)]

26. J.E. Furneaux, D.A. Syphers, J.S. Brooks, G.M. Schmiedeshoff, R.G. Wheeler, G. Dorda, R.P. Smith and P.J. Stiles, **in:** "Proc. of the 18th Conf. on the Physics of Semiconductors", O. Engström, ed., World Scientific, Singapore (1987)

27. V.M. Pudalov and S.G. Semenchinskii, **Pis'ma Zh. Eksp. Teor. Fiz.**, 39:143 (1984), [**JETP Lett.**, 39:170 (1984)]

28. V.M. Pudalov, S.G. Semenchinskii and V.S. Edelman, **Solid State Comm.**, 51:713 (1984)

29. S.G. Semenchinskii, **Zh. Eksp. Teor. Fiz.**, 91:1804 (1986), [**Sov. Phys. JETP**, 64:1068 (1988)]

30. P.C. van Son, G.H. Kruithof and T.M. Klapwijk, **Surf. Sci.**, 229:57 (1990)

THEORY OF ELECTRONIC TRANSPORT IN LOW DIMENSIONAL SEMICONDUCTOR

MICROSTRUCTURES

S. Das Sarma

Center for Theoretical Physics and Joint Program
for Advanced Electronic Materials
Department of Physics, University of Maryland
College Park, Maryland 20742-4111, USA

ABSTRACT

I discuss complementary theoretical aspects of electronic
transport in low dimensional semiconductor microstructures
emphasizing, in particular, novel transport mechanisms and
quantitative comparisons with experiments. The two specific
topics discussed are: a) the temperature dependence of low
temperature (4-40 K) ohmic mobility in two-dimensional
modulation-doped ultra high mobility GaAs heterostructures; and,
b) the conductance quantization phenomenon in the **ballistic**
transport properties of quasi-one-dimensional constrictions or
quantum point contacts. These two examples, respectively, bring
out characteristic features of the physical aspects of diffusive
and ballistic transport in low dimensional microstructures. I
also present some calculated numerical results for elastic and
inelastic mean free paths in low dimensional GaAs structures.

1.   INTRODUCTION

    Electronic transport in low dimensional semiconductor
systems has been studied extensively during the last twenty-five
years ever since the discovery of the quantized two-dimensional
electron gas in silicon inversion layers[1,2]. In the last ten
years attention has mostly been focused on electronic transport
in very high mobility two-dimensional structures based on
modulation-doped GaAs-$Al_xGa_{1-x}As$ (or, other related III-V
semiconductor systems) heterojunctions, quantum wells, and
superlattices. Recently, improved growth and fabrication
techniques have allowed creation of various types of one-
dimensional quantum wire or zero-dimensional quantum dot
structures. Experimental studies of electronic transport in
such novel systems is a very active current area of research.

    In this paper we will discuss a number of theoretical
aspects of electronic transport in low dimensional semiconductor
systems. Since the subject of transport in semiconductor
microstructures is truly a vast one, we concentrate here on two

Condensed Systems of Low Dimensionality
Edited by J. L. Beeby *et al.*, Plenum Press, New York, 1991

interesting and complementary issues involving fundamentally different aspects of low dimensional electronic transport, namely diffusive and ballistic conduction at low temperatures in very high mobility GaAs microstructures.

In the following we discuss theoretical issues pertaining to the following aspects of electronic transport properties of low dimensional systems: a) Acoustic phonon scattering effect[3-7] on the ohmic mobility of a two-dimensional electron gas (2DEG) in high mobility, modulation-doped GaAs heterostructures (discussed in Section 2); and, b) the effects of temperature, disorder, and constriction geometry on the ballistic conductance quantization phenomenon[8-12] in one-dimensional quantum point contact structures (discussed in Section 3).

The above two topics, to be discussed below in Sections 2 and 3 respectively, represent totally different aspects of low dimensional electronic transport. The first topic deals with the temperature dependence of the low temperature (4-40 K) two-dimensional mobility of a high mobility modulation-doped GaAs heterojunction. We get excellent quantitative agreement with experimental results and calculate the absolute upper bound on the maximum achievable mobility in high mobility GaAs hetero-structures. This topic (a), discussed in Section 2, is thus an example of conventional diffusive transport in a high mobility two-dimensional system with the emphasis on a particular scattering mechanism (namely, the acoustic phonon scattering) limiting the two-dimensional mobility and its complete quantitative understanding.

The second topic (b), discussed in Section 3, deals with ballistic transport in narrow constrictions or quantum point contacts made by using ultramicrolithography (or, even nano-lithography) from high mobility two-dimensional electron gas systems in GaAs heterostructures. Our main emphasis here is to theoretically investigate how various physical effects such as disorder and temperature affect the recently observed ballistic conductance quantization phenomenon in narrow constrictions. We also study numerically the effect of the constriction geometry (i.e. its shape and size) on the quantization. This topic is an example of coherent quantum transport in low dimensional semiconductor structures and emphasizes qualitative and semi-quantitative understanding of the problem.

Finally, in Section 4 we give some numerical results for two important length scales in GaAs heterostructures, both of which have some bearing on topics (a) and (b) above. These lengths are the inelastic and the elastic mean free paths in GaAs heterostructure. Both electron-electron and electron-LO phonon interaction contribute to inelastic scattering and an accurate knowledge of inelastic scattering length is important in understanding ballistic hot electron transistor operation and electron focusing experiments. The elastic mean free path arises mainly from impurity scattering and is the ultimate mobility limiting mechanism in high mobility structures at the lowest temperatures.

This brief introduction serves to show that there are many interesting and complementary aspects to electronic transport in semiconductor microstructures, not all of which can obviously be mentioned here. We restrict ourselves to some of the very

262

recent theoretical work done in this general subject area in our group. For lack of space we do not consider the situation in the presence of an external magnetic field which, of course, is an interesting subject by itself.

2. ACOUSTIC PHONON SCATTERING IN 2DEG

Ever since the invention of the modulation doping technique, transport properties of two-dimensional electron gas (2DEG) systems in high mobility, modulation doped $Al_xGa_{1-x}As/$ GaAs heterojunctions have been a subject of great interest, from both the technological and fundamental points of view. Electron mobilities, μ, higher than 10^6 cm^2/Vs are now routinely obtained and, very recently, mobilities in excess of 10^7 cm^3/Vs have been achieved[3,7]. Scattering mechanisms limiting mobility in such structures are of both intrinsic and extrinsic types. Extrinsic effects associated with charged impurity scattering can be reduced by improving growth and fabrication techniques and the steady improvement in mobility over the last decade is a result of improving the purity of the GaAs material and of using increasingly wide spacer thicknesses so as to reduce scattering by the remote dopants. The ultimate intrinsic limit on the highest achievable mobility at a particular temperature is, however, set by acoustic phonon scattering which, unlike Coulomb scattering by impurities, cannot be eliminated. Our focus in this paper is on a thorough understanding of the acoustic phonon scattering limited mobility, μ_{ac}, of a GaAs based 2DEG.

In the process of developing an accurate quantitative theory for phonon scattering, we resolve a number of controversies[13-15] in the literature regarding the correct values of various electron-phonon coupling constants in GaAs heterostructures and obtain[16] excellent agreement with the available experimental results for high mobility samples, in the temperature range of 4-40 K, and the electron density range of $(0.3-6) \times 10^{11}$ cm^{-2}. Since we do not take into acount impurity scattering effects, we are necessarily dealing with the temperature dependence in rather high mobility structures for which the zero temperature mobilities are $\mu_0 > 10^5$ cm^2/Vs, and, at intermediate temperature (T ~ 4-40 K) which are not too low or high. Above 40 K, scattering by LO phonons starts becoming important. More details of our theory can be found in Ref. 16.

The reason that such a great deal of experimental and theoretical attention has been focused on the ohmic mobility of a 2DEG is that the temperature dependent mobility has important information about scattering mechanisms operative in the system. For carrier densities, $N_s = (1-6) \times 10^{11}$ cm^{-2}, it is experiment-ally[3-7] observed that in the temperature range 4-40 K the reciprocal mobility increases linearly with temperature, $\mu^{-1} = \mu_0^{-1} + \alpha T$. The slope, α, increases with N_s independently of μ_0. At low temperatures the electron mobility is limited by the deformation potential and piezoelectric coupled acoustic mode phonon scattering, together with ionized impurity scattering. In a typical $Al_xGa_{1-x}As/GaAs$ modulation doped heterojunction the electron wavefunction is mainly confined in GaAs so that the influence of alloy disorder scattering can be considered to be negligible. Interface roughness scattering also plays a rather minor role due to the smoothness of the interface in high quality heterojunctions.

Undoubtedly, the current extensive investigation of the mobility in a 2DEG is partly motivated by the controversy surrounding the exact value of the deformation potential constant, D, in GaAs. In particular, it is well established that for T < 40 K scattering from ionized impurities is independent of temperature in *ultra-high* mobility samples. This eliminates the need to quantify the ionized impurity scattering, and means that any temperature dependence of the mobility in ultra-high mobility samples can be solely attributed to acoustic mode phonons. Reasonable agreement of the calculated temperature dependence of mobility with experimental results[4-7] is obtained only by assuming a value of D = 12-13.5 eV, which is approximately 50-100% more than the generally accepted value in bulk GaAs (D = 7 eV). Enhanced values of D have also been independently inferred from an analysis of the energy relaxation data in GaAs heterojunctions. However, in some other somewhat more approximate[14] studies, which neglect screening effects, the temperature dependence of mobility is satisfactorily explained using a value of D = 7 eV. We show our calculated linear temperature coefficient α as a function of electron density N_s in Fig. 1 and compare it to existing experimental results.

We calculate the acoustic phonon scattering limited mobility μ_{ac} within the following model and approximations: (1) The electrons are assumed to occupy only the lowest quantum subband (which is a good approximation for $N_s \leq 6 \times 10^{11}$ cm^{-2} and T ≤ 40 K). (2) The subband wavefunction is approximated by

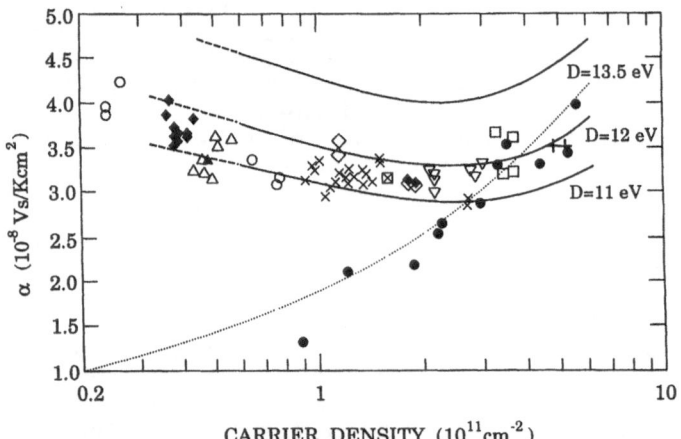

Fig. 1. Temperature coefficient, α as a function of the electron sheet density N_s. The solid lines are the theoretical curves (Ref. 16) for three different values of the deformation potential constant D. These curves are extrapolated to lower densities by the dashed lines. The dotted line is the theoretical (Ref. 13) curve for D = 13.5 eV, and a vanishing depletion density without any temperature dependent effects. Results reported by Mendez, Price and Heiblum (Ref. 4) are plotted by full circles; other data points are from Harris et al. (Ref. 3).

a standard variational wavefunction which should work well for the ground state of interest here. (3) We assume, consistent with all the previous work in the field, the bulk GaAs acoustic phonons to be the only phonons contributing to scattering. (4) The electron acoustic phonon scattering is treated in the usual quasi-elastic, equipartition approximation which should be an excellent approximation above the Bloch-Grüneisen range (i.e. as long as $k_B T$ is larger than the energy of an acoustic phonon with $2k_F$ wavevector). (5) The deformation potential and piezo-electric coupled acoustic phonon scattering rates are calculated in the standard momentum relaxation time approximation[13] with full wavevector and temperature dependent static screening[2] of the electron-phonon interaction included in RPA.

Once the individual scattering rates, τ_{DP}^{-1} and τ_{PE}^{-1} respectively, for the deformation potential and piezoelectric coupling are obtained from Fermi's golden rule, the mobility can be calculated from its definition[2] (without making the Matthiessen rule approximation). The ohmic mobility is $\mu_{ac} = e<\tau>/m^*$, where $1/\tau = 1/\tau_{DP} + 1/\tau_{PE}$, and the thermal average, using the Fermi distribution $f_0(E)$, is given by,

$$<\tau> = \int_0^\infty \tau(E)E\left[\frac{df_0(E)}{dE}\right] dE / \int_0^\infty E\left[\frac{df_0(E)}{dE}\right]dE. \qquad (1)$$

We numerically evaluate μ_{ac} in the temperature range $T = 4-40$ K for a range of carrier densities $N_s = (0.3-6) \times 10^{11}$ cm^{-2}. We keep both the temperature dependence of screening and the effect of thermal smearing of the electron energy. We also estimate their relative importance at various densities. If both the temperature dependent effects in the evaluation of $<\tau>$ are suppressed, it follows from the equipartition of electron-phonon scattering that μ_{ac}^{-1} is strictly linearly proportional to T. It turns out that an approximate linear dependence is preserved even in the presence of these two effects in the temperature and density range of interest in this paper.

The combined and isolated effects of the temperature dependence of $\mu_{ac}(T)$ are shown in Fig. 2 for $N_s = 1 \times 10^{11}$ cm^{-2} and 6×10^{11} cm^{-2}. The solid lines represent the theoretical calculations which include both the temperature dependent screening and thermal averaging over energy whereas the thin solid lines exclude both effects and are, therefore, strictly linear in T. As expected, at the higher density $N_s = 6 \times 10^{11}$ cm^{-2}, the temperature dependent effects are not very significant. At high N_s thermal averaging over energy is completely negligible and the effect of the temperature dependence of screening is noticeable only at higher temperatures $T > 20$ K. At the lower density, both the temperature dependent effects are rather important, with the temperature dependent screening effect being stronger than the thermal averaging effect.

We make a linear fit of the thick solid lines (which include both the temperature dependent effects) in the temperature range, $T = 20-40$ K, where we expect equipartition to be valid, and identify its slope with the experimentally

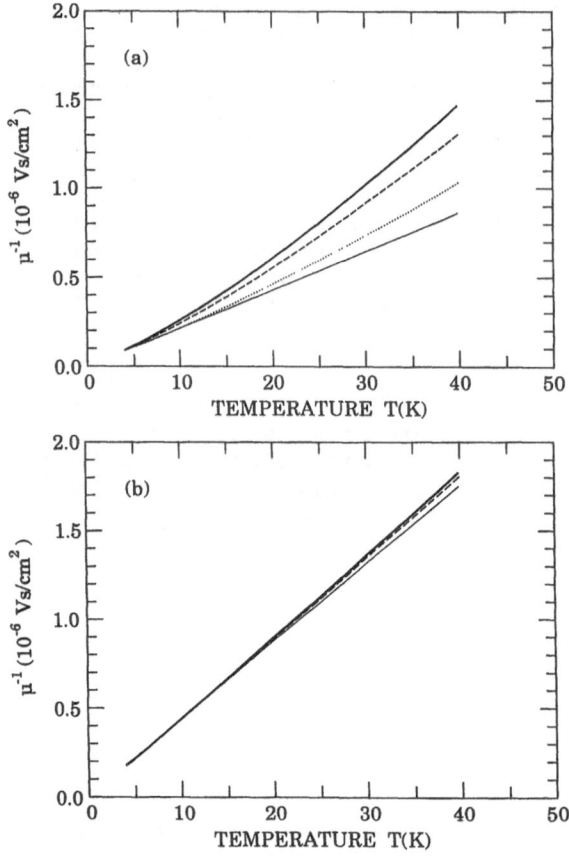

Fig. 2. Calculated temperature dependence of the reciprocal mobility for two different electron sheet densities. (a) $N_s = 1 \times 10^{11}$ cm^{-2} and (b) $N_s = 6 \times 10^{11}$ cm^{-2}. The results given by the thick solid lines include temperature dependent screening and thermal averaging over energy. The dashed lines represent calculations which do not have the averaging over energy. The dotted lines represent calculations which ignore the temperature dependence of screening. The thin lines correspond to calculations without any temperature dependent effects. In (b) the dotted line coincides with the thin line. $N_{depl} = 5 \times 10^{10}$ cm^{-3} and D = 13.5 eV. (From Ref. 16).

measured temperature coefficient, $\alpha = d\mu_{ac}^{-1}(T)/dT$. The results of our theoretical calculation of $\alpha(N_s)$ for the density range $N_s = (0.6-6) \times 10^{11}$ cm^{-2} are given in Fig. 1 for three different values of D. Assuming a value of $N_{depl} = 5 \times 10^{10}$ cm^{-2} for the depletion density, the optimal value for D is slightly less than 12 eV. Although the theoretical curve for $\alpha(N_s)$ which ignores both the temperature dependent effects agrees fairly well with the experimental data at high densities above $N_s \sim 3 \times 10^{11}$ cm^{-2}, the fit becomes increasingly poor at lower densities. The temperature dependencies arising from screening and thermal averaging significantly augment μ_{ac}^{-1} at high temperatures,

266

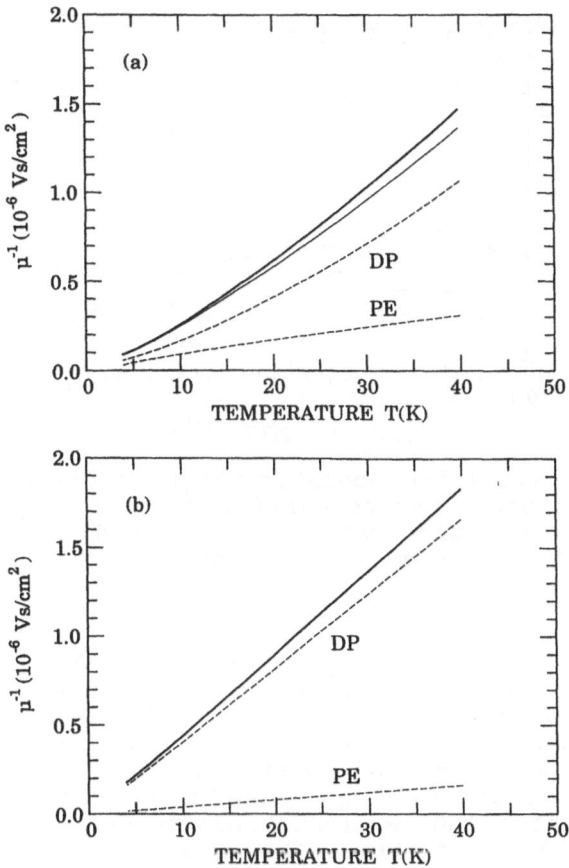

Fig. 3. Calculated temperature dependence of the
reciprocal mobility for two different electron
sheet densities. (a) $N_s = 1 \times 10^{11}$ cm^{-2} and
(b) $N_s = 6 \times 10^{11}$ cm^{-2}. Results given by the
dashed lines are for the deformation potential
and peizoelectric coupled acoustic-mode phonons.
Their sum is given by the thin solid lines, its
discrepancy from the thick solid lines shows the
departure from Matthiessen's rule. In (b) the
thin and thick solid lines coincide. $N_{depl} =$
5×10^{10} cm^{-2} and D = 13.5 eV. (From Ref. 16).

increasing the temperature coefficient α as the density is
decreased, giving rise to the nonmonotonic behavior of $\alpha(N_s)$
seen in Fig. 1. Our claim is that this explains the rather
curious minimum in $\alpha(N_s)$ observed in the experimental data[3] of
Harris et al.

In Fig. 3 we show the individual temperature dependencies
of the reciprocal mobility limited by the deformation potential
and piezoelectric coupled acoustic mode scattering. Their sum
is compared to the total acoustic phonon scattering limited
reciprocal mobility to check the validity of Matthiessen's rule.
At high densities and low temperatures, Matthiessen's rule holds
extremely well, but at lower densities and high temperatures it
does not work so well, although the effect on $\alpha = d\mu_{ac}^{-1}(T)/dT$ is

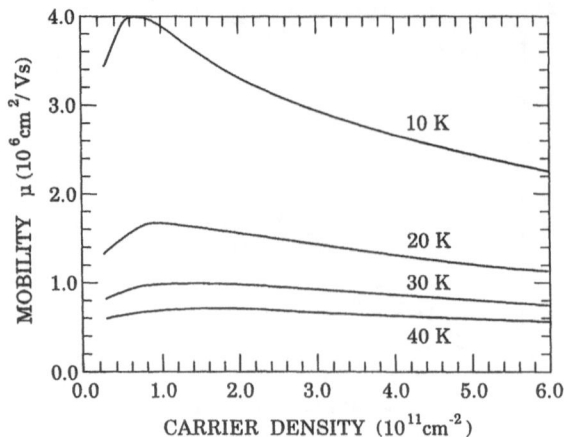

Fig. 4. The calculated acoustic phonon limited mobility,
μ_{ac} plotted versus carrier density N_s for four
different temperatures. $N_{depl} = 5 \times 10^{10}$ cm^{-2}
and D = 13.5 eV. (From Ref. 16.)

small (less than 10%). Finally in Fig. 4 we show the density
dependence of the acoustic phonon scattering limited mobility
itself for four different temperatures. The maximum in $\mu_{ac}(N_s)$
for a given temperature arises from a competition between the
screening effect which dominated at lower N_s (and tends to
increase μ_{ac} with increasing N_s), and the matrix element effect
which dominates at higher N_s (and tends to decrease μ_{ac} with
increasing N_s). Both of these effects go down with increasing
temperature, producing a very weak density dependent $\mu_{ac}(N_s)$ at
higher T. The approximate maximum intrinisic mobility at
T = 10 K is 4×10^6 cm^2/Vs occurring around $N_s \sim 8.5 \times 10^{10}$ cm^{-2}.
We point out that the results shown in Fig. 4 apply only when
impurity scattering has been completely eliminated. In real
systems, there will be an additional N_s dependence of μ arising
from screening effects associated with the charged impurity
scattering.

In conclusion, we have shown[16] that a careful consideration of the temperature dependent effects of screening and
thermal averaging can quantitatively explain the temperature and
the density dependence of the acoustic phonon scattering limited
mobility in the high mobility 2DEG in GaAs heterostructure. In
particular, we show that careful consideration of temperature
dependent screening and thermal averaging explains quantitatively the experimentally observed minimum[3] in the
temperature coefficient, $\alpha(N_s)$, of the electron mobility in GaAs
heterolayer. We also provide[16] an absolute upper limit on the
maximum possible intrinsic mobility in a GaAs based 2DEG in the
4-40 K temperature regime.

3. QUANTUM BALLISTIC CONDUCTION IN POINT CONTACTS

It is straightforward to show that for ballistic conduction
in a strictly quasi-one-dimensional (1D) system, conductance
will be quantized as G = n(2e^2/h) where n is the number of
occupied 1D subbands (and the factor of 2 arises from spin).

The total current J in 1D ballistic transport at T = 0 is given by

$$J = \sum_n J_n = \sum_n e v_{nk} D_n (E_F) e \, V \qquad (2)$$

where J_n is the current in individual subbands, v_{nk} is the electronic velocity, D_n is the 1D subband density of states at the Fermi energy E_F and V is the applied voltage. The density of states D_n is defined to be

$$D_n = \frac{2}{2\pi} \left(\frac{\partial E_{nk}}{\partial k} \right)^{-1} \Bigg|_{E_F = E_{nk}} , \qquad (3)$$

where

$$E_{nk} = E_n + \frac{\hbar^2 k^2}{2m} \qquad (4)$$

is the subband energy. Using the fact that

$$v_{nk} = \hbar^{-1} \frac{\partial E_{nk}}{\partial k} , \qquad (5)$$

$$v_{nk} D_n = \frac{2}{h} , \qquad (6)$$

which leads to

$$J = \left[\sum_n \frac{2e^2}{h} \right] V$$

i.e.

$$G = J/V = 2ne^2/h. \qquad (7)$$

An equivalent description of the quantization is to use the two probe Landauer formula

$$G = \frac{2e^2}{h} \, \mathrm{Tr} \, tt^+ , \qquad (8)$$

where t is the transmission matrix of the constriction. In the absence of any scattering, t is given trivially for the system by

$$t_{ij} = \delta_{ij} , \qquad (9)$$

leading to the quantized conductance $G = 2ne^2/h$ given by Eqn. 7.

The above discussion shows that the conductance quantization phenomenon should occur in a one-dimensional system provided the transport is truly ballistic in nature. This complete absence of any scattering is, of course, an idealization because in real systems there is always some disorder.

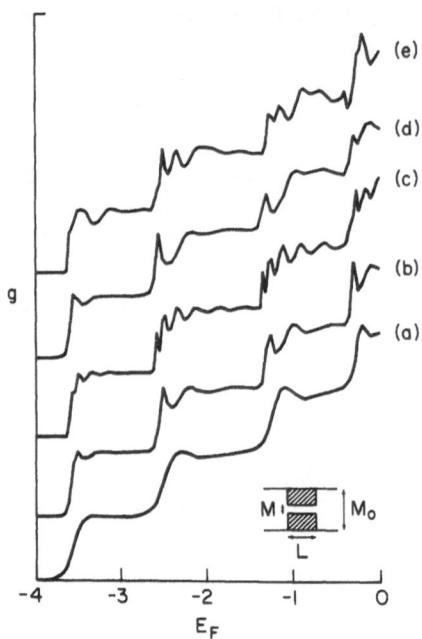

Fig. 5. Shows the calculated dimensionless conductance g
for a sharp and rectangular wide-narrow-wide
geometry constriction (the geometry shown in
bottom right hand corner) as a function of the
chemical potential E_F: (a) L = 4, W = 0.2; (b) L
= 8, W = 0.2; (c) L= 16, W = 0.2; (d) L = 128, W
= 0.2; (e) L = 16, W = 0.0. All have M = 4.
(From Ref. 19.)

In Figs. 5-10 we study numerically the effect of disorder,
temperature, inelastic scattering, and of the finite
constriction size on the conductance quantization phenomenon.
Our calculation is based on a two-dimensional, nearest neighbor,
tight-binding Anderson Hamiltonian on a square lattice[17]. We
take the band width of the model to be 4 so the hopping term for
kinetic energy is being chosen to be unity. Disorder enters as
usual through the diagonal term and is characterized by the
disorder strength parameter W which denotes the width of the
rectangular distribution over which the site diagonal energies
are randomly distributed. We have an additional nonrandom
diagonal term which produces the confinement of the electrons
and models the constriction geometry, mimicking the gate
potential in some suitable way. Our unit of length is the
lattice constant of the tight-binding model and our unit of
energy is the nearest neighbor hopping energy.

     To obtain the system conductance we calculate directly the
conductance of the system by evaluating the Kubo formula using
the recursive Green's function technique following the well
known procedure[18] of Fisher and Lee. This theoretical
technique is standard and has been extensively used and
discussed in the literature. Our calculation[19] is equivalent
to using the two probe Landauer formula for the dimensionless
conductance g:

     $g = G/(2e^2/h) = Tr(tt^+)$.

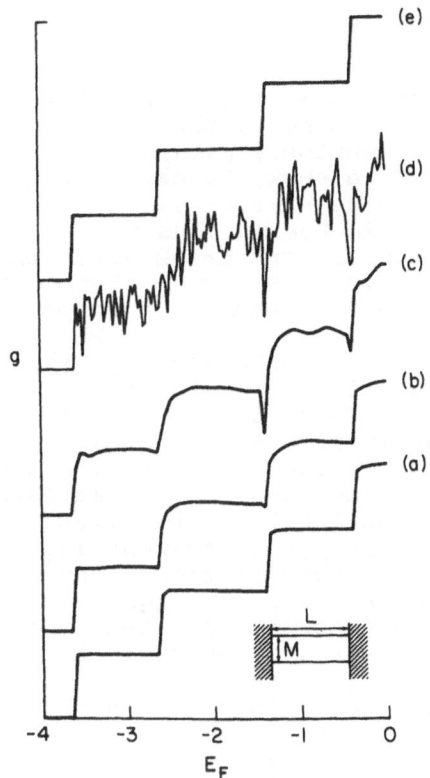

Fig. 6. The same as in Fig. 1 for a narrow one-
dimensional strip geometry (as shown in the
bottom right hand corner) with M = 32 and M_0 =
4: (a) L = 4, W = 0.0; (b) L = 8, W = 0.0;
(c) L = 16, W = 0.0; (d) L = 8, W = 0.2;
(e) L = 16, W = 0.2. (From Ref. 19.)

Temperature effects are introduced in the theory[19] by
convoluting the single particle Green's function (calculated
recursively) with the finite temperature Fermi function and
inelastic scattering effects are introduced by adding a small
imaginary part γ to the energy E at which the Green's function
is calculated (W is the parameter for elastic scattering in the
system). Thus, our calculated conductance, $g(E_F)$, as a function
of the chemical potential E_F depends on five parameters W
(disorder induced elastic scattering strength), T (the
temperature), γ (the inelastic broadening due to, for example,
electron-electron and electron-phonon interactions), L (the
length of the constriction along the direction of transport),
and M (width of the constriction along the transverse dimension
perpendicular to the current flow - the third direction is
neglected here assuming that the electron gas is two-
dimensional). In principle, γ is a function of T, but we
parametrize them independently here. Finally, the shape of the
constriction is the other important consideration which has been
discussed extensively in the literature. Since the shape effect
has been a subject of study in a number of recent theoretical
investigations, we leave that out of this paper for the sake of
brevity, concentrating here instead on the effects of

scattering, system size and finite temperature on the basic quantization phenomenon.

We restrict ourselves in this paper to two idealized constriction geometries, namely, the wide-narrow-wide (inset of Fig. 5) and the one-dimensional strip (inset of Fig. 6) geometries. In Fig. 5 we show the calculated dimensionless conductance g for the wide-narrow-wide geometry as a function of the chemical potential E_F for various values of the constriction length (L) and the elastic scattering strength (W) with T and γ both being zero. The widths M and M_0 of the (narrow) constriction and the (wide) reservoirs respectively are kept fixed for the results shown in Fig. 5. The curves (a) - (c) are truly ballistic with W = 0 (no disorder) whereas (d) and (e) have finite elastic scattering as shown. The structure on the conductance plateaus in Figs. 5a - c arises entirely from quantum resonances or interference associated with the sharp edges of the wide-narrow-wide structure (this quantum resonance structure on the quantized conductance plateaus has been extensively discussed in the theoretical[20] literature on the subject, but has not been yet experimentally seen, presumably because the experimental constriction is smooth and the transport through it is adiabatic). In Figs. 5d - e, the quantum resonance structure is somewhat softened by disorder, but eventually for large disorder, the conductance fluctuation effect destroys the quantization phenomenon. In the intermediate region, structure could arise on the conductance plateaus due to disorder effects.

In Fig. 6 we show the calculated conductance for a strip geometry for $\gamma = T = 0$ and a fixed width M = 4 but for different values of W and L. For the curve (e), $W = \gamma = T = 0$ and, therefore, the quantization is perfect in this geometry as discussed before. For a fixed W = 0.2, the quantization is destroyed progressively as L increases from L = 4 (a) through L = 8 (b) to L = 16 (c). Eventually for L = 128 (d), the system length being much larger than the mean free path, the quantized conductance phenomenon is replaced by the conductance fluctuation phenomenon as one would expect. Thus, one expects best quantization for short constrictions or quantum point contacts, keeping the elastic mean free path much longer than the constriction length.

In Figs. 7 - 10 we show various aspects of our calculated conductance for the strip geometry (of Fig. 6). In Figs. 7, 8, 9, 10 we depict, respectively, the dependence of $g(E_F)$ on variations in elastic (W) and inelastic (γ) scattering strength, the constriction width (M), the constriction length (L), and, the temperature (T), keeping all the other parameters fixed in each figure. one can see that larger values of L, M, W, γ, and T all lead to inferior quantization. The general behavior of our numerical results is in very good qualitative agreement with experimental results.

In this paper we have only considered two extreme geometries for the quantum constriction, namely, the wide-narrow-wide abrupt geometry and the smooth strip geometry. Numerical results for the calculated conductance using the recursive Green's function technique are qualitatively the same in both except that the quantum resonances associated with the sharp edges of the constriction show up only in the wide-narrow-wide structure. We have studied a number of other more

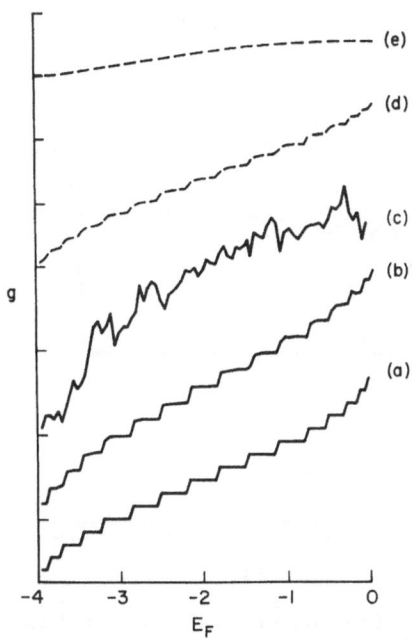

Fig. 7. The dimensionless conductance g as a function of
the chemical potential E_F for fixed L, M and T,
and for various values of elastic and inelastic
scattering strengths (a) W = 0.01, $\gamma = 10^{-6}$; (b)
W = 0.1, $\gamma = 10^{-6}$; (c) W = 1.0, $\gamma = 10^{-6}$; (d) W =
0.01, $\gamma = 10^{-3}$; (e) W = 0.01, $\gamma = 10^{-2}$. (From
Ref. 19.)

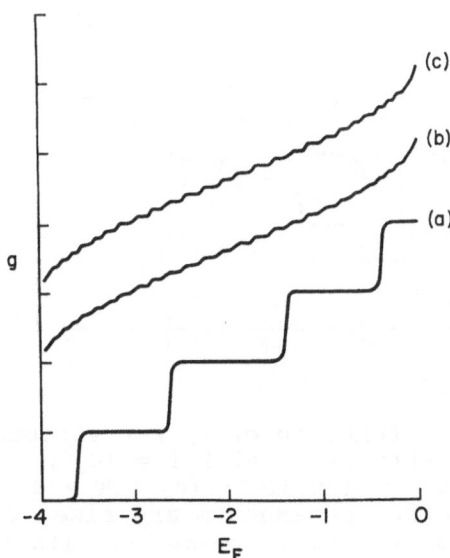

Fig. 8. g as a function of E_F for fixed L, W and γ, and
for various constriction widths: (a) M = 4, $k_B T$
= 10^{-2}; (b) M =32, $k_B T = 10^{-2}$; (c) M = 32, $k_B T =$
5 x 10^{-3}. (From Ref. 19.)

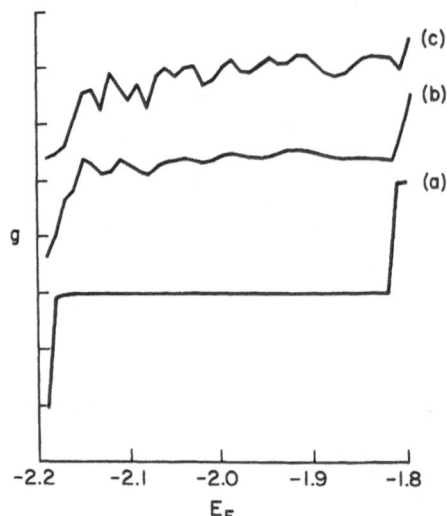

Fig. 9.   g as a function of E_F on a fixed step (n
          = 8) for various constriction lengths:
          (a) l = 4, (b) L = 256, (c) L = 512.
          All other parameters (M = 16, W = 0.1,
          $\gamma = 10^{-6}$) are fixed.   (From Ref. 19.)

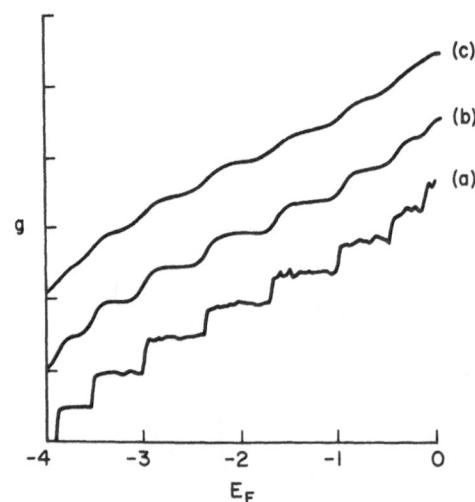

Fig. 10. g as a function of E_F for various
          temperatures: (a) $k_B T = 10^{-3}$;
          (b) $k_B T = 4 \times 10^{-2}$; (c) $k_B T = 8 \times 10^{-2}$.
          All other parameters are fixed (L = 256,
          M = 8, W = 0.1, $\gamma = 10^{-6}$). (From Ref.
          19.)

realistic (and smoother) constriction geometries (e.g. horn-shaped, triangular, concave, etc.) using the same technique. The details for those will be published[19] elsewhere. The main result is that the smoother and more adiabatic the constriction becomes, the quantization becomes softer with additional structures smoothed out and the transition from one constriction to another more gradual. This conclusion is in agreement with that of a number of other groups[20]. The experimentally studied constrictions are thought to be rather smooth due to screening effects, producing rather smooth quantized conductance consistent with our numerical results.

4.   ELECTRON MEAN FREE PATH IN MICROSTRUCTURES

An important quantity in connection with electronic transport in microstructures is the electronic mean free path (MFP) in the system, which, in principle, determines whether the system is in the diffusive or the ballistic transport regime. In particular, when the system size (in the direction of transport or current propagation) is much longer than the MFP, transport is diffusive, whereas for systems shorter than the MFP, transport is ballistic (strictly speaking, the length to be compared with the MFP is *not* the system size, but the distance between source and drain where the electrons are injected into and removed from the system). Unfortunately, depending on one's definition, there are several *different* MFPs in a system, which has caused some confusion in the literature. In this section of the paper we discuss these different MFPs and provide some numerical results for GaAs microstructures.

One must fundamentally distinguish between the elastic (energy-conserving) and the inelastic (energy-nonconserving) MFP. The inelastic MFP is associated with the loss of quantum phase coherence and is produced by two-particle interaction effects such as electron-electron and electron-phonon interactions. At $T = 0$ and at the Fermi energy ($E = E_F$) inelastic scattering necessarily vanishes and the inelastic MFP is infinite. But, at finite temperatures and/or for energies away from the Fermi energy inelastic scattering limits the MFP. The elastic MFP arises from elastic scattering with disorder, arising from the static impurities, imperfections, interface roughness and alloy disorder. The interaction is a one electron interaction and quantum phase coherence is not lost in elastic scattering. Elastic MFP defines the distance over which wavevector is a good quantum number and is connected with the momentum relaxation (rather than the energy relaxation which is determined by the inelastic MFP) determining the ohmic mobility.

Each type of MFP is associated with a lifetime (a scattering time or a decay time) τ (where τ^{-1} defines the corresponding scattering rate) and the MFP is given by

$$\text{MFP} = v_F \tau \tag{10}$$

where v_F is the Fermi velocity or the relevant electron velocity defined as

$$v = (2E/m)^{1/2}, \tag{11}$$

where E is the electron energy (which may be measured from the Fermi energy E_F for inelastic scattering). It turns out that the scattering time τ_t associated with transport (the so-called transport "relaxation time") which enters into the definition of the mobility by the formula

$$\mu = \frac{e\tau_t}{m} \, , \tag{12}$$

is not the same as the single particle lifetime τ_s arising from the imaginary part of the single particle self energy $\Sigma(k,\omega)$:

$$\tau_s = \tfrac{1}{2} \left[|Im\Sigma(k,\omega)| \right]^{-1} . \tag{13}$$

For elastic scattering by remote charged impurities in a modulation doped heterostructure $\tau_{t,s}$ are given by[21]

$$[\tau_{t,s}]^{-1} = \frac{2\pi m N_i}{\hbar^3} \int \frac{d^2k}{(2\pi)^2} f_{t,s} \left| u\left(2k_F \sin\frac{\theta}{2}\right) \right|^2 \frac{\delta(k-k_F)}{k} e^{-4k_F z_i \sin\frac{\theta}{2}} \tag{14}$$

where N_i is the density of the random charged impurity centers, $u(k)$ is the screened electron impurity Coulomb interaction, z_i is the separation between the electron plane and the impurity plane (for simplicity, we model modulation doping by considering the charged impurities to be randomly distributed in a plane separated from the 2DEG by a distance z_i). The function $f_{t,s}$ is given by,

$$\left. \begin{aligned} f_t &= 1 - \cos\theta \\ f_s &= 1 \end{aligned} \right\} \, , \tag{15}$$

where θ is the scattering angle. The difference between τ_t and τ_s is that τ_t^{-1} does not include contributions from forward scattering whereas τ_s^{-1} does. For isotropic scatterers (i.e. short range scattering), $\tau_t = \tau_s$ because the vertex correction arising from the $\cos\theta$ factor vanishes. In modulation doped GaAs heterostructures, however, the potential due to remote dopants is smooth and long ranged, making most of the scattering small angle scattering so that $\tau_t > \tau_s$. We have calculated[21], in the Born approximation, $\tau_{t,s}$ for a GaAs heterostructure in the modulation doped situation taking into account scattering by (screened) ionized impurities and the results are shown in Fig. 11. Note that for the realistic situation of Fig. 11, the actual density of the 2DEG depends somewhat on the spacer thickness because the remote dopants themselves release the electrons to produce the 2DEG. In Fig. 12 we show our calculated ratio of τ_t/τ_s for the simple model defined by Eqn. 14 for various values of the spacer thickness z_i as a function of the ratio k_F/q_{TF} where k_F and q_{TF} are the Fermi and Thomas-Fermi screening wavevectors given in a GaAs 2DEG by

$$k_F = (2\pi N_s)^{1/2} = 1.772 \times 10^6 \ (2N_s/10^{12} cm^{-2})^{1/2} \tag{16}$$

$$q_{TF} = \frac{2me^2}{\kappa\hbar^2} \simeq 2 \times 10^6 cm^{-1} \ .$$

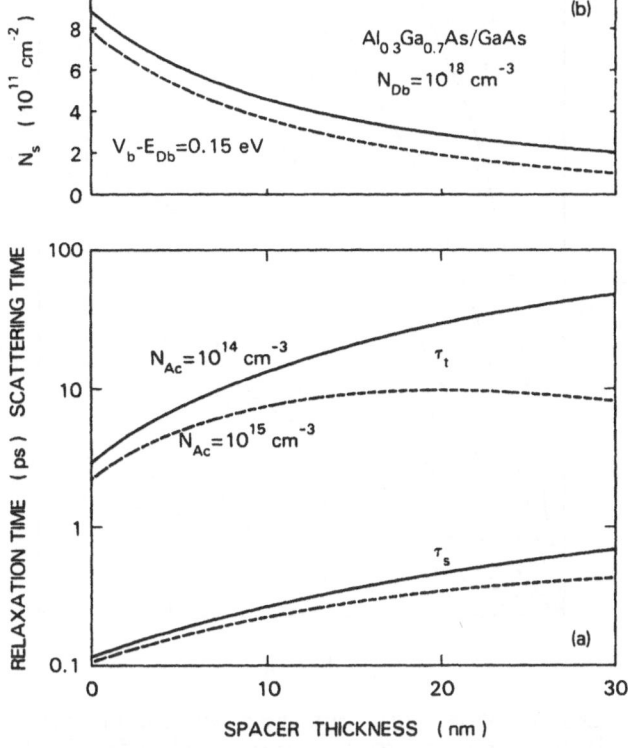

Fig. 11. (a) Calculated values of τ_t and τ_s versus
spacer thickness for electrons in $Al_{0.3}Ga_{0.7}As/$
GaAs heterojunctions with GaAs doping levels of
10^{14} cm^{-3} (full curves) and 10^{15} cm^{-3} (dashed
curves) and a donor doping level of 10^{18} cm^{-3}
in the $Al_{0.3}Ga_{0.7}As$. Note that the electron
mobility is $\mu[cm^2/Vs] = 2.5 \times 10^4 \tau_t[ps]$ and
that the level broadening is $\Gamma[meV] =$
$0.33/\tau_s[ps]$. (b) Corresponding values of the
channel electron density N_s. The difference
between the conduction band offset V_b and the
donor binding energy in the barrier, E_{Db}, is
taken to be 0.15 eV. All values are for
absolute zero. (From Ref. 21.)

In a GaAs based 2DEG, k_F/q_{TF} is mostly between 0.4 and 1 in the
range of $N_s = 10^{11}$-10^{12} cm^{-2}, and, as one can see from Fig. 11,
in this weak screening situation $\tau_t/\tau_s \gg 1$ for $z_i > q_{TF}^{-1} \approx 50$ Å.
For larger spacer thicknesses, $\tau_t \sim 10$-100 τ_s even for
moderately low values of N_s. This is due to the fact in
modulation doped GaAs heterostructure, the scattering potential
due to the distant dopant is smooth and most of the scattering
is in the forward direction which greatly enhances τ_s^{-1} over
τ_t^{-1}. In real systems, the actual value of τ_t/τ_s will depend
crucially on which impurities are causing the dominant
scattering - if the unintentional impurities in GaAs are the
main scatterers ($z_i = 0$), then τ_t/τ_s is not large (at most a
factor of 2) whereas, if the remote dopants are the main
scatterers, τ_t/τ_s could be very large indeed.

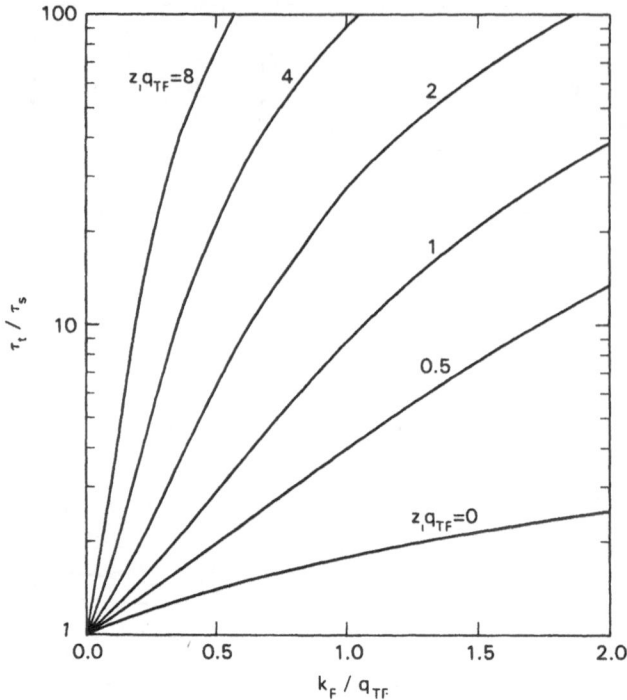

Fig. 12. Same as Fig. 11, but for an ideal two-
dimensional electron gas with six different
values of the separation z_i between the
electron layer and the impurity layer. Note
that the random phase approximation and the
long-wavelength limit give the same results
here. The value of q_{TF} for two-dimensional
electrons in GaAs is 2.0×10^6 cm^{-1} and the
corresponding value for a Si(001) inversion
layer is 1.9×10^7 cm^{-1} if the average
dielectric constant of Si and SiO$_2$ is used.
(From Ref. 21.)

There have been many experimental confirmations of the fact
that in modulation doped GaAs heterostructures, $\tau_t > \tau_s$. The
proper MFP which determines whether transport is ballistic or
not should be determined by using τ_s in Eqn. 10 and, therefore,
conductance fluctuation effects typically show up in GaAs
constrictions way below the length scale determined by the
mobility MFP.

For inelastic scattering, the situation with respect to
transport is more complex[22] because electron-electron
scattering, an important inelastic process, does not directly
contribute to resistivity. Acoustic phonon scattering, on the
other hand, can be thought of as quasi-elastic (even though,
strictly speaking, it is an inelastic process) and the transport
MFP can be obtained from our calculated results in Section 2 by
using the following relationship between the MFP and mobility
(μ) in a 2DEG system:

$$\text{MFP(\AA)} = 1167 \left[\frac{\mu}{10^4 \, cm^2/V \cdot s} \right] \left[\frac{N_s}{10^{12} \, cm^{-2}} \right]^{1/2}. \qquad (17)$$

Thus, at low temperatures (< 10 K), one concludes from Fig. 4, that the mobility MFP in GaAs heterostructure arising solely from acoustic phonon scattering is in the 100 μm range, going down inversely linearly in temperature. Acoustic phonon scattering is also a fairly short range scattering, making transport and single particle MFPs approximately equal unlike the situation for impurity scattering described above.

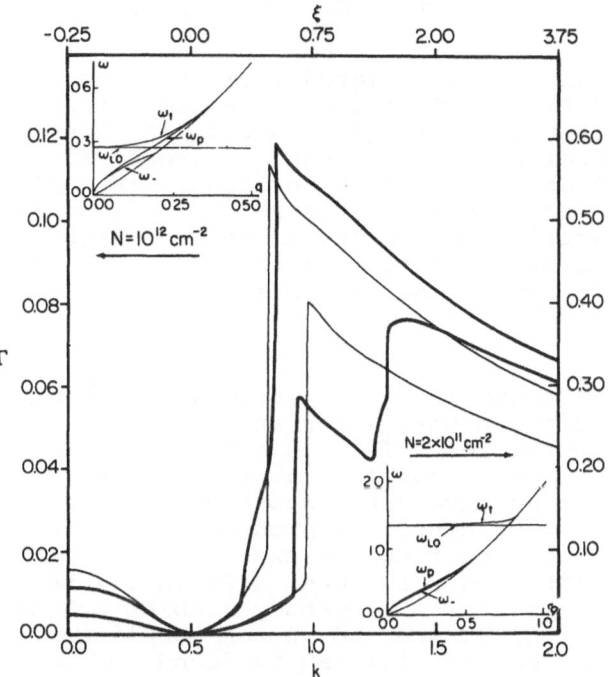

Fig. 13. Shows the calculated inelastic scattering rate $\Gamma(k)$ corresponding to lateral two-dimensional motion for the uncoupled (thin lines) and coupled (thick lines) two-dimensional electron gas systems for two electron densities. The insertions describe the mode coupling in the q-ω plane. The two upper curves, the upper insertion and the vertical scale on the left correspond to n = 10^{12} cm^{-2}. The two lower curves, the lower insertion and the vertical scale on the right correspond to n = 2 x 10^{11} cm^{-2}. Wavevectors are measured in units of $2k_F$, energies and damping rates are measured in units of $4E_F$. In the upper horizontal scale some values of the electron kinetic energy $\xi(k) = k^2 - 0.25$ are indicated. For n = 2 x 10^{11} cm^{-2} and k < 0.5 the coupled and uncoupled results are not distinguishable in the scale of the figure. For n = 10^{12} cm^{-2}: $2k_F$ = 5 x 10^6 cm^{-1} and $4E_F$ = 137 meV and for n = 2 x 10^{11} cm^{-2}: $2k_F$ = 2.2 x 10^6 cm^{-1} and $4E_F$ = 27.4 meV. The uncoupled and coupled results refer to the only electron-electron scattering and electron-electron plus electron-LO phonon scattering cases respectively. (From Ref. 22.)

279

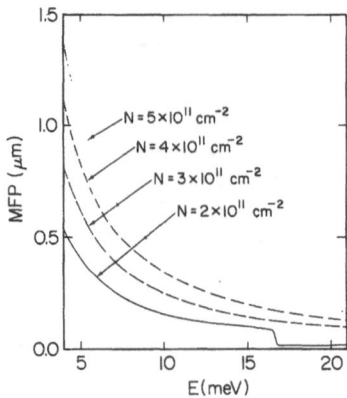

Fig. 14. Shows the calculated inelastic mean-free-path (MFP) in a GaAs-based 2DEG as a function of the electron energy E measured with respect to the Fermi energy for four different values of the electron density N as shown. The electron energy range here is rather low, making quasi-particle scattering associated with electron-electron interaction the main inelastic process (the result here corresponds to the regime just above E_F in Fig. 13).

For hot electrons ($E > E_F$), electron-electron and electron-LO phonon scattering become important inelastic channels[22] even at low temperatures. This is not an important consideration for ohmic mobility at low temperatures, but these inelastic proces- ses are very important factors in a number of low temperature experiments[23] such as ballistic hot electron transistor and electron focusing experiments. The calculation of inelastic scattering rate (and, MFP) in a 2DEG is complicated because it is manifestly a many-body problem[22] involving dynamical screening, plasmon-LO phonon coupling and Landau damping. Theoretical details can be found in the literature[22]. In Figs. 13 and 14 we show some numerical results[22] for our calculated damping rate Γ ($= \hbar[2\tau]^{-1}$) and the MFP respectively at T = 0 including both electron-electron and electron-LO phonon scattering. The results are shown as a function of energy. At low ($E - E_F$), electron-electron scattering is the main scattering mechanism and in a 2DEG, $\Gamma \sim (E - E_F)^2 \ln|E - E_F|$ with $\Gamma \to 0$ as $E \to E_F$. When $E - E_F = \hbar\omega_{LO}$, the optical phonon emission channel opens up and Γ becomes large (and the MFP small). We find that for GaAs based 2DEG, the inelastic MFP becomes the shortest length scale in the problem once $|E - E_F| \geq 1$ meV. At large values of ($E - E_F$), the inelastic MFP is only a few hundred Å. The numerical results of Figs. 13 and 14 are for a strict 2DEG interacting with the bulk LO-phonons of GaAs - when finite thickness effect (and slab and interface phonon modes) are included in the theory[22], the results change by less than 40%.

CONCLUSION

We have provided theoretical results for two complementary aspects of electronic transport in low dimensional GaAs microstructures. These are the ohmic, diffusive transport (Section 2) in high mobility 2DEG occurring in modulation doped GaAs heterostructures, and, ballistic, quantum transport (Section 3) in narrow constrictions (or point contacts) in high mobility GaAs 2DEG. We also provide a critical discussion (Section 4) of the various electron mean free paths, both elastic and inelastic, that enter into transport considerations of low dimensional systems.

Acknowledgment

The author acknowledges collaboration with his graduate students Song He, Rodolfo Jalabert and Tetsuo Kawamura whose work is reviewed here. The work has been supported by the US-ARO, US-ONR, and US-DoD.

REFERENCES

1. A.B. Fowler, F.F. Fang, W.E. Howard and P.J. Stiles, **Phys. Rev. Lett.**, 16:901 (1966)
2. T. Ando, A.B. Fowler and F. Stern, **Rev. Mod. Phys.**, 54:437 (1982) and references therein.
3. J.J. Harris, C.T. Foxon, D. Hilton, J. Hewett, C. Roberts and S. Auzoux, **Surf. Sci.**, 229:113 (1990)
4. E.E. Mendez, P.J. Price and M. Heiblum, **Appl. Phys. Lett.**, 45:294 (1984)
5. K. Hirakawa and H. Sakaki, **Phys. Rev. B**, 33:8291 (1986)
6. B.J.F. Lin, D.C. Tsui and G. Weimann, **Solid State Commun.**, 56:287 (1986)
7. H.L. Stormer, L.N. Pfeiffer, K.W. Baldwin and K.W. West, **Phys. Rev. B**, 41:1278 (1990)
8. B.J. van Wees, H. van Houten, C.W.J. Beenakker, J.G. Williamson, L.P. Kouwenhoven, D. van der Marel and C.T. Foxon, **Phys. Rev. Lett.**, 60:848 (1988)
9. D.A. Wharam, T.J. Thornton, R. Newbury, M. Pepper, H. Ahmed, J.E.F. Frost, D.G. Hasko, D.C. Peacock, D.A. Ritchie and G.A.C. Jones, **J. Phys. C**, 21:L209 (1988)
10. B.J. van Wees, L.P. Kouwenhoven, E.M.M. Willems, C.J.P.M. Harmans, J.E. Mooij, H. van Houten, C.W.J. Beenakker, J.G. Williamson and C.T. Foxon, **Phys. Rev. B**, in press.
11. Y. Hirayama, T. Saku and Y. Horikoshi, **Phys. Rev. B**, 39:5535 (1989)
12. G. Timp, R.E. Behringer, J.E. Cunningham and E.H. Westerwick, p.347 of this volume.
13. P.J. Price, **Phys. Rev. B**, 32:2643 (1985)
14. W. Walukiewicz, H.E. Ruda, J. Lagowski and H.C. Gatos, **Phys. Rev. B**, 32:9645 (1985)
15. B. Vinter, **Phys. Rev. B**, 33:5904 (1987), and **Surf. Sci.**, 170:445 (1986)
16. T. Kawamura and S. Das Sarma, **Phys. Rev. B**, 42:3725 (1990)
17. X.C. Xie and S. Das Sarma, **Phys. Rev. B**, 38:3529 (1988)
18. D.S. Fisher and P.A. Lee, **Phys. Rev. B**, 23:6851 (1981)
19. Song He and S. Das Sarma, **Phys. Rev. B**, 40:3379 (1989); **Solid State Electronics**, 32:1695 (1989), and, to be published.
20. A. Szafer and A.D. Stone, **Phys. Rev. Lett.**, 62:300 (1989); G. Kirczenow, **Solid State Commun.**, 68:715 (1988); E.G. Haanappel and D. van der Marel, **Phys. Rev. B**, 39:5484 (1989); E. Tekman and S. Ciraci, **Phys. Rev. B**, 39:8772 (1989); N. Carcia and L. Escapa, **Appl. Phys. Lett.**, 54:1418

(1989); A. Matulis and D. Segzda, **J. Phys. Condens. Matter**, 1:2289 (1989)

21. S. Das Sarma and F. Stern, **Phys. Rev. B**, 33:8442 (1985)
22. R. Jalabert and S. Das Sarma, **Phys. Rev. B**, 39:5542 (1989); **Solid State Electronics**, 32:1259 (1989); **Phys. Rev. B**, 40:9793 (1989)
23. U. Sivan, M. Heiblum and C.P. Umbach, **Phys. Rev. Lett.**, 63:992 (1989); J.R. Hayes, A.F.J. Levi and W. Wiegmann, **Phys. Rev. Lett.**, 54:1570 (1985); J. Spector, H.L. Stormer, K.W. Baldwin, L.N. Pfeiffer and K.W. West, **Appl. Phys. Lett.**, 56:1290 (1990); H. van Houten, B.J. van Wees, J.E. Mooij, C.W.J. Beenakker, J.G. Williamson and C.T. Foxon, **Europhys. Lett.**, 5:721 (1988)

TRANSPORT PROPERTIES OF $Al_xGa_{1-x}As$/GaAs HETEROSTRUCTURES:

THEORETICAL RESULTS

A. Gold

Department of Physics
Massachusetts Institute of Technology
Cambridge, 02139 Massachusetts, USA

ABSTRACT

We discuss unconventional electronic properties of disordered electron gases in $Al_xGa_{1-x}As$/GaAs heterostructures. The unconventional electronic properties arise because of the long-range random potential (for remote doping and background doping), and the existance of plasmons in an interacting electron gas. The scattering time, the single-particle relaxation time, and the frequency-dependent scattering time are calculated. Localization phenomena are discussed. The theoretical results are compared with experiments.

1.   INTRODUCTION

The transport properties of low dimensional electron gases are a subject of continuous interest. For a review on the properties of silicon metal-oxide-semiconductor (MOS) structures, see Ref. 1. Transport properties of electrons in disordered systems are usually discussed by considering a short-range random potential and noninteracting electrons. The scattering time τ_t determines the static and the dynamical conductivity within the Drude formula. In this model only one parameter, the scattering time τ_t, described the effects of disorder.

Weak localization phenomena were also described within this model and the quantum interference effects received large attention during the last ten years[2]. The quantum interference effects[3] give rise to conductivity changes of a few percent only.

In this paper I review a more realistic model for the effects of disorder in $Al_xGa_{1-x}As$/GaAs heterostructures by considering a long-range random potential due to a large spacer width α in these structures and by considering the effects of the electron-electron interaction potential. It turns out that the system has to be described by three different scattering times:

Condensed Systems of Low Dimensionality
Edited by J. L. Beeby *et al.*, Plenum Press, New York, 1991

(a)   the scattering time τ_t,
(b)   the dynamical scattering time τ_p: $\tau_t/\tau_p = 4.60(\alpha/a*)^{1/2}$,
(c)   the single-particle scattering time τ_s: $\tau_t/\tau_s = (2k_F\alpha)^2$.

a* is the effective Bohr radius and k_F the Fermi wave number. We derive the relation: $\tau_s \ll \tau_p \ll \tau_t$. For $Al_xGa_{1-x}As/GaAs$ heterostructures with a* = 100 Å we get $\tau_s = \tau_t/63$ and $\tau_p = \tau_t/10$ for $\alpha = 5a*$ and an electron density $N = 1 \times 10^{11}$ cm^{-2}. From these considerations it is evident that a model of non-interacting electrons and a short-range random potential is quite unable to describe the effects of disorder in the real systems as for instance $Al_xGa_{1-x}As/GaAs$ heterostructures with large spacer width.

In this paper we review our theoretical work on the transport properties of electrons in $Al_xGa_{1-x}As/GaAs$ heterostructures. Our theoretical results are for zero temperature. Additional theoretical work concerning the mobility of heterostructures can be found in Refs. 4-7 and in a recent review[8]. For further experimental results, see Ref. 9.

The paper is organized as follows. In Section 2 we describe the model. The results for the scattering time and the single-particle relaxation time are given in Section 3. Results for the dynamical scattering time are presented in Section 4. A comparison between silicon metal-oxide-semiconductor structures and $Al_xGa_{1-x}As/GaAs$ heterostructures is made in Section 5. We conclude in Section 6.

2.   MODEL: CHARGED IMPURITIES

We consider two models for the distribution of charged impurities in the $Al_xGa_{1-x}As/GaAs$ heterostructures. Both models turn out to be important for the understanding of transport properties in these structures. In the first model we assume that the dopant atoms are located in the $Al_xGa_{1-x}As$ and are separated from the $Al_xGa_{1-x}As/GaAs$ interface by a spacer of width α (remote doping). In this case the scattering is due to intentional doping. The two-dimensional electron gas is located near the $Al_xGa_{1-x}As/GaAs$ interface in the GaAs. For simplicity we assume that the dopants are located in a plane (which is parallel to the interface) and are characterized by a (two-dimensional) impurity density N_i. The random potential $<|U(\mathbf{q})|^2>$ in the Fourier space with $\mathbf{q}$ as a two-dimensional wave vector is written as

$$<|U(\mathbf{q})|^2> = N_i\left(\frac{2\pi e^2}{\epsilon_L q}\right)^2 F_R(q,\alpha,1/b) . \qquad (1)$$

$F_R(q,\alpha\ 1/b)$ is the form factor. 1/b describes the extension of the envelope wavefunction of the electron gas into the GaAs. For $1/b \ll \alpha$ one can use $F_R(q,\alpha,1/b = 0) = \exp(-2q\alpha)$[1]. For large α only small wave numbers (momentum transfers) are important and the random potential is long-ranged. The form factor for 1/b > 0 can be found in Ref. 10.

Unintentional doping due to residual impurities introduced during the growth of the sample can be considered as a homo-geneous distribution of impurities with a (three-dimensional) background doping density N_B. The random potential is expressed as

$$<|U(\mathbf{q})|^2> = N_B \left[\frac{2\pi e^2}{\epsilon_L q} \right]^2 F_B(q,b)/q . \qquad (2)$$

The form factor $F_B(q,b)$ was derived in Ref. 11. For small wave numbers one finds the anomalous behavior $<|U(\mathbf{q})|^2> \propto 1/q^3$.

For a rough interface the well known interface-roughness scattering[1,4] could be an important scattering mechanism. The mobility in structures with a "normal interface" ($Al_xGa_{1-x}As$/GaAs means $Al_xGa_{1-x}As$ on top of GaAs) seems not to be determined by interface roughness scattering. The "inverted interface" (GaAs on top of $Al_xGa_{1-x}As$) might be more sensitive to interface roughness scattering.

3.   STATICS: τ_t AND τ_s

The scattering time τ_t of a two-dimensional electron gas can be calculated according to the theory of Stern and Howard[12]. τ_t determines the mobility μ via the effective mass m: $\mu = e\tau_t/m$. For large electron densities the mobility of $Al_xGa_{1-x}As$/GaAs heterostructures with a large spacer width is determined by background impurities. Recent experimental results[13-15] showed a weak density dependence of the mobility. The characteristic density dependence of the mobility for homogeneous background doping found in the calculation is

$$\mu_B \propto N^\beta/N_B \qquad (3)$$

with $\beta \sim 0.7$[11,16,17]. The numerical value for β depends weakly on the electron density. A theoretical calculation of the mobility for impurities in the GaAs suggested that the density dependence of the mobility is weaker than the density dependence of the mobility for impurities in the $Al_xGa_{1-x}As$[16].

The density dependence of the mobility for remote impurity scattering is given by[7,18]

$$\mu_R \propto N^{3/2}\alpha^3/N_i \qquad (4)$$

and shows for fixed impurity density a much stronger dependence on the electron density than the mobility for homogeneous background doping. The background doping density for the samples with $\mu \sim 1 \times 10^7$ cm^2/Vs at $N = 1 \times 10^{11}$ cm^{-2} is about $N_B \sim 2 \times 10^{13}$ cm^{-3}[19]. For the comparison of recent experiments[13-15] with the theory for background doping, see Fig. 1.

The results given above have been derived within the lowest-order Born approximation and multiple-scattering effects are neglected in the calculations. Multiple-scattering effects can be included within a "self-consistent" Born approximation[20]. In such a calculation one finds a metal-insulator transition at a certain electron density N_c. Only for $N > N_c$ the mobility is finite. Crossed diagrams, which give rise to weak localization effects, are neglected in this approach but the ladder diagrams are included. For details, see Ref. 20.

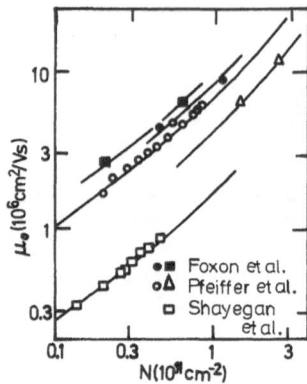

Fig. 1. Mobility μ_0 (lowest-order Born approximation) versus electron density N for homogeneous background doping and various background doping densities N_B as the solid lines (1.9×10^{13} cm^{-3} < N_B < 1.1×10^{14} cm^{-3})[19]. The experimental results of Refs. 13-15 are also shown.

For remote impurity scattering one gets[21]

$$N_c = (N_i/16\pi\alpha^2)^{1/2} \ . \tag{5}$$

This criterion is equivalent to the criterion derived for neutral impurities and a long-range random potential of range q_0 ($q_0 \sim 1/\alpha$)[22]. A similar criterion as in Eqn. 5 has been derived within a percolation approach[23]. We would like to mention that the mean free path l_c, calculated at the transition point and in lowest order Born approximation, is given by $l_c = 2\alpha$ and increases with increasing spacer width. For homogeneous background doping an approximative expression for N_c is given by[24]

$$N_c = (N_B{}^2/2\pi)^{1/3} \ . \tag{6}$$

One can show that for most experimental situations N_c for remote doping is much larger than N_c for homogeneous background doping. For N_c < N < $2N_c$ strong deviations from the lowest order result are observed in the self-consistent calculation of the mobility and at N = N_c the mobility drops to zero. In Fig. 2 we show recent experimental results for the mobility versus the electron density for sample M73 and measured at 40 mK, see Fig. 2 of Ref. 25. For large electron densities the mobility is determined by background doping. We interpret the strong decrease of the mobility for sample M73 (the spacer width is about 3000 Å) at low electron densities as a metal-insulator transition due to the remote doping. Our theory describes the mobility in a quite satisfactory way, see the solid line in Fig. 2[26]. The fit parameter used for the calculation is the background doping density N_B. For the remote doping we used the doping structure as given in Ref. 25. More details concerning this experiment and analytical results for the mobility can be found in Ref. 26. Similar experimental results as seen in the experiment of Ref. 25 have been reported in structures with smaller spacer width: 350 Å < α < 650 Å[17]. The comparison between experiment and our theory is satisfactory, too[11].

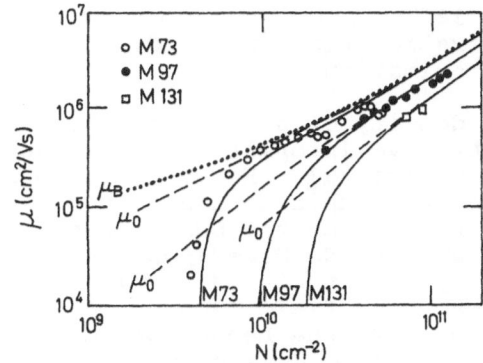

Fig. 2. Mobility μ (self-consistent) and μ_0
(lowest-order Born approximation) for
remote doping (for doping structures
corresponding to M73, M97 and M131, see
Ref. 25) and homogeneous background doping
(μ_B for $N_B = 6 \times 10^{13}$ cm^{-3})[26]. The open
circles, full circles, and the open
squares represent the experimental results
of Ref. 25.

Disorder gives rise to a modification of the density of
states (DOS) and the characteristic time is the single-particle
relaxation time τ_s. For weak disorder the DOS (ρ) is expressed
in terms of τ_s, the Fermi energy, and the DOS (ρ_0) of the free
electron gas as[18]

$$\rho/\rho_0 = 1 - 1/2\pi\epsilon_F\tau_s .$$ (7)

For remote doping τ_s was calculated in Ref. 27,28. Analytical
results for various scattering mechanisms and for weak and
strong disorder were derived in Ref. 18. For weak disorder and
for remote doping we got

$$\tau_t/\tau_s = \begin{cases} 1 & \text{for } 2k_F\alpha \ll 1 \\ (2k_F\alpha)^2 & \text{for } 2k_F\alpha \gg 1 \end{cases} .$$ (8)

The difference between τ_t and τ_s can be explained as
follows. Backward scattering contributes most to the resistance
(the inverse scattering time) while forward scattering does not
contribute. For the single-particle relaxation time forward
scattering and backward scattering contribute both and therefore
one expects that $\tau_t > \tau_s$. For a long-range random potential
($2k_F\alpha \gg 1$) backward scattering (with momentum transfer $q \sim 2k_F$)
is strongly suppressed and $\tau_t \gg \tau_s$. For homogeneous background
doping an infrared singularity was found for the single-particle
relaxation rate (calculated in lowest order Born approximation)
and the ratio τ_t/τ_s depends on the background doping density.
For small background doping density and small electron density
we found[18]

$$\tau_t/\tau_{sr} = (1 + k_F a^*)^2 \ln(4\epsilon_F\tau_{sr}) \gg 1 ,$$ (9)

see Eqn. 39 of Ref. 18. τ_{sr} in Eqn. 9 is the renormalized (r) single-particle relaxation time and calculated within the self-consistent Born approximation.

The electronic density of states is difficult to measure. Tunneling experiments gave information on the subband structure in two-dimensional systems[29]. However, it appears that the resolution obtainable with this method is not large enough to derive detailed information on the DOS. From the temperature dependence of the Shubnikov-de Haas (SdH) oscillations one can determine a scattering time τ_{SdH}. If one assumes that for low magnetic field the magnetic field dependence of the DOS can be neglected (which means that the Landau bands are overlapping) one can argue that $\tau_{SdH} \sim \tau_s$. Measurements of τ_{SdH} showed that the ratio τ_t/τ_{SdH} can be strongly enhanced[30]. For a recent experimental and theoretical analysis for $\tau_t/\tau_{SdH} \sim \tau_t/\tau_s$ (with τ_s calculated), see Ref. 31.

4.   DYNAMICS: $\tau(\omega)$ AND τ_p

The scattering time has to be generalized to a frequency-dependent scattering time $\tau(\omega)$ if the dynamical conductivity is considered. The presence of plasmon excitations in an interacting electron gas enhances the density of final states for the decay of current modes into density modes[20]. For remote doping $1/\tau(\omega)$ can be calculated in analytical form and we obtained[32]

$$\tau_t/\tau(\omega) = 1 + (2\pi^2\alpha/a\ast)^{1/2}|\Omega|^5\exp(-\Omega^2) \tag{10}$$

with $\Omega = \omega(\alpha a\ast k_F^2/2\epsilon_F^2)^{1/2}$. In Fig. 3 we show $\tau_t/\tau(\omega)$ versus frequency. The maximum of $\tau_t/\tau(\omega)$ occurs at the frequency

$$\omega_0 = \epsilon_F(5/k_F^2\alpha a\ast)^{1/2} . \tag{11}$$

The enhancement factor is expressed as[32]

$$\tau_t/\tau_p = 4.60(\alpha/a\ast)^{1/2} \tag{12a}$$

with

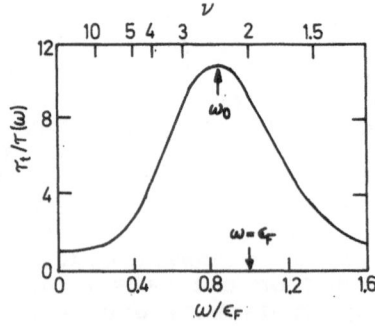

Fig. 3. $\tau_t/\tau(\omega)$ versus frequency according to Eqn. 10 for $N = 1.5 \times 10^{11}$ cm^{-2} and $\alpha = 750$ Å [32]. ν is the Landau filling factor.

$$\tau_p = \tau(\omega_0) \ . \tag{12b}$$

For $N = 1.5 \times 10^{11}$ cm^{-2} and $\alpha = 750$ Å as in Fig. 3 we get $\omega_0 \sim \epsilon_F$. We conclude that in measurements of dynamical properties one must be very careful in determining the scattering time and one should not expect that τ_t is the relevant time scale in these experiments. Only for $\omega \ll \omega_0$ one finds that $\tau_t \sim \tau(\omega)$. For $\omega \sim \omega_0$ the dissipation is mainly determined by plasmon dynamics: $\tau(\omega) \sim \tau_p \ll \tau_t$. For $\omega \gg \omega_0$ one finds that $\tau(\omega) \gg \tau_t$ and the relaxation time is (as for $\omega \ll \omega_0$) determined by the electron-hole spectrum[32].

A frequency-dependent scattering time has been found in experiments in silicon MOS structures with impurities at the interface ($\alpha = 0$)[33]. These experimental results have been interpreted as plasmon dynamics. Strong disorder can also account for a frequency-dependent scattering time, see Ref. 34. In heterostructures with large spacer width the effect of the plasmon dynamics is strongly enhanced compared to silicon MOS systems.

Due to the strong frequency dependence of $\tau(\omega)$ we expect in the dynamical conductivity a peak at $\omega = 0$ (the Drude peak) and near $\omega = \omega_0$ (the plasmon peak)[32]. A similar result was found for disordered three-dimensional electron gases[35]. The reduction of the scattering time due to the plasmon dynamics (calculated in perturbation theory and for an interacting electron gas) is a much stronger effect than the reduction of the scattering time due to weak localization effects (calculated in perturbation theory and for a noninteracting electron gas)[36].

For background doping the frequency dependence of the scattering time is weaker than for remote doping[37]. In case of two scattering mechanisms the enhancement factor of the relaxation rate is reduced compared to the enhancement factor for remote doping[32].

5.   DISCUSSION

Our calculation of τ_t/τ_s indicates that a large mobility as found in certain samples of Al$_x$Ga$_{1-x}$As/GaAs heterostructures is not a sufficient criterion to argue for weak disorder. For $\tau_t/\tau_s \gg 1$ and $2k_F\alpha \gg 1$ the effect of the disorder on the density of states can be quite large even if the mobility is high. Ultra-high mobility samples of silicon MOS structures have a mobility of about 5×10^4 cm^2/Vs[38] and $\tau_t/\tau_s \sim 1$. This corresponds to $\tau_s = 5.4 \times 10^{-12}$s. For Al$_xGa_{1-x}$As/GaAs heterostructures with $\mu = 5 \times 10^6$ cm^2/Vs and $2k_F\alpha = 5$ we get $\tau_s = 7.6 \times 10^{-12}$s, which is of the same order of magnitude as τ_s for the silicon MOS structure.

Our calculations of the frequency-dependent scattering time show that for a long-range random potential a Drude picture with a frequency-independent scattering time only makes sense for $\omega \ll \omega_0$. In far-infrared experiments with $\omega \sim \omega_0$ the frequency dependence of the scattering time must be included and in this regime the Drude formula for the conductivity makes no sense. We expect that plasmon-excitation spectroscopy via a grating coupler or Raman spectroscopy could be used to examine $\tau(\omega)$. For a more detailed discussion, see Ref. 32.

The strong decrease of the mobility at a certain low electron density and the good agreement of our calculation with the experiments leads us to propose that a metal-insulator transition occurs in these structures. The samples with an ultra-high mobility ($\mu > 1 \times 10^6$ cm^2/Vs) should be considered as good metals. In spite of these experimental results it seems that the "absence of diffusion" in two-dimensional systems as claimed in Ref. 3 should be considered with some caution. Our theory with realistic scattering potentials and with multiple scattering effects included seems to describe the experimental results in the regime of weak **and** strong disorder. Our theory can be used to get information on the strength of the disorder and the relevant scattering mechanisms[11,26]. Such information is not obtainable from the theory generally described as "weak localization theory"[2].

In the presence of two scattering mechanisms the dominant scattering mechanism might be different for different scattering times as measured in experiments. This result has already been found in experiments (τ_t and τ_s have been measured) on thin quantum wells with impurity scattering and interface roughness scattering[31].

τ_t, τ_s and τ_p in Eqns. 8 and 10 have been calculated in the lowest order Born approximation. If multiple-scattering effects are taken into account one finds $\tau_{tr} < \tau_t$ [20], $\tau_{sr} > \tau_s$ [18] and $\tau_{pr} < \tau_p$ [20]; r means renormalized due to multiple-scattering effects. We conclude that

$$\tau_{tr}/\tau_{sr} < \tau_t/\tau_s \ . \tag{13}$$

For strong disorder ($N \sim N_c$) we expect $\tau_{tr}/\tau_{sr} < 1$.

CONCLUSION

The long-range random potential and the existence of plasmons give rise to three different scattering times in a disordered two-dimensional electron gas: for remote doping one finds $\tau_s \ll \tau_p \ll \tau_t$. Background doping is characterized by $\tau_s \ll \tau_p < \tau_t$. For the interpretation of experimental results it is crucial to know which scattering time is measured and which scattering mechanism is relevant.

In GaAs/Al$_x$Ga$_{1-x}$As heterostructures with high mobilities one finds that for low electron densities the electrons are scattered by remote impurities located in the Al$_x$Ga$_{1-x}$As (remote doping). For large electron densities the scattering by impurities located in the GaAs (background doping) represents the dominant scattering mechanism. The density for the cross-over from scattering by remote doping to scattering by background doping depends on the spacer width and the impurity concentrations.

Acknowledgment

I thank A. Ghazali for a critical reading of the manuscript and P.A. Lee for discussions. This work was supported by the US Joint Services Electronics Program (under Contract No. DAAL03-89-C-0001).

REFERENCES

1. T. Ando, A.B. Fowler and F. Stern, **Rev. Mod. Phys.**, 54:437 (1982)
2. P.A. Lee and T.V. Ramakrishnan, **Rev. Mod. Phys.**, 57:287 (1985)
3. E. Abrahams, P.W. Anderson, D.C. Licciardello and T.V. Ramakrishnan, **Phys. Rev. Lett.**, 42:673 (1979)
4. T. Ando, **J. Phys. Soc. Jpn.**, 51:3900 (1982)
5. F. Stern, **Appl. Phys. Lett.**, 43:974 (1983)
6. W. Walukiewicz, H.E. Ruda, J. Lagowski and H.C. Gatos, **Phys. Rev. B**, 30:4571 (1984)
7. P.J. Price, **Surf. Sci.**, 143:145 (1984)
8. J. Harris, J.A. Pals and R. Woltjers, **Rep. Prog. Phys.**, 52:1217 (1989)
9. K. Hirakawa and H. Sakaki, **Phys. Rev. B.**, 33:8291 (1986)
10. A. Gold, **Z. Phys. B**, 63:1 (1986)
11. A. Gold, **Appl. Phys. Lett.**, 54:2100 (1989)
12. F. Stern and W.E. Howard, **Phys. Rev.**, 163:816 (1967)
13. M. Shayegan, V.J. Goldman, M. Santos, T. Sajoto, L. Engel, and D.C. Tsui, **Appl. Phys. Lett.**, 53:2080 (1988)
14. C.T. Foxon, J.J. Harris, D. Hilton, J. Hewett and C. Roberts, **Semicond. Sci. Technol.**, 4:582 (1989)
15. L. Pfeiffer, K.W. West, H.L. Stormer and K.W. Baldwin, **Appl. Phys. Lett.**, 55:1888 (1989)
16. A. Gold, **J. Phys. (Paris)**, C5:255 (1987)
17. C. Jiang, D.C. Tsui and G. Weimann, **Appl. Phys. Lett.**, 53:1533 (1988)
18. A. Gold, **Phys. Rev. B**, 38:10798 (1988)
19. A. Gold, **Phys. Rev. B**, 41:8537 (1990)
20. A. Gold and W. Götze, **Phys. Rev. B**, 33:2495 (1986)
21. A. Gold, **Phys. Rev. B**, 42 submitted (1990)
22. A. Gold and W. Götze, **J. Phys. C**, 14:4049 (1981)
23. A.L. Efros, **Solid State Commun.**, 70:253 (1989)
24. A. Gold, **Solid State Commun.**, 60:531 (1986)
25. T. Sajoto, Y.W. Suen, L.W. Engel, M.B. Santos and M. Shayegan, **Phys. Rev. B**, 41:8449 (1990)
26. A. Gold, **Phys. Rev. B**, 42, submitted (1990)
27. J.P. Harrang, R.J. Higgins, R.K. Goodall, P.R. Jay, M. Laviron and P. Delescluse, **Phys. Rev. B**, 32:8126 (1985)
28. S. Das Sarma and F. Stern, **Phys. Rev. B**, 32:8442 (1985)
29. M. Zachau, F. Koch, K. Ploog, P. Roentgen and H. Beneking, **Solid State Commun.**, 59:591 (1986)
30. M.A. Paalanen, D.C. Tsui and C.J.M. Hwang, **Phys. Rev. Lett.**, 51:2226 (1983)
31. U. Bockelmann, G. Abstreiter, G. Weimann and W. Schlapp, **Phys. Rev. B**, 41:7864 (1990)
32. A. Gold, **Phys. Rev. B**, 41:3608 (1990)
33. A. Gold, W. Götze, C. Mazure and F. Koch, **Solid State Commun.**, 49:1085 (1984)
34. A. Gold, S.J. Allen, B.A. Wilson and D.C. Tsui, **Phys. Rev. B**, 25:3519 (1982); S.J. Allen, D.C. Tsui and F. de Rosa, **Phys. Rev. Lett.**, 35:1359 (1975)
35. W. Götze and P. Wölfle, **Phys. Rev . B**, 6:1226 (1972)
36. L.P. Gorkov, A.I. Larkin and D.E. Khmelnitzkii, **JETP Lett.**, 30:228 (1979)
37. A. Gold, **Phys. Rev. B**, 35:723 (1987)
38. M.G. Gavrilov, Z.D. Kvon, I.V. Kukushkin and V.B. Timofeev, **JETP Lett.**, 39:507 (1984)

ELECTRON ENERGY RELAXATION AND CHARGE BUILDUP EFFECTS IN

RESONANT TUNNELING DEVICES

L. Eaves

Department of Physics
University of Nottingham
Nottingham NG7 2RD, UK

ABSTRACT

The use of magnetocapacitance, magnetoconductance and photoluminescence spectroscopy measurements on asymmetric double barrier resonant tunneling structures is described. Fourier analysis of the magnetocapacitance oscillations can be used to measure the electron space-charge buildup in the quantum well and in the emitter accumulation layer. Thermalization of the resonantly tunneling electrons is observed. The variation of the photoluminescence linewidth with applied voltage also provides a direct signature of the space-charge buildup. Both the photoluminescence and electrical measurements provide clear evidence for sequential tunneling processes.

1.   INTRODUCTION

Although resonant tunneling in double barrier semiconductor heterostructure devices was proposed and demonstrated in the early 1970s[1,2], the subject remains one of intense experimental and theoretical interest. In this article we shall examine two topical and controversial aspects of resonant tunneling: first, the question of whether the process is coherent or sequential[3-5]; secondly, the process of space-charge buildup in the quantum well and the intrinsic bistability effect in the current-voltage characteristics to which this process gives rise[7-13].

We will focus attention on asymmetric double barrier resonant tunneling structures in which the two tunnel barriers have different widths. In such structures, the space-charge buildup can greatly exceed the values which occur in more conventional double barrier structures of the type used in high frequency applications. The paper will describe a series of electrical[14] and optical spectroscopy measurements[15,16] which provide quantitative information about (a) the sheet density of electrons that builds up in the quantum well under resonant tunneling conditions and (b) the energy relaxation processes which resonantly-tunneling electrons undergo.

Condensed Systems of Low Dimensionality
Edited by J. L. Beeby *et al.*, Plenum Press, New York, 1991

2.   DETAILS OF STRUCTURE OF THE DEVICE

Our asymmetric structure is shown schematically below:

0.5 μm GaAs, 2 x 10^{18} cm^{-3}, top contact
50 nm GaAs, 10^{17} cm^{-3}
50 nm GaAs, 10^{16} cm^{-3}
3.3 nm GaAs, undoped
11.1 nm Al$_{0.4}$Ga$_{0.6}$As, undoped (thick barrier)
5.8 nm GaAs, undoped (well)
8.3 nm Al$_{0.4}$Ga$_{0.6}$As, undoped (thin barrier)
3.3 nm GaAs, undoped
50 nm GaAs, 10^{16} cm^{-3}
50 nm GaAs, 10^{17} cm^{-3}
2 μm GaAs, doped to 2 x 10^{18} cm^{-3}
n^{+}GaAs substrate

The layers were processed into mesas of diameter 200 μm.   The
conduction band profile, with the top contact biased positively,
is shown in Fig. 1.   Undoped GaAs spacer layers separate the
heavily doped n^{+} contact regions from the barriers.   The spacer
layers prevent dopant diffusion into the barriers and thus
reduce scattering which can degrade the characteristics of a
double barrier structure[17].   Under an applied voltage a bound
state is formed in the accumulation region adjacent to the
emitter barrier and, at liquid helium temperature, the
associated two-dimensional electron gas (2DEG) is degenerate.
Resonant tunneling occurs when the energy ϵ_0 of the 2DEG quasi-
bound state coincides with that of a quasi-bound state of the

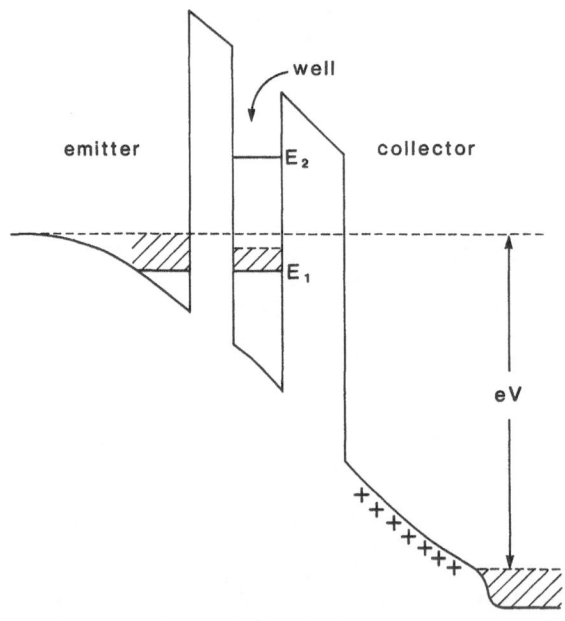

Fig. 1. Conduction band profile under applied voltage V
        showing bound-state levels (solid lines) in
        emitter and well and quasi-Fermi levels (dashed
        lines).

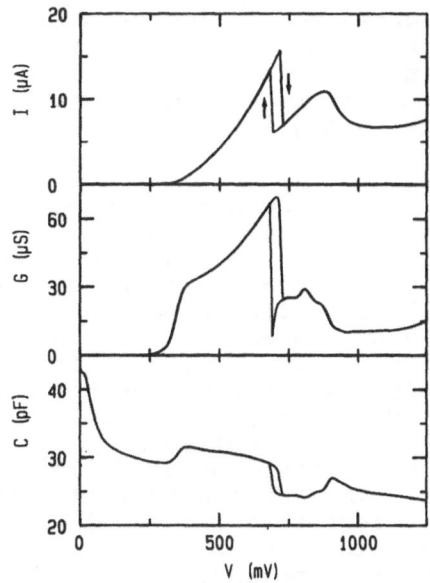

Fig. 2. Voltage dependence of dc current I, ac
capacitance C and parallel conductance G
measured at 1 MHz and 4 K.

quantum well (ϵ_1 or ϵ_2). In the remainder of the lightly doped
region between the emitter barrier and the n^+ layer the Fermi
level is close to the conduction band edge since the doping
density (10^{16} cm^{-3}) is close to the Mott metal-insulator
transition. However, this does not give rise to an important
series resistance.

3. MAGNETOTUNNELING AND CAPACITANCE MEASUREMENTS

The current voltage characteristic for our asymmetric DBS
at 4 K is shown in Fig. 2 for the region corresponding to
resonant tunneling into the first quasi-bound state ϵ_1 of the
quantum well. The region of the second resonance is considered
later. The figure shows clearly the threshold for resonant
tunneling at V_{th} = 330 mV, the turn off at V_p = 725 mV where the
current peaks, and the region of intrinsic bistability. The
transverse components of momentum and hence transverse kinetic
energy are conserved in the resonant tunneling process. Total
energy is also conserved so resonant tunneling occurs when the
energies of the bound states in the accumulation layer and
quantum well essentially coincide. However, because of finite
level widths, the tunneling rate from emitter into well will be
a sharply peaked resonant function of the voltage drop across
the emitter barrier. The observed resonant tunneling range V_{th}
to V_p corresponds to moving up this resonance peak. It is the
screening effect of the charge buildup in the quantum well
(electrostatic feedback) that is responsible for the extended
voltage range over which resonant tunneling is observed in the
I(V) curve and the appearance of current bistability. When the
applied voltage is increased, only a very small voltage change
across the emitter barrier is needed to charge up the well and,
due to the screening effect of this charge, almost all the extra

voltage drop occurs across the collector barrier and depletion layer. When biased in the opposite direction a very sharp maximum in the I(V) curve is observed[10,11] since in this case (thin collector barrier) there is little charge buildup and electrostatic feedback is negligible. The bistability does not then appear.

Conservation of transverse kinetic energy in resonant tunneling requires the Fermi levels in the accumulation layer and quantum well to coincide. But the theory of resonant tunneling[9] shows that the states of transverse motion in the well are only partially occupied, since the occupancy is determined dynamically by the balance between the transition rates into and out of the well. However, if the energy relaxation time is much less than the charge storage time in the well, the electron distribution will be able to thermalize to the lattice temperature and establish a well defined quasi-Fermi level below that in the emitter contact. This is the situation illustrated in Fig. 1. The Fermi energy of a thermalized 2DEG in the quantum well is less that the Fermi energy of the 2DEG in the accumulation layer since the bound-state levels essentially coincide during resonant tunneling.

We have investigated this charge buildup and thermalization of electrons in the quantum well by studying magneto-oscillations in the capacitance of the structure for a magnetic field $\underline{B} \parallel \underline{J}$. In this geometry, the states of transverse motion of a degenerate 2DEG (in emitter or well) are quantized into Landau levels. Theoretically, oscillations with a definite period, $\Delta(1/B)$, in $1/B$ arise from a modulation of the charge when Landau levels pass through the quasi-Fermi level. This charge modulation affects the distribution of electric potential and screening lengths and hence modulates the capacitance of the device. The frequency of the oscillations $B_f = \{\Delta(1/B)\}^{-1}$ is thus related to the Fermi energy E_F by $B_f = m*E_F/e\hbar$ where $m*$ is the effective mass. For fully, as opposed to partially, occupied states $E_F = \hbar^2 \pi n/m*$, where n is the areal electron density. This gives $n = 2eB_f/h$.

The voltage dependence of the differential capacitance C and parallel conductance $G = 1/R$ are also plotted in Fig. 2. The differential parameters were measured at 1 MHz with a modulation of 3 mV using a Hewlett-Packard 4275A LCR meter in the mode which analyses the impedance of a device as a capacitor C and parallel resistor R. Between 10 kHz and 2 MHz these parameters are also independent of measurement frequency. Figure 3 shows typical oscillatory structure in the variation of capacitance with magnetic field. Similar oscillations are also observed in the current I and conductance G but are less well defined. To reveal more clearly multi-periodic behavior we have Fourier analyzed the experimental capacitance traces. The distributions of magneto-oscillation frequencies B_f thus obtained are shown in Fig. 4 for different applied voltages.

Below the threshold voltage V_{th}, there is a single peak in the magneto-oscillation spectrum. This peak is clearly associated with the 2DEG in the emitter accumulation layer, since the corresponding areal density n_a increases steadily with voltage up to V_{th} as shown in Fig. 5. Moreover, the static capacitance values, given by the ratio of total accumulation charge to applied voltage, are in good agreement with the ac

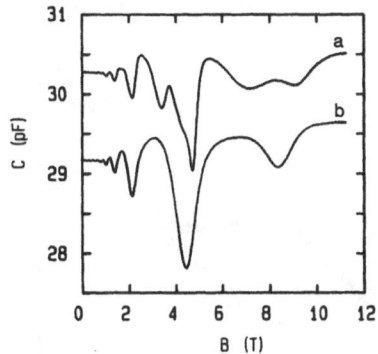

Fig. 3. Magneto-oscillations in differential
capacitance C vs. magnetic field B for
applied voltage (a) 600 mV and (b) 300 mV.

values shown in Fig. 2 and are consistent with our theoretical
modeling[1] of the potential distribution across the heterostructure. We have also investigated the magneto-oscillations
when the magnetic field is tilted at an angle θ to the normal to
the layers. Up to θ = 40° the form of the oscillations as a
function of $B\cos\theta$ is largely unchanged. This confirms the two-
dimensional nature of the electrons in the emitter accumulation
layer.

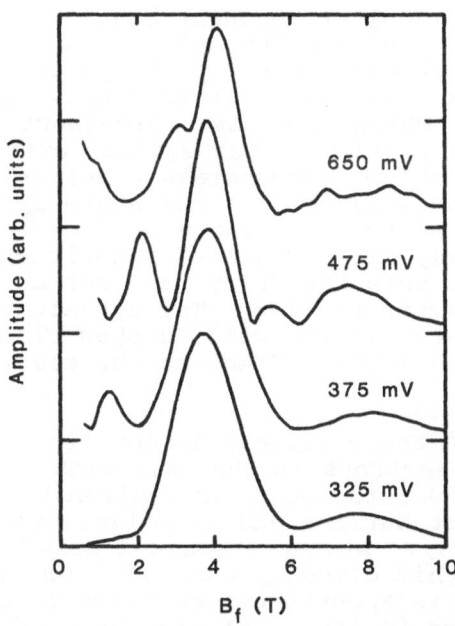

Fig. 4. Spectrum of the magneto-oscillation frequencies
B_f obtained by Fourier transforming them in $1/B$
space. The peak position B_f for each voltage
gives the inverse period. The weak structure at
around 8 T corresponds to the second harmonic of
the series due to the charge in the accumulation
layer.

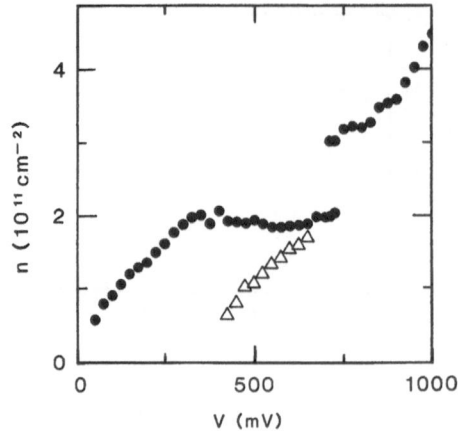

Fig. 5. Areal density n versus voltage V for charge in the accumulation layer n_a (circles) and well n_w (triangles). The values of n are deduced from the peaks in the Fourier spectrum.

Between V_{th} and V_p, this magneto-oscillation frequency is independent of voltage (Fig. 4) and, within experimental error, the accumulation density n_a remains constant at 2.2×10^{11} cm^{-2}. Hence the electric field and voltage drop across the emitter barrier remain virtually unchanged in this range, which is consistent with our model of resonant tunneling between bound states in the emitter and quantum well. Also in the resonant tunneling region, a second, weaker peak appears in the magneto-oscillation spectrum (Fig. 4). The frequency of this peak gives an areal density n_w, which increases throughout this range and approaches n_a at the voltage V_p for maximum current (Fig. 5). We attribute this peak to a degenerate electron distribution stored in the quantum well whose temperature $T_e << \hbar\omega_c/k = 15$ K at B = 1 T. The equality of n_a and n_w at V = V_p is to be expected theoretically since the peak transition rate into the well is much greater than the decay rate out of the well throughout the collector barrier. The dynamically determined occupancy of the states in the well is then close to unity and energy relaxation has little effect on the electron distribution.

This cooling of the electrons is confirmed by comparing the lifetime τ_c of the electrons in the well with the energy relaxation rate. The lifetime τ_c is limited by tunneling through the collector barrier and is related to the current density by $J = n_w e/\tau_c$. At the resonance peak (J $\simeq$ 0.06 Acm^{-2}, $n_w \simeq 2 \times 10^{11}$ cm^{-2}) this gives $\tau_c \sim 0.6$ µs. The energy relaxation must be via spontaneous emission of acoustic phonons since the temperature is low (4 K) and the electron kinetic energies ($E_F \sim 7$ meV) are too small for optic-phonon processes. An estimate of the emission rate τ^{-1} from the deformation potential gives $\tau_{ph} \sim 10^{-9}$ s for a well width of 5.8 nm. This is indeed much shorter than τ_c. We note that electron-electron scattering is not important here since the phonon emission rate is sufficiently large to thermalize the electron distribution to the lattice temperature within the available time τ_c.

When V increases above V_p a transition occurs in which charge is expelled from the well with a consequent redistribution of potential, and resonant tunneling can no longer occur. This results in a step-wise increase in the accumulation layer density to $n_a \simeq 3 \times 10^{11}$ cm^{-2}, as shown in Fig. 5. For V > V_p only a single magneto-oscillation period is observed. The different charge states of the device on the high and low current parts of the hysteresis loop are also clearly shown by the different magneto-oscillation frequencies observed. The corresponding sheet density increases smoothly with voltage showing that, in this case, there is little charge buildup in the well. We note here that the broad maximum in I(V) above the current bistability is due to inelastic tunneling processes involving the emission of a longitudinal optic phonon[17] by an electron which tunnels from the emitter into the quantum well.

The principal features of the C(V) curve of Fig. 2 can also be understood in terms of our model. We may regard the DBS as two parallel plate capacitors C_1 (emitter barrier and accumulation layer) and C_2 (collector barrier and depletion layer) connected in series. The charge on the common central plate corresponds to the charge stored in the quantum well. The steep fall in capacitance at low voltages is due to the rapid increase in depletion length in the lightly doped (10^{16} cm^{-3}) collector layer which decreases C_2. The slower decrease for V > 50 mV is due to the less rapid rate of depletion of the layer doped to 10^{17} cm^{-3}. However, the most notable features are the sharp increase in capacitance at V_{th} and subsequent fall at V_p. When the bound states of the accumulation layer and quantum well are on resonance, the voltage drop across C_1 is essentially constant. An increase in applied voltage therefore appears almost entirely across C_2. The measured differential capacitance is thus $C \simeq C_2$ in the resonant tunneling region, whereas off-resonance $C = C_1 C_2 / (C_1 + C_2) < C_2$.

The quasi-static argument is not obviously applicable at 1 MHz, where the measurements of differential capacitance were made. In a small signal analysis[19] based on the sequential theory of resonant tunneling, C_1 and C_2 have parallel resistors R_1 and R_2 respectively. R_1 allows electrons tunneling from the emitter to charge the quantum well whilst R_2 allows the stored charge to leak through the collector barrier. Our previous argument is equivalent to the assumption that, during resonant tunneling, R_1 becomes very small and effectively short-circuits C_1 so that the measured capacitance $C \simeq C_2$. By identifying the time constant $R_2 C_2$ with the storage time τ_c we have $\tau_c = R_2 C_2 \sim$ RC during resonance. The values shown in Fig. 2 give $\tau_c \sim 0.4$ μs at V_p, which is consistent with our previous estimate from the dc current and charge. An approximate theoretical estimate of τ_c can be made by calculating the attempt rate (obtained from the energy of the bound state above the bottom of the well ~ 68 meV) and the transmission coefficient of a rectangular barrier (conduction band offset ~ 320 meV). This gives $\tau_c \sim 1$ μs. Under bias a smaller value is expected since the average collector-barrier height is then decreased. Measurements of the tunneling escape time from the quantum well of a DBS have also been made by studying the temporal decay of photoluminescence [19]. However, the optical determination of τ_c is limited to thin barrier structures for which τ_c is less than the radiative recombination time $\tau_r \sim 0.5$ ns. In our asymmetric structure it is the thick collector barrier which gives rise to the long τ_c.

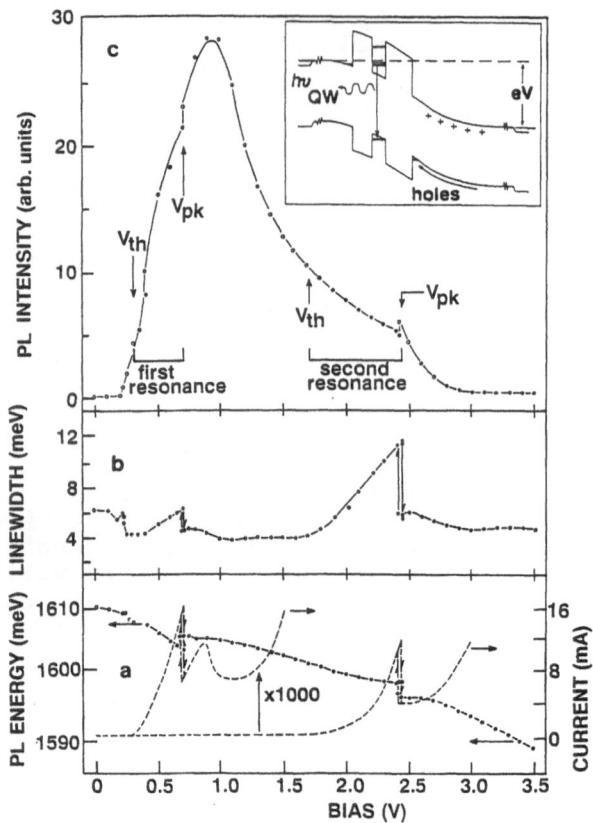

Fig. 6. Variation with bias voltage: current (Fig. 6a dashed); PL peak position (Fig. 6a circles); PL linewidth (full width half maximum) (Fig. 6b); integrated PL intensity (Fig. 6c). Open circles - low I state, full circles - high I state. The bistable, hysteresis loops are exhibited in the bias dependence of the current, PL peak energy and linewidth. The inset shows a diagram of the energy bands and quasi-Fermi levels in the structure at the bias voltage of the first resonance. [After Refs. 15 and 16.]

4.   PHOTOLUMINESCENCE MEASUREMENTS

The above picture of the resonant tunneling process in our asymmetric resonant tunneling device is complemented by low temperature photoluminescence (PL) measurements. These were performed at 2 K using 1.96 eV incident radiation of low power density (< 1 W cm^{-2}) with the device in forward bias. Figure 6 compares the I(V) characteristics, taken this time under weak illumination, with the key features of the PL data: PL intensity, linewidth and photon energy at the PL peak. The peak in I(V) at 2.4 V, not shown in Fig. 2, corresponds to resonant tunneling into the second quasi-bound state (ϵ_2) of the quantum well. This also shows a region of intrinsic bistability[10,11]. Detailed modeling of the device shows that on the second peak in I(V), the electronic space-charge is mainly in the lower state of the electron quantum well whereas a significant proportion of the total tunnel current arises from electrons directly tunneling out of the upper state[20].

The PL arises from electron-hole recombination between the lowest electron and heavy hole states in the quantum well. On resonance, the electrons present in the well arise predominantly from the charge buildup which occurs during the tunneling, whereas the holes are photocreated principally in the GaAs contact regions. The holes then diffuse and drift to the collector barrier where they accumulate and tunnel into the well (inset to Fig. 6).

Figure 6a shows the variation of PL peak energy with V. The PL peak energy corresponds to recombination of holes with electrons in the lower bound state of the (electron) quantum well, even when the device is biased for resonant tunneling of electrons into the second bound state. This indicates the importance of intersubband scattering from ϵ_2 to ϵ_1 in relaxing the energy of resonantly tunneling electrons. The inter-subband scattering rate due to the emission of longitudinal optic phonons is around 10^{12} s^{-1}, which is close to our estimated value[16] from the rate of tunneling out of the upper subband. A large shift to lower energy of 20 meV at 3.3 V, due to the Stark shift of the electron and hole quasi-bound states, is observed. The shift is close to that expected for a ~ 60 Å well in an electric field of ~ 2 x 10^5 V cm^{-1}. There are two discontinuities and bistable regions in the variation of peak energy with bias (expanded version, Fig. 7), corresponding to those in I(V). These arise from the changes in the energies of the quasi-bound states in the well, and of the band gap renormalization, between the on and off resonant states.

The variation of PL linewidth versus bias is shown in Fig. 6b, and on an expanded scale in Fig. 7. As V is swept, the

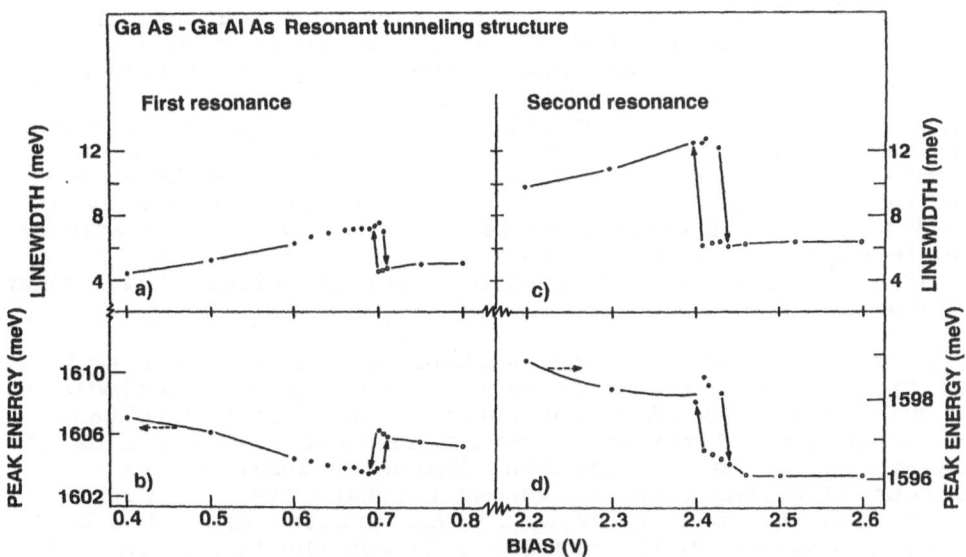

Fig. 7. Expanded plots of the variation of the PL peak energy and linewidth around the bistable regions of the first and second resonances. [After Refs. 15 and 16.]

linewidth increases up to the switch-off of the first resonance
(0.7 V) where it falls sharply and remains fairly constant until
the threshold voltage for the second resonance at ~ 1.7 V. Here
the linewidth steadily increases with tunneling current up to
the switch-off, where it again falls sharply and then stays
constant in a similar manner to the first resonance. The
variation of the linewidth with bias is very similar to that of
the current, see Fig. 6a. Bistability and accompanying
hysteresis loops are again observed. The increase in PL linewidth with resonant current is due to free carrier broadening
and indicates filling of electron states in the well, as
confirmed by magnetic field measurements[15,16]. The PL spectra
on and off resonance are displayed in Fig. 8. When on resonance
(Figs. 8a and 8c), the PL is broad and peaks on the low energy
side with a tail to higher energy, as expected for a high
density 2D electron gas with weak disorder. The two spectra off
resonance (Figs. 8b and 8d), where there is relatively little
charge in the well, are much narrower.

The change in linewidth between the on and off resonant
states can be used to determine n_w at the resonances. For the
first resonance, by convolving the lineshape observed off
resonance (where $n_w \approx 0$) with a Fermi function (including a
decreasing oscillator strength towards E_F), a good fit to the
experimental spectrum can be obtained, and values of $E_F = 7.5 \pm$
1.0 meV and $n_w = (2.2 \pm 0.3) \times 10^{11}$ cm^{-2} deduced. This is in
excellent agreement with that given by the electrical
measurements (see Fig. 5). For the second resonance, the
fitting is less reliable due to the absence of a sharp Fermi
cut-off in the spectrum, but values of $E_F = 12 \pm 2$ meV and $n_w =$
$(3.5 \pm 0.5) \times 10^{11}$ cm^{-2} are estimated. More accurate values for
n_w, ($2.2 \pm 0.1 \times 10^{11}$ and $3.2 \pm 0.1 \times 10^{11}$ cm^{-2} for the first and
second resonances respectively) without any fitting procedure,
are obtained from the magneto-PL measurements[15,16]. Detailed
analysis of the photoluminescence spectra[16] also provides a
value for the electron temperature in the quantum well. On the
first resonance, the electrons which have tunneled into the well
undergo almost complete thermalization to the lattice
temperature (~ 4 K), in agreement with the magnetocapitance
measurements. On the second resonance, the effective
temperature of the electrons which have scattered into the lower
bound state ϵ_1 is estimated to be around 30 K. The high
electron temperature on the second resonance arises because the
tunneling out rate from the ϵ_1 subband (estimated to be about
10^9 s^{-1} at 2.4 V) is comparable to or faster than the electron-
acoustic phonon relaxation rate[24], $2\tau_{ac}^{-1} \sim 10^8 - 10^9$ s^{-1}.
Both of these rates are, of course, about 10^3 times slower than
the inter-subband scattering rate.

Direct evidence for inter-subband transitions from ϵ_2 to ϵ_1
comes from the high frequency magneto-oscillations, periodic in
1/B, which are observed in the tunnel current in the voltage
range between the first and second resonances. This series of
oscillations is due to scattering induced transitions of
electrons tunneling into the second resonant state ϵ_2 to the
upper Landau levels of ϵ_1[22,23]. This process modulates both
the charge density in the quantum well and the tunnel current.
Two series of high frequency oscillations are observed, as shown
in Fig. 9 which plots the frequency $B_f = [\Delta(1/B)]^{-1}$ of both
series against bias. The higher frequency series corresponds to
elastic scattering from the emitter to the well and the lower

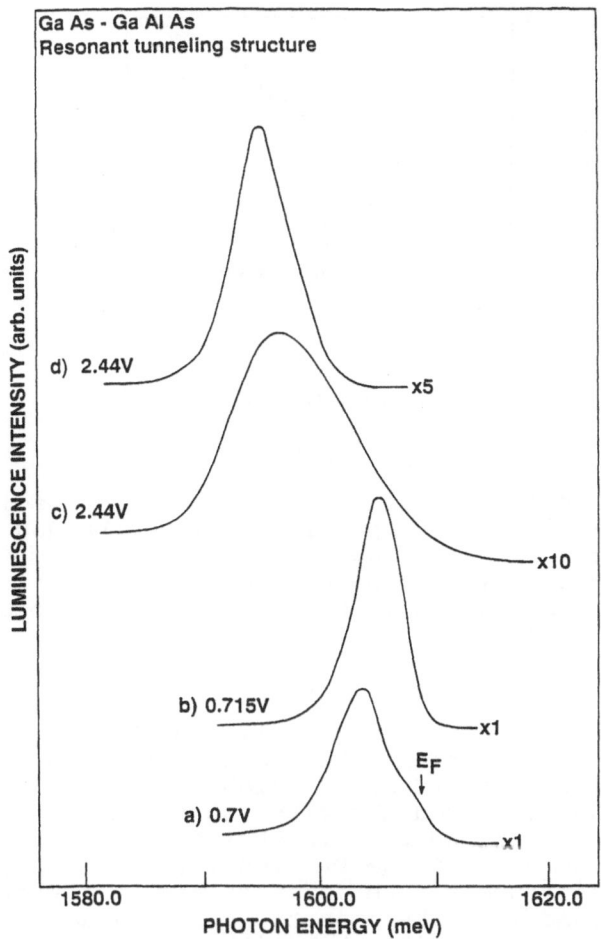

Fig. 8. PL spectra in the high and low current states at
the first resonance: (a) 0.7 V, 18.9 μA and (b)
0.715 V, 10.46 μA; and for the second resonance;
(c) 2.44 V, 12.33 mA and (d) 2.44 V, 3.93 mA.
The Fermi energy cut-off is indicated in (a).
[After Refs.15 and 16.]

series to a similar transition but with the emission of a
longitudinal optic phonon of energy $\hbar\omega_L$. The periods of the two
series are given by the equation

$$\epsilon_2 \simeq \epsilon_0 + eV_1 = \epsilon_1 + i\hbar\omega_c$$

where V_1 (< V) is the voltage drop between emitter and quantum
well, $\hbar\omega_c$ is the cyclotron frequency and i is the index of the
ith Landau level of the ϵ_1 quantum well state.

Figure 6c shows the PL intensity as a function of V. The
maximum of PL intensity occurs between the two resonances at 1.0
V where there is no electron charge buildup. In contrast to the
model of Young and co-workers[21] we therefore deduce that the
PL intensity is not, in general, simply proportional to n_w in
the quantum well, but also depends strongly on the population of

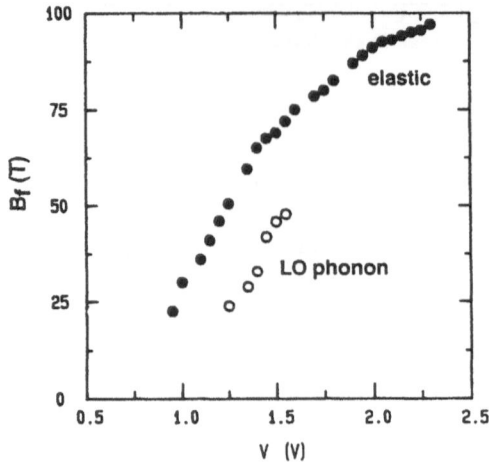

Fig. 9. Periodicity of the high frequency oscillations corresponding to inter-Landau
level scattering transitions; full circles
- elastic and open circles - inelastic
series of oscillations. (M.L. Leadbeater,
unpublished.)

the minority carrier holes in the well. Most of the holes which
participate in the PL tunnel into the well through the thicker
collector barrier. The effective height of the collector
barrier to the holes is reduced with increasing bias, thus
accounting in part for the rise in PL intensity between 0 and 1
V. The strong decrease of the PL signal from 1 to 3.3 V can be
understood by reference to the inset of Fig. 6. When the bias
is increased beyond 1 V, the injected holes will tunnel through
the first barrier more easily, but with an increasing number
passing straight over the second barrier. The dynamic hole
population in the well will then fall with a consequent decrease
in PL intensity as the bias is increased further (1.0 V to 3.5
V), in agreement with experiment.

CONCLUSION

     We have described the use of high magnetic fields and
photoluminescence techniques to reveal new phenomena in resonant
tunneling structures. Particular emphasis has been placed on
the importance of space-charge buildup and its role in intrinsic
bistability and on energy relaxation effects.

Acknowledgment

     This work is supported in part by the Science and
Engineering Research Council. The author gratefully
acknowledges the collaboration of Drs. M.L. Leadbeater, D.G.
Hayes, E.S. Alves, G.A. Toombs, P.E. Simmonds, M. Henini, O.H.
Hughes (Nottingham) and Dr. M.S. Skolnick (RSRE, Great Malvern)
in the work described in this paper. A full discussion of the
use of photoluminescence and magnetophotoluminescence spectroscopy for the study of resonant tunneling devices is given in
the paper by Skolnick et al.[16].

REFERENCES

1.   R. Tsu and L. Esaki, **Appl. Phys. Lett.**, 22:562 (1973)
2.   L.L. Chang, L. Esaki and R. Tsu, **Appl. Phys. Lett.**, 24:593
      (1974)
3.   M.C. Payne, **J. Phys. C: Solid State Phys.**, 19:1145 (1986)
4.   T. Weil and B. Vinter, **Appl. Phys. Lett.**, 50:1281 (1987)
5.   S. Luryi, **Appl. Phys. Lett.**, 47:490 (1985)
6.   B. Ricco and M. Ya Azbel, **Phys. Rev. B**, 29:1970 (1984)
7.   V.J. Goldman, D.C. Tsui and J.E. Cunningham, **Phys. Rev.
      Lett.**, 58:1257 (1987a) and V.J. Goldman, D.C. Tsui and J.E.
      Cunningham, **Phys. Rev. B**, 35:9387 (1987b)
8.   C.A. Payling, E.S. Alves, L. Eaves, T.J. Foster, M. Henini,
      O.H. Hughes, P.E. Simmonds, F.W. Sheard and G.A. Toombs,
      Surf. Science, 196:404 (1988)
9.   F.W. Sheard and G.A. Toombs, **Appl. Phys. Lett.**, 52:1228
      (1988)
10.   E.S. Alves, L. Eaves, M. Henini, O.H. Hughes, M.L.
      Leadbeater, F.W. Sheard, G.A. Toombs, G. Hill and M.A.
      Pate, **Electronics Lett.**, 24:1190 (1988). *In this paper the
      bias direction for occurrence of intrinsic bistability was
      referred to as "reverse bias".*
11.   E.S. Alves, L. Eaves, M. Henini, O.H. Hughes, M.L.
      Leadbeater, F.W. Sheard and G.A. Toombs, JSAP-MRS Int.
      Conf. on Elec. Mats., Tokyo, 1988, p.165 **in**: "Proceedings,
      Materials Research Society", (1989)
12.   A. Zaslavsky, V.J. Goldman and D.C. Tsui, **Appl. Phys.
      Lett.**, 53:1408 (1989)
13.   M.L. Leadbeater, E.S. Alves, L. Eaves, M. Henini, O.H.
      Hughes, F.W. Sheard and G.A. Toombs, **Semicond. Sci.
      Technol.**, 3:1060 (1988)
14.   M.L. Leadbeater, E.S. Alves, F.W. Sheard, L. Eaves, M.
      Henini. O.H. Hughes and G.A. Toombs, **J. Phys.: Conds.
      Matter**, 1:10605 (1989)
15.   D.G. Hayes, M.S. Skolnick, P.E. Simmonds, L. Eaves, D.P.
      Halliday, M.L. Leadbeater, M. Henini and O.H. Hughes, Proc.
      4th Int. Conf. on Modulated Semiconductor Structures 1989,
      Surf. Sci., 228:373 (1990)
16.   M.S. Skolnick, D.G. Hayes, P.E. Simmonds, A.W. Higgs, G.W.
      Smith, H.J. Hutchinson, C.R. Whitehouse, L. Eaves, M.
      Henini, O.H. Hughes, M.L. Leadbeater and D.P. Halliday,
      Phys. Rev. B, 41:10754 (1990)
17.   M.L. Leadbeater, E.S. Alves, L. Eaves, M. Henini, O.H.
      Hughes, A. Celeste, J.C. Portal, G. Hill and M.A. Pate,
      Phys. Rev. B, 39:3438 (1989)
18.   F.W. Sheard and G.A. Toombs, **Sol. St. Electronics**, 32:1443
      (1989)
19.   M. Tsuchiya, T. Matsusue and H. Sakaki, **Phys. Rev. Lett.**,
      59:2356 (1987)
20.   T.J. Foster, M.L. Leadbeater, D.K. Maude, E.S. Alves, L.
      Eaves, M. Henini, O.H. Hughes, A. Celeste, J.C. Portal, D.
      Lancefield and A.R. Adams, **Sol. St. Electronics**, 32:1731
      (1989)
21.   J.F. Young, B.M. Wood, G.C. Aers, R.L.S. Devine, H.C. Liu,
      D. Landheer, M. Buchanan, A.L. Springthorpe and P.
      Mandeville, **Phys. Rev. Lett.**, 60:2085 (1988)
22.   L. Eaves, G.A. Toombs, F.W. Sheard, C.A. Payling, M.L.
      Leadbeater, E.S. Alves, T.J. Foster, P.E. Simmonds, M.
      Henini, O.H. Hughes, J.C. Portal, G. Hill and M.A. Pate,
      Appl. Phys. Lett., 52:212 (1988)
23.   M.L. Leadbeater and L. Eaves, to be published

24. J.F. Ryan, R.A. Taylor, A.J. Turberfield, A. Mauel, J.M. Worlock, A.C. Gossard and W. Weigmann, **Phys. Rev. Lett.**, 53:1841 (1984)

SATURATED MINI-BAND TRANSPORT IN SEMICONDUCTOR SUPERLATTICES

G. Brozak,† M. Helm, R. Bhat, F. DeRosa, P. Grabbe,
D.M. Hwang, M. Koza, T. Duffield, C.H. Perry* and
S.J. Allen, Jr.

Bell Communications Research, Inc.
Redbank, NJ 07701, USA
†Physics Department, Northeastern University
Boston, MA 02115, USA

1. INTRODUCTION

Strong nonlinear response is an essential feature of narrow band transport. These potentially strong nonlinearities derive from the fact that a perturbation applied to a narrow band will exhaust the spectrum of states and the response will be saturated. The most exciting example is the long sought for but never achieved Bloch oscillation [1]. Here, we describe two different experiments that explore the saturated response of a narrow conduction band. First we discuss the effect of magnetic fields that drive the cyclotron resonance above the top of the mini-band and then, secondly, we document quenching of the narrow band conductivity at temperatures that exceed the mini-band width.

2. BAND SATURATED CYCLOTRON RESONANCE

Cyclotron resonance is a classic band structure probe that is able to measure without ambiguity the transport mass and scattering rate in any system that is sufficiently clean that the carrier is able to circle the applied magnetic field at least a few times before scattering. As a result, in a superlattice, when the magnetic field is oriented perpendicular to the growth direction, the observation of cyclotron resonance has the power to determine the vertical transport mass as well as the scattering rate. It is important to point out that cyclotron resonance requires that the electron execute coherent transport around the magnetic field and the probe is specific to quantum, coherent tunneling transport in the vertical direction. Phonon assisted hopping through barriers or thermally activated transport over barriers can not contribute to the resonance since they are basically scattering processes. The simple

---

*Guest Scientist at the Francis Bitter National Magnet Laboratory.

Condensed Systems of Low Dimensionality
Edited by J. L. Beeby *et al.*, Plenum Press, New York, 1991

observation of cyclotron resonance in the tunneling direction is unambiguous proof of quantum or coherent transport along the superlattice direction[2].

Implicit in the discussion of cyclotron resonance in normal solids is the assumption that the cyclotron diameter is large compared to the period of the periodic potential, and that the cyclotron frequency is much less than the band width. Unlike normal solids the length scale of the periodic potential in a semiconductor superlattice can become comparable to the magnetic length or cyclotron diameter and more important the magnetic energy can exceed the mini-band width. This is rarely, if ever, encountered in conventional solids, but under these conditions the cyclotron resonance will begin to "sense" the "graininess" or microscopic character of the periodic potential. These effects were anticipated as early as 1955 by Harper[3,4] but not observed until 1985 by Maan and co-workers in interband magneto-luminenscence[5-8] and by Duffield and co-workers[9,10] in tunneling cyclotron resonance. Interband magnetoluminescence experiments revealed that the Landau level fan of interband transitions disappeared when the cyclotron energy exceeded the mini-band width. In cyclotron resonance measurements, the low field resonance evolves at high magnetic fields into a field independent absorption controlled by the single barrier

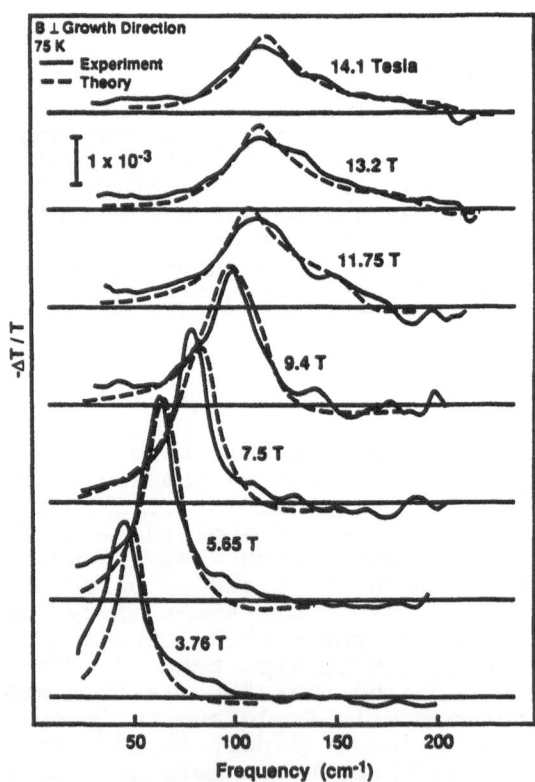

Fig. 1.   Fractional change in transmission as a function
of frequency with magnetic field, oriented
perpendicular to the growth direction, as a
parameter. Dashed curves are the result of the
model calculation.

tunneling rate. The frequency at which the cyclotron resonance saturates is determined by the mini-band width. The saturation of cyclotron resonance is unique to superlattice transport in narrow mini-bands.

The saturation effect in tunneling cyclotron resonance is shown in Fig. 1. These experiments were performed on a superlattice with a period of 10.4 nm and barriers 2 nm wide with 25% Al[10]. At low magnetic fields the resonance increases linearly and can be characterized by a cyclotron mass given by the geometric mean of the mini-band mass, m_s, and the in-plane GaAs mass, m_0,

$$\omega_c = eB/(m_0 m_s)^{1/2}. \tag{1}$$

At large magnetic fields the resonance ceases to increase and saturates around 120 cm^{-1}

If we assume that the superlattice mini-band is described by the tight binding approximation then the dispersion is given by

$$E(k) = -\Delta \cos(k z_0), \tag{2}$$

where z_0 is the superlattice period.

The low magnetic field resonance can be used to determine the mini-band mass, m_s, which is directly related to the mini-band width by

$$m_s^{-1} = \Delta z_0^2 / \hbar^2 . \tag{3}$$

From Fig. 1, we deduce that $m_s = .1 m_e$ and with Eqn. 3 we estimate that the mini-band width is ~ 120 cm^{-1}. This is where the cyclotron resonance in Fig. 1 is seen to saturate and broaden to high frequencies.

Heuristically speaking the resonance has exhausted all the states in the mini-band and can not increase in frequency. From another point of view as the electron rotates in a magnetic field it finds itself propagating down the superlattice in a forbidden band and therefore resonance can not proceed.

The phenomenon is amenable to a precise analytical description by diagonalizing Schrödinger's equation in a magnetic field in a one-dimensional periodic potential. Following Maan [5] we calculate the eigenstates of the following Hamiltonian

$$H = -\frac{\hbar^2}{2m} \frac{\partial^2}{\partial z^2} + \frac{1}{2} m \omega_c^2 z^2 + V(z + z_1) \tag{4}$$

where $\hbar$ and m are Planck's constant and the electron mass in GaAs, ω_c is the cyclotron frequency and $V(z)$ is the periodic potential. The origin has been chosen to be the orbit center which is displaced by z_1 from the origin of the periodic potential. We approximate the 2 nm wide barriers by delta functions with strength given by the barrier height-width product.

The eigenvalues that result from this calculation are shown in Fig. 2. In the Landau gauge the orbit center or center of

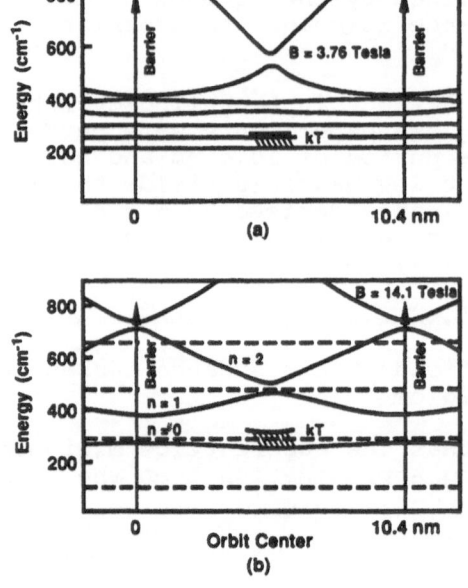

Fig. 2. Landau levels versus orbit position with respect
to the barriers for a magnetic field of 3.76 T,
(a) and 14.1 T, (b). The dashed lines indicate
where the levels would be without the barriers.
The cross hatched area indicates the scale of kT
for the experiments.

the wavefunction remains a good quantum number and Fig. 2
displays the Landau levels as a function of the orbit center
with respect to the periodic potential. In a uniform system
these would be a series of straight lines separated by $\hbar\omega_c$.

At modest magnetic fields, ~ 4 T, the lowest levels still
retain the appearance of a conventional Landau level diagram
with the energy independent of orbit center. The energy level
separation is given by the geometric mean of the in-plane and
mini-band mass, (1). As we rise in Landau level index the
levels develop considerable dispersion when the energy exceeds
the mini-band width.

At this point it is useful to refine the conditions for no
Landau level dispersion. Earlier we suggested that the relevant
parameter could be taken to be either the ratio of magnetic
length to superlattice period or cyclotron frequency to mini-
band width. These two ratios are effectively the same measure
only for the lowest Landau levels. The higher Landau levels
have in fact larger cyclotron diameter and hence have larger
ratio of cyclotron diameter to superlattice period yet show more
dispersion. Although the extent of the wavefunction is larger
for the higher Landau levels the wavefunctions contain a larger
number of nodes reflecting the higher kinetic energy of the
particle and it is the oscillatory features of the higher Landau
levels that sense the relative position of the wavefunction and
the periodic potential. As a result the critical parameter is
in fact the energy relative to the mini-band width. This fact

is readily apparent in the original interband magneto-luminescence of Maan[5]. There, the high lying Landau levels disappeared as they passed through the top of the mini-band. In the experiments we are discussing here, however, we are always probing the lowest Landau levels and the ratio of magnetic length/period is essentially equivalent to magnetic energy/mini-band width in controlling the dispersion of the low lying levels.

In Fig. 2, at ~ 14 T, all Landau levels suffer dispersion due to the superlattice periodic potential. The resonance will be given by a superposition of vertical transitions from the ground state to the first excited state. It is apparent that the lowest transition frequency is found at the barrier. At that point the electron may be thought, in classical terms, to execute cyclotron motion but tunnel twice per period through the barrier. The highest transition frequency is found for orbits centered on the quantum well. Here the electron is restrained not only by the magnetic field but also by the barriers on either side of the quantum well.

The full conductivity tensor may be derived for this system and the resulting far-infrared absorption calculated. The results are shown as a dashed lines in Fig. 1. It is apparent that the model reproduces in detail the saturating behavior of the cyclotron resonance accompanied by a broadening to high frequencies. The strongest feature in the inhomogeneously broadened line is the low frequency shoulder which corresponds to the barrier bound resonance, that is to say, resonance between states impaled on the barrier. For these orbits the time the electron takes to circle the magnetic field is determined predominantly by tunneling through the barrier and not by the magnetic field.

The relation of this frequency to the mini-band width can also be readily understood. The mini-band width is controlled by the tunneling rate between adjacent quantum wells. (Brozak et al.[11] have recently used these measurements to determine the tunneling rate through pure AlAs barriers deep in the barrier band gap). Since the saturated cyclotron resonance is also controlled by this tunneling rate it is not surprising that cyclotron resonance saturates at this same value and experimentally defines the mini-band width.

3. THERMAL QUENCHING OF MINI-BAND TRANSPORT

A striking prediction of narrow mini-band transport is that the conductivity should vanish in the high temperature limit. This is readily apparent in the case of a narrow three-dimensional band[12]. Conceptually all states in momentum space are equally populated, but not completely filled, and an applied electric field cannot bias this distribution to cause current flow. Although the effect may not be as robust in a super-lattice where the narrow band controls transport in the growth direction only, the conductivity in this direction should also be quenched as the temperature rises to uniformly populate all the states in directed momentum space.

To experimentally test this concept we measured the Drude conductivity along the growth direction of several superlattices

as a function of temperature from 50 to 200 K[11,13,14]. It is
important to measure the complete intra-mini-band response. A
measurement of the d.c. value alone is rendered ambiguous since
it can be altered by changes in mobility. By measuring the full
Drude response we are free of this ambiguity.

Three different superlattice structures were investigated
and all three demonstrated thermal quenching of the mini-band
transport. We focus here on a superlattice with 22 nm period
and 2 nm barriers. Low n doping of the order 2×10^{15} cm^{-3}
minimizes the effect of impurity scattering. The mini-band
width in this superlattice is only 33 K. The next highest mini-
band is separated by more than 250 K, so we can, to good
approximation, concern ourselves only with transport in the
lowest mini-band.

To couple the radiation field to the vertical motion we
chose to use a grating coupler as decribed by Heitmann et al.
[15,16] and Helm et al.[17]. This is depicted in Fig. 3. For
radiation wavelengths in the GaAs long compared to the period of
the grating, diffraction is ignored, and we focus on the near
field pattern imposed on the electric field. A substantial
component is developed under the grating lines in a direction
along the superlattice and will sense the conductivity of
interest. However, the component of the electric field parallel
to the quantum wells and barriers will still sense the in-plane
motion and render the experiment useless. To remove this defect
we apply a strong magnetic field along the superlattice
direction. This has the effect of lifting all of the oscillator
strength for the in-plane motion to the cyclotron frequency and
the low frequency response reveals the pure superlattice mini-
band conductivity.

From a conceptual point of view, it is important to point
out that despite the fact that the strong magnetic field

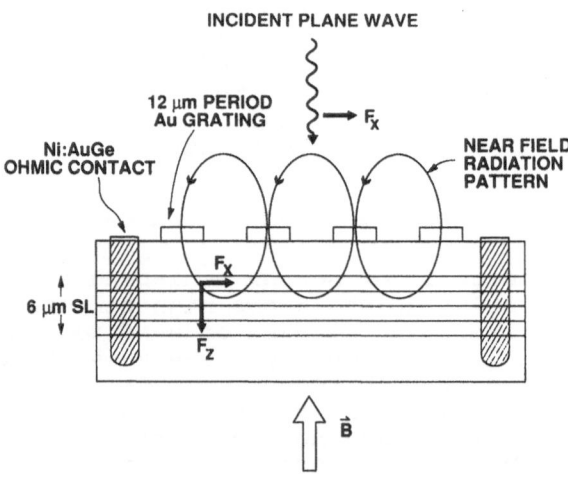

Fig. 3. The near field of the grating coupler
couples to both the perpendicular and in-
plane electron motion. However, the
perpendicular magnetic field lifts the
absorption due to the in-plane motion to
the cyclotron frequency.

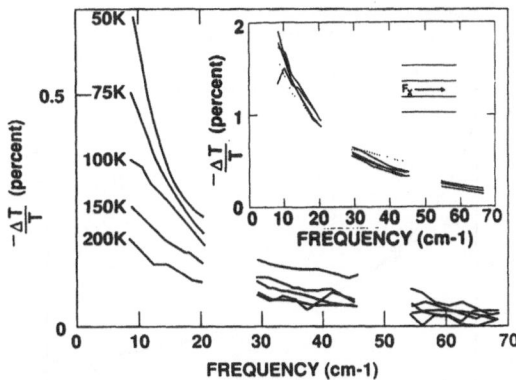

Fig. 4. Frequency dependent conductivity,
(arbitrary units), in the superlattice
direction with temperature as a parameter.
The inset shows the corresponding in-plane
conductivity with no substantial temperature dependence. The dotted curve in the
inset is the result at 50 K which begins
to be affected by carrier freeze out.

parallel to the superlattice appears to produce a quasi-one-
dimensional mini-band, this is not a critical issue. Even
though the excited Landau levels be populated their respective
quasi-one-dimensional mini-bands will be saturated provided that
kT is larger than their mini-band widths and the conductivity
will be quenched. The most important issue to produce quenching
of the conductance along the superlattice direction is that the
excited mini-bands in the quantum well which have larger band
widths must not be appreciably populated. The strong magnetic
field serves only to remove the in-plane response which is not
rejected by the grating from the part of the spectrum which is
dominated by the superlattice mini-band Drude.

Although we expect the absorption of the radiation under
these conditions to be proportional to the superlattice
conductivity[16], the proportionality factor is not known. In
Fig. 4 we show the mini-band Drude conductivity in arbitrary
units, at temperatures from 50 to 200 K. A striking drop in the
response is observed as the temperature is increased. As an
inset, we display the in-plane Drude conductivity, in zero
magnetic field. There is no significant change in the strength
of the in-plane Drude as is expected.

Unfortunately, we are restricted to temperatures above 50 K
to avoid freeze out of the carriers on the donors. Indeed, in
the inset, the dotted curve, which represents the in-plane Drude
at 50 K, begins to develop an increase in the high frequency
tail, at the expense of the low frequency response, which is due
to the appearance of donor bound electrons.

In Fig. 5 we display the conductivity at 10 cm^{-1} as a
function of temperature as well as the predictions of the theory
of Ktitorov et al.[12]. We fit to the theory at 75 K and let
the rest of the data fall where they may. The fit is excellent
considering the fact that the depression of the 50 K data is

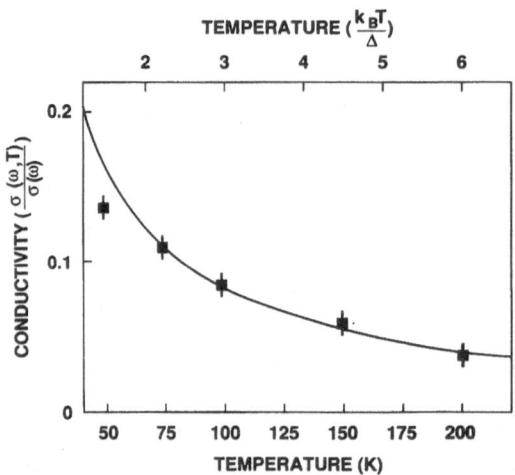

Fig. 5. Mini-band conductivity, (arbitrary units),
at 10 cm^{-1} as a function of temperature.
Solid line is the theory of Ktitorov
(1972) fixed to the results at 75 K.

probably due to the fact that the electrons are beginning to
freeze out on the donors. On the other hand, the confrontation
between experiment and theory is not complete since we are in
the high temperature limit where the conductivity is expected to
simply fall as 1/kT. Nevertheless, it is a gratifying and
substantial confirmation of thermal saturation and quenching of
the conductivity in a mini-band in a semiconductor superlattice.

SUMMARY

        We have reviewed two experiments that expose in a graphic
way the effect of conduction band saturation on electron
dynamics in superlattice mini-bands. In the first case we show
how the cyclotron resonance literally saturates when the
cyclotron energy exceeds the mini-band width. In this limit, a
real space description is required and the resonance frequency
is dominated by barrier bound resonances. In the second case
the superlattice Drude conductivity is shown to be quenched as
the thermal energy unformly populates the mini-band. This
result is particularly important for it is a precursor to Bloch
oscillation. If the band can be saturated by electron heating
in an applied electric field then conditions should be correct
for differential negative resistance controlled by Bloch
oscillations. Falling short of this important goal it is
apparent that this system should be rich in a variety of nonlinear phonomena excited by intense microwave and millimeter
wave fields.

REFERENCES

1.   L. Esaki and R. Tsu, **IBM J. Res. Develop.**, 14:61 (1970)
2.   T. Duffield, R. Bhat, M. Koza, F. DeRosa, D.M. Hwang, P.
      Grabbe and S.J. Allen Jr., **Phys. Rev. Lett.**, 56:2724 (1986)
3.   P.G. Harper, **Proc. Phys. Soc. London**, A58:879 (1955)

4.   P.G. Harper, **J. Phys. Chem. Solids**, 82:495 (1967)
5.   J.C. Maan, **Springer Series in Solid State Sciences**, 53:183 (1984)
6.   J.C. Maan, **Festkorperprobleme**, 27:137 (1987)
7.   G. Belle, J.C. Maan and G. Weimann, **Solid State Commun.**, 56:65 (1985)
8.   G. Belle, J.C. Maan and G. Weimann, **Surf. Sci.**, 170:611 (1986)
9.   T. Duffield, R. Bhat, M. Koza, F. DeRosa, K.M. Rush and S.J. Allen Jr., **Phys. Rev. Lett.**, 59:2693 (1987)
10. T. Duffield, R. Bhat, M. Koza, D.M. Hwang, F. DeRoas, P. Grabbe and S.J. Allen Jr., **Solid State Commun.**, 65:1483 (1988)
11. G. Brozak, E.A. de Andrada e Silva, L.J. Sham, F. DeRosa, P. Miceli, S.A. Schwarz, J.P. Harbison, L.T. Florez and S.J. Allen, Jr., **Phys. Rev. Lett.**, 64:471 (1990)
12. S.A. Ktitorov, G.S. Simin and V.Ya. Sindalovski, **Sov. Phys. Sol. St.**, 13:1872 (1972)
13. G. Brozak, F. DeRosa, D.M. Hwang, P. Micelli, S.A. Schwarz, J.P. Harbison, L.T. Florez and S.J. Allen, Jr., **Surf. Sci.**, to be published.
14. G. Brozak, M. Helm, F. DeRosa, C.H. Perry, M. Koza, R. Bhat, S.J. Allen, Jr., **Phys. Rev. Lett.**, 64:3163 (1990)
15. D. Heitmann and U. Mackens, **Phys. Rev. B**, 33:8269 (1986)
16. D. Heitmann, Kotthaus and E.G. Mohr, **Solid State Commun.**, 44:715 (1987)
17. M. Helm, P. England, E. Colas, F. DeRosa and S.J. Allen, Jr., **Phys. Rev. Lett.**, 63:74 (1989)

INTERSUBBAND CARRIER SCATTERING IN MULTIPLE QUANTUM WELL STRUCTURES

A. Seilmeier

Physikalisches Institut
Universität Bayreuth
D-8580 Bayreuth, FRG

1.   INTRODUCTION

Carrier transport and carrier relaxation in semiconductors are determined by several scattering mechanisms. A mechanism specific for heterostructures is the scattering between the subbands which are generated by carrier confinement in the layers. Intersubband scattering is also of importance for the performance of infrared detectors and modulators based on quantum well structures.

Intersubband relaxation has attracted a lot of experimental and theoretical attention in recent years. Various experimental techniques have been introduced: i) the Raman technique probes the occupation of an excited subband via resonantly enhanced Raman scattering[1,2]. ii) The infrared bleaching technique measures the recovery of intersubband absorption as a function of time[3]. iii) The determination of the saturation intensity of an intersubband transition gives information on the corresponding relaxation time[4]. iv) Interband absorption changes at special wavelengths are connected with intersubband scattering processes[5,6]. v) Resonant tunneling times in coupled quantum wells may be analogous to intersubband scattering times[7,8]. Some of the methods (e.g. the Raman technique and the IR bleaching technique) give direct information on the occupation of subbands. Other techniques (e.g. probing via interband absorption changes) are more indirect; simultaneous carrier heating processes and simultaneous relaxation processes in the hole states have to be taken into account.

In this paper the intersubband relaxation in n-type modulation doped $GaAs/Al_xGa_{1-x}As$ quantum well structures is discussed. The experimental data are obtained from infrared bleaching experiments[3]. Surprisingly long time constants of the order of 10ps are found. The long time constants may be attributed to a transfer of electrons to the potential minimum in the barrier generated by modulation doping of the samples[9].

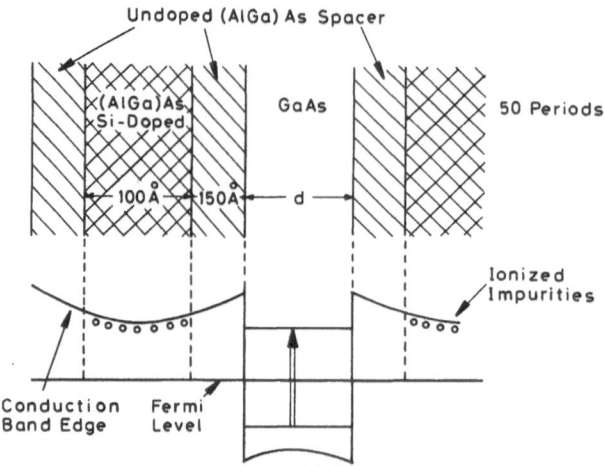

Fig. 1. Structure and conduction band of the n-type
modulation doped GaAs/Al$_x$Ga$_{1-x}$As quantum well
structures investigated.

2. EXPERIMENTAL

The structure and the conduction band of the n-type
modulation doped GaAs/Al$_x$Ga$_{1-x}$As samples investigated are shown
in Fig. 1. The samples are grown by molecular beam epitaxy on
(100) semi-insulating GaAs substrates of 350 μm thickness and
consist of fifty thin, undoped GaAs layers with individual
thicknesses between $d_{GaAs} = 47$ Å and 73 Å. They are embedded in
400 Å thick Al$_x$Ga$_{1-x}$As layers, in which the central 100 Å are
doped with Si. This leads to a two-dimensional carrier
concentration of about 6×10^{11} cm^{-2} per GaAs quantum well as
determined from Hall effect and Shubnikov-de Haas experiments.
The total multi-quantum well structures are clad between 0.2 μm
thick Al$_x$Ga$_{1-x}$As layers to avoid surface depletion and substrate
effects.

The energy of the subbands depends critically on the well
width and the aluminum concentration x of the barrier.
Modulation doping provides a large number of electrons in the
lowest subband and generates shallow potential minima in the
barriers (see Fig. 1).

The infrared bleaching technique requires ultrashort
infrared pulses tunable at wavelengths longer than 7 μm. An
intense infrared pulse resonantly excites electrons from the
lowest subband to the first excited subband resulting in a
bleaching of the intersubband transitions (see Fig. 1). A
second weaker light pulse of the same frequency monitors the
absorption change as a function of time. The absorption
recovery reflects the repopulation of the lowest subband.

The picosecond laser system used for the infrared bleaching
experiments is shown in Fig. 2. It is based on a parametric
amplification process. We start with a modelocked Nd-doped
glass laser system which generates single pulses of 5 ps
duration. The pulses pump a traveling wave infrared dye laser
(TWDL) producing tunable pulses between 1.15 μm and 1.40 μm[10].

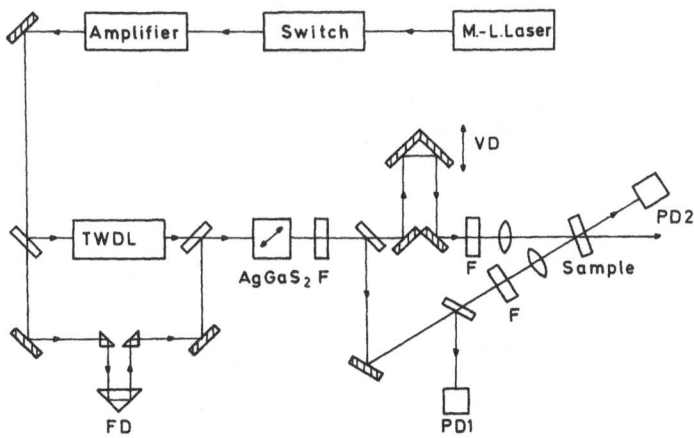

Fig. 2. Experimental system for infrared bleaching
experiments with picosecond time resolution.

A second part of the glass laser pulse and the dye laser pulse
are mixed in an $AgGaS_2$ crystal. In this way pulses at the
difference frequency tunable between 4 and 12 μm are generated
[11]. The pulses exhibit an energy of several microjoules, a
spectral width of about 10 cm^{-1}, and a pulse duration of 2 ps as
determined by a background-free autocorrelation technique.

The tunable infrared pulses are divided into pump and
weaker probe pulses which are focused on the sample. A
noncolinear geometry with an angle of ~ 30° between pump and
probe beam is used. The change of transmission of the probe
beam is studied as a function of delay time.

3.   RESULTS

In the following data are presented for thin ($d_{GaAs} < 75$ Å)
n-type modulation doped GaAs/Al$_x$Ga$_{1-x}$As quantum well structures
with a subband separation larger than the LO phonon energy. The
investigations were performed at room temperature. Prior to a
discussion of time resolved results, the infrared absorption of
the intersubband transitions is discussed.

It is well known that transitions between subbands exhibit
large oscillator strengths. Strong absorption is, however, only
observed for an electric field vector of the infrared light
normal to the layers. Samples are prepared with special prism
geometry leading to a large component of the electric field
vector normal to the layers (see inset of Fig. 3a). In this way
strong infrared absorption is achieved.

Figure 3 compares the intersubband absorption a) of a
sample with 75 Å well thickness and b) of a sample with 52 Å
well thickness both consisting of 50 wells. The spectra are
measured with the sample geometry shown in the inset. The
intersubband absorption is in the order of A = 1. This value is
sufficiently large to measure reliable data on transient
transmission changes. In Fig. 3 we observe a shift of the
absorption bands to higher energies and an increase of the
linewidth with decreasing well thickness. The larger linewidth

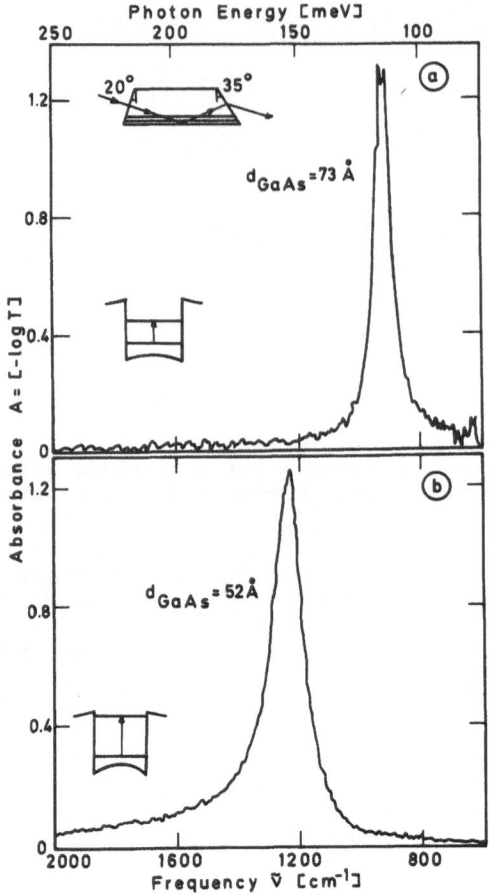

Fig. 3. Infrared absorption of multi-quantum well samples of
a) 73 Å well thickness, and of b) 52 Å well thickness. The absorption is due to intersubband
transitions. Inset: sample geometry.

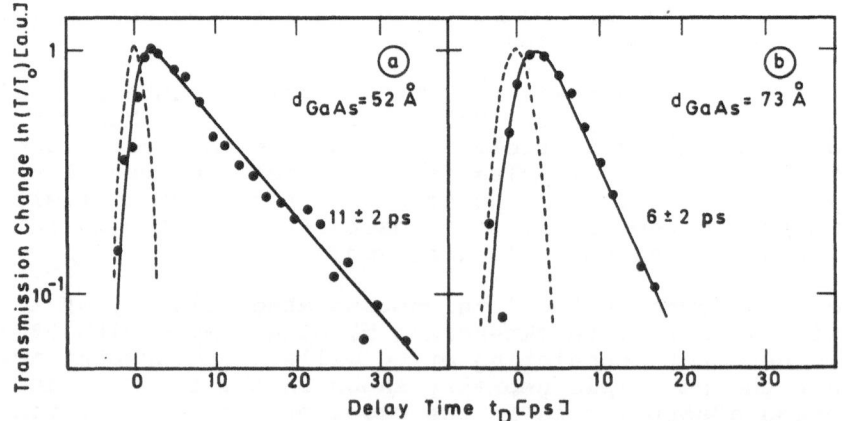

Fig. 4. Bleaching of the probe pulse as a function of
the delay time for 52 Å and 73 Å well widths
measured at room temperature. T_0 is the
transmission without excitation.

Table 1. Survey of intersubband time constants in GaAs/ $Al_x Ga_{1-x} As$ multiple quantum well structures.

Well thickness d (Å)	Al concentration x	Carrier density/well $n(10^{11} cm^{-2})$	Subband separation ΔE (cm^{-1})	Intersubband relaxation τ (ps)
Undoped				
116	0.3	-	518	< 10 [a]
120	0.3	-	520	< 12 [b]
146	0.3	-	416	≤ 1 [c]
215	0.3	-	216	~ 500 [a]
240	0.3	-	128	~ 20 [b]
n-type Modulation Doped				
47 ± 3	0.35	6.0	1300	14 ± 2
52 ± 2	0.35	6.1	1225	11 ± 2
59 ± 2	0.33	6.8	1140	8 ± 2
59 ± 1	0.29	6.6	1015	8 ± 2
73 ± 1	0.31	5.7	920	6 ± 2

a: Oberli et al.[1]; b: Levenson et al.[6]; c: Tatham et al.[2].

and the broad high energy tail in Fig. 3b indicate that the energy of the excited subband is close to the top of the well. The broadening is believed to be due to mixing with barrier states (see below) and continuum states.

Time-resolved data taken at room temperature are shown in Fig. 4a for the 52 Å sample and in Fig. 4b for the 73 Å sample. The broken curve represents the autocorrelation of the infrared pulses. The signal probed in Fig. 4

$$\ln(T/T_0) = \sigma l (N_2 - N_1 + N_0)$$

depends on the instantaneous number of carriers in the upper state N_2 and on the depletion of the lower state $(N_0 - N_1)$. (N_0 and N_1 are the total number of electrons and the number of electrons in the lower state, respectively; σ is the absorption cross section, l is the absorption length). We observe a rapid rise of the transmission and an absorption recovery on a longer time scale. Decay times of 11 ps and 6 ps are found for the 52 Å and for the 73 Å sample, respectively. The absolute value of the observed increase in transmission approximately corresponds to a complete bleaching of the absorption band.

Experimental data on undoped and modulation doped GaAs/ $Al_x Ga_{1-x} As$ multiple quantum well structures are summarized in Table 1. The shortest time constants are observed for undoped structures with a well thickness smaller than 200 Å. Longer time constants are observed in n-type modulation doped samples and, in particular, in samples with a well thickness larger than 200 Å.

4.   DISCUSSION

The detailed mechanism of intersubband relaxation is of special interest. In this section we focus on intersubband scattering in n-modulation doped multiple quantum well structures.

Electron-polar LO phonon interaction is generally accepted as the dominant relaxation mechanism for intersubband transitions in thin quantum wells (subband separation > $\hbar\omega_{LO}$). This fact is confirmed by the results on thicker quantum wells with $\hbar\omega_{LO}$ > subband separation (d_{GaAs} > 200 Å in Table 1) in which intersubband transitions via electron-polar LO phonon are not possible. In this case longer time constants are observed[1,6].

A first estimate of the intersubband relaxation times is obtained from an analytic expression derived by Ridley[15], assuming infinitely deep potential wells, and bulk phonon modes. In this model the momentum conservation approximation is used, i.e. the k-vector perpendicular to the layers is supposed to be conserved during the scattering process.

The scattering time τ_{21} of an intersubband transition between the first excited subband and the lowest subband is given by

$$\tau_{21} = \left[\frac{1}{2} W_0 \sqrt{\hbar\omega_{LO}/E_1} \{1+n(\hbar\omega_{LO})\} \left(\frac{1}{4+\hbar\omega_{LO}/E_1} + \frac{1}{12+\hbar\omega_{LO}/E_1} \right) \right]^{-1}$$

In this equation only the dominant processes which emit a phonon are taken into account.

$$W_0 = \frac{e^2}{4\pi\hbar} \left[\frac{2m^*\omega_{LO}}{\hbar} \right]^{1/2} \left[\frac{1}{\epsilon_\infty} - \frac{1}{\epsilon_s} \right] = 7.7 \times 10^{12} \text{ s}^{-1}$$

is the basic rate for electron-polar LO-phonon scattering, ω_{LO} the phonon frequency, $E_1 = \hbar^2\pi^2/2m^*L^2$ the energy of the lowest subband, and $n(\hbar\omega_{LO})$ the phonon population density. For a well width of L = 50 Å a time constant τ_{21} = 2 ps is estimated which is considerably shorter than the experimentally determined values for the n-type modulation doped samples. For an undoped sample, however, a time constant close to the calculated value τ_{21} has been observed for a well thickness of 146 Å[2].

Treating intersubband scattering in more detail shows that the process is highly involved. Several mechanisms have to be included which may increase the intersubband relaxation times: i) screening of the electron phonon interaction, ii) non-equilibrium phonon populations, iii) finite potential well depths, and iv) processes occurring with a small mismatch Δk of the momentum perpendicular to the layers, which are allowed due to an uncertainty on account of the finite well thickness. The four mechanisms considered increase the time constants by approximately 30%. Consequently, the calculated time constants are still shorter than 3 ps and do not fit the experimentally determined time constants in the order of 10 ps found for n-type modulation doped samples.

322

The calculations discussed so far are done with bulk phonon modes and not with two-dimensional phonon modes. Models taking into account quasi-2D phonon modes have been published by Ridley,[12], and Jain and Das Sarma[13]. In the two papers different approaches are used to describe phonons in quantum well structures, but in both models the time constants for infinite deep wells are not changed significantly by introducing 2D phonon modes. Jain and Das Sarma[13], give also some numbers for finite quantum well depths. For a 50 Å sample intersubband scattering times of 5 to 10 ps are estimated.

In the following the influence of the specific potential structure of n-type modulation doped quantum well structures on the intersubband relaxation is discussed which is believed to be of particular importance for the relatively long time constants observed in these samples. The ionized doping atoms (donors) in the center of the barrier give rise to a shallow potential minimum as shown in Figs. 1 and 5 containing several bound states. The eigenstates and wavefunctions of the entire heterostructure potential consisting of quantum wells and potential minima in the barriers have to be determined by a self-consistent band structure calculation. In Fig. 5 calcu- lated eigenstates and the corresponding wavefunctions are shown for a structure with a well width of 50 Å and a doping concen- tration of 6×10^{11} cm^{-2} [14].

The two bands of lowest energy are the lowest subband of the well and the ground state of the potential minimum in the barrier (solid and dotted line in Fig. 5). The wavefunctions indicate a strong confinement of the carriers in the well and the barrier, respectively. The higher states are mixed states made up from excited subband states and excited barrier states which are close to resonance. These states exhibit a large probability of finding electrons both in the well and in the barrier. An efficient transfer of electrons from the wells to the potential minima of the barriers and vice versa is expected. In Fig. 5 the energy levels are drawn in the well and in the barrier minimum, depending where they are arising from.

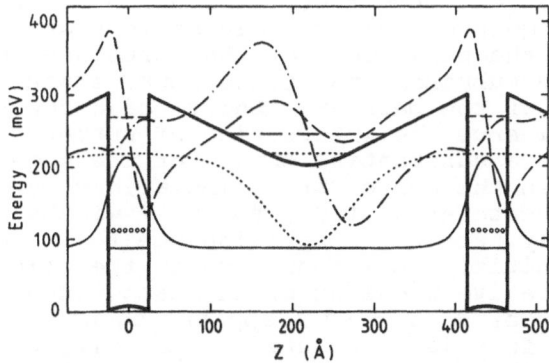

Fig. 5. Potential, eigenstates and wavefunctions for a n-modulation doped GaAs/Al$_x$Ga$_{1-x}$As quantum well structure: well width 50 Å; doping concentration 6×10^{11} cm^{-2} [14].

In the time resolved experiment electrons are excited to the first excited subband of the well (broken line in Fig. 5). The corresponding wavefunction exhibits a considerable probability for the electrons staying in the doping region of the barrier. In this way electrons are transferred to the barrier; there they relax to the lowest state of the potential minimum via LO phonon emission. This state is strongly confined in the potential minimum. The electrons can only return to the lowest subband of the well via excited states with wavefunctions exhibiting again a large amplitude within both the barrier and the well (e.g. the dash-dotted band). Such a relaxation path requires absorption of a phonon. Taking into account the electron transfer from the well to the barrier and the way back via excited states time constants in the order of 10 ps are estimated for the 52 Å sample. Moreover, the model qualitatively reproduces the observed dependence of the time constants on the well thickness.

More accurate calculations using Monte Carlo simulations are reported by Goodnick and Lugli[16]; Goodnick[17], and Educato et al.[18]. The calculations clearly show a delay of the recovery of intersubband absorption due to the transfer of electrons to the doping region of the barrier. Time constants are calculated which are even longer than the experimentally determined values. The numbers obtained from the Monte Carlo simulation are reduced again by taking into account a dynamic change of the self-consistent potential due to the transfer of carriers into the barriers[17].

The model calculations give strong evidence that inter-action between the subbands of the well and barrier states is the most relevant process which delays the return of electrons to the lowest subband of the well. This result is corroborated by additional measurements in samples with the same well width of 59 Å but different doping concentrations. In highly doped samples the potential minimum in the barrier is more pronounced leading to a stronger interaction of the upper subband of the well with barrier states. At low doping concentrations the barrier states are located at considerably higher energies than the upper subband of the well resulting in a reduced interaction and a substantially faster intersubband relaxation time.

It should be mentioned that intersubband scattering in modulation doped quantum well structures is highly sensitive to small changes in the parameters of the structure which determine the energy of the subbands and of the bound states of the potential minima of the barrier. Shifts of a few meV may considerably influence the interaction of subbands with barrier states leading to a substantial change of the eigenfunctions. The differences in the Monte Carlo simulations by Goodnick and Lugli[16] and Educato et al.[18], for well widths of 50 Å, and 47 Å, respectively, are not surprising due to the different parameters and initial conditions used in the calculations. In any case, quantitative modeling of the data taken for structures with narrow wells of $d_{GaAs} = 50$ Å appears to be extremely difficult. This fact is attributed to the difficult determination of the exact parameters of very thin modulation doped quantum well structures.

CONCLUSIONS

Progress has been made in the understanding of intersubband scattering processes. Modulation doping of the structure turned out to be of considerable influence on the intersubband relaxation. However, there are still open questions which have to be addressed in future experimental and theoretical work. Experiments at different sample temperatures should give more detailed information on the transfer channels of electrons from the wells to the barrier states and vice versa and their respective mechanisms. Quantitative agreement between model calculations and experimental results is expected from improved models which take into account the transient changes of the subband energies and wavefunctions with high accuracy.

Acknowledgment

The author thanks R. Berger, H.-J. Hübner and M. Wörner for their contributions. The work was done in close collaboration with G. Abstreiter and G. Weimann, Walter Schottky Institut, Technische Universität München. The high quality quantum well structures were grown by G. Weimann. The work is supported by the Deutsche Forschungsgemeinschaft.

REFERENCES

1. D.J. Oberli, D.R. Wake, M.V. Klein, J. Klem, T. Henderson and H. Morkoc, **Phys. Rev. Lett.**, 59:696 (1987)
2. M.C. Tatham, J.F. Ryan and C.T. Foxon, **Phys. Rev. Lett.**, 63:1637 (1989)
3. A. Seilmeier, H.-J. Hübner, G. Abstreiter, G. Weimann and W. Schlapp, **Phys. Rev. Lett.**, 59:1345 (1987)
4. F.H. Julien, J.-M. Lourtioz, N. Herschkorn, D. Delacourt, J.-P. Pocholle, M. Papuchon, R. Planel and G. LeRoux, **Appl. Phys. Lett.**, 53:116 (1988)
5. R.J. Bäuerle, T. Elsaesser, W. Kaiser, H. Lobentanzer, W. Stolz and K. Ploog, **Phys. Rev. B**, 38:4307 (1988)
6. J.A. Levenson, G. Dolique, J.L. Oudar and I. Abram, **Solid State Electronics**, 32:1869 (1989)
7. B. Deveaud, F. Clerot, A. Chomette, A. Regreny, R. Ferreira, G. Bastard and B. Sermage, **Europhys. Lett.**, 11:367 (1990)
8. G. Livescu, A.M. Fox, D.A.B. Miller, T. Sizer and W.H. Knox, **Phys. Rev. Lett.**, 63:438 (1989)
9. A. Seilmeier, M. Wörner, G. Abstreiter, G. Weimann and W. Schlapp, **Superlattices and Microstructures**, 5:569 (1989)
10. T. Elsaesser, H.-J. Polland, A. Seilmeier and W. Kaiser, **IEEE J. Quant. Electron.**, 20:191 (1984)
11. T. Elsaesser, H. Lobentanzer and A. Seilmeier, **Opt. Commun.**, 52:355 (1985)
12. B.K. Ridley, **Phys. Rev. B**, 39:5282 (1989)
13. J.K. Jain and S. Das Sarma, **Phys. Rev. Lett.**, 62:2305 (1989)
14. G. Abstreiter, M. Besson, K. Heinrich, A. Köck, G. Weimann and R. Zachai, **in**: Proceedings of the NATO Workshop on Resonant Tunneling in Semiconductors: Physics and Applications, L.L. Chang, ed., Plenum Press, to be published.

15. B.K. Ridley, **J. Phys. C**, 15:5899 (1982)
16. S.M. Goodnick and P. Lugli, **in:** OSA Proc. on **Picosecond Electronics and Optoelectronics**, 4:158 (1989)
17. S.M. Goodnick, **in:** Proc. of NATO ARW on Spectroscopy of Semiconductor Microstructures, G. Fasol, A. Fasolino, P. Lugli, eds., Plenum Press, (1990)
18. J.L. Educato, D.W. Bailey, A. Sugg, K. Hess and J.P. Leburton, **Solid State Electron.**, 32:1615 (1989)

ELECTRON-PHONON INTERACTION IN TWO-DIMENSIONAL ELECTRON GASES

W. Dietsche

Max-Planck-Institut für Festkörperforschung
7000 Stuttgart 80
Federal Republic of Germany

ABSTRACT

The interaction of a 2DEG with acoustic phonons has been studied in detail including the conservation laws for energy and momentum and the elastic anisotropy of the semiconductor materials. Most significant is the occurrence of a maximum in-plane phonon momentum of $2k_F$ up to which phonon scattering is possible ($2k_F$ cut-off). This effect was demonstrated by analyzing acoustical phonons emitted from a heated 2DEG and from absorption measurements of monochromatic phonons. In the latter case a newly developed technique based on the phonon-drag effect was employed. With the same technique the anisotropy of the electron-phonon interaction was made visible in a phonon imaging experiment.

1.   INTRODUCTION

The interaction of a two-dimensional electron gas (2DEG)[1] with acoustic phonons is important because almost any excess energy in the 2DEG is relaxed by phonon emission. Furthermore, the scattering by acoustic phonons forms the ultimate limit for the transport mobility[2]. The absorption and emission of phonons is also of fundamental interest because of the different dimensionalities of the two types of excitations.

In this contribution I report on our work on the emission and absorption of acoustic phonons by 2DEGs[3-5]. Since our experiments were largely based on the use of superconducting tunnel junctions, we were limited to study effects in zero magnetic field. Phonon experiments in high magnetic fields were performed by L.J. Challis and co-workers[6].

An important feature of our work is the spectral information about the phonon absorption and the emission process. This was achieved by using superconducting tunnel junctions as detectors and generators of phonons. In addition, with the help of a phonon imaging technique, we were able to observe the angular dependence of the phonon absorption.

Condensed Systems of Low Dimensionality
Edited by J. L. Beeby *et al.*, Plenum Press, New York, 1991

2.   PHONON EMISSION IN MOS STRUCTURES

We studied the phonon emission in MOS structures by passing a current through the 2DEG and observing the emitted phonons. The lowest subband in (100) Si is circular in k-space and is filled up to a Fermi vector $k_F = (2\pi N_s/g_V)^{1/2}$ where N_s is the electron density and $g_V = 2$ is the valley degeneracy. The probability of phonon absorption or emission can be calculated using Fermi's golden rule:

$$\Gamma(q) = \frac{2\pi}{\hbar} \sum_{k_i} |M|^2 \delta(\epsilon_i + \hbar\omega - \epsilon_f) f(k_i)[1 - f(k_f)]$$

where f is the Fermi distribution function which includes a drift momentum $\mathbf{K}_d = m^*\mathbf{v}_D/h$ due to a transport current, $\mathbf{k}_i$ and $\mathbf{k}_f$ are the two-dimensional wave wave vectors of the initial and final electron states, respectively, and M is the matrix element. This matrix element is nonzero only if $\mathbf{k}_i - \mathbf{k}_f = \mathbf{q}_p$ where $\mathbf{q}_p$ is the component of the phonon wavevector in the electron plane. Since the electrons are localized in the perpendicular direction there is a large uncertainty in their perpendicular momentum leading to coupling to phonons with perpendicular components of less than 1/d. Here d is the "thickness" of the 2DEG. Quantitatively this is described by a form factor.

An interesting feature at low temperatures is the existence of a maximum q_p value of the emitted phonon spectrum of about $2k_F$ originating from electron transitions along the diameter of the Fermi circle. Thus the spectrum of emitted phonons extends to maximum values of $q = 2k_F/\sin\theta$ or of frequency $\Omega = v_s 2k_F/\sin\theta$ where v_s is the sound velocity. Calculated spectra of transverse phonons emitted in the [111] direction are shown in Fig. 1 for different k_F values (traces marked 1 to 6). They show a distinct maximum just below $2k_F$ and then a rapid fall-off.

The experiments were performed with (100) Si samples of about 3 mm thickness. On one side of the crystal a MOS

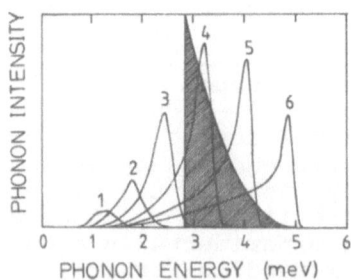

Fig. 1. Theoretical phonon-emission spectra for transverse acoustic (TA) phonons. The numbered traces correspond to k_F-values in units of 10^6 cm^{-1}. The spectra are tuned across the detector response function with increasing k_F.

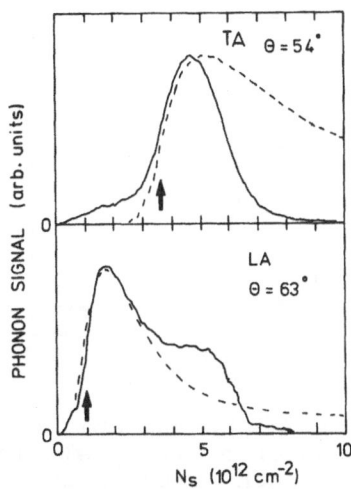

Fig. 2. Experimental phonon emission as function
of electron density for two phonon modes.

structure was prepared[3]. Current pulses of about 100 ns
duration were passed through the 2DEG resulting in phonon pulses
which were detected at the opposite surface. As detector a
superconducting Pb tunnel junction was generally used. This
detector has a cut-on frequency in the detection sensitivity
which corresponds to the superconducting energy gap (650 GHz).
At higher phonon frequencies there is an effective cut-off due
to the isotope scattering in Si. The combined effects lead to a
detector response function as indicated by the hatched area in
Fig. 1. In the course of the experiment N_s was varied by
changing the gate voltage attracting more or less electrons
toward the Si-SiO$_2$ interface. Consequently a sharp rise of the
detector signal was expected when the phonon emission spectra
were tuned across the detector response function.

Experimental results are shown in Fig. 2 (solid lines).
Here the TA and LA signals are plotted versus N_s. The emission
direction was in both cases near [111]. Steep increases were
indeed observed where expected (arrows). Since the LA phonon
spectrum extends to higher frequencies (they have a higher sound
velocity) the LA signal increase occurs already at smaller
densities. Comparison with theory (dashed lines) shows good
agreement for small N_s values but considerable deviations at
larger N_s. Similar agreement with theory at small and
disagreement at large N_s was observed if the emission angle was
varied[3].

We believe that the deviations are due to electron
transitions within and between higher subbands which lie close
to E_F at large N_s. Higher subbands were completely neglected in
the theory. To test this assumption uniaxial stress was applied
to the sample. This shifts the different subbands with respect
to each other and particularly the E_0' band which lies near the
zone edge starts to be populated. The effect on the phonon
emission is seen in Fig. 3: the anomalous signal fall-off occurs
already at smaller electron densities and the phonon emission
disappears altogether at higher stresses. In the region of the

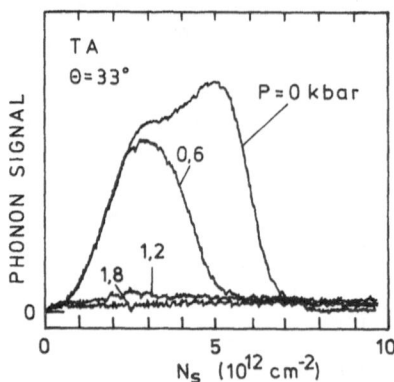

Fig. 3. As Fig. 2 for TA phonons emitted under a smaller
angle. Stress effectively eliminates the
emission of phonons with frequencies in the
detection band.

signal fall-off in Figs. 2 and 3 there is actually an excess of
low frequency phonons as is apparent if an Al-junction detector
with 100 GHz cut-on is used[7].

This excess phonon intensity also shifts to smaller
electron densities under uniaxial stress. Thus it seems as if
the upper subbands lead to the opening of new emission channels
of low frequency phonons.

3. PHONON ABSORPTION EXPERIMENTS IN GaAs HETEROSTRUCTURES

The most straightforward way to measure phonon absorption
is to monitor the phonon flux through a semiconductor interface
and record the change of signal with and without a 2DEG. This
was indeed done by Hensel et al.[8] in a reflection experiment.
In their situation the phonons did not only pass through the
2DEG but were also reflected off it. Other phonons were
reflected off the Si-SiO$_2$ interface near the 2DEG and also
reached the detector. The intensities of the two reflected

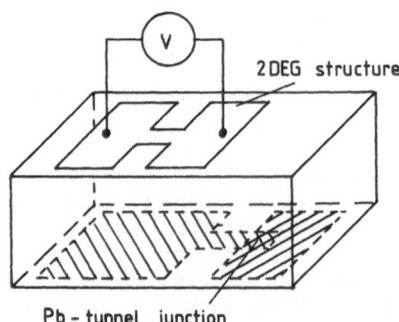

Fig. 4. Experimental set-up for a phonon-drag
experiment with a tunnel junction
generator.

beams happened to be about the same. Interferences between the
two beams made interpretation of the results very difficult. In
our experiment we measured the response of the 2DEG to the
phonons which was directly proportional to the phonon
absorption. Interferences affect the result only very little
because the intensity of the phonons at the 2DEG is almost not
affected by the small reflectivity off the semiconductor
interface. The reflection from the free surface of the crystal
is also unimportant because such surfaces are sufficiently
contaminated to absorb the phonons.

As experimental probe we utilised the phonon drag
effect[4,5]. The experimental setup is shown in Fig. 4. This
time the 2DEG device was prepared on a GaAs-AlGaAs heterostructure grown by molecular beam epitaxy on GaAs wafers of
0.6 mm thickness and (100) orientation. The 2DEG had a charge
density of 5.3 x 10^{11} cm^{-2}. A structure was etched out
consisting of two contact areas (0.5 x 0.8 mm^2 each) connected
by a 50 x 80 μm^2 bridge. A voltmeter was attached to the
contacts. Monochromatic phonons were generated by a Pb tunnel
junction (0.2 x 0.2 mm^2) placed on the opposite surface of the
sample. Phonons from the generator hit the 2DEG and, if
absorbed, they transfer their parallel momentum to the 2DEG.
This additional momentum will rapidly be distributed over all
electrons leading to a drift velocity of the electrons. Let the
channel of the 2DEG structure point into the x-direction, then

$$\dot{Q}_x = \int q_x [\Gamma(\mathbf{q})/v] F(\mathbf{q}) d\mathbf{q}$$

is the total transferred momentum along the channel in unit
time. Here $F(\mathbf{q})d\mathbf{q}$ is the phonon flux hitting the 2DEG, v is the
sound velocity, $\Gamma(\mathbf{q})$ is the absorption probability of the
phonons which has the same form as the emission probability
described earlier except that the $2k_F$ cut-off should be sharper
because the electron temperature is not raised by a transport
current. Under steady state conditions the phonon drag voltage
along the channel will be $V = \hbar\dot{Q}_x l/e$ where l is the channel
length and e is the unit charge.

Experimental results are shown in Fig. 5. Data obtained
with the phonon generator being placed in three different
directions from the channel are presented. The phonon drag
voltages are plotted versus frequency $f = \Omega/2\pi$. Absolute
intensity values can not be compared because of different
generator efficiencies. In each trace, we find a steady rise of
the signal until a cut-off is reached which we relate to the $2k_F$
condition. In one direction two cut-offs were observed; in this
case two phonon modes contributed to the signal. The spectral
width of the cut-offs is presently limited by the angular
resolution of the experiment. If we read the cut-off
frequencies from the experimental data (dashed lines) and use
average focusing corrected absorption angles we find k_F = (1.85
$\pm$ 0.15) x 10^6 cm^{-1} which is in excellent agreement with the 1.82
x 10^6 cm^{-1} value calculated from the electron density.

The phonon drag technique was also used by us to measure
the angular dependence of the phonon imaging technique[9]. The
Pb-junction generator was replaced by an Al film onto which a
laser (HeNe, 1 mW) was focused. The laser was raster scanned

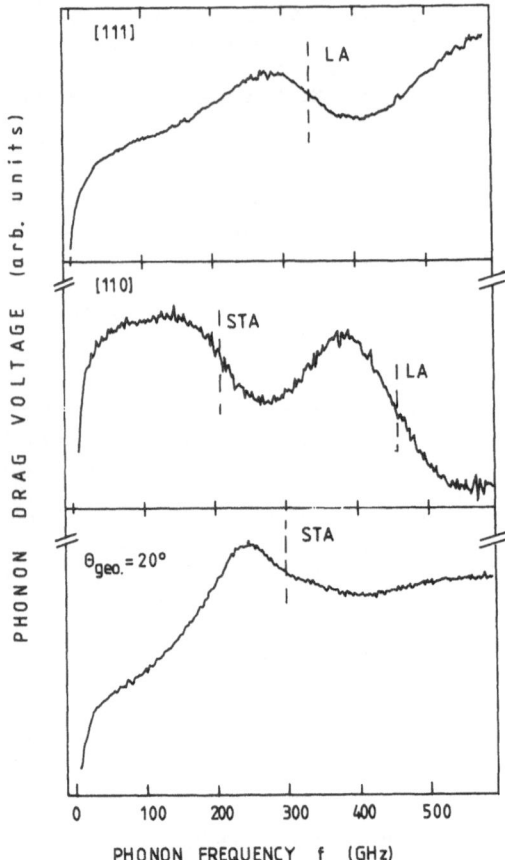

Fig. 5. Phonon-drag voltages as function of phonon
frequency for three different directions.

across the surface and phonons emanating from the heated spot
hit the 2DEG channel from different directions. Simultaneously
the phonon drag voltage was recorded and displayed as grey tones
on a TV monitor. The phonon flux in such an experiment is
modulated by the phonon focusing. The focusing pattern expected
in a (100) GaAs sample is shown in Fig. 6a. Bright areas
correspond to high phonon fluxes.

The measured phonon drag image is shown in Fig. 6b. Zero
voltage corresponds to an average grey tone while positive and
negative values are displayed brighter and darker, respectively.
The maximum voltages were of the order of 0.5 μV. Sharp
features are visible which coincide with some of those of the
focusing pattern. As expected for the phonon drag effect the
voltages changed sign whenever the q_x component of the incident
phonons reversed its sign. Most prominent are the FTA ridges
along the {100} planes. On the other hand, the STA structures
along the {110} planes do not show up in this measurement. The
two broad structures which are centered at [111] directions are
LA phonons as revealed by time resolved measurements. The
sharpness of the structures demonstrates that only the bridge in
the 2DEG device contributed to the observed pattern while phonon
drag voltages in the relatively large contact areas seemed to
average out.

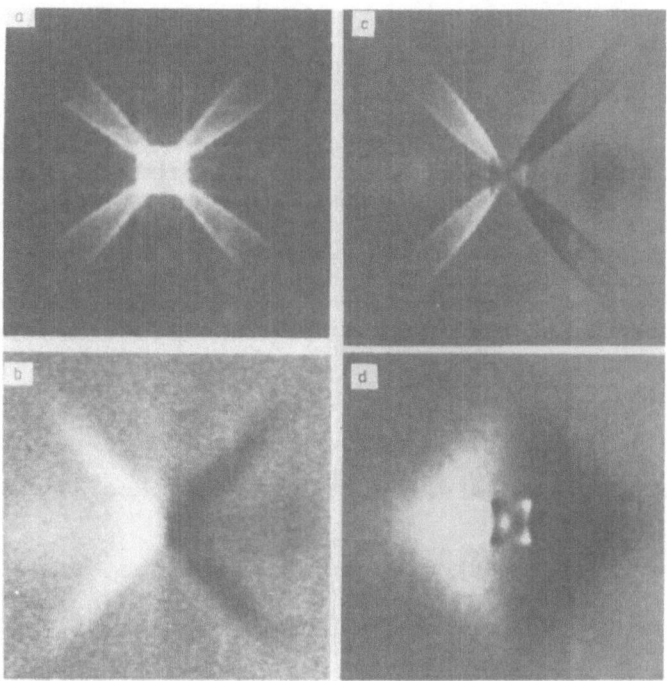

Fig. 6. (a) Phonon focusing pattering in GaAs
(simulated); (b) measured image of the phonon-
drag voltage; (c) calculated image of the
phonon-drag pattern using piezoelectric interaction; (d) same as (c) but using deformation
potential interaction.

The absence of the STA modes and the relatively strong
appearance of the LA phonons is indicative of a strong phonon
polarization dependence of the electron-phonon interaction. In
GaAs there are two types of coupling possible: deformation
potential and piezoelectric. In the first case the square of
the matrix element M is proportional to $(\Xi_d \; \mathbf{q} \cdot \mathbf{a})^2$ where Ξ_d is
the deformation potential and $\mathbf{a}$ is the unit polarization vector
of the phonon with wavevector $\mathbf{q}$. For the piezoelectric
interaction, on the other hand, the corresponding expression is
$e_{14}^2 (a_x q_y q_z + a_y q_x q_z + a_z q_x q_y)^2$ where e_{14} is a piezoelectric
constant[10]. In both cases screening is neglected. For a
comparison with experiment the phonon focusing calculations were
repeated but the resulting phonon fluxes were multiplied by a
number proportional to q_x times the phonon absorption
probability. In Figs. 6c and 6d the resulting theoretical
phonon drag images are shown if piezoelectric and deformation
potential interaction is assumed, respectively.

Obviously, the piezoelectric pattern agrees much better
with the experimental data than the deformation potential one.
Therefore one can conclude that the coupling in GaAs
heterostructures is of the piezoelectric type. There is one
area in the experimental image which does not agree with the
theoretical one: near the [100] direction we observed a much
higher signal than expected. This phenomenon is probably due to

thermal voltages which develop at the boundary of the 2DEG and the highly doped contact area[11]. In more recent experiments this effect could be removed by better contact geometries.

In conclusion, it was shown that information about the emission and absorption spectra of 2DEGs can be obtained using phonon spectroscopy. The most interesting feature found was the $2k_F$ cut-off. The different dimensionalities of electron gases and phonons led to cut-off frequencies which were dependent on emission angle and phonon mode. Using the phonon imaging technique the dependence of the phonon interaction on direction or, more precisely, on polarization was measured. In GaAs heterostructures good agreement with piezoelectric interaction was found.

REFERENCES

1. T. Ando, A.B. Fowler and F. Stern, **Rev. Mod. Phys.**, 54:437 (1982)
2. S. Das Sarma, p.261 of this volume
3. M. Rothenfusser, L. Köster and W. Dietsche, **Phys. Rev. B**, 34:5518 (1986)
4. H. Karl, W. Dietsche, A. Fischer and K. Ploog, **Phys. Rev. Lett.**, 61:2360 (1988)
5. A. Lega, H. Karl, W. Dietsche, A. Fischer and K. Ploog, **Surf. Science**, 229:116 (1990)
6. L.J. Challis, A.J. Kent and V.W. Rampton, p.967 **in:** "Phonons 89", S. Hunklinger, W. Ludwig and G. Weiss, eds., World Scientific, Singapore
7. W. Dietsche, M. Rothenfusser and L. Köster, **Physica Scripta**, T25:194 (1989)
8. J.C. Hensel, R.C. Dynes and D.C. Tsui, **J. Phys. (Coll.)**, 42:C6-308 (1981)
9. G.A. Northrop and J.P. Wolfe, p.165 **in:** "Nonequilibrium Phonon Dynamics", W.E. Bron, ed., Plenum
10. B.K. Ridley, "Quantum Processes in Semiconductors", Clarendon, Oxford. (1982)
11. F. Dietzel, W. Dietsche and K. Ploog, **Physica B**, 165:877 (1990)

OSCILLATING TRANSVERSE VOLTAGE IN A CHANNEL WITH QUANTUM POINT

CONTACT VOLTAGE PROBES

L.W. Molenkamp, H. van Houten, C.W.J. Beenakker,
R. Eppenga and C.T. Foxon*

Philips Research Laboratories
5600 JA Eindhoven, The Netherlands
* Philips Research Laboratories
Redhill, Surrey, RH1 5HA, UK

ABSTRACT

We have observed a transverse voltage on passing a current
through a narrow channel, electrostatically defined in a two-
dimensional electron gas, at *zero* magnetic field. The channel
is fitted with two opposite quantum point contact voltage
probes, and the voltage occurs when these probes are differently
adjusted, so that the transmission probabilities through the
probes have a different energy dependence. The transverse
voltage occurs only in the nonlinear response regime, and is
even in the applied current; the driving force of the effect is
the current-heating of the electrons in the channel. We observe
strong oscillations in the transverse voltage as the number of
occupied subbands in one of the voltage probes is varied by
means of electrostatic or magnetic depopulation. Model
calculations show that this novel effect is a manifestation of
the oscillatory thermopower of a quantum point contact predicted
by Streda. The effect can thus be used to obtain information on
electron heating.

1.   INTRODUCTION

The quantized conductance of a short and narrow
constriction (quantum point contact) in a two-dimensional
electron gas (2DEG) was discovered only recently[1], but has
already been utilized advantageously for the experimental study
of a variety of fundamental transport phenomena[2]. Part of
this research was aimed directly at the transport properties of
the point contacts themselves. In the regime of linear response
this work included, e.g., detailed studies of the ubiquity of
the quantization at zero magnetic field[3] and experiments on
the magnetic depopulation[4] of one-dimensional (1D) subbands at
finite fields. In a later stage, the research was extended to
the nonlinear regime, involving measurements and interpretation
of the nonlinear I-V characteristics of a quantum point contact
[5]. Subsequently, electronic instabilities occurring at high
voltages across the point contact were reported[6].

Condensed Systems of Low Dimensionality
Edited by J. L. Beeby *et al.*, Plenum Press, New York, 1991

Another class of experiments includes studies where two adjacent point contacts are used as voltage and current probes of ballistic transport phenomena. This approach was first used to study coherent electron focusing[7]. The electron focusing technique was subsequently used to study scattering processes[8] and to detect the injection of hot electrons through a voltage biased point contact[9]. At high magnetic fields, adjacent point contacts have been used for selective excitation and detection of the edge channels that are responsible for the transport in this regime, resulting in the observation of an anomalous quantization of the quantum Hall plateaus[10]. A geometry with two opposite point contacts was used to study the ballistic series resistance[11] and to detect electron beam collimation[12].

In this paper we demonstrate how one can use two opposite point contact voltage probes to detect electron heating in a current carrying channel. We will show that this detection technique constitutes a measurement of the thermopower of a quantum point contact, thereby enabling an estimate of the amount of electron heating in the channel. In addition, we observe strong oscillations that are related to the depopulation of 1D electric subbands in the quantum point contacts. Our main results have been briefly reported elsewhere[13].

2. TRANSVERSE VOLTAGE QUADRATIC IN THE CURRENT

Using electron beam lithography, we have fabricated split-gate microstructures on a GaAs/(Al,Ga)As heterojunction wafer with a 2DEG mobility of about 100 $m^2V^{-1}s^{-1}$ and a carrier density $n_s = 3.5 \times 10^{15}$ m^{-2}. The structures consist[12] of a narrow channel of 18 μm length and a width W_0 of 4 μm. On both sides of the channel two point contacts are defined, with 3 μm separation. A schematic layout of the gates and ohmic contacts is given in Fig. 1. Unless stated otherwise, the samples are kept at a temperature of 1.65 K and no magnetic field is applied.

We have performed a number of experiments to study the electron heating in the channel. In a first experiment, a dc

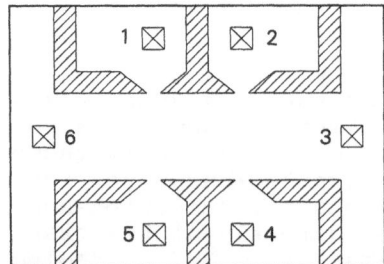

Fig. 1. Schematical layout of the samples used in these experiments. The hatched areas indicate the top gates, the crosses depict the ohmic contacts. In the actual devices, the channel has a length of 18 μm and a width W_0 of 4 μm. The separation between two adjacent point contacts is 3μm.

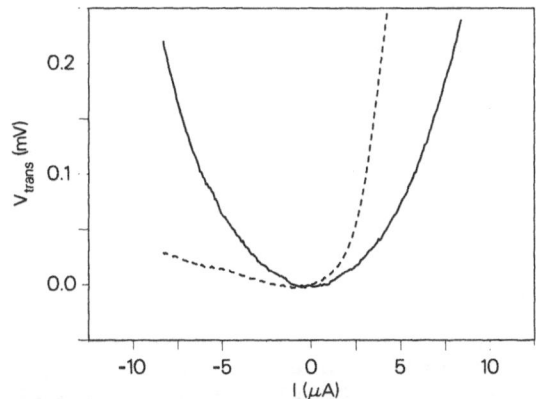

Fig. 2. The dependence of V_{trans} on the current in the channel. In this plot, $V_{gate\ 1}$ = -2.97 V, corresponding to $R_{pc\ 1}$ ≈ 12.9 kΩ, and $V_{gate\ 5}$ = -0.7 V ($R_{pc\ 5}$ ≈ 0.5 kΩ). The lattice temperature T_0 = 1.65 K. The drawn (dashed) curve was obtained using ohmic contacts 6 and 3(2) as current source and drain, respectively. (In these experiments we have not averaged over both current directions.)

current I is passed through the channel, using 2DEG contacts 6 and 3 as source and drain, respectively. We now measure the voltage on contact 5 relative to that on contact 1, i.e. a transverse voltage $V_{trans} \equiv V_5 - V_1$, where we have adjusted the gates such that the resistance of point contact 1 (leading to voltage probe 1) is much higher than that of point contact 5. As discussed below, electron heating leads to a positive value of V_{trans} due to accumulation of electrons in the 2DEG region behind the most strongly pinched-off point contact (1). The curve in Fig. 2 is a plot of V_{trans} as a function of the current I, for $V_{gate\ 1}$ = -2.97 V, $V_{gate\ 5}$ = -0.7 V and I ranging between -10 and +10 μA. The observation of a nonzero V_{trans} in the absence of a magnetic field is by symmetry not allowed in the linear response regime. The observed V_{trans} exhibits, to a very good approximation, a quadratic dependence upon I. It is one of the very few examples of even-order nonlinear behavior reported so far in semiconductor microstructures. Previously, a second order nonlinearity was observed as second harmonic generation in the quantum diffusive transport regime at mK temperatures[14-16]. This effect saturated at very low current levels - four orders of magnitude lower than those relevant for our structures. We have observed the quadratic current dependence in this configuration for currents up to 20 μA and temperatures up to 30 K.

Since our signal is relatively large and robust, it is evident that the observed behavior is not related to quantum coherence in the channel. We attribute the effect to current heating of the electrons in the channel. In a homogeneous, isotropic material, Joule heating can only lead to odd-order nonlinearities in the longitudinal voltage response[16,17]. The reason is one of symmetry: a longitudinal voltage has to change sign on reversing the current. The even-order nonlinearity

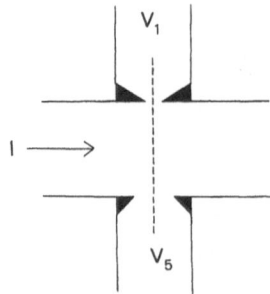

Fig. 3. An arrangement for transverse voltage
measurements in a channel. The point
contact voltage probes are indicated in
black. The dashed line is a line of
mirror symmetry; its presence demands that
the transverse voltage should be even in
the current.

observed here is due to the effect of the voltage probes
themselves on the measured voltage difference. The measurement
geometry is shown schematically in Fig. 3. As is evident from
this figure, symmetry now requires that any transverse voltage
$V_5 - V_1$ should be even in the current.

Assuming Joule heating as the driving force of our effect,
the simplest classical mechanism leading to a transverse voltage
emerges when one realizes that in a point contact the bottom of
the conduction band is raised with respect to the 2DEG regions
leading to the point contact[2]. Hot electrons in the channel
will easily overcome this barrier and enter the cold regions
behind the point contacts. Since we use the point contacts as
voltage probes, this flux must be compensated by a flux of cold
electrons back into the channel. These cold electrons can only
cross the energy barrier of the point contact provided the cold
2DEG region is sufficiently charged up. We thus expect a
voltage to develop across the point contacts which depends on
the energy barrier height, and thus on the resistance of the
point contacts. To the extent that the heated electrons in the
channel can be described by a Fermi-Dirac distribution, our
transverse voltage becomes simply the difference in thermo-
voltage of two differently adjusted point contacts. While the
classical mechanism described above is qualitatively correct, it
cannot explain the oscillations in the transverse voltage
discussed in the following section. A quantum mechanical
treatment is necessary to account for all our observations, and
will be given below.

In view of our claim that Joule heating causes the
transverse voltage, it is important to estimate the actual
temperature of the electrons in the channel. We assume that the
lattice temperature T_0 is unaffected and uniform over the
sample. A rough indication of the electron temperature T in the
channel can be obtained from the heat-balance equation

$$c_v (T - T_0) = (I/W_0)^2 \rho \tau_\epsilon \tag{1}$$

where $c_v = (\pi^2/3)(k_B T/E_F) n_s k_B$ is the heat capacity per unit area
of the 2DEG, ρ is the resistivity in the channel, and τ_ϵ an

energy relaxation time associated with energy transfer from the electron gas to the lattice. For I = 5 μA and an estimated[18] $\tau_\epsilon \approx 10^{-10}$ s this yields T - $T_0 \approx$ 1 K. A quantitative calculation of the electron temperature should account for the temperature dependence of c_v and τ_ϵ, and should also include the heat conduction in the 2DEG for the actual device geometry. An additional contribution of comparable magnitude to T - T_0 results from the contact resistance R_{pc} of about 200 Ω. This result can be readily obtained from the expression for the Sharvin resistance[2], R_{pc} = $(h/2e^2)(\pi/k_F W_0)$, which leads to a voltage drop $IR_{pc}/2$ at the entrance of the channel. (Note that this mechanism allows for the observation of a transverse voltage in a channel that is much shorter than the mean free path of the electrons). The magnitude of V_{trans} in the present geometry appears to saturate for currents larger than 20 μA. This is probably due to the temperature dependence of c_v and τ_ϵ (cf. Eqn. 1), in addition to nonnegligible lattice heating.

A larger V_{trans} for a given current can be realized by injecting the current over a barrier. This leads to a large voltage drop close to the detecting voltage probes; in other words, the contact resistance contribution to the electron heating process becomes dominant. The dashed curve in Fig. 2 was obtained for the same gate voltages as used for the full curve, but now using contacts 6 and 2 as current source and drain, respectively. The enhancement of V_{trans} when the electrons are accelerated by the voltage drop over point contact 6 (which has a resistance of approximately 10 kΩ) is quite dramatic. Of course, in this case the hot electron distribution may differ appreciably from a heated Fermi-Dirac distribution, in which case V_{trans} is not simply related to the thermopower of the voltage probes. Note also that V_{trans} is not even in I in this particular measurement geometry. The transverse voltage is only enhanced for electrons that first are accelerated over the barrier and then detected by the point contact voltage probes. In the reverse current direction, the electrons are heated after passing the voltage probes and V_{trans} does not increase.

3.   QUANTUM OSCILLATIONS IN V_{trans}

To further elucidate the origin of V_{trans} we have performed a series of experiments using the same geometry as described above (the current I is passed through ohmic contacts 6 and 3, and the voltage measured is V_{trans} = V_5 - V_1), but now with a fixed current of I of 5 μA. We scan the voltage of the gates leading to contact 1, while keeping the voltage on the other gates fixed at -2.0 V. To eliminate any spurious longitudinal resistance contributions (which occur in some samples due to a small misalignment of the voltage probes) the signal is averaged over both current directions. The full curve in Fig. 4a is the result of such an experiment. Dramatic oscillations in V_{trans} are observed, that are not expected from the above classical description. For comparison, we have plotted the gate voltage dependence of the (two-terminal) resistance of the scanned point contact (1) in the same figure (dashed line). The oscillations in V_{trans} peak whenever the resistance of the point contact changes from one plateau to another. From the expression[2] for the quantized resistance of a point contact, R_{pc} = $h/2e^2 N(E_F)$, we see that this occurs whenever the number $N(E_F)$ of 1D electric subbands in the quantum point contact that are available for

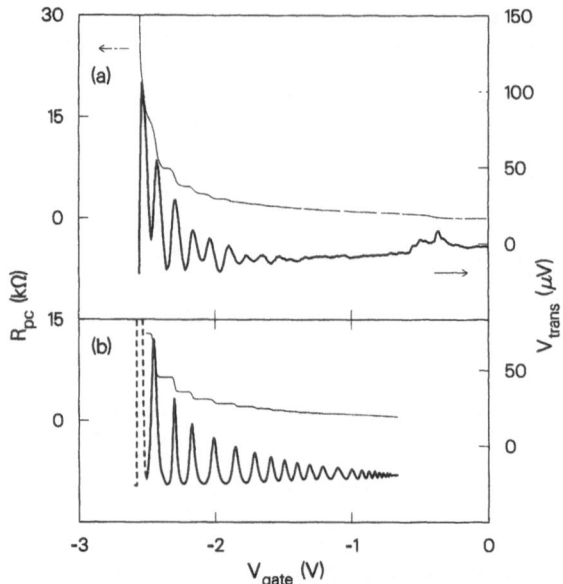

Fig. 4. (a) Experimental traces of V_{trans} (thick curve) and R_{pc} (thin curve) as a function of V_{gate} at a lattice temperature $T_0 = 1.65$ K for $I = 5$ μA and $V_{gate\ 5} = -2.0$ V. These data were obtained from a different sample than in Fig. 2. (b) Calculation of the transverse voltage (thick curve) using Eqns. 1-3 with electron temperature $T = 4$ K, $T_0 = 1.65$ K, and $E_F = 13$ meV. The thin curve gives the dependence of R_{pc} on V_{gate}, calculated for a temperature T_0 using experimental values of W and E_0.

electrons at the Fermi energy E_F, changes by one. Below we will present a quantitative model, which will demonstrate explicitly the close relation between V_{trans} and the thermopower of a quantum point contact.

To model our experiment we straightforwardly extend Streda's calculation[19] of the thermopower of a quantum point contact to finite temperature and voltage differences. (The method to calculate the thermopower from the transmission probabilities was developed by Sivan and Imry[20], along the lines of Landauer's formula for the conductance[21].) As the voltage probes draw no net current, the influx of hot electrons $I_{hot \to cold}$ from the channel into the voltage probe should exactly cancel the flux of cold electrons in the reverse direction, $I_{cold \to hot}$. We thus have $I_{hot \to cold} - I_{cold \to hot} = 0$, or

$$2\frac{e}{h} \int_0^\infty t(E)[f_{hot}(E) - f_{cold}(E)]dE = 0 \tag{2}$$

where $t(E)$ is the transmission probability summed over the 1D subbands that propagate through the point contact at energy E, and f_{hot} and f_{cold} are the distribution functions of the electrons in the channel and in the cold 2DEG region. For the present, we assume that these functions are well represented by Fermi-Dirac distributions at chemical potentials E_F and $E_F + \Delta\mu$, and at temperatures T and T_0, respectively. Note that Eqn. 2 holds for any hot electron distribution function which depends on energy only, so that it can also be used, e.g., to describe the effects of injection over a barrier, as discussed in the previous section. The quantum point contact is modeled by a square well lateral confinement potential of width W and well bottom at energy E_0 (measured with respect to the conduction band bottom in the channel). Assuming a transmission of unity for each of the N(E) subbands in the point contact, we have

$$t(E) = N(E) = \text{Int } [(2m/\hbar^2)^{1/2}(E - E_0)^{1/2}W/\pi]\theta(E - E_0), \qquad (3)$$

where Int denotes truncation to an integer, and $\theta(_x)$ is the unit step function. From Eqns. 2 and 3 we can obtain $\Delta\mu$ numerically. Since V_{trans} is the difference of the voltage measured by two differently adjusted point contacts, the above calculation should be repeated for the reference point contact, to obtain $\Delta\mu^{ref}$, which has a constant value. The transverse voltage is then found from $V_{trans} = (\Delta\mu - \Delta\mu^{ref})/e$. For comparison with the experiment, we have treated $\Delta\mu^{ref}$ as an adjustable baseline.

The result of our calculation, for $T_0 = 1.65$ K and assuming T = 4 K (consistent with the estimate from the heat balance equation discussed above for a current I = 5 μA), is given in Fig. 4b. Experimentally determined values for W and E_0 were used. The good agreement of the calculated curve with the experimental data of Fig. 4a indicates that our theoretical understanding of the effect is basically correct. No detailed quantitative agreement is obtained, and was not to be expected. For example, the peaks in the experiment are broader than the theoretical ones. Similarly, the experimental point contact resistance R_{pc} shows less pronounced steps than the calculation. Since both R_{pc} and V_{trans} depend on the detailed behavior of t(E) near E_F, both discrepancies may be ascribed, at least partly, to uncertainties in the transmission probability. The experiments show additional structure around threshold ($V_{gate} = -0.5$ V) where the point contact (and the channel) is just defined. This is explained by the associated abrupt change[2] in $t(E_F)$. The voltage peak near $V_{gate} \approx -2.6$ V (just beyond the R = $h/2e^2$ resistance plateau), turns out much weaker in the experiment than in our calculations (dashed part). The size of this peak is very sensitive to the (unknown) details of the dependence of t(E) on V_{gate} in the pinch-off regime, and we have not attempted to achieve a better agreement.

The connection of the above calculation with Streda's result on the thermopower of a quantum point contact can be made explicit on considering the limiting behavior at low lattice temperatures and small electron heating ($k_B T_0$ and $k_B T$ both much smaller than the subband separation at the Fermi energy). In this limit, Eqns. 2 and 3 yield the result[19] that the peak in $\Delta\mu$ when the (N + 1)-th subband is depleted has amplitude $\Delta\mu \approx (\ln 2)k_B(T - T_0)/N$. The transverse voltage experiment employing current heating is a very convenient way to measure the thermopower in the quantum ballistic transport regime, where only temperature differences on the scale of a mean free path are important. Application of an external temperature difference to a semiconductor sample containing a quantum point contact would only result in a very small temperature drop over the point contact, and the signal would be obscured by spurious contributions from the thermopower of the bulk 2DEG, enhanced by phonon drag contributions[22]. Since the lattice temperature T_0 is kept uniform in the present experiments, phonon drag does not play a role. Current heating was very recently also used for the study of universal thermopower fluctuations in the phase coherent diffusive transport regime[23].

4. QUANTUM OSCILLATIONS AT MODERATE MAGNETIC FIELD

Using the same measurement geometry as described above, we have also performed measurements of the quantum oscillations in moderate magnetic fields. The results of such experiments, for magnetic fields of 0, 0.75, 1.5 and 3.0 T are shown in Fig. 5. In a moderate perpendicular magnetic field, the number of 1D subbands in a point contact is determined both by the electrostatic potential of the gates and by the magnetic field[2]. The intersubband energy spacing of these magnetoelectric subbands increases with increasing magnetic field. Consequently, the number of occupied 1D subbands $N(E_F)$ for a given Fermi energy E_F

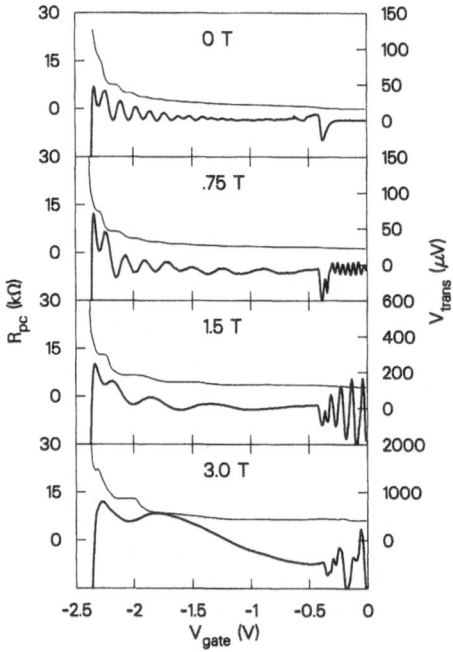

Fig. 5. The effects of magnetic depopulation on V_{trans}. Again, the thick curves give V_{trans} and the thin curves R_{pc}. The magnetic fields at which the data were taken are indicated in the figures. Other experimental conditions: $T_0 = 1.65$ K, $I = 5$ μA and $V_{gate\ 5} = -1.0$ V. The rapid oscillations at small gate voltages are due to depopulation of the Landau levels in the 2DEG area underneath the gate. The data were taken from the same sample as in Fig. 4; however, due to thermal cycling the zero-field dependence of R_{pc} on V_{gate} has changed.

decreases with increasing magnetic field. This magnetic depopulation effect can be very directly observed by measuring the conductance of a quantum point contact as a function of gate voltage, for various magnetic fields. The magnetic depopulation manifests itself by a reduction of the number of plateaus in a given gate voltage interval[4].

Magnetic depopulation of 1D subbands in the quantum point contact affects the oscillations in V_{trans} as a function of V_{gate} in a very similar manner: in a magnetic field, V_{trans} still exhibits peaks at gate voltages where the number of magnetoelectric subbands changes by one, but the oscillations are more widely spaced in gate voltage. The increase in magnitude of V_{trans} at the magnetic field values used in Fig. 5 is caused by the enhancement of the longitudinal resistivity ρ_{xx} due to Shubnikov-de Haas oscillations and, hence, of the power dissipation in the channel.

In the traces taken at finite magnetic fields, additional rapid oscillations are seen in Fig. 5 at small gate voltages where one point contact and one of the channel boundaries are not yet fully defined. For $V_{gate}^{def} < V_{gate} < 0$, where V_{gate}^{def} is the depletion threshold under the gate, the partially depleted 2DEG area underneath the gates acts as a weak potential barrier for the electrons moving from the channel region to the 2DEG region surrounding the ohmic contacts. Thus, a transverse voltage will develop on passing a current through the channel, analogous to the situation discussed in the previous paragraphs where the point contact acts as potential barrier. The rapid oscillations as a function of gate voltage can thus be explained by the stepwise decrease in the number of Landau levels in the barrier region. (Because of the absence of lateral confinement, the relevant quantum states in the wide regions under the gate are bulk Landau levels, rather than 1D magnetoelectric subbands[2].) Note that the trace taken at 3.0 T shows a doublet structure, presumably due to spin splitting.

The rapid oscillations are linear in V_{gate}, which implies that the density of the 2DEG underneath the gate is also linear in V_{gate} (as expected when the capacitance between gate and 2DEG does not depend on V_{gate}). We can estimate the carrier density under the gate at zero gate voltage, n_s^{gate}, by measuring the period ΔV_{gate} of the oscillations in this gate voltage region. We use

$$\nu = \frac{h n_s^{gate}}{2eB} = \frac{V_{gate}^{def}}{\Delta V_{gate}}, \tag{4}$$

where ν is the filling factor (the factor of two accounts for the two spin states). From Fig. 5 we find that for all three fields $n_s^{gate} \approx 3.3 \times 10^{15}$ m^{-2}, which is only slightly smaller than the density of the bulk 2DEG ($n_s \approx 3.5 \times 10^{15}$ m^{-2}), which was determined from Shubnikov-de Haas measurements on the same sample. The electron concentration in the 2DEG is thus hardly affected by the presence of a grounded top gate.

CONCLUSIONS AND OUTLOOK

We have demonstrated a novel type of nonlinearity in a semiconductor nanostructure, caused by electron heating. In contrast to thermal effects in bulk material, where only odd-order nonlinearities in the longitudinal voltage are expected, we find a transverse voltage that is even in the current. The essential requirement to observe such a voltage is a difference in transmission probability for hot and cold electrons in at least one of the voltage probes. The origin of this difference can be classical (a potential barrier in the probes), or quantum mechanical (discrete 1D subbands in the probes). The presence of point contact voltage probes in our channel has enabled us to observe the quantum oscillations in the thermopower of the point contacts[19]. Other applications in physics might include the determination of hot electron distributions in spatial directions that are not accessible to conventional hot electron spectroscopy[24]. Finally, one can anticipate that this method may be used as a sensitive probe for the study of magnetothermal effects, such as magnetophonon resonances. Since the effect is not fundamentally limited to very low temperatures, its application in devices, e.g., high frequency mixers and parametric amplifiers is well worth pursuing.

Acknowledgment

We acknowledge C.E. Timmering for sample preparation, M.A.A. Mabesoone for technical assistance during the experiments, and A.A.M. Staring, J.G. Williamson, and M.F.H. Schuurmans for valuable discussions.

REFERENCES

1. B.J. van Wees, H. van Houten, C.W.J. Beenakker, J.G. Williamson, L.P. Kouwenhoven, D. van der Marel and C.T. Foxon, **Phys. Rev. Lett.**, 60:848 (1988); D.A. Wharam, T.J. Thornton, R. Newbury, M. Pepper, H. Ahmed, J.E.F. Frost, D.G. Hasko, D.C. Peacock, D.A. Ritchie and G.A.C. Jones, **J. Phys. C**, 21:L209 (1988)
2. H. van Houten, C.W.J. Beenakker and B.J. van Wees, **in:** "Semiconductors and Semimetals", M. Reed, ed., Academic Press, New York, to be published.
3. G. Timp, **in:** "Semiconductors and Semimetals", M. Reed, ed., Academic Press, New York, to be published.
4. B.J. van Wees, L.P. Kouwenhoven, H. van Houten, C.W.J. Beenakker, J.E. Mooij, C.T. Foxon and J.J. Harris, **Phys. Rev. B**, 38:3625 (1988)
5. L.P. Kouwenhoven, B.J. van Wees, C.J.P.M. Harmans, J.G. Williamson, H. van Houten, C.W.J. Beenakker, C.T. Foxon and J.J. Harris, **Phys. Rev. B**, 39:8040 (1989)
6. R.J. Brown, M.J. Kelly, M. Pepper, H. Ahmed, D.G. Hasko, D.C. Peacock, J.E.F. Frost, D.A. Ritchie and G.A.C. Jones, **J. Phys. Condens. Matter**, 1:6285 (1989)
7. H. van Houten, B.J. van Wees, J.E. Mooij, C.W.J. Beenakker, J.G. Williamson and C.T. Foxon, **Europhys. Lett.**, 5:721 (1988); H. van Houten, C.W.J. Beenakker, J.G. Williamson, M.E.I. Broekaart, P.H.M. van Loosdrecht, B.J. van Wees, J.E. Mooij, C.T. Foxon and J.J. Harris, **Phys. Rev. B**, 39:8556 (1989)
8. J. Spector, H.L. Stormer, K.W. Baldwin, L.N. Pfeiffer and

K.W. West, **Surf. Sci.**, 228:283 (1990)

9.   J.G. Williamson, H. van Houten, C.W.J. Beenakker, M.E.I. Broekaart, L.I.A. Spendeler, B.J. van Wees and C.T. Foxon, **Phys. Rev. B**, 41:1207 (1990)

10.   B.J. van Wees, E.M.M. Willems, C.J.P.M. Harmans, C.W.J. Beenakker, H. van Houten, J.G. Williamson, C.T. Foxon and J.J. Harris, **Phys. Rev. Lett.**, 62:1181 (1989)

11.   D.A. Wharam, M. Pepper, H. Ahmed, J.E.F. Frost, D.G. Hasko, D.C. Peacock, D.A. Ritchie and G.A.C. Jones, **J. Phys. C**, 21:L887 (1988)

12.   L.W. Molenkamp, A.A.M. Staring, C.W.J. Beenakker, R. Eppenga, C.E. Timmering, J.G. Williamson, C.J.P.M. Harmans and C.T. Foxon, **Phys. Rev. B**, 41:1274 (1990)

13.   L.W. Molenkamp, H. van Houten, C.W.J. Beenakker, R. Eppenga and C.T. Foxon, **Phys. Rev. Lett.**, 65:1052 (1990)

14.   R.A. Webb, S. Washburn and C.P. Umbach, **Phys. Rev. B**, 37:8455 (1988)

15.   S.B. Kaplan, **Surf. Sci.**, 196:93 (1988)

16.   P.G.N. de Vegvar, G. Timp, P.M. Mankiewich, J.E. Cunningham, R. Behringer and R.E. Howard, **Phys. Rev. B**, 38:4326 (1988)

17.   R. Landauer, **in**: "Nonlinearity in Condensed Matter", A.R. Bishop, D.K. Campbell, P. Kumar and S.E. Trullinger, eds., Springer, Berlin (1987)

18.   D.R. Leadley, R.J. Nicholas, J.J. Harris and C.T. Foxon, **Solid State Electronics**, 32:1473 (1989)

19.   P. Streda, **J. Phys. Condens. Matter**, 1:1025 (1989)

20.   U. Sivan and Y. Imry, **Phys. Rev. B**, 33:551 (1986)

21.   R. Landauer, **IBM J. Res. Dev.**, 1:223 (1957)

22.   R. Fletcher, M. D'Iorio, A.S. Sachrajda, R. Stoner, C.T. Foxon and J.J. Harris, **Phys. Rev. B**, 37:3137 (1988); C. Ruf, H. Oblow, B. Junge, E. Gmelin, K. Ploog and G. Weimann, **Phys. Rev. B**, 37:6377 (1988)

23.   B.L. Gallagher, T. Galloway, P. Beton, J.P. Oxley, S.P. Beaumont, S. Thoms and C.D.W. Wilkinson, **Phys. Rev. Lett.**, 64:2058 (1990)

24.   See, e.g. J.R. Hayes, A.F.J. Levi and W. Wiegman, **Phys. Rev. Lett.**, 54:1570 (1985); A. Palevski, M. Heiblum, C.P. Umbach, C.M. Knoedler, A.N. Broers and R.H. Koch, **Phys. Rev. Lett.**, 62:1776 (1989)

THE SUPPRESSION OF RANDOM TELEGRAPH NOISE IN A POINT CONTACT

G. Timp, R.E. Behringer, J.E. Cunningham and
E.H. Westerwick

AT & T Bell Laboratories
Holmdel, New Jersey 07733, USA

ABSTRACT

The conductance of a one-dimensional (1D) constriction in a
two-dimensional electron gas (2DEG) can be quantized in steps of
$2e^2/h$ as a function of the width of the constriction,
corresponding to the depopulation of 1D subbands, provided the
constriction is short enough so that backscattering from
impurities is improbable. Generally, impurities or disorder
have a deleterious effect on the quantization of the conductance
of a one-dimensional constriction, but not always. We
demonstrate that scattering from a defect can be suppressed in a
modulation doped, 1D constriction through an examination of
discrete changes in the conductance of the wire which resemble a
random telegraph signal (RTS). The discrete changes in the
conductance develop from time-dependent fluctuations in the
scattering potential of a single defect in close proximity to
the 2DEG. We show that the switching associated with RTS is
quenched whenever an integral number of 1D subbands are
occupied, but reappears near the threshold for the occupation of
a subband. The suppression of the switching is attributed to
the quenching of small angle impurity scattering in a one-
dimensional wire.

1.   RANDOM TELEGRAPH NOISE

Discrete changes in the conductance of a wire, which
resemble a random telegraph signal (RTS), are prevalent in the
drain-source conductance of small area ($\leq 1 \ \mu m^2$) field effect
transistors (FET) near threshold[1-4]. RTS is supposed to
develop from time-dependent fluctuations in the scattering
potential of a single defect in close proximity to the two-
dimensional electron gas (2DEG) in a FET. The discrete
modulation may develop from either the emission or capture of a
single electron by a trap[1,2]. RTS is supposed to be a
constituent of the (1/f) noise observed in large devices
[2,3,5,6]. Because it represents a time-dependent change in the
threshold voltage for a FET, it is the bane of submicron
semiconductor electronics. The intensive investigation of RTS

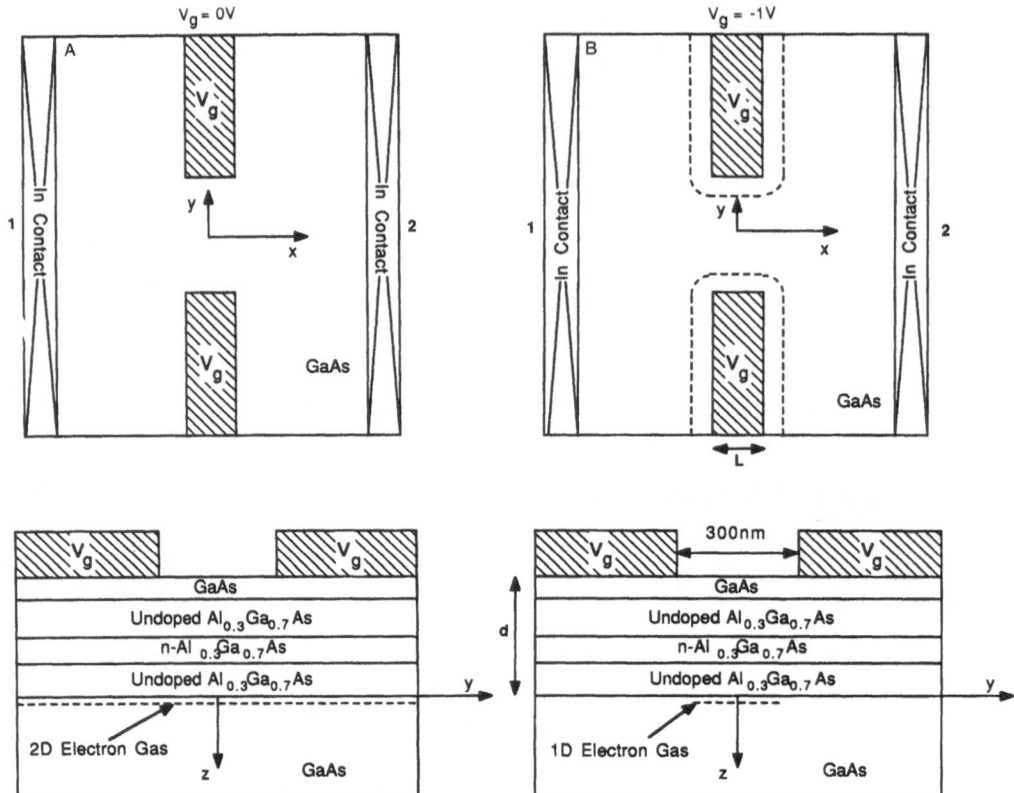

Fig. 1. Figure 1(a) and (b) respectively show the
geometry of the split-gate used to investigate
the two terminal resistance of a constriction in
a 2DEG for two different bias voltages: $V_g = 0$ V
and $V_g = 1$ V. The top portions of the figures
show schematically the top plan of the device
and the location of indium contacts. (Actually,
the contacts are approximately 300 μm from the
gate electrodes and there are intervening
voltage terminals.) The thickness, d, is
usually < 100 nm while the gap between the
electrodes is typically 300 nm wide. The
depletion around the gate electrodes is indicated by the dashed lines in Fig. 1(b). The
lower portions of the figures represent a cross-
section through the device. The 2DEG is
localized within 10 nm of the lower AlGaAs/GaAs
heterointerface. For gate voltages $V_g < -0.5$ V
the 2D electron gas is constrained within the
gap to the region indicated by the dashed lines
in Fig. 1(b).

has so far focused on the dependence of the capture and emission
kinetics of the trap on gate votage, drain current and
temperature[2,7].

We report here generally on the effect of disorder on the
conductance of split-gate FETs used to make quantum point
contacts[8-10], and specifically on the effect of RTS on the

conductance. We show that RTS can be suppressed when the channel of a small FET becomes comparable to the wavelength of an electron at the Fermi energy. The scattering is quenched when an integral number of one-dimensional subbands (1D) are occupied. After Sakaki[11], we attribute the suppression of RTS to the absence of small angle scattering in one dimension.

A typical split-gate FET used to make a quantum point contact is represented schematically in Fig. 1. The electrostatic potential provided by the split gate is used to laterally constrain a high mobility 2DEG in a modulation-doped AlGaAs/GaAs heterostructure to the region within the gap between the gate electrodes. The heterostructure consists of a GaAs substrate, an $Al_xGa_{1-x}As$ (x = 0.3) spacer layer, a δ-doped Si layer, an AlGaAs buffer layer and a doped GaAs cap. The dimensions for the heterostructures examined in this work, along with the mobility and carrier density of the 2DEG, are summarized in Table 1. The split gates are fabricated using electron beam lithography to prepare a mask for lift-off. A Ti/Au or Ti/AuPd film approximately 7.5/50 nm thick is evaporated onto the mask and lifted off to give the split gate electrodes. The split gate electrodes are nominally 200 nm wide with an intervening gap of 300 nm. The disposition of the gate electrodes on top of the heterostructure is shown in top of Fig.2.

By applying a negative voltage to the split gates, the 2DEG gas at the AlGaAs/GaAs interface, immediately beneath the electrodes, is depleted and so the 2DEG is laterally constrained within the gap between the electrodes. Figure 2 shows the

Table 1. A list of n (the 2D carrier carrier density), μ (the 2D mobility), d_s (the $Al_{0.3}Ga_{0.7}As$ spacer layer thickness), d_b (the $Al_{0.3}Ga_{0.7}As$ buffer layer thickness), d_c (the thickness of the GaAs cap layer), the low temperature mean free path L_e estimated from the mobility, and the electron wavelength λ_F for the six different materials examined in this work. The concentration of Si in the delta-doped layer is approximately 4×10^{16} m^{-2} except for heterostructure #6 which has a doping concentration of 5×10^{16} m^{-2}.

| | Heterostructure | | | | | |
	#1	#2	#3	#4	#5	#6
$n/10^{15}$m^{-2}	1.0±0.07	1.6±0.05	2.0±0.08	2.75±0.05	4.4±0.2	8.0±0.2
μ(m^2/Vs)	85±4.	93±2.	105.±5.	115.±5.	87±2	5.±1.0
d_s (nm)	78	50	47	42	27	7.7
d_b (nm)	52	40	28	24	53	70.
d_c (nm)	6	9	9	6	7	6
L_e (μm)	3.0	4.3	5.6	6.7	6.6	0.5
λ_F	84.	63.	46.	48.	39.	28.

resistance (conductance), $R_{12,12}$ ($G_{12,12}$), measured as a function of the applied gate voltage found in devices fabricated on heterostructures #4 and #5. The notation $R_{kl,mn}$ denotes a resistance measurement in which there is a positive current from lead k to l, and a positive voltage is detected between leads m and n. The series resistance, found at $V_g = 0$ V, was subtracted from the measured resistance obtained as a function of gate voltage to give the date plotted in Fig. 2. The depletion of the 2DEG immediately beneath the gate electrodes occurs when $V_g \approx -0.30$ V, and is not shown in Fig. 2. As the gate voltage decreases, the constriction within the gap between the electrodes narrows; the carrier density within the constriction decreases; and plateaus are observed in the resistance[9,10]. The plateaus found in the resistance of constrictions with a lithographic length of 200 nm are clearly evident in Fig. 2a. The conductance obtained by inverting the resistance versus gate voltage is also shown in Fig. 2a; i.e. $G_{12,12} \equiv 1/R_{12,12}$. $G_{12,12}$ is evidently quantized in steps of $2e^2/h$ with about 1-5% accuracy as an increasingly negative gate voltage makes the constriction narrower[12]. The conductance of Fig. 2a is well represented by: $G_{12,12} = 2e^2N/h$, with N an integer ranging from 1 to about 11 depending on the device.

Each step found in the conductance as a function of V_g corresponds to the depopulation of one 1D subband due to a change in the width of the constriction, W, relative to λ_F. According to the Landauer formula[13], the conductance is:

$$G = (2e^2/h) \sum_{j,k=1}^{N} t_{jk}$$

where t_{jk} is the probability intensity for transmission from subband k into subband j, and N is the number of subbands in the constriction. If we assume that: (1) the temperature is much smaller than the separation between transverse energy levels, (2) the length of the perfect wire is much longer than the decay length for evanescent modes, and (3) there is no scattering (i.e. $t_{ij} = \delta_{ij}$); then the conductance of the constriction with N occupied subbands is approximately: $G = (2e^2/h)N$. Thus, the ideal conductance of a constriction is a quantized function which measures the number of occupied 1D subbands.

Impurity scattering can have a deleterious effect on the quantization of the conductance. For example, we find that, as the length of the constriction increases, the quantization deteriorates. Figure 2b shows the resistance and conductance determined as above for a constriction with gate electrodes of length L = 600 nm, spaced with a 300 nm gap on the same heterostructures. The deviations of the conductance from $2e^2N/h$ shown in Fig. 2b are reproducible and time-independent. While the quantization of the resistance is still apparent in the top of Fig. 2b; the quantization is not evident at all in the lower portion of Fig. 2b. The accuracy of the quantization is only about 90-95% for the first three steps in the upper portion of Fig. 2b, with no quantization observed for higher N. The quantization in constrictions with L = 600 nm has deteriorated drastically from that shown in Fig. 2a where L = 200 nm, yet the mean free path estimated using the 2D mobility and carrier

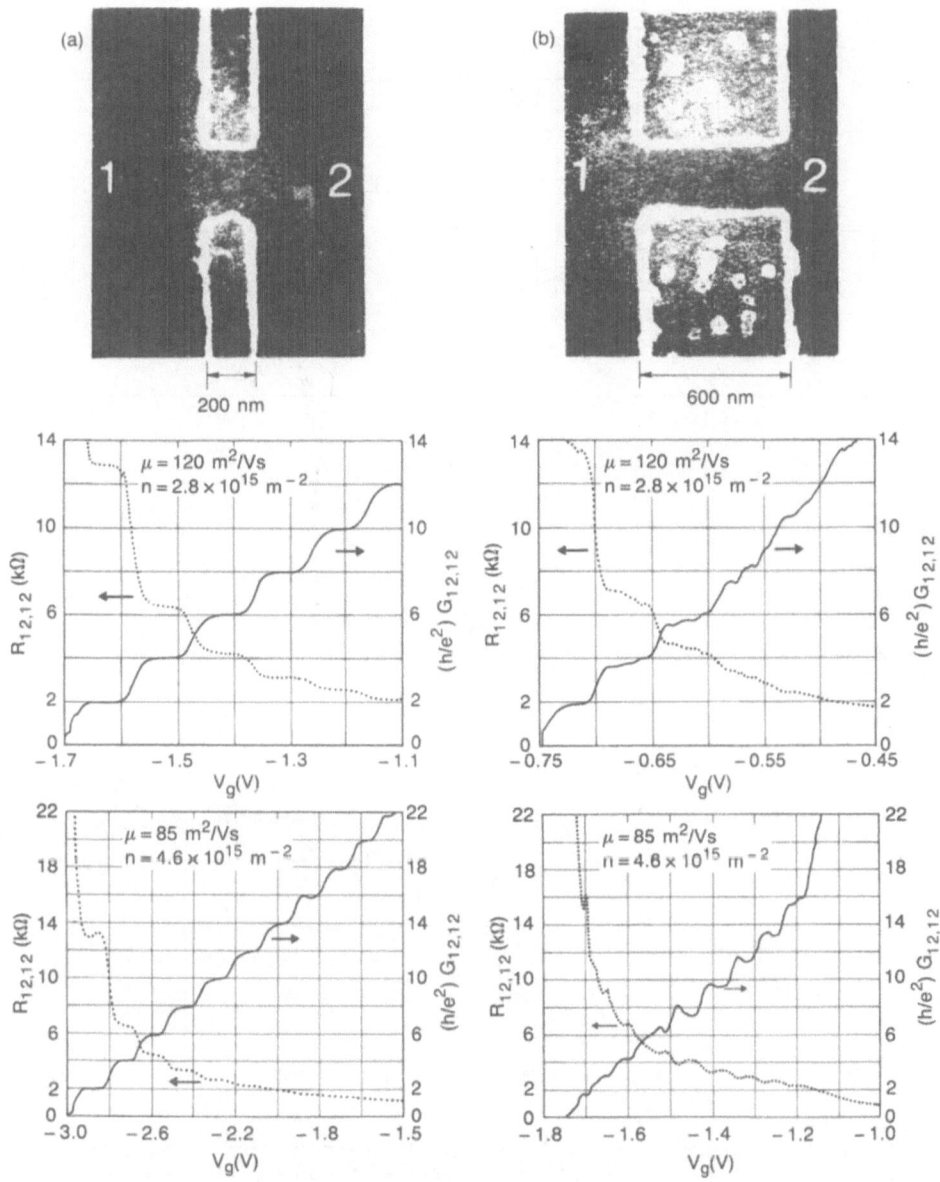

Fig. 2. The two terminal resistances (conductance), $R_{12,12}$ ($G_{12,12}$), of a constriction in a 2DEG as a function of gate voltage, V_g, obtained at 280 mK for two different heterostructures with mean free paths, estimated from the mobility, of about 7 μm. The data shown in the top of Figs. 2(a) and (b) were obtained from devices in heterostructure #4, while the data of the bottom figures were obtained from devices in heterostructure #5 (see Table 1). In Fig. 2(a) the split-gate is 200 nm long with a 300 nm gap between the electrodes. Figure 2(a) shows that the conductance, $G_{12,12}$, is quantized in steps of $2e^2/h$ as a function of V_g. In Fig. 2(b) the split-gate electrodes are 600 nm long with a 300 nm gap between electrodes. Although the low temperature mean free path in the 2DEG is much longer than the length of the constriction, Fig. 2(b) shows deterioration of the quantization of the two terminal resistance.

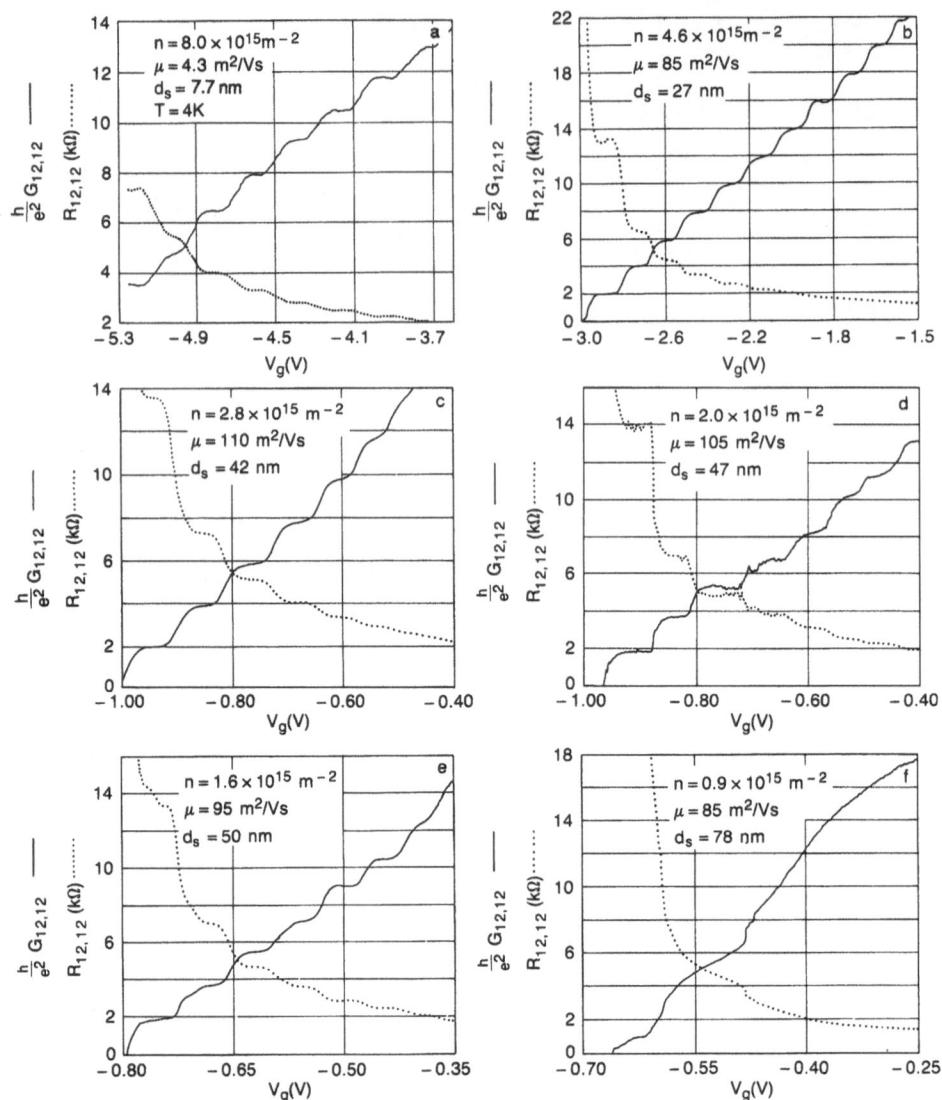

Fig. 3. Figure 3 shows the two terminal resistance (conductance) $R_{12,12}$ ($G_{12,12}$) measured in devices made in the six different heterostructures of Table 1. The data of Fig. 3(a),(b),(c),(d),(e) and (f) were obtained from devices made in heterostructure #6,#5,#4,#3,#2 and #1 of Table 1 respectively using the same split-gate electrode geometry. The split-gate electrodes were 200 nm long with a 300 nm gap. The resistance (conductance) of Figs. 3(b-f) was measured at T = 280 mK, while the resistance (conductance) of Fig. 3(a) was obtained at T = 4 K.

density is still a factor of ten larger than the lithographic length of the constriction!

Apparently, the electronic mean free path in the 2DEG is not a realistic estimate for the mean free path in a 1D constriction[12]. It may actually be irrelevant. For example,

Fig. 3 represents two terminal measurements of the conductance of constrictions in the six different heterostructures characterized in Table 1. Generally, as the carrier density of the 2DEG and the elastic mean free path estimated from the mobility of the 2DEG decrease, the quantization deteriorates. However, when the 2D carrier density is large, as in Fig. 3a, the quantization of the conductance is accurate to 85% in a heterostructure which has a mean free path of only 0.5 μm at T = 285 mK. (The data of Fig. 3a was obtained at a temperature of 4 K, but the accuracy of the quantization did not improve when measured at T = 285 mK.) The mean free path in the constriction may be different than in the 2DEG because electrons in a constriction comparable in width to the Fermi wavelength do not effectively screen the remote ionized impurities in the doped layer. The large potential fluctuations which result probably increase backscattering in the constriction[14].

Depending on the position, scattering from disorder may reduce or enhance the conductance because of backscattering[15]. For example, if an impurity is in the constriction, and backscatters or reflects an electron, the transmission probabilities, t_{jk} become less than unity and thus the conductance is reduced from the quantized value[12]. Alternatively, if the impurity is outside the constriction, the conductance may be either enhanced or reduced[15]. In either case, backscattering from an impurity destroys the quantization of the conductance. Small angle scattering, which is prevalent in GaAs/AlGaAs heterostructures[16,17], does not necessarily destroy quantization. Since the quantization is determined by a sum over the transmission probabilities associated with all of the subbands in the constriction, it is resilient with respect to small angle scattering provided the scattering only changes the distribution of the current among the subbands in the constriction[18]. Figure 4 shows the two scattering events which are detrimental to conductance quantization: (1) a large angle backscattering event away from threshold, and (2) a small angle backscattering event near threshold. As illustrated by Fig. 4, the quantization is sensitive to small angle scattering only near the threshold for occupation of subband.

According to our interpretation, Fig. 5 illustrates the effects of small angle scattering on the quantization of the conductance. Figure 5 shows the four terminal conductance. $G_{12,43}$ found when electrode pair A-C of the device shown in the inset is used to define a constriction. (Note that using this measurement scheme there is no series resistance when $V_A = V_C = 0$ V.) When either electrode pair A-C or B-D is used to define the constriction, time-dependent fluctuations are observed near the threshold for a step in the conductance for $N \leq 5$, indicative of RTS. Although the fluctuations are time-dependent, the statistics associated with the fluctuations at a particular gate voltage are not[2]. As shown in Fig. 5, the conductance as a function of time measured when $V_g = V_A = V_C = -2.250$ V, -2.103 V, and -1.905 V fluctuates with an amplitude of $\Delta G_{12,43} \approx e^2/h$. The largest change in conductance observed is $1.2e^2/h$. The conductance generally fluctuates between predominantly three different values; the time spent at each value, which is supposed to correspond to a particular configuration of the defect and the captured electron[3], varies depending upon the gate voltages[2]. In stark contrast, for $V_g = V_A = V_D = -2.163$ V and -2.037 V, the value of $G_{12,43}$ is relatively

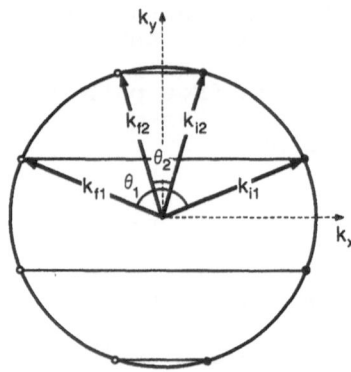

Fig. 4. A schematic representation of the Fermi surface
of a narrow constriction in a 2DEG. The circle
represents the Fermi surface of a 2DEG in
reciprocal space; the horizontal lines reflect
the allowed transverse wavevectors in the 1D
constriction. Current is applied along the k_x
direction. Two scattering events occurring at
the Fermi energy are depicted in the figure. In
the first (unlikely) event k_{i1} associated with
an occupied subband is scattered through a large
angle θ to k_{f1}, resulting in the deterioration
of the quantization. Alternatively, in the
second event, k_{i2} is backscattered through a
small angle < 60° to k_{f2}, and the quantization
deteriorates.

independent of time, i.e. the fluctuations are suppressed at the
center of the N = 1 and N = 2 plateaus.

We suppose that the suppression of the RTS is due to the
suppression of small angle scattering in the 1D constriction.
Sakaki[11,19] first proposed that small angle scattering from an
impurity could be suppressed in a one-dimensional wire. When
the width of the wire becomes comparable to the Fermi
wavelength, the quantization of the transverse momentum due to
the confinement allows scattering only through certain discrete
angles. If a modulation-doped wire is made so narrow that only
one 1D subband is occupied, then only improbable, large angle
backscattering is possible and a dramatic increase in the
mobility (μ)[20], the electron lifetime[16], and the low
temperature mean free path is expected. If the defect
associated with the RTS is in the δ-doped layer, for example,
which is set back from the 2DEG, then large angle backscattering
is improbable but small angle scattering is still possible and
could destroy the quantization especially near the threshold for
the occupation of a subband. The change in the conductance
associated with a reflection[13] near threshold is constrained
by unitarity to be less than $2e^2/h$.

We contend that the defect associated with the RTS is near
the constriction, somewhere in the gap between the electrodes A-
G and B-D, because it is observed when the pairs A-C or B-D are
biased independently or all together. (RTS is not observed when
the pairs A-B, B-C, C-D, OR A-D are biased, but the gate
voltages used are a factor of 2 smaller than those used to bias

the A-C pair, and no RTS is observed when A-C is biased at such low V_g.) Furthermore, we contend that the defect is in the AlGaAs, separated from the constriction, since the RTS depends on V_g as shown in Fig. 5, and is suppressed when an integral number of 1D subbands are occupied independent of the lateral position of the constriction with the gap between A-C.

In an attempt to ascertain the position of the defect, we laterally changed the position of the constriction within the gap between A-C by independently biasing the split-gates. Independent control of the electrodes provides the opportunity to laterally change the position of the constriction while maintaining the same number of 1D subbands[21]. It has been shown for a split gate configuration similar to the inset to Fig. 5 that, in the absence of scattering, the maximal value of the potential measured from the center of the gap between A-C corresponds to the position, $y_0 = D\Delta V/2V^p$, where $\Delta V \equiv V_A - V_C$, D = 425 nm is the width of the gap between A-C, and $V^p \approx -0.35$ V for the device of Fig. 5.) While the lateral position of the constriction within the gap depends linearly on ΔV, the changes

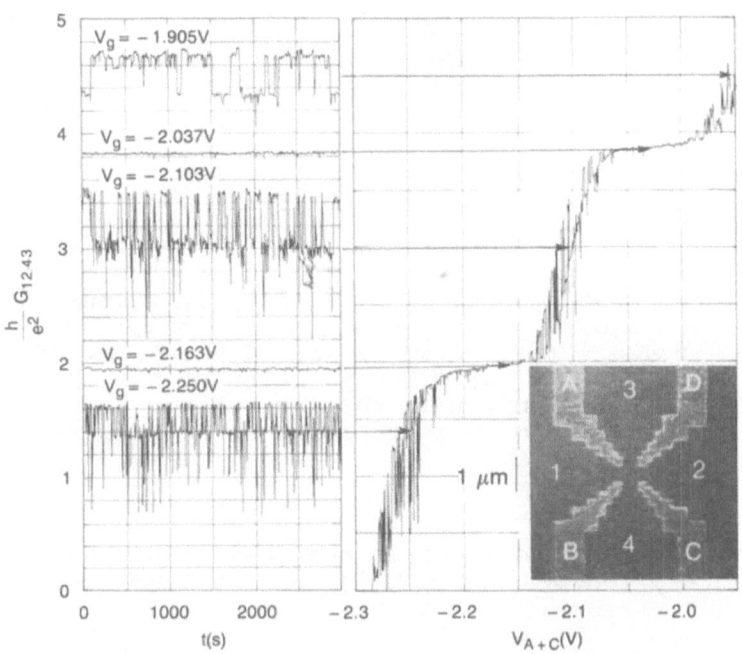

Fig. 5. The four terminal resistance (conductance), $R_{12,43}$ ($G_{12,43}$), of a constriction in a 2DEG as a function of gate voltage (right), $V_g = V_A = V_C$, and as a function of time (left), t, found at 280 mK. The resistance is quantized as a function of V_g corresponding to the number of 1D subbands occupied within the constriction. However, the resistance fluctuates wildly at the threshold of a step which is indicative of RTS. The inset shows the disposition of the gate electrodes and the convention used to denote the leads and the gate electrodes. The device was fabricated on heterostructure #4.

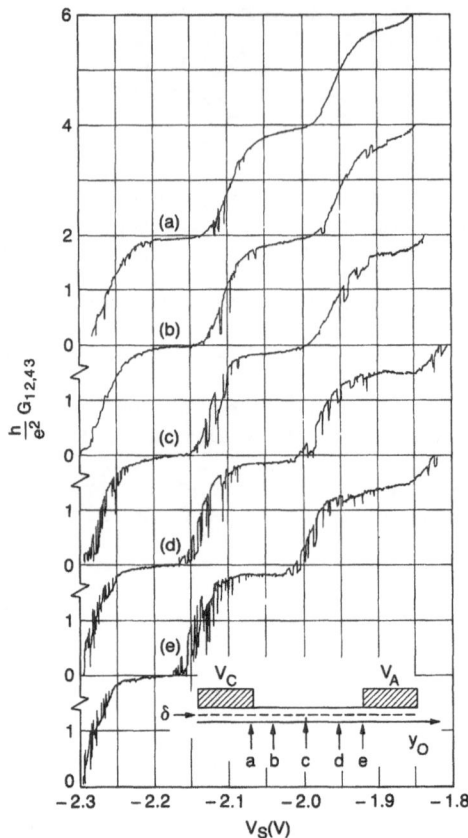

Fig. 6. The conductance, $G_{12,43} \equiv 1/R_{12,43}$, versus gate
voltage, $V_s = \frac{1}{2}(V_A + V_C)$, for $\Delta V = V_A - V_C =$
-0.425 V (a), -0.275 V (b), 0 V (c), 0.275 V (d)
and 0.425 V (e). The position of the 1D
constriction within the gap between the electrodes of Fig. 5 is supposed to be linearly
related to ΔV.

in the threshold for the steps in the conductance as a function
of $V_s = \frac{1}{2}(V_A + V_C)$ depend on $(\Delta V/V^p)^2$ and are supposed to be
small[21].

In Fig. 6, $G_{12,43}$ is shown as a function of V_s for $\Delta V =$
-0.425 V, -0.275 V, 0 V, 0.275 V, 0.425 V ideally corresponding
to $y_0 = -260$ nm, -170 nm, 0, 170 nm and 260 nm. The
quantization of the conductance is preserved when $\Delta V \neq 0$, and
the thresholds are not sensitive to changes in ΔV as anticipated. We observe RTS at the threshold for occupation of a
subband for each ΔV, but the magnitude and characteristic
frequency of the switching noise changes dramatically with the
lateral position of the constriction. RTS is reduced when
$\Delta V < 0$ compared to $\Delta V \geq 0$ which may indicate that the trap is in
the AlGaAs near electrode A.

We can estimate the vertical separation between the defect
and the constriction from the width of the transition between
plateaus where RTS is observed. For example, at the threshold

for occupation of the N = 2 subband, where RTS is first observed (V_g = -2.14 V), the longitudinal component of the wavevector, k_x in Fig. 5, vanishes. The RTS vanishes on the N = 2 plateau at V_g = -2.06 V where the Fermi wavevector k_F = 7.93 x 10^7 m^{-1}. (k_F is determined by the carrier density in the constriction, n = 1 x 10^{15} m^{-2}, which we deduced from magnetoresistance measurements.) At the same gate voltage we deduce that W ≈ 100 nm using a hard wall confinement potential. Thus, k_x = {k_F^2 - $(2\pi/W)^2$}$^{1/2}$ ≈ 5 x 10^7 m^{-1} when RTS is suppressed at the N = 2 conductance plateau. The absence of scattering for -1.98 V > V_g > -2.06 V, where k_x > 5 x 10^7 m^{-1}, implies a vertical separation between defect and constriction[22] of about z = k_x^{-1} > 21 nm, a distance comparable to the spacer layer thickness.

Our data is consistent with a switching defect in the donor layer changing the potential at the AlGaAs/GaAs interface. If the RTS found as a function of ΔV is all due to the same trap, that trap affects a region at least 260 nm wide, half as large as the size of the gap. This deduction is consistent with inadequate screening of potential fluctuations at the AlGaAs/GaAs interface in the gap[14]. Alternatively, if different traps are activated with each ΔV, the relevant trap must always be removed from the constriction because RTS is always suppressed whenever an integral number of subbands is occupied.

In sum, while impurity scattering can cause the quantization of the conductance of a point contact to deteriorate[12,15], it does not always have a deleterious effect. Specifically, we have shown that RTS is quenched when the conductance corresponds to an integral number of occupied 1D subbands. We attribute the suppression of RTS to the suppression of small angle scattering as predicted to occur in a narrow wire by Sakaki[11].

REFERENCES

1. K.S. Ralls, W.J. Skocpol, L.D. Jackel, R.E. Howard, L.A. Fetter, R.W. Epworth and D.M. Tennant, **Phys. Rev. Lett.**, 52:228 (1984)
2. R.E. Howard, et al., **IEEE Trans. Electron Devic.**, ED-32:1669 (1985)
3. M.J. Uren, D.J. Day and M.J. Kirton, **Appl. Phys. Lett.**, 47:1195 (1985)
4. C.T. Rogers and R.A. Burhrman, **Phys. Rev. Lett.**, 55:859 (1985)
5. C. Surya and T.Y. Hsiang, **Phys. Rev. B**, 35:6342 (1987)
6. A. van der Ziel, "Noise in Solid State Devices and Circuits", Wiley, New York (1986)
7. K.K. Hung, P.K. Ko, Chunming Hu and Yiu Chung Cheng, **IEEE Electron Dev. Lett.**, EDL-11:90 (1990)
8. K.F. Berggren, T.J. Thornton, D.J. Newson and M. Pepper, **Phys. Rev. Lett.**, 57:1769 (1986)
9. B.J. van Wees, H. van Houten, C.W.J. Beenakker, J.W. Williamson, L.P. Kouwenhoven, D. van der Marel and C.T. Foxon, **Phys. Rev. Lett.**, 60:848 (1988)
10. D.A. Wharam, T.J. Thornton, R. Newbury, M. Pepper, J.E.F. Frost, D.G. Hasko, D.C. Peacock, D.A. Ritchie and G.A.C. Jones, **J. Phys. C**, 21:L209 (1988)
11. H. Sakaki, **Jpn. J. Appl. Phys.**, 19:L735 (1980)

12. G. Timp, R. Behringer, S. Sampere, J.E. Cunningham and R.E. Howard, p.331 **in**: "Proc. Int. Symp. on Nanostructure Physics and Fabrication", W.P. Kirk and M. Reed, eds., Academic Press, New York (1989)
13. Y. Imry, Physics of Mesoscopic Systems, p.101 **in**: "Directions in Condensed Matter Physics", G. Grinstein and G. Mazenko, eds., World Scientific Press, Singapore (1986)
14. J.H. Davies and J.A. Nixon, **Phys. Rev. B**, 39:3423 (1988)
15. E.G. Haanappel and D. van der Marel, **Phys. Rev. B**, 39:5484 (1989); C.S. Chu and R.S. Sorbello, **Phys. Rev. B**, 40:5941 (1989); and A. Szafer and A.D. Stone, **Phys. Rev. Lett.**, 62:300 (1989)
16. S. Das Sarma and F. Stern, **Phys. Rev. B**, 32:8442 (1985)
17. F. Fang, T.P. Smith and S.L. Wright, **Surf. Sci.**, 196:1988 (1988)
18. J.H. Davies, private communication, and J.H. Davies, J.A. Nixon and H.U. Baranger, preprint.
19. G.B. Lesovik, **JETP Lett.**, 49:592 (1989)
20. K. Ismail, D.A. Antoniadis and H.I. Smith, **Appl. Phys. Lett.**, 54:1130 (1989)
21. L.I. Glazman and I.A. Larkin, preprint ICTP (1989)
22. P.J. Price, **Surf. Sci.** 143:145 (1984)

QUANTIZATION AND RESONANT STRUCTURE IN LOW DIMENSIONAL

STRUCTURES

D.A. Wharam

Sektion Physik, Ludwig-Maximilians-Universität
Geschwister-Scholl-Platz 1
8000 München 22, FRG

1.   INTRODUCTION

The discovery of the quantization of resistance in narrow
ballistic channels defined in the two-dimensional electron gas
(2DEG) of a GaAs-AlGaAs heterojunction has prompted a large
number of theoretical and experimental studies. Whilst the
origin of the quantization itself is now well understood there
remain a number of outstanding problems associated with such
constrictions and it is these problems that are here addressed.
In particular it has been predicted that resonant structure may
be observed in the conductance of such channels as a consequence
of standing waves along the channel; the observation of such
resonances is here discussed within the context of which
theoretical model best describes the observed conductance.

2.   RESISTANCE MEASUREMENTS

In Fig. 1 the low temperature (T ~ 100 mK) resistance of a
typical narrow device is plotted as a function of the negative
bias applied to the gate electrodes. The gate configuration is
shown in Fig. 2, the defined channel length is 300 nm and width
500 nm. The resistance was measured between optically defined
voltage probes situated at the edge of a 25 μm wide mesa and
separated by 100 μm. Hence at zero gate bias the measured
resistance was merely that of four squares of 2D material, and
was used to determine the low tempeature mean free path of the
electrons. The measured mobility of the sample was 83 m^2 $V^{-1}s^{-1}$
which combined with the observed carrier density of 4.2 10^{15} m^{-2}
yielded an electronic mean free path of 8.8 μm. At a negative
gate bias of -0.5 V there was a step in resistance associated
with the depletion of the electrons from the 2DEG beneath the
gate and the consequent definition of the narrow channel. As
the negative bias was further increased towards pinch-off, the
width of the channel narrowed and steps between well defined
resistance plateaus were observed. The resistance of the
channel itself was derived by subtracting the resistance of the
surrounding 2DEG and the values of plateaus resistance thereby
derived were found to be quantized according to R = $h/2ie^2$ where

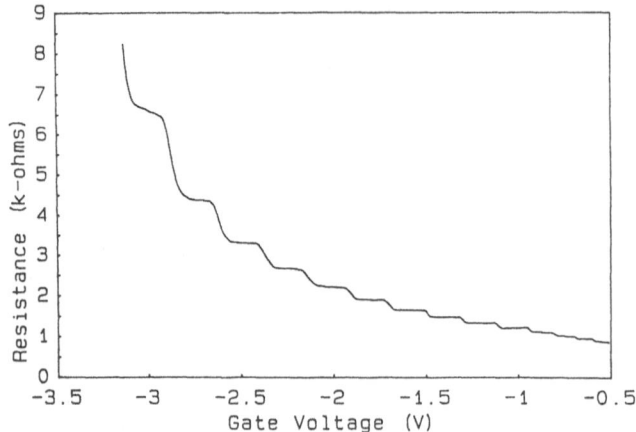

Fig. 1. The resistance of a typical split-gate device is
plotted as a function of the gate bias applied
to the electrodes. The observed quantized steps
in resistance are quantized according to R =
$h/2ie^2$ (Wharam et al. 1988[2]).

i is an integer. This quantization of resistance has been
observed in a wide range of fabricated narrow channels of
different lengths and different material parameters. However,
identical devices fabricated on the same wafer could have
markedly different gate characteristics suggesting that the

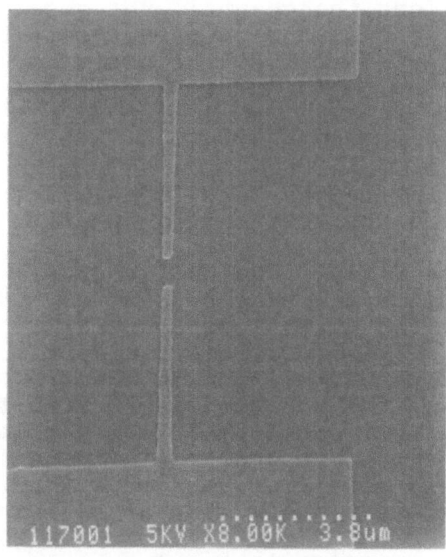

Fig. 2. An SE Micrograph showing the electron beam
defined gate structure of a typical split-
gate device. The lighter area is gold-
palladium which is thermally evaporated
onto the surface of the GaAs-AlGaAs
heterostructure.

particular distribution of scattering centers in the vicinity of
the channel played a crucial role in the observation of this
phenomenon.

The quantized result is most easily understood from the
exact cancellation that exists in one dimension between the
density of states at the Fermi energy and the Fermi velocity.
The resulting current per channel, or subband, is then
determined purely by the product of the voltage dropped and the
fundamental ratio $e^2/\pi\hbar$. If there are N occupied subbands then
the total current is given by $I = (Ne^2/\pi\hbar)\Delta V$ which leads
directly to the quantized result. The assumption of one-
dimensional statistics used to derive this result is entirely
equivalent to the assumption of a quantized transverse momentum
as used by van Wees et al.[1]. This assumption requires, in
principle at least, that the channel be considerably longer than
it is wide. Close to pinch-off, when the channel becomes very
narrow, this requirement is satisfied, but when the channel is
first defined this is clearly not the case and the validity of
the above assumptions has to be questioned. Within the framework of the one-dimensional model states are occupied up to a
well-defined energy, the Fermi energy, such that the required
carrier density in the constriction is achieved. As the channel
width narrows the energy separation between the one-dimensional
subbands increases and eventually an infinity in the density of
states associated with the ground state energy of a particular
subband passes through the Fermi energy. At this point the
subband no longer contributes to the current flowing through the
constriction and transition between two quantized resistance
plateaus is observed.

The transition between quantized plateaus is, in practice,
not discontinuous, due to both the thermal broadening of the
electronic energies at the Fermi energy and the finite lifetime
of the momentum eigenstates. The former leads to a thermal
spreading in the occupancy of states at the Fermi energy whilst
the latter leads to the formation of a new density of states
which is no longer truly one-dimensional. At temperatures below
$T = \hbar/k_B\tau$, where k_B is Boltzmann's constant and τ the elastic
lifetime, the lifetime broadening of the states dominates; in a
typical high-mobility heterostructure this corresponds to a
temperature of approximately 0.3 K. For a typical device the
plateaus were not clearly resolved at 4.2 K although there was
some evidence of quantum structure, and below 500 mK there was
no significant improvement in the quality of the plateaus as the
temperature was further decreased. From the data of Fig. 1 the
effective broadening has been extracted approximately by
assuming that the variation of channel width with gate voltage
is effectively linear. In addition it was assumed that the
ground state energies were given by those of a particle in a
box. Thus the effective broadening was determined from the
variation in measured gate voltage over a typical transition
between plateaus. This yielded energy broadening between
plateaus of the order of 0.3 ± 0.1 meV corresponding to thermal
broadening of ~ 3 K, or, alternatively, lifetime broadening
associated with a mean free path of ~ 0.6 μm. Although this
estimate of the broadening cannot be regarded as accurate it is
clear that under the experimental conditions of the measurement
neither of the above conditions was satisfied and the magnitude
of the broadening must be explained by some alternative
mechanism.

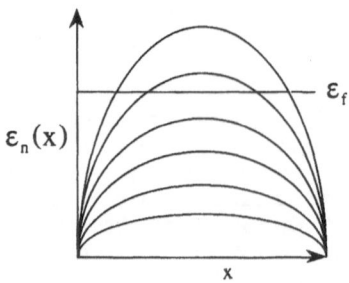

Fig. 3. The transverse energies of the lowest
modes in a narrow constriction are
schematically plotted as a function of
position along the channel.

3. THE ADIABATIC APPROXIMATION

Whilst the one-dimensional subband model is capable of
explaining the quantization of resistance as well as the
qualitative features of the effects of broadening it is not at
all clear why the assumption of one-dimensional statistics is
valid for a constriction of small aspect ratio. Furthermore it
does not address the effects of quantum coherence between states
at the entrance and exit of the constriction. In general the
confining potential produced by the application of a negative
bias to a split-gate structure is a highly complex function of
position whose exact solution is only possible in cases of high
symmetry, for example in the case of an infinitely long
channel[3,4]. When the aspect ratio of the constriction is
small the width of the confining potential is no longer
independent of the position along the channel. There is a
smooth variation in the width of the confining potential which
is enhanced by the limitations of the fabrication technique; in
Fig. 1 a rounding of the split-gate electrodes is clearly
visible. This expectation is confirmed by the recent depletion
calculations of Davies[5] and Kumar et al.[6]

Glazman et al.[7] have shown that provided the variation of
the width, $d(x)$, of the confining potential is smooth at the
scale of the Fermi wavelength then an adiabatic separation of
the variables in the Schrödinger equation is possible. The
transverse energies of the modes within the channel transform
smoothly with the position along the channel, the precise form
of this dependence being determined by the variation of the
channel width and the shape of the confining potential. This
behavior is illustrated qualitatively in Fig. 3.

As the transverse energies pass successively through the
Fermi energy the motion along the channel becomes evanescent and
the mode no longer contributes to the conductance of the
channel. Thus the number of subbands contributing to the
current is determined by the narrowest point of the constriction. However, the finite variation in channel width about the
narrowest point leads to a correction to the conductance of the
form;

$$\delta G = \frac{2e^2}{h} [1 + \exp(-z\pi^2 \sqrt{2R/d})]^{-1} \tag{1}$$

where R is the radius of curvature at the narrowest point and z is a parameter defining the position on each plateau ($z = k_f d/\pi - n$). This correction, which allows for the finite transmission probabilities of classically forbidden states, leads to a broadening of the transition regions between plateaus which would otherwise be expected to be theta function type discontinuities. This correction is expected to dominate for temperatures below $k_B T = n\hbar^2/m(2Rd^3)^{1/2}$. As discussed earlier, the observed broadening was considerably larger than that expected from either thermal or lifetime broadening. Under the assumption that the additional broadening was given by the conductance correction of Eqn. 1 the data of Fig. 1 were used to extract the radius of curvature of the confining potential. A linear variation of k_f with gate voltage was assumed and the derived radius of curvature was found to be 110 ± 10 nm and showed no systematic variation as the width of the channel decreased. The calculated magnitude of the radius of curvature is of the same order as that derived from the depletion calculations of Davies[5].

4.   THE WIDE-NARROW-WIDE GEOMETRY

The adiabatic model discussed in the previous section assumes that the channel width varies smoothly at the scale of the Fermi wavelength. Whilst this assumption is valid at the center of the constriction it is not clear that it holds for the constriction entrance and exit where the potential varies rapidly along the device length. A number of papers[8-11], have addressed the transmission of states through a constriction with abrupt edges using the so-called Wide-Narrow-Wide (WNW) geometry, which is illustrated in Fig. 4.

Szafer and Stone[9] considered a WNW geometry and showed that the quantized result can be readily derived by using both an exact recursive method as well as a mean field approximation (MFA). Furthermore, they have shown that the quantization accuracy, which according to their model can be as good as 0.01%, is perturbed by two effects. First, they predict a series of resonant peaks as a consequence of successively constructive and destructive internal reflections within the channel; such resonant structure becomes important for $k_n L > 1$

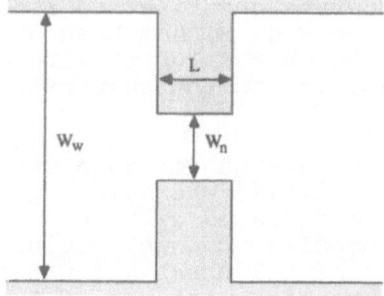

Fig. 4. A schematic illustration of the so-call wide-
        narrow-wide geometry. The shaded area rep-
        resents the depletion region defined by the gate
        electrodes.

where k_n is the wavevector of the n-th mode propagating at the Fermi energy. Secondly, they predict that evanescent modes within the channel contribute to the conductance when $\kappa_n L < 1$ where κ_n is the magnitude of the imaginary wavevector of the evanescent mode. Thus if the channel length, L, is sufficiently long to damp out any evanescent wave it is also long enough to give rise to resonant peaks. Szafer and Stone have also considered the effect of the constriction tapering upon such resonant structure and have found that when the confining potential is no longer abrupt, as must be the case in any realistic device, there is decreased reflection at the ends of the constriction leading to smaller resonances. However, they expect that such resonances should still occur but might be broadened by the temperature of the experiments. They model the temperature dependence of the van Wees[1] device and show that resonant structure should persist up to a temperature $T_o \sim 2.8$ K although the experimental data showed no resonant structure at the temperature of observation, 600 mK.

The effect of constriction tapering has also been considered in a WNW geometry by Escapa and Garcia[12] who have shown that the introduction of tapered horns to the abrupt junction the resonant structure in the calculated transmission coefficients rapidly disappeared. They conclude that for any practical device the variation of the depletion region in the vicinity of the constriction is sufficiently gradual for transport to be adiabatic.

5.   RESONANT STRUCTURE IN NARROW CONSTRICTIONS

Oscillatory structure in the conductance of a narrow constriction has been observed by Brown et al.[13] and was attributed to length dependent resonances as discussed in the previous section. However, the suggestion that such resonances were due to impurity states could not be definitively discounted. Significantly, the length of the narrow constriction considered was 0.8 μm and, whilst it might be argued that increasing the length of the channel leads to a more well-defined channel with sharper corners, it is also clear that the statistical probability of incorporating an impurity within the channel is also greatly enhanced. In the experiments of Brown et al. reducing the temperature led to an improvement in the observed structure but interestingly at no temperature were good well-defined plateaus observed. By contrast in the "best" structures studied accurate quantization (to within 0.1% of the theoretical values - see Fig. 5) has been observed at temperatures between 500 mK and 100 mK with no indication of resonant structure evolving as the temperature of observation was decreased.

An alternative explanation for such structure has been suggested by McEuen et al.[14] who have suggested that such resonant structure is associated with the presence of an impurity within the immediate channel vicinity, a conclusion supported by the experimental observation that thermal cycling of the sample led to the disappearance of the observed resonant structure. Furthermore, McEuen et al. used the resonance height and width to derive the energy and decay length of the bound state associated with the impurity and found excellent agreement with the "hydrogenic" donor atom model.

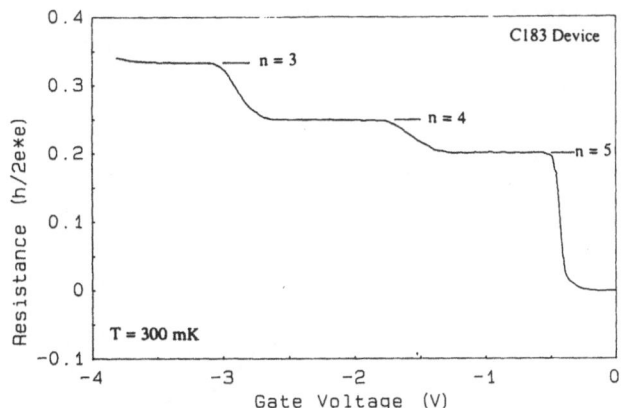

Fig. 5. The quantization accuracy is shown for a good
sample at a measurement temperature of 300 mK.
Note in particular the quality of the plateaus'
structure and the absence of any resonant
structure.

6.   PARALLEL FIELD MEASUREMENTS

     A magnetic field affects both the spin state of an electron
as well as its orbital motion and it is convenient to
investigate such spin effects with the field parallel to the
plane of the 2DEG.   In this orientation orbital effects are
negligible due to the spatial quantization of 2D systems[15],
and it is therefore expected that the dominant effect is to lift
the spin degeneracy which has been previously assumed.   This
gives rise to additional plateaus at quantized values of
resistance $R = h/2e^2(i + 1/2)$.   Typical parallel field data are
presented for a narrow ballistic constriction in Fig. 6; the
measurement temperature being 100 mK.   The additional plateaus
are resolved for fields in excess of 10 T but only for the high
index subbands.   The high value of field required to observe
such spin-splitting is in contrast to the comparatively low
fields, typically ~ 2 T, necessary for spin-splitting of the
Landau levels in the 2DEG of a GaAs-AlGaAs heterostructure.
Such low field splitting is a consequence of the enhancement of
the Landé g factor, first proposed by Janak[16] to explain
measurements in Si MOSFETs; g factors as high as 6.2 have been
reported by Nicholas et al.[17] in tilted GaAs-AlGaAs heterojunctions (cf. for bulk GaAs g = -0.44).   However, recent
differential capacitance measurements performed upon quasi-one-
dimensional grating structures have shown a novel anisotropy[18]
with respect to the orientation of the magnetic field.   With the
magnetic field perpendicular to the plane of the 2DEG spin-
splitting of the Landau levels was observed, but with the field
parallel to the plane of the 2DEG and perpendicular to the
grating no such spin-splitting of the states was resolved even
with fields as large as 20 T.   The enhancement of the g factor
has been explained in 2D systems in terms of an imbalance in the
number of spin-up and spin-down electrons[16] and requires that
the Landau level spacing is much greater than the level
broadening.   Smith et al.[18] suggest that the same mechanism
may be ineffective in the parallel orientation since the 1D
subband separations are small, typically ~ a few meV, and that

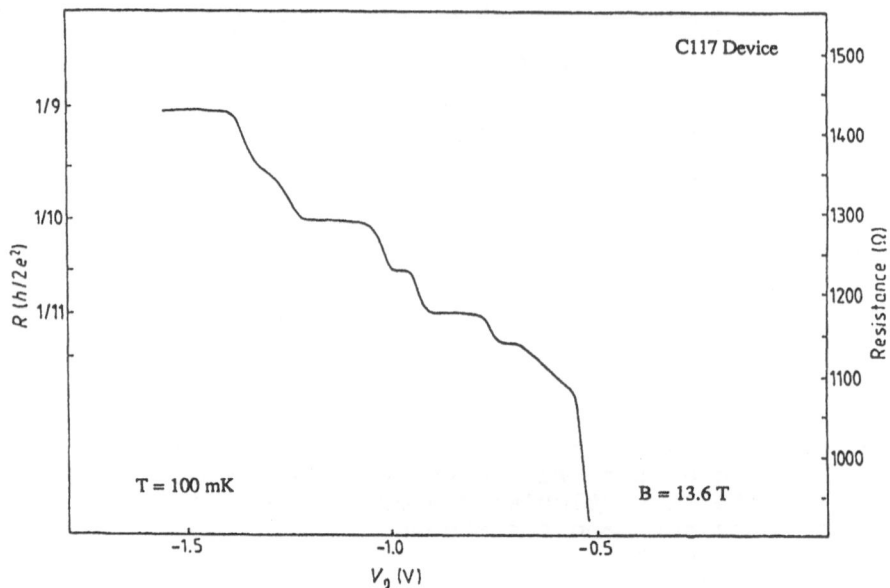

Fig. 6. The device resistance is plotted as a function
of the applied gate voltage for an applied
magnetic field of 13.6 T. The spin-split
plateaus at $h/21e^2$ and $h/23e^2$ are clearly
resolved.

the g factor is close to its bulk value. At 20 T in GaAs this
implies a spin-splitting of approximately 0.5m eV and would not
be resolvable. Smith et al. also observed an additional
anisotropy when the field was parallel to both the plane of the
2DEG and the 1D grating. With the field in this orientation the
capacitance oscillations reflecting the 1D subband structure
were rapidly damped as the field strength increased.

The adiabatic model of quantum transport in a narrow
constriction, discussed in section 3, has been extended by
Glazman and Khaetskii[19] to incorporate the effects of a
magnetic field in the parallel orientation. Considering only
the spin-splitting of the states Glazman and Khaetskii have
shown that the spin-split subband modes switch into the
conductance of the channel for different values of their
dimensionless parameter $z = k_f d/\pi$ (see Section 3). When the
magnitude of spin-splitting exceeds the broadening associated
with the transmission of evanescent states as discussed earlier
aditional plateaus are observed. Since the spin-splitting
scales with plateau index, n, whilst the broadening is
proportional to $\sqrt{n}$ the additional plateaus are only observed
above a certain critical value of plateau index given by;

$$n_c = \frac{\lambda_f}{R}\left[\frac{2\epsilon_f}{\pi^2 g\mu_B B}\right]^2 . \tag{2}$$

For the device considered in the data of Fig. 6 the values of
Fermi energy and wavelength are $\epsilon_f = 14$ meV and $\lambda_f = 40$ nm. In
addition it is observed empirically that the spin-splitting

becomes apparent for subbands with index n > 9. Thus the product Rg^2 can be directly evaluated and is equal to $1.1 \ 10^{-7}$ m. An accurate determination of R, the curvature at the center of the constriction, is difficult, however the smallest possible value, and hence the largest g factor, would be given by half the lithographically defined channel length. Such an estimate is not unreasonable since the spin-split plateaus were observed very close to device definition when the lateral spread of the confining potential is minimal. Thus assuming R ~ 1500 Å as the smallest possible curvature gives an upper limit for g of 0.86. As an upper limit this value is sufficiently close to the bulk g factor of 0.44 to suggest that in the parallel field orientation there is little or no enhancement of the g factor in such one-dimensional structures.

CONCLUSIONS

The observed quantization of resistance is now well understood and has been derived from a variety of theoretical models. However, the predicted resonant structure in single narrow constrictions has yet to be conclusively observed and it seems likely that in any realistic device inevitable smoothing of the defining potential will lead to an absence of resonant structure. The observation of resonant structure requires the presence of an additional scattering potential which may be introduced accidentally, as in the case of undesired impurities located within the channel, or deliberately using lithographic techniques as has already been illustrated in the zero dimensional structures of Smith et al.[20] and also in the closely related Fabry-Perot resonators[21].

REFERENCES

1. B.J. van Wees, H. van Houten, C.W.J. Beenakker, J.G. Williamson, L.P. Kouwenhoven, D. van der Marel and C.T. Foxon, **Phys. Rev. Lett.**, 60:848 (1988)
2. D.A. Wharam, T.J. Thornton, R. Newbury, M. Pepper, H. Ahmed, J.E.F. Frost, D.G. Hasko, D.C. Peacock, D.A. Ritchie and G.A.C. Jones, **J. Phys. C**, 21:L209 (1988)
3. S.E. Laux and F. Stern, **Appl. Phys. Lett.**, 49:91 (1986)
4. S.E. Laux, D.J. Frank and F. Stern, **Surf. Science**, 196:101 (1988)
5. J.H. Davies, private communication (1989)
6. A. Kumar, S.E. Laux and F. Stern, Preprint from APS (St. Louis) meeting, March 1989.
7. L.I. Glazman, G.B. Lesovik, D.E. Khmel'nitskii and R.E. Shekhter, **JETP Lett.**, 48:218 (1988)
8. I.B. Levinson, **JETP Lett.**, 48:273 (1988)
9. A. Szafer and A.D. Stone, **Phys. Rev. Lett.**, 62:300 (1989)
10. G. Kirczenow, **Solid State Comm.**, 68:715 (1988)
11. G. Kirczenow, **J. Phys. Condens. Matter**, 1:305 (1989)
12. L. Escapa and N. Garcia, **J. Phys. Condens. Matter**, 1:2125 (1989)
13. R.J. Brown, M.J. Kelly, R. Newbury, M. Pepper, B. Miller, H. Ahmed, D.G. Hasko, D.C. Peacock, D.A. Ritchie, J.E.F. Frost and G.A.C. Jones, Preprint (1989)
14. P.L. McEuen, B.W. Alphenaar, R.G. Wheeler and R.N. Sacks, **Surf. Sci.**, 229:312 (1990)
15. T. Ando, A.B. Fowler and F. Stern, **Rev. Mod. Phys.**, 54:437 (1982)

16. J.F. Janak, **Phys. Rev.**, 178:1416 (1969)
17. R.J. Nicholas, R.J. Haug, K. von Klitzing and G. Weimann, **Phys. Rev. B**, 37:1294 (1988)
18. T.P. Smith III, J.A. Brum, J.M. Hong, C.M. Knoedler, H. Arnot and L. Esaki, **Phys. Rev. Lett.**, 61:585 (1988)
19. L.I. Glazman and A.V. Khaetskii, **J. Phys. Condens. Matter**, 1:5005 (1989)
20. C.G. Smith, M. Pepper, H. Ahmed, J.E.F. Frost, D.G. Hasko, D.C. Peacock, D.A. Ritchie and G.A.C. Jones, **J. Phys. C**, 21:L893 (1988)
21. C.G. Smith, M. Pepper, H. Ahmed, J.E.F. Frost, D.G. Hasko, D.C. Peacock, D.A. Ritchie and G.A.C. Jones, **J. Phys. Condens. Matter**, 1:9035 (1989)

BALLISTIC TRANSPORT THROUGH A QUANTUM POINT CONTACT

E. Tekman and S. Ciraci

Department of Physics, Bilkent University
Bilkent 06533, Ankara, Turkey

ABSTRACT

The ballistic transport of electrons through narrow
constrictions is investigated for a variety of configurations.
It is found that for a uniform constriction the conductance is
quantized in units of $2e^2/h$ for long channels. The interference
of waves in the constriction gives rise to the resonance
structure superimposed on the quantized steps. The lack of the
resonance structure in the experimental results are attributed
to temperature effects and/or adiabatic transport due to
tapering of the constriction. It is shown that elastic
scattering by an impurity distorts the quantization of
conductance. Novel resonant tunneling effects due to formation
of quasi-zero-dimensional states are predicted for an attractive
impurity or a local widening at the center of the constriction.
Possible applications of the method in scanning tunneling
microscopy are discussed.

1.   INTRODUCTION

Owing to the advances in growth and lithography techniques
during the last two decades quasi-one-dimensional (quasi-1D)
electron devices have been fabricated, and novel quantum effects
have been discovered thereof[1]. In particular, the devices in
which electrons move ballistically (i.e. without being
scattered) have been developed and used in several recent
experiments. The appealing feature of ballistic devices is that
they provide experimental tests for textbook problems of quantum
mechanics. A large number of papers appeared on the subject
which became a part of the *mesoscopics*, a term that is used to
describe a new length scale for physical events, between
microscopic and macroscopic scales.

Recently the theoretical investigations of electrical
transport on quantum scales received wide interest owing to the
transport measurements at the *mesoscopic* scale. Part of the
experimental results were explained using the appropriate
Landauer formula[2]. On the other hand, in several works the

physical implications of the Landauer formulae and their connections to the standard linear response theory were analyzed and questioned[3]. There are, however, still open problems and controversial issues in the basic theory of transport in *mesoscopic* systems. Nevertheless, as we show in this study, simple linear response arguments together with the solutions of the Schrödinger equation are sufficient to understand most of the events related to the ballistic transport.

Recently, the Delft-Philips collaboration[4] and the Cambridge group[5] reported the quantization of conductance for a ballistic quantum point contact (QPC) in a 2D electron gas (EG). Using the high mobility GaAs-AlGaAs heterojunctions and split gate technique they produced a small constriction on the sample. A channel was obtained from this constriction by applying a negative bias to the split gate, and thus by causing the depletion of electrons beneath the gate. This way, the portion of the 2DEG lying below the gap of the split gate electrode remains conducting. In the experiment[4,5] the length of the constriction d, is smaller than the electron free path l_e, so that electrons are prevented from being scattered in and around the constriction. Also the width of the constriction w, is comparable to the Fermi wavlength λ_F, so that the quantum size effects become important. At very low temperature they found that the two terminal conductance of the QPC, G_c, changes as a function of the gate voltage V_g (or equivalently the width w) approximately in units of $2e^2/h$. This observation was interpreted as the quantization of conductance.

Earlier, the quantum of conductance $2e^2/h$ was attributed to the contact resistance at the junctions between the ballistic device and the reservoirs[6,7]. However, the exact (or accurate) quantization was not considered as a possibility since the various geometrical and material parameters were thought to interfere with the quantum size effects. The quantization, which is achievable (with an accuracy of one percent) for a variety of system parameters in the experiment, is far from being coincidental. On the other hand, since the quantum of conductance can routinely be measured with an accuracy of 1 part in 10^7 in quantum Hall effect experiments, the quantization occurring in the QPC is highly distorted.

Initially the quantization of conductance was explained [4,5] by assuming that the 2DEG regions act as ideal reservoirs and the QPC can be perceived as a uniform electron waveguide connected to reservoirs at each end. The current carrying states in the waveguide have the subband structure because they are confined in the transverse direction, leading to the quantization of the lateral momentum. The propagation in the nth subband takes place whenever the minimum energy of the subband ϵ_n, is smaller than the Fermi level E_F. When one of the reservoirs has an infinitesimally larger chemical potential than the other, the difference given by $\Delta\mu$, each occupied subband contributes to the current an amount

$$I_n = ev_n(E_F)D_n(E_F)\Delta\mu, \tag{1}$$

where $v_n(E)$ is the velocity of electrons in the nth subband with energy E and $D_n(E)$ is the density of states. Since the system is effectively 1D as far as the flow of the carriers is concerned, the product $v_n(E)D_n(E)$ is equal to $e/\pi\hbar$ and is independent of energy and subband index n. Hence each occupied subband contributes an equal amount to the current. This means that the conductance of the QPC is given by

$$G_c = (2e^2/h)N_c, \qquad (2)$$

where N_c is the number of occupied subbands. The conductance is quantized and increases by a quantum of conductance $2e^2/h$ whenever the energy of a subband ϵ_n, dips in the Fermi level. This can be achieved either by widening the QPC[4,5] or by increasing the density of electrons[4]. It is noted that this simple model neglects any quantum mechanical effect arising from the 2DEG portions and their boundaries with the QPC, and it conjectures an ideal step structure. The experimental results, on the other hand, display some deviations from this ideal result which are accurately reproducible and device dependent. Consequently, a quest for a better theoretical explanation for the quantized conductance has been started just after the experiments were reported.

Subsequent to the observation of quantization of conductance several groups dealt with the development of a realistic theory for ballistic transport through QPC[8-13]. Kirczenow[8], Szafer and Stone[9], and Tekman and Ciraci[12] used a boundary matching technique which makes use of the subband structure explicitly. On the other hand, the tight-binding model of Haanapel and van der Marel[10], the scattering model of García and co-workers[11] and the Anderson model of He and Das Sarma[13] incorporate the subband structure implicitly. The existing studies covered a wide range of physical aspects such as geometry effects[9,11,12] temperature averaging [9,12,14], *adiabatic* evolution of states[12,15,16] quasi-0D states and resonant tunneling[12], channel roughness[12], and impurity scattering[10,17-19]. Currently most of the problems related to ballistic transport through a QPC at zero magnetic field are solved and conceptually understood.

As expected several factors affect the transport in a QPC and may lead to deviations from the exact quantization. Our goal is to present a detailed analysis within the framework of the theory developed by the authors[12]. In Section 2 we explain the essential points of the theory. Several effects which may influence ballistic transport will be dealt with in Section 3. The concluding remarks are in Section 4.

2. FORMULATION

2.1 Conductance of a Uniform Channel

The subject matter of our study is to solve the Schrödinger equation and to obtain the conductance for various QPC geometries. Normally, the potential due to the split-gate is obtained self-consistently using the Poisson equation. Since the self-consistent solutions are tedious and require extensive computations, we use a realistic model potential to represent the QPC. The existence of the subband structure and thus the quantization of conductance is not affected in any essential manner by the detailed shape of the potential as long as the QPC is quasi-1D. To this end, the extent of wave function perpendicular to the 2DEG is neglected and the system is assumed to be strictly 2D. For a uniform QPC of length d (with the propagation axis along the z-direction and the transverse axis along the y-direction) the potential can be written as

$$V(y,z) = V_c(y)\theta(z)\theta(d - z), \tag{3}$$

where the confining potential $V_c(y)$ has eigenfunctions $\phi_n(y)$ satisfying

$$\{-\frac{\hbar^2}{2m*}\frac{d^2}{dy^2} + V_c(y)\}\,\phi_n(y) = \epsilon_n\phi_n(y). \tag{4}$$

The subband structure can be obtained by assuming free propagation along z-axis so that $\exp(\pm i\gamma_n z)\phi_n(y)$ are the current carrying solutions for the 2D Schrödinger equation in the constriction ($0 < z < d$) with energy $\epsilon_n + \hbar^2\gamma_n^2/2m*$.

Clearly the form of the confining potential plays an important role as far as the quantitative results are concerned. It was determined both self-consistently[20] and empirically [26]. Laux and co-workers[20] calculated the self-consistent potential due to a split gate and found that when the number of the occupied subbands is small $V_c(y)$ is parabolic, and as N_c increases it becomes flat near the center in the y-direction. This result was verified for the QPC used in the experiments [26], and it was shown that the most appropriate form for $V_c(y)$ is a quantum well with *smooth* parabolic walls. We use both parabolic and infinite well confinement since they provide significant simplifications in calculations. Nevertheless, any form of the confining potential may be used in the following, since the formulation does not explicitly depend on it. It is also shown that[20] the minimum of the potential is changing with the gate voltage. This implies that the QPC between two 2DEG regions actually has a saddle point structure. Although such a feature is neglected in Eqn. 3, the effect of the saddle point potential is analyzed later. Note that the electron density[26] or equivalently Fermi wavelength λ_F is a function of V_g. Therefore, the results presented in Section 3 can not be compared directly to the experiments. A study similar to that given in Ref. 26 is necessary to obtain the conductance as a function of V_g. The saddle point structure, on the other hand, can be found by including the effects of the fringing fields as well.

With the knowledge of ϵ_n and $\phi_n(y)$ one can solve the 2D Schrödinger equation for the potential given in Eqn. 3. Since the linear response theory is employed, it is sufficient to find the solutions $\psi(y,z)$ with energy $E = E_F$. In the 2DEG ($z < 0$ and $z > d$) $\psi(y,z)$ is a linear combination of 2D plane waves, each of which constitutes a solution. In the constriction ($0 < z < d$) a linear combination of the subband states is used. Note that, for a finite length constriction one has to include also the states lying below E_F, which have decaying wave functions along the z-direction. As shown below, they become essential when tunneling through the QPC is concerned. The solution for an incident right-going plane wave with the wave vector $k = (\kappa_o, k_o)$ can be written as

$$\psi_k(y,z) = e^{i(\kappa_0 y + k_0 z)} + \int_{-\infty}^{\infty} d\kappa\, B_k(\kappa) e^{-i(\kappa y + k_z(\kappa)z)} \quad \text{for } z < 0$$

$$= \sum_n [e^{i\gamma_n z}\theta_n(k) + e^{-i\gamma_n z}\Delta_n(k)]\phi_n(y) \qquad \text{for } 0 < z < d \tag{5}$$

$$= \int_{-\infty}^{\infty} d\kappa\, A_k(\kappa) e^{i(\kappa y + k_z(\kappa)z)} \qquad \text{for } d < z$$

where $k_z^2(\kappa) + \kappa^2 = 2m^*E_F/\hbar^2$ and $k_0 = k_z(\kappa_0)$. The coefficients A_k, B_k, $\theta(k)$ and $\Delta(k)$ have to be determined using the boundary conditions at the edges of the constriction (i.e., $z = 0$ and $z = d$). Making use of the transverse Fourier transform

$$\Phi(q) = (2\pi)^{-1/2} \int_{-\infty}^{\infty} dy\, e^{-iqy}\phi(y), \tag{6}$$

the continuity of the wave function and its derivative at the edges can be incorporated in the problem. The Fourier transform of Eqn. 5 and its derivative along the z-direction at the edges can be calculated easily in terms of $\Phi(q)$. The coefficients A_k and B_k in Eqn. 5 are eliminated from these equations and one gets

$$\sum_n [(k_z(\kappa)+\gamma_n)\theta_n(k)+(k_z(\kappa)-\gamma_n)\Delta_n(k)]\phi_n(\kappa) = (2\pi)^{1/2} 2k_0 \delta(\kappa-\kappa_0), \tag{7}$$

$$\sum_n [(k_z(\kappa) - \gamma_n)e^{i\gamma_n d}\theta_n(k) + (k_z(\kappa) - \gamma_n)e^{-i\gamma_n d}\Delta_n(k)]\phi_n(\kappa) = 0. \tag{8}$$

Equations 7 and 8 can be further simplified by multiplying from left by $\Phi_m^*(\kappa)$ and integrating over κ. The equations obtained therefrom can be cast in a matrix form and are solved for the coefficient vectors $\bar{\theta}$ and $\bar{\Delta}$ to get

$$\bar{\theta}(k) = [1 - \{\bar{r}(k)\exp(i\bar{\Gamma}d)\}^2]^{-1}\bar{t}(k), \tag{9}$$

$$\bar{\Delta}(k) = \exp(\bar{\Gamma}d)\bar{r}(k)\exp(i\bar{\Gamma}d)\bar{\theta}(k), \tag{10}$$

where the matrix $\bar{\Gamma}$ is diagonal with elements $\Gamma_{ij} = \delta_{ij}\lambda_j$. The reflection matrix $\bar{r}$ and transmission vector $\bar{t}$ are given by

$$\bar{r}(k) = (\bar{\Gamma} + \bar{\kappa})^{-1}(\bar{\Gamma} - \bar{\kappa}), \tag{11}$$

$$\bar{t}(k) = (2\pi)^{1/2}(\bar{\Gamma} + \bar{\kappa})^{-1} 2k_0 \bar{\Phi}^\dagger(\kappa_0). \tag{12}$$

Here $\bar{\Phi}$ is the vector of transverse Fourier transforms of the wave functions and the momentum matrix $\bar{\kappa}$ is defined via

$$K_{mn} = \int\limits_{-\infty}^{\infty} d\kappa \; \Phi_m^*(\kappa) k_z(\kappa) \Phi_n(\kappa). \tag{13}$$

It was shown[12] that the matrix $\bar{r}$ and the vector $\bar{t}$ consist of reflection and transmission coeficients for a semi-infinite constriction connected to a 2DEG for incidence from the constriction and 2DEG, respectively. The analogy between the quasi-1D system and the strictly 1D system was also established by the authors[12] based on Eqns. 9-12.

Finally, the conductance of the QPC is calculated within the linear response theory using the expectation value of the current operator $\hat{j}$, with respect to the current carrying solutions $\psi_k(y,z)$

$$G = \int \frac{dk}{(2\pi)^2} < \Psi_k |j| \Psi_k > \delta \left[\frac{\hbar^2 k^2}{2m^*} - E_F \right] \theta(k_z). \tag{14}$$

Here the δ-function selects the states on the Fermi circle and $\theta(k_z)$ guarantees that the incident wave vector k is pointing at the constriction. Expanding the integrand for the wave function given in Eqn. 5 one gets[12]

$$G = \frac{e^2}{\pi h} \int\limits_{-k_F}^{k_F} \frac{d\kappa}{k_z(\kappa)} \left\{ \bar{\theta}^\dagger(k)\bar{\Gamma}_R\bar{\theta}(k) - \bar{\Lambda}^\dagger(k)\bar{\Gamma}_R\bar{\Lambda}(k) + 2\text{Im}[\bar{\theta}^\dagger(k)\bar{\Gamma}_I\bar{\Lambda}(k)] \right\} \tag{15}$$

where the matrix $\bar{\Gamma}$ is decomposed into its real and imaginary parts, $\bar{\Gamma} = \bar{\Gamma}_R + i\bar{\Gamma}_I$. The first and second terms in this expression correspond to right- and left-going waves in the constriction, respectively. Their interference effects are responsible for resonance structure which is described below. The third term, on the other hand, corresponds to evanescent states lying above E_F and yields tunneling effects. The advantage of the present formalism and that developed by Kirczenow[8] lies in the fact that the size of the matrices used in the calculations is independent of the length of the constriction. In addition to that, this size does not exceed 20 x 20 for observing more than 15 steps in the conductance curve. Thus, the explicit use of the subband structure makes the computations more tractable than those for tight-binding[10,13] and scattering[11] methods.

2.2 Transfer Matrix Method

In order to include several geometrical effects which may play crucial roles in quantized conductance and the saddle point structure of the potential we extend the above formalism. To this end, we seek solutions to the 2D Schrödinger equation for a potential

$$V(y,z) = V_s(z) + V_c(y,z)\theta(z)\theta(d - z), \tag{16}$$

374

where the potential $V(y,z)$ is separated into two parts, one for the saddle point structure $V_s(z)$, and the second for the geometry of the constriction $V_c(y,z)$. Note that $V_s(z)$ is the minimum value of the potential $V(y,z)$ along the y-direction and $V_c(y,z)$ is changed according to the variation of the width of the constriction. In order to use a wave function similar to that given in Eqn. 5 in the 2DEG (i.e., $z < 0$ and $z > d$), it is assumed that the saddle point potential $V_s(z)$ is vanishing out of the constriction. The 2D Schrödinger equation for the potential given by Eqn. 16 can not be solved exactly. A transfer matrix method is used to obtain an approximate solution as described below.

First, the constriction is divided into a large number of segments. In each segment $V_s(z)$ and $V_c(y,z)$ are assumed to be constant. Thus, the solutions in the ith segment are the same as that of a uniform constriction with confining potential $V_c(y,z_i)$ and the zero of energy is shifted by $V_s(z_i)$. Hence, the solution in the constriction can be written as

$$\psi_k(y,z) = \sum_n [\exp\{i\gamma_n(z_i)z\}\theta_n(k,z_i) + \exp\{-i\gamma_n(z_i)z\}\Delta_n(k,z_i)]\phi_n(y,z_i)$$

(17)

for $z_{i-1} < z < z_i$.

For the nonuniform constriction the boundary conditions at $\{z_i\}$ have to be taken into account together with those at $z = 0$ and $z = d$. This is done using the transfer matrix method. Consider an infinite constriction which has two uniform semi-infinite parts joining at $z = z_1$. The wave function can be written as

$$\bar{\phi}_1(y)[\exp(i\bar{\Gamma}_1 z)\bar{\theta}_1 + \exp(-i\bar{\Gamma}_1 z)\bar{\Delta}_1] \quad \text{for } z < z_1$$

$$\bar{\phi}_2(y)[\exp(i\bar{\Gamma}_2 z)\bar{\theta}_2 + \exp(-i\bar{\Gamma}_2 z)\bar{\Delta}_2] \quad \text{for } z_1 < z .$$

(18)

Using the continuity of the wave function and its derivative at $z = z_1$ one gets the matrix equation

$$\begin{bmatrix} \bar{\theta}_1 \\ \bar{\Delta}_1 \end{bmatrix} = \frac{1}{2} \begin{bmatrix} e^{-i\bar{\Gamma}_1 z_1}(\bar{S}+\bar{\Gamma}_1^{-1}\bar{S}\bar{\Gamma}_2)e^{i\bar{\Gamma}_2 z_1} & e^{-i\bar{\Gamma}_1 z_1}(\bar{S}-\bar{\Gamma}_1^{-1}\bar{S}\bar{\Gamma}_2)e^{-i\bar{\Gamma}_2 z_1} \\ e^{i\bar{\Gamma}_1 z_1}(\bar{S}-\bar{\Gamma}_1^{-1}\bar{S}\bar{\Gamma}_2)e^{i\bar{\Gamma}_2 z_1} & e^{i\bar{\Gamma}_1 z_1}(\bar{S}+\bar{\Gamma}_1^{-1}\bar{S}\bar{\Gamma}_2)e^{-i\bar{\Gamma}_2 z_1} \end{bmatrix} \begin{bmatrix} \bar{\theta}_2 \\ \bar{\Delta}_2 \end{bmatrix}$$

(19)

$\bar{S}$ being the overlap matrix,

$$S_{mn} = \int_{-\infty}^{\infty} dy\ \phi_{1m}^*(y)\phi_{2n}(y).$$

It can be shown that the transfer matrix given in Eqn. 19 is analogous to the strictly 1D transfer matrix. The full solution can be found by using this transfer relation and the boundary conditions at the edges. The conductance is obtained by evaluating the expression in Eqn. 15 in any one of the segments. In the following section the transfer matrix method is used to investigate several problems related to QPC.

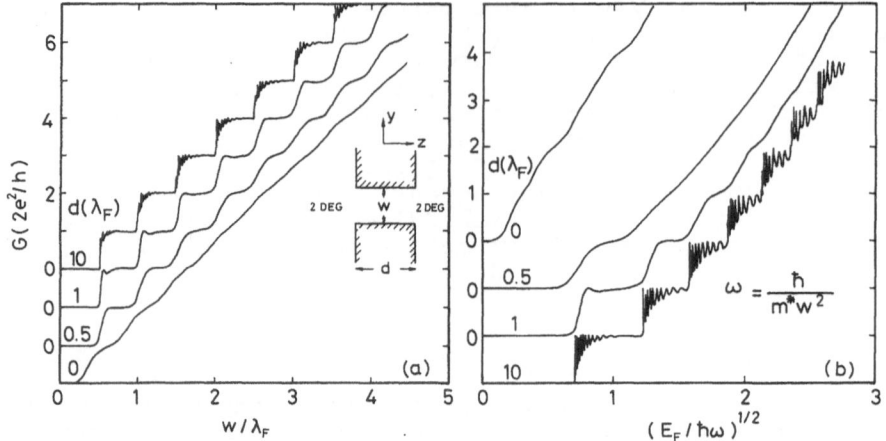

Fig. 1. Conductance G of the QPC as a function of width
w for (a) infinite well, (b) parabolic confinement in the transverse direction.

3.   RESULTS

The formalism outlined in the preceding section is used to
calculate the ballistic conductance in QPC with different
structural parameters. In addition, the effects of the
variation of potential in the constriction, as well as those of
finite temperature and bias are explored. In view of these
results we attempt to develop a better understanding of the
underlying physical events leading to the deviation from the
exact quantization.

3.1 Uniform Constriction

First we consider a constriction of uniform width and
finite length which is connected to a 2DEG at both ends. We
consider two types of confinement, namely an infinite well of
width w and a parabolic potential of width $w = (\hbar/m*\omega)^{1/2}$. The
essential differences between the two are the energy separation
between the subbands [$\epsilon_n = \hbar^2(2\pi/w)^2/2m*$ for the infinite well
and $\epsilon_n = \hbar\omega(n + 1/2)$ for the parabolic potential] and the
lateral wave function [for the infinite well $\phi_n(y)$ is zero for
$|y| > w/2$, however for the parabolic potential $\phi \alpha \exp\{-y^2/2w^2\}$].

The conductance curves G(w) of uniform QPC for infinite
well and parabolic potential are illustrated in Figs. 1a and 1b,
respectively. It is seen that G(w) exhibits different behavior
for different confining potentials. For very short constrictions the semiclassical conductance was calculated first by
Sharvin[22], who showed that the conductance of the QPC is
independent of any material properties and solely determined by
the geometry of the contact and electron density. This conductance is referred to as the Sharvin conductance and is given by
$G_s = (2e^2/h)2w/\lambda_F$. One observes in Fig. 1a that for constrictions of vanishing length the whole conductance curve is
displaced with respect to G_s and weak oscillations are superimposed on top of the straight line. The displacement is due to
the Heisenberg uncertainty principle and the oscillations are
related to the quantum interference effects. For longer

constrictions a step structure begins to emerge from this
quantum Sharvin conductance.

The larger is d, the sharper are the quantum jumps and the
closer are the steps to integer multiples of $2e^2/h$. The most
important feature of G(w) curves for uniform constrictions is
the resonance structure superimposed on the plateaus, which
originates from the interference of the reflected waves from the
ends of the constriction. That is, actually the conductance
does not exhibit a staircase structure, but oscillates between
the quantized value (resonances) and minima (antiresonances).
The number of resonances on each plateau increases with
increasing d but decreases with increasing w (or increasing
subband index n). The important point to note is that the
experimentally observed conductance curves are lacking the
resonance structure. Moreover, the sharp corners of the step
structure are rounded. Consequently, the experimental data and
theoretical predictions[8-12] obtained for uniform constrictions
are at variance.

Since the sharp Fermi distribution is smeared out at finite
but low temperature, the quantum conductance is expected to
deviate from that at T = 0 K. Assuming the decrease in the
carrier mobility is not substantial and transport through the
QPC is still ballistic, the use of the Fermi-Dirac distribution
suffices to include the effect of finite temperature[9,12,14].
It is found that for experimentally relevant temperature scales
(T ≃ 1 K) the resonance structure is destroyed for long
constrictions (d ≃ $10\lambda_F$), but not for the shorter ones (d ≃ $2\lambda_F$).
For higher temperatures (T ≃ 5 K) the resonance structure is
totally eliminated even for short constrictions, but the step
structure is still apparent for small subband index n. Similar
effects were obtained for a finite bias[14,23]. For example, a
bias of eV = $0.05E_F$ is sufficient to destroy the resonance
structure, and for a bias of Ev = $0.5E_F$ the quantized steps
begin to disappear for high index subbands. Note that in the

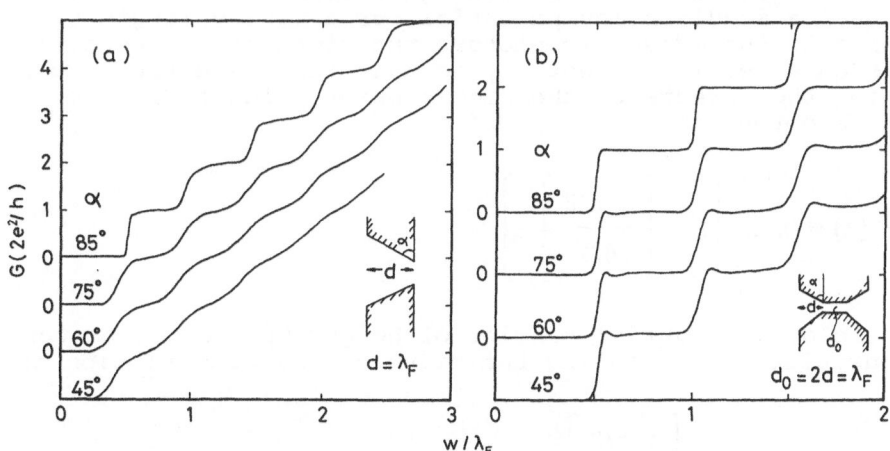

Fig. 2. Conductance G of the QPC as a function of width
w for (a) wedge-like constriction, (b) flared
constriction yielding adiabatic transport. The
confinement in the transverse direction is
infinite well.

experiments[4,5] the bias eV is taken to be smaller than kT so that the temperature effects are more important than those of the finite bias. Experiments carried out to investigate the nonlinear conductance of the QPC showed that[24] the linear voltage dependence on the plateaus is invalidated above a critical bias.

3.2 Flared Constriction and Adiabatic Evolution

A uniform constriction with abrupt connections to the 2DEG regions as described in Fig. 1 is a very idealistic situation. First of all, the openings to 2DEG are either flared or smooth due to limitations in fabrication. In addition, the electrostatic depletion region created by the split gate cannot be as sharp as the lithographic geometry. Thus, inevitably there has to be a flaring at the entrance and exit of the QPC. Calculations of the conductance $G(w)$ for the tapered and the wedge-like opening indicates a pronounced effect of the opening geometry. In Fig. 2a the step structure becomes apparent only for wedge angle $\alpha > 75°$, and for $\alpha \simeq 45°$ the conduction is still in the *quantum Sharvin* regime.

A constriction with a finite uniform part and flared openings to 2DEG with large wedge angle is reminiscent of the constriction considered by Glazman and co-workers[15]. Figure 2b shows that the resonance structure and smooth steps evolves to flat plateaus with sharp steps as α increases. This is explained by the *adiabatic* evolution of the current carrying states where the subband quantum number n is conserved throughout the constriction, and subband energy $\epsilon_n(z)$ acts as an effective 1D potential.

The saddle point potential also affects the conductance curve. Such a case was analyzed analytically by Büttiker[25]. Using the *adiabatic* approximation[15,16] he showed that the conductance of a saddle point constriction of infinite length is quantized under certain conditions. We used the transfer matrix approach described in the preceding section to investigate the effects of a saddle point potential for a finite length QPC. According to the *adiabatic* picture the width of the constriction acts as an effective potential, so that there are two ways of analyzing the effects of the saddle point. The first one is to include a potential

$$V_s(z) = \eta \ E_F \left[1 - \left[\frac{z-d/2}{d/2} \right]^2 \right] , \tag{20}$$

which yields a potential barrier of height ηE_F at the center of the constriction. Changing the width of the constriction in the form

$$w(z) = w_o + \xi\lambda_F \left[\frac{z-d/2}{d/2} \right]^2 , \tag{21}$$

on the other hand, produces an effective potential barrier which has the saddle point structure. The results obtained by using

378

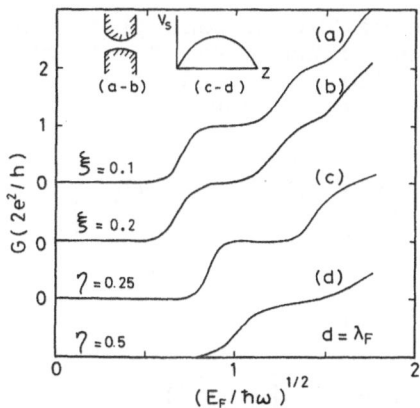

Fig. 3. Conductance G of the QPC as a function of
width w for saddle point structures given
by (a-b) Eqn. 21, (c-d) Eqn. 20. The
confinement in the transverse direction is
parabolic. The width is given by
$w = (\hbar/m^*\omega)^{1/2}$.

both kinds of constriction are shown in Fig. 3. Comparing these
conductance curves with that given in Fig. 1b one sees that the
saddle point potential used in Fig. 3 leads to *adiabatic*
evolution of states so that the resonance structure is suppressed. However, as η gets larger the real potential starts to
dominate the effective potential and the steps between the
plateaus become wider and the quantization disappears.
Similarly, as ξ gets larger the length of the effective
potential barrier gets smaller so that the evanescent states
become more effective, and yields narrowing of the plateaus and
distortion of quantization.

3.3 Elastic Scattering by Impurities

One of the important problems related to the quantization
of conductance is the effect of the scattering events due to
impurities[25] and potential fluctuations[28] in the channel.
We investigated the effects of elastic scattering by impurities
by performing exact calculation of conductance[17]. The eigenstates of a uniform constriction in the presence of a scattering
potential $v_I(y,z)$ can be found by solving the Lippman-Schwinger
equation

$$\psi_j(y,z) = e^{i\gamma_j z}\phi_j(y) + \int dy'dz' g(z-z';y,y')v_I(y',z')\psi_j(y'z'). \tag{22}$$

We adopt a model impurity potential $v_I(y,z)$ since the solution
of Eqn. 22 for an arbitrary potential is quite complicated. To
this end we approximated the scatterer by

$$v_I(y,z) = \frac{\hbar^2\beta}{m^*}\exp(-q|y - y_I|)\,\delta(z - z_I), \tag{23}$$

which is δ-function in the z-direction, and has the exponentially decaying form in the y-direction with a decay length

of q^{-1}. The strength of this potential is set by the magnitude of β, which may be both attractive ($\beta < 0$) and repulsive ($\beta > 0$). For this form of the potential the conductance of an infinite constriction can be found as

$$G_\infty = \frac{2e^2}{h} \sum_j^{\epsilon_j < E_F} \left\{ 1 + 2\frac{\text{Im}(\bar{\Omega})_{jj}}{\gamma_j} + \frac{\text{Re}(\bar{\Omega}^\dagger \bar{\Gamma}^{-1} \bar{\Omega})_{jj}}{\gamma_j} \right\},$$ (24)

where $\bar{\Omega} = \bar{u}(\bar{\Gamma} + i\bar{u})^{-1}\bar{\Gamma}$ with

$$u_{ij} = \beta \int dy\, \phi_i^*(y)\phi_j(y)\exp(-q|y - y_I|).$$ (25)

The calculation of the conductance for a finite length QPC is described elsewhere[17]. The results obtained from this formalism are in overall agreement with the earlier ones[10,18,19]. In addition to that, the present approach has the advantage of investigating several features of the scattering potential in a systematic way[17].

In Fig. 4 we present the results obtained for attractive impurities with various β and q values. The ratio β/q is kept constant in order to minimize the effect of the varying integrated strength[17]. In what follows we focus our attention on a novel effect of the attractive impurities which recently found experimental evidence[29]. This is the enhanced back-scattering due to quasi-bound states. Ignoring the intersubband interaction which is included via the nondiagonal elements of $\bar{u}$ given in Eqn. 25, one is left with the strictly 1D problem for each subband. Clearly, an attractive δ-function potential has a bound state in 1D. Since the bound states do not carry any current for infinitely long systems, the effects of these bound

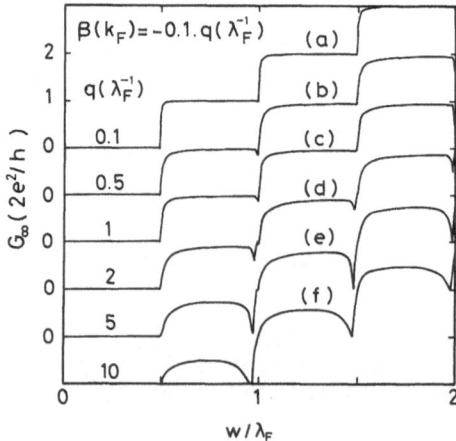

Fig. 4. Conductance G of the infinite constriction as a function of width w in the presence of an attractive impurity with $y_I = 0.13\,\lambda_F$. The confinement in the transverse direction is infinite well.

states would not appear in the conductance curve. This is the case for Fig. 4a, since the intersubband interaction is negligible due to long decay length q^{-1}. Decreasing q [Fig. 4b-c], some dips begin to emerge just below the propagation threshold of the subbands $n \geq 1$. For intermediate values of q [Fig. 4d-e] there are both dips and peaks, and for large q [Fig. 4f] the peaks merge to the propagation threshold leaving only the dips. These dips are the signs of enhanced backscattering resulting from the intersubband interaction. When the coupling between the subbands is not negligible, one has to deal with the full quasi-1D Schrödinger equation. The 1D bound states described above are degenerate with the continuum of propagating states from lower lying subbands and they have to be mixed. Such a mixing gives rise to two types of states. The first one is the quasi-bound states with no net current which are reminiscent of the true bound states. These do not contribute to the current for infinite constriction. The second type of states are complete backscattering states, which have unity reflection probability. The wave function for these states has a standing wave form in the lower lying subband and a quasi-bound state contribution from the next subband. For very large values of q [Fig. 4f], the first type of states disappears since the 2D δ-function has no bound states. Recently Faist and co-workers[29] observed the dips and absence of plateaus in conductance curves of highly disordered samples. Although the argument given above does not directly apply to the experiment, the underlying physics is just the enhancement in backscattering due to the presence of localized states.

3.4 Quasi-0D States and Resonant Tunneling

According to the *adiabatic* picture of ballistic transport, the subband energy $\epsilon_n(z)$ acts as a 1D effective potential. Thus by widening the constriction at the center, $z = d/2$, it should be possible to form a cavity (effective potential of which resembles to a quantum well as illustrated in Fig. 5) along the channel. A similar effect can be observed when the potential along the constriction, $V_s(z)$ in Eqn. 16, has a minimum at the center. Thus, the constriction effectively acts like a quantum well between two potential barriers for certain QPC configurations. Accordingly, such a structure leads to formation of quasi-0D states which are bound to the cavity and give rise to resonant tunneling (RT) effects for the finite QPC. It was shown that[12,30] even for sharp QPC geometries (that is, the *adiabatic* approximation is not valid) RT peaks are observable in the conductance curve G(w). Typical results obtained with QPC geometries which are adequate for RT are shown in Fig. 5. Clearly, just below the steps corresponding to dipping of a subband below the Fermi level, i.e., $w < n\lambda_F/2$, there are sharp peaks with amplitude $2e^2/h$. These RT features are analogous to the double barrier RT structures obtained by *band gap engineering* methods. Namely, in addition to ordinary tunneling through the channel, there is a resonance event taking place as evidenced from the peaks in the conductance curve. For a certain width of the constriction, resonance takes place and unity transmission through the channel is achieved. Destroying the resonance condition (by increasing the width) the conductance rapidly drops to its ordinary tunneling value, however. It was also shown that an attractive impurity at the center of the constriction yields RT characteristics[17,30]. A detailed analysis of RT effects in QPC is given elsewhere[30].

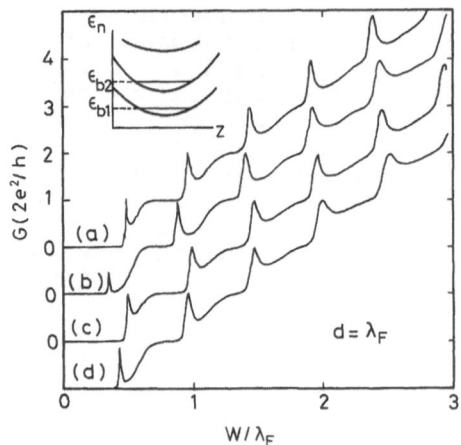

Fig. 5. Conductance G as a function of width w of the channel for configurations appropriate to observe RT effects. At the center (a) the channel gets wider by an amount 0.2 λ_F; (b) there is a potential well of height 0.2 E_F, both of length 0.5 λ_F. The channel is modulated by a cosine-wave so that at the center (c) it is wider by an amount 0.3 λ_F; (d) there is a potential well of height 0.3 E_F. The confinement in the transverse direction is infinite well.

Although observation of Q0D states was reported[26], it is not clear so far whether these are really RT events since the resistance, but not the conductance curve, has peaks.

3.5 Applications in Scanning Tunneling Microscopy

The electron transport occurs normally via tunneling between the sharp tip and sample surface in scanning tunneling microscopy (STM). A significant potential barrier between the electrodes characterizes the normal tunneling regime. As the tip approaches the sample, the potential barrier lowers gradually and finally collapses due to tip-sample interaction. However, ballistic transport sets in when the lowest energy state (which is the solution of the Schrödinger equation between the electrodes) dips in the Fermi level, i.e., the effective barrier collapses. This way a channel of atomic dimensions opens in the potential barrier between the electrodes. Since the size of this channel is comparable with λ_F of the electrodes quantum effects similar to those discussed before become pronounced. In particular the conductance of the point contact can be quantized. In fact Gimzewski and Möller[31], who measured the tunneling current as a function of the tip-sample distance, observed oscillations as the tip pushed further to the sample after the hysteric deformation. Whether the observed oscillations have some bearing on the quantization of the conductance has been the subject of interest[32,33]. We calculated the current as a function of distance covering a wide range[34] (~ 2-10 Å) by using a 3D potential similar to that given in Eqn. 16. Our results provide a thorough understanding of several features of point contact such as appearance of the first plateau, oscillations, and the possibility of having decreased conductance with decreasing tip-sample distance. In particular our results indicate that the point contact through a short constriction prevents the quantization of the conductance and pronounced quantum oscillations arising therefrom, and favors the ballistic transport taking place in the *quantum Sharvin* regime. Then, the observed oscillations are attributed rather to the irregular enlargement of the contact area.

Field emission of the electrons from an atomically sharp tip happens to be an interesting event relevant to STM. Fink[35] produced such sharp and stable tips and obtained a focused low energy electron beam with substantial current density. The potential in the vicinity of such a tip can be described by a funnel[36], the minimum of which lies above the Fermi level for intermediate electric fields. The lateral confinement of the states in this funnel imposes an effective potential which is higher than the actual potential barrier. This funnel is connected to the outer vacuum region with a horn-like opening. In such a system, electrons which enter the funnel tunnel through the effective barrier, and then evolve in the horn-like opening. Having left the horn they finally move semiclassically in the outer vacuum region with laterally uniform electric field[37]. The variation of potential energy near the tip is reminiscent of the model used in the ballistic transport through a constriction, except that the electrons are tunneling the potential barrier before the vacuum region. Earlier theories[36,37] have controversial explanations for the main effect responsible for focusing.

We have studied the field emission from an atomically sharp tip and investigated the origin of collimation[38]. Our calculations indicate that during the tunneling through the potential barrier, low transverse momentum states are selected due to lower effective potential. This filtering of the low subband quantum number states is found to be most essential for the collimation of the beam. For the typical horn-like openings the *adiabatic* evolution of states (i.e., transmission without intersubband scattering) does not take place. Nevertheless, the horn-like opening may improve the collimation owing to the dominant transverse momentum reduction effect. That is, the effect of the widening of the orifice is superior to the intersubband scattering which increases the lateral momentum, even though the evolution is not *adiabatic*. In addition, such field emitting structures were shown[30,38] to exhibit novel resonance effects.

CONCLUSIONS

We have presented a formalism based on the linear response theory and studied the ballistic transport in quasi-1D systems and in the quantum regime. We showed that a uniform constriction of length $d \sim 5\lambda_F$ gives rise to sharp steps. However, due to the interference of waves a resonance structure is superimposed on the plateaus. Such a resonance structure is lacking in the experimental conductance curve. We showed that the resonance structure can be eliminated by several effects. In particular, smooth entrance followed by a uniform part give rise to step structure without resonance due to *adiabatic* evolution of subband states. Using the same formalism, we conjectured new resonant tunneling effects. We used our formalism to study the STM at small tip-sample distance, and developed an understanding for point contact and focusing of electron beams emitted from a point source.

Acknowledgment

This work was partially supported by a Joint Project Agreement between Bilkent University and IBM Zurich Research Laboratory. The authors acknowledge fruitful discussions and collaboration with A. Baratoff in the study of focused electron emission.

REFERENCES

1. For a review see: "Nanostructure Physics and Fabrication", M.A. Reed and W.P. Kirk, eds., Academic Press, New York (198)
2. R. Landauer, **IBM J. Res. Develop.**, 1:223 (1957)
3. H.U. Baranger and A.D. Stone, **Phys. Rev.**, 40:8169 (1989) and references therein
4. B.J. van Wees, H. van Houten, C.W.J. Beenakker, J.G. Williamson, L.P. Kouwenhoven, D. van der Marel and C.T. Foxon, **Phys. Rev. Lett.**, 60:848 (1988)
5. D.A. Wharam, M. Pepper, H. Ahmed, J.E.F. Frost, D.G. Hasko, D.C. Peacock, D.A. Ritchie and G.A.C. Jones, **J. Phys. C**, 21:L209 (1988)
6. Y. Imry, **in**: "Directions in Condensed Matter Physics", vol. 1, G. Grinstein and G. Mazenko, eds., World Scientific, Singapore (1986)
7. R. Landauer, **Z. Phys. B**, 68:217 (1987)
8. G. Kirczenow, **Solid State Commun.**, 68:715 (1988); **Phys. Rev. B**, 39:10452 (1989)
9. A. Szafer and A.D. Stone, **Phys. Rev. Lett.**, 62:300 (1989)
10. E.G. Haanapel and D. van der Marel, **Phys. Rev. B**, 39:5484; D. van der Marel and E. G. Haanapel, *ibid.* B39:7911 (1989)
11. L. Escapa and N. García, **J. Phys. Condens. Matter**, 1:2125 (1989); N. García and L. Escapa, **Appl. Phys. Lett.**, 54:1418 (1989)
12. E. Tekman and S. Ciraci, **Phys. Rev. B**, 39:8722 (1989); *ibid.* 40:8559 (1989)
13. S. He and S. Das Sarma, **Phys. Rev. B**, 40:3379 (1989)
14. P.F. Bagwell and T.P. Orlando, **Phys. Rev. B**, 40:1456 (1989)
15. L.I. Glazman, G.B. Lesovik, D.E. Khmel'nitskii, R.I. Shekhter, **Pis'ma Zh. Tear. Fiz.**, 48:218 (1988) [**JETP Lett.**, 48:238 (1988)]
16. A. Yacoby and Y. Imry, **Phys. Rev. B**, 41:5341 (1990)
17. E. Tekman and S. Ciraci, **Phys. Rev. B**, 42: Nov. 15 (1990)
18. C.S. Chu and R.S. Sorbello, **Phys. Rev. B**, 40:1456 (1989)
19. P.F. Bagwell, **Phys. Rev. B**, 41:10354 (1990)
20. S.E. Laux, D.J. Frank and F. Stern, **Surf. Sci.**, 196:101 (1988)
21. D.A. Wharam, U. Ekenberg, M. Pepper, D.G. Hasko, H. Ahmed, J.E.F. Frost, D.A. Ritchi, D.C. Peacock and G.A.C. Jones, **Phys. Rev. B**, 39:6283 (1989)
22. Yu. V. Sharvin, **Zh. Eksp. Teor. Fiz.**, 48:984 (1965) [**Sov. Phys. - JETP**, 21:655 (1965)]
23. L.I. Glazman and A.V. Khaetski, **Pis'ma Zh. Eksp. Teor. Fiz.**, 48:546 (1988) [**JETP Lett.**, 48:592 (1988)]
24. L.P. Kouwenhoven, B.J. van Wees, C.J.P. Harmans, J.G.

Williamson, H. van Houten, C.W.J. Beenakker, C.T. Foxon and J.J. Harris, **Phys. Rev. B**, 39:8040 (1989)

25. M. Büttiker, **Phys. Rev. B**, 41:7906 (1990)
26. C.G. Smith, M. Pepper, H. Ahmed, J.E.F. Frost, D.G. Hasko, D.C. Peacock, D.A. Ritchie and G.A.C. Jones, **J. Phys. C**, 21:L893 (1988)
27. G. Timp, R.E. Behringer, J.E. Cunningham and E.H. Westerwick, p. 347 of this volume
28. J.A. Davies, J.A. Nixon and H.U. Baranger, p. 387 of this volume
29. J. Faist, P. Guéret and H. Rothuizen, **Phys. Rev. B**, 42:3217 (1990)
30. E. Tekman and S. Ciraci, **in**: "Resonant Tunneling of Semiconductors: Physics and Applications", L.L. Chang, C. Tejedor and E.E. Mendez, eds., Plenum Press, New York (in press)
31. J.K. Gimzewski and R. Möller, **Phys. Rev. B**, 36:1284 (1987)
32. N.D. Lang, **Phys. Rev. B**, 36:8173 (1987)
33. J. Ferrer, A. Martin-Rodero and F. Flores, **Phys. Rev. B**, 38:10113 (1988)
34. S. Ciraci and E. Tekman, **Phys. Rev. B**, 40:11969 (1989)
35. H.W. Fink, **Physica Scripta**, 38:260 (1988)
36. N.D. Lang, A. Yacoby and Y. Imry, **Phys. Rev. Lett.**, 63:1499 (1989)
37. H. de Raedt, N. García and J.J. Sáenz, **Phys. Rev. Lett.**, 63:2260 (1989)
38. E. Tekman, S. Ciraci and A. Baratoff, **Phys. Rev. B**, 42: Oct. 15 (1990)

POTENTIAL FLUCTUATIONS IN HETEROSTRUCTURE DEVICES

John H. Davies, John A. Nixon and Harold U. Baranger†

Department of Electronics & Electrical Engineering
Glasgow University, Glasgow G12 8QQ, UK
† AT & T Bell Laboratories, Crawford Corner Road
Holmdel, NJ 07733, USA

ABSTRACT

Ionized donors in a doped heterostructure have random positions, which gives rise to a random potential with long-ranged fluctuations. This has a small effect on fast electrons, allowing high mobilities in two-dimensional electron gases, but becomes much more significant when the density of electrons is low. We have analysed the effect of this random potential using a semi-classical, self-consistent method to calculate the density of electrons. This includes a realistic potential from a patterned gate, and is followed by a quantum-mechanical calculation of the conductance.

A nominally uniform two-dimensional electron gas becomes inhomogeneous as its average density is lowered, eventually breaking into isolated puddles and undergoing a classical metal-insulator transition. A quantum wire is strongly distorted by the random potential, which impedes ballistic propagation. The wire becomes discontinuous before it is narrow enough to support monomode propagation. The conductance of a short constriction or 'point contact' (0.2 μm) is well quantized. The guiding potential is rounded, suppressing resonances within the constriction. Quantization in a longer constriction (0.6 μm) is spoiled by scattering from the random potential. These results agree well with experiments by G.L. Timp et al.

1. INTRODUCTION

Recent advances in quantum ballistic transport, such as the observation of the quantized conductance of a narrow constriction[1,2], rely heavily on the low scattering rate in a two-dimensional electron gas (2DEG) trapped at a heterojunction. Mobilities exceeding 10^3 $m^2V^{-1}s^{-1}$ have been reported[3], for which a straightforward calculation of the mean free path gives a value approaching 0.1 mm. It appears that scattering is very weak in these structures and that the random potential from the ionized donors must be very small. We shall show that this

Condensed Systems of Low Dimensionality
Edited by J. L. Beeby *et al.*, Plenum Press, New York, 1991

picture changes drastically if the average density of electrons is lowered, either by reducing the concentration of the 2DEG uniformly or by patterning the 2DEG into narrow 'quantum wires'. There are two principal reasons for this change. First, the screening becomes weaker as the density is lowered, so the random potential increases. Secondly, the random potential has long-ranged fluctuations with most of its weight as small wavevectors. Energetic electrons are only scattered weakly by such a potential, but it becomes more effective as the average density and Fermi energy are lowered.

We find that a nominally uniform 2DEG undergoes a metal-insulator transition and breaks up into isolated puddles as its average density is lowered. The edges of a quantum wire are roughened by the random potential, and the wire becomes discontinuous as its average width is reduced. This will frustrate the goal of making wires behave like monomode waveguides, which is desirable to increase the interference of electron waves. A short 'point contact' (0.2 μm) is little affected by the random potential, because the active area is smaller than the coherence area of the fluctuations, and quantization remains good. The fluctuations induce scattering within a longer contact (0.6 μm) and quantization is spoiled. Our calculations are based on a semi-classical self-consistent model of a heterostructure in which we treat the donors explicitly as discrete random charges[4]. The conductance of electrons moving through this potential is then calculated quantum mechanically.

Classical potential fluctuations arising from the random positions of ionized impurities in impurity bands have been extensively studied[5], as have the effect of discrete dopants on ohmic contacts[6] and planar doped barriers[7]. These were three-dimensional systems: 2DEGs differ in two important aspects. First, the screening is much less effective[8]. Secondly, the electrons are usually separated from the ionized donors by a spacer layer. This cuts off the core of the Coulomb potential, leaving only the tails. Many tails overlap in space, giving rise to a random potential with long-ranged fluctuations in which the contributions from individual donors cannot be distinguished. The details of our model are described in Section 2. Statistical properties of the potential are calculated in Section 3 for two limits, the bare potential (no electrons) and when there is a high density of electrons so that screening is linear. The random potential and the Fermi energy of the 2DEG are comparable between these limits and screening is highly nonlinear. Efros[9] has considered this in connection with the width of plateaus in the integral quantum Hall effect. All such calculations are restricted to nominally uniform 2DEGs and cannot treat a patterned gate. Our previous work on a MODFET with a short gate[10] included the gate potential, but treated screening in a simplistic way. Our numerical method removes these restrictions. It is described in Section 4, and results for nominally uniform 2DEGs, quantum wires and point contacts are given in Section 5. We start by describing the model in detail.

2.   MODEL

Our model of a 2DEG in a heterostructure includes random donors, the potential from a gate and screening. A typical

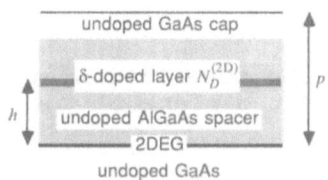

Fig. 1. A typical δ-doped heterostructure,
showing our notation.

δ-doped structure is shown in Fig. 1. Note that $N_D^{(2D)}$ is a
density of donors per unit *area*. We have also considered doped
layers of nonzero thickness, but this paper is restricted to δ-
doping for simplicity. We have assumed that the other layers
are undoped but this is not vital. The main features of the
model are as follows.

i)   Donors are assumed to be fully ionized and distributed at
      random through the δ-doped layer. We ignore parallel
      conduction, DX centers and possible correlation between the
      positions of donors.

ii)   The electrons are treated as strictly two-dimensional, a
      distance p below the free surface (or gate). A 2DEG is
      typically about 10 nm thick, which is much smaller than
      other relevant length scales.

iii) Surface states pin the chemical potential 0.8 eV below the
      conduction band on free surfaces of GaAs. The chemical
      potential is assumed to be flat throughout the structure,
      and the surface can therefore be treated as an
      equipotential with any gate voltages superposed[11]. This
      simplifies the calculations greatly, but it is difficult to
      believe that the surface states can remain in equilibrium
      with the 2DEG at the low temperatures typically used for
      experiments. We propose to investigate the effect of
      different boundary conditions in the future, but the
      experimental position is unclear.

iv)   The density of electrons at any point is given by a
      semiclassical Thomas-Fermi approximation, and depends only
      on the potential at that point[18,19]. This will be a good
      approximation provided that the Fermi wavelength is much
      smaller than the scale on which the potential varies. This
      is accurate when the density of electrons is high, but
      becomes less good at low densities or if the electrons
      occupy isolated 'puddles'. The alternative would be to
      treat the electrons quantum mechanically, and solve
      Schrödinger's equation on a large 3D grid - a major
      computational task.

v)   After the density of electrons and the self-consistent
      potential have been computed semi-classically, the
      conductance is calculated in a purely quantum-mechanical
      way.

3. ANALYTICAL RESULTS

Within this model, the potential at **r** contributed by a donor i, at a point $\mathbf{r}_i$ in the plane of the 2DEG and a height h above it, is given by

$$\phi(\mathbf{r},\mathbf{r}_i) = \frac{e}{4\pi\epsilon\epsilon_0}\left[\frac{1}{\sqrt{|\mathbf{r}-\mathbf{r}_i|^2 + h^2}} - \frac{1}{\sqrt{|\mathbf{r}-\mathbf{r}_i|^2 + (2p-h)^2}}\right], \tag{1}$$

where the second term is the image contributed by the surface states, and all vectors are two-dimensional in the plane of the 2DEG. The total (bare) potential from the donors ϕ_{imp} is given by summing Eqn. 1. The variance σ^2 of ϕ_{imp} is given by

$$\sigma^2 = N_D^{(2D)}\int d\mathbf{r}[\phi(0,\mathbf{r})]^2. \tag{2}$$

There would be an additional integration over h for a thick layer. Substituting Eqn. 1 into Eqn. 2 gives

$$\sigma^2 = \left[\frac{e}{4\pi\epsilon\epsilon_0}\right]^2 2\pi N_D^{(2D)}\log\frac{p^2}{h(2p-h)}. \tag{3}$$

The most significant feature of Eqn. 3 is probably the weak dependence on h, the spacer between the δ-doped layer and the 2DEG. As an example, consider the layer used in reference[14] which had p = 72 nm, h = 42 nm and $N_D^{(2D)} = 5 \times 10^{16}$ m^{-2}. This gives σ = 27 meV, which is much larger than the Fermi energy of about 10 meV.

This potential is screened linearly if there is a high density of electrons in the 2DEG. The Thomas-Fermi dielectric function is $\epsilon(q) = 1 + q_{TF}/q$ in 2D, which is the limit as q $\rightarrow$ 0 of the Stern dielectric function[8]. The Thomas-Fermi screening wavevector $q_{TF} \approx (5$ nm$)^{-1}$ in GaAs. Including this gives

$$\sigma^2 = \left[\frac{e}{4\pi\epsilon\epsilon_0}\right]^2 2\pi N_D^{(2D)}\left[g(2q_{TF}h) + g(2q_{TF}[2p-h]) - 2g(2q_{TF}p)\right], \tag{4}$$

where

$$g(z) = (1+z)e^z E_1(z) - 1 \sim \frac{1}{z^2} \text{ as } z \rightarrow \infty, \tag{5}$$

and $E_1(x)$ is the exponential integral (reference [15], Section 5.1.1). Again this can be extended to thick doped layers, which have been studied previously[16]. It gives σ = 2.2 meV for the example above, much smaller than the unscreened value but far from negligible. The random potential is only slightly reduced on changing from a thick doped layer to δ-doping, assuming that the distribution of impurities remains random. We have also

390

calculated the correlation function. It is long-ranged, and decays as a power law at large distances for both the unscreened and screened potentials, so a correlation length cannot be rigorously defined.

The active regions of typical devices have a rapidly varying density of electrons and lie between the two limits considered above, with strongly nonlinear screening. Further progress therefore requires numerical modeling.

4. NUMERICAL METHOD

The numerical calculations were performed for a square sample with 1-2 μm sides and a grid spacing of about 10 nm. Donors were distributed at random and their potential calculated by summing Eqn. 1 to give ϕ_{imp}, using nearest-image boundary conditions[17]. Three more terms must be added to get the total self-consistent potential ϕ_{tot}. These are ϕ_s from the surface states (-0.8 V, from the energy of the conduction band above the chemical potential), ϕ_g from the gate and ϕ_e from the electrons in the 2DEG. The density of electrons in the 2DEG is given in terms of ϕ_{tot} by the local equation

$$ n(r) = \frac{m}{\pi\hbar^2} \left[\mu - e\phi_{tot}(r) \right] \theta \left[\mu - e\phi_{tot}(r) \right] \tag{6} $$

according to the Thomas-Fermi approximation, where μ is the chemical potential. The discontinuities in the conduction band cancel because both the surface and the 2DEG are in GaAs. The electronic contribution ϕ_e to the total potential was found from $n(r)$ by integrating Poisson's equation over the plane of the 2DEG, including image charges. The system was solved self-consistently by under-relaxation with an estimated error of less than 10 μV in ϕ_{tot}.

The conductance of electrons in this potential was then found by a purely quantum mechanical method[18-20]. Perfect leads are attached to the system by extending the potential profile at the left and right-hand edges outwards to infinity. The Green's function for electrons is then calculated by building up the system in slices from left to right. The transmission matrix and conductance are then derived. A magnetic field can be included but is not considered here. We have taken the temperature to be zero throughout.

5. NUMERICAL RESULTS

We shall describe numerical results for three systems: nominally uniform two-dimensional electron gases, narrow 'quantum wires' and short constrictions that behave as point contacts.

5.1 Nominally Uniform Two-dimensional Electron Gas

We start by describing the results for a 2DEG whose average density can be controlled by the voltage on a large gate, and which would be uniform if there were no fluctuations. The layers match those of references [12] and [13]. A contour map

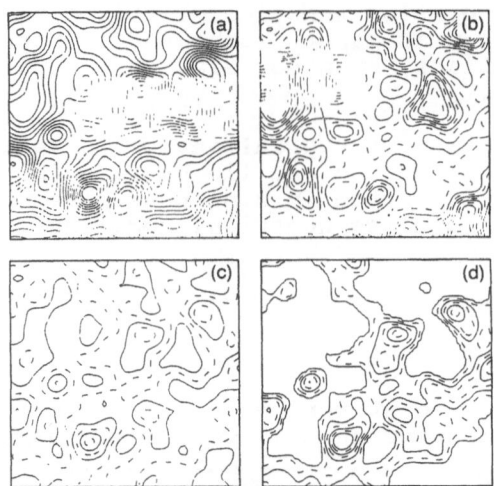

Fig. 2. Contours of self-consistent potential ϕ_{tot} for a
0.5 μm square taken from a 1 μm square sample
with a uniform gate. This sample had p = 70 nm,
h = 28 nm and $N_D^{(2D)}$ = 2.16 x 10^{16} m^{-2}, following
reference[12]. Full contours are 6 mV apart
with dashed lines half-way between. The
electron coverage fractions f are (a) 0.00; (b)
0.57; (c) 1.00. (d) shows the electron density
corresponding to (b), with contours 6 x 10^{14} m^{-2}
apart.

of the bare potential is shown in Fig. 2a. It is smooth, with
features on a typical length scale of 0.1 μm which is much
larger than the separation of impurities (< 0.01 μm). This
emphasizes that the potential is a 'collective' one, with
features arising from fluctuations in the density of donors
rather than from individual ions. The measured standard
deviation is 18 mV, rather less than the analytic estimate of 27
mV for this layer because of large finite-size effects when
there are no electrons present to provide screening.

Electrons were introduced by lowering the voltage on the
gate V_g in steps from a large negative value. The results are
shown in Fig. 2. The coverage fraction f is defined as the
fraction of the area of the sample where electrons are present.
Electrons first fill the deep, isolated pockets in the
potential. These electrons are classically localized if we
ignore tunneling through the broad potential barriers that
separate the pockets.

The standard deviation σ of the self-consistent potential
ϕ_{tot} falls as more electrons enter and screening improves, but
the effect is not uniform over the whole sample. The
fluctuations in ϕ_{tot} are greatly weakened where electrons are
present, and much less affected elsewhere. This is illustrated
well by Figs. 2b and 2d, the self-consistent potential and
corresponding electron density. The minimum value of σ,
corresponding to Fig. 2c, is 2.6 mV which agrees with the
analytic estimate of 2.8 mV.

There is a classical metal-insulator transition in a 2D system when f reaches the percolation threshold, 0.5[21]. This occurs at an average density of $\bar{n}_c = 4 \times 10^{14}$ m^{-2} in our sample, in good agreement with the estimate of Efros[22] who also discusses the experimental evidence for this transition. There is no further improvement in screening in our semiclassical model if more electrons are added after f has reached 1, so σ remains constant (this would not be true in a quantum mechanical calculation). Even at very high densities of electrons, about 3.5×10^{14} m^{-2} localized electrons remain in the band tail of our sample.

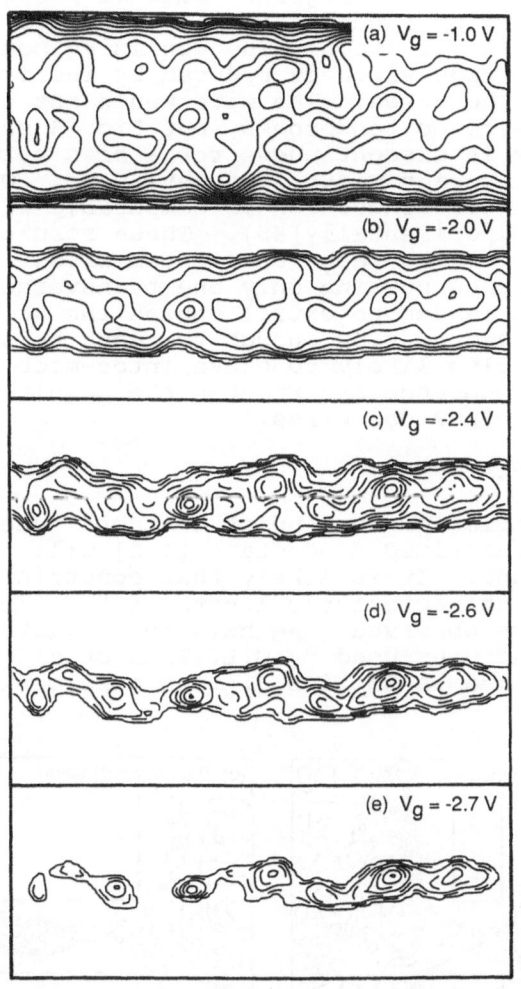

Fig. 3. Density of electrons in a 1 μm length of a wire confined by a split gate with a 0.4 μm gap, for several gate biases. Full contours start from zero and are 6×10^{14} m^{-2} apart, with dashed contours halfway between. The fluctuations become more severe as the average width decreases, and the wire becomes discontinuous for a gate bias between -2.6 V and -2.7 V; it would pinch off at -2.8 V if there were no fluctuations.

5.2   Quantum Wires

The results for quantum wires are more spectacular.   We considered a wire confined by a pair of negatively biased gates[23,24] separated by a gap of 0.4 μm[12].   The bias repels electrons from the regions under the gates, leaving a narrow channel under the gap.   The channel can be squeezed further by making the gates more negative, until all electrons are removed from the 2DEG.   The average density of electrons in these structures is very small, so screening is feeble.

The density of electrons is shown in Fig. 3.   The boundaries of the occupied regions would be parallel straight lines if there were no random potential.   Clearly the wire is far from this ideal situation.   The confining potential gets rougher as the average width is reduced by increasing the negative bias on the gates.   The wire would not pinch off until –2.8 V if there were no randomness, but Fig. 3 shows that the fluctuations make it discontinuous for a bias between –2.6 V and –2.7 V where its average width is about 80 nm.   Our estimate of the minimum width for conduction is comparable with the smallest widths observed experimentally[25].   These results have major implications for devices based on quantum wires.   The ideal wire would be monomode, supporting only one transverse state. Unfortunately the random potential causes the wire to break up before it can be made narrow enough to reach this limit.   The fluctuations are also likely to cause inter-mode scattering which restricts coherence and reduces the magnitude of inter-ference phenomena in wider wires.

5.3   Point Contacts

Narrow constrictions or 'point contacts' used to investigate the quantized conductance[1,2] will also be affected by the fluctuations.   It is likely that constrictions must be shorter than the characteristic length of the fluctuations for quantization to be observed.   We have investigated this by modeling the structures used by G.L. Timp et al.[14], based on the layers described in Section 3.   The gaps between the gates

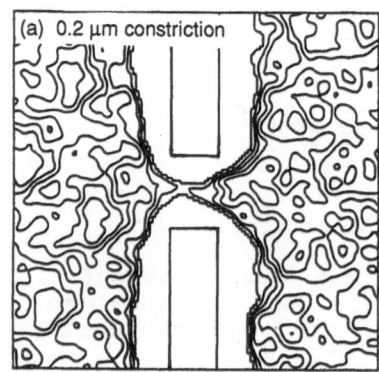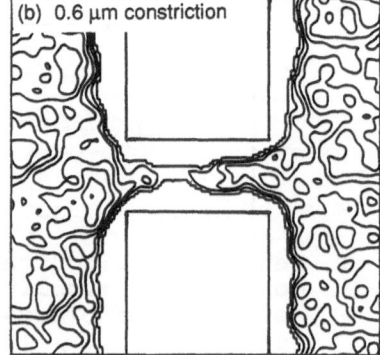

Fig. 4. Gate pattern on surface and density of electrons
in 2DEG for a 1.5 μm square around point
contacts: (a) length 0.2 μm at a bias of –1.8 V;
(b) length 0.6 μm at –0.8 V.   Contours start
from zero and are 6×10^{14} m^{-2} apart.

are 0.3 μm wide, and the constrictions are 0.2 μm and 0.6 μm long in the two devices studied. Typical electron densities are shown in Fig. 4, with conductance in Fig. 5.

Figure 4 shows that the saddle point of the potential in the constriction is shifted slightly in the 0.2 μm sample, but not greatly distorted. The 'throat' of the 0.6 μm constriction is much longer than the length scale of the fluctuations and is likely to induce scattering. The calculations of conductance shown in Fig. 5 confirm this: the short constriction shows at least six good steps in conductance, while quantization is clearly much poorer in the longer constriction which gives only one good plateau. This agrees well with the experiments (Figs. 2b and 4b in reference [14]). A surprising feature is that quantization in the 0.2 μm constriction is *better* in the experiment than the theory, while it is *worse* for the 0.6 μm constriction. Part of this may arise from the specific distribution of impurities: we have found large differences between individual samples with the same average parameters. It is also probable that our assumption that the impurities are fully ionized is incorrect, and this would lead to an over-estimate of the random potential.

Another important feature is the rounded shape of the guiding potential. This suppresses resonances within the constriction of the form that have been widely reported in the theoretical literature[26-29]. We have verified this by repeating our calculations, omitting the random potential but retaining the realistic gate potential. Unfortunately there are large finite-size effects in our calculations at present, which mean that the gate voltage cannot be directly compared with experiment, but these are being eliminated. We are also modeling devices used by the Dutch group[1], which have gates of different shape.

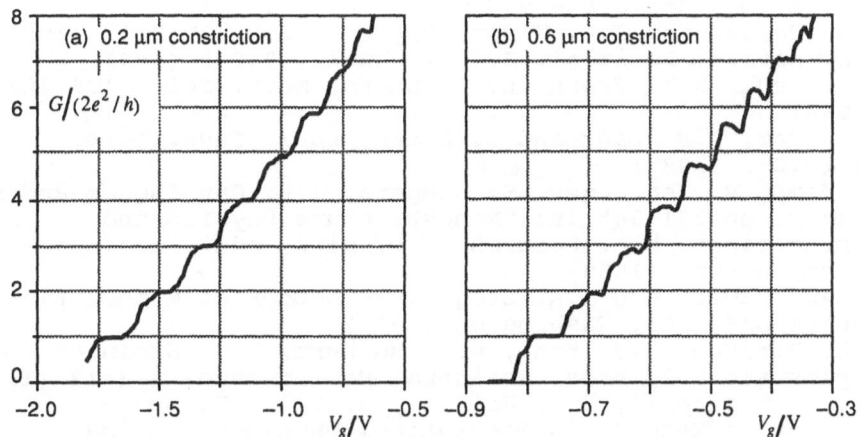

Fig. 5. Normalized conductance G/(2e²/h) as a function of gate voltage V_g for the two devices shown in Fig. 4.

CONCLUSIONS

Our calculations show that the traditional approximation of treating doped regions as uniform charge densities is inadequate in small low-dimensional devices. This randomness is unavoidable, and contrasts with the problem studied by Kumar et al.[13], who considered the effect of fabrication errors in the gate. Electrons in highly constrained 2DEGs - point contacts or narrow wires - are severely scattered by potential fluctuations on a length scale of less than 0.5 μm. It appears that structures like conventional modulation-doped layers, with a high density of unscreened ionized impurities, will have to be abandoned if devices based on monomode quantum wires are to be fabricated reliably.

Acknowledgment

J.H. Davies and J.A. Nixon were supported by the UK Science and Engineering Research Council through grants GR/E73192, GR/E18186 and GR/F80890

REFERENCES

1. B.J. van Wees, H. van Houten, C.W.J. Beenakker, J.G. Williamson, L.P. Kouwenhoven, D. van der Marel and C.T. Foxon, **Phys. Rev. Lett.**, 60:848 (1988)
2. D.A. Wharam, T.J. Thornton, R. Newbury, M. Pepper, H. Ahmed, J.E.F. Frost, D.G. Hasko, D.C. Peacock, D.A. Ritchie and G.A.C. Jones, **J. Phys. C**, 21:L209 (1988)
3. C.T. Foxon, J.J. Harris, D. Hilton, J. Hewitt and C. Roberts, **Semicond. Sci. Technol.**, 4:582 (1989)
4. J.A. Nixon and J.H. Davies, **Phys. Rev. B**, 41:15 April 1990
5. B.I. Shklovskii and A.L. Efros, "Electronic Properties of Doped Semiconductors", Springer-Verlag, Berlin (1984)
6. W.J. Boudville and T.C. McGill, **Appl. Phys. Lett.**, 48:791 (1986)
7. D. Arnold and K. Hess, **J. Appl. Phys.**, 61:5178 (1987)
8. T. Ando, A.B. Fowler and F. Stern, **Rev. Mod. Phys.**, 54:437 (1982)
9. A.L. Efros, **Solid State Commun.**, 67:1019 (1988)
10. J.H. Davies and J.A. Nixon, **Phys. Rev. B**, 39:3423 (1989)
11. J.H. Davies, **Semicond. Sci. Technol.**, 3:995 (1988)
12. S.E. Laux, D.J. Frank and F. Stern, **Surf. Sci.**, 196:101 (1988)
13. A. Kumar, S.E. Laux and F. Stern, **Appl. Phys. Lett.**, 54:1270-1 (1989)
14. G. Timp, R. Behringer, S. Sampere, J.E. Cunningham and R.E. Howard, pp.331-345 **in:** "Nanostructure Physics and Fabrication", M.A. Reed and W.P. Kirk, eds., Academic Press, Boston (1989)
15. M. Abramowitz and I.A. Stegun, "Handbook of Mathematical Functions", NBS, Washington (1970)
16. J.M. Rorison, M.J. Kane, D.C. Herbert, M.S. Skolnick, L.L. Taylor and S.J. Bass, **Semicond. Sci. Technol.**, 3:12 (1988)
17. See, for example, D.W. Heerman, p.17 **in:** "Computer Simulation Methods in Theoretical Physics", Springer-Verlag, Berlin (1986)
18. D.J. Thouless and S. Kirkpatrick, **J. Phys. C**, 14:235 (1981)
19. P.A. Lee and D.S. Fisher, **Phys. Rev. Lett.**, 47:882 (1981)
20. A.D. Stone, **Phys. Rev. Lett.**, 54:2692 (1985)

21. R. Zallen and H. Scher, **Phys. Rev. B**, 4:4471 (1971)
22. A.L. Efros, **Solid State Commun.**, 70:253 (1989)
23. H.Z. Zheng, H.P. Wei, D.C. Tsui and G. Weimann, **Phys. Rev. B**, 34:5635 (1986)
24. T.J. Thornton, M. Pepper, H. Ahmed, D. Andrews and G.J. Davies, **Phys. Rev. Lett.**, 56:1198 (1987)
25. D.A. Wharam, U. Ekenberg, M. Pepper, D.G. Hasko, H. Ahmed, J.E.F. Frost, D.A. Ritchie, D.C. Peacock and G.A.C. Jones, **Phys. Rev.B**, 39:6283 (1989)
26. L.I. Glazman, G.B. Lesovick, D.E. Kmelnitskii and R.I. Shakhter, **Pis'ma Zh. Teor. Fiz.**, 48:218 (1988) [**JETP Lett.**, 48:238 (19880]
27. G. Kirczenow, **Solid State Commun.**, 68:715 (1988)
28. A. Szafer and A.D. Stone, **Phys. Rev. Lett.**, 62:300 (1989)
29. E.G. Haanappel and D. van der Marel, **Phys. Rev. B**, 39:5484 (1989)

SEMICONDUCTOR STRAINED LAYERS

SYNTHESIS, PROPERTIES AND APPLICATIONS OF STRAINED LAYERS AND HETEROSTRUCTURES

Pallab Bhattacharya

Solid State Electronics Laboratory
Department of Electrical Engineering & Computer
Science, The University of Michigan, Ann Arbor
Michigan 48109-2122, USA

ABSTRACT

Strained layers and their heterostructures have emerged as important materials for application to high-speed electronic and optoelectronic devices. They provide an additional degree of freedom in band gap engineering. More recently, it has become clear that in the pseudomorphic regime, large changes can be made in the valence band structure of III-V compounds, resulting in dramatic changes in their electronic and optical properties. The growth of pseudomorphic materials by techniques such as MBE and the associated growth modes are also areas where a lot of understanding has developed in the recent past. It is clear, for example, that for large misfits ($\epsilon \geq 1.5\%$) pseudomorphic layers prefer to grow in well ordered three-dimensional islands instead of two-dimensional layers as in the lattice-matched case.   In this paper, a brief review is made of the growth of In-bearing III-V compounds, their electrical and optical properties and their application to the design of high performance MODFETs, lasers, detectors and electro-optic devices.

1.   INTRODUCTION

Strained semiconductors and their heterostructures have merged as important materials for application to high speed electronic and optoelectronic devices. They provide an additional degree of freedom in band gap engineering. More recently it has become clear that in the pseudomorphic regime, large changes can be made in the valence band structure of III-V compounds, resulting in dramatic changes in their electronic and optical properties[1,2]. Essentially, the coupling between the light and heavy hole subbands are altered, producing altered hole effective masses, transport properties and optical absorption. Currently pseudomorphic heterostructures and quantum well systems are being developed for device applications and the quality of the heterointerfaces are extremely important. The latter depends, to a great extent on the growth modes. In other words, it is important to establish whether strained layers grow in a layer-by-layer mode, as for lattice-matched systems, or in an island mode. Such a growth mode would produce a rough growth front and a rough interface in heterostructures. Very recently new understanding has been gained on these issues and it is apparent that techniques such as migration enhanced epitaxy (MEE)[3,4] may alleviate the problem. Increased understanding of the growth and properties of strained layers and heterostructures have led to electronic and optoelectronic devices with performance better than in their lattice-matched versions. In what follows, I will briefly review the growth, properties and applications of In-bearing III-V strained single and multi-layered structures.

2. GROWTH OF STRAINED LAYERS BY MOLECULAR BEAM EPITAXY

In lattice-matched systems, the general nature of growth in MBE is fairly well understood (although a number of details are not). The growth front during MBE can be described as being in between two extreme growth modes: (a) a layer-by-layer growth mode where the growth front essentially consists of at most two exposed (cation) monolayers and, even though atoms impinge randomly from the vapor phase, they move rapidly on the exposed surface and attach themselves (get incorporated) at step edges and (b) a three-dimensional growth mode where the impinging atoms have essentially no surface kinetics and the surface is rough due to the intrinsic statistical fluctuations. These growth modes can be readily observed by in situ reflection high energy electron diffraction (RHEED). The above picture is a good description of the lattice-matched systems if the free energy minimum for the growth surface favors an atomically smooth surface. For the (100) surface, one has cation or anion layers and energetically the difference between a perfectly smooth (Fig. 1a) and a rough (Fig. 1b) surface is essentially due to in-plane bonds which are second neighbor bonds. If the

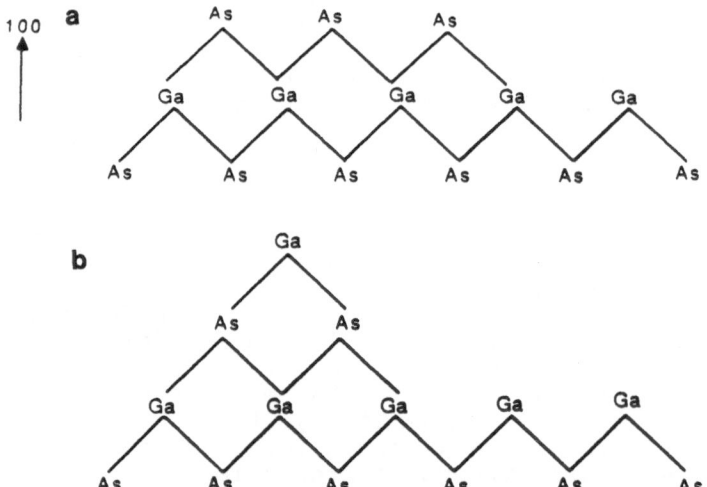

Fig. 1. (a) Atomically abrupt and (b) rough (100) surface. The atoms on the rough surface have the same number of nearest neighbor bonds but fewer second neighbor bonds.

second neighbor bond strength was zero, the rough surface would
be the lower free energy surface (due to entropy considerations). The second neighbor bonds are expected to be an order
of magnitude smaller than nearest neighbor bonds, but although
small, at usual MBE temperatures $|kT| < |W_2|$ (= second neighbor
bond energy), and one has the possibility of a two-dimensional
layer-by-layer growth mode.

In the case of strained epitaxy, in addition to
considerations arising from second neighbor bond strengths, one
has to contend also with the strain energy. The problem of
strained layers under equilibrium has been studied for several
decades[5,6], although only recently the predictions on critical
thicknesses in MBE grown III-V systems have been verified[7-9].
According to these theories (based on energy balance between
strain energy and chemical energy due to dislocation generation)
for any lattice mismatch ΔR, as long as the epilayer grown on a
thick substrate is below a certain thickness (the critical
thickness), the strain is absorbed coherently and no
dislocations are produced. However, above the critical
thickness, misfit dislocations are produced at the epilayer-
substrate interface. Such an assertion has been verified
experimentally in a number of systems[8,9]. It is, however,
important to know how strain affects the growth mode below
critical thickness.

Detailed RHEED oscillation measurements and measurement of
the temporal change of lattice spacing during the growth of
InGaAs on GaAs (compressively strained) clearly show that growth
changes from a layer-by-layer mode to an island mode as the
misfit exceeds 1.5%. The same is true for InGaAs on InP
substrates. From simple thermodynamic considerations we have
shown[10] that the minimum free energy for a strained system
favors a three-dimensional (3D) surface.

It is clear that the growth of strained systems cannot be
improved simply by enhancing the surface migration rate of
atoms. One expects the free energy to depend upon the surface
reconstruction during growth. In MBE the surface is anion
stabilized with a c(2 x 4) reconstruction. It is therefore
important to examine other possible surface reconstructions
which might change the surface chemical energy. Since MBE of
III-V semiconductors cannot be carried out under cation-rich
conditions because the excess cation causes nonstoichiometric
growth we have examined the MEE approach[6-8] where a few
monolayers (up to 4) of cations and a few monolayers of the
anion are alternately deposited by shutter control. The surface

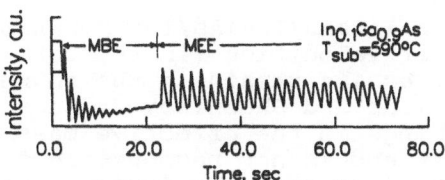

Fig. 2. RHEED oscillation data showing changes in
growth modes during pseudomorphic epitaxy.
The oscillations die during normal MBE and
are restored during subsequent MEE.

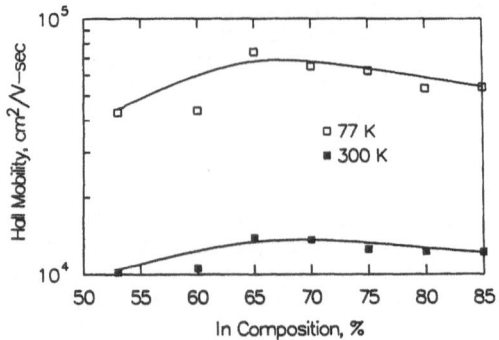

Fig. 3. Measured Hall mobilities of pseudomorphic
$In_xGa_{1-x}As/In_{0.52}Al_{0.48}As$ MODFETs with increasing
In content in the channel at 77 and 300 K. The
lines are guides to the eye.

recombination thus alternates between cation and anion
stabilized. Figure 2 shows the growth of InGaAs on GaAs. After
the oscillations die during normal MBE, growth is continued by
MEE, and it is seen that oscillations are restored. The latter
indicates the restoration of layer-by-layer growth.

As stated earlier, the growth modes and interface quality
directly affect the properties of heterostructure devices. As
an example, Fig. 3 shows the measured mobilities in pseudomorphic $In_{0.53+x}Ga_{0.47-x}As/In_{0.52}Al_{0.48}As$ heterostructures for
increasing x. The decrease in mobility at large values of x is
attributed to island growth in the channel region and have been
analyzed taking into account interface roughness scattering and
other relevant scattering mechanisms. A rough interface would
affect the performance of almost every active device made with
strained heterostructures.

3. ELECTRONIC AND OPTICAL PROPERTIES OF STRAINED
HETEROSTRUCTURES AND DEVICE APPLICATIONS

Singh, Bhattacharya and co-workers[1,2] and other authors
[11] have made detailed studies of the electronic and optical
properties of strained heterostructures and quantum wells.
Biaxial strain alters the band structure and the electronic and
optical properties in these materials. Some specific examples
are briefly mentioned.

a) n- and p-type pseudomorphic MODFETs

Experiments with InGaAs/InAlAs/InP MODFETs show that
mobilities and carrier velocities (in the 2DEG) steadily improve
with increasing compressive strain in the channel region.
Theoretical calculations and Shubnikov-de Haas measurements
indicate that the change in the effective mass with increasing
In is not significant and is not responsible for the enhancement
in mobilities. We believe that the improvement results from
reduced alloy scattering, reduced intersubband scattering, and
reduced impurity scattering, all of which result from a higher
conduction band offset and increased carrier confinement in the
two-dimensional electron gas. Highest frequency and lowest

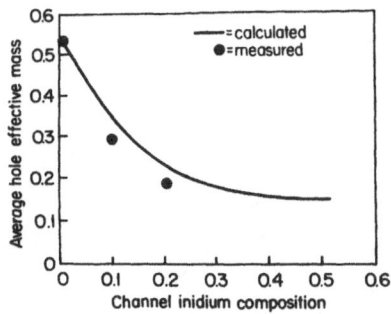

Fig. 4. Measured hole effective mass in
In$_x$Ga$_{1-x}$As/AlGaAs modulation-doped
heterostructures.

noise performance to date have been demonstrated by InP-based
pseudomorphic MODFETs[12-14].

In p-MODFETs, however, due to a lifting of the light hole
(LH) - heavy hole (HH) degeneracy in the valence bands, the hole
effective masses can be made very light[15]. Measured hole
masses in InGaAs/AlGaAs/GaAs p-MODFET structures, with
increasing In content in the p-type channel, are shown in
Fig. 4. Such results have prompted the development of p-MODFETs
for complimentary logic and other circuits applications[16].

b)   Strained Quantum Well Lasers

Strained layer InGaAs/GaAs quantum well lasers are
attractive light sources for GaAs and InP-based optoelectronic
circuits. The transparency of the substrates implies an
increased flexibility in circuit design. To minimize heat
dissipation and to facilitate a high packing density these
semiconductor lasers have to be highly efficient. A high
electrical to optical power conversion efficiency requires a low
threshold current density and a high external differential
quantum efficiency. Quantum well lasers have demonstrated lower
threshold currents mainly because of the altered density of
states function in the quantum well. However, further
improvements are desirable, particularly for integrated lasers,
where the standard schemes for heat sinking are not applicable.

We have recently calculated[17] the modification of the
hole dispersion relations in biaxial compressive pseudomorphic
quantum well laser structures. Several important conclusions
emerge from these studies: i) the modal purity of the laser
output increases dramatically as the strain increases. This
results from the strain-induced splitting between the $|\frac{3}{2}$, $\pm\frac{3}{2}>$
(heavy hole) states and the $|\frac{3}{2}$, $\pm\frac{1}{2}>$ (light hole) states which
greatly reduces the occupation of the light hole states. For
example, it is seen that in the lattice matched MQW laser there
is a significant fraction of TM mode material gain but in the
strained system the TM mode gain remains essentially zero even
at very high injection values; ii) the carrier injection
required to obtain a particular TE mode peak material gain (or
threshold current) drops rapidly as the strain increases as
shown in Fig. 5; iii) as a consequence of the lower threshold,
the high temperature behavior is also improved.

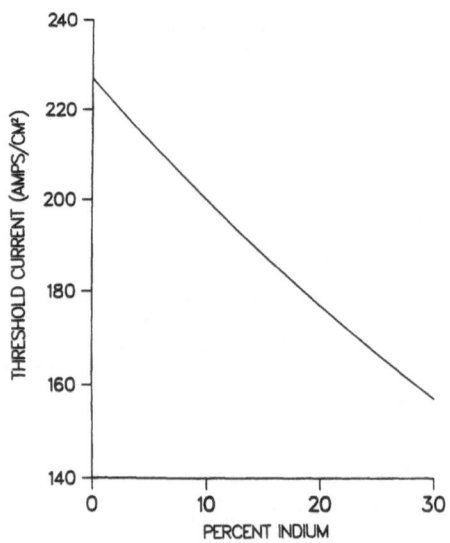

Fig. 5. Variation of threshold current density
with excess in In, x, in
$In_{0.53+x}Ga_{0.47-x}As/In_{0.52}Al_{0.48}As$.

Preliminary experiments have been made by us on 60 μm
striped coupled quantum well $In_xGa_{1-x}As/AlGaAs$ lasers, with
x = 0.1, 0.2 and 0.3. It is clear that the threshold currents
are lower for increasing x. Lower thresholds in strained QW
lasers are recently being reported by other workers[18-20]. It
was also confirmed experimentally that the emission from the
laser with x = 0.3 consists purely of TE modes, while for the
lattice-matched GaAs/AlGaAs QW lasers, the output consists of a
mixture of TE and TM modes.

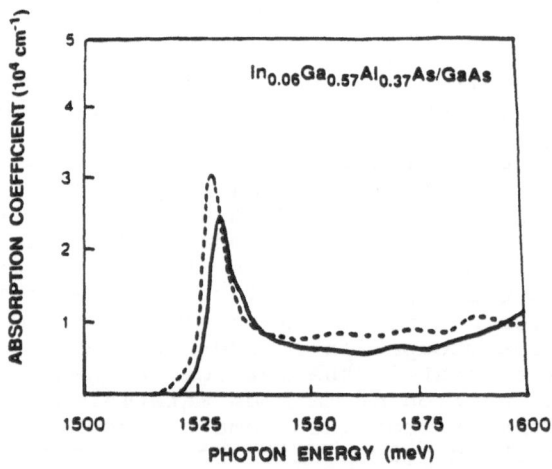

Fig. 6 Experimental (solid lines) and calculated
(dashed line) absorption spectra in a biaxially
tensile strained QW at 11 K showing a merger of
the light and heavy hole states.

c)   Detectors and Modulators

In the design of high-speed photodiodes, it would be helpful if the absorption coefficient of the absorbing region could be enhanced. It has been shown that it is possible to merge the heavy and light hole excitonic resonances through use of biaxial tensile strain in the well regions[2]. Calculations and experiments show a strong increase in optical absorption, by almost a factor of 2, due to this merger. The calculations include the heavy hole - light hole coupling which is essential to understand this phenomenon. Calculated and experimental data are shown in Fig. 6. It may be remembered that in the case of biaxial compressive strain in the well region, the separation of the HH and LH resonances increase with strain.

The merger of the excitonic resonances and increase in absorption is also useful for quantum well Stark-effect modulators, where the modulation ratio can be enhanced, or lower voltage operation can be obtained. The increase in absorption may also be important for nonlinear devices or quantum well electro-optic devices[21] operating at optical wavelengths close to the excitonic absorption edge.

CONCLUSION

In this talk and paper I have reviewed briefly the current status on the understanding of strained layer epitaxy by MBE and the electrical and optical properties in strained hetero-structures and quantum wells. It is clear that some of the unique properties are leading to a large enhancement in the performance characteristics of pseudomorphic electronic and optoelectronic devices. An area of concern, where some work needs to be done is the stability of such structures under long-term operation and harsh environments.

Acknowledgment

The author would like to thank his co-workers J. Singh, D. Pavlidis, P.R. Berger, J. Pamulapati, D. Biswas, N. Debbar, J. Loehr, J.E. Oh, M. Jaffe and S.C. Hong. The work was partly supported by the Army Research Office (URI Program) under contract DAAL03-87-K0007.

REFERENCES

1.   M. Jaffe and J. Singh, **J. Appl. Phys.**, 65:329 (1989)
2.   S-C Hong, G.P. Kothiyal, N. Debbar, P. Bhattacharya and J. Singh, **Phys. Rev. B**,37:878 (1988)
3.   J.M. Gerard and J.Y. Marzin, **Appl. Phys. Lett.**, 53:568 (1988)
4.   R. Katsumi, H. Ohno, H. Ishii, K. Matsuzaki, Y. Akatsu and H. Hasegawa, **J. Vac. Sci. Technol.**, B6:593 (1988)
5.   F.C. Frank and J.H. van der Merwe, **Proc. R. Soc. London**, A198:205 (1949); 198:216 (1949); C.A.B. Ball and J.H. van der Merwe, **Dislocations in Solids**, 6:121 (1983)
6.   J.W. Matthews and A.E. Blakeslee, **J. Cryst. Growth**, 27:118 (1974)
7.   T.G. Anderson, Z.G. Chen, V.D. Kulakovski, A. Uddin and J.T. Vallin, **Appl. Phys. Lett.**, 51:752 (1987)

8. P.L. Gourley, I.J. Fritz and L.R. Dawson, **Appl. Phys. Lett.**, 52:377 (1988)

9. Y. Yokomo, Y. Fukuda and M. Seki, **Appl. Phys. Lett.**, 52:380 (1988)

10. P.R. Berger, K. Chang, P. Bhattacharya, J. Singh and K.K. Bajaj, **Appl. Phys. Lett.**

11. E.P. O'Reilly, **Semic. Sci. Technol.**, 4:121 (1989)

12. G.I. Ng, W-P. Hong, D. Pavlidis, M. Tutt and P.K. Bhattacharya, **Electron. Dev. Lett.**, 9:439 (1988)

13. U.K. Mishra, A.S. Brown, S.E. Rosenbaum, C.E. Hooper, M.W. Pierce, M.J. Delaney, S. Vaughn and K. White, **Electron Dev. Lett.**, 9:647 (1988)

14. P.C. Chao, A.J. Tessmer, K-H.G. Duh, P. Ho, M-Y. Kao, P.M. Smith, J.M. Ballingall, S-M.J. Liu and A.A. Jabra, **Electron. Dev. Lett.**, 11:59 (1990)

15. M. Jaffe, J.E. Oh, J. Pamulapati, J. Singh and P. Bhattacharya, **Appl. Phys. Lett.**, 54:2345 (1989)

16. T.J. Drummond, T.E. Zipperian, I.J.H. Fritz, J.E. Schirber and T.A. Plut, **Appl. Phys. Lett.**, 49:461 (1986)

17. G. Feak, D. Nichols, D. Biswas, J.P. Loehr, P.K. Bhattacharya and J. Singh, presented at the Biennial IEEE/Cornell Conference on Active Semiconductor Devices, Ithaca, August 1989.

18. K.J. Beermink, P.K. York and J.J. Coleman, **Appl. Phys. Lett.**, 55:2585 (1989)

19. D.F. Welch, W. Streifer, C.F. Schaus, S.Sun and P.L. Gourley, **Appl. Phys. Lett.**, 56:11 (1990)

20. D.P. Bour, R. Martenelli, F.Z. Hawrylo, G.A. Evans, N.W. Carlson and D.B. Gilbert, **Appl. Phys. Lett.**, 56:318 (1990)

21. U. Das, Y. Chen, P. Bhattacharya and P.R. Berger, **Appl. Phys. Lett.**, 53:2129 (1988)

IN SITU INVESTIGATION OF THE LOW PRESSURE MOCVD GROWTH OF
LATTICE-MISMATCHED SEMICONDUCTORS USING REFLECTANCE ANISOTROPY
MEASUREMENTS

O. Acher, S.M. Koch, F. Omnes,
M. Defour, B. Drévillon* and M. Razeghi

Thomson-CSF LCR
Domain de Corbeville, F 91404, Orsay, France
*LPICM, Ecole Polytechnique
F 91128 Palaiseau, France

ABSTRACT

    The growth of InAs on InP and InP on GaAs is investigated
using Reflectance Anisotropy (RA) measurements. Very large
optical anisotropies are observed, related to the three-
dimensional growth mode of these materials. A model is proposed
to account for the optical properties of the samples, using
effective medium theories to describe the roughness. Good
quantitative agreement is obtained for small roughness
thickness, and a qualitative description is found for larger
roughness features. The RA technique is found to be very useful
to monitor the growth of lattice-mismatched materials,
particularly at the nucleation stage.

1.   INTRODUCTION

    There has been recently a great deal of interest in the
growth of III-V compounds on alternative substrates. GaAs on
silicon and InP on GaAs on silicon structures are very
attractive for device applications, and much work has been done
using these materials systems[1,2]. Our group recently reported
the first operation of a GaInAsP-InP laser emitting at 1.3 μm on
a silicon substrate[3], and the operation of a GaInAs-InP on
GaAs on silicon photodiode[4]. The growth of lattice-mismatched
III-V layers on other III-V substrates also has important
applications. For example, since semi-insulating InAs
substrates are not commercially available, GaAs and InP semi-
insulating substrates can be used for InAs growth.

    However, lattice mismatched growth is difficult and depends
critically on the growth conditions, particularly during the
nucleation stage. For this reason, it would be extremely
important to have a convenient in situ characterization tool to
monitor such growths.

    Until recently, very few techniques were available for in
situ characterization of MOCVD growth. A technique recently

developed[5,6], called Reflectance Difference Spectroscopy
(RDS)[7,8] or Reflectance Anisotropy (RA)[9,10], is a means of
obtaining chemical and structural information about the layer
during the growth. It can be used on MOCVD reactors, despite
the deposition of reaction products on the reactor walls during
the growth[11,12]. This technique can be used to perform
spectroscopic or kinetic studies. Since the present study
focuses essentially on the kinetic use of this technique, and
also to mark the difference with another technique called
Differential Reflectivity[13], the term RA will be used in
preference to RDS. RA involves illuminating the sample with
polarization-modulated light under normal incidence and
detecting the polarization of the reflected light. This is also
done in ellipsometry; RA differs in that perpendicular incidence
is used so that the optical anisotropy of the sample is
measured. Our group has shown that the RA signature of heterojunctions is very sensitive to the growth conditions and could
be related to heterojunction quality[12]. The growth of InAs on
InAs was also investigated, and the RA signal was found to be
very sensitive to growth conditions and also to flow or pressure
transients[14]. This last point is very attractive for gas flow
control and reactor geometry optimization.

Our group reported the observation of large RA signals
recorded during the first stages of the growth of lattice-
mismatched layers[12]. This was observed both for InP/GaAs/
silicon and for InAs/InP, using different wavelengths. In this
paper, we describe in more detail the occurance of these large
anisotropies during the growth of the first 4000 Å. It is shown
that anisotropic three-dimensional growth is responsible for
these observations. A model is proposed and satisfactory
agreement is obtained with experiment for the first stages of
the growth. Quantitative and qualitative information can be
extracted from RA measurements. As the roughness dimensions
increase, precise modeling is not possible, but the broad
features of RA behavior are well understood. The use of RA for
in situ monitoring of the growth of lattice-mismatched layers is
illustrated.

2.   EXPERIMENTAL DETAILS

2.1 Growth Procedure

The present study was performed using a Low-Pressure
Metalorganic Chemical Vapor Deposition (LP-MOCVD) reactor
described elsewhere[15]. Two types of structures are
considered: InAs/InP, and InP/GaAs on Si. In both cases the
lattice parameter of the epilayer is different from that of the
substrate. The lattice mismatch is 3.2% in the case of the
InAs-InP system and 3.8% for the InP-GaAs pair.

The growth of InAs was done directly on InP substrates.
The substrate orientation was (100) with a 2° tilt toward [01$\bar{1}$],
unless otherwise stated. After introduction into the reactor,
the substrates were heated under AsH_3 until they reached the
growth temperature of 480°C. The H_2 flow through TMI bubbler
was typically 100 cm^3/min, and AsH_3 flow was between 5 and 20
cm^3/min. Under these conditions, the growth rate of InAs,
deduced from thickness measurement performed on thick layers,
was about 270 Å/min.

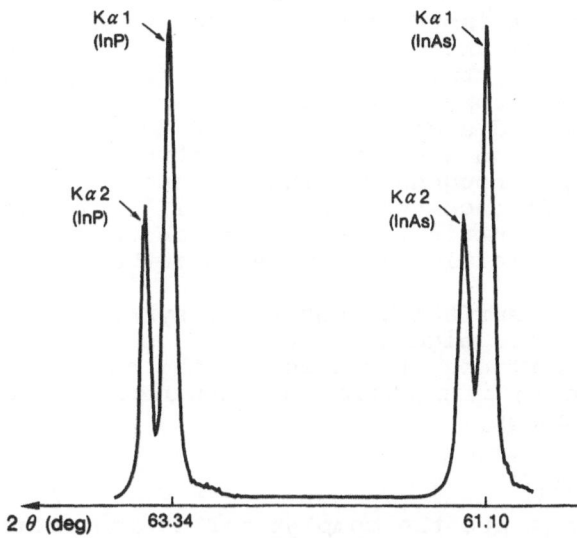

Fig. 1. X-ray simple diffraction pattern of 1 μm thick
InAs layer grown on an InP substrate.

Layer quality was assessed by X-ray simple diffraction and
Hall electron mobility on 1 μm thick InAs/InP layers. The X-ray
diffraction pattern has well-separated $K_{\alpha 1}$ and $K_{\alpha 2}$ diffraction
peaks (Fig. 1), indicative of good structural quality. The
typical Hall electron mobility at room temperature was about
11,500 $cm^2/V/s$.

The growth of InP/GaAs/silicon structures was detailed
elsewhere[16]. Some InP-GaInAs superlattices can be included in
the InP to reduce threading dislocation propagation. Lasers
emitting at 1.3 μm in cw mode were fabricated using these
structures, indicative of excellent material quality[3].

RA records of InP-GaInAs superlattices grown on GaAs/Si are
reported elsewhere[12,17]. The present study only deals with
the simpler case of InP grown directly on a smooth thick GaAs
top layer, without any superlattice. The growth rate of InP is
180 Å/min.

Both in the case of InAs/InP and InP/GaAs/Si, one can
observe that the sample can become hazy during the first minutes
of the growth. This is indicative of three-dimensional (3D)
growth. If the growth is conducted properly, this haze
disappears after 10 or 20 minutes and after one hour of growth
the sample surface is mirror-like.

2.2 The RA Technique

RA measures the reflectance anisotropy of a sample under
normal incidence. It can be described either as a normal
incidence ellipsometer, or as a dichroism setup working in
reflection instead of transmission. Since bulk III-V semiconductors are nearly optically isotropic, because of their
cubic symmetry, a major contribution to reflectance anisotropy
is expected to arise from lower symmetry regions: the surface
and the interface. Various structural and chemical contri-

411

butions to optical anisotropies measured by RA have been studied by Aspnes[5,6,11,18] and others[10,19,20]. For (100)-oriented III-V semiconductors, bulk contributions are expected to be null, and the anisotropy is expected to arise from the surface. Chemi- or physisorbed species contribute to the RA signal. Optical anisotropy can also arise from structural contributions, such as anisotropic roughness. The presence of a buried interface also influences the RA signal[17]. In the present study, structural contributions will be shown to play a prominent role for lattice-mismatched growths.

For the (100) surface orientation, symmetry considerations show that the optical eigenaxes are [001] and [01$\bar{1}$]. This holds for exact (100) surfaces, but also for the substrates tilted toward [01$\bar{1}$] used in this study. RA measures the relative reflectance anisotropy:

$$(r_{01\bar{1}} - r_{001})/r$$

where r_{011} and $r_{01\bar{1}}$ are the complex reflection coefficients for light polarized along [011] and [01$\bar{1}$] respectively, and r is the average value. For a small anisotropy, the real part of the relative reflectance anisotropy corresponds to the relative anisotropy of the amplitude of the reflection coefficient and the imaginary part is the phase anisotropy. With $r_{011} = |r_{011}| \exp(i\Delta_{011})$ and similar notation for $r_{01\bar{1}}$:

$$\text{Re}\{(r_{01\bar{1}} - r_{011})/r\} = (|r_{01\bar{1}}| - |r_{011}|)/|r| \tag{1}$$

$$\text{Im}\{(r_{01\bar{1}} - r_{011})/r\} = \Delta_{01\bar{1}} - \Delta_{011}. \tag{2}$$

For larger anisotropies, this correspondance is no longer valid. We chose to work with $(r_{01\bar{1}} - r_{011})/r_{011}$ when the anisotropy exceeds 20%.

The RA setup used for the present study has been described elsewhere[12,14]. It was derived from a phase-modulated ellipsometer[21,22] modified to work at near normal incidence. The light sources were either a 75 W xenon lamp, or a low power (<0.5 mW) He-Ne laser emitting at $\lambda = 5435$ Å or $\lambda = 6328$ Å. Briefly, the polarization of the incoming light is modulated at $\omega = 50$ kHz using a photoelastic modulator, and RA detects the intensity modulation of the reflected light. The optical anisotropy of the sample can be determined from measurement of the reflected light using equations established for phase-modulated ellipsometers[22]. The reflection coefficients r_p and r_s corresponding to eigenpolarizations in ellipsometry are replaced by r_{011} and $r_{01\bar{1}}$ associated with eigenpolarizations in RA. Using a convenient orientation of the different optical elements[7], and for small anisotropies, the first harmonic is proportional to $\text{Im}\{(r_{01\bar{1}} - r_{011})/r\}$, and the second harmonic to $\text{Re}\{(r_{01\bar{1}} - r_{011})/r\}$. The proportionality coefficients are determined using a calibration procedure similar to that described in Ref. 22. For large anisotropies (in the case of InP/GaAs/Si for example), the signal analysis is performed using equations established for phase-modulated ellipsometers. It is convenient to introduce the angles ψ and Δ, defined by $(r_{01\bar{1}}/r_{011}) = \tan\psi \exp(i\Delta)$. ψ and Δ are related to the signal by the same relationships as the ellipsometric angles[22].

The birefringence of the reactor tube modifies the polarization of the beam. This affects mainly the measurements

of Im$\{(r_{01\bar{1}} - r_{011})/r\}$[18], the coupling to the real part being
only a second order-effect[7]. For that reason, the noise level
for Re$\{(r_{01\bar{1}} - r_{011})/r\}$ is generally lower than 10^{-4}, while it is
one order of magnitude higher for Im$\{(r_{01\bar{1}} - r_{011})/r\}$.
Therefore, the variations of the imaginary part associated with
surface chemistry modifications are concealed by noise. In
contrast, it will be shown that for lattice-mismatched growth
the signals measured are much larger than those due to surface
chemistry changes: it is possible to record both real and
imaginary part of the reflectance anisotropy with a small level
of noise. The perturbation due to the reactor tube also adds an
offset to the RA signal, so it is possible to measure only
variations of optical anisotropy. The offset can be evaluated
with an uncertainty of about $4 \cdot 10^{-3}$, which is large compared to
surface chemistry changes, but negligible compared to typical
variations during the growth of lattice-mismatched materials.

There is a simple way to determine that the measured RA
signal does not arise from parasitic effects. Depending on the
substrate orientation, the measured signal is either
$(r_{011} - r_{01\bar{1}})/r$ or its opposite, $(r_{01\bar{1}} - r_{011})/r$. Two identical
experiments with the substrate orientation differing by 90°
should change the sign of the signal. Conversely, when a
material system is well known, the measured sign of the RA
signal indicates the orientation[12].

3. RESULTS

3.1 Overview

Figure 2 shows the RA record of the growth of InAs on an
InP substrate for 15 minutes, using 6328 Å light wavelength.
One striking feature is the order of magnitude of the measured
anisotropies, on the order of 5%. In contrast, the RA signals
measured for lattice-mismatched growth are generally about
10^{-3}[7,12,14]. The occurrence of these large RA signals is
correlated with the occurrence of 3D growth, to a hazy surface

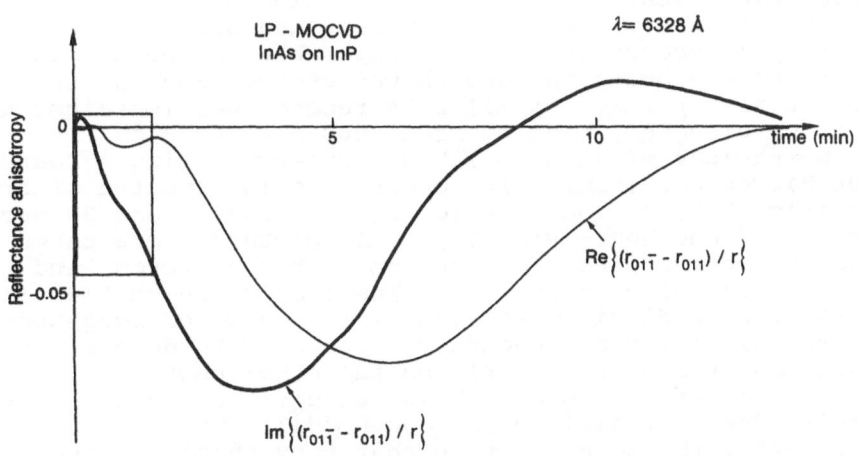

Fig. 2. RA record of the growth of an InAs layer on an
InP substrate, using λ = 6328 Å light
wavelength.

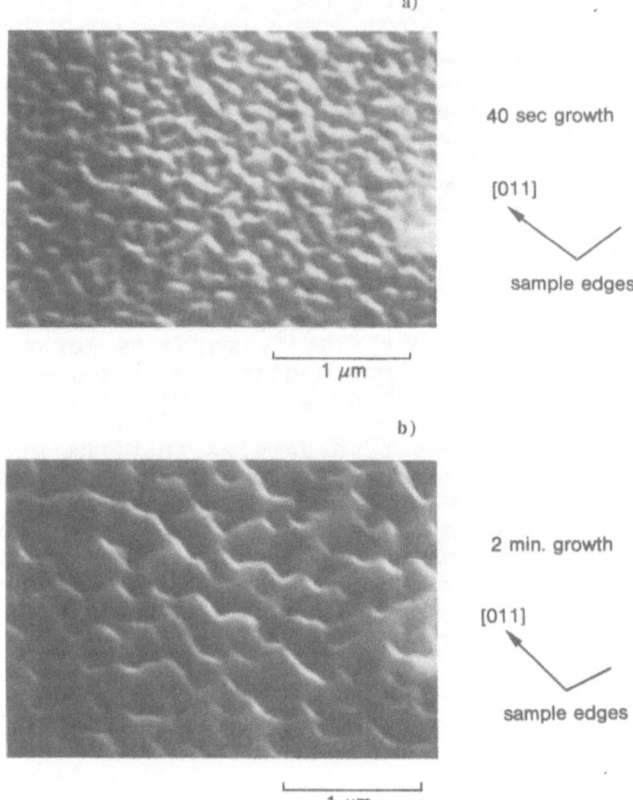

a)

40 sec growth

[011]

sample edges

1 μm

b)

2 min. growth

[011]

sample edges

1 μm

Fig. 3. Scanning electron microscope pictures of InAs on
InP after: a) 40 sec growth; b) 2 minutes
growth.

appearance, and to a decrease in the measured reflectivity. The
timescale of these anisotropy changes is also suggestive of
structural evolution. By the end of 15 minutes growth, the RA
signal had returned to zero, the surface had smoothed, and the
reflectivity increased to a level comparable to the initial
value. After one hour, the growth was stopped and the sample
surface was mirror-like. Similar RA records were obtained using
different wavelengths, ranging from 3800 Å to 6328 Å. The
surface morphology of InAs on InP was observed using a Scanning
Electron Microscope (SEM). The growth was interrupted after 40
seconds (Fig. 3a), 2 minutes (Fig. 3b), 3 minutes and 30 seconds
(Fig. 3c), and one hour (Fig. 3d). All pictures were taken with
the electron beam coming from the top of the pictures, and the
samples were tilted by about 45°. The resolution on Fig. 3a is
not sufficient to distinguish clearly the shape of roughness.
The lateral dimension of roughness increase with deposition time
and its geometry appears clearly on the other photos. It
consists mainly of rectangular holes of different sizes, but all
with their edges parallel to [011] and [01Ī]. This observation
is consistent with the prediction that they should be the
privileged symmetry directions. The dispersion in size and
shape is quite important on Figs. 3a to 3c. These pictures do
not reveal clear anisotropic patterns. The anisotropy may arise

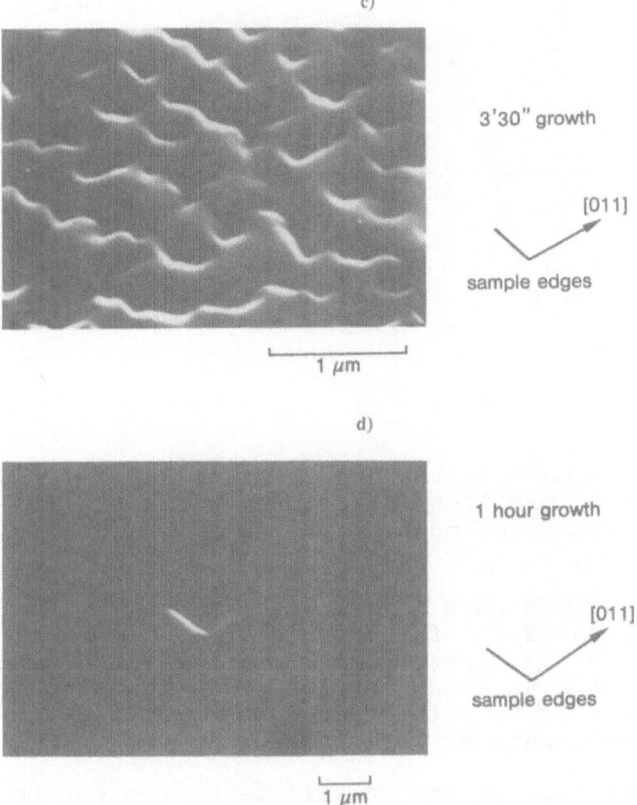

c)

3'30" growth

[011]

sample edges

⊢——— 1 μm ———⊣

d)

1 hour growth

[011]

sample edges

⊢— 1 μm —⊣

Fig. 3. Scanning electron microscope pictures of InAs on
InP after: c) 3 minutes 30 seconds' growth; d)
one hour's growth.

from the lateral dimensions of the holes, but also from their
slope, or their spacing. After one hour's growth, very few
roughness patterns can be seen on the surface; one of these
features is shown in Fig. 3d. This is consistent with the fact
that the sample is mirror-like after one hour's growth. The
dimensions of roughness features increase with time (note the
factor 3 difference between the magnifications of Figs. 3c and
3d). One should point out that when the growth of the samples
represented on Figs. 3a to 3c was stopped, the RA signal
continued to have an evolution as long as the sample was
maintained at growth temperature. This suggests that some mass
transport occurs on the surface at that temperature. As a
consequence, Figs. 3a to 3c may not reflect exactly the shape of
roughness under growth conditions. For the same reason, it is
difficult to conduct spectroscopic studies on such samples.
There is an evolution of RA properties during the stopping of
the growth and it makes it difficult to compare kinetic and
spectroscopic measurements. A fast cooling down may be required
to freeze the optical anisotropy.

Figure 4 shows the RA record, using 5435 Å, light of the
growth of InP on a GaAs on silicon substrate. In this case the
optical anisotropy is extremely large, with at some times more

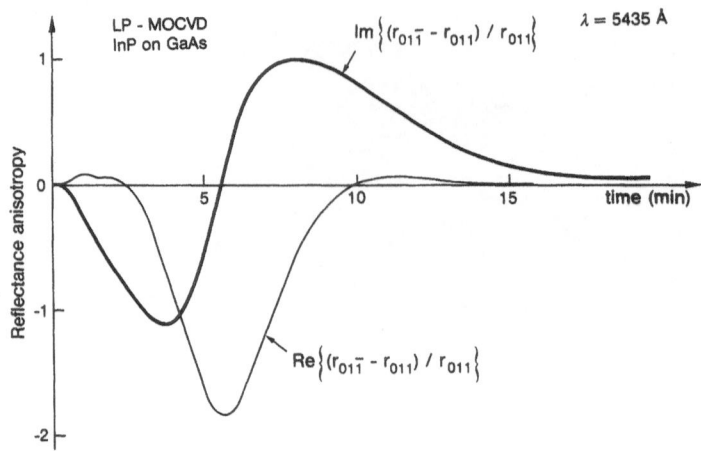

Fig. 4. RA record of the growth of an InP layer on a
         GaAs on Silicon structure, using λ = 5435 Å
         light wavelength.

than a factor 2 difference between r_{011} and $r_{01\bar{1}}$. However, the
broad features, the sign of the variations, are similar in Fig.
2 and Fig. 4. This suggests that the underlying mechanism
responsible for the RA signal evolution is the same, and related
to 3D growth, rather than to effects associated with the
specific material, such as chemical bonding.

For further discussion, it will be convenient to
distinguish various stages in the RA records. The initial
transient corresponds to the first 5 to 10 seconds of the
growth. The experimental curve shown on Fig. 5 corresponds to
the inset of Fig. 2, and it appears clearly that the initial
behavior of RA signal differs from the behavior after 10
seconds' growth. During this initial transient, the dominant
contribution to RA does not arise from surface roughness, as
will be discussed further. In the first stage, corresponding to
the first minute, the RA record exhibits a rapid, nearly linear
variation of $Im\{(r_{01\bar{1}} - r_{011})/r\}$. In contrast,
$Re\{(r_{01\bar{1}} - r_{011})/r\}$ remains quite small. In Fig. 2, it shows a
kind of oscillation with a time constant clearly different from
the rest of the variations. The second stage corresponds to the
rest of the growth, where the reflectance anisotropy shows a
kind of damped oscillating behavior with large period and
amplitude.

3.2 Optical Models

3.2.1 Presentation. It is clear from direct observation
and SEM pictures (Fig. 3) that the surface of the growing
lattice-mismatched layer is rough during the first stages of the
growth. It is suspected that this roughness accounts for the
large reflectance anisotropies observed. In this section, we
present models of the effect of surface roughness on the optical
properties of the sample.

Optical properties of rough surfaces are commonly treated
using Effective Medium Theories (EMTs). Briefly, a rough
surface can be approximated as an homogeneous film, with a

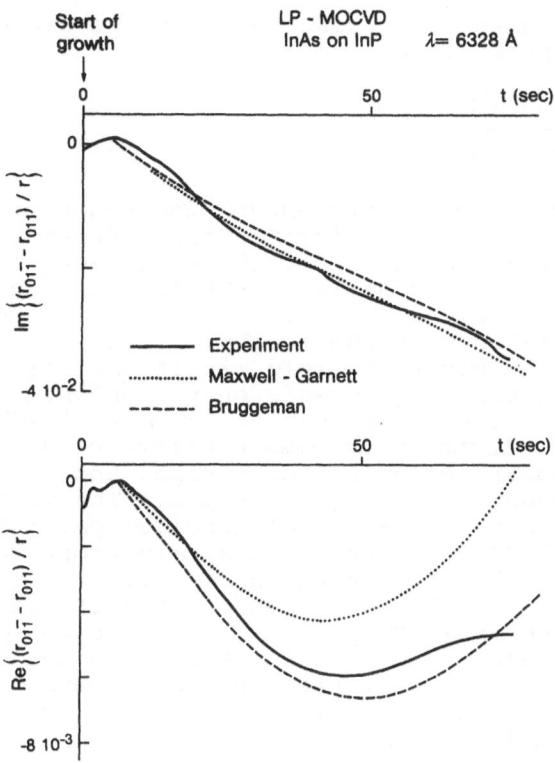

Fig. 5. Comparison between experiment (same as Fig. 1)
and theory for the RA record at $\lambda = 6328$ Å of
the first minute of growth of InAs on InP. a)
$\text{Im}\{(r_{01\bar{1}} - r_{011})/r\}$; b) $\text{Re}\{(r_{01\bar{1}} - r_{011})/r\}$. Note
the change in scale between the two figures.
Maxwell-Garnett fit was obtained with $F\delta qde/dt =$
17 Å/min. Bruggeman fit corresponds to $f = 0.7$,
$q = 0.23$ and $de/dt = 100$ Å/min.

thickness corresponding to the height of roughness and with a
dielectric constant beween that of the dense medium and that of
the ambient. Accordingly, a layer with a rough surface can be
modeled as two layers: a dense layer covered by a rough layer.
A rough InAs/InP sample can be described as a stack of three
homogeneous layers: InP substrate, InAs dense layer and rough
layer. The reflection coefficients of such a multilayer
structure and its associated optical anisotropy can easily be
calculated using the Abeles formalism[23], provided that the
dielectric constant and thickness of all layers are known.
Dielectric constants of GaAs, InP and InAs are tabulated[24],
and that of rough layer evaluated using EMTs.

However, no reflectance anisotropy will arise from that
model, unless the shape of the roughness is anisotropic. It is
well known that oblong particles may affect light polarization;
this effect is known as form birefringence[25]. It is possible
to get from EMTs a quantitative evaluation of the form
birefringence associated with the anisotropic roughness, as
shown below.

417

All EMTs yield the following form for the dielectric function of the effective medium[26]:

$$\epsilon_{eff} = \frac{q\epsilon_0\epsilon_1 + (1-q)\epsilon_h[f\epsilon_1 + (1-f)\epsilon_0]}{(1-q)\epsilon_h + q[f\epsilon_0 + (1-f)\epsilon_1]} \qquad (3)$$

where ϵ_1 is the dielectric function of the dense medium and ϵ_0 that of vacuum. f is the filling factor, corresponding to the proportion of medium 1 making up this roughness. If f<0.5, the roughness can be viewed as hills. If f>0.5, the roughness can be viewed as a layer with depressions in it. The depolarization factor q is associated with the roughness shape and accounts for the screening of the electric field within the effective medium. For anisotropic patterns, q depends on the orientation of the polarization, and one expects that $q_{011} \neq q_{01\bar{1}}$, leading to $\epsilon_{eff,011} \neq \epsilon_{eff,01\bar{1}}$. If the roughness consists of long stripes, q = 0 for light polarized along the stripes and q = 1 for light polarization perpendicular to the stripes. The value of q associated with the main axes of ellipsoidal particles is reported by Kittel[27]. There is no straightforward derivation of q for patterns like that shown in Fig. 3. ϵ_h is the dielectric function of the background or "host" medium. In the Maxwell-Garnett theory (M-G), $\epsilon_h = \epsilon_1$ or $\epsilon_h = \epsilon_0$, depending on the choice of prevalent medium. The Bruggeman theory takes $\epsilon_h = \epsilon_{eff}$. This is often preferred to M-G, because it is self-consistent and does not attribute a particular role to either medium 0 or medium 1.

The present approach leads to a description of a sample using four media: the ambient, a rough anisotropic layer, a dense layer and the substrate. With the Abeles matrix formulation it is very easy to compute the optical properties of such multilayer systems. However, in order to get a better understanding of the observations, it is advantageous to use simpler models, if possible with algebraic solutions. Our SEM observations and our thickness measurement suggest a growth process where both the dense layer and the top rough layer are growing.

One may try to use separate models for the different stages of the RDS records. During the first stage (Fig. 6a), the roughness thickness e is very small compared to the light wavelength λ, allowing a first order expansion in e/λ. In the second part of the growth (Fig. 6c), the roughness should be treated without the small thickness approximation. The effect of the substrate is ignored, because the dense layer is thick enough to absorb the light before it reaches the layer/substrate interface. This assumption is fully valid provided that the dense layer thickness exceeds the penetration depth of light in the material.

3.2.2 Optical Model for Small Roughness. The model related to the first stage of the growth is developed in the following way. We begin with a model developed for the RA study of heterojunctions, describing the RA signal for the structure shown schematically in Fig. 6b: an isotropic layer growing on an isotropic substrate, with anisotropic surface and interface. Surface and interface are modeled as anisotropic layers with respective thickness $e_s, e_i \ll \lambda$. The Abeles matrix formulation

418

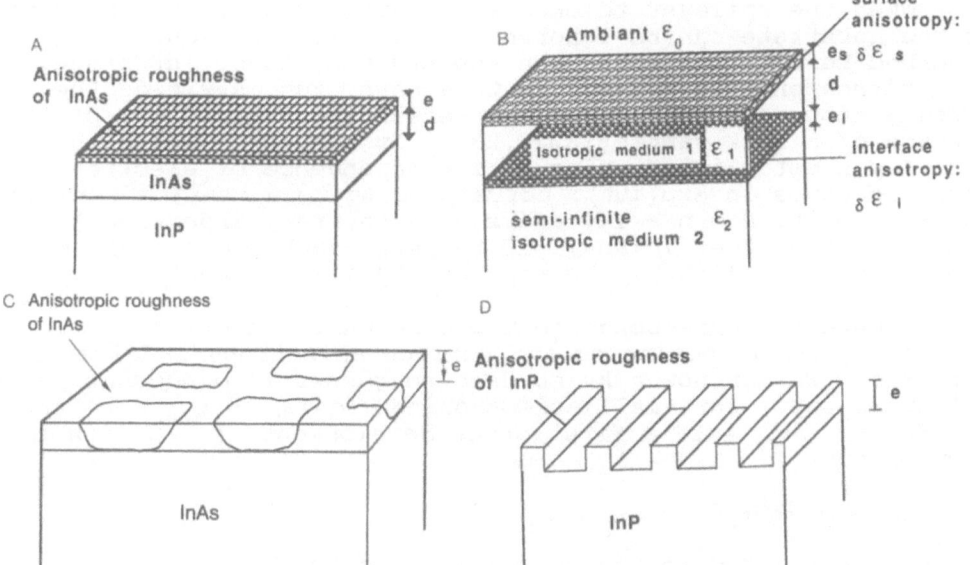

Fig. 6. Schematic view of the 4 models used to account
for the observed RA signal: a) model with semi-
infinite substrate, InAs dense layer, and InAs
rough anisotropic layer with thickness e << λ;
b) model with semi-infinite substrate, isotropic
epilayer, anisotropic surface and interface; c)
model with semi-infinite InAs dense layer and
InAs rough anisotropic layer; d) model with
semi-infinite layer and roughness consisting of
infinite stripes. This corresponds to a
particular case of model c).

and a first-order development in e/λ yield the reflectance
anisotropy:

$$\frac{(r_{011} - r_{0\bar{1}1})}{r} = \frac{4\pi i}{\lambda} \frac{N_0}{\epsilon_1 - \epsilon_0} e_s \delta\epsilon_s + \frac{16\pi i}{\lambda} \frac{N_0 \epsilon_1}{(\epsilon_1 - \epsilon_0)(N_1 + N_2)^2}$$

$$\times \left[\frac{X}{[1 + (r_{12}/r_{01})X](1 + r_{01} r_{12} X)} \right] \left[e_i \delta\epsilon_i + \frac{\epsilon_1 - \epsilon_2}{\epsilon_1 - \epsilon_0} e_s \delta\epsilon_s \right] \quad (4)$$

with

$$X = \exp(-4\pi i N_1 d/\lambda), \quad r_{01} = (N_0 - N_1)/(N_0 + N_1),$$

$$r_{12} = (N_1 - N_2)/(N_1 + N_2),$$

$$\delta\epsilon_s = (\epsilon_{01\bar{1}} - \epsilon_{011})_{surface}, \quad \delta\epsilon_i = (\epsilon_{01\bar{1}} - \epsilon_{011})_{interface}$$

N_0, N_1, and N_2 are the optical indices of the ambient, the dense
layer, and the substrate, respectively; d is the thickness of
the layer.

When the epilayer thickness is large enough, the influence of the substrate can be ignored because all the light is absorbed before it reaches the substrate-epilayer interface. This corresponds to X = 0. In this case, Eqn. 4 yields the already known expression for the reflectance anisotropy of the surface of a bulk sample[18,28,29]. It should be noted that Eqn. 4 was established using a time dependence of electromagnetic waves in exp(iωt), corresponding to a complex optical index given by N = n - ik, where n is the real index and k is the extinction coefficient. If the exp(-iωt) convention is preferred, Eqn. 4 should be replaced by its conjugate.

Figure 6a corresponds to the case where interface anisotropy $\delta\epsilon_i$ is neglected and surface anisotropy is given by the EMTs. We can now take surface roughness into account in the following way. For small anisotropies ($\delta q = (q_{01\bar{1}} - q_{011}) \ll 1$), and for $\epsilon_1 \gg \epsilon_0 = 1$, a first-order development of M-G with $\epsilon_h = \epsilon_1$ gives:

$$\delta\epsilon_s = -\delta q F \epsilon_1 \qquad (5)$$

where F is related to the roughness geometry by:

$$F = f(1 - f)/(1 - qf)^2. \qquad (6)$$

The RA signal corresponding to case of Fig. 6a with the Maxwell-Garnett description of roughness is obtained using Eqns. 4 and 5, with $\epsilon_1 \gg \epsilon_0$; e is the roughness thickness.

$$(r_{01\bar{1}} = r_{011})/r = -(4\pi i/\lambda) Fe\delta q [1 + 4X'/(1 - X')^2] \qquad (7)$$

with

$$X' = r_{12} \exp(-4iN_1 d/\lambda), \quad r_{12} = (N_1 - N_2)/(N_1 + N_2),$$

r_{12} being the reflection coefficient of the layer-substrate interface.

It is clear from expression (7) that the anisotropic roughness affects mainly the imaginary part of reflectance anisotropy, as it is observed experimentally during the first stages of the growth shown on Figs. 2 and 5. The real part arises only from the term in X', which is a perturbation. It corresponds to light reflected at the substrate-dense layer interface. As the dense layer thickness increases, interference conditions between the light reflected at the surface and at the interface are modified. This leads to an oscillating behavior of Re{$(r_{011} - r_{01\bar{1}})/r$} as d increases. The characteristics of roughness appear in the factor Feδq . It is clearly not possible to independently obtain values for f, q, e, δq from the experimental data. Indeed, the product Feδq is the only variable parameter that need be introduced in a fit to the experimental measurement. It is important to emphasize that the relationship between real and imaginary parts deduced from Eqn. 7 is independent of roughness characteristics. For a given substrate and layer, if the growth rate of the dense layer is known, the model predicts the value of the real part of reflectance anisotropy as a function of the imaginary part only. Therefore, the ability of the model to account for both components of the RA signal is a good test of its validity.

If the Bruggeman theory is used instead of M-G to calculate the value of $\delta\epsilon_s$, one no longer obtains a simple analytic expression similar to Eqn. 5. $\delta\epsilon_s$ has to be evaluated numerically using Eqn. 3 and the RA signal is computed using Eqn. 4. Bruggeman is expected to be more accurate, but both theories lead to similar results, as will be shown in the next section. The discussion based on Eqn. 7 is expected to remain essentially valid.

3.2.3 Optical Model for Large Roughness Thickness. After both the rough and the dense layer have grown for some time, another model (Fig. 6c) should be used to describe the second stage. This is indeed a three-medium model, with an anisotropic intermediate medium treated using the M-G or Bruggeman theory. The RA signal associated with such a sample is easily computed as a function of roughness characteristics. The main features expected from such a model are the possibilities to account for large anisotropies, and the interference-like features measured by RA (Fig. 2, Fig. 4). Clearly, the thicker the anisotropic medium, the greater the interaction with light, and therefore a larger optical anisotropy. Constructive or destructive interference within this layer is possible only if the material is not too absorbing. Indeed, effective medium described by the M-G or Bruggeman theory is significantly less absorbing than the dense medium. The penetration depth of light in the rough layer exceeds the wavelength in this material. One can therefore expect to observe oscillating interference-like features on the RA signal when the thickness of the roughness increases, even for light wavelengths strongly absorbed in the dense medium.

One should point out that the validity of EMT depends on the dimensions of the rough features compared to the light wavelength. If the roughness consists of a series of parallel stripes of infinite length (Fig. 6d), the ratio between the lateral dimensions (stripe width and spacing) and the wavelength is the relevant criterion. This ratio should be small in order for the theories to be valid; discrepancies appear when this ratio exceed 0.1. However, SEM images (Fig. 3) show that in fact roughness dimensions are comparable with light wavelength after four minutes growth of InAs on InP. However, even if EMTs are used somewhat outside of their range of validity, they can still give a qualitative idea of the optical properties of a rough layer; one cannot expect to get quantitative information from RA measurements after a few minutes growth.

3.3 The First Stage of 3D Growth

Figures 5,7,8 present the RA record of the first minute of InAs/InP growth and the corresponding fit using the model shown in Fig. 6a. Figure 5 is a detail of Fig. 2. After an initial transient of 5 to 10 seconds, $\text{Im}\{(r_{01\bar{1}} - r_{001})/r\}$ shows a nearly linear variation. This suggests that the roughness thickness e increases linearly with time. Equation 7 shows that the slope of the imaginary part of reflectance anisotropy is $F\delta qde/dt$. This is the only adjustable parameter used in the M-G model to fit both the real and the imaginary part of $s (r_{01\bar{1}} - r_{011})/r$. The evolution of the thickness d of the dense layer as a function of time has to be known in order to evaluate X' in Eqn. 7. The growth rate of the dense InAs layer dd/dt was taken as 250 Å/min for Figs. 5 and 7. This value was chosen to be slightly lower than the 270Å/min measured experimentally on

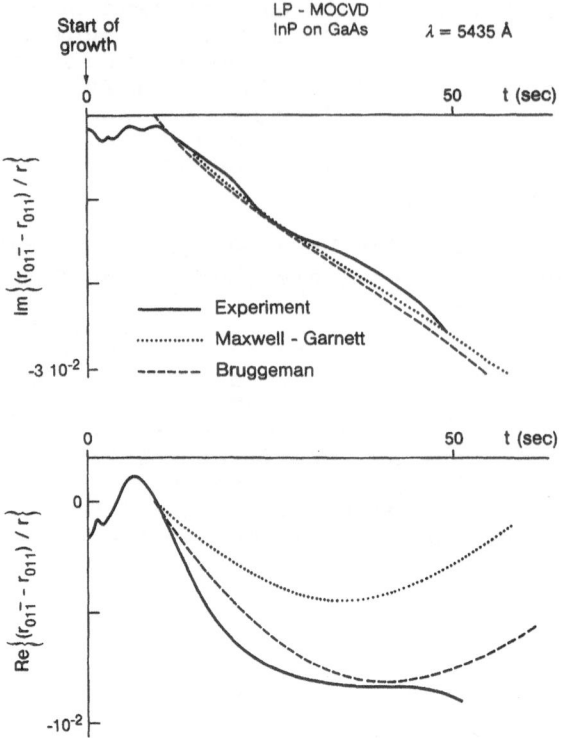

Fig. 7. Comparison between experiment and theory for the
RA record at $\lambda = 5435$ Å of the first minute of
growth of InAs on InP. Maxwell-Garnett fit was
obtained with $F\delta q de/dt = 19.5$ Å/min. Bruggeman
fit corresponds to $f = 0.7$, $q = 0.5$, $\delta q = 0.28$
and $de/dt = 100$ Å/min.

thick samples. This is because only part of the growth
contributes to the dense layer, another part contributes to
increase roughness thickness. In the case of Fig. 8, where the
growth rate was doubled, dd/dt was taken as 500 Å/min. It can
be noticed that a small error in the evaluation of dd/dt would
slightly change the period of the oscillation of $Re\{(r_{01\bar{1}} -
r_{011})/r\}$ but would not significantly affect the agreement
observed between the experiment and the model.

The simulations shown on Figs. 5,7,8 using the Bruggeman
theory were made using $f = 0.7$, $q = 0.5$, and by varying shape
anisotropy δq and roughness growth rate de/dt. No attempt was
made to find the best fit by allowing all parameters to vary.
As discussed previously, it is not possible to obtain reliable
evaluation for each parameter individually, but only for
combinations of them. Using the M-G theory, the model yields
17 Å/min, 19.5 Å/min and 35.5 Å/min for the $F\delta q de/dt$ product for
Figs. 5,7,8 respectively. The corresponding values using
Bruggeman theory are 11.5 Å/min, 14 Å/min and 24 Å/min. This
shows that the M-G and Bruggeman theory yield comparable
$F\delta q de/dt$ products for a given sample.

As can be seen from the figures, both models give
comparable predictions. The model based on Bruggeman theory

422

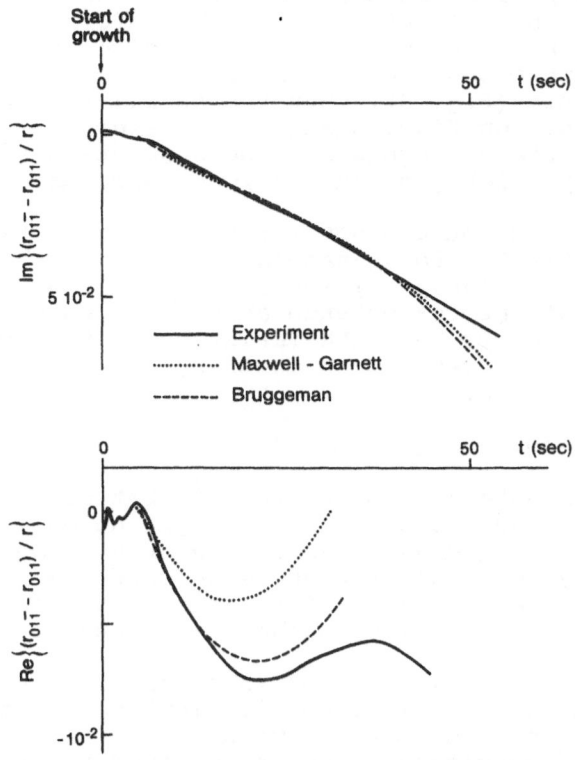

Fig. 8. Comparison between experiment and theory for the
RA record at λ = 5435 Å of the first minute of
growth of InAs on InP, for twice the usual
growth rate. Maxwell-Garnett fit was obtained
with F δqde/dt = 35.5 Å/min. Bruggeman fit
corresponds to f = 0.7, q = 0.5, δq = 0.28 and
de/dt = 170 Å/min.

accounts very well for the behavior of the real part of the
reflectance anisotropy, and for the one order of magnitude
difference between the real and the imaginary parts. The model
agrees well with the measurements both at λ = 5435 Å (Fig. 7)
and at λ = 6328 Å (Fig. 5). The Fδqde/dt products found in both
cases are similar. Doubling the growth rate (Fig. 8) yields a
nearly doubled Fδqde/dt product. The period of the oscillation
of Re$\{(r_{01\bar{1}} - r_{011})/r\}$ is shorter, in accordance with the higher
growth rate of the dense layer. Figures 5,7,8 show that the
model proposed gives a satisfactory description of our
observations. The slope of Im$\{(r_{01\bar{1}} - r_{011})/r\}$ gives a
quantitative indication of the anisotropic thickness growth
rate. The oscillatory behavior of Re$\{(r_{01\bar{1}} - r_{011})/r\}$ as a
function of dense layer thickness predicted by Eqn. 7 is
actually observed. Other experiments have shown that this
feature is observed also at 5000 Å wavelength, but is no longer
seen with UV light. This is because the InAs epilayer is too
absorbing in the UV, and the quantity X' in Eqn. 7 vanishes even
for small epilayer thickness.

The same model was also used for the growth of InP on GaAs. The sign of the reflection coefficient between InP and GaAs is the opposite of that of InAs on InP. Therefore, the sign of initial variation of $Re\{(r_{01\bar{1}} - r_{011})/r\}$ (given by Eqn. 7) should be opposite of that observed in the case of InAs on InP. This is indeed the case (Fig. 5). But the oscillatory behavior especially visible on Figs. 6 and 8 is masked here by the fast variations of surface roughness. The assumption that $e \ll \lambda$ breaks down very quickly on the InP/GaAs samples.

The model does not account for the initial stage, corresponding to the first 5 to 10 seconds of the growth. It takes between 4 seconds (Figs. 5,8) and 9 seconds (Fig. 7) for the imaginary part of the reflectance anisotropy to begin its linear variation. This shows that the variation of RA signal during this initial stage comes from a contribution other than anisotropic 3D growth.

3.4 Subsequent Growth

After one minute of growth, the fit between the calculated and measured RA signal evolution no longer holds for $Re\{(r_{01\bar{1}} - r_{011})/r\}$. This is expected, because the model displayed in Fig. 6a is valid only for small roughness thickness $e \ll \lambda$. One should use the model developed for the second stage (Figs. 6c,6d) that works for finite roughness thickness.

This is shown on Fig. 9, for InP/GaAs/Si growth, where the calculated reflectance anisotropy is plotted as a function of roughness thickness and compared with the RA versus time record of Fig. 4. The calculated curve is obtained by approximating the roughness to be stripes (Fig. 6d), which is the most anisotropic shape ($\delta q = 1$). Both M-G and Bruggeman models are equivalent in that case. The filling factor f was taken to be 0.5. The theoretical curve as a function of roughness thickness e exhibits the same general features as the variations of measured signal as a function of time. There is agreement for the sign of the variation, the order of magnitude, the sign change of the imaginary part, and the correspondance between the inflection points and the extrema of both components. This similarity between the theoretical curves as a function of e and the experimental record as a function of time suggests that the roughness thickness e increases linearly with time. Differences exist between the observation and the model, but many reasons may account for this. The roughness shape, the filling factor and the growth rate of e may vary with time. Besides, as previously mentioned, the EMT is not used in its range of strict validity.

The growth of InAs on InP was investigated at different wavelengths, ranging from 3800 Å to 6328 Å. The broad features and the sign of the variations are the same at all wavelengths, except the initial oscillation of $Re\{(r_{01\bar{1}} - r_{011})/r\}$ as mentioned before. This is consistent with a structural origin of RA behavior. The model of Fig. 6c was used to account for a RA record of InAs/InP. Figure 10 represents the calculated RA signal for the growth of rough InAs/InP as a function of roughness thickness e. The roughness shape is kept constant ($f = 0.7$, $q = 0.5$, $\delta q = 0.16$). The model agrees quite well with the experimental record of the RA signal as a function of time (same record as the one detailed in Fig. 7). While it was not possible to separately determine the value of de/dt and the

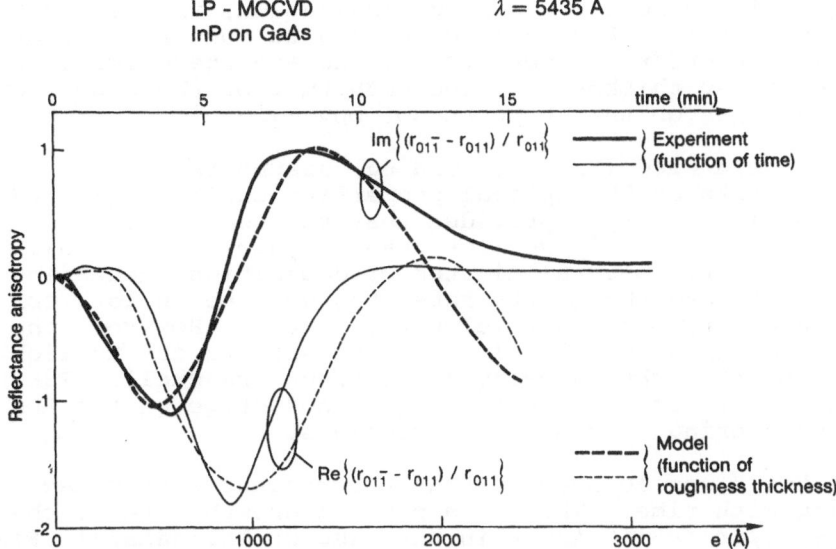

Fig. 9. Comparison between experimental RA record as a
function of time (same as Fig. 4), and
theoretical prediction of model (Fig. 6d) as a
function of roughness thickness. The filling
factor was taken as f = 0.5.

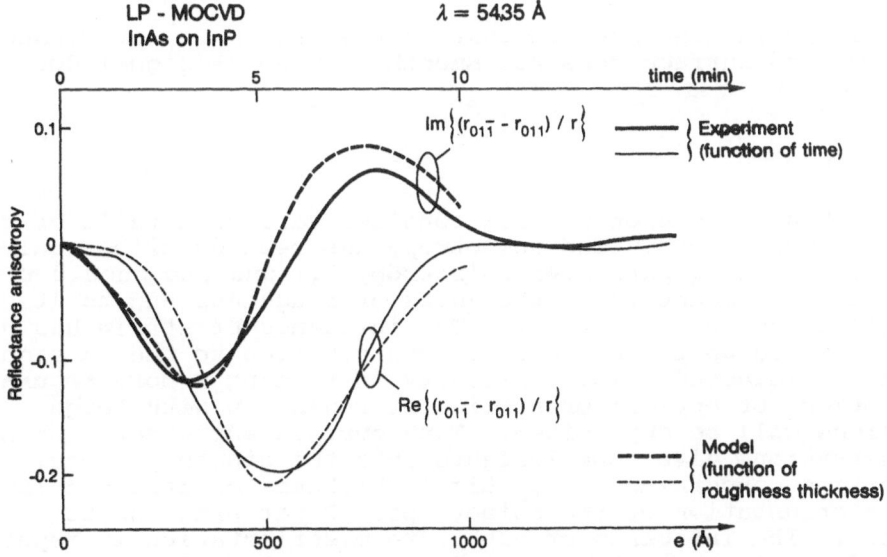

Fig. 10. Comparison between experimental RA record as a
function of time of InAs/InP, and theoretical
prediction of model (Fig. 6c) as a function of
roughness thickness. The roughness was modeled
using Bruggeman theory, with f = 0.7, $\delta q = 0.16$
and q = 0.5.

shape anisotropy δq during the first minute of the growth, these two parameters have quite different effects on the RA features for longer timescales. As a rough indication, the period of the oscillations of the RA signal correspond to constructive or destructive interference conditions, and are therefore related to the roughness thickness e; the magnitude of the signal at the exrtrema depends on the shape anisotropy δq.

The models used (Figs. 6c and 6d) assume that the influence of the substrate on the optical properties can be neglected. This assumption is valid provided that the penetration length of light in the material is smaller than the dense layer thickness. This is true after several minutes of deposition, depending on the wavelength and the growth rate, but does not hold after only one minute growth for the wavelengths we use. However, the former paragraph shows clearly that the effects due to light reflected by the substrate-layer interface are small. Taking this effect into account during the second stage of the RDS signal would bring only small corrections.

Figures 9 and 10 suggest that the roughness thickness keeps increasing with time. The corresponding growth rate of the roughness de/dt is 160 Å/min in the case of InP/GaAs/Si (Fig. 9), and 100 Å/min in the case of InAs/InP (Fig. 10). However, after 10 or 15 minutes growth, the surface improves, leading to a mirror-like surface after one hour's growth. This is consistent with the fact that RA signal returns to zero, instead of exhibiting more oscillations as the models would predict. The mechanisms that could lead both to increase in apparent roughness thickness and in the disappearance of roughness have not been investigated. They are beyond the scope of present modeling.

One should also mention that if the growth is not conducted properly, the surface does not smooth and the RA signal does not return to zero.

4.   DISCUSSION

In this discussion we will consider various details of the models in more depth. The anisotropy has been described only with the screening parameter anisotropy δq, and one should try to relate this quantity to the shape of roughness and to its lateral dimensions anisotropy. The influence of strain has not been mentioned up to now, and its contribution to the RA signal should be discussed. The difficulty in finding a more accurate description for optical properties of roughness with large dimensions will be emphasized. With our present understanding, RA observations give some insights into the growth process. Information concerning the critical thickness of strained layers and the orientation of the anisotropic 3D patterns can be obtained. The influence of substrate misorientation is reported and evidence for mass-transport during annealing is presented. With our present understanding, the information obtained using RA can be useful for the optimization of the growth of lattice-mismatched semiconductors.

4.1 Discussion of Optical Models

The first stages of the growth have been described successfully using the model of Fig. 6a, as the agreements between experiment and theory show in Figs. 5,7,8. The roughness features are small, and EMTs are clearly used within their range of validity. It is possible to give an idea of the correspondance between this parameter and the shape of roughness in some cases. The sign of δq indicates that the screening is less effective along the [011] direction than along [01$\bar{1}$]. This suggests that the roughness patterns are elongated in the [011] direction. While it is not absolutely clear from SEM pictures, it is plausible. For oblate spheroids, δq depends on the ratio between axial length c/a[27]; for example, $\delta q = 0.3$ corresponds to $c/a \approx 1.4$. This elongation is moderate, but can account for the RA behavior observed on InAs on InP (Fig. 10). The determination of the depolarization factor associated with a given particle shape is not straightforward, and no attempt was made to relate δq to SEM observations. The dispersion of roughness shape has not been taken into account in our calculations. It is possible to include it in a model[30], but it would add more parameters.

In the absence, to our knowledge, of available data concerning high temperature values of the optical indices of InAs and InP, room-temperature values have been used. The main parameter affected by this simplification is the reflection coefficient at the substrate-layer interface, r_{12}, which relates the variation of the real part of the RA signal to the variation of the imaginary part (Eqn. 7). However, r_{12} depends mainly on the difference between the indices, and should not be too affected by temperature since the indices of both InAs and InP are expected to have similar variation with temperature.

No effects other than structural contributions have been taken into account in our model. The contribution of chemisorbed species to the RA signal is known to be in the 10^{-3} range, in particular for InAs as shown in other studies[12,14]. It may account for the variations of the RA signal in the first seconds of the growth, but clearly not for the larger variations observed afterwards. The effect of the strain, on the other hand, should be discussed. It is known that a lattice-mismatched layer is very strained at the beginning of the growth and gradually relaxes. If the deformation of the epilayer is purely tetragonal, no RA contribution is expected to arise from the strain. If the relaxation process is different in the [011] and [01$\bar{1}$] directions, as has been reported in the case of GaAs on silicon[31], it is expected to have an influence on the RA signal. The effect of strain on the optical properties of semiconductors has been studied with piezoreflectance measurements[32,33]. The effect is extremely wavelength-dependent, increases by one order of magnitude and has a sign change near the critical points E_0 and E_1. This clearly rules out anisotropic strain as a major contribution to the RA records of lattice-mismatched materials, because the observed features are essentially similar at all wavelengths. The influence of strain may however be a key feature in understanding the evolution of the RA signal in the first 5 or 10 seconds. Using available data on GaAs[32], a strain anisotropy in the growth plane of 10^{-3} would lead to a RA contribution less than 10^{-3} far below E_1. It would reach 5×10^{-3} in the E_1, $E_1 + \Delta_1$ region. Careful spectroscopic measurements would be necessary to assess this effect.

As previously mentioned, EMTs are based on the assumption that the dimensions of the roughness are very small compared to

427

the light wavelength. This assumption clearly holds during the first minute, but it is no longer true after 5 minutes. The use of EMT to account for RA behavior after 5 minutes of growth (in Figs. 9 and 10) requires some further discussion. The reflectivity of the sample decreases by a factor 2 to 6 after 10 minutes' of 3D growth, before returning to a level comparable to its original value. EMTs fail to account for this evolution of the reflectivity. One reason is that they do not take into account light diffusion. When the typical size of roughness patterns approaches the light wavelength, part of the light is diffused, leading to a decrease in reflectivity and to a hazy appearance. But the reflectance anisotropy is expected to be not much affected by the diffused light, since it contributes to decrease both r_{001} and $r_{01\bar{1}}$ individually.

A more accurate model would be able to account for the optical properties of roughness with characteristic dimensions comparable to the wavelength. Much effort has been devoted to the study of scattering by anisotropic particles[34-37], leading very often to numerical solutions requiring long computations. The modeling of anisotropic patterns on a surface may be more difficult than that of isolated particles[38]. Besides, SEM pictures show that the roughness is not uniform, and a distribution of roughness characteristics should be used. This inhomogeneity may also induce partial depolarization of reflected light. A model based on the Fresnel-Kirchhoff theory of diffraction accounts for the decrease in reflectivity of rough surfaces with lateral roughness dimensions large compared to the wavelength, and small roughness thickness[39]. This model is valid only if the slopes of the roughness features are very small, which does not seem to hold in the present case and is unable to account for reflectance anisotropy under normal incidence. The difficulty of finding a model that would account both for reflectance measurements and characterization in polarized light has often been mentioned[40].

4.2 Insights on the Growth Process

Using this understanding of the RA technique, it is now possible to discuss the aspects of growth where information was obtained. Among the crystal growth issues that can be dealt with, using our present understanding of RA, is the assessment of critical thickness where 3D growth begins. During the initial stage of the growth, RA behavior does not correspond to anisotropic 3D growth. This could mean that the growth is two-dimensional during the first 5 to 10 seconds. This corresponds to Stranski-Krastanov growth: growth of several layers of a complete InAs film followed by island formation. After this initial stage, both the dense and the rough part of the film grow. The RA signal evolution observed during the initial transient corresponding to the first seconds of growth may arise from a variety of contributions. The contribution of chemi- and physisorbed species are expected to be different in the case of InP or in the case of growing InAs. Atomic rearrangements at the surface or interface can affect the signal. Another contribution may arise from anisotropic strain in the layer, as discussed in the last paragraph.

The magnitude of the RA signal for InAs/InP is very dependent on growth conditions, but also on substrate cleaning and heating procedure. It makes the quantitative RA assessment of Fδqde/dt more appreciable for in situ monitoring. The small difference between the values deduced from Figs. 5 and 7 corresponding to two different growths at two different wavelengths may not arise from unexpected spectroscopic dependence, but simply because the growth conditions were slightly different between the two experiments. The choice of wavelength for performing such in situ studies is not crucial. The same broad features are found at all wavelengths, except the oscillation on $Re\{(r_{01\bar{1}} - r_{011})/r\}$ during the first minute. It is not visible for wavelengths too strongly absorbed in the dense medium.

Our observations show that during the growth of both InAs/InP and InP/GaAs/Si, anisotropic 3D growth occurs. The magnitude of this effect is different for the two systems. Modeling suggests that roughness patterns are more anisotropic and grow faster in the case of InP/GaAs than in the case of InAs/InP.

In the preliminary study, the growth of InAs on InP was stopped, while keeping AsH_3 flux and temperature constant. Figure 11 reports the RA record of a 3'30" growth of InAs on InP, followed by about one minute under AsH_3 without growth, at the same temperature. This is the same sample as that on Fig. 3c. The RA signal changes significantly during this annealing, suggesting that there was mass transport and change in roughness characteristics. The change during the first 30 seconds is by about 5% for $Re\{(r_{01\bar{1}} - r_{011})/r\}$, and 20% for $Im\{(r_{01\bar{1}} - r_{011})/r\}$, which corresponds to a variation rate larger than during the growth! Then, the optical anisotropy reaches a steady-state value. The change in RA is expected to come from a change in roughness shape and characteristics, since only structural contributions can account for this order of magni- tude. The mechanism for such structural evolutions has not been investigated. It is important information that stopping the growth and keeping the sample at the same temperature does not stop surface evolution. As previously mentioned, it makes a comparison between kinetic RA data and spectroscopic RA recorded after growth has been stopped more difficult. This is because the evolution of the optical properties of a sample during the process of stopping the growth and cooling down is not negligible.

The models proposed for the first and second stages of the RA records of roughness suggest that roughness thickness continuously increases with time. It is also clear that after some time the roughness contribution disappears, indicating that the surface has smoothed. These two aspects are not contra-dictory if one supposes that the roughness thickness increases while its filling factor increases toward 1. Figures 3a to 3d are consistent with roughness features increasing in size but becoming less and less numerous as a function of time. The mechanism for this has not been further investigated.

Two factors contribute to the difference in optical proper-ties of the [011] and [01$\bar{1}$] directions. One is the chemical asymmetry between the two directions: the dangling bonds of the group V element are aligned along the [01$\bar{1}$] direction. The other is the structural asymmetry due to the presence of steps on the surface. The steps are due to the slight tilt of the surface of the substrate relative to (100). For the samples

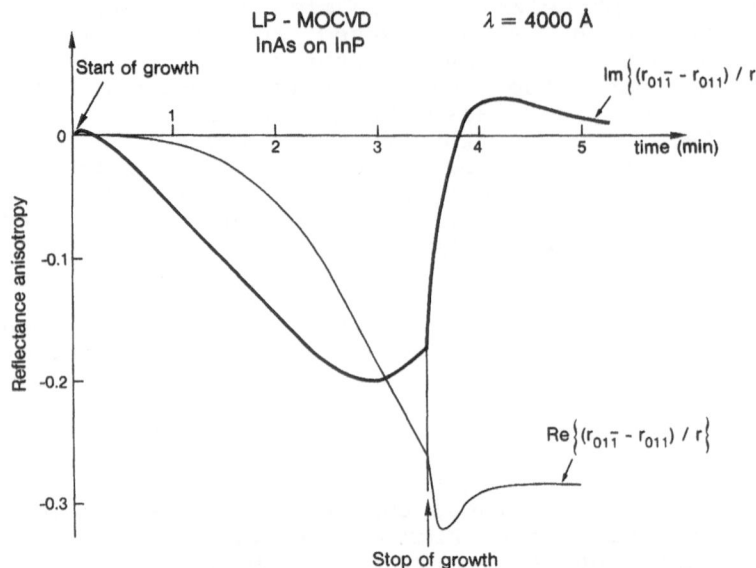

Fig. 11. RA record of the growth of an InAs on an InP substrate, using $\lambda = 4000$ Å light wavelength. The growth is stopped after 3 minutes and 30 seconds, the AsH_3 flux and temperature being unchanged. Note the important change of RA signal in the seconds following the growth interruption.

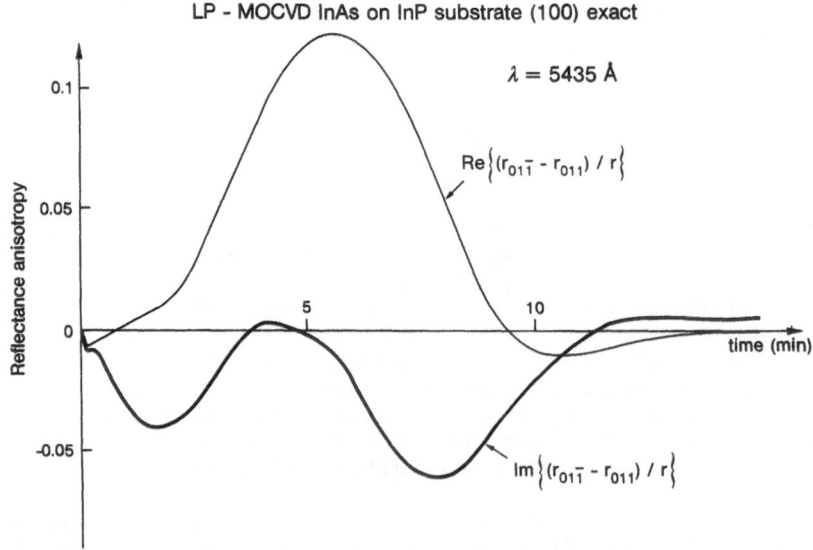

Fig. 12. RA record of the growth of an InAs layer on an InP substrate with (100) exact surface orientation, using $\lambda = 5435$ Å light wavelength.

we use, misoriented toward [01$\bar{1}$], the steps are along the [011] direction. Steps are known to play an important role in the first stages of the growth of GaAs/Si[31]: nucleation occurs at the step edges and GaAs islands are elongated along the step. In the present study, the sign of δq tells us that roughness is elongated along [011], which is also the direction of the steps. Preliminary experiments were conducted with (100) exact substrates, in order to assess the importance of the structural asymmetry of the substrate on the RDS behavior (Fig. 12). The RA features related to 3D growth are quite different from that observed on substrates with 2° misorientation toward [01$\bar{1}$]. It suggests that the structural rather than the chemical anisotropy of the surface of the substrate determines the privileged directions of anisotropic 3D growth. It shows the importance of substrate misorientation for the growth of lattice-mismatched materials. With our present understanding, the RA record of Fig. 12 is difficult to comment in more detail. It is not even clear that [011] and [01$\bar{1}$] are the optical eigenaxes. The orientation of our commercially available substrates is not specified with a precision better than 0.5°, and optical eigenaxes may correspond to the orientation of the residual tilt. Further studies with different substrate misorientations are required to separate the influence of the chemical and structural asymmetry of the substrate on the roughness characteristics.

CONCLUSION

The observation of large reflectance anisotropies during the LP-MOCVD growth of lattice-mismatched semiconductors has been reported for the first time. The case of InAs on InP and InP on GaAs have been investigated. The difference between the reflection coefficients r_{001} and $r_{01\bar{1}}$ can exceed 5% in the case of InAs/InP, and 100% in the case of InP/GaAs/Si. It has been shown that these anisotropies are related to three-dimensional growth. For small roughness thickness, it is possible to get quantitative information from RA observation. The fact that both real part and imaginary part of the reflectance anisotropy are in agreement with our model is a good confirmation of its validity. For larger roughness thickness, only a qualitative description was found. It has been shown that substrate mis-orientation has a large influence on the way 3D growth occurs. We found that mass transport takes place at the growth temperature after the growth is stopped. The growth of lattice-mismatched materials is difficult and very sensitive to growth conditions. The information provided by RA is therefore very attractive for the optimization of the growth of lattice-mismatched semiconductors. They can help in assessing the reproducibility and optimizing growth procedure. If the growth is not conducted properly optical anisotropy does not return to zero, and RA indicates it. In many cases, one should avoid 3D growth, and the measure of $Im\{(r_{01\bar{1}} - r_{011})/r\}$ is a good indication for that. There is one limitation, however: RA is not sensitive to roughness, but to roughness anisotropy. RA can be very useful to study annealing. More generally, RA should be convenient to assess the efficiency of procedures to prevent or cancel 3D growth. This is especially appreciable in the severe MOCVD environment, where very few investigation techniques can be used in situ during the growth.

REFERENCES

1. M. Razeghi, M. Defour, F. Omnes, P. Maurel, E. Bigan,
    O. Acher, J. Nagle, F. Brillouet and J.C. Portal, **Journ.
    Cryst. Growth**, 93:776 (1988)
2. M. Razeghi, M. Defour, F. Omnes, J. Nagle, P. Maurel,
    O. Acher, A. Huber and D. Mijuin, **Mat. Res. Soc. Symp.
    Proc.**, 126:143 (1988)
3. M. Razeghi, M. Defour, R. Blondeau, F. Omnes, P. Maurel,
    O. Acher, F. Brillouet, J.C. Fan and J. Salerno, **Appl.
    Phys. Lett.**, 53:2389 (1988)
4. M. Razeghi, F. Omnes, R. Blondeau, P. Maurel, M. Defour,
    O. Acher, E. Vassilakis, G. Mesquida, J.C. Fan and
    J. Salerno, **Appl. Phys. Lett.**, 65:4066 (1989)
5. D.E. Aspnes and A.A. Studna, **Phys. Rev. Lett.**, 54:1956
    (1985)
6. D.E. Aspnes, **J. Vac. Sci. Technol.**, B3:1498 (1985)
7. D.E. Aspnes, **Appl. Phys. Lett.**, 52:957 (1988)
8. D.E. Aspnes and A.A. Studna, **J. Vac. Sci. Technol.**, A5:546
    (1987)
9. R. Del Sole, **Journ. of Mat. Sci. and Engineering.**, B5:
    (1990)
10. S.E. Acosta-Ortiz and A. Lastras-Martinez, **Solid State
    Comm.**, 64:809 (1987)
11. D.E. Aspnes, E. Colas, A.A. Studna, R. Bhat, M.A. Koza,
    V.G. Keramidas, **Phys. Rev. Lett.**, 61:2782 (1988)
12. O. Acher, F. Omnes, M. Razeghi and B. Drévillon, **Journ. of
    Mat. Sci. and Engineering**, B5:223 (1990)
13. T. Lopez-Rios, Y. Borensztein and G. Vuye, **Phys. Rev.B.**,
    30:659 (1984)
14. S.M. Koch, O Acher, F. Omnes, M. Defour, B. Drévillon and
    M. Razeghi, to be published in **Journ. of Appl. Phys.**
15. M. Razeghi, "The MOCVD Challenge", Adam Hilger, (1989)
16. M. Razeghi, M. Defour, F. Omnes, P. Maurel, J. Chazelas,
    F. Brillouet, **Appl. Phys. Lett.**, 53:725 (1988)
17. M. Razeghi and O. Acher, NATO ASI Series (1989), in press
18. D.E. Aspnes, A.A. Studna, L.T. Florez, Y.C. Chang,
    J.P. Harbinson, M.K. Kelly and H.H. Farrell, **J. Vac. Sci.
    Technol.**, B7:901 (1989)
19. F. Manghi, R. Del Sole, E. Molinari and A. Selloni, **Surf.
    Sci.**, 211/212:518 (1989)
20. V.L. Berkovits, L.F. Ivantsov, I.V. Makarenko, T.A.
    Minashvili and V.I. Safarov, **Solid State Commun.**, 64:767
    (1987)
21. B. Drévillon, J. Perrin, R. Marbot, A. Violet and J.L.
    Dalby, **Rev. Sci. Instrum.**, 53:969 (1982)
22. O. Acher, E. Bigan, B. Drévillon, **Rev. Sci. Instrum.**, 60:65
    (1989)
23. See for example R.M.A. Azzam and N.M. Bashara,
    "Ellipsometry and Polarized Light", North-Holland,
    Amsterdam (1987)
24. "Handbook of Optical Constants of Solids",   E.D. Palik,
    ed. Academic Press (1985)
25. M. Born and Wolf, p.705 **in**: "Principles of Optics",
    Pergammon Press, Oxford (1980)
26. D.E. Aspnes, **Journal de Physique**, C10:3 (1983)
27. C. Kittel, p.405 **in**: "Introduction to Solid State Physics",
    Wiley (1976)
28. J.D.E. McIntyre and D.E. Aspnes, **Surf. Sci.**, 24:417 (1971)
29. R. Del Sole and A. Selloni, **Solid State Comm.**, 9:825 (1984)
30. Y. Borenszstein, M. Jebari, T. Lopez-Rios and G. Vuye,

432

Europhys. Let., 7:617 (1988)

31. S.M. Koch, PhD Thesis, Stanford University (1988)
32. D.D. Sell, **Surf. Sci.**, 37:876 (1973)
33. J. Camassel, D. Auvergne and H. Mathieu, **J. Appl. Phys.**, 46:2683 (1975)
34. S. Asano and G. Yamamoto, **Appl. Optics**, 14:29 (1975)
35. S. Asano, **Appl. Optics.**, 18:712 (1979)
36. P.W. Barber and D.-S. Wang. **Appl. Optics.**, 17:797 (1978)
37. N. Uzunoglu, B.G. Evans and A.R. Holt, **Electron. Lett.**, 12:312 (1976)
38. D.W. Berreman, **Journ. Opt. Soc. Am.**, 60:499 (1970)
39. I. Ohlidal and F. Lukes, **Optica Acta**, 19:817 (1972)
40. C. Pickering, R. Greef and A.M. Hodge, **Semicond. Sci. Technol.**, 4:574 (1989)

EPITAXIAL GROWTH OF STRAINED III/V SEMICONDUCTOR ALLOYS –

COMPOSITION AND MICROSTRUCTURE

G.B. Stringfellow

University of Utah
Salt Lake City, Utah 84112, USA

ABSTRACT

Microscopic and macroscopic strain are enormously important factors in the energetics of III/V alloys. The enthalpy of mixing, previously thought to be due to so-called "chemical" effects, is in fact a result of the microscopic strain energy of the system due to distorting the bonds. The attempts of the system to minimize the strain energy have profound effects on the epitaxial growth of semiconductor alloys. This paper will begin with a review of the thermodynamics of mixing of III/V alloys. This will be the basis of a discussion of the effects of lattice mismatch on solid composition for epitaxial growth by liquid phase epitaxy, molecular beam epitaxy, and organometallic vapor phase epitaxy. The macroscopic strain energy also stabilizes alloys which would be unstable in the bulk, thus allowing the growth of pseudomorphic alloys. Nonrandom atomic arrangements are adopted by the system in an attempt to minimize strain energy, including clustering and phase separation as well as the formation of atomically ordered structures.

1.   INTRODUCTION

A wide range of III/V semiconductor compounds and alloys is required for modern optoelectronic and electronic devices. For example, photon detectors are required over an enormous wavelength range from the visible, < 0.5 μm, to as long as 12 μm. This requires a range of alloys from AlGaInP, for visible optoelectronic devices, to InAsSb for devices operating at long wavelengths. Both photonic and electronic devices capable of ultra-high frequency operation require complex structures with atomic dimensions. For some of these materials and structures, no substrate is available with the ideal lattice constant. Thus, the layers are unavoidably strained. Frequently, the thin layers in superlattice and quantum well structures are intentionally strained to produce desirable changes in the energy band gap as well as the detailed band structure[1]. Strain-layer superlattices are even used as dislocation traps[1] for the removal of threading dislocations

which would otherwise enter the device active layers with
undesirable consequences.

As our understanding of the basic characteristics of III/V
semiconductor alloys has unfolded, it has become increasingly
apparent that the strain energy is a major factor governing the
positioning and arrangement of the atoms. The microscopic
strain energy, i.e., that associated with the stretching and
bending of individual bonds, determines how the atoms are
positioned in a disordered lattice[2,3]: the atomic positions do
not correspond to those in the virtual crystal model. This
microscopic strain energy also determines the mixing enthalpy in
III/V alloys[4-6]. This means that the enthalpy of mixing is
always positive and increases with the difference in lattice
constant of the constituent compounds. In systems where the
lattice size difference is large, this leads to regions of solid
immiscibility, i.e. regions where the free energy of the system
can be decreased by separating the 50/50 alloy into macroscopic
regions having compositions nearer the pure end components.

In addition, macroscopic strain effects associated with
differences in lattice parameter between a coherently strained
epitaxial layer and the substrate, or another epitaxial layer,
may have a profound effect on the composition of the solid[7].
This is to be expected, since the mixing enthalpy and the
macroscopic elastic strain energy have the same physical origin.
The effects of macroscopic strain energy play a particularly
important role for immiscible alloys. The large coherency
strain energy increases the energy associated with fluctuations
in solid composition. This prevents phase separation via
spinodal decomposition for a homogeneous alloy with a
composition inside the region of solid immiscibility at a
particular temperature[8]. Even though phase separation is
prevented by the macroscopic coherency strain energy, the total
energy of the immiscible alloy can be decreased by the formation
of ordered structures composed of alternating layers of the
large and small atoms[9].

These diverse phenomena, related to the effects of strain
energy on the thermodynamics of mixing and the detailed atomic
arrangements in alloys, will all be discussed in this paper
dealing with the epitaxial growth of strained III/V
semiconductor alloys.

2.   THERMODYNAMICS OF MIXING (microscopic strain energy)

The first thermodynamic investigation of a semiconductor
alloy was for the Si-Ge system[10]. The liquid-solid phase
diagram indicates that, surprisingly, both the liquid and solid
phases are nearly ideal solutions, i.e. the enthalpy (energy) of
mixing is nearly zero and the entropy of mixing is nearly the
configurational value for a random solution. The deviations
from ideal behavior were accounted for using the regular
solution model[11]. In this model the entropy of mixing is
ideal and the enthalpy of mixing is given by the symmetric
function. $\Delta H^M = x(1 - x)\Omega$. Ω is the interaction parameter, which
is purely an adjustable parameter to fit the data in this model.

An attempt to produce a more physical model of the enthalpy
of mixing of III/V alloys led to the delta-lattice-parameter

(DLP) model[12] based on the dielectric theory of electro-negativity of Phillips and van Vechten[13,14]. This resulted in an expression for the interaction parameter depending only on the equilibrium atomic spacings of the end component binary compounds, a_{AC} and a_{BC} for an alloy of AC and BC,

$$\Omega = K(a_{AC} - a_{BC})^2/(a_{AC} + a_{BC})^{4.5} \quad . \tag{1}$$

The value of K was obtained by making a least squares fit of Eqn. 1 to available experimental values of Ω. The result was a surprisingly good agreement between experimental and calculated results. The model explains the experimental observations that the interaction parameter increases with increasing size difference for the constituent compounds and is always positive. The form of Eqn. 1 led Fedders and Muller[5] to suggest that the enthalpy of mixing is due to strain energy effects involved in stretching the AC and BC bonds from their equilibrium values. Using the virtual crystal model, they obtained values of Ω approximately 4 times larger than the experimental values. This disparity was explained as due to the failure of the virtual crystal model in III/V alloys, as shown by the EXAFS data of Mikkelson and Boyce for the GaAs-InAs alloy system[2]. More recently both Fukui and Martins and Zunger[3] have produced accurate calculations of Ω with no adjustable parameters. The interactions between atoms are entirely due to strain.

A feature of both the DLP and strain energy models is that they predict a large positive enthalpy of mixing for alloys composed of binary constituents having a large difference in lattice parameter. This means that the total energy of the system can be reduced by allowing an alloy within a certain composition range to separate into two distinct phases with compositions near the end-point binary compounds. This will be true only at temperatures below a specific critical temperature, T_c, since at high temperatures entropy dominates the free energy expression and favors formation of a disordered alloy. The values of T_c for various III/V pseudobinary alloys are listed in Table 1.

The existence of a solid phase miscibility gap causes grave problems for the growth of III/V alloys by near-equilibrium growth techniques such as liquid phase epitaxy (LPE), where the constituents can readily pass in either direction across the growth interface. For example, attempts to grow GaAsSb alloys within the region of solid immiscibility by LPE yield crystals of two solid compositions at the edge of the miscibility gap rather than a single alloy[15]. The GaAs-rich and GaSb-rich crystals produced simultaneously by LPE lie on either side of the binodal curve, connected by a horizontal tie line. Similar problems have been reported for the LPE growth of other alloys such as GaInAsP[16] and AlGaAsSb[17], useful for optoelectronic devices operating at 1.3-1.6 μm and 2.0-2.5 μm wavelength ranges, respectively. In some cases compositional fluctuations are observed rather than separation into two distinct phases. This is indicative of the initial stages of spinodal decomposition, a term used to describe the gradual phase separation process beginning in a homogeneous alloy with a composition well within the miscibility gap where the curvature of the free energy versus composition curve is negative, i.e. $d^2G/dx^2 < 0$. In this case phase separation can occur gradually with no activation barrier.

Table 1. Calculated values of critical temperature with, T_s, and without, T_c, the inclusion of the coherency strain energy (data from Refs. 8 and 37).

System	a_{AC} (Å)	a_{AD} or a_{BC} (Å)	T_C (K)	T_S (K)
TERNARY ALLOYS				
$GaAs_ySb_{1-y}$	5.6536	6.0950	856	*
$Ga_xIn_{1-x}P$	5.4512	5.8696	908	*
$InAs_ySb_{1-y}$	6.0590	6.4794	572	*
$Ga_xIn_{1-x}AS$	5.6536	6.059	729	*
GaP_yN_{1-y}	5.4512	4.49**	8591	1820
GaP_ySb_{1-y}	4.4512	6.095	1965	*
InP_ySb_{1-y}	5.8696	6.4794	1306	*
QUATERNARY ALLOYS				
$Ga_xIn_{1-x}As_yP_{1-y}$			1081	*
$Ga_xIn_{1-x}As_ySb_{1-y}$			1428	*
$Ga_xIn_{1-x}P_ySb_{1-y}$			2470	*
$InP_xAs_ySb_{1-x-y}$			1319	*
$GaP_xAs_ySb_{1-x-y}$			1996	*
$Al_xGa_yIn_{1-x-y}P$			973	*
$Al_xGa_yIn_{1-x-y}As$			735	*
$Al_xGa_yIn_{1-x-y}Sb$			462	*

* $T_s < 0$ Alloys stable at all temperatures and concentrations.
** Estimated, for cubic GaN.

3. EFFECTS OF LATTICE MISMATCH ON EPITAXIAL GROWTH (macroscopic strain energy)

It has been recognized since early epitaxial growth experiments that high quality thick (> 0.5 μm) epitaxial layers can be produced only when the lattice mismatch ($\Delta a/a$) between epitaxial layer and substrate is smaller than a fraction of a percent. Otherwise, mismatch dislocations are generated which have deleterious effects on the material quality and prevent the fabrication of certain devices with desired performance characteristics. For larger values of lattice mismatch, even

more detrimental effects are observed due to the inability of
growth to occur via the layer-by-layer Frank van der Merwe (FM)
mechanism[18] which is desirable for the growth of high quality
epitaxial layers with smooth interfaces. For large values of
lattice mismatch the strain energy of the epitaxial layer can be
reduced by the growth of local islands rather than covering the
entire substrate. This 3-dimensional mode of growth, termed
Volmer-Weber growth[18], is, of course, extremely detrimental
for material used in the fabrication of quantum well, modulation
doped, and superlattice structures.

Very thin layers cover the substrate and remain coherently
strained even at fairly large values of lattice mismatch. This
is the basis of the epitaxial growth of strain-layer superlattices and intentionally strained layers[1]. We are less
interested in systems where the lattice parameter mismatch is
large enough to prevent FM growth.

3.1 Lattice Latching

The alloy has ways of minimizing the macroscopic strain
associated with lattice parameter mismatch which were initially
surprising and somewhat controversial. Today, however, it is
well accepted that the composition of the epitaxial layer grown
by LPE may be perturbed in a measurable way due to the elastic
energy associated with the lattice mismatch. This lattice
"latching" or "pulling" effect was first observed for the LPE
growth of GaInP in 1972[7]. It was observed that as the
composition of the liquid was slowly changed, from run to run,
the composition of unconstrained, noncoherent, platelets growing
around the edge of the substrate changed as expected from the
phase diagram. However, the composition of the epitaxial layer
remained nearly constant, as seen in Fig. 1. It should be noted
here that both coordinates are experimental solid compositions.
The nearly horizontal solid line was calculated thermodynamically with inclusion of the strain energy of the coherent
epitaxial layer. Based on the above discussion of the enthalpy
of mixing for III/V alloys it should not be surprising that the
macroscopic strain energy is large enough to have a significant
effect on the solid composition. The effect will be miniscule
for systems composed of compounds having nearly the same lattice
constants, such as AlGaAs and GaAsP, where entropy plays the
dominant role. In systems near the critical temperature, such
as GaInP, the free energy versus composition curve is nearly
flat, so a perturbation due to the elastic energy will have a
dominant effect. The lines intersecting the main line at nearly
right angles, labeled 1, 0.5, 0.2 and 0.1 μm, represent the
calculated solid compositions for layers of various thicknesses
which are not completely coherent, i.e. they contain mismatch
dislocations. This situation is described more completely by
Hirth and Stringfellow[19].

The lattice latching effect has now been repeatedly
observed in the GaInP system by many authors[20], although the
magnitude of the effect is dependent on kinetic parameters which
affect nucleation and the growth process as well as dislocation
generation[21]. It is also reliably seen as several other
systems including GaInAs[22], GaInAsP[23] and AlGaAsSb[24]. The
phenomenon has been put on an improved theoretical footing by
the work of Larche and Cahn[25] and Wood and Zunger[26].

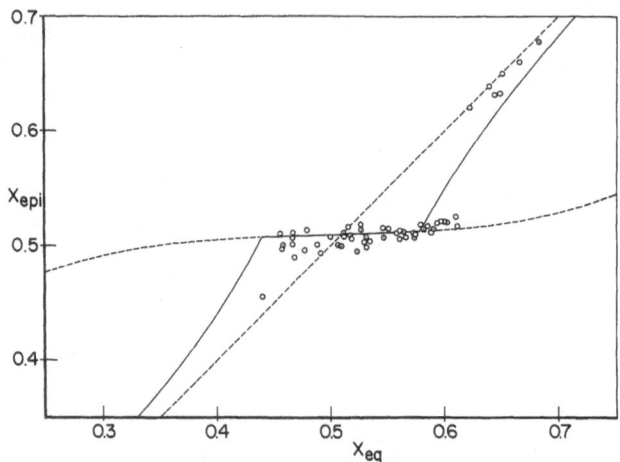

Fig. 1. Composition of liquid phase epitaxial $Ga_x In_{1-x}P$
layers (x_{epi}) versus that of unconstrained
platelets (x_{eq}). Lattice matching occurs for
x = 0.5 (after Stringfellow[7]).

Effects which resemble lattice latching have been observed
for molecular beam epitaxial (MBE) and organometallic vapor
phase epitaxial (OMVPE) growth of III/V alloys. Turco and
Massies[27] observed that the In incorporation into MBE-grown
AlInAs was increased for an InP, as opposed to a GaAs,
substrate. The effect was attributed to the influence of strain
on the thermodynamic driving force for In incorporation. It is
important to note that the effect is only observed at high
temperatures where In can re-evaporate from the surface if not
incorporated. Allovon et al.[28] observed a significant lattice
latching effect, similar to that seen in Fig. 1, in AlInAs grown
by MBE. The observed effect was found to increase with
increasing temperature. However, the magnitude of the latching
was found to considerably exceed that predicted from
thermodynamic equilibrium calculations, leading to the
suggestion that strain energy may affect the kinetics of surface
processes.

A similar conclusion was reached by Leys et al.[29] for the
OMVPE growth of GaAsP strained layers. They found that As
incorporation was markedly suppressed for growth of thin layers
on a GaP substrate. The magnitude of the effect was again found
to exceed the thermodynamic predictions. They suggested that
the strain and/or steric effects produced a change in the
adsorption and surface kinetic processes. For OMVPE growth of
GaInP at 675°C, Schaus et al.[30] reported two phenomena
indicative of lattice latching. The solid composition was found
to be nearly independent of the trimethylgallium flow rate and
to be dependent on the lattice constant of the substrate for the
same growth conditions. The authors explained their results in
terms of the thermodynamic latching effect described above, but
included no thermodynamic calculations. Again, it is worth
stressing that such an effect is unexpected unless the
temperature is high enough to allow evaporation of In produced
by the pyrolysis of trimethylindium. Such an effect at lower

temperatures or for Al-Ga alloys would be surprising indeed unless the temperature were sufficiently low to retain volatile organometallic species on the surface which could re-evaporate. Finally, Duchemin et al.[31] reported data indicative of lattice latching for the OMVPE growth of GaInAsP layers on InP substrates, although they did not comment on the observed independence of solid composition of the gas flow rate around the conditions for lattice parameter match.

3.2 Stabilization of Metastable Alloys

An effect similar to that described above, also caused by the macroscopic strain energy in a coherent system, is the stabilization of metastable alloys. We discussed above the miscibility gap existing for alloys with large differences in the lattice constants of the constituent compounds. The miscibility gap prevents the growth of many alloys, such as GaAsSb, by LPE. However, Nahory et al.[17] demonstrated that AlGaAsSb alloys within the miscibility gap could be grown by LPE provided they were lattice matched to the substrate. The explanation is simply that a coherency strain energy is associated with the growth of the two alloys at either end of the solid-solid tie line connecting points on the binodal curve, while the energy of the lattice matched alloy contains no strain energy term. Similar effects were demonstrated by Quillec et al.[32] for the growth of GaInAsP alloys by LPE on an InP substrate. This alloy has been the subject of numerous investigations due to its use for emitters and detectors in the commercially important 1.3-1.6 μm wavelength range. An interesting observation is that the substrate stabilization is more effective for growth on (111)-oriented substrates than for the (100) orientation[33], in agreement with thermodynamic predictions[25,34] which include the strain energy, since the (100) direction is elastically softer than (111).

The coherency elastic energy also stabilizes a homogeneous alloy within the miscibility gap[35]. The initial stages of spinodal decomposition of AB_xC_{1-x} involve the formation of coherent clusters of AC- and BC-rich material. This creates an elastic energy comparable in magnitude to the thermodynamic driving force for clustering from the positive enthalpy of mixing. The entire spinodal can be calculated including the elastic energy terms[35]. The values of critical temperature including strain, T_s, are included in Table 1. It can be seen that nearly all III/V alloys are stable at room temperature. This has important implications for the use of metastable alloys for device applications. They should not phase separate during either processing or use, except very near the surface, where the coherency strain energy is reduced[36].

In recent years the growth of a wide range of metastable alloys by OMVPE and MBE has been demonstrated. Both processes operate with a very high amount of supersaturation in the vapor. For growth conditions commonly used, the vapor and solid at the interface are nearly in equilibrium[37]. This requires that nearly all of the nutrient atoms arriving at the solid/vapor interface are incorporated. To be specific, during the OMVPE growth of an alloy such as AlGaAs with a value of V/III ratio much greater than unity and low growth temperatures, the concentrations of the Al and Ga species are reduced by orders of magnitude near the interface. The growth process is controlled

441

by diffusion of these species through the vapor to the interface. In this case a nearly random arrangement of Al and Ga atoms is deposited on the surface, without the possibility of leaving. If this random atomic arrangement is covered by the next layer before the atoms can diffuse significant distances on the surface, the random arrangement is "frozen" into the solid. Once in the solid the atoms migrate very little due to the extremely low diffusion coefficients. The AlGaAs system was chosen for this example because the stability of the solid is extremely high, i.e. the vapor pressures of Al and Ga are extremely low at the interface at all useful growth temperatures. This alloy also happens to have a small enthalpy of mixing, so phase separation would be unlikely to occur. However, the same basic process occurs during low temperature OMVPE growth of metastable GaInP (or GaInAs) alloys with a high V/III ratio. Although GaP- and InP-rich clusters might form at the surface if they had time, the very short time before the following layer of GaInP is deposited effectively prevents this process. As we will see below, a small amount of atomic rearrangement may occur at the surface when the diffusion distances required are small. To form ordered structures the atoms need diffuse only a few Ångstroms.

Control of the V/III ratio to encourage depletion of the species on the sublattice where mixing occurs is vital to the successful growth of alloys in the miscibility gap. This is illustrated by the growth of GaAsSb alloys by OMVPE. In this case, a high value of V/III ratio depletes the Ga at the interface, but not the As and Sb. Since As and Sb can leave the interface, the alloy grown will be near the edge of the region of solid immiscibility[37,38]. However, when the V/III ratio is near unity, the As and Sb are depleted at the interface and growth of the metastable alloy occurs via the process described above. In this way layers of GaAsSb[39], as well as even less stable alloys such as GaInAsSb[40], InPSb[41], GaPSb[42], and GaInPSb[43] have been produced for the first time, by OMVPE

The ability to grow metastable alloys by OMVPE has recently been extended to InAsSbBi alloys where the second phase is a liquid[44], although the success has been limited except at very low temperatures where formation of the liquid phase is suppressed.

3.3 Ordering

Even though the immiscible alloys are prevented from phase separation by the coherency strain energy, the system can relax somewhat the microscopic strain energy of the random alloy by forming ordered structures where the large and small atoms arrange themselves into alternating layers. Different ordered structures are formed depending on the crystallographic planes involved. The lowest energy[45] chalcopyrite or $E1_1$ structure involves ordering on (210) planes while the much less stable, but frequently observed, CuPt or $L1_1$ structure involves ordering on (111) planes. The other commonly observed ordered structure, CuAu or $L1_0$, involves ordering on (100) planes. Each ordered structure has several variants depending only on which of the {h,k,l} planes is ordered. Thus, for example, the CuPt structure has four variants and the CuAu structure has three. Typically, not all of the possible variants are observed, which offers a clue that kinetic factors play a significant role in

Table 2. Summary of Ordering in III/V Alloys

Material/Technique	$\Delta a/a$(%)	Substrate Orientation	Ordered Structure	Reference
ANION SUBLATTICE				
GaAsSb OMVPE	7.54	(100),(110), (221)and (311)	$L1_0$, $E1_1$	Jen et al[52]
GaAsSb MBE	7.54	(100)	$L1_1$	Murgatroid et al[53]
GaAsSb MBE	7.54	(100)	$L1_1$	Ihm et al[54]
InAsSb OMVPE	6.72	(100)	$L1_1$	Jen et al[55]
GaPSb OMVPE	10.7	(100)	$L1_1$ (Very Weak)	Jen et al[56]
InPSb OMVPE	9.9	(100)	$L1_1$	Jen et al[56]
GaAsP OMVPE	3.42	(100)	$L1_1$	Jen et al[57]
CATION SUBLATTICE				
AlGaAs OMVPE	0.16	(100), (110)	$L1_0$	Kuan et al[46]
GaInAs LPE	6.92	(100)	$L1_0$, $E1_1$ (DO_{22}?)[1]	Nakayama et al[58]
GaInAs MBE	6.92	(110)	$L1_0$	Kuan et al[59]
GaInAs(Sb) OMVPE		(100)	$L1_0$, $E1_1$	Jen et al[60]
GaInAs(P) VLE		(100)	$L1_1$	Shahid et al[61]
GaInP OMVPE	7.39	(100)	$L1_1$	Goral et al[62]
GaInP OMVPE	7.39	(100)	$L1_1$	Gomyo et al[48,51]
GaInP OMVPE	7.39	(100)	$L1_1$	Ueda et al[63]
AlInAs OMVPE	6.76	(100)	$L1_1$	Norman et al[64]

their formation. The ordered structures observed by different research groups using various epitaxial growth techniques are summarized in Table 2. It is seen that the chalcopyrite structure, which has the lowest energy in the bulk, is seldom observed. Also unexpected is that the AlGaAs system orders[46], even though there is no elastic energy driving force since AlAs and GaAs have nearly equal lattice parameters. Conversely, GaPSb alloys, with a large size difference, appear to contain compositional modulations due to clustering but no ordering[47].

The CuPt ordered structure is most commonly observed. This is explained in terms of the kinetic processes involved in the formation of the ordered structure on the surface during growth. The surface structure of alternating rows of large and small atoms along the [110] and [1$\bar{1}$0] directions on the (001) surface gives the most favorable surface energy[48]. When the relative displacements from layer to layer occur in a regular pattern, this leads to two variants of the CuPt structure. Suzuki et al.[48] have suggested the involvement of {111} facets on the surface during growth, which have never been observed, to explain the development of the CuPt structure. The substrate misorientation from (001) toward either of the two perpendicular <110> directions will favor the formation of a single variant.

The above explanations may not be sophisticated enough to explain the kinetic effects involved in the formation of ordered structures since the surface is expected to reconstruct. Murgatroid et al.[49] have suggested that the (2x4) reconstructed surface observed for III/V surfaces under MBE conditions may explain the CuPt ordering in MBE grown GaAsSb alloys. The As/Sb-rich surface is suggested to be covered with rows of As-Sb dimers oriented in the same sense. Such a model may explain the CuPt variants observed for the MBE grown GaAsSb, which are the same as those observed by Suzuki et al.[48] for GaInP grown by OMVPE. This is an important observation, which complicates the development of a single compelling model. Either of the existing models predicts the formation of different variants for ordering on the group III and group V sublattices[49]. Thus, the observation of the same variants for GaAsSb and GaInP seems to necessitate two different models.

The formation of ordered structures appears to have significant effects on the electrical and optical properties of III/V alloys. Perhaps the most significant effect is the large change in band gap predicted for various ordered structures. For example, the band gap of $GaAs_{0.5}Sb_{0.5}$ with (111) ordering is predicted to be nearly 0.8 eV lower than that for the (210) ordered structure[50]. A reduction in band gap of nearly 100 meV has been attributed to (111) ordering in $Ga_{0.5}In_{0.5}P$[51].

CONCLUSIONS

Elastic strain energy is seen to be extremely important for the epitaxial growth of III/V alloys. The microscopic strain energy involved in distorting individual bonds to form a disordered alloy is seen to be the principal factor influencing the enthalpy of mixing. This gives rise to positive values of enthalpy of mixing for all III/V alloys with the magnitude increasing with the square of the difference in lattice parameters of the constituent compounds. This gives rise to miscibility gaps in several III/V alloys. It also has a strong influence on the atomic arrangements formed in III/V alloys during growth, leading in many cases to the formation of ordered structures. On a macroscopic scale the strain energy involved in the growth of III/V alloys on a nearly lattice matched substrate has a large effect on solid composition. So-called lattice latching or pulling leads to the growth of nearly lattice matched epitaxial layers over a range of nutrient concentrations which would produce a sizeable variation in the composition of unconstrained crystals. This phenomenon is widely observed for LPE growth and appears to occur to a lesser extent for the MBE and OMVPE processes. The macroscopic strain energy also has the effect of stabilizing nonequilibrium, pseudomorphic phases. One example is the ability to grow alloys within a solid phase miscibility gap by LPE when the alloy is lattice matched to the substrate while the separate, equilibrium phases are not. Similarly, when a nearly random arrangement of the constituents can be trapped on the surface during OMVPE or MBE, leading to the growth of alloys within the miscibility gap, the macroscopic strain energy prevents phase separation. This has led to the growth and characterization of a number of highly metastable III/V alloys.

Acknowledgments

The authors wish to acknowledge the Department of Energy for financial support of much of the work described in this paper.

REFERENCES

1. L.R. Dawson, **J. Cryst. Growth**, 98:220 (1989)
2. J.C. Mikkelson and J.B. Boyce, **Phys. Rev. Letters**, 49:1412 (1982)
3. J.L. Martins and A. Zunger, **Phys. Rev.**, B30:6217 (1984); T. Fukui, **J. Appl. Phys.**, 57:5188 (1985)
4. G.B. Stringfellow, **J. Cryst. Growth**, 27:21 (1974)
5. P.A. Fedders and M.W. Muller, **J. Phys. Chem. Solids**, 45:685 (1984)
6. D.M. Wood and A. Zunger, **Phys. Rev.B**, 40:4062 (1989)
7. G.B. Stringfellow, **J. Appl. Phys.**, 43:3455 (1972)
8. G.B. Stringellow, **J. Cryst. Growth**, 65:454 (1983)
9. G.B. Stringfellow, **J. Cryst. Growth**, 98:108 (1989)
10. C.D. Thurmond, **J. Electrochem. Soc.**, 57:827 (1953)
11. F.A. Trumbore, C.R. Isenberg and E.M. Porbansky, **J. Phys. Chem. Solids**, 9:60 (1958)
12. G.B. Stringfellow, **J. Cryst. Growth**, 27:21 (1974)
13. J.A. van Vechten, **Phys. Rev.**, 170:773 (1968)
14. J.C. Philips, "Bonds and Bands in Semiconducturs", Academic Press, New York, (1973)
15. J.R. Pessetto and G.B. Stringfellow, **J. Cryst. Growth**, 62:1 (1983)
16. T. Kato, T. Matsumoto, T. Kobatake and T. Ishida, **Japan. J. Appl. Phys.**, 26:L1161 (1987)
17. R.E. Nahory, M.A. Pollack, E.D. Beebe, J.C. DeWinter and M. Ilegems, **J. Electrochem. Soc.**, 125:1053 (1978)
18. R. Bruinsma and A. Zangwill, **J. Physique**, 47:2055 (1986)
19. J.P. Hirth and G.B. Stringellow, **J. Appl. Phys.**, 48:1813 (1977)
20. see, for example, J. Ohta, M. Ishikawa, R. Ito, and N. Ogasawara, **Japan. J. Appl. Phys.**, 22:L136 (1983); V.V. Kuznetsov, P.P. Moskvin, V.S. Sorokin, **J. Cryst. Growth**, 88:241 (1988)
21. Yu. B. Bolkhovityanov, **Phys. Stat. Sol.**, (a)76:85 (1983)
22 Y. Takeda and A. Sasaki, **J. Cryst. Growth**, 45:257 (1978); M. Quillec, H. Launois and M.C. Joncour, **J. Vac. Sci. and Tech.**, B1:238 (1983); K. Nakajima, T. Tanahashi, K. Akita and T. Yamaoka, **J. Appl. Phys.**, 50:4975 (1979); P.K. Bhattacharya and S. Srinivasa, **J. Appl. Phys.**, 54:5090 (1983)
23. S. Fukui, M. Tobita, S. Furuta, S. Sakai and M. Umeno, **Japan. J. Appl. Phys.**, 27:379 (1988)

24. A.N. Baranov, S.G. Konnikov, T.B. Popova, V.E. Umansky and Yu. P. Takovlev, **J. Cryst. Growth**, 66:547 (1984)
25. F.C. Larche and J.W. Cahn, **J. Appl. Phys.**, 62:1232 (1987)
26. D.M. Wood and A. Zunger, **Phys. Rev. B**, 38:12756 (1988)
27. F. Turco and J. Massies, **Appl. Phys. Lett.**, 51:1989 (1987)
28. M. Allovon, J. Primot, Y. Gao and M. Quillec, **J. Electron. Mater.**, 18:505 (1989)
29. M.R. Leys, H. Titze, L. Samuelson and J. Petruzzello, **J. Cryst. Growth**, 93:504 (1988)
30. C.F. Schaus, W.J. Schaff and J.R. Schealy, **J. Cryst. Growth**, 77:360 (1986)
31. J.P. Duchemin, J.F. Hirtz, M. Razeghi, M. Bonnet and S.D. Hersee, **J. Cryst. Growth**, 55:64 (1981)
32. M. Quillec, C. Daguet, J.L. Benchimol and H. Launois, **Appl. Phys. Lett.**, 40:325 (1982)
33. M.C. Joncour, J.L. Benchimol, J. Burgeat and M. Quillec, **J. Physique Colloq. C5**, 43:3 (1982); S. Tanaka, K. Hiramatsu, Y. Habu and I. Akasaki, **J. Cryst. Growth**, 87:446 (1988)
34. B. de Cremous, **J. Physique Colloq.**, 43:3455 (1982)
35. G.B. Stringfellow, **J. Electron. Mater.**, 11:903 (1982)
36. R.M. Cohen, M.J. Cherng, R.E. Benner and G.B. Stringfellow, **J. Appl. Phys.**, 57:4817 (1985)
37. G.B. Stringfellow, Chapter 3 **in**: "Organometallic Vapor Phase Epitaxy: Theory and Practice", Academic Press, Boston, (1989)
38. C.B. Cooper, R.R. Saxena and M.J. Ludowise, **J. Electron. Mater.**, 11:1001 (1982)
39. G.B. Stringfellow and M.J. Cherng, **J. Cryst. Growth**, 64:413 (1983)
40. M.J. Cherng, Y.T. Cherng, H.R. Jen, P. Harper, R.M. Cohen and G.B. Stringfellow, **J. Cryst. Growth**, 77:408 (1986)
41. M.J. Jou, Y.T. Cherng and G.B. Stringfellow. **J. Appl. Phys.**, 64:1472 (1988)
42. M.J. Jou, Y.T. Cherng, J.R. Jen and G.B. Stringfellow, **Appl. Phys. Lett.**, 52:549 (1988)
43. M.J. Jou, D.H. Jaw, Z.M. Fang and G.B. Stringfellow, **J. Cryst. Growth**, (to be published)
44. K.Y. Ma, Z.M. Fang, D.H. Jaw, R.M. Cohen, G.B. Stringfellow, W.P. Kosar and D.W. Brown, **Appl. Phys. Lett.**, 55:2420 (1989)
45. J.E. Bernard, R.G. Dandrea, L.G. Ferreira, S. Froyen, S.H. Wei and A. Zunger, **Appl. Phys. Lett.**, 56:731 (1990)
46. T.S. Kuan, T.F. Kuech, W.I. Wang and E.L. Wilkie, **Phys. Rev. Lett.**, 54:201 (1985)
47. H.R. Jen, M.J. Jou and G.B. Stringfellow (unpublished results)
48. T. Suzuki, A. Gomyo and S. Iijima, **J. Cryst. Growth**, 93:396 (1988)
49. I.J. Murgatroyd, A.G. Norman and G.R. Booker, **J. Appl. Phys.**, 67:2310 (1990)
50. S.H. Wei and A. Zunger, **Phys. Rev. B.**, 39:3279 (1989)
51. A. Gomyo, T. Suzuki and S. Iijima, **Phys. Rev. Lett.**, 60:2645 (1988)
52. H.R. Jen, M.J. Cherng and G.B. Stringfellow, **Appl. Phys. Lett.**, 48:1603 (1986)
53. J. Murgatroid, A.G. Norman and G.R. Booker, Materials Research Society Spring Meeting, Palo Alto, CA, April 1986
54. Y.E. Ihm, N. Otsuka, J. Klem and H. Morkoc, **Appl. Phys. Lett.**, 51:2013 (1987)
55. H.R. Jen, K.Y. Ma and G.B. Stringfellow, **Appl. Phys. Lett.**, 54:1154 (1989)

56. H.R. Jen, M.J. Jou and G.B. Stringfellow (unpublished results)

57. H.R. Jen, D.S. Cao and G.B. Stringfellow, **Appl. Phys. Lett.**, 54:1890 (1989)

58. H. Nakayama and H. Fujita, GaAs and Related Compounds, **Inst. of Phys. Conf. Ser.**, 79:289 (1986)

59. T.S. Kuan, W.I. Wang and E.L. Wilkie, **Appl. Phys. Lett.**, 51:51 (1987)

60. H.R. Jen, M.J. Cherng and G.B. Stringfellow, GaAs and Related Compounds, **Inst. of Phys. Conf. Ser.**, 83:159 (1987)

61. M.A. Shahid and S. Mahajan, **Phys. Rev. Lett.**, B38:1344 (1988)

62. J.P. Goral, M.M. Al-Jassim, J.M. Olsen and A. Kibbler, p.583 **in**: "Epitaxy of Semiconductor Layered Structures", Materials Research Soc., Pittsburgh, PA. (1988)

63. O. Ueda, M. Takikawa, J. Komeno and I. Umebu, **Japan. J. Appl. Phys.**, 26:L1824 (1987)

64. A.G. Norman, R.E. Mallard, I.J. Burgatroyd, G.R. Booker, A.H. Moore and M.D. Scott, Microscopy of Semiconducting Materials 1987, **Inst. of Phys. Conf. Ser.**, 87:77 (1987)

STRAIN RELAXATION AND INFLUENCE OF RESIDUAL STRAIN IN GaAs AND
(AlGa)As/GaAs HETEROSTRUCTURES GROWN ON (100) Si BY MOLECULAR
BEAM EPITAXY

W. Stolz,†‡ K. Nozawa,† Y. Horikoshi,†
L. Tapfer‡ and K. Ploog‡

Wiss. Zentrum für Materialwissenschaften und
Fachbereich Physik, Philipps-Universität
' D-3550 Marburg, Federal Republic of Germany
† NTT Basic Research Laboratory
Musashino-shi, Tokyo 180, Japan
‡ Max-Planck-Institut für Festkörperforschung
D-7000 Stuttgart 80, Federal Republic of Germany

ABSTRACT

The optimization of the low temperature growth start in the
heteroepitaxial growth of GaAs on (100) Si by using low As_4/Ga
flux ratios and migration-enhanced epitaxy (MEE) yields smooth
epitaxial layer surfaces. A model for a more controlled and
homogeneous strain relaxation by nucleation of misfit
dislocations at surface steps in a subsequent annealing step is
discussed. Inhomogeneous residual strain deteriorates strongly
the quality of (AlGa)As epitaxial layers. These residual strain
fluctuations as well as the residual density of threading
dislocations are reduced by thermal cycle annealing of 1.5 μm
thick GaAs buffer layers. The above presented different steps
for optimized GaAs buffer layers result in improved quality of
GaAs/AlAs superlattices (SL) as indicated by narrow X-ray
diffraction linewidths of 580 μrad (FWHM) for the GaAs buffer
peak, the GaAs/AlAs SL main peak and SL satellite peaks. Strain
effects resulting from the difference in thermal expansion
coefficients in the GaAs/Si material system for the described
process steps are discussed.

1.   INTRODUCTION

The heteroepitaxial growth of GaAs on Si substrate has
attracted great attention in recent years[1,2], because of the
advantages offered by integrating Si and GaAs based device
structures. In addition to this great potential for device
applications, the GaAs/Si material system has become a model
system for studying the influence of strain properties on the
epitaxial growth mechanism. The understanding of this influence
is of key importance both for the strain-released heteroepitaxy
(as in the GaAs/Si system) as well as for the pseudomorphic
heteroepitaxy (as in the (GaIn)As/GaAs or SiGe/Si material

systems). In the former system the epitaxial growth process should lead to a strain free GaAs layer with a small number of residual threading dislocations, while the strain in the latter systems should be maintained to alter the physical properties of the epitaxial layer structure.

For the GaAs/Si material system additional problems exist despite the difference in lattice constants. These problems are the polar on nonpolar character of the GaAs/Si interface, the difference in thermal expansion coefficients and the possible cross-doping of the two materials[3]. In this article, we will briefly describe optimized growth start conditions with respect to a realization of a charge neutral GaAs/Si interface and of smooth epitaxial layer surfaces. The smoothness of the GaAs epitaxial layer surface is a prerequisite for a more controlled strain relaxation by formation of misfit dislocations at steps on the epitaxial layer surface. This behavior is discussed in the second section of our paper. Experiments for a reduction of residual threading dislocations and of residual strain fluctuations by thermal annealing and thermal cycle annealing steps are described in the third part, leading to an improvement especially of GaAs/AlAs heterostructures on Si substrate. Finally, the influence of the difference in thermal expansion of GaAs and Si on specific process steps is discussed, where one can take advantage of this difference.

All of the above described steps are designed in a way to avoid 'unnecessary' defects being created during epitaxial layer growth. 'Unnecessary' defects are, for example, stacking faults/microtwins and threading dislocations, created by the coalescence of 3D islands during the beginning of the epitaxial growth process[4], if this islanding can be avoided by specific process steps. Another avoidable source of defects are point defects and other structural defects, which are formed by using standard molecular beam epitaxy (MBE) instead of the more suitable migration-enhanced epitaxy (MEE)[5] technique for the growth of the low temperature GaAs prelayer.

2. GROWTH START AND STRAIN RELAXATION FOR GaAs/Si LAYERS GROWN BY MIGRATION-ENHANCED EPITAXY (MEE)

In the beginning of the investigation of the hetero-epitaxial growth of GaAs on Si, standard As-rich growth start conditions were applied[6,7,8]. Often even the high temperature annealing step to clean the Si substrate surface was performed in an As atmosphere. These high temperature (> 600°C) Si annealing steps lead to a stable As layer on the Si surface[9,10]. This stable As layer, covalently bonded to the Si substrate, hinders if not makes it impossible to realize a charge-neutral GaAs/Si interface by the required rearrangement of the interface atoms[11]. This becomes especially true for the low temperature prelayer growth, where, because of the existing energetic barrier, the Si-As bonds can not be broken up. If GaAs growth is started under these conditions immediate islanding of the GaAs epilayer even for layer thicknesses below one monolayer is observed. If the Si substrate is heat treated in an As free atmosphere and exposed to As_4 at low substrate temperatures (< 450°C), however, a different metastable As surface structure exists[10].

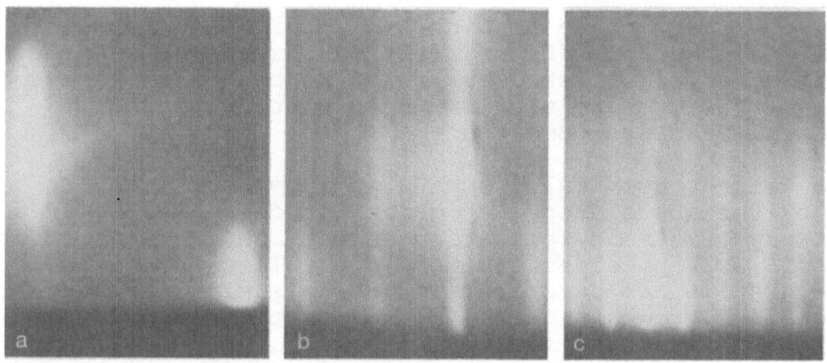

Fig. 1. RHEED pattern of GaAs on Si grown by MEE at
300°C: a) after deposition of two monolayers of
GaAs ([100] azimuth at 300°C), b) and c) after
deposition of 50 nm of GaAs ([1$\bar{1}$0] azimuth and
[1$\bar{1}$0] azimuth, respectively recorded at 580°C).

For nucleation of GaAs on this metastable As-Si surface
structure we observe a clear improvement of the RHEED pattern,
where streaky patterns are maintained after growth starts
(Fig. 1a)[12]. This optimized growth start at low temperature
in conjunction with the use of the MEE growth technique under
small As_4/Ga flux ratios instead of standard MBE yields smoother
GaAs low temperature prelayers with better structural and
optical properties[12]. The improved quality by using MEE[13]
and low As_4/Ga flux ratios[14,15] has also been established by
other groups. As we would like to point out, the thin GaAs
layer (2 monolayers) on Si is not in an equilibrium state. This
becomes evident when this layer is heated up to 600°C. A
transition of this smooth layer to an island-like structure
takes place with a reappearance of the Si bulk RHEED pattern.
Therefore, the observation of the metastable, but smooth low
temperature GaAs layer is not in contradiction to recent
theoretical equilibrium descriptions of the growth mechanism of
GaAs on Si[16]

The interface charge neutrality under the above described
growth start is thought to be realized in a simple picture by
formation of both Ga-Si and As-Si bonds at the interface,
whereas the last Si plane remains unbroken[17]. Evidence for
the formation of both Ga-Si and As-Si bonds at the interface
comes from photoemission core-level spectroscopy[18]. A recent
ab initio total energy calculation[19] of the stability of
various configurations of the first monolayers of GaAs on Si
comes to the conclusion that the lowest energy configuration is
a mixed GaAs bilayer, which completely wets the Si surface.
Further growth on this mixed GaAs layer is modeled by 3D island
nucleation at surface steps[20]. However, the number of nuclei
formed on this surface is found to be a function of the As_4/Ga
flux ratio[14]. In the growth model presented the mobility of
Ga adatoms under low As_4 overpressure is increased and, thus, is
the number of growth nuclei and the lateral growth rate of the
nucleated islands, as also verified experimentally for standard
MBE growth[14]. The enhancement of Ga adatom mobility is even
larger when using the MEE growth technique instead, leading to
smooth epilayer surfaces under a quasi-2D growth mechanism[12].

The absence of anti-phase domains, which might be formed because of this mixed structure of the first monolayers of GaAs on Si, is established by RHEED studies for 50 nm thin GaAs layers at 600°C (Fig. 1b,c), where similar RHEED patterns as for homoepitaxial GaAs are observed.

These almost atomically smooth epilayer surfaces, as realized by the above described growth process, result in a different way of strain relaxation as compared to GaAs 3D island structures[4,22]. Besides the more controlled and homogeneous strain relaxation, as described in the following, the advantage of the quasi-2D growth process is the drastic reduction in micro-twins and stacking faults, when compared to 3D island nucleation[14]. The strain relaxation for the quasi-2D grown GaAs epilayers is investigated by wafer bending, determined by reflection of a laser beam from the surface, X-ray diffraction (XRD) and transmission electron microscopy (TEM) investigations[21].

There exist two possible mechanisms in the GaAs/Si system for post-growth wafer bending. One originates in the difference of the thermal expansion coefficients. The second is an incompletely relaxed GaAs epilayer. These two mechanisms cause opposite ways of wafer bending, as schematically indicated in Fig. 2. The results of the wafer bending experiments are summarized in Fig. 3. Here the inverse of the curvature radius R is shown as a function of the growth temperature for samples, grown on different substrate off-orientations. The expected dependence of 1/R according to the bimetal strip model[23] is also included as a straight line, estimating the bending caused by the difference in thermal expansion alone. The respective bending of the wafer is schematically included in the left part of Fig. 3. The total thickness is 1 μm for all epilayers shown. One series of samples was grown at a fixed temperature (solid

Wafer Bending by
Difference in Thermal Expansion
$\alpha_{GaAs} > \alpha_{Si}$

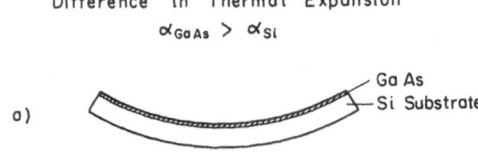

a)

Wafer Bending by
Not Completed Strain Relaxation
$a_{GaAs} > a_{Si}$

b)

Fig. 2. Schematics of wafer bending for the GaAs/Si material system caused by a) difference in thermal expansion after cooling from the growth temperature and b) incomplete strain relaxation in the GaAs epilayer.

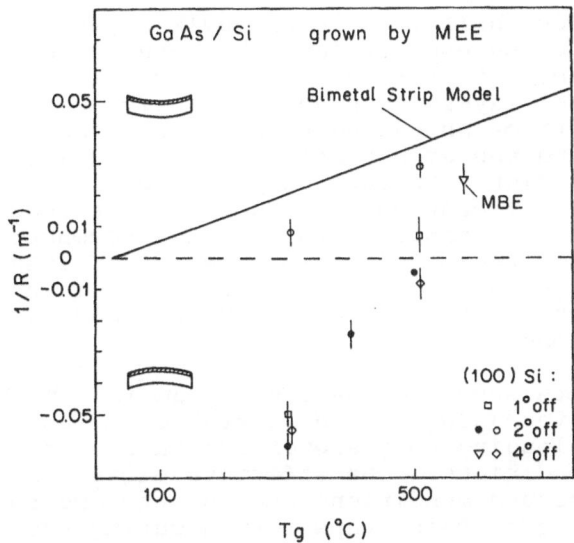

Fig. 3. Inverse of the wafer curvature R as a function
of the growth temperature for different substrate off-orientations. Solid marks indicate
nonannealed samples, open marks samples annealed
at 580°C for 15 min after deposition of a 50 nm
GaAs prelayer at 300°C.

marks), while for the second series, growth was initiated at
300°C. After deposition of 50 nm of GaAs the samples were
heated to 580°C for 15 min. Then, growth was continued at the
respective temperature in the range of 300°C to 580°C. These
samples are represented by open marks in Fig. 3.

The most important results from this investigation are the
reproducible realization of opposite wafer bending, and the
dependence of strain relaxation on the substrate off-
orientation. XRD investigations confirm that GaAs layers are
under compressive strain for wafers with negative curvature
radius. Furthermore, when etching off the GaAs epilayer
completely, the wafer bending is removed. Thus, the observed
opposite wafer bending originates from the still compressively
strained GaAs layer, caused by the incomplete strain relaxation
under the respective growth and/or annealing conditions. The
observation, that strain relaxation in epilayers having
thicknesses exceeding the equilibrium critical thickness, can be
hindered at low substrate temperature is now well established
for strained material systems. More interesting is that
different substrate off-orientations lead to a difference in
strain relaxation. GaAs epilayers, grown on nominally 2° off-
oriented (real value 1.9°) are more relaxed than layers grown on
nominally 1° and 4° off-oriented substrates (real 1.0° and 4.1°
respectively). This demonstrates that surface steps as well as
their spacing influence the strain relaxation mechanism.

Misfit dislocations in epitaxial layers grown on high
quality substrates, having a low density of built-in threading
dislocations, have to be nucleated on the epitaxial layer
surface[24]. On the surface the required energy for dislocation

nucleation is about half that in the bulk crystal. This energy barrier can, thus, be used at low substrate temperatures to inhibit the dislocation formation and to realize thicker strained epitaxial layers than under equilibrium conditions. Besides the growth or annealing temperature, the actual surface reconstruction and the step configuration also influence the dislocation formation. Surface steps could act as nucleation sites for misfit dislocations because of their topological preference. The importance of the surface reconstruction has been shown by the strain relaxation results of InAs grown on GaAs[25], where pseudomorphic growth could be maintained for thicker epilayers under III-stable surface conditions than under V-stable conditions.

The alternate supply of Ga and As_4 during the MEE growth mode leads to alternating Ga and As stable surface conditions, thus eventually leading to a slower relaxation than for standard As stable MBE conditions. The effective migration of Ga atoms under low As pressure conditions smooths the growth surface. The number of GaAs surface steps formed during the subsequent As_4-deposition period should therefore be small. However, the substrate misorientation creates surface steps on the GaAs surface. For the lattice mismatch of GaAs on Si, the average spacing between misfit dislocations in the [011] direction is 9.6 nm. If the dislocations are to be formed at surface steps resulting from the substrate off-orientation, the optimal value for the off-orientation in this crystallographic direction would be 1.7°.

This gives an explanation for the preference of the 1.9° off-oriented wafers. Because of the near matching of the average terrace width and the average dislocation spacing, almost all dislocations can nucleate at steps on the GaAs surface for this off-orientation, thus leading to a locally more homogeneous strain relaxation at an earlier stage of the growth process, as compared to the other off-orientations. The precise microscopic mechanism for this nucleation of misfit dislocations at surface steps is still under investigation. A differing model to that of Maree et al.[24] for SiGe/Si strained layers has recently been published[26]. A discussion of this controversy for the GaAs/Si material system is not attempted here.

However, this more homogeneous strain relaxation can be used to form a regular dislocation network at the GaAs/Si interface in a two-step temperature process for GaAs epilayers grown on properly off-oriented substrates. After growth of 50 nm of GaAs at 300°C by MEE under the above discussed growth start conditions, the sample is heated to standard substrate temperatures of 580°C. During these process steps, misfit dislocations nucleate on the epitaxial layer surface at surface steps and glide on (111) planes to the GaAs/Si interface, thus, forming a dislocation network there. The high resolution TEM micrograph (Fig. 4) shows that the average spacing of dislocation formed at the interface is approximately 10 nm, agreeing favorably with the estimated value of 9.6 nm from the lattice mismatch of GaAs on Si. The formation of this regular dislocation network at the interface explains the decrease in density of residual dislocations in the GaAs epilayer. We would like to point out that this dislocation network is obtained under normal growth conditions, requiring no post-growth high temperature annealing step[27].

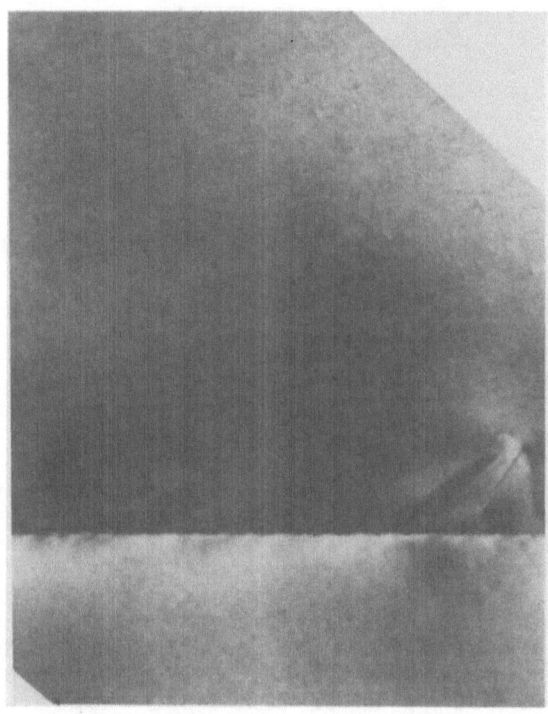

Fig. 4. High resolution cross-sectional transmission
electron microscopy micrograph of the GaAs/Si
interface region.

3.   THERMAL ANNEALING OF THE GaAs BUFFER LAYER

The annealing of the GaAs buffer layer is performed in
order to reduce the number of residual threading dislocations
and the inhomogeneous strain fluctuations still present in the
epitaxial buffer layer. The different annealing steps were
performed after growth of a 1.5 μm thick GaAs buffer layer.
A 1 μm thick GaAs/AlAs SL layer was grown at 600°C to serve as
the active layer used for testing the achieved quality of this
buffer layer. The GaAs buffer was annealed either for 15
minutes at a constant temperature between 600°C and 700°C or by
cyclic thermal annealing between 650°C and 300°C. During the
cyclic annealing the temperature was kept at its maximum for 5
minutes. The XRD pattern recorded in the vicinity of the (400)
GaAs reflection for three of these samples are shown in Fig. 5.
The upper trace is recorded for a sample without annealing or
thermal cycling of the GaAs buffer layer, the central trace for
a sample annealed for 15 minutes at 700°C and the lower one for
a sample which was cycled three times between 300°C and 650°C.

It is obvious from inspection of Fig. 5, that thermal
annealing of the GaAs buffer layer results in improved epitaxial
layer quality, as also observed in the literature (see for
example [27]). However, it is misleading, that the annealing
procedure would lead to identical layer qualities independent of
the number of defects present in the epitaxial layer before the
annealing step. The samples with a higher defect density before

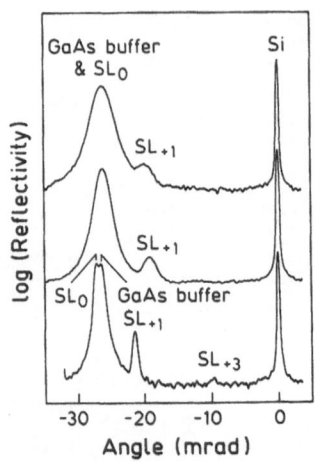

Fig. 5. X-ray diffraction pattern (logarithmic scale) of
1 μm thick GaAs/AlAs superlattices with 1.5 μm
thick GaAs buffer layers grown on (100) Si,
recorded in the vicinity of the (400) GaAs
reflection. Upper trace recorded for sample
without thermal treatment of the GaAs buffer
layer, middle trace for sample with constant
temperature annealing of the GaAs buffer for 15
min. at 700°C, lower trace for sample with
cyclic thermal annealing of the GaAs buffer
layer (three times 650°C-300°C).

the annealing step exhibit also higher defect densities after
the annealing. Thus, the avoidance of the creation of any
unnecessary defects in the growth start, prelayer and buffer
layer growth results in lower defect density material also after
the specific annealing processes.

Marked differences in the structural properties of the
GaAs/AlAs SL structures for the two annealing procedures are
observed: i) the thermally cycled samples exhibit narrower
linewidths, and ii) the SL main peak (SL_0) and the SL satellite
peak (SL_{+1}) have the same linewidth for these samples. In
contrast, for the samples annealed at a constant temperature the
SL satellite reflection is much broader than the main SL peak.
This behavior, although improved in the present unannealed and
annealed samples, has been observed earlier[28] and was
interpreted as an effect of interface roughness and of inferior
quality of the constituent AlAs layer in the SL structure. This
indicates that cyclic thermal annealing conditions are more
efficient in realizing GaAs buffer layers with reduced threading
dislocation densities and residual strain fluctuations than
constant temperature annealing conditions. Similar behavior has
also been observed for MOVPE-grown GaAs/Si layers[29], although
much higher annealing temperatures were used in that case. An
explanation, however, for this difference for constant
temperature and cyclic thermal annealing conditions is still
lacking.

The observed narrow XRD linewidths of 580 to 630 μrad
(FWHM) of the GaAs buffer layer peak and the SL main peak and SL

satellite peak for the thermal cycle samples are to our knowledge the best values reported so far for such thin epitaxial layers, although only three thermal cycles were performed for the samples reported here. The identical linewidths for the SL main peak and the GaAs buffer peak in the XRD pattern for these samples indicate comparable threading dislocation densities in the GaAs buffer and the SL layer. Because of the contribution of dislocations, particle size broadening, and residual strain fluctuations to the linewidth of the XRD pattern[28 and references therein], it is difficult to distinguish between these contributions and to establish precise dislocation densities in these layers. Assuming that the XRD linewidth broadening is caused entirely by dislocations, we obtain an upper limit of $1.5 \ 10^7 \ cm^{-2}$ for the average dislocation density in the 1.5 μm thick GaAs buffer layer. This upper limit for the average dislocation density is a clear improvement over values reported in the literature for such thin epitaxial layers. The identical linewidths of the SL main peak and SL satellite peak indicate improved structural properties, especially of the AlAs layers and the heterointerfaces in the GaAs/AlAs SL grown on Si substrate.

4.   EFFECTS OF THE DIFFERENCE IN THERMAL EXPANSION COEFFICIENTS OF GaAs AND Si

The difference of thermal expansion of GaAs and Si[30, and references therein] is considered to be a problem, because of the straining of the GaAs layer during cooling after growth. This behavior leads to a bending of the Si wafer and is critical to post-growth processing steps for integrated circuits. In the following we would like to briefly discuss two situations, where one can take advantage of the difference in thermal expansion. The first advantage is the additional straining of the GaAs layer during a high temperature annealing step in order to relax more strain than correlated to the epilayer thickness. The second one is the overall smaller amount of stress for certain GaAs/(AlGa)As heterostructures, because AlAs layers, having a larger lattice constant than GaAs, are less strained when grown on thermally strained GaAs/Si than on GaAs substrates.

Because of the smaller thermal expansion of Si, any GaAs layer becomes thermally strained when the substrate temperature is raised. This thermal strain adds to the residual elastic strain in the epilayer and, thus, can be used to relax more strain than that corresponding to the present epilayer thickness. By cooling back to the normal growth temperature this thermal strain is again removed. This leads to the situation that during the subsequent growth of the active device structure at normal substrate temperature the critical thickness for the then smaller amount of strain present in the epilayer may never be reached for proper design of the annealing step. Therefore, no additional dislocations are formed during the growth of the active device structure.

The second advantage of the difference in thermal expansion can be drawn for (AlGa)As/GaAs heterostructures on GaAs/Si. This we would like to briefly discuss using the example of a (AlGa)As/GaAs heterostructure laser. In Fig. 6 the lattice constants of GaAs, the parallel lattice constant of GaAs on Si, and the perpendicular lattice constant of AlAs on GaAs are shown

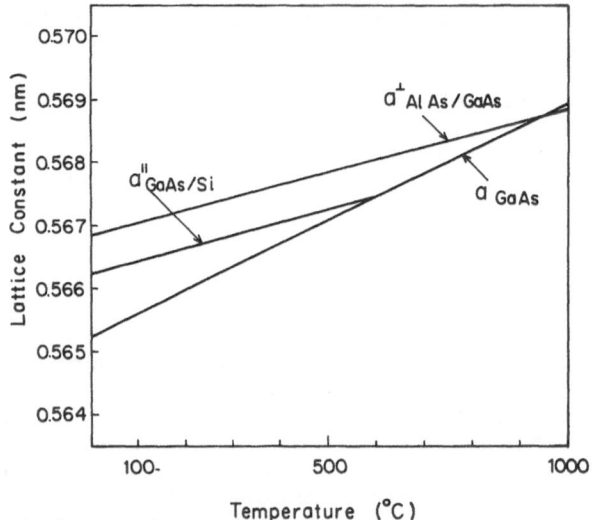

Fig. 6. Lattice parameters of GaAs, of AlAs
(perpendicular lattice constant) on GaAs and of
GaAs (parallel lattice constant) on Si as a
function of the temperature. The AlAs layer on
GaAs is assumed to be elastically strained,
while the GaAs epitaxial layer on Si is assumed
to be completely relaxed at the growth
temperature (600°C).

as a function of the temperature. The AlAs epilayer on GaAs
substrate is assumed to be elastically strained, while the GaAs
layer on Si is assumed to be completely relaxed at the growth
temperature of 600°C and becomes thermally strained upon cooling
to room temperature. From inspection of Fig. 6, it is obvious
that the (AlGa)As confinement layers are under strain at room
temperature for laser structures on GaAs substrate, whereas the
GaAs active region is strain-free. For the same structure on
GaAs/Si the situation is reversed. In this case the GaAs active
region is strained and the lattice constant parallel to the
surface is almost identical to the unstrained lattice constant
of AlAs at room temperature. This implies that the (AlGa)As
optical confinement layers are almost unstrained. As the active
layer thickness is small compared to the thickness of the
(AlGa)As confinement layers this could lead to a smaller amount
of stress for the same structure grown on GaAs/Si than on GaAs
substrate and, thus could lead to an improved stability of the
heterostructure at room temperature.

CONCLUSIONS

     Optimized substrate preparation and low temperature growth
start by using MEE leads to an improved quasi-2D growth
mechanism. This results in smooth epitaxial layer surfaces with
a smaller amount of stacking faults and point defects in the
initial stage of growth. Strain relaxation is found to depend
on the substrate off-orientation. A model for nucleation of
misfit dislocations at epilayer surface steps and subsequent
glide to the GaAs/Si interface is discussed. Formation of a

regular dislocation network at the interface under standard growth conditions is realized without a post-growth high temperature annealing step. Different thermal annealing processes for reduction of residual threading dislocations and strain fluctuations are presented, leading to a clear improvement especially of (AlGa)As/GaAs heterostructures on Si substrates. Finally, two positive aspects of the difference of thermal expansion of GaAs and Si are discussed.

Acknowledgment

Expert technical support by A. Fischer and J. Knecht is gratefully acknowledged. Part of this work has been sponsored by the Bundesministerium für Forschung und Technologie of the Federal Republic of Germany.

REFERENCES

1. Mat. Res. Soc. Symp. Proc. Vol. 67, "Heteroepitaxy on Silicon", J.C.C. Fan and J.M. Poate, eds., Materials Research Society, Pittsburgh (1986)
2. Mat. Res. Soc. Symp. Proc. Vol. 91, "Heteroepitaxy on Silicon", J.C.C. Fan, J.M. Phillips and B.Y. Tsaur, eds., Materials Research Society, Pittsburgh (1987)
3. H. Kroemer, **J. Cryst. Growth**, 81:193 and references therein (1987)
4. H.L. Tsai and Y.C. Kao, **J. Appl. Phys.**, 67:2862 (1990)
5. Y. Horikoshi, M. Kawashima and H. Yamaguchi, **Japan. J. Appl. Phys.**, 27:169 (1988)
6. W.I. Wang, **Appl. Phys. Lett.**, 44:1149 (1984)
7. R. Fischer, W.T. Masselink, J. Klem, T. Henderson, T.C. McGlinn, N.V. Klein, J.H. Masur and J. Washburn, **J. Appl. Phys.**, 58:374 (1985)
8. T.C. Chong, C.G. Fonstad, **J. Vac. Sci. Techn.**, B5:815 (1987)
9. R.I.G. Uhrberg, R.D. Bringans, R.Z. Bacharach and J.E. Northrup, **Phys. Rev. Lett.**, 56:520 (1986)
10. M. Kawabe and T. Ueda, **Japan. J. Appl. Phys.**, 26:L114 (1987)
11. W.A. Harrison, E.A. Kraut, J.R. Waldrop and R.W. Grant, **Phys. Rev. B**, 18:4402 (1978)
12. W. Stolz, M. Naganuma and Y. Horikoshi, **Japan. J. Appl. Phys.**, 27:L283 (1988)
13. J. Varrio, H. Asonen, A. Salokatve, M. Pessa, E. Ranhala and J. Kleinonen, **Appl. Phys. Lett.**, 51:1801 (1987)
14. J.E. Palmer, G. Burns, C.G. Fonstad and C.V. Thompson, **Appl. Phys. Lett.**, 55:990 (1989)
15. N. Chand, F. Ren, A.T. Macrander, J.P. van der Ziel, A.M. Sergent, R. Hull, S.N.G. Chu, Y.K. Chen and D.V. Lang, **J. Appl. Phys.**, 67:2343 (1990)
16. J.E. Northrup, **Phys. Rev. Lett.**, 62:2487 (1989)
17. W. Stolz, Y. Horikoshi, M. Naganuma and K. Nozawa, Proc. 5th Int. Conf. Molecular Beam Epitaxy, Sapporo (1988), **J. Cryst. Growth**, 95:87 (1989)
18. R.D. Bringans, M.A. Olmstead, R.I.G. Uhrberg and R.Z. Bachrach, **Phys. Rev.B**, 36:9569 (1987)
19. E. Kaxiras and J.D. Joannopoulos, **Surface Science**, 224:515 (1989)
20. E. Kaxiras, O.L. Alerhand, J.D. Joannopoulos and G.W. Turner, **Phys. Rev. Lett.**, 62:2484 (1989)

21. W. Stolz, Y. Horikoshi and M. Naganuma, **Japan. J. Appl. Phys.**, 27:L1140 (1988)
22. R. Hull and A. Fischer-Colbrie, **Apply. Phys. Lett.**, 50:851 (1987)
23. S. Timoshenko, **J. Opt. Soc. Am.**, 11:233 (1925)
24. P.M.J. Maree, J.C. Barbour, J.F. van der Veen, K.L. Kavanagh, C.W.T. Bulle-Lieuwma and M.P.A. Viegers, **J. Appl. Phys.**, 62:4413 and references therein (1987)
25. M. Munekata, L.L. Chang, S.C. Woronick and Y.H. Kao, **J. Cryst. Growth**, 81:237 (1987)
26. D.J. Eaglesham, E.P. Kvam, D.M. Maher, C.J. Humphreys and J.C. Bean, **Phil. Mag.**, A59:1059 (1989)
27. J.W. Lee, H. Shichiijo, H.L. Tsai and R.J. Matyi, **Appl. Phys. Lett.**, 50:31 (1987)
28. L. Tapfer, J.R. Martinez and K. Ploog, **Semicond. Sci. Technol.**, 4:617 (1989)
29. H. Okamoto, Y. Watanabe, Y. Kadota and Y. Ohmachi, **Japan. J. Appl. Phys.**, 26:L1950 (1987)
30. Thermophysical Properties of Matter Vol. 13, Thermal Expansion Nonmetallic Solids, Y.S. Touloukian, R.K. Kirby, R.E. Taylor and T.Y.R. Lee, eds., Plenum Pbl. Corp., New York (1977)

STRAIN RELAXATION BY MISFIT DISLOCATIONS

M. Mazzer

Department of Physics
University of Padova
Via Marzolo 8, 35131 Padova, Italy

1. INTRODUCTION

The mechanism of strain release in heteroepitaxial mismatched structures of semiconductors by misfit dislocations (MD) is still not well known. The so called equilibrium theories[1,2] correctly predict the critical film thickness above which pseudomorphic structures are no longer stable and misfit dislocations appear at the interface between film and substrate. However, above the critical thickness value the experimental rate of strain release as a function of the thickness is much lower than predicted by these theories. Some authors suggest that the measured strain may be greater than the predicted one since the system may be in a metastable equilibrium state because of the low rate of strain release at room temperature[3,4] or because the activation energy for the nucleation of new dislocations is not available[5]. However, experiments with samples having different misfit values and growth rates show that the residual strain is likely only to depend on the film thickness[6], so the samples should be close to a stable equilibrium state which should be different from the one depicted by the equilibrium theories.

An accurate description of the effects of the MDs on the strain field in the film is necessary in order to get more insight into the mechanism of strain relaxation. Recent experiments have detected asymmetries in the residual strain field, i.e. MDs may induce deformations which can be appreciably different from the simple tetragonal distortion[7]. This fact should be related to asymmetries in the distributions of the dislocations and/or of their Burgers vectors. In fact it is known that the MDs are not homogeneously distributed in particular at the first stage of the generation process[8].

In the first part of this paper the experimental data concerning a set of $In_xGa_{1-x}As/GaAs$ samples analyzed by means of both RBS-channeling and Double Crystal X-ray Diffractometry (DCD) is presented in order to study the effective dependence of the tetragonal part of the strain field on the layer thickness. Then by means of a model based on the linear elasticity theory

Condensed Systems of Low Dimensionality
Edited by J. L. Beeby *et al.*, Plenum Press, New York, 1991

the calculation of the strain tensor is performed for small misfit systems without the assumption of simple tetragonal distortion. In this way it is possible to get some interesting information about the link between the dislocation distribution and geometry and the structure of the film primitive cell. For this purpose a method for RBS-channeling and DCD data reduction which allows one to obtain all the parameters describing the strain tensor is proposed. The results of Kavanagh et al.[7] are then interpreted in the frame of this model and important indications about the possible evolution of the layer lattice structure as the MD generation process goes on are obtained.

2. TETRAGONAL STRAIN RELEASE AS A FUNCTION OF THE LAYER THICKNESS

One of the more important conclusions of the large amount of experimental work on the strained heterostructures of semiconductors is that the rate of tetragonal strain release as a function of the layer thickness, above the critical value predicted by the equilibrium theories, is lower than it is expected to be. Furthermore, there is evidence that the same curve of strain release is followed by samples with different growth rate and initial misfit, i.e. the residual strain must be a function of the thickness only. This behavior is confirmed by a set of $In_xGa_{1-x}As/GaAs$ samples grown by MBE at CSELT Laboratory (Torino) and analyzed at Laboratori Nazionali di Legnaro (Padova) and at MASPEC (Parma) by means of RBS-channeling and DCD respectively[6]. Channeling and DCD data were elaborated under the assumption of tetragonal distortion so that the information obtained (Table 1) was substantially the so-called parallel strain $\epsilon_\parallel$ in the layer, that is the component $\epsilon_{xx} = \epsilon_{yy}$ of the strain tensor in a frame of reference with the z axis perpendicular to the interface. These data, together with the results of other authors[9-11], are then reported in Fig. 1 in order to stress that we can hardly speak of metastable equilibrium state for this structure.

If the samples are in equilibrium then the Matthews models for the MDs generation mechanism underlying the equilibrium theories is likely to neglect something important such as, for example, the interaction between the dislocation lines. It has been shown that the direction of a MD may change abruptly because of the interaction with other dislocations[12]. In this way the equilibrium between the self-energy of the dislocation lines and the elastic strain energy per unit area in the layer at a given thickness may be reached at higher values of the strain. The same effects may also be responsible for the asymmetries in the strain release in InGaAs/GaAs hetero-structures detected by Kavanagh et al. which are related to the distribution of the MDs at the interface. As a first step toward a deeper understanding of these phenomena it is then necessary to study how the film lattice may be distorted in the presence of MDs beyond the hypotheses of simple tetragonal distortion.

3. ELASTIC DISTORTION FIELD IN THE PRESENCE OF MISFIT DISLOCATIONS

When a strained heteroepitaxial structure is pseudomorphic the film is distorted in such a way that only one parameter, the

Table 1. Summary of the strain measurements. The film
thickness and the In composition are those
measured by RBS. The alloy misfit is
calculated from the measured composition and
Vergard's law (see Ref. 6).

Sample	Thickness (nm)	Growth rate (nm/sec)	Composition (%)	Misfit (%)	$-\epsilon_{\parallel}$ (Chann) (%)	$-\epsilon_{\parallel}$ (DCD) (%)
S2	408	0.135	3.5	0.25±0.02	0.25±0.02	
S4	314	0.114	10.3	0.74±0.04	0.38±0.03	
S6	26	0.114	11.0	0.79±0.04	0.77±0.04	
S7	68	0.114	10.4	0.75±0.04	0.68±0.04	0.66±0.04
S8	310	0.114	8.5	0.61±0.04	0.37±0.03	0.31±0.04
S9	90	0.114	7.5	0.54±0.04	0.57±0.04	0.48±0.03
S10	36	0.114	8.0	0.57±0.07	0.55±0.03	
S11	20	0.114	8.0	0.57±0.04	0.57±0.05	
S12	810	0.281	8.0	0.57±0.04	0.19±0.02	
S13	1156	0.281	8.0	0.57±0.04	0.17±0.02	0.21±0.03
S14	200	0.281	8.0	0.57±0.04	0.40±0.02	
S15	345	0.330	15.0	1.07±0.04	0.40±0.03	0.33±0.04
S16	45	0.330	15.0	1.07±0.04	1.16±0.06	
S17	23	0.330	15.0	1.07±0.04	1.11±0.08	

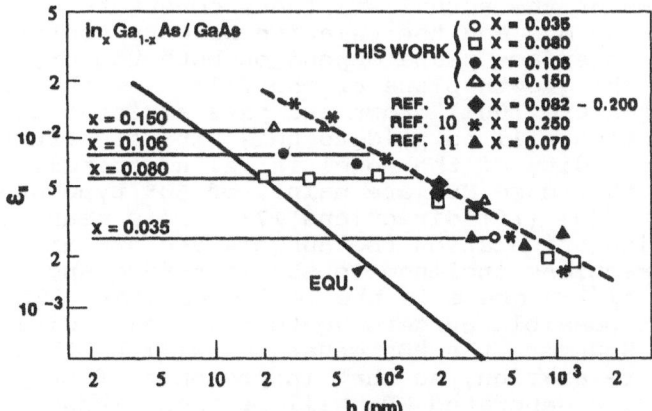

Fig. 1. Parallel strain measured by ion channeling for
all samples is reported as a function of the
epilayer thickness. Horizontal lines are drawn
at the value corresponding to the misfit of the
various In compositions. Also drawn is the
prediction of the equilibrium theory (see Ref.
2) for single layers. The results of other
authors are reported for comparison (see Ref.6).
The dashed line through the data is drawn to
show the lower slope with respect to the
equilibrium theory.

misfit, is sufficient to describe the system. In other words, the strain tensor is diagonal and the primitive cell of the film is tetragonally distorted. MDs induce a plastic distortion in the film which can be described in terms of a distortion tensor $u_{ij}^p = \partial u_i^p / \partial x_j$, [13] where u^p is the displacement field, superimposed on the pre-existing elastic strain field f_{ij}. In general u_{ij}^p is neither diagonal nor homogeneous[14] as the field generated by a single dislocation depends strongly on the distance from the dislocation core. For large misfit systems, growth modes are three-dimensional and the strain relaxation mechanism may not be easy to describe but in low misfit single layer structures growth modes are two-dimensional and MDs lie at the interface between film and substrate[15]. In the following only small misfit systems will be considered. In this case the greater the distance from the interface plane the smaller the amplitude of the field oscillation due to discreteness of the dislocation distribution. It can be shown that the field dependence on the spatial coordinates can be neglected outside a "transition region" extending from the dislocation network up to a distance comparable to the dislocation spacing[14]. In this limit, as will be shown in the following, the total distortion field $u_{ij} = f_{ij} + u_{ij}^p$ depends on six parameters, three of which enter into the symmetrical part ϵ_{ij} of the tensor (the strain tensor) describing the primitive cell deformations, while the others are related to three independent small rotations of the film lattice as a whole with respect to the substrate. All these parameters turn out to depend only on the distribution of the dislocation Burgers vectors.

The calculation may be performed with the help of a model based on the linear elasticity theory where the layer is treated as a semi-infinite elastic continuum. The anisotropy of the real structure can be accounted for by expressing the isotropic elastic constants μ and ν (the shear modulus and the Poisson's ratio) in terms of the actual stiffness constants of the crystal[16]. The MDs and the relative Burgers vectors distribution on the interface plane depend on both the crystal structure and the growth plane of the film. As the most common heterostructures are (001) grown and have a zinc-blende type lattice, attention will be paid to this kind of structure although the validity of the model is quite general. It is well known that in this case MDs are mainly of 60° type and lie in $[\bar{1}10]$ (t') and $[1\bar{1}0]$ (t'') directions[17]. This means that for each dislocation orientation the Burgers vectors are along the four <110> directions inclined to the interface and have magnitude $B = a/\sqrt{2}$ where a is the cell parameter. This makes a total of eight possible Burgers vectors for each dislocation orientation. However, the MDs generated will be those inducing strain energy relaxation, so that in presence of compressive misfit strain the generated MDs will be those inducing tensile strain and conversely. The Burgers vectors can be divided into two sets as illustrated in Fig. 2[14].

The first set contains the vectors $\mathbf{B}'_{1,2} = 1/2\ a(1,0,\pm1)$, $\mathbf{B}'_{3,4} = 1/2\ a(0,-1,\pm1)$ for direction t' and $\mathbf{B}''_{1,2} = 1/2\ a(-1,0\ \pm1)$, $\mathbf{B}''_{3,4} = 1/2\ a(0,-1,\pm1)$ for direction t'' and is associated with MDs generating tensile strain. The other set contains the vectors opposite to those of the first set and corresponds to MDs generating compressive strain. For both t' and t'' directions MDs are supposed to make up a superposition of many arrays containing an infinite number of equally spaced

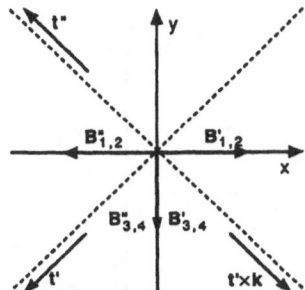

Fig. 2. The projection on the (001) interface of
the eight possible Burgers vectors
relative to dislocation lines generating
tensile strain in a zinc-blende type
crystal. The dashed lines represent the
dislocation orientations.

dislocations having the same Burgers vectors. If the Burgers
vectors are distributed among the four possible values B'_i
(B''_i), i = 1,.,4 with density n(B'_i) (n(B''_i)) we can define

$$\mathbf{b'} = \sum_i \mathbf{B'}_i n(\mathbf{B'}_i) \text{ and } \mathbf{b''} = \sum_i \mathbf{B''}_i n(\mathbf{B''}_i)$$

as the mean Burgers vectors per unit length in the two
directions. Summing up the contributions of all the dislocation
arrays the total plastic distortion field u^p_{ij} outside the
transition region in the film is found to be[14]:

$$\overline{\overline{u}}^p = \frac{1}{2} \begin{bmatrix} b_\perp + \delta_\parallel & \delta b_\perp & 0 \\ \delta b_\perp & b_\perp - \delta b_\parallel & 0 \\ 0 & 0 & -\alpha b_\perp \end{bmatrix} + \frac{1}{2} \begin{bmatrix} 0 & b_\parallel & \sqrt{2}\delta b_z \\ -b_\parallel & 0 & \sqrt{2} b_z \\ -\sqrt{2}\delta b_z & -\sqrt{2} b_z & 0 \end{bmatrix} \quad (1)$$

where $\alpha = 2\nu/(1 - \nu)$. If $\mathbf{z}$ is the normal to the interface
plane oriented as the growth direction the six parameters in the
previous equation are defined as follows:

$$b_\parallel = \mathbf{b''} \cdot \mathbf{t''} + \mathbf{b'} \cdot \mathbf{t'}$$

$$\delta b_\parallel = \mathbf{b''} \cdot \mathbf{t''} - \mathbf{b'} \cdot \mathbf{t'}$$

$$b_\perp = \mathbf{b''} \cdot (\mathbf{z} \wedge \mathbf{t''}) + \mathbf{b'} \cdot (\mathbf{z} \wedge \mathbf{t'}) \quad (2)$$

$$\delta b_\perp = \mathbf{b''} \cdot (\mathbf{z} \wedge \mathbf{t''}) - \mathbf{b'} \cdot (\mathbf{z} \wedge \mathbf{t'})$$

$$b_z = (\mathbf{b''} + \mathbf{b'}) \cdot \mathbf{z}$$

$$\delta b_z = (\mathbf{b''} - \mathbf{b'}) \cdot \mathbf{z} \quad .$$

The geometrical meaning of the parameters in Eqn. 1 may be
understood with the help of Fig. 3a where the contribution of
the $\mathbf{t'}$ oriented dislocations is shown.

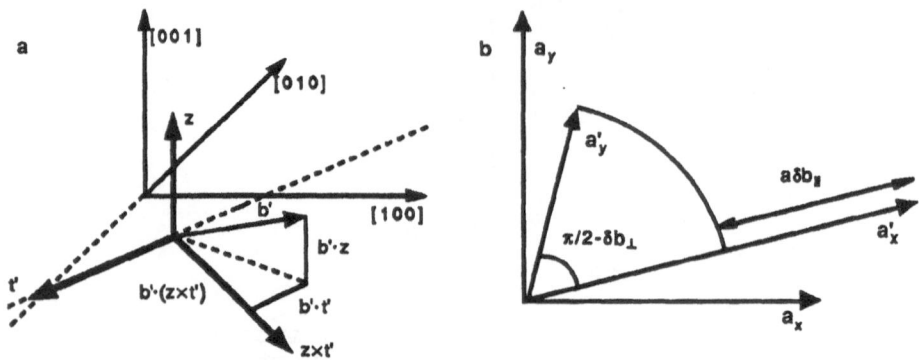

Fig. 3. (a) The mean Burgers vector for **t'** oriented
dislocations is decomposed into three components
in order to show the different contributions to
the elastic distortion field. (b) The
deformation of the base of a simple cubic
lattice primitive cell due to a nonbiaxial
strain field. The difference between the screw
component of the total Burgers vector per unit
length relative to the dislocation lines
oriented as **t'** and those oriented as **t"** is
proportional to the difference between the
moduli of the transformed primitive vectors $\mathbf{a'}_x$
and $\mathbf{a'}_y$ differs from a right angle by an amount
$\delta b_\perp$ (the effects of the deformation are
exaggerated for the sake of clarity).

The first term in Eqn. 1 is the plastic strain $\bar{\epsilon}^p$ which
describes the MDs contribution to the primitive cell
deformations. In addition to the well known tetragonal
distortion, described by the parameter $b_\perp$, two additional
independent deformations appear. In fact if $\mathbf{a}_x = a\mathbf{x}$, $\mathbf{a}_y = a\mathbf{y}$
and $\mathbf{a}_z = a\mathbf{z}$ are the primitive vectors of a simple cubic cell in
an undeformed lattice and $\mathbf{a'}_x$, $\mathbf{a'}_y$, $\mathbf{a'}_z$ are the same vectors
under the action of the total strain $\epsilon_{ij} = f_{ij} + \epsilon^p_{ij}$, then (see
Fig. 3b) it is easy to show that the angle θ between $\mathbf{a'}_x$ and
$\mathbf{a'}_y$ and the difference $\mathbf{a'}_x - \mathbf{a'}_y$ between the norms of the same
vectors are given by $\pi/2 - \delta b_\perp$ and $a\delta b_\parallel$ respectively.

It is interesting to note that these two parameters are
simply related to the dislocation densities. In fact, let δn be
the difference between the MDs density in **t"** and in **t'**
directions and δ^*n the difference between the total number of
dislocations per unit length (no matter what their direction)
having the Burgers vector perpendicular to the y axis and to the
x axis respectively. Bearing in mind the sets of possible
Burgers vectors given above, we find:

$$\delta n = n' - n" = \sum_{i=1}^{4} n(\mathbf{B"}_i) - \sum_{i=1}^{4} n(\mathbf{B'}_i) = \frac{2}{B}\delta b_\perp \qquad (3)$$

and:

$$\delta*n = n_{1,2} - n_{3,4} = \sum_{i=1}^{2} [n(B'_i) + n(B''_i)] - \sum_{i=3}^{4} [n(B'_i) + n(B''_i)] = \frac{2}{B} \delta b_\parallel \ . \qquad (4)$$

The connection between the deformations of the film primitive cell and the dislocations distribution is completed by the tetragonal distortion which depends of course on the total MDs density; i.e.:

$$n = n' + n'' = n_{1,2} + n_{3,4} = \sum_{i=1}^{4} [n(B'_i) + n(B''_i)] = \frac{2}{B} b_\perp \ . \qquad (5)$$

The parameters b_z, δb_z, and $b_\parallel$ in the second term of Eqn. 1 are proportional to the rotation angles of the whole film around the undistorted crystallographic directions [100], [010] and [001] respectively. Therefore the effects of the plastic distortion described by the antisymmetric part of the tensor $\bar{u}^p$ are the same that could be obtained by applying a proper external torque to the film.

4. CHARACTERIZATION OF HETEROSTRUCTURES IN THE PRESENCE OF A NONTETRAGONAL DISTORTION

For a complete characterization of this kind of structure by means of techniques such as RBS-channeling and DCD, it is necessary to perform a minimum number of independent measurements in order to determine an equal number of significative parameters.

The channeling technique allows one to know the structure of the strained film primitive cell by determining a set of lattice axial or planar directions[18]. These directions cannot be referred to the underlying structure of the substrate. In fact, because of the steering effect, the channeling condition for the incident beam can be determined with sufficient precision for the film structure but not for the substrate. Therefore only the symmetric part ϵ_{ij} of the elastic distortion tensor (i.e. $b_\perp$, $\delta b_\perp$ and $\delta b_\parallel$) can be determined. If the strain field is not biaxial, this requires at least three independent measurements but actually a greater number of experimental points (dip minima) are collected and properly fitted in order to improve the precision.

Two physical quantities which can be directly related to the elastic distortion field can be measured by means of DCD for each set of reflection planes. They are the difference $\Delta d/d$ of the lattice plane spacing between the substrate and the film and the angle $\Delta\chi$ between the normal γ' to the reflection plane in the epitaxial layer and that in the substrate (γ)[19]. The first quantity depends only on the strain field, i.e. on the parameters f (the misfit), $b_\perp$, $\delta b_\perp$ and $\delta b_\parallel$, while the other one depends on the whole elastic distortion field. In polar coordinates we have $\gamma \equiv (\cos\phi\sin\theta, \sin\phi\sin\theta, \cos\theta)$, and then we find[14]:

$$\frac{\Delta d}{d}(\gamma) = - (\alpha + 1) \left(f + \frac{b_\perp}{2} \right) \cos^2\theta + \frac{1}{2} [b_\perp + \sin^2\theta \; \zeta(\phi)] \qquad (6)$$

where

$$\zeta(\phi) = \delta b_\parallel \cos(2\phi) + \delta b_\perp \sin(2\phi) \; . \qquad (7)$$

From Eqns. 6,7 we can see that (nnm) or (nn̄m) type reflections (n and m integers) where ϕ is $\pm \pi/4$ or $\pm 3\pi/4$, allow the determination of $\delta b_\perp$ as follows:

$$\left[\frac{\Delta d}{d} \right]_{\phi = \frac{\pi}{4} + k\frac{\pi}{2}} - \left[\frac{\Delta d}{d} \right]_{\phi = \frac{3\pi}{4} + k\frac{\pi}{2}} = \sin^2\theta \; \delta b_\perp \qquad (8)$$

while for (0nm) or (n0m) reflections we have:

$$\left[\frac{\Delta d}{d} \right]_{\phi = k\frac{\pi}{2}} - \left[\frac{\Delta d}{d} \right]_{\phi = \frac{\pi}{2} + k\frac{\pi}{2}} = \sin^2\theta \; \delta b_\parallel \qquad (9)$$

and then $\delta b_\parallel$ can be easily obtained.

All these reflections together with a symmetric reflection ($\theta = 0$) then allow the calculation of both f and $b_\perp$, i.e. the tetragonal part of the deformation since:

$$\left[\frac{\Delta d}{d} \right]_{\phi} + \left[\frac{\Delta d}{d} \right]_{\phi + \frac{\pi}{2}} = - 2(\alpha + 1) \left(f + \frac{b_\perp}{2} \right) \cos^2\theta + b_\perp \; . \qquad (10)$$

Finally, the other three parameters may be determined only by using the $\Delta\chi$ data.

5. DISCUSSION

Experimental data showing a nonbiaxial strain release are presented in a recent paper by Kavanagh et al.[7]. They detected large in-plane strain asymmetries in $In_xGa_{1-x}As/GaAs$ films thicker than 300 nm (with x = .07) where the dislocation densities are rather large and equal, within the experimental error, in each <110> direction. As was shown in the previous paragraph the condition $\delta n \approx 0$ implies that $\delta b_\perp \approx 0$ then asymmetries are possible only if $\delta b_\parallel \neq 0$. It follows that the base of the simple cubic primitive cell should be of rectangular shape and in terms of dislocation densities we should have:

$$\begin{cases} n'_{12} + n'_{34} \approx n''_{12} + n''_{34} \\ n'_{12} + n''_{12} \neq n'_{34} + n''_{34} \end{cases} \qquad (11)$$

showing that the component of the Burgers vectors parallel to the interface plane should be oriented mainly in one direction.

The same authors reported that thinner films, having the same composition, behave in the exactly opposite way. Despite the dislocation density being highly different for the two directions, this was not so for the strain relief. This can be explained within our model. Since the dislocations lie predominantly in one direction (for instance t') then $\delta b_\perp \approx -b_\perp \neq 0$ so that the base of the cubic cell becomes of rhomboid shape. However, its sides may have the same length if $\delta b_\parallel \approx -b_\parallel \approx 0$ or equivalently:

$$n'_{12} \approx n'_{34} \tag{12}$$

that is, there should be no asymmetry in the Burgers vectors distribution.

We have looked for an analogous behavior among our samples. A further set of channeling and DCD measurements was made on samples S7 and S14, presented above in this paper, in order to determine the nondiagonal components of the strain tensor. However, despite these samples being expected to have the highest asymmetries, the measured strain fields turned out to be essentially biaxial within the experimental errors[14].

As shown in Ref. 14, it is worth noting that if the dislocation interaction is neglected the elastic energy density per unit volume in the layer is a minimum when the distortion field is purely tetragonal. Then an equilibrium state characterized by a nonbiaxial strain field may be justified only by taking into consideration the energy contribution from the dislocation interaction.

CONCLUSION

A model for an accurate description of the strain relaxation by misfit dislocations in heteroepitaxial structures has been presented. For (001) grown single layers having zinc-blende structure and low misfit it has been shown how the distortion of the layer lattice depends on both the MD density at the interface and the mean Burgers vectors per unit length in the two <110> directions. The plastic distortion field generated by the dislocations depends on six parameters, three of which enter into the symmetric part of the distortion tensor (the strain tensor) describing the primitive cell deformations. The film deformation is simply tetragonal if, and only if, the mean Burgers vectors per unit length have the same screw component and the same projection of the edge component to the interface. These conditions are not generally satisfied so the base of the deformed primitive cell may be rectangular or rhomboid in shape. These "asymmetries" in the strain relaxation may be detected by means of the usual characterization techniques and appropriate methods for data reduction in the case of RBS-channeling and DCD have been suggested.

From the experimental point of view there are two important things. Kavanagh et al. detected relevant strain asymmetries in some of their samples. Furthermore, Drigo et al., in agreement with other authors, found that the tetragonal part of the residual strain is always greater than the predicted equilibrium value without any evidence that the samples were in a metastable equilibrium state. The interaction between the

469

dislocations may be responsible for this behavior as it may
raise the energy value necessary to generate a new MD. Work is
in progress in order to obtain further information about the
mechanism of the MDs generation from the first stage just beyond
the critical thickness until the strain relaxation is almost
completed. The dislocation and the relative Burgers vectors
distributions will be analyzed independently of the strain
measurements by means of Transmission Electron Microscopy (TEM)
and by the Dechanneling technique.

REFERENCES

1. J.H. Van der Merwe and C.A. Ball, in: "Epitaxial Growth",
    part B ch.6, J.W. Matthews, ed., Academic, New York,
    (1975)
2. J.W. Matthews and A.E. Blakeslee, **J. Cryst Growth**, 27:118
    (1974)
3. B.W. Dodson and J.V. Tsao, **Appl. Phys. Lett.**, 51:1325
    (1987)
4. B.W. Dodson and J.V. Tsao, **Appl. Phys. Lett.**, 52:852 (1988)
5. B.W. Dodson, **Phys. Rev. Lett.**, 35:5558 (1987)
6. A.V. Drigo, A. Aydinly, A. Carnera, F. Genova, C. Rigo,
    C. Ferrari, P. Franzosi and G. Salviati, **J. Appl. Phys.**,
    66:1975 (1989)
7. K.L. Kavanagh, M.A. Capano, L.W. Hobbs, J.C. Barbour,
    P.M.J. Marée, W. Schaff, J.W. Mayer, D. Pettit,
    J.M. Woodall, J.A. Stroscio and R.M. Feenstra, **J. Appl.
    Phys.**, 64:4843 (1988)
8. H. Strunk, W. Hagen and E. Bauser, **Appl. Phys.**, 18:67
    (1979)
9. K. Kamigaki, H. Sakashita, H. Kato, M. Nakayama, N. Sano
    and H. Terauchi, **Appl. Phys. Lett.**, 49:1071 (1986)
10. P.J. Orders and B.F. Uscher, **Appl. Phys. Lett.**, 50:980
    (1987)
11. P.M.J. Marée, J.C. Barbour, J.F. Van de Veen, **J. Appl.
    Phys.**, 62:4413 (1987)
12. V.I. Ddovin, L.A. Matveeva, G.N. Semonova, M. Ya. Skorohod,
    Yu. A. Tkhorik and L.S. Khazan, **Phys. Stat. Sol. (a),**
    92:379 (1985)
13. A.M. Kosevich in: "Dislocations in Solids" (vol.1), F.R.N.
    Nabarro, ed., North Holland, Amsterdam (1978)
14. M. Mazzer, A. Carnera, A.V. Drigo and C. Ferrari, **J. Appl.
    Phys.**, 68:531 (1990)
15. G. Salviati, C. Ferrari, A.V. Drigo, F. Romanato and F.
    Genova, Proceed. 17th Congress of the Italian Electronic
    Microscopy Society, Lecce 4-7 Oct. 1989
16. J. Hornstra and W.J. Bartels, **J. Cryst. Growth**, 44:513
    (1978)
17. S. Amelinckx, in: "Dislocations in Solids (II)", F.R.N.
    Nabarro, ed., North Holland, Amsterdam (1979)
18. A. Carnera and A.V. Drigo, **Nucl. Instr. Meth.**, B44:357
    (1990)
19. K. Ishida, J. Matsui, T. Kamejima and I. Sakuma, **Phys.
    Stat. Sol. (a)**, 31:255 (1975)

STRAINED LAYER HETEROSTRUCTURES AND SUPERLATTICES BASED ON GROUP

IV ELEMENTS

G. Abstreiter, K. Eberl, E. Friess, U. Menczigar and
W. Wegscheider

Walter Schottky Institut
Technische Universität, München, D-8046 Garching, FRG

ABSTRACT

Heterostructures and superlattices based on the group IV
elements Si, Ge and α-Sn are fabricated by low temperature
molecular beam epitaxy. The main problems of high quality
growth are the large lattice mismatch and the strong
interdiffusion and segregation. We have performed detailed
investigations of growth mode, interface sharpness, critical
thickness of the Si/Ge and Ge/α-Sn systems. LEED, TEM,
Auger- and Raman spectroscopy are used as "in situ" and "ex-
situ" analytical tools. The achievement of short period superlattices with nearly atomically sharp interfaces is possible.

1.   INTRODUCTION

Si based heterostructures and superlattices are of
considerable interest both due to their fundamental physical
properties and their potential future applications. A
combination of Si, Ge, α-Sn and perhaps even SiC and diamond
like C opens various possibilities for band structure
engineering which eventually might lead to applications in
ultrafast electronics, optoelectronics and optics. Hetero-
bipolar transistors, high electron mobility transistors,
resonant tunneling devices, infrared- and far-infrared detectors
and perhaps even light emitting devices are among the most
promising candidates[1-3]. Most of these novel devices require
high quality interfaces and sharp doping profiles. One of the
major problems for heteroepitaxy of the group IV elements is the
large lattice mismatch. This is shown in Fig. 1 where the
energy gaps and lattice constant of C, SiC, Si, Ge and α-Sn are
given. Also included in the figure are some of the most
important III-V semiconductors. The full circles indicate
semiconductors with a direct band gap, the open ones have an
indirect fundamental gap. Apart from the large lattice mismatch, which leads to extremely small critical layer thicknesses, the large difference of the melting points and the
tendency of segregation and interdiffusion require nonequilibrium epitaxial growth conditions and in some cases also a

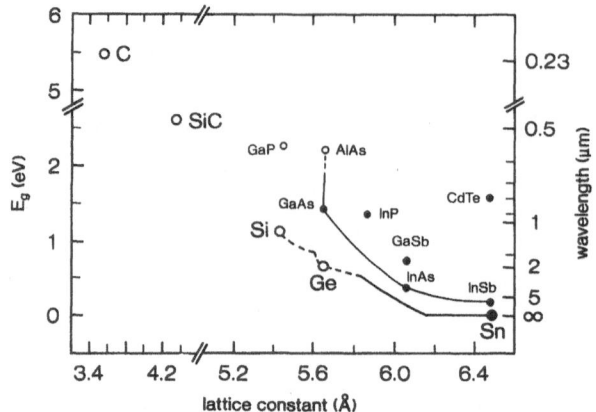

Fig. 1. Energy gaps and lattice constant of C,
SiC, Si, Ge, α-Sn and some of the most
important III-V semiconductors.

large temperature variation during growth. We have used a
special low temperature molecular beam epitaxy with in situ
analysis to optimize the growth conditions. Selected results
are presented in the main part of this contribution. The phonon
properties of ultra short period superlattices are studied by
Raman spectroscopy. They are discussed briefly in the last
section.

2.   LOW TEMPERATURE MOLECULAR BEAM EPITAXY

A modified molecular beam epitaxy technique is used for
nonequilibrium growth of group IV heterostructures and
superlattices. The growth chamber is equipped with LEED and AES
in order to obtain in situ information on the crystalline
quality and the chemical composition of the deposited films.
The substrates can be heated by radiation and cooled alternatively with water or liquid nitrogen which allows rapid
temperature (T_g) variations during growth. This turned out to
be an essential feature especially for the synthesis of Sn/Ge
superlattices. Typical growth temperatures for Si/Ge and Sn/Ge
superlattices were in the range of 250°C - 400°C and 50°C -
350°C respectively. The pressure during growth is within the
10^{-11} mbar region at growth rates as low as 5 Å/min. Si(001)
and Ge(001) as well as SiGe buffer layers which provide a
lattice constant intermediate between that of Si and Ge are used
as substrates.

The formation of the Si/Ge(001) and Ge/Si(001) heterostructures was investigated with LEED and AES to find the
optimum growth temperature[4]. Figure 2 shows the condensate/
substrate ratio of the AES intensities of the Si(92 eV) and
Ge(47 eV) lines as a function of the growth temperature for a)
one ML of Ge on Si(001) and b) one ML of Si on Ge(001). The
escape length of the Auger electrons is 5.5 Å for Si(92 eV) and
7.5 Å for Ge(47 eV). The dashed and the pointed arrows mark the
expected Auger ratio of ideal atomically sharp overlayers. For
one ML Ge on Si(001) substrate we observe a deviation from the
ideal ratio at T_g > 330°C. At higher temperatures, intermixing

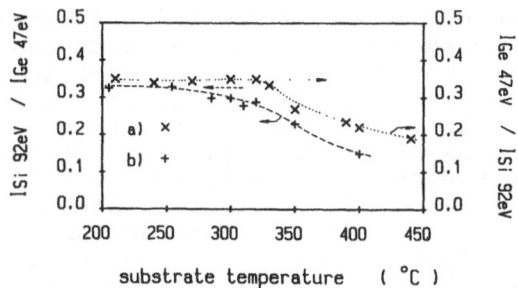

Fig. 2. Condensate/substrate ratio of the normalized AES
intensities of the Si(92 eV) and Ge(47 eV) lines
as a function of the growth temperature for a)
one ML of Ge on Si(001) and b) one ML of Si on
Ge(001). The dashed and the pointed arrow mark
the expected Auger ratios for the related ideal
heterostructures.

or three-dimensional growth occurs. A quantitative evaluation
of the energy dependence of the LEED spots shows that there is
an optimum in surface flatness between T_g = 310°C and T_g = 360°C
for one ML Ge on Si(001). For growth temperatures higher than
400°C there is enhanced spot broadening in the LEED pattern
which indicates increased surface roughness. The AES data for
one ML Si on Ge(001) substrate in Fig. 2 show a deviation from
the ideal curve (dashed arrow) already at T_g = 280°C. This is
mainly due to Ge segregation on the Si layer. At higher
temperatures considerable intermixing must also be taken into
account. The mean terrace width on the surface on one ML Si on
Ge(001) is considerably larger than in the inverse case of one
ML Ge on Si[4,5]. The results in Fig. 2 and similar AES
measurements within pseudomorphic Si/Ge superlattices on Si and
Ge substrates indicate, that in order to achieve atomically flat
interfaces, T_g should be kept slightly below 280°C during the
formation of the Si/Ge interface, whereas a temperature of about
310°C to 320°C leads to the best Ge/Si interfaces.

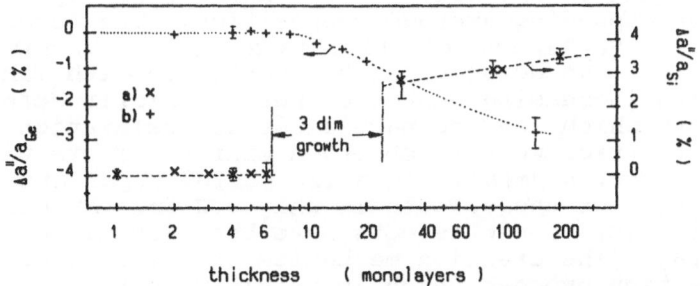

Fig. 3. Relative change of the lateral lattice constant
as a function of thickness for a) Ge on Si(001)
and b) for Si on Ge(001). The measurement was
performed using a vidicon based LEED system.
The growth temperature was T_g = 310°C.

LEED can also be used as a sensitive instrument to follow the relaxation process with increasing thickness. Figure 3 shows the relative change of the lateral lattice constant as a function of thickness for a) Ge on Si(001) and b) for Si on Ge(001). The Ge film is lattice matched to the Si substrate up to 6 ML. But already after the deposition of the first ML Ge an increased surface roughness is observed which leads finally to three-dimensional (3D) growth (for more than 6 ML Ge). For thicker Ge layers ($\geq$ 30 ML Ge) we find an improvement of the surface quality at T_g = 310°C. For 30 ML Ge we observe a lateral lattice constant which is about 3% larger than that of the Si substrate, as determined from the decrease of the LEED spot separation. With increasing thickness of the Ge layer the surface quality improves as indicated by the error bars in Fig. 3. Simultaneously the lattice constant changes towards that of Ge (complete relaxation). The width of the region of 3D growth and the relaxation level $\delta a/a_{Si}$ depends strongly on the substrate temperature[6].

The behavior of $\delta a/a_{Ge}$ for Si overlayers is demonstrated in Fig. 3b. Up to about 8 ML the Si film is laterally extended by 4.2% (lattice matched to the Ge substrate). For higher coverages the Si overlayer begins to relax by the formation of misfit defects. In contrast to Ge on Si there is no pronounced 3D growth and faceting of the Si surface at T_g = 310°C. Reasonable surface flatness is obtained up to three ML of Si. There was no detectable additional broadening of the LEED spots, contrary to the case of Ge on Si, where after 1 ML already a slight broadening of the LEED spots is observed. In both cases (Si on Ge and Ge on Si) an analysis of the LEED spot profile shows that the surface of a 6 ML film exhibits many random steps of different atomic heights, but it still adapts to the lattice constant of the substrate.

Based on such detailed LEED and AES investigations we prepared short period $Si_m Ge_n$ superlattices on Si and Ge substrates (sequence of alternating m MLs Si and n MLs Ge). Figure 4a shows a TEM micrograph of a 20 period $Si_3 Ge_9$ superlattice of high structural quality[7]. In this sample the 3 ML Si layers are 4.2% laterally extended, whereas the Ge layers are not strained. Besides the critical thickness for lattice-matched growth of the individual Si layers there is a second critical thickness for the superlattice as a whole, which is roughly equal to the critical thickness of a $Si_x Ge_{1-x}$ alloy with the corresponding average composition. The total thickness of a 20 period $Si_3 Ge_9$ superlattice is about 330 Å, far below the critical value. No defects can be identified with TEM in this sample. With increasing number of periods misfit defects appear in the layers which lead to partial strain relaxation. In a 40 period superlattice we have observed misfit defects with a mean distance of about 5 μm[8]. In a 120 period superlattice the mean separation of the defects is only 0.2 μm. An example is shown in Fig. 4b. These defects have been identified as microtwins[8]. The creation mechanism of these defects provides a new relaxation process which is peculiar to tensile strain[8]. The equilibrium lattice constant of the $Si_3 Ge_9$ structure is 1% less than that of the Ge substrate. In the 120 period $Si_3 Ge_9$ superlattice relaxation $\delta a/a_{Ge}$ is between 0.44 to 0.88% as determined by LEED and TEM[5]. Thus the superlattice is not yet fully relaxed. In the inset of Fig. 4 we show also a selected area electron diffraction pattern. Additional spots appear

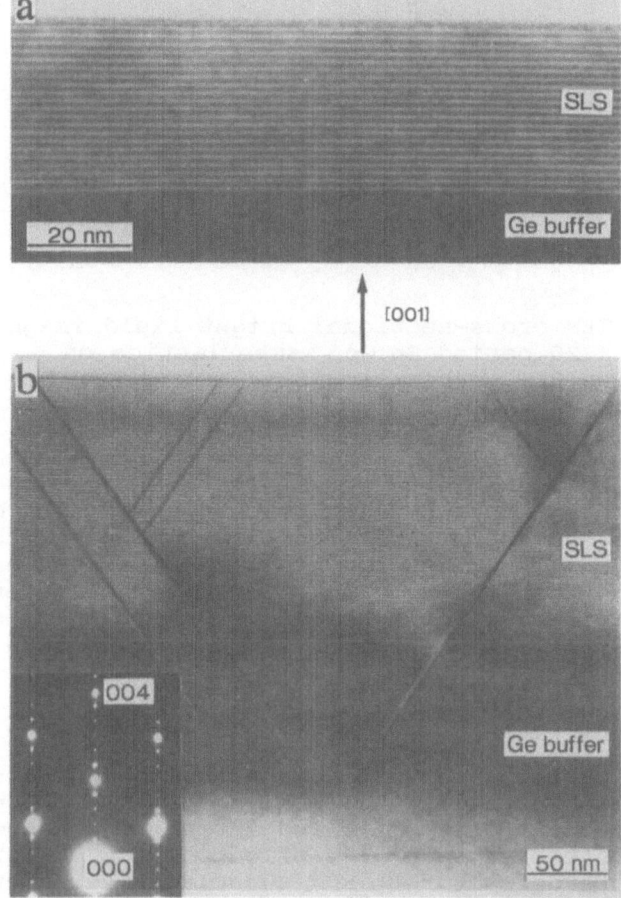

Fig. 4. TEM cross-sectional bright field images of
Si$_3$Ge$_9$ superlattices on Ge(001); a) 20 periods
thick, lattice matched to Ge; b) 120 periods
thick and partly relaxed superlattice. The
inset shows a selected area diffraction pattern
of the superlattice in the [110] pole.

in-between the main diffraction points which are due to the
superlattice periodicity of 12 ML.

Apart from the Si/Ge system we have also tried to achieve
high quality α-Sn/Ge superlattices[9]. Besides the problems
caused by the large lattice mismatch there exist considerable
other difficulties which must be overcome in preparing Sn/Ge
structures on Ge substrates. A basic problem is for example the
strong tendency of Sn to segregate on the surface of the Ge film
at temperatures even as low as 150°C[10]. Figure 5 shows a
cross sectional TEM micrograph of a 20 period Sn$_2$Ge$_{20}$
superlattice on Ge(001) substrate. During the deposition of the
individual layers, the substrate temperature was varied between
50°C to 230°C in order to avoid Sn segregation. The clearly
distinguishable dark and bright lines correspond to the Sn and
Ge layers respectively. The structure is lattice matched to Ge.

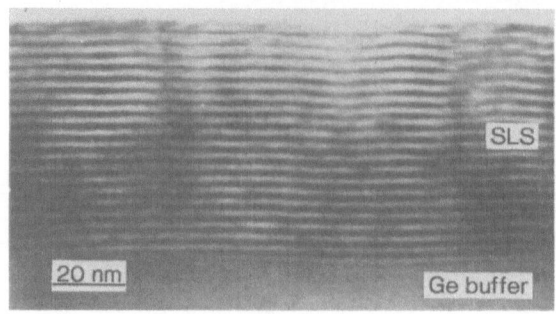

Fig. 5. TEM cross-sectional bright field image of
a 20 period Sn_2Ge_{20} superlattice on
Ge(001) substrate.

The realization of such Sn/Ge short period strained layer
superlattices was possible by a modified MBE technique with
large modulation of the substrate temperature during growth
which can be reached by a very effective cooling of the
substrate holder. Important features are also the very low
growth rates and the extremely low growth temperatures. This
opens new possibilities for nonequilibrium growth of semiconductors or metals with widely different properties.

3.   RAMAN SCATTERING

     The short period superlattices have also characteristic
phonon properties which lead to detailed information on superlattice period, built-in strain, and interface roughness. Thus
phonon Raman scattering is a versatile tool for the analysis of
thin layers, heterostructures, and superlattices. Raman spectra
of Si_nGe_n superlattices with nominal values of n = 3,6,12 are
compared with that of a corresponding $Si_{0.5}Ge_{0.5}$ alloy in Fig. 6.
New features appear in the wave number range below 250 cm^{-1} due
to so-called folded acoustic phonons[11]. The enlargement of
the primitive unit cell of the superlattices in growth direction
leads to a reduced Brillouin zone in the corresponding direction
in k-space. This results effectively in a folding of the
acoustic branch with the average sound velocity of Si and Ge.
These modes can be treated in an elastic continuum model. In
backscattering from (100) surfaces mainly the folded LA modes
are observed. The folded acoustic modes shift to higher
energies with decreasing period, but they can be clearly
resolved even for the sample with an individual layer thickness
of only 3 ML. For the larger period superlattices additional
higher order folded acoustic modes are observed whose
intensities depend on the quality of the interface and the exact
ratio of the layer thicknesses.

     In the wavenumber range above 250 cm^{-1} optical phonons are
observed due to Si-Si, Ge-Ge and Si-Ge vibrations. The
intensity of the latter one is very sensitive to intermixing and
interface roughness. The Si-Ge mode is weakest in the Si_6Ge_6
superlattice. The considerable increase of this mode in the
Si_3Ge_3 superlattice reflects the limitations of interface sharpness which is achievable in this system even at the low growth

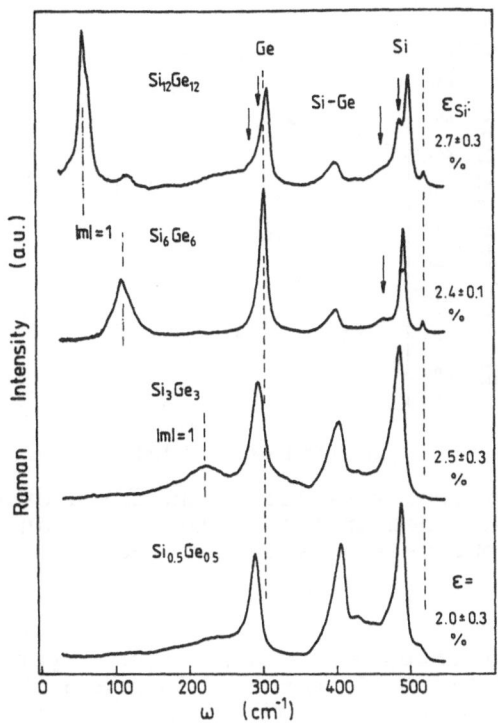

Fig. 6. Raman spectra of three Si_nGe_n superlatices with
n = 3,6,12 and an alloy with the same Si and Ge
content.

temperature used in our studies. There are, however, still
clear differences in the optical phonon spectra as compared to
the random alloy. In the $Si_{12}Ge_{12}$ superlattice the Si-Ge mode
is also slightly increased with respect to the Si_6Ge_6 sample.
This is due to the increased interface roughness when the
individual layer thickness gets too large. The modes around 500
cm^{-1} are confined optical modes in the Si slabs. They are well
confined because there is no overlap of the Si and Ge LO phonon
branches. At the low energy side of the main structure there
appear for the 12 ML Si two additional peaks and for the 6 ML Si
one additional peak. These optical phonon modes reflect
standing waves which fit into the Si layers[12]. The splitting
of the modes depends mainly on the thickness of the layers and
reflects the bulk dispersion of the optical phonon branches. It
is more difficult to resolve these modes in Ge because of the
weaker LO phonon dispersion. In addition the Ge LO phonon modes
overlap with the LA modes of Si. The interaction between these
modes results in a finite dispersion of the Ge LO modes[13]. It
has been shown, however, that these effects are small and they
follow roughly also the bulk LO phonon dispersion, although
details of the observed higher order confined modes (marked by
arrows in Fig. 6) indicate deviations due to the finite
dispersion[12].

In addition to these confinement effects a further energy
shift has to be taken into account due to the built-in lateral
strain. The samples whose Raman spectra are shown in Fig. 6

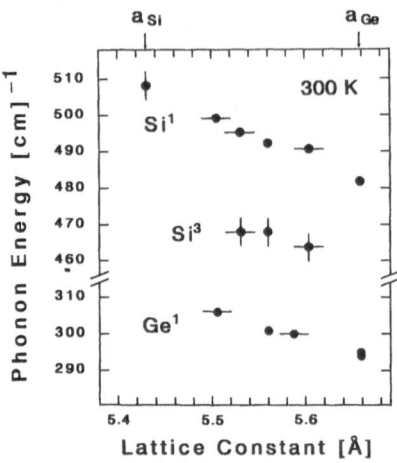

Fig. 7. Phonon energies of first and third order
        confined modes in 6 ML thick slabs as a function
        of the built-in lateral lattice constant.

were grown on a strain symmetrizing buffer layer. Therefore ϵ_{Si}
is of the order of 2% which causes a downward shift of the Si
optical modes and an upward shift of the Ge modes with respect
to the unstrained situation. This is shown more clearly in Fig.
7 where the energies of the first and third order confined modes
of 6 ML Si slabs and the first order quasi-confined mode of 6ML
Ge slabs are plotted versus lateral lattice constant which can
be adjusted by a buffer layer and the overall thickness of the
superlattices. A linear relationship is observed for all modes
within the experimental resolution in reasonable agreement with
recent calculations[14]. A different slope as predicted
theoretically for the Si^3 and Si^1 mode, and which reflects the

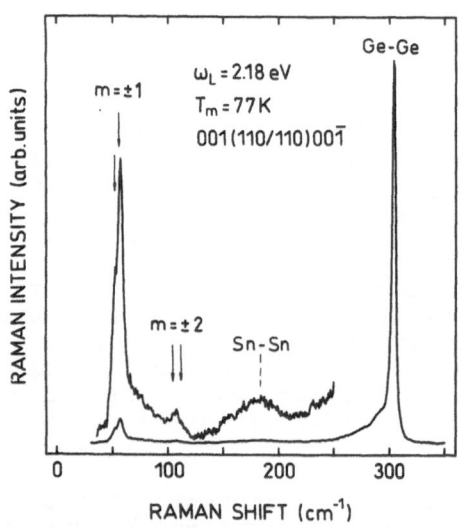

Fig. 8. Raman spectrum of a 20 period Sn_2Ge_{20}
        superlattice grown on a Ge(001) substrate.

strain dependent Si LO phonon dispersion, cannot be resolved unambiguously within the uncertainty of our experimental results.

Finally we show as a first result a Raman spectrum of a nominally Sn_2Ge_{20} superlattice (Fig. 8) which exhibits very similar features as discussed for the short period Si/Ge superlattices. The most prominent peak at about 300 cm^{-1} is attributed to the Ge LO phonon in the Ge slabs. The folded LA doublets (m = ±1, m = ±2) confirm the achievement of a periodic compositional modulation of this structure as shown already by TEM (see Fig. 5). The arrows mark the expected positions according to the elastic continuum theory and the assumption of a period length of 22 ML and a weighted averaged sound velocity of Ge and α-Sn. The agreement between theory and experiment is excellent. The broad peak slightly below 200 cm^{-1} is assumed to originate from Sn-Sn optical vibrations. Its position and width reflect the built-in strain in the Sn layers, but also the intermixing with the Ge layers which is also evident from the TEM analysis of these superlattices.

CONCLUDING REMARKS

We have demonstrated that short period superlattices of lattice-mismatched group IV elements can be achieved with nearly atomically flat interfaces by low temperature molecular beam epitaxy. This opens a wide field of band structure engineering which eventually might lead also to novel future device applications.

Acknowledgment

We are grateful to H. Cerva and H. Oppolzer of the Siemens research laboratories for the excellent collaboration and help in the TEM analysis of the short period superlattices and to M. Bichler for his expert help in the MBE technology. The work benefitted from the financial support of the Siemens AG and the Deutsche Forschungsgemeinschaft.

REFERENCES

1. E. Kasper **in**: "Heterostructures on Silicon: One Step Further with Silicon", Y.I. Nissim and E. Rosencher, eds., Kluwer Academic Publishers, Dordrecht, NATO Series E, **Applied Sciences**, 160:101 (1988) and references therein.
2. S. Luryi and S.M. Sze, pp182-240 **in**: "Silicon Molecular Beam Epitaxy", E. Kasper and J.C. Bean, eds., CRC Press, Boca Raton (1988), and references therein.
3. T.P. Pearsall, pp 451-459 **in**: "Silicon Molecular Beam Epitaxy", E. Kasper and E.H.C. Parker, eds., North-Holland, Amsterdam (1990) and references therein.
4. K. Eberl, W. Wegscheider, G. Abstreiter, to be published in **J. Cryst. Growth**
5. K. Eberl, W. Wegscheider, E. Friess and G. Abstreiter, p 153 **in**: "Heterostructures on Silicon: One Step Further with Silicon", Y.I. Nissim, E. Rosencher, eds., Kluwer Academic Publishers, (1989)
6. K. Eberl, E. Friess, W. Wegscheider, U. Menczigar and G. Abstreiter, **Thin Solid Films**, 183:95 (1989)

7. W. Wegscheider, K. Eberl, H. Cerva, H. Oppolzer, **Appl. Phys. Lett.**, 55:448 (1989)

8. W. Wegscheider, K. Eberl, G. Abstreiter, H. Cerva, H. Oppolzer, Proc. of the MRS Spring Meeting 1990, San Francisco.

9. W. Wegscheider, K. Eberl, U. Menczigar and G. Abstreiter, **Appl. Phys. Lett.**, 57:875 (1990)

10. P.R. Pukite, A. Harwit and S.S. Iyer, **Appl. Phys. Lett.**, 54:2142 (1989)

11. See for example, H. Brugger, E. Friess, G. Abstreiter, E. Kasper and H. Kibbel, **Semicond. Sci. Technol.**, 3:1166 (1988); also, D.J. Lockwood, p.481 of this volume and references therein.

12. E. Friess, K. Eberl, U. Menczigar and G. Abstreiter, **Solid State Commun.**, 69:899 (1989)

13. E. Molinari and A. Fasolino, **Appl. Phys. Lett.**, 54:1220 (1989)

14. A. Fasolino, E. Molinari and A. Qteish, p.495 of this volume.

RAMAN SPECTROSCOPY OF Si-Ge ATOMIC LAYER SUPERLATTICES

D.J. Lockwood

Physics Division
National Research Council
Ottawa K1A 0R6, Canada

ABSTRACT

Raman spectroscopy has proved to be a valuable technique for investigating the dynamical and structural properties of thin layer heterostructures. We review here the application of the technique to $(Ge_m Si_n)_p$ atomic layer superlattices with $m,n \leq 12$ and $p < 100$. A series of strong broad peaks are seen in the low frequency spectra of these structures that are due to resonant acoustic-phonon modes. The positions and widths of the peaks are sensitive to the superstructure total thickness and the surface boundary condition. The superlattice folded-acoustic and optic phonon peak parameters vary considerably with m,n and the substrate/buffer layer material. Modeling of the spectra reveals the differences between the actual and target structures.

1. INTRODUCTION

Raman spectroscopy has proved to be a versatile non-destructive technique for investigating the dynamical and structural properties of thick layer heterostructures. In the case of superlattices, for example, the folded acoustic-phonon spectrum can be used to obtain information on the individual layer thicknesses, the superlattice period, the alloy composition, and the interface quality, while the optic mode spectrum provides information on the local structure and the strain within the layers[1,2]. Theoretical models used to interpret the Raman spectra are generally in good accord with experiment.

Recently, attention has turned to studying ultrathin or atomic layer superlattices (ALS) of the form $(A_m B_n)_p$ comprising p repeats of m,n monolayers of material A,B. Such ALSs are also different from the thick layer case, because, in essence, for small m,n new crystallographic unit cells are being formed along the growth direction[3]. There have been numerous Raman studies of $[(GaAs)_m (AlAs)_n]_p$ and related systems, especially of the optic modes under resonance excitation conditions[4]. Here, I

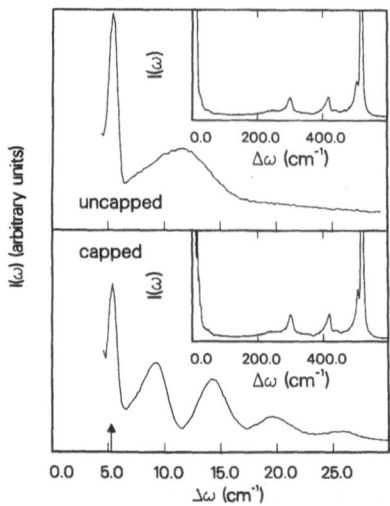

Fig. 1. Experimental Raman spectrum of $(Ge_2Si_2)_{10}$ on
Si(100) without and with a thin Si capping
layer. The arrow indicates the Brillouin peak
(after Ref. 7).

review recent developments in Raman studies of the somewhat
simpler (monatomic) system $(Ge_mSi_n)_p$ with emphasis on
experimental results. Theoretical treatments of the vibrational
properties of such ALSs have been dealt with in greater detail
in the accompanying paper by A. Fasolino et al.[5] and
elsewhere[6].

2.   ACOUSTIC PHONONS

     Consider, for example, a $(Ge_2Si_2)_{15}$ ALS grown on Si. A
Ge_2Si_2 unit cell replaces the substrate Si_4 cell along the (100)
growth direction, and the folded acoustic modes of such a thin
structure are expected at optical mode frequencies ($\omega > 100$ cm^{-1}
say). Somewhat surprisingly, Dharma-wardana et al.[7] have
observed a number of intense low frequency modes (apart from the
usual Brillouin line) with $\omega < 100$ cm^{-1} whose frequency and
width depend on the presence or absence of a Si capping layer
(see Fig. 1). These resonant phonon modes arise from the
interaction between the continuum of acoustic modes in the Si
substrate and the quasi-localized (slab) modes in the
superlattice (plus cap).

     A simple linear-chain dynamical model including only
nearest neighbor interactions proved sufficient to calculate the
observed spectrum[7]. The Si substrate was modeled sufficiently
well by 3000 atom layers in the calculation with the ALS (plus
cap) on top. The force constant between Si substrate layers
spaced 1.36 Å apart was chosen to be $f_{Si} = 1.62 \times 10^5$ dynes/cm
to give the observed Brillouin frequency of 5.3 cm^{-1}. The three
force constants corresponding to the interactions between
adjacent Ge-Ge, Ge-Si and Si-Si layers spaced 1.46, 1.41 and
1.36 Å apart, respectively, in the ALS were equated to one force
constant, f_{ALS}, since the long-wavelength acoustic modes are

482

insensitive to the microscopic details. The equation of motion for the surface atom layer, ν, is

$$m_{Si} \ddot{z}_\nu = f_{ALS}(z_{\nu-1} - z_\nu) - f_{ALS}\sigma z_\nu \tag{1}$$

where m_{Si} is the Si mass, z the displacement in the growth direction, and parameter σ is zero and one for free and anchored surfaces, respectively. The Raman intensity in the photoelastic coupling model for acoustic phonons[8] is given by

$$I(\omega) \ \alpha \ \sum_i \frac{\Gamma}{(\omega-\omega_i)^2+\Gamma^2} \left[\frac{n(\omega_i)+1}{\omega_i} \right] \left| \int_{-\infty}^{+\infty} e^{-iqz} P(z) \frac{\partial u_i(z)}{\partial z} \, dz \right|^2 \tag{2}$$

where ω_i is the mode frequency, u_i the mode eigenvector, $P(z)$ the photoelastic constant, $n(\omega_i)$ the Bose factor, and Γ a damping parameter set at 0.5 cm^{-1}. The calculated spectra shown in Fig. 2 were obtained with $f_{ALS} = 0.44 f_{Si}$ and are in good agreement with experiment. To obtain agreement with the observed intensities the photoelastic constant of the ALS, P_{ALS}, had to be increased over that of the substrate value P_{Si}: for the uncapped ALS, P_{ALS} was set at $10P_{Si}$, whereas for the capped ALS values of $P_{ALS} = 50P_{Si}$ (!) and $P_{cap} = 10P_{Si}$ were needed. The calculated spectrum is also quite sensitive to the surface boundary conditions. The best fit with experiment was obtained with a nearly free surface ($\sigma = 0.1$).

3.   FOLDED ACOUSTIC MODES

Folded acoustic modes in Si-Ge ALSs have been observed in a wide variety of samples by a number of laboratories[9]. A typical spectrum is shown in Fig. 3, where several orders of the folded acoustic modes of a $(Ge_8Si_{12})_p$ ALS grown on a strain-compensating buffer layer of $Ge_{0.6}Si_{0.4}$ are visible[10]. The total thickness of the ALS was 0.2 μm corresponding to $p \approx 70$. According to the elastic continuum theory of Rytov[11], the phonon dispersion is given approximately by[8]

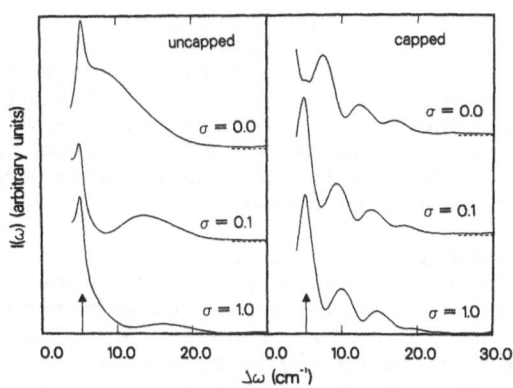

Fig. 2. Calculated Raman spectrum of $(Ge_2Si_2)_{10}$ on Si(100) without and with a Si cap (after Ref. 7).

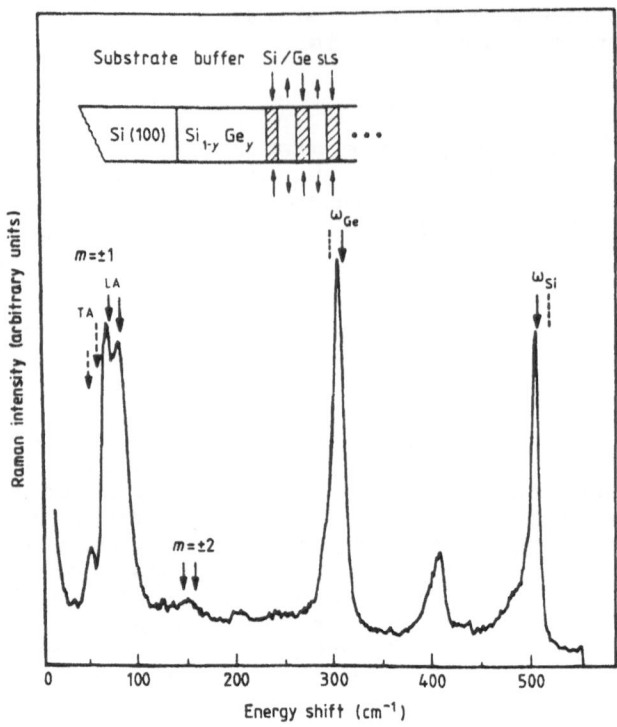

Fig. 3. Raman spectrum of $(Ge_8Si_{12})_{\sim 70}$ on $Ge_{0.6}Si_{0.4}$ (100) (after Ref. 10).

$$\omega(q) = v_{ALS} |(2\pi m/d) \pm q| \qquad (3)$$

where $m = 0,1,2...$ is the zone folding index, d is the ALS periodicity, and v_{ALS} is the ALS sound velocity, which is a function of the Si and Ge layer thicknesses, densities and sound velocities. In the exact expression[11], gaps appear in the dispersion at $q = 0$ and π/d, but Eqn. 3 is usually satisfactory provided the light scattering wavevector is not close to $q = 0$ or π/d (modulo $2\pi/d$). Figure 3 shows folded longitudinal acoustic (LA) doublets for $m = 1,2$ and a weak (symmetry forbidden in true backscattering) transverse acoustic (TA) doublet corresponding to $m = 1$. The predictions of the full Rytov model, given by the arrows in Fig. 3, are in excellent agreement with experiment. However, for ultrathin layer systems (with $m,n < 6$ say), where the folded modes are at much higher frequencies (≥ 150 cm^{-1}), the Rytov model is no longer satisfactory and a model including the phonons from the optic as well as acoustic branches is needed.

The folded acoustic mode intensities are a sensitive indicator of ALS interface quality[12]. Substantial interface blurring due to interdiffusion during growth and/or surface roughness greater than plus or minus one monolayer results in considerably weaker and often broader Raman peaks. In samples with poor quality interfaces no folded modes are visible. Likewise, the diffusion resulting from annealing the ALSs at high temperatures will greatly affect the Raman intensities [10,13]. An example of such a study is shown in Fig. 4 for the

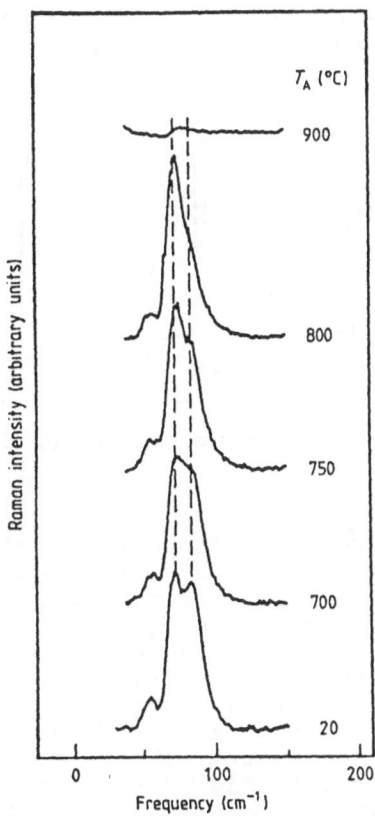

Fig. 4. Raman spectrum of the $m = \pm 1$ folded acoustic phonons of the (Ge_8Si_{12}) ALS of Fig. 3 after incremental annealing for 15 min. at various temperatures (after Ref. 10).

ALS considered above. The period of the ALS is not greatly affected by annealing and thus the LA peak frequencies scarcely alter with annealing temperature. However, the line intensities are affected (see Fig. 4) and the $m = 2$ doublet intensity decreases at a faster rate than the $m = 1$ case consistent with a gradual transition from the original square well profile in the atomic mass to a more sinusoidal profile due to atomic diffusion across the Si-Ge interfaces[10].

4.   OPTIC PHONONS

The most widely studied features in Si-Ge ALSs have been the optic phonons, from both the theoretical and experimental points of view. Theoretical results for these phonons in ASLs for various values of m and n are given in Ref. 5 and elsewhere[14].

Experimentally, three stronger features are typically observed in the Raman spectrum at frequencies near 300, 400 and 500 cm^{-1}, respectively. A typical ALS spectrum[15] showing these three features is given in Fig. 5. The peaks labeled Ge and Si are due to optic modes confined largely in the Ge and Si

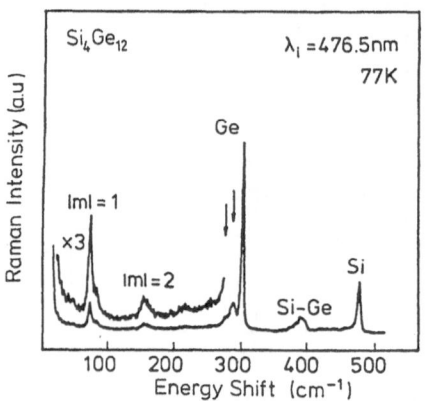

Fig. 5. Raman spectrum of $(Ge_{12}Si_4)_{15}$ on Ge(100) (after
Ref. 15).

layers respectively. The ubiquitous Si-Ge peak near 400 cm^{-1} is
commonly attributed to vibrations at the interfaces between Ge
and Si layers and the intensity of this peak is correlated with
interface blurring/defects/disorder[10,12,15]. Theoretical
calculations have shown that the Si-Ge peak does not appear in
the Raman spectrum of the longitudinal phonons for ALSs with
perfectly abrupt interfaces[16]. The frequencies of the Ge and
Si peaks are sensitive to the degree of confinement of their
vibrational modes and the amount of strain in the layers. For
example, the solid arrows in Fig. 3 mark the positions expected
for the Ge and Si peaks, which have been shifted by strain from

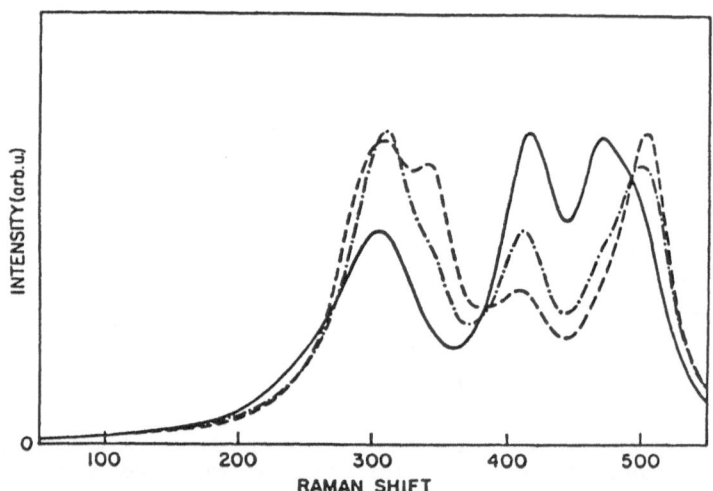

Fig. 6. Calculated Raman spectra of a $(Ge_4Si_4)_{150}$ ALS
with disorder parameters $p_1 = p_2 = 0.9$ (dashed
line) and $p_1 = 0.9$, $p_2 = 0.6$ (dashed-dotted
line), and a random $Si_{0.5}Ge_{0.5}$ alloy $[p_1 = p_2 =
0.5]$ (solid line). p_1 and p_2 are the
probabilities that the central and interfacial
planes of atoms in the ALS are the expected ones
(after Ref. 17).

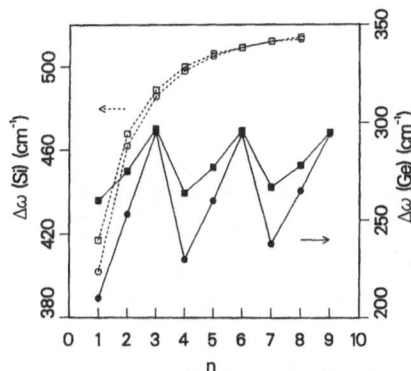

Fig. 7. Quasi-confined modes of Si-Ge ALSs: the highest
Si optical-phonon frequency in $(Ge_1Si_n)_{24}$ (open
square) and $(Ge_3Si_n)_{24}$ (open circle) and the
highest Ge optical-phonon frequency in
$(Ge_3Si_n)_{24}$ (full square) and $(Ge_2Si_n)_{24}$ (full
circle) as a function of n (after Ref. 13).

their respective bulk frequencies (marked by the dashed lines in
Fig. 3). However, the observed peak positions lie at slightly
lower frequencies as a result of confinement. Higher-order
confined optical modes occur within thicker Ge or Si layers, as
is shown, for example, in Fig. 5 where the two arrows mark
additional confined modes associated with the Ge layers in the
$(Ge_{12}Si_4)_{15}$ ALS.

A Raman study of order and disorder in Si-Ge ALSs[17] has
shown that disorder greatly affects the intensities and, to a
lesser extent, the positions of the three characteristic Raman
peaks. The effects of disorder are shown in Fig. 6, where it

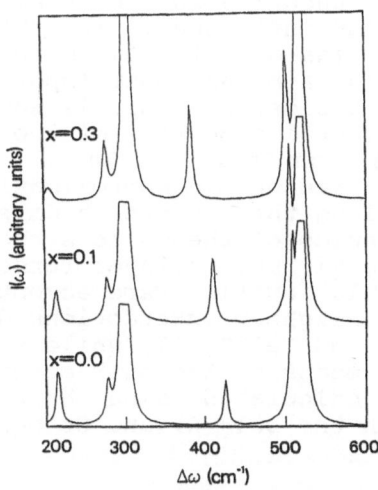

Fig. 8. Result of interface blurring (parameter x) on
the Raman spectrum of $(Ge_6Si_6)_{24}$. The strong
peak at 520 cm^{-1} arises from the substrate and
cap Si layers (after Ref. 13).

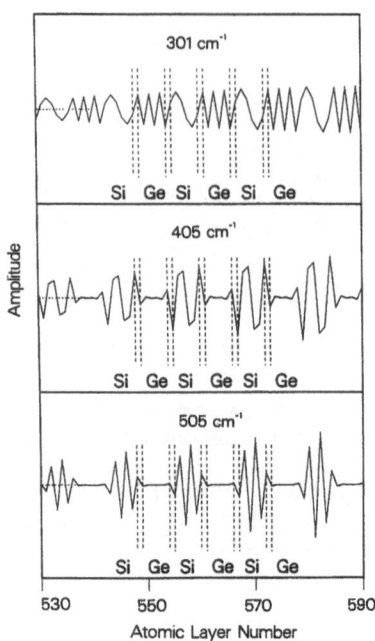

Fig. 9. The vibration amplitudes of the three optical
phonon modes of $(hGe_4h1Si_41)_{24}$ with x = 0.15.
The dashed vertical lines indicate the blurred
interface layers (after Ref. 13).

should be noted that the intensity of the Si-Ge peak near 400
cm^{-1} is very sensitive to the degree of disorder. For this
reason, the Si-Ge peak intensity is a good indicator of the
degree of interface sharpness and intralayer purity[12].

So far, nearly all theories for the vibrational modes in
ALSs have assumed perfect interfaces. In attempting to model
Raman spectra of actual (as-grown) ALSs, however, all three
factors - confinement, strain, and most importantly, interface
blurring - need to be considered. This is shown from a
theoretical and experimental study[13] of two ALSs: $(Ge_2Si_6)_{48}$
on Si(100) and $(Ge_8Si_4)_{24}$ on Ge(100). In this study a linear
chain model, which is a reasonable approximation for the
longitudinal phonons along the ALS growth direction[5], was used
to investigate the dynamics of the whole structure: substrate
plus ALS plus cap. The Si- and Ge-layer force constants, k,
used in the model calculations[13] were essentially the values
including up to fourth neighbor interactions determined
previously for bulk Si and Ge[18,19], while for Si-Ge interactions the arithmetic means of the bulk constants were used.
Raman intensities were calculated using the bond polarizability
model[20,21], for which the bulk Si and Ge polarizabilities, α,
are known[20].

Theoretical calculations for a $(Ge_mSi_n)_{24}$ structure[13]
with m,n < 10 demonstrate the systematics of quasi-confinement,
strain and interface roughness in such ALSs. For example, Fig.
7 shows the n dependence of the highest frequency modes in the
Si and Ge layers for small m values. From this figure it is

Table 1. Effects of confinement, strain and interface blurring on the optical phonon frequencies of bulk Si and Ge when incorporated into an ALS grown on Si or Ge. A +/- implies a shift up/down in frequency.

	Confinement	Strain	Rough interface
ALS on Si			
Si frequency	-	+	-
Ge frequency	-	+	+
ALS on Ge			
Si frequency	-	-	-
Ge frequency	-	-	+

evident that the Si mode is essentially confined to the Si layers for n > 2, and that the mode frequency is still below that of bulk Si(520 cm^{-1}) even for large n, due to the confinement. The behavior for the partially-confined Ge mode is quite different, since these vibrations are buried in the acoustic continuum of Si vibrations. A cyclical variation of the Ge mode frequency is found (see Fig. 7), and it approaches, but never reaches, the bulk Ge value (301 cm^{-1}) for n = 3,6,9, etc. This behavior simply reflects the fact that three silicon atoms have nearly the same mass as one Ge atom[13].

The effect of interface blurring on the Raman spectrum is shown in Fig. 8. Here, the $(Ge_m Si_n)_p$ ALS has been replaced by a smudged (h Ge_{m-2} h 1 Si_{n-2}1)$_p$ structure, where the previously perfect Ge-Si and Si-Ge interfaces have been replaced by h-1 and 1-h, respectively. The interfacial Ge atom has, on average, been replaced with another atom of mass

$$m_h = (1 - x)m_{Ge} + xm_{Si} \quad (x \leq 0.5) \tag{4}$$

due to an admixture x of Si atoms into the Ge layer. Likewise the 1 atom has mass

$$m_1 = xm_{Ge} + (1 - x)m_{Si} . \tag{5}$$

The force constants and polarizabilities of the smudged layers are also linearly intermixed. In Fig. 8 it can be seen that the frequencies and intensities of the optic phonon near 400 cm^{-1} and the folded mode near 200 cm^{-1} are quite sensitive to the admixture x, whereas the main Si and Ge peaks are relatively unaffected. As noted earlier, the peak at ~ 400 cm^{-1} is usually referred to as the Si-Ge line. However, the vibrational amplitude of this mode in (hGe$_4$h1Si$_4$1)$_{24}$ (see Fig. 9) shows that it is really a 1-Si-Si-Si-Si-1 antisymmetric vibration with little penetration into the Ge layers. The Si-Ge label is thus a misnomer in this case and, in general, care should be taken in the proper use of the "Si-Ge" label in ALS spectra.

The effects of confinement, strain and interface blurring on the frequencies of the highest-frequency Si and Ge modes are summarized in Table 1. This table shows that the three factors can act in quite different ways on the mode frequency depending

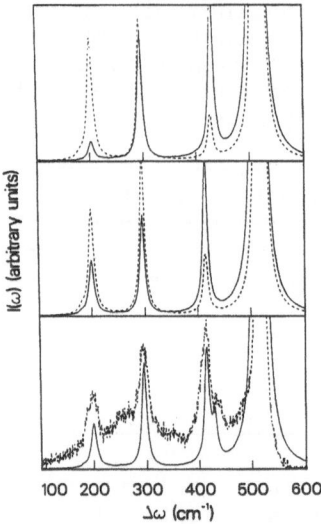

Fig. 10. Theoretical simulation (solid line, bottom
panel) of the experimental Raman spectrum
(dashed line, bottom panel) of $(Ge_2Si_6)_{48}$ on
Si(100). See text for details (after Ref. 13).

on the type of mode and the substrate material. Thus, a priori,
it is difficult to, say, estimate the strain in the ALS Ge/Si
layer from the shift in the Ge/Si frequency from the bulk value,
as is done for thick layer superlattices and epilayers[12]. All
three factors need to be considered in any analysis of the ALS
Raman spectrum, as the following examples demonstrate.

The experimental Raman spectrum[13] of a nominal $(Ge_2Si_6)_{48}$
ALS grown on Si(100) is given in the bottom panel of Fig. 10.
Trial model calculations with a variety of structures having m
and n close to m = 2 and n = 6, respectively, showed the
spectrum to comprise two independent components arising from the
target structure and a blurred interface structure. The
calculated Raman spectrum of the target structure is shown in
the top panel of Fig. 10 (dashed line), but α_{Si} had to be
approximately doubled to reproduce the experimental intensities
and it was also necessary to increase k_{Si} by 1.4% (solid line).
The middle panel of Fig. 10 (dashed line) shows the calculated
spectrum of the smudged structure $(h_2lSi_4l)_{48}$ with x = 0.075.
Again α_{Si} had to be nearly doubled and $k_{Si}/k_l,k_h$ enhanced/
diminished by ~ 3% to match experiment (solid line). The Ge
layers, having become intermixed with Si, carry little or no
strain. A reconstructed spectrum comprising 25% of the target
structure spectrum plus 75% of the blurred-interface structure
spectrum is in good agreement with experiment (see the bottom
panel of Fig. 10).

Similar results were obtained from a simulation of the
Raman spectrum of a nominal $(Ge_8Si_4)_{24}$ ALS grown on Ge(100)[13].
Again, two structures provided the major contribution: one with
mixed Ge-Si interfaces - $(hGe_6hlSi_2l)_{24}$ with x = 0.05; and the
other with all Si layers mixed - $(hGe_6hl_4)_{24}$ with m_l 69% Si and
m_h 50% Si. The top panel of Fig. 11 shows the calculated

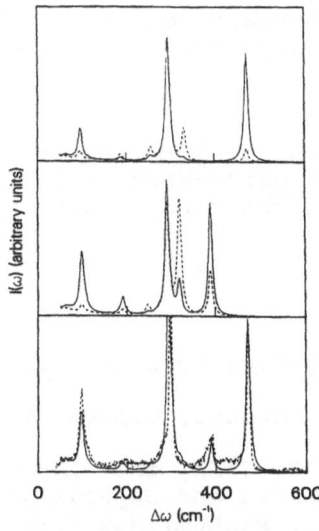

Fig. 11. Theoretical simulation (solid line, bottom
        panel) of the experimental Raman spectrum
        (dashed line, bottom panel) of $(Ge_6Si_4)_{24}$ on
        Ge(100). See text for details (after Ref. 13).

spectrum of the first structure assuming the h and Ge layers are
unstrained and with k_{Si} and k_l reduced by 9% from the bulk Si
value (dashed line), and with revised polarizabilities of $4\alpha_{Si}$
for the Si layers, $2\alpha_{Si}$ for the l layers, and $\alpha_{Ge}/4$ for the h
layers (solid line). For the second structure, the spectrum
given in the middle panel of Fig. 11 has been calculated with k_l
reduced by 11% from bulk Si (dashed line), and with the modified
polarizabilities of $2\alpha_{Si}$ for the l layers and $\alpha_{Ge}/4$ for the h
layers (solid line). Taking 80% of the first structure spectrum
plus 20% of the second spectrum yields the solid line in the
bottom panel of Fig. 11 in good agreement with the actual
spectrum (dashed line)

CONCLUSIONS

     The Raman spectrum of folded acoustic modes in Si-Ge ALSs
is a reliable indicator of sample quality: the positions of the
peaks are sensitive to the respective layer thicknesses and
their intensities are sensitive to layer interface sharpness.
The Rytov elastic continuum model accurately reproduces the
folded mode results for thicker layer ALSs.

     Modeling the Raman spectra of ALSs has shown that the
positions of the optical phonon peaks cannot simply be used to
determine intralayer strain. The effects of confinement and,
most importantly, the interface blurring that occurs in real
structures must also be considered. Such modeling can lead to
useful characterization of these structures, providing estimates
of the interface mixing, intralayer strain, and differences from
the target structure.

     More experimental and theoretical work needs to be done to
explain the enhanced bond polarizabilities (or photoelastic

coupling coefficients) that are required to reproduce the phonon intensities in ALS Raman spectra.

Acknowledgment

The essential contributions of J.-M. Baribeau (MBE growth and X-ray analysis) and M.W.C. Dharma-Wardana and G.C. Aers (lattice dynamics calculations) in the work carried out at NRC are gratefully acknowledged.

REFERENCES

1. J. Sapriel and B. Djafari Rouhani, **Surface Science Reports**, 10:189 (1989)
2. B. Jusserand and M. Cardona, **Topics in Applied Physics**, 66:49 (1989)
3. See for example, M. Cardona, in: "Light Scattering from Semiconductor Structures and Superlattices", D.J. Lockwood and J.F. Young, eds., Plenum, New York (1991)
4. See for example, Z.V. Popovic, M. Cardona, E. Richter, D. Strauch, L. Tapfer and K. Ploog, **Phys. Rev. B**, 40:3040 (1989)
5. A. Fasolino, E. Molinari and A. Qteish, p.495 of this volume
6. See for example, G. Kanellis, p.207 in: "Spectroscopy of Semiconductor Microstructures", G. Fasol et al., eds., Plenum, New York (1989); E. Molinari and A. Fasolino, p. 195 of the same book; E.A. Montie, G.F.A. van de Walle, D.J. Gravesteijn, A.A. van Gorkum and W.J.O. Teesselink, **Semicond. Sci. Technol.**, 4:889 (1989); J. White, G. Fasol, R.A. Ghanbari, C.J. Gibbings and C.G. Tuppen, **Thin Solid Films**, 183:71 (1989)
7. M.W.C. Dharma-wardana, D.J. Lockwood, G.C. Aers and J.-M. Baribeau, **J. Phys. Condens. Matter**, 1:2445 (1989) and **Appl. Phys. Lett.**, 52:2040 (1988)
8. C. Colvard, T.A. Gant, M.V. Klein, R. Merlin, R. Fisher, H. Morkoç and A.C. Gossard, **Phys. Rev.B**, 31:2080 (1985)
9. Some representative references are: G. Abstreiter, **Thin Solid Films**, 183:1 (1989); J.-M. Baribeau, D.J. Lockwood, N.L. Rowell, M.W.C. Dharma-wardana, D.C. Houghton and Song Kechang, **Mat. Res. Soc. Symp. Proc.**, 160:95 (1990) S.J. Chang, C.F. Huang, M.A. Kallel, K.L. Wang, R.C. Bowman, Jr., and P.M. Adams, **Appl. Phys. Lett.**, 53:1835 (1988); H. Okumura, K. Miki, K. Sakamoto, T. Sakamoto, K. Endo and S. Yoshida, **Appl. Surf. Sci.**, 41-42:548 (1989); M. Ospelt, W. Bacsa, J. Henz, K.A. Mäder and H. von Känel, **Superlattices Microstruct.**, 5:71 (1989)
10. H. Brugger, E. Friess, G. Abstreiter, E. Kasper and H. Kibbel, **Semicond. Sci. Technol.**, 3:1166 (1988)
11. S.M. Rytov, **Acoust. Zh.**, 2:71 (1956) [**Sov. Phys. - Acoust.**, 2:67 (1956)]
12. D.J. Lockwood and J.-M. Baribeau, in: "Light Scattering from Semiconductor Structures and Superlattices", D.J. Lockwood and J.F. Young, eds., Plenum, New York (1990)
13. M.W.C. Dharma-wardana, G.C. Aers, D.J. Lockwood and J.-M. Baribeau, **Phys. Rev. B**, 41:5319 (1990)
14. See, for example, the references quoted in Ref. 6 above.
15. G. Abstreiter, K. Eberl, E. Friess, W. Wegscheider and R. Zachai, **J. Crystal Growth**, 95:431 (1989)
16. A. Fasolino and E. Molinari, **Appl. Phys. Lett.**, 54:1220 (1989)

17. J. Menéndez, A. Pinczuk, J. Bevk and J.P. Mannaerts, **J. Vac. Sci. Technol. B**, 6:1306 (1988)
18. A. Fleszar and R. Resta, **Phys. Rev.B**, 34:7140 (1986)
19. K. Kunc and P. Gomes DaCosta, **Phys. Rev.B**, 32:2010 (1985)
20. S. Go. H. Bilz and M. Cardona, **Phys. Rev. Lett.**, 34:580 (1975)
21. R.J. Bell, **Methods in Computational Phys.**, 15:215 (1976)

COMBINED EFFECT OF STRAIN AND CONFINEMENT ON THE PHONON SPECTRA

OF IV/IV SUPERLATTICES: TOWARDS A QUANTITATIVE DESCRIPTION

A. Fasolino, E. Molinari† and A. Qteish‡

International School for Advanced Studies
Strada Costiera 11, I-34100 Trieste, Italy
† CNR, Istituto di Acustica "O.M. Corbino"
Via Cassia 1216, I-00189 Roma, Italy
‡ Cavendish Laboratory, Madingley Road
Cambridge, CB3 OHE, UK

ABSTRACT

The lattice dynamical properties of Si/Ge superlattices are studied by means of the interplanar force constants recently calculated *ab-initio* for strained bulk Si and Ge. The longitudinal modes of (001)-grown superlattices, with different strain distributions, can be accurately studied. We compare our results with available experimental data. We discuss the possible extension of our conclusions to other IV/IV superlattices, particularly to α-Sn/Ge.

1.   INTRODUCTION

Strained layer Si/Ge superlattices (SLs) have recently attracted much interest because of their novel basic properties and because of the obvious interest of Si-based materials for device applications. Despite the large lattice mismatch (~ 4%), short period Si/Ge SLs can be grown by MBE, either on Si or Ge or $Si_{1-x}Ge_x$ substrates, with different strain distributions in the Si and Ge layers.

Much experimental[1-8] and theoretical[9-15] work has been devoted to their lattice dynamical properties. Raman studies of the vibrational properties of Si/Ge SLs have been performed for layer and interface characterization[1-8]. The interpretation of these spectra requires to deal at the same time with the strain, due to the lattice mismatch, and with confinement effects, due to the layering. Qualitatively, strain is the dominant contribution for the shift of the highest, Si-like and Ge-like, optical mode while confinement effects dominate the frequency spacing of successive modes.

Studies on GaAs/AlAs SLs have shown that phonons in SLs are related to the bulk dispersion of the constituents and that an approximate correspondence between the optical SL modes and the

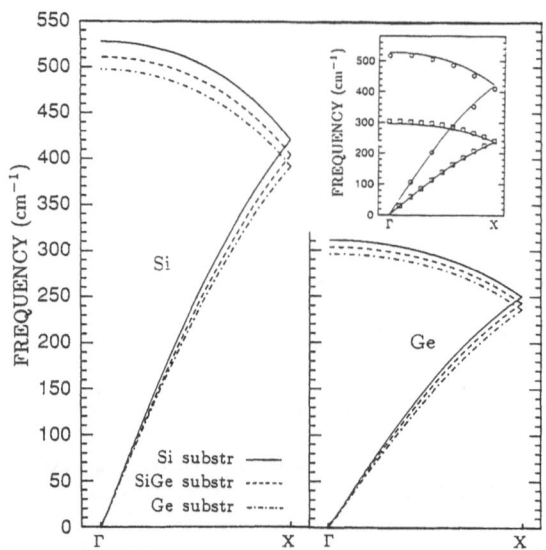

Fig. 1. Longitudinal phonon dispersions for bulk Si and Ge along the (001) direction, at three values of strain corresponding to lattice-matching to a Si-substrate (solid line), $Si_{0.5}Ge_{0.5}$ substrate (dashed line), and Ge substrate (dashed dotted line), calculated with the interplanar force constants of Ref. 15. The calculated $\omega^{LO}(\Gamma)$ for unstrained Si and Ge are 528.0 and 296.2 cm^{-1} respectively. In the inset, the unstrained dispersions of Si and Ge are compared to the experimental values (circles, for Si from Ref. 22, T = 296 K, and squares for Ge from Ref. 23, T = 80 K).

bulk phonons at selected wavevectors q in the growth direction can be made[16,17,14]. In order to take strain into account it is then, in principle, necessary to know the strain dependence of the full bulk dispersion of the constituents. Such knowledge is not experimentally available; only the shift of the Γ-point bulk phonons is known[18,19]. Therefore it has been, up to now, assumed that the whole optical branches follow rigidly the shift of the modes at Γ.

Force constants, recently calculated *ab-initio* for strained bulk Si and Ge[15], yield instead a nonrigid shift of the longitudinal (L) dispersion along the (001) growth direction. Although the major effect of strain is the large shift of $\omega^{LO}(\Gamma)$, the predicted modification of the bulk *dispersion* has a quantitatively appreciable effect on the frequency spacing of the SL optical modes[15]. Qualitatively, the inclusion of strain in this fashion does not change the previous under- standing of the vibrational properties of this system[9-14], but we believe that it can improve the quantitative agreement between theory and experiments which is crucial for a detailed characterization.

In the following we calculate the L modes of (001) SLs along the growth direction by use of such interplanar force

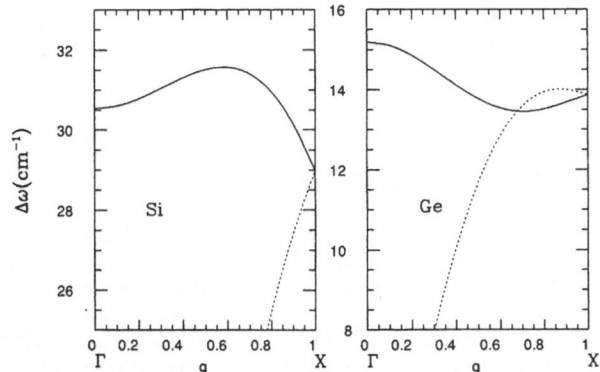

Fig. 2. Frequency shift induced by strain in going from a Ge-substrate to a Si-substrate of the optical (solid line) and acoustical (dotted line) bulk modes as a function of wavevector in the (001) direction. If the shift of the optical dispersion were rigid, the solid curves would be a constant, equal to 30.6 cm^{-1} and 15.2 cm^{-1} for Si and Ge respectively.

constants[15]. In Section 2 the results of Ref. 15 for bulk Si and Ge are reviewed. In Section 3 we discuss the effect of layering and compare the SL Si-like confined modes and Ge-like quasi-confined modes[11] with the corresponding bulk dispersions. Section 4 is instead focused on the effect of

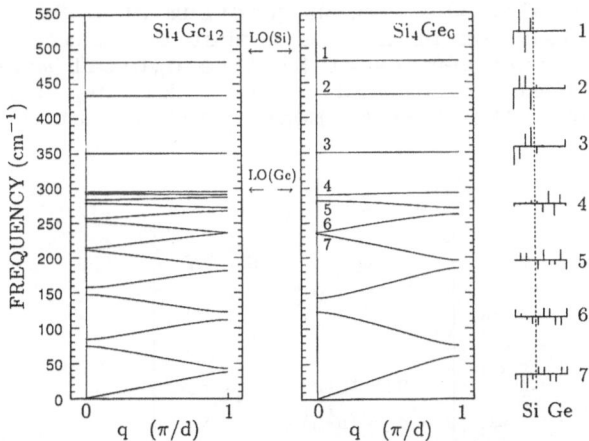

Fig. 3. Phonon spectrum of a Si_4Ge_{12} SL (left) and of a Si_4Ge_6 SL (center) on a Ge substrate in the SL Brillouin Zone (d is the SL period). On the right the displacement of the modes 1-7 at the Γ-point are displayed. Mode 1-3 are the optical Si-like confined modes. Mode 4-7 are Ge-like quasi-confined mode. They have the typical displacement of optical modes in the Ge-layer (compare e.g. mode 4 with mode 1) and that of acoustical modes in the Si-layer. The amplitude in the Si-layer is not evanescent like in true confined modes.

strain. In Section 5 we discuss the comparison with experiments and in Section 6 we extend the conclusions drawn for Si/Ge SLs to other IV/IV SLs, particularly α-Sn/Ge. Finally, in Section 7 we summarize our results.

2. EFFECT OF STRAIN ON BULK Si AND Ge

We show in Fig. 1 the bulk phonon dispersion of Si and Ge along the (001) direction, as obtained with the interplanar force constants calculated *ab-initio* by Qteish and Molinari[15] when the in-plane lattice constant is matched to that of either a Si or a Ge or a $Si_{0.5}Ge_{0.5}$ substrate. While the absolute values may be slightly at variance with the experimental determination, depending also on the measurement temperature, the frequency shift of 30.6 and 15.2 cm^{-1} of the Γ-point frequencies, obtained for Si and Ge respectively, in going from a Ge-substrate to a Si-substrate, is in good agreement with those of 31 and 16 cm^{-1} extracted from experiments[18,19].

The changes in the *dispersion* for different substrates (i.e. for different strain) are better seen in Fig. 2 where we plot, for Si and Ge, the difference between $\omega^{LO}(q)$ on a Ge-substrate and that on Si-substrate as a function of q along (001). If the shifts of the optical dispersion were rigid, these curves would be a constant, equal to 30.6 cm^{-1} and 15.2 cm^{-1} for Si and Ge respectively.

3. CONFINEMENT OF SL MODES AND RELATIONSHIP TO BULK DISPERSIONS

In Fig. 3 we plot the phonon spectrum of a Si_4Ge_{12} and of a Si_4Ge_6 SL, the subscript being the number of atomic (001) planes. The different nature of Si-like optical modes (between $\omega^{LO}(\Gamma)$ of Si and that of Ge) and the Ge-like ones (below $\omega^{LO}(\Gamma)$ of Ge) is evident. The Si-like modes have a flat dispersion

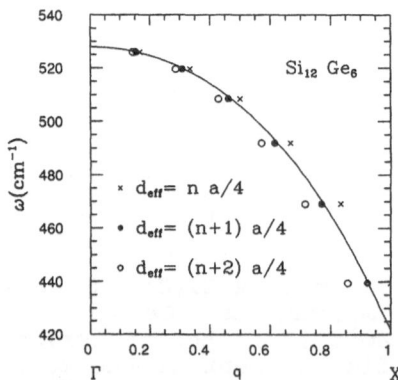

Fig. 4. Si-like ω_m^{LO} confined modes of a $Si_{12}Ge_6$ SL on a Si-substrate plotted onto the unstrained Si bulk dispersion at wavevectors $q_{eff} = m\pi/d_{eff}$. We remind the reader that X = 2π/a. An effective confinement length $d_{eff} = (n + 1) a/4$ (n being the number of Si *atomic* planes) is appropriate to describe all Si-like modes.

since they are confined in the Si layer, being at frequencies
where no bulk Ge modes exist. The modes of Ge fall instead in
the region of overlap between the Ge "optical" continuum and the
Si "acoustical continuum". Quasi-confined modes[11], non-
exponentially damped in the Si layer, appear just below $\omega^{LO}(\Gamma)$
of Ge. The similarity to true confined modes is enhanced for
thicker Ge layers, where the dispersion of a few modes becomes
almost flat and the amplitude of vibration is mostly concentrated in the Ge layer. Even for thinner Ge layers, however,
the displacements associated with these modes are those of
optical modes in the Ge layer (see e.g. mode 4 in Fig. 3 and
compare to the Si-like mode 1), and of acoustical modes in the
Si layer. Therefore they are expected to have Raman efficiency
comparable to that of true confined modes.

We address now the relationship between the SL modes and
the bulk dispersion[16,17,14,13,2]. We plot in Fig. 4 the
calculated Si-like ω_m^{LO} confined modes of a $Si_{12}Ge_6$ SL on a Si-
substrate onto the appropriate unstrained Si bulk dispersion at
wavevectors $q_{eff} = m\pi/d_{eff}$. We find that an effective
confinement length $d_{eff} = (n + 1)a/4$ (n being the number of Si
atomic planes and a/4 being the interplanar spacing) is
appropriate to describe all Si-like modes as proposed in Ref. 2.

This is no longer true for the Ge-like modes. In Fig. 5a
we plot, with an asterisk, the Ge-like modes of a Si_4Ge_{18} SL on
a Ge-substrate onto the unstrained Ge bulk dispersion at the
wavevectors where the bulk mode frequencies equal those of the
SL modes. Clearly, the SL modes are not at all equally spaced
in wavevector, and, except for the first two modes, they are
paired into doublets, due to their dispersion (see also Fig. 3).

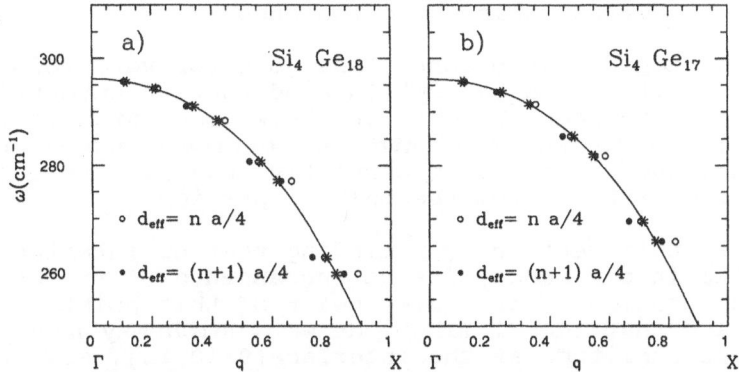

Fig. 5. a) Ge-like modes of a Si_4Ge_{18} SL on a Ge-
substrate onto the unstrained Ge bulk
dispersion. The asterisks are at the wavevectors where the bulk mode frequencies equal
those of the SL modes. Empty and solid circles
are at wavevectors $q_{eff} = \pi/d_{eff}$, with $d_{eff} =$
na/4 and $d_{eff} = (n + 1) a/4$ respectively. By
comparing with panel b), where the SL modes of a
Si_4Ge_{17} are shown in the same fashion as in a),
it can be seen that one or the other
correspondence, represents better the even or
odd modes, depending on the (even or odd) number
of Ge layers.

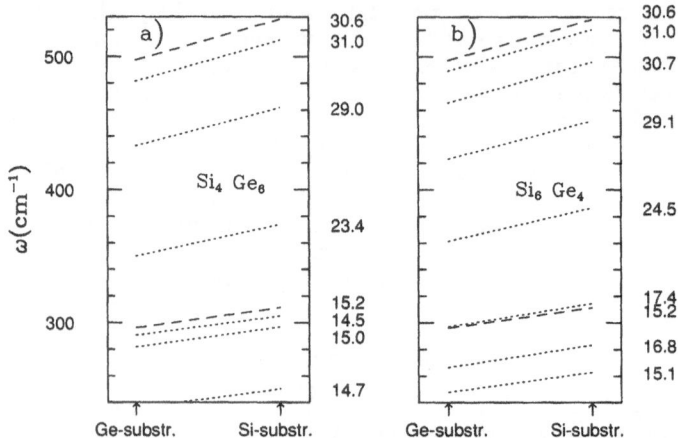

Fig. 6. Dotted line: variation of the SL mode
frequencies of a Si_4Ge_6 SL as a function of
strain, going from a Ge to a Si substrate.
Dashed line: same as before for the bulk Si and
Ge mode at Γ. The numbers on the right are the
differences in cm^{-1} for each mode.

Thus, the definition of an effective confinement length is
somewhat ill-defined. In fact, an effective length $d_{eff} = na/4$
is the best to describe the odd modes, while $d_{eff} = (n + 1)a/4$
is more appropriate for the even ones. The situation is
reversed for an odd number of Ge planes, as shown in Fig. 5b.
Thus, depending on the number (even or odd) of Ge planes, by use
of one or the other d_{eff} either even or odd modes fall
reasonably well onto the bulk dispersion. The number of Si
planes does not affect this correspondence.

This correspondence becomes less good for very thin Ge
layers. Thus, it should be kept in mind that an automatic
correspondence, between SL modes and bulk modes at selected
wavevector q_{eff}'s is more delicate for Ge-like modes than for
Si-like modes and, if not done correctly, may lead to over-
estimate the departure from the bulk dispersion.

We close this Section by reminding that no interface modes
are predicted in our model, due to the absence of frequency gaps
in the Si dispersion. For a discussion of this point see Ref.
10. We stress that the Si-like mode ω_3^{LO}, which may have a
nonnegligible amplitude at the interface[9-10,12], may fall
around 400 cm^{-1} for particularly thin Si layers, that is,
approximately at the frequency where a structure, which we
attribute to interface alloying or disorder, appears in most
experimental Raman spectra. However, the mode ω_3^{LO} shifts very
rapidly with varying Si thickness[9], so that a layer thickness
independent mode, such as experimentally observed, cannot be
attributed to it.

4.   EFFECT OF STRAIN ON SL MODES

As far as the layer thickness is kept below a critical
value[3], SLs can be grown, without dislocations, on several
types of substrates, ranging from pure Si to pure Ge. In this

$Si_{12}Ge_{12}$

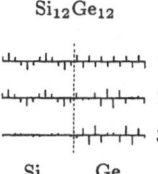

Si   Ge

Fig.7.   Displacements of selected modes of a $Si_{12}Ge_{12}$
          SL. Mode 1 is the Si-like mode ω_9^{LO} which falls
          in frequency just above $\omega^{LO}(\Gamma)$ of Ge. The
          displacement in the Ge layer is the one of bulk
          optical modes at Γ, evanescently damped.
          Therefore it has appreciable Raman strength,
          even though it is a high index mode. Modes 2
          and 3 are the first and second Ge-like modes ω_1^{LO}
          and ω_2^{LO}. The displacement of mode 2 is slightly
          deformed with respect to a typical first Ge-like
          mode. This fact occurs whenever the number of
          Si planes is a multiple of three, i.e. whenever
          the last Si-like mode falls very close to $\omega^{LO}(\Gamma)$
          of Ge.

case the effect of strain can be clearly looked at by studying a
given SL on different substrates. We plot in Fig. 6 the optical
modes of a Si_4Ge_6 and of a Si_6Ge_4 SLs, calculated on a Si and on
a Ge substrate. The lines between the two sets of values are
guides to the eye although a linear interpolation is appropriate
[15]. It can be seen that the highest Si-like mode and the
highest Ge-like one follow almost rigidly the behavior of the
bulk $\omega^{LO}(\Gamma)$, also shown in the figure. This fact compares very
well with the strain dependence reported in Ref. 2. The slope
of successive modes may instead be different. They, in fact,
follow the strain dependence of bulk modes at the wavevectors
q_{eff}'s discussed in the previous Section. For instance, the Si-
like mode ω_3^{LO} of Si_4Ge_6 varies only by 23.4 cm^{-1} in going from a
Ge to a Si substrate, against the 31.0 cm^{-1} of ω_1^{LO}. This fact
might be the reason for the large spreading and deviations from
the bulk dispersion of the Si-like modes observed in Ref. 2,
where the Si-like modes have been rigidly shifted for strain and
reported on the unstrained Si bulk dispersion. Moreover, the
change of slope may also be nonmonotonic for higher index modes,
as is shown in Fig. 6 for Ge like modes. This is due to the
rather complicated dependence of bulk modes at different q's
shown in Fig. 2 and also to the interplay between the different
Si and Ge dependence on strain, which becomes important for
nonstrictly confined modes. We note in passing, that in our
calculations for Si_6Ge_4, there is a Si-like mode which falls
just above $\omega^{LO}(\Gamma)$ of Ge. This fact happens whenever the number
of Si planes is a multiple of three. The displacements of such
modes are rather peculiar, as shown in Fig. 7 for a $Si_{12}Ge_{12}$ SL
(mode 1). In the Ge layer, it consists in fact of the typical
bulk LO displacements at Γ, but evanescently damped. This is
why these modes have an appreciable Raman intensity, even if
they are high index Si-like modes.

5.   COMPARISON WITH EXPERIMENTS

     In order to compare our results with experimental data, we
need to know both the thicknesses of the Si and Ge layers and

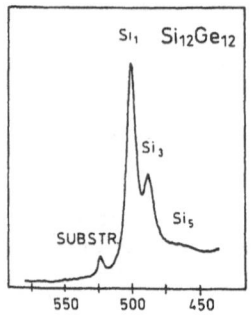

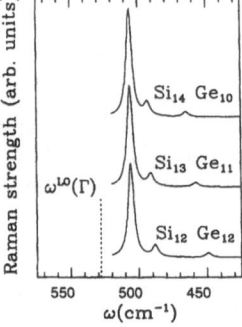

Fig. 8. Comparison of the experimental spectrum of a
Si$_{12}$Ge$_{12}$ SL (top) from Ref. 2 with the
calculated spectrum of a Si$_{12}$Ge$_{12}$, Si$_{13}$Ge$_{11}$ and
of a Si$_{14}$Ge$_{10}$ SL on a Si$_{0.36}$Ge$_{0.64}$ substrate. The
latter reproduces the experimental spectrum,
quite well once all frequencies are shifted by
8 cm^{-1}. Also the experimental peak labeled as
substrate which comes from an unstrained Si
buffer compares very well with our $\omega^{LO}(\Gamma)$ of
unstrained Si, reported as a dashed line in the
figure.

the composition of the substrate. Most of the results are
instead given in the literature, already corrected for the
strain, according to the hypothesis of a rigid shift of phonon
branches with strain. Consequently, many details of the samples
are not given. Moreover, as discussed in Section 2, there may
be small variations of the absolute values of the frequencies,
due to different temperature conditions. These differences can
be in the range 1 - 10 cm^{-1}. Last but not least, the results of
an *ab-initio* parameter-free calculation can be trusted with a
precision of a few cm^{-1}. They should however be reliable, even
on the scale of 1 - 2 cm^{-1}, if *differences* of phonon frequencies
are considered. This is why we attempt a comparison with
experiments only for cases where more than one confined or
quasi-confined mode is observed. In Fig. 8 we compare our
results with the experimental ones[2] for the Si-like modes of a
Si$_{12}$Ge$_{12}$ on a Si$_{.36}$Ge$_{.64}$ substrate. We simulate the observed
spectrum by calculating the Raman strength with a simple bond
polarizability model[20], assuming equal bond polarizability for
both materials. The calculated strength is dressed with a
5 cm^{-1} wide Lorentzian. This broadening makes the comparison
easier, but not much physical meaning should be attached to it.

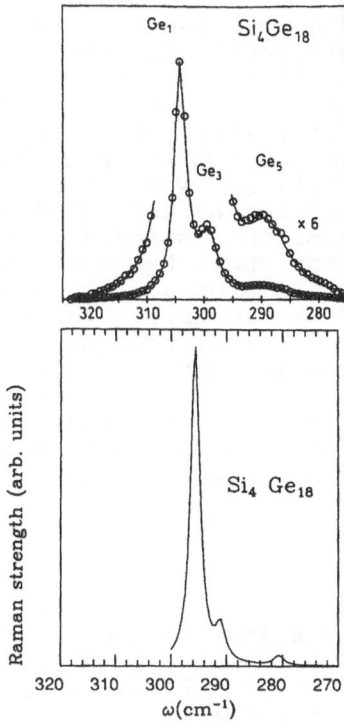

Fig. 9. Comparison of the experimental spectrum of a
Si_4Ge_{18} SL (top) from Ref. 2 with the calculated
spectrum of the same SL on a Ge substrate. They
compare very well once frequencies are shifted
by 8 cm^{-1}. See also the text.

The observed spacing of frequencies cannot be explained by our
model with a thickness of 12 Si planes, but is very well
reproduced by 14 Si atomic planes. It is interesting that also
the experimental peak labeled as substrate, which comes from an
unstrained Si buffer compares very well with the position of our
$\omega^{LO}(\Gamma)$ of unstrained Si. In Fig. 9 we compare, in the same way
but with a 2 cm^{-1} wide Lorentzian, our results for the Ge-like
modes of a Si_4Ge_{18} with the experimental ones[2]. The agreement
is remarkable.

As far as the comparison of behavior with strain is
concerned, we are not aware of published experimental studies of
the strain dependence of SL modes which are not the highest Si-
like or Ge-like one, as reported in Ref. 2. Such a study is in
progress[21] and will constitute a crucial test for our model.

We wish to make a last comment concerning the Ge-like
quasi-confined modes. We have seen that, for thick layers, they
behave more or less as true confined modes. For thinner layers,
where they should have a certain dispersion in the SL Brillouin
Zone, it is however, difficult to try and follow this dispersion
because the range of available laser lines spans a very minor
portion of the SL Brillouin Zone. The different nature of Ge-
like modes could be ascertained experimentally if the frequency
separation of successive modes could be observed to be a non-

monotonic function of the mode number. In other words, high
index modes should be grouped in doublets close in energy as
shown in Figs. 3 and 5. However, this is also difficult to do
because of Raman selection rules which, off-resonance, make only
odd modes observable. Therefore, we consider the study of the
Ge-like modes of a Si_3Ge_9 SL versus laser frequency, reported in
Ref. 2, extremely interesting. The authors observe, under
resonant conditions, both second and third Ge-like modes. If
this is the explanation of the observed double peak, it is very
interesting to notice that these two modes occur much closer in
energy than expected from a simple picture of confinement. We
show in Fig. 10 what we should expect for a Si_4Ge_9 SL on a Ge
substrate, and it is quite clear that the second and third Ge-
like modes occur in a doublet and at a frequency separation very
similar to that of Ref. 2. In the case of a Si_3Ge_9 SL we obtain
an even smaller separation between second and third Ge-like
modes. This is in our opinion the first clear-cut experimental
evidence of the dispersive nature of the Ge-like optical quasi-
confined modes.

6.   OTHER IV/IV SLs

The progress in the epitaxial growth of strained materials
means that several groups are starting to grow other IV/IV SLs
even more strained that Si/Ge. Preliminary Raman results have
been obtained for α-Sn/Ge SLs[21].

Qualitatively, the considerations illustrated for the Si/Ge
system[9-12] can be extended also to the other possible pairs.
Confined modes are expected for the lighter material and quasi-

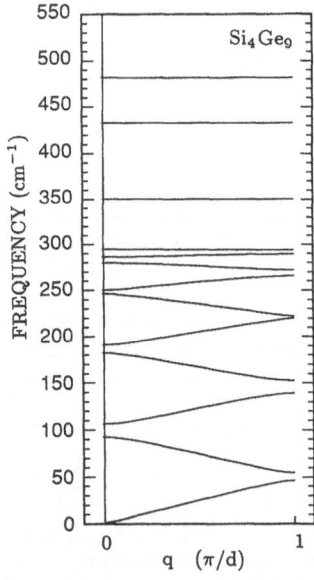

Fig. 10. Phonon dispersion of a Si_4Ge_9 SL on a Ge
         substrate. The first three Ge-like modes are
         at 294.5, 286.2 and 279.7 cm^{-1}. Notice that
         the second and third Ge-like modes occur in a
         doublet (see text).

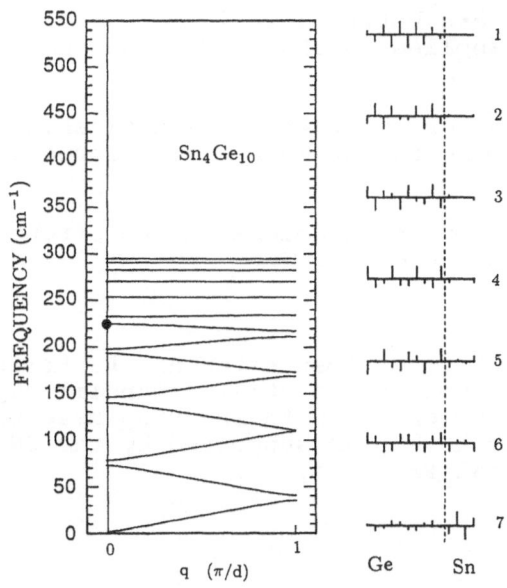

Fig. 11. Phonon dispersion of a α-Sn_4Ge_{10} SL on a Ge
substrate and corresponding displacement
patterns of the modes at Γ in order of
decreasing frequency. Modes 1-6 are Ge-like
confined modes while mode 7 (indicated by a
full dot in the dispersion) is quasi-confined
in the α-Sn layer.

confined modes for the heavier one. No L interface modes are
expected. We have done a preliminary calculation for the system
α-Sn/Ge and assumed a mass approximation. This leads for the
bulk to a much higher $\omega^{LO}(\Gamma)$ (230 instead of the experimental
200 cm^{-1}) but strain is expected to stiffen the phonons of α-Sn
so that this approximation becomes less bad. Results for a
α-Sn_4Ge_{10} SL are shown in Fig. 11. It is interesting to notice
that a quasi-confined mode appears for α-Sn. Such a mode, with
nonzero Raman activity, is present even with only two planes of
a α-Sn. Moreover the Ge-like optical modes are now truly
confined and they are well described by $d_{eff} = (n + 1)a/4$ like
the Si-like modes in Si/Ge SLs.

SUMMARY AND CONCLUSIONS

We have studied the longitudinal phonons of (001) Si/Ge
strained layer SLs along the growth direction by means of
interplanar force constants calculated *ab-initio* for strained Si
and Ge.

For Si/Ge SLs we find that: i) the strain dependence of the
first Si-like confined mode and of the first Ge-like quasi-
confined mode follows that of $\omega^{LO}(\Gamma)$ and can be used for strain
determination. Successive modes may deviate appreciably from
this dependence for thin layers; ii) a well defined confinement
length $d_{eff} = (n + 1)a/4$ can be associated with Si-like modes.
Conversely, in SLs with an even number of Ge planes, the odd Ge-
like modes are satisfactorily described by $d_{eff} = na/4$ while the

even Ge-like modes are better described by $d_{eff} = (n + 1)a/4$. The opposite rule applies to SLs with an odd number of Ge planes.

We find a good agreement with the available experiments and believe that our approach can make a step forward for sample characterization.

We discuss how our conclusions can be extended to other IV-IV SLs, particularly to αSn-Ge SLs.

Acknowledgment

We are grateful to G. Abstreiter and E. Friess for useful discussions. We acknowledge partial financial support by INFM (Italy) and by the Science and Engineering Research Council (UK). Calculations have been supported by the CNR-CINECA and SISSA-CINECA National Projects.

REFERENCES

1. K. Eberl, G. Krötz, R. Zachai and G. Abstreiter, **Journal de Physique**, 5:329 (1987); E. Friess, H. Brugger, K. Eberl, G. Krötz and G. Abstreiter, **Solid State Commun.**, 69:899 (1989)
2. E. Freiss, K. Eberl, U. Menczigar and G. Abstreiter, **Solid State Commun.**, 73:203 (1990)
3. see e.g. K. Eberl, E. Friess, W. Wegscheider, U. Menczigar and G. Abstreiter, **Thin Solid Films**, 183:95 (1989)
4. J.C. Tsang, S.S. Iyer and S.L. Delage, **Appl. Phys. Lett.**, 51:1732 (1987); S.S. Iyer, J.C. Tsang, M.W. Copel, P.R. Pukite and R.M. Tromp, **Appl. Phys. Lett.**, 54:219 (1989)
5. D.J. Lockwood, M.W.C. Dharma-wardana, G.C. Aers and J.M. Baribeau, **Appl. Phys. Lett.**, 52:2040 (1988); M.W.C. Dharma-wardana, G.C. Aers, D.J. Lockwood and J.M. Baribeau, **Phys. Rev.B**, 41:5319 (1990)
6. W. Bacsa, H. von Känel, K.A. Mäder, M. Ospelt and P. Wachter, **Superlattices and Microst.**, 4:717 (1988); M. Ospelt, W. Bacsa, J. Henz, K.A. Mäder and H. von Känel, **Superlattes and Microstr.**, 5:71 (1989)
7. J. Menéndez, A. Pinczuk, J. Bevk and J.P. Mannaerts, **J. Vac. Sci. Technol.**, B6:1306 (1988)
8. M.I. Alonso, F. Cerdeira, D. Niles, M. Cardona, E. Kasper and H. Kibbel, **Journ. Appl. Phys.**, 66:5645 (1989)
9. A. Fasolino and E. Molinari, **Journal de Physique**, C5:569 (1987)
10. E. Molinari and A. Fasolino, **Appl. Phys. Lett.**, 54:1220 (1989)
11. A. Fasolino, E. Molinari and J.C. Maan, **Phys. Rev.**, B39:3923 (1989)
12. M.I. Alonso, M. Cardona and G. Kanellis, **Solid State Commun.**, 69:479 (1989)
13. R.A. Ghanbari and G. Fasol, **Solid State Commun.**, 70:1025 (1989)
14. E. Molinari, A. Fasolino and K. Kunc, p.663 **in**: "Proc. 18th Int. Conf. on the Physics of Semiconductors", O. Engstrom, ed., World Scientific (1987); A. Fasolino and E. Molinari, **Surf. Sci.**, 228:112 (1990)
15. A. Qteish and E. Molinari, **Phys. Rev.B**, to be published.
16. A.K. Sood, J. Menéndez, M. Cardona and K. Ploog, **Phys. Rev. Lett.**, 54:2111 (1985)

17. B. Jusserand and D. Paquet, **Phys. Rev. Lett.**, 56:1752 (1986)
18. E. Anastassakis, A. Pinczuk, E. Burstein, F.H. Pollak and M. Cardona, **Solid State Commun.**, 8:133 (1970)
19. F. Cerdeira, C.J. Buchenauer, F.H. Pollak and M. Cardona, **Phys. Rev.B**, 5:580 (1972)
20. B. Jusserand and D. Paquet **in**: "Semiconductor Heterojunctions and Superlattices", N.Boccara, G. Allan, G. Bastard, M. Lannoo and M. Voos, eds., Springer (1986)
21. G. Abstreiter, private communication.
22. G. Dolling **in**: "Symposium on Inelastic Scattering of Neutrons in Solids and Liquids", IAEA, Wien, II:37 (1963); see also: G. Nilsson and G. Nelin, **Phys. Rev.B**, 6:3777 (1972)
23. G. Nilsson and G. Nelin, **Phys. Rev.B**, 3:364 (1970)

ELECTRONIC STRUCTURE AND OPTICAL PROPERTIES OF STRAINED LAYER

SUPERLATTICES

G. Duggan, K.J. Moore, K. Woodbridge, C. Roberts,
N.J. Pulsford* and R.J. Nicholas*

Philips Research Laboratories
Redhill, Surrey RG1 5HA, UK
* The Clarendon Laboratory
Parks Road, Oxford OX1 3PU, UK

ABSTRACT

The essential elements of strain effects on the electronic structure of $In_x Ga_{1-x}As$-(AlGa)As on GaAs are elucidated. Attention is focused on the optical properties of quantum wells and superlattices (SL) with In fractions < 0.12. Photo-luminescence and photoluminescence excitation spectroscopy at ~ 4 K has revealed sharp exciton features and allowed us to identify $\Delta n = 0$ and $\Delta n \neq 0$ transitions and follow the development of SL minibands.

1. INTRODUCTION

A desire to achieve lattice matching limits the possible number of interesting and exploitable III-V heterojunction combinations that can be grown on available, device quality substrates. However, lattice-mismatched systems, again with commercially exploitable and technologically interesting properties, could be grown on readily available, device grade, substrate materials of GaAs and InP. Uniquely, these mismatched materials offer the opportunity of tailoring two of the fundamental properties of semiconductors i.e. the energy gap and lattice parameter independently of one another[1]. Here, attention is restricted to the moderately strained III-V combination of (InGa)As-GaAs deposited on GaAs substrates. For modest amounts of indium in the alloy the one-dimensional potential steps in a layered combination are relatively small. For example, the conduction band step is of the order of 70-80 meV for an indium fraction of ~ 0.12, making the system an ideal one to study superlattice effects.

2. THEORETICAL BACKGROUND - STRESS EFFECTS ON THE BAND STRUCTURE

In considering the description of (InGa)As-GaAs QWs grown on a (001) substrate (and buffer) the (InGa)As layers will be

Condensed Systems of Low Dimensionality
Edited by J. L. Beeby *et al.*, Plenum Press, New York, 1991

under biaxial compression in the layer planes (x-y) and subjected to a shear, uniaxial stress in the growth direction (z). The GaAs substrate and buffer regions are far thicker than the epitaxially deposited (InGa)As layers and provided we do not exceed the critical layer thickness[1] then we are justified in assuming that it is only the (InGa)As layers that are elastically strained.

The biaxial strain, ϵ, is given by

$$\epsilon = (a_2 - a_1)/a_1 \tag{1}$$

where a_1 and a_2 are the lattice constants of (InGa)As and GaAs respectively. The lattice constant of the alloy is calculated by assuming a linear variation between the binary end-members.

As noted above, the (InGa)As is under compression in the layer planes ($\epsilon < 0$) and the net effect of the hydrostatic and uniaxial components of the strain is to increase the band gap relative to the unstrained value and remove the degeneracy of the light and heavy hole valence bands. The application of stress radically alters the in-plane (x-y) dispersion of the light and heavy holes, with the $m_j = \frac{3}{2}$ (heavy hole) band acquiring a "light" mass in the x-y plane whilst the opposite is true for the $m_j = \frac{1}{2}$ (light hole) band. This so-called "mass reversal" also occurs in unstrained quantum well systems and is a reflection of the lowering of the symmetry of the crystal. The strain does not couple the $m_j = \frac{3}{2}$ and $m_j = \frac{1}{2}$ bands in the z-direction, however the changes in the magnitudes of the fundamental heavy and light hole energy gaps do mean that there are some changes to the effective masses in this direction. These changes can be estimated using the Kane model and for moderately strained materials they are almost negligible. We therefore ignore this effect in our analysis.

If we denote the unstrained band gap of the $In_xGa_{1-x}As$ by $E_0(x)$ the strained conduction to heavy hole (C-HH) and strained conduction to light hole band gaps are given by[2]

$$E_0^{C-HH} = E_0(x) + \delta E_H - \delta E_S \tag{2}$$

$$E_0^{C-LH} = E_0(x) + \delta E_H + \delta E_S - (\delta E_S)^2/2\Delta_0 \tag{3}$$

where Δ_0 is the spin-orbit splitting in the (InGa)As and δE_H and δE_S are the hydrostatic and shear components of the strain splittings respectively. These last two quantities are expressed in terms of the elastic stiffness constants (C_{ij}) and appropriate deformation potentials as[2]

$$\delta E_H = 2a[(C_{11} - C_{12})/C_{11}]\epsilon \tag{4}$$

$$\delta E_S = 2b[(C_{11} + 2C_{12})/C_{11}]\epsilon. \tag{5}$$

The parameters a and b are the interband hydrostatic and uniaxial deformation potentials respectively. Values of a and b for the ternary are determined by a linear interpolation between the values for the binary end members. The values used in our calculations are gathered together in Table 1.

Of course, in the systems we are concerned with we not only have to account correctly for strain induced effects but also

Table 1. Deformation potentials and elastic stiffness
constants used in the calculation of the
strained (InGa)As band gaps. Values for the
ternary are found from linear interpolation.

Material	a(eV)	b(eV)	C_{11} (10^{11}dyn/cm^2)	C_{12} (10^{11}dyn/cm^2)
GaAs	-7.1	-1.7	11.88	5.38
InAs	-5.9	-1.8	8.33	4.53

for quantum confinement effects. We have calculated the
position and width of the appropriate electron and hole subbands
within the envelope function approximation; matching conditions
used ensured continuity of the particle wavefunction and
"current" at each interface. Necessary input parameters into
the calculations are the variations in effective masses as a
function of In fraction, the variation of the unstrained gap
$E_0(x)$, the change in Δ_0 with In fraction and, of course, the
appropriate band offsets. The variations in the physical
parameters with In fraction are gathered together in Table 2.

In all our comparisons we have used a band offset, assumed
to be independent of the In fraction and equal to a value of Q_c
of 0.67[3,4,5]. Our definition of Q_c is such that this quantity
corresponds to that fraction of the energy gap difference
between the GaAs barrier and the *strained* conduction to heavy
hole band gap of (InGa)As that appears in the conduction band.
This particular value means that for GaAs barriers we always
have a mixed situation for the heavy and light holes; the heavy
holes being confined in the (InGa)As wells whilst the highest
energy light hole state is in the GaAs. Clearly the values of
some other parameters will also be important in comparing the
calculated and measured transition energies, and it is fair to
say that many of these remain uncertain. This seems to be
particularly true of the GaAs deformation potential, a. The
value we are currently using is -7.1 eV, which is the value
adopted by Gershoni and co-workers[6,7] to fit optical data on
both (InGa)As-InP and (InGa)As-GaAs quantum wells.

Table 2. Variation of materials parameters as a function
of In fraction, x. All masses are in units of
m_0 and all energies in eV.

Quantity	Variation with In fraction, x
Unstrained gap (4K)	$E_0(x)=1.519-1.5387x+0.475x^2$
electron mass	$m_e^*=0.0665-0.0435x$
heavy hole mass	$m_{hh}^*=0.34$
light hole mass	$m_{lh}^*=0.094-0.062x$
spin-orbit splitting	$\Delta_0=0.341-0.09x+0.14x^2$

3.   GROWTH AND SAMPLE DETAILS

The layers were all deposited by molecular beam epitaxy in a Varian modular Gen II growth system.   The samples were deposited on undoped (001) GaAs substrates mounted in indium free holders and rotated at ~ 20 rpm during growth to ensure lateral uniformity of the layers.   The substrate temperature during growth of the samples with an indium mole fraction of near to 0.12 was nominally 580°C, which is in the temperature range where significant re-evaporation of In occurs.   To compensate for the loss, an increased In flux has to be used. Growth rates for both GaAs and (InGa)As were measured using the RHEED oscillation technique on a GaAs monitor slice prior to growth of the quantum well samples.   The growth rate for GaAs was 1 μm/hr and the As_4 flux supplied to the surface was just sufficient to maintain an arsenic stabilized surface for both GaAs and (InGa)As.   None of the samples was intentionally doped.

The growth sequence always commenced with the deposition of a 1 μm GaAs buffer layer followed by either 2, 5 or 20 repeats of an (InGa)As-GaAs well and barrier combination where the wells were nominally 25 Å wide and barriers 100 Å thick.   Each structure was terminated with a GaAs capping layer.   We have no evidence to suggest that any of the samples exceeds the critical thickness limit for the relaxation of the compressive strain via dislocation generation.   High resolution X-ray diffraction measurements on these samples is in support of this assertion[8].

4.   EXPERIMENTAL DETAILS

The photoluminescence (PL), photoluminescence excitation (PLE) and circularly polarized PLE (PPLE) measurements were made at ~ 4 K.   The PL measurements were made using an Ar^+ pumped dye (styryl-9) laser set at an energy above the GaAs band gap.   The same dye laser provided the tunable excitation source for the PLE measurements.   PPLE is a useful technique for distinguishing luminescence associated with either light or heavy holes. Details of the technique can be found in Ref. 9.   In our PPLE experiments, the linearly polarized laser light was chopped by an oscillating stress plate to produce alternating σ^+ and σ^- excitation.   The spectra were recorded at the e1-hh1 exciton position, selectively detecting only changes in one sense of the circularly polarized emission.   The phase of the detection system was arranged so that this heavy hole signal gave rise to a peak in the PPLE.   Higher energy transitions involving heavy hole states will therefore result in an increase in the PL intensity and hence a peak in the PPLE spectrum.   Conversely the participation of light hole states will decrease the PL intensity, producing dips.

5.   RESULTS AND DISCUSSION

5.1 Coupled Wells - Evolution into Superlattices

Little work exists which catalogues the changes of the electronic structure of quantum wells into their SL counterparts as one alters not the dimensions of the component parts but the number of repeats of the well and barrier "building block".   An exception is the work of Morris and co-workers[10] who characterized, both optically and structurally, (InGa)As-GaAs "superlattices", as a function of the number of periods in the structure. These authors studied samples grown by low pressure metal-organic chemical vapor deposition on (001) oriented GaAs substrates, the nominal In fraction was close to 0.16 and the well/barrier ratio was approximately constant at ~ 70 Å/70 Å. The number of repeats of the well/barrier structure was varied from 5 through 10 to 20. If one assumes that the band offset ratio between the strained electron to heavy hole gaps of (InGa)As and GaAs is 67:33 at this indium fraction[3,5] ($x \sim 0.16$) and calculates the predicted band widths for the $n = 1$ electron states then one arrives at values of ~ 7.5 meV. Clearly, the nonzero band width indicates that the system is a coupled one and even for the 5-repeat structure the separation between the hybridized, isolated well electron states of ~ 1.5 meV suggests that this structure will be quite a good approximation to a superlattice for these carriers.

We have designed (InGa)As-GaAs samples so that the $n = 1$ electron miniband width for an infinite periodic structure would be ~ 20 meV. We have studied samples that have 2, 5 and 20 (InGa)As layers, all with a nominal indium fraction of ~ 0.12. In contrast to other work, this particular set of samples gives us the opportunity to observe the hybridization of the individual, isolated quantum well eigenfunctions into the extended miniband states of the SL. Furthermore we can follow the evolution of states above the barriers as continuum resonances are lost and forbidden gaps and unconfined minibands appear in the carrier dispersion[11].

The PLE spectra are presented in Figs. 1a - 1c for the samples with 2, 5 and 20 wells respectively. The circularly polarized PLE spectra for the samples are represented by the dashed lines in this figure.

5.1.1. The 2-well sample. Even though we refer to this sample as having two wells, with our supposed strained electron to heavy hole band offset ratio of 67:33 the sample is only a symmetric, double quantum well system for the electrons and heavy holes. The electrons and light holes form a type-II configuration so that the light hole valence band edge is exactly out of phase with that of the heavy holes, making the (InGa)As layers barriers to light hole motion in the growth direction (z-direction). A schematic of the conduction and valence band edge configurations for this sample is shown in Fig. 2a. The structure is simultaneously a symmetric double well system for electrons and heavy holes and a double barrier system for the light holes.

Given this band structure arrangement and our assumptions about the magnitude of the various potential steps, we can make use of some straightforward quantum mechanics[11] to predict the allowed optical transitions. Taking transitions involving electrons and heavy holes first: the eigenfunctions for electrons and heavy holes in this DQW structure are simply the symmetric and antisymmetric combinations of the eigenfunctions of the isolated QWs, the symmetric combination having the lower energy eigenvalue. Allowed optical transitions should only occur between those energy levels whose envelope functions have the same parity. In addition to the parity allowed transitions

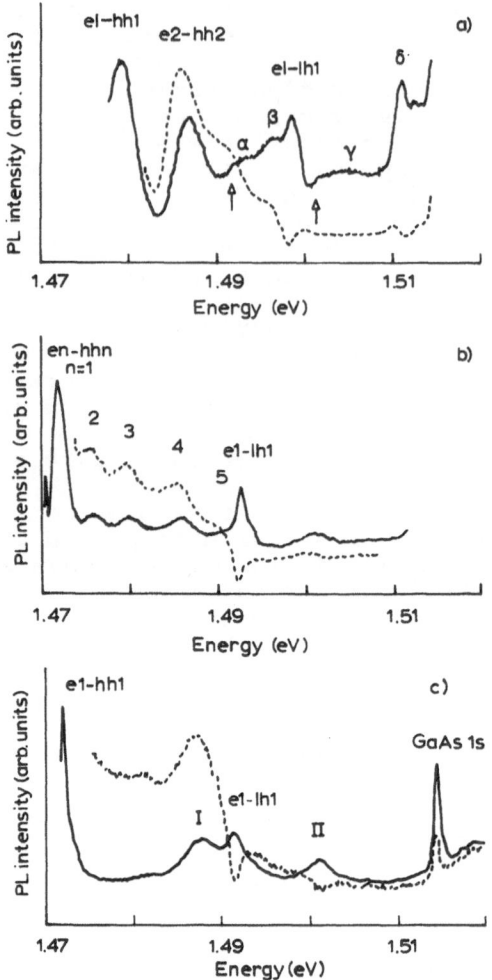

Fig. 1. Low temperature (~ 4 K) photoluminescence
excitation (PLE) spectrum (full curve) and
circularly polarized photoluminescence excitation (PPLE) spectrum (dashed curve) from
(InGa)As-GaAs quantum well samples with
(a) 2 (InGa)As layers, (b) 5 (InGa)As layers and
(c) 20 (InGa)As layers.

between confined states of the DQW there also exists the
possibility of allowed optical transitions between confined
electron (heavy hole) states and the heavy hole (electron)
continuum in the GaAs. The situation for optical transitions
involving light holes is somewhat different. The (InGa)As
layers are barriers for the light holes. With an offset ratio
of 67:33 and using the deformation potentials in Table 1, we
estimate that for an indium fraction of ~ 0.11, the light hole
barriers are only ~ 4.5 meV high. We have calculated the
transmission coefficient through the nominal 25 Å/100 Å/25 Å
double barrier structure and find, as expected, no sharp peak in
the transmission coefficient. The combined effects of a light
mass particle, low potential steps and thin barriers smear out

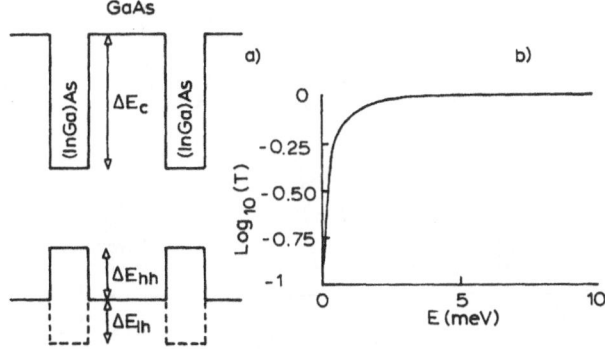

Fig. 2. (a) Schematic of the conduction and valence band
         edges for the '2-well' sample assuming a
         strained conduction band offset fraction of 0.67
         and (b) the energy spectrum for the probability
         of transmission of a light hole through the
         structure. For an indium fraction of 0.11 the
         light hole barrier is ~ 5 meV in magnitude.

any sharp features, making the influence of the (InGa)As only
the weakest of perturbations on the light hole motion. It is
reasonable to expect that the light holes in the system will
behave in an almost three-dimensional manner and we therefore
expect to find optical transitions involving confined electrons
and quasi-3D light holes in the GaAs.

     Returning to Fig. 1a, we identify the lowest energy
transition with the creation of the e1-hh1 free exciton, where
e1 and hh1 label the levels corresponding to the symmetric
combinations of the isolated well eigenfunctions. The next peak
is assigned to absorption between the higher energy,
antisymmetric heavy hole and antisymmetric electron states of
the DQW, which we call e2-hh2. The PPLE spectrum for this
sample allows us to confirm this assignment and to reveal the
character of further transitions. The dip at 1.4986 eV
indicates that this transition involves a light hole state.
This we assign to the e1-lh1 free exciton.

     Further features are clearly visible in the spectrum. The
most obvious of these are labeled by the lower case Greek
characters in the figure. To aid in the identification of these
features we have "fitted" the positions of the $\Delta n = 0$
transitions that we have identified. We allow the layer
thicknesses and indium fraction to vary between limits set by
high resolution X-ray diffraction on similar samples and the
relatively small, but inevitable, errors associated with flux
variations etc. during growth at the substrate temperature of
580°C. Our "best fit" to this particular data is with an indium
fraction of 0.105, a GaAs barrier of 110 Å and (InGa)As wells
25 Å wide. To make the comparison we assumed a binding energy
of ~ 8 meV for the e1-hh1 exciton[12] and a slightly smaller
value for those transitions which we think involve extended or
quasi-three-dimensional states e.g. e1-lh1. With the aid of the
calculations we assign peaks α and β to the exciton transitions
e1-hhc and ec-hh1 respectively. The notation hhc(ec) indicates
that the state involved is that at (or very close to) the onset

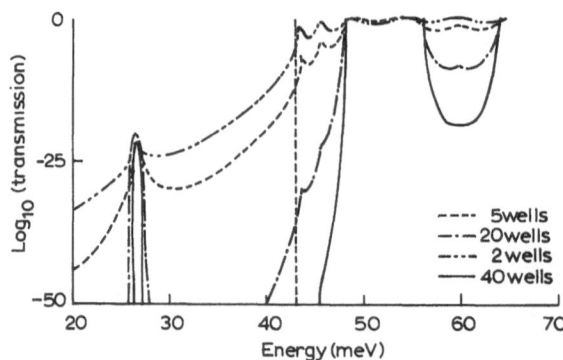

Fig. 3. Calculated transmission coefficient for the
heavy holes through samples with 2, 5, 20 and 40
wells. Note that the resonances centered near
27 meV which are below the top of the barrier
should reach unity, the fact that they do not is
an artefact of the number of energy points used
in the calculation. All other features are
significant. The dashed vertical line at ~ 43
meV correspond to the height of the heavy hole
potential step. The energy zero is at the
position of the strained (InGa)As heavy hole
valence band maximum.

of the heavy hole (electron) continuum states at the top of the
heavy hole (electron) barrier. In fact the "continuum" states
involved can be calculated by examining the transmission
resonances in the energy range above the top of the barriers
(see Fig. 3). Depending on the precise details of effective
masses, barrier heights and well widths, these transmission
resonances can give rise to sharp peaks that manifest themselves
as exciton transitions in the PLE or absorption spectrum[11,12].
Approximately 2 meV below the peak α we see a distinct rise in
the PLE signal. We cannot be definite about the origin of this
rise, but speculate that it may be due to the excited states of
the e2-hh2 exciton and its attendant continuum. The rather
broad feature labeled γ we think is probably associated with the
transition e2-hhc. On the low energy side of γ (marked by an
arrow) we again note a distinct rise in the PLE signal. Once
more we cannot be categorical about the origin of the feature,
but given the quasi-3D nature of the e1-lh1 exciton (and
therefore its smaller binding energy), this feature could well
originate from excited states of the e1-lh1 exciton, unresolved
from the continuum. Finally, consider the origin of the sharp
peak δ. The PPLE in this region of the spectrum is a little
ambiguous, but the increase at 1.51 eV suggests that the peak δ
probably involves some heavy hole contribution. The
transmission resonances for the heavy hole DQW system are shown
in Fig. 3. We find a sharp resonance at the top of the heavy
hole well, a further slightly stronger resonance a couple of meV
higher and then almost unity transmission thereafter. The onset
of this region of unity transmission in fact marks the energy at
which a miniband will be formed in the SL which has the same
well and barrier dimensions as the 2-well sample. We therefore
assign the peak δ to an exciton transition involving the state
in the heavy hole continuum at which unity transmission is

reached and, most likely from its energy position, the confined electron state, e2.

5.1.2. The 5-well sample. Figure 1b shows the PLE and PPLE spectra from the 5-well sample. Again strong e1-hh1 and e1-lh1 peaks are evident. Between these two sharp features there are three clear peaks and we note an asymmetry of the light hole peak on its low energy side. Three peaks are also seen in the PPLE in this energy range and a distinct change of slope can also be discerned prior to the light hole dip. All these features are of heavy hole nature and we assign them to free excitons built from the optically allowed combinations of the confined excited states of the coupled five-well sample i.e. en-hhn where n runs from 2 to 5. The 5 electron and hole states are once more the linear combinations of the isolated well eigenstates. The calculated separation between the individual electron levels ranges between 2 and 5 meV. This separation is too large to talk about this structure in terms of being a superlattice. Indeed, to analyse the spectral data in a proper manner the individual electron states have to be considered. Once more we have adjusted the nominal values of well and barrier width and indium fraction so that the envelope function calculations (including correction for exciton binding energies as in the 2-well case) reasonably reproduce the strong exciton features. To describe this sample we used x = 0.11, and (InGa)As width of 27 Å and a GaAs barrier width of 110 Å. To complete the analysis of the spectrum we need to interpret the peak at ~ 1.501 eV. Again we have calculated the transmission probability of the various carriers through this structure and in Fig. 3 we show the calculations for the heavy holes. If we compare the 5-well (dashed) calculation to that of the 2-well (dotted) then we find they are remarkably similar. Below the heavy hole barrier, ΔE_{hh}, which has a magnitude of 43 meV, we find strong resonances associated with the 5 bound states of the system. These are centered at an energy about 27 meV above the strained (InGa)As valence band edge and all contained in a width of < 2 meV. Just above the top of the barrier we again see a sharp resonance, however now a little weaker than in the 2-well spectrum, this is followed by a stronger resonance about 2 meV higher in energy. Above this point this is again unity trans-mission probability until one reaches an energy of ~ 57 meV, then the coefficient drops below unity and remains constant until ~ 65 meV where it regains its unity value. This behavior in the region between 57 meV and 65 meV indicates that part of the energy spectrum where a "forbidden" gap will appear in the eigenstates above the barrier as the number of periods increases. These calculations again indicate that there is a fairly high probability of finding hole states at or near to the top of the well whose wavefunctions will have some overlap with the confined electron states in the system. We therefore attribute the peak at ~ 1.501 eV to an exciton transition involving a confined electron state (most likely, e5) and a resonant state in the heavy hole continuum. A contributing factor to this transition being weaker than the equivalent one in the 2-well sample is the diminished heavy hole transmission probability near the top of the well, indicating a reduced amplitude of the heavy hole wavefunction in the well regions.

5.1.3 The 20-well sample. The evolution of the system to a superlattice can be followed by further increasing the number of wells in the system. The final sample we have studied has a total of 20-wells (20-barriers for the light holes) and the PLE and PPLE spectra from this sample are shown in Fig. 1c. Given the nominal well and barrier widths we calculate that a SL with this period would have an n = 1 electron miniband width of ~ 20 meV, and that the top of the band would be at an energy smaller than the value of the 1D SL potential step, ΔE_c. This means that the separation between individual electron states would be of the order of 1 meV - intuitively, a value small enough to talk about the electrons having formed a miniband rather than having to consider 20 individual eigenstates. The transition to a superlattice means that there now exists a 1D dispersion relation for the carriers in the growth direction. The wavevector in this direction, denoted by q, has now become a good quantum number. The periodic variation in the potential in the growth direction will open up forbidden energy gaps in this E-q dispersion at the values of $q = n(\pi/d)$, where n is an integer and d is the period of the SL. In a reduced zone scheme, forbidden gaps appear between bands alternately at $q = (\pi/d)$ and at $q = 0$, i.e. at the 1D mini-Brillouin zone-edge and zone-center.

Further calculation supports the idea that the 20-well sample is a good approximation to a SL. In Fig. 3 we show the energy spectrum of the transmission coefficient for the heavy holes through the 20-well structure. A clear band of states appears (centered at ~ 27 meV) below the heavy hole barrier (again 43 meV in magnitude). However, in contrast to the previous cases the transmission probability at the top of the well (near 43 meV) has dropped by ~ 25 orders of magnitude. The transmission probability only reaches unity 5 meV or so above ΔE_{hh}, and an allowed band of states extends up to ~ 57 meV. This is followed by a "forbidden" gap of ~ 8 meV in magnitude until the next band of allowed states is encountered. For illustrative purposes we have increased the number of wells to 40 and repeated the calculation. Qualitatively there is no change in the spectrum from the 20-well calculation. Quantitatively the only difference is the increase in the contrast between the allowed and forbidden regions of the spectrum, but, for example, there is no discernible change in the width of the forbidden gap between the allowed heavy hole states. From the calculations presented above, showing the existence of forbidden gaps in the transmission through the 20-well sample, it is clear that the number of repeats is sufficient for the structure to be a good approximation to a superlattice.

Let us now return to Fig. 1c. The sharp peak at 1.4721 eV we assign to the recombination of free excitons at an energy smaller than the energy gap between the bottom of the n = 1 electron miniband and the top of the n = 1 heavy hole miniband at the center of the 1D mini-Brillouin zone i.e. q = 0. We refer to this transition as e1-hh1(Γ). At 1.4914 eV the PPLE spectrum shows a distinct dip - this transition we ascribe to the e1-lh1(Γ) free exciton transition. The further sharp peak at 1.5146 eV we assign to the 1s state of the bulk GaAs free exciton. Two further, somewhat broader, features appear at 1.4879 eV and 1.5011 eV which we refer to as peaks I and II respectively.

In our analysis of the spectral features of the 2-well and 5-well samples we explained the additional features in the spectrum by invoking transitions between confined states and resonances in the heavy hole or electron continuum. To explain peaks I and II it seems that it is also necessary to invoke transitions to higher states, but, as we have shown above, we can no longer talk about resonances in the continuum. The number of wells in the structure means we have to formulate an explanation that properly takes into account the 1D periodicity of the structure and the evolution of the miniband dispersion.

The strongest direct optical transitions are expected at $q = 0$ or at $q = (\pi/d)$. Furthermore, the parity of the hybridized, individual-well envelope functions in the z-direction suggests that strong transitions will only occur between subbands with the same index, n. An intriguing feature of the SL band structure is the existence of M_1 critical points at the top (bottom) of the nth electron (hole) subband at $q = n(\pi/d)$. Zone-folding back to the first 1D minizone means that the M_1 points alternate between $q = (\pi/d)$ and $q = 0$. An M_1 critical point (often called a saddle point) means that the effective mass in the region of the critical point is negative. The Coulomb interaction between electrons and holes associated with these saddle points can correlate their motion leading to the formation of excitonic resonances below the energy of the saddle point. In bulk materials Kane[13] for example showed theoretically that these resonances manifest themselves as rather broad, asymmetric features. In Kane's treatment the resonance appears centered at an energy below the M_1 critical point which is almost identical to the energy at which the ground state exciton appears below the minimum M_0. Recently saddle point excitons in SLs have been studied theoretically[14] and been seen experimentally in PLE studies of GaAs-(AlGa)As samples[15,16].

Peak I in the PLE spectrum of the 20-well samples is the exciton resonance associated with the M_1 critical points of the n = 1 electron and heavy hole minibands at $q = (\pi/d)$. We denote this transition as e1-hh1(π). The calculated sum of the n = 1 electron and heavy hole miniband widths for this sample is ~ 16 meV and we note that the separation between e1-hh1(Γ) and e1-hh1(π) is also ~ 16 meV. For this particular sample therefore the e1-hh(π) saddle point exciton resonance appears at an energy below the saddle point equal to the energy at which the e1-hh1(Γ) exciton ground state appears bound below the fundamental gap. This is quantitatively what is expected from Kane's analysis and lends added weight to our assignment. All that remains now is to assign peak II.

We think that the feature at 1.501 eV again has its origins associated with an M_1 critical point. This time the saddle point involved is the M_1 critical point of the n = 2 heavy hole miniband at $q = 0$, and we assign the peak to the parity forbidden e1-hh2(Γ) exciton transition. This $\Delta n \neq 0$ transition may become allowed for a variety of reasons: (i) asymmetry in the potential profile caused by extraneous electric fields, grading of the indium profile or local departures from a random alloy, (ii) some deviation from the exact periodicity of the SL or (iii) by analogy with the GaAs-(AlGa)As systems, in-plane mixing of the light and heavy holes[17]. A-priori we cannot rule out any of these possibilities, and (i) and (ii) are distinctly possible given the state of knowledge about the MBE growth of strained, indium-containing alloys, particularly at elevated growth temperatures. Point (iii) is the principal reason that $\Delta n \neq 0$ transitions are seen in the GaAs-AlAs system and we suggest it is the most probable reason for the observation of this feature here. The in-plane mixing of the light and heavy holes may also explain the somewhat ambiguous nature of the PPLE signal in the vicinity of the e1-hh2(Γ) exciton. One further point remains to be made about this saddle point feature. Unlike the e1-hh1(π) exciton, which is a resonance, it is possible (because only the hole effective mass is negative) that the e1-hh2(Γ) exciton will be a bound state solution of the corresponding effective mass equation that describes the exciton.

SUMMARY

We have concentrated on elucidating some of the electronic and optical properties of moderately strained (InGa)As-GaAs isolated quantum wells, coupled wells and superlattices. By varying the number of repeats of the well and barrier building-block of an (InGa)As-GaAs strained layer structure we have illustrated the evolution of the system from coupled quantum wells to a superlattice. The PLE spectra of the 2-well and 5-well samples have shown strong $\Delta n = 0$ exciton transitions and further features associated with transitions from confined states to continuum states. When a SL is formed in the 20-well sample, q becomes a good quantum number and we see momentum conserving optical transitions at $q = 0$ and $q = (\pi/d)$. We have also identified transitions associated with M_1 critical points in the SL band structure. One of these is a $\Delta n = 0$ exciton resonance below the saddle point whilst the other is most likely a bound state of a $\Delta n \neq 0$ exciton, e1-hh2(Γ), which has become allowed due to mixing of the light and heavy hole bands away from $k = 0$.

REFERENCES

1. G.C. Osbourn, **Superlattices and Microstructures**, 1:223 (1985)
2. F.H. Pollak and M. Cardona, **Phys. Rev.**, 172:816 (1968)
3. J-Y. Marzin, M-N Charasse and B. Sermage, **Phys. Rev. B**, 31:8298 (1985)
4. J. Menendez, A. Pinczuk, D.J. Werder, S.K. Sputz, R.C. Miller, D.L. Sivco and A.Y. Cho, **Phys. Rev. B**, 36:8165 (1987)
5. M.J. Joyce, M.J. Johnson, M. Gal and B.F. Usher, **Phys. Rev. B**, 38:10978 (1988)
6. D. Gershoni, H. Temkin, M.B. Panish and R.A. Hamm, **Phys. Rev. B**, 39:5531 (1989)
7. D. Gershoni, J.M. Vandenberg, S.N.G. Chu, H. Temkin, T. Tanbun-Ek and R.A. Logan, **Phys. Rev. B**, 40:10017 (1989)
8. P.F. Fewster - private communication
9. C. Weisbuch, R.Dingle, A.C. Gossard and W. Weigmann, **Sol. St. Comm.**, 38:709 (1981)
10. D. Morris, C. Lacelle, A.P. Roth, P. Maigne and J.L. Brebner, in: "Proceedings of 4th Int. Conf. on Modulated Semiconductor Structures", Ann Arbor, Michigan (July 1989) **Surf. Sci.**, 228:347 (1990)
11. G. Bastard, "Wave Mechanics Applied to Semiconductor Heterostructures", Les Editions de Physique, France (1989)

12. J.J. Song, Y.S. Yoon, A. Fedotowsky, Y.B. Kim, J.N. Schulman, C.W. Tu, D. Huang and H. Morkoç, **Phys. Rev. B,** 34:8958 (1986)
13. E.O. Kane, **Phys. Rev.,** 180:852 (1969)
14. H. Chu and Y-C. Chang, **Phys. Rev. B,** 36:2946 (1987)
15. J.J. Song, P.S. Jung, Y.S. Yoon H. Chu, Y-C. Chang and C.W. Tu, **Phys. Rev. B,** 39:5562 (1989)
16. B. Deveaud, A. Chomette, F. Clerot, A. Regreny, J.C. Maan, R. Romestain, G. Bastard, H. Chu and Y-C. Chang, **Phys. Rev. B,** 40:5802 (1989)
17. J.N. Schulman and Y-C. Chang, **Phys. Rev. B,** 31:2056 (1985), Y-C. Chang and J.N. Schulman, **Phys. Rev. B,** 31:2069 (1985)
18. K.J. Moore, G. Duggan, K. Woodbridge and C. Roberts, **Phys. Rev. B,** 41:1090 (1990)
19. K.J. Moore, G. Duggan, K. Woodbridge and C. Roberts, **Phys. Rev. B,** 41:1095 (1990)

MAGNETOSPECTROSCOPY OF MOCVD GROWN GaInSb/GaSb STRAINED LAYER

QUANTUM WELLS

G. Rees,* S.K. Haywood,† R.W. Martin, N.J. Mason,
R.J. Nicholas, G.M. Sundaram, P.J. Walker,
and R.J. Warburton

Clarendon Laboratory, Oxford, UK
† Presently at Department of Electronic Engineering
University College, London, UK

1. INTRODUCTION

Biaxial strain in a pseudomorphically grown heterostructure provides a further item in the toolkit of the band structure engineer, complementing alloying and quantum confinement. It is a valuable asset for constructing new materials offering improved optoelectronic device performance. Modeling and design of devices is an important step[1-4] in the fabrication process and demands a knowledge of these factors of strain, confinement and material composition on electronic structure. The effective Hamiltonian provides an economical and accurate means of describing band structure for device modeling purposes which automatically includes simplifications due to symmetry.

In this paper we show how measurements of cyclotron resonance[5], interband-magnetophotoconductivity and Shubnikov-de Haas effect[6] supply complementary information on the electronic structure of a strained quantum well which can be interpreted simply within an effective Hamiltonian model for device design. The symmetry reduction caused by quantum confinement and biaxial strain is known to split the light and heavy hole levels and to simplify the associated dispersion curves[7]. We show how this simplification operates in magnetospectroscopy and in particular modifies the spin structure, leading to a 2D spin system for $m_J = \pm\frac{3}{2}$ 'heavy' holes and complementary behavior for $m_J = \pm\frac{1}{2}$ 'light' holes.

In the following we describe our effective Hamiltonian calculations and the simplifications arising from level splitting. These calculations provide a framework for interpreting measurements on p-type, strained layer quantum wells. We describe how tilted field Shubnikov-de Haas measurements provide evidence of a 2D spin system for heavy holes and we show how this comes about from the reduction to tetragonal symmetry caused by confinement and strain.

---

* On leave from Plessey Research, Caswell, Towcester, UK.

Condensed Systems of Low Dimensionality
Edited by J. L. Beeby *et al.*, Plenum Press, New York, 1991

2.   EFFECTIVE HAMILTONIAN CALCULATIONS

We consider the 4 x 4 effective Hamiltonian[8-12] describing the $m_J = \pm\frac{3}{2}$ and $\pm\frac{1}{2}$ valence band levels in a magnetic field in the axial approximation[13], partitioned into 2 x 2 blocks:

$$H = \begin{bmatrix} P_+ - 3\kappa\mu_B B_z & 0 & Q - \sqrt{3}(\gamma_3 + \kappa)\mu_B B_- & R \\ 0 & P_+ + 3\kappa\mu_B B_z & R^\dagger & -Q^\dagger + \sqrt{3}(\gamma_3 - \kappa)\mu_B B_+ \\ Q^\dagger - \sqrt{3}(\gamma_3 + \kappa)\mu_B B_+ & R & P_- - \kappa\mu_B B_z & -2\kappa\mu_B B_- \\ R^\dagger & -Q + \sqrt{3}(\gamma_3 - \kappa)\mu_B B_- & -2\kappa\mu_B B_+ & P_- + \kappa\mu_B B_z \end{bmatrix} \qquad (1)$$

The Hamiltonian is expressed in the basis $m_J = \frac{3}{2}, -\frac{3}{2}, \frac{1}{2}, -\frac{1}{2}$ using the following terminology:

$$P_\pm = -p_z^2(\gamma_1 \mp 2\gamma_2)/2m - 2B_z\mu_B(\gamma_1 \pm \gamma_2)(a_+a_- + \tfrac{1}{2}) \mp \zeta + V$$

$$Q = \gamma_3 p_z ((12\mu_B B_z)/m)^{1/2} a_-$$

$$R = \sqrt{3}(\gamma_2 + \gamma_3)\mu_B B_z a_-^2$$

$$B_\pm = B_x \pm iB_y.$$

If this Hamiltonian is evaluated in the gauge $A = -(yB_z, zB_x, xB_y)$ then the Landau ladder operators for cyclotron motion about B_z, the field in the growth direction, are given by

$$a_\pm = (2\hbar eB_z/c)^{-1/2}(\hbar(\partial/\partial x \pm i\partial/\partial y)/i - e(yB_x \pm izB_y)/c).$$

For most of the measurements, and for the calculations reported here, the magnetic field is perpendicular to the well so that $B_\pm = 0$ and the in-plane field also disappears from $a_\pm$.

The Luttinger parameters γ_1, γ_2, γ_3 and κ are material and therefore position dependent, as are the strain splitting, 2ζ and the mean position of the valence band edge, V. Only the well is strained and the energy zero is chosen so that V = 0 in the cladding also. Band edge configurations are calculated using model solid theory[14] which predict a well for the $m_J = \pm\frac{3}{2}$ heavy holes and a barrier for the $m_J = \pm\frac{1}{2}$ light holes. Luttinger parameters are taken from magneto-optic measurements on the bulk $Ga_{1-x}In_xSb$ alloy[15]. These are modified to account for the hydrostatic component of strain in the well which increases the band gap, E_g, the conduction band effective mass, m_c^*, and the light hole mass, $(\gamma_1 + 2\gamma_2)^{-1}$ by 11% (appropriate to the 1% strain in our measured samples, see later), leaving the heavy hole mass $(\gamma_1 - 2\gamma_2)^{-1}$ unchanged. The spin-magnetic field coupling parameter κ, for the well is adjusted to fit Shubnikov-de Haas measurements (see later). The final set of band parameters is given in Table 1.

The Hamiltonian of Eqn. 1 has eigensolutions of the form

$$\psi = (\phi_{n-1}f_1(z), \phi_{n+2}f_2(z), \phi_n f_3(z), \phi_{n+1}f_4(z))^\dagger \qquad (2)$$

where the ϕ_n are Landau wavefunctions describing the motion in

Table 1. Electronic structure parameters used in
effective Hamiltonian calculations

	$\gamma 1$	$\gamma 2$	$\gamma 3$	κ	ζ(meV)	V(meV)	E_g(meV)	g*	m_c*
GaSb	14.3	5.3	4.0	4.0	——	——	——	——	——
$Ga_{0.85}In_{0.15}Sb$	14.4	5.3	6.3	5.7	37.5	16.2	698	-10.4	0.04

the plane of the well about the field B_z and n = -2,-1,0,...,
with appropriate elements set equal to zero to avoid negative
Landau quantum numbers.

Using the form Eqn. 2 as an eigenfunction of the
Hamiltonian Eqn. 1 yields an eigenvalue equation for $(f_i(z))$,
the column vector formed from the envelope functions describing
the z-motion:

$$(H_n(z) - E)(f_i(z)) = 0 \qquad (3)$$

where $H_n(z)$ is a 4 x 4 differential matrix operator acting in
the z-domain. Equation 3 is solved by linear combinations[16]
of bulk solutions for the well and cladding regions. Interface
conditions produce relations between the expansion coefficients
whose consistency provides a secular equation for the energy
eigenvalue. By suitable manipulation of the component matrices
the secular determinant can be made real definite instead of
complex. Eigenvalues can then be found by seeking a simple sign
change as a function of energy. It is well to be aware that the
determinant contains poles as well as zeros, analogous to those
which interlace the zeros in a 1D square well for a free
electron. In the present case these zeros and poles can cross
and hybridize with magnetic field. Foreknowledge helps to
unravel complex numerical behavior in tracing the eigenvalues
along the vein of a fan diagram. This method constitutes a
rapid and economical procedure for evaluating the eigenvalues of

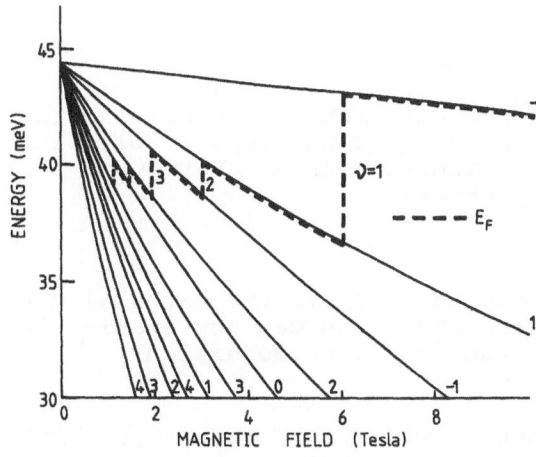

Fig. 1. Calculated $m_J = \frac{3}{2}$ energy levels versus magnetic
field. The dotted lines indicate the Fermi
level at 0 K for sample 449.

a model Hamiltonian for comparison with experiment. A typical
fan diagram is shown in Fig. 1.

2.1 Simplifications due to Level Splitting

Both biaxial strain and quantum confinement are known to
remove the degeneracy between the $m_J = \pm\frac{3}{2}$ and $\pm\frac{1}{2}$ levels at the
band edge[7]. This reduction to tetragonal symmetry also splits
degeneracy in the magnetic spectrum and can be used to simplify,
at least qualitatively, our picture of magnetic spectra.

The terms $P_\pm$ on the diagonals of Eqn. 1 are the source of
this splitting. If we ignore the off-diagonal 2 x 2 blocks then
provided the field lies in the growth direction the Hamiltonian
becomes diagonal (the effects of the in-plane field are
discussed later). The nonmagnetic components of the remaining
diagonal terms provide the splitting, ΔE, between the $m_J = \pm\frac{3}{2}$
and $\pm\frac{1}{2}$ levels. If this splitting is large then the off-diagonal
2 x 2 blocks can be treated as a perturbation, in which they
contribute only in second order, and this justifies our block-
diagonalization.

The eigenvalues then have the simple form:

$$E_{n,\pm 3/2} = E_H - \hbar\omega_c (\gamma_1 + \gamma_2)(n_{\pm 3/2} + 1/2) \mp 3\kappa\mu_B B_z \qquad (4a)$$

$$E_{n,\pm 1/2} = E_L - \hbar\omega_c (\gamma_1 - \gamma_2)(n_{\pm 1/2} + 1/2) \mp \kappa\mu_B B_z \qquad (4b)$$

where E_H and E_L are the zero field band edges for $m_J = \pm\frac{3}{2}$ and $\pm\frac{1}{2}$
being the eigenvalues of $-p_z^2 (\gamma_1 \mp \gamma_2)/2m \mp \zeta + V$, ω_c is the free
electron cyclotron frequency and $n_{\pm 3/2}$ and $n_{\pm 1/2} = 0,1,2,3...$

Thus our simple picture for strong symmetry breaking,
caused by strain or quantum confinement, is of a fan of Landau
levels, depending linearly on B_z, with cyclotron masses
intermediate between those of light and heavy holes, split by
spin and associated with pure $m_J = \pm\frac{3}{2}$ and $\pm\frac{1}{2}$ states. The extent
to which the veins of Fig. 1 are curved or hybridized is a
measure of the importance of higher order corrections introduced
by off-diagonal elements and of the failure of our simple
picture.

2.2 Interface Conditions

We have used our calculations to examine the dependence of
energy levels on interface conditions. Following Ref. 9 we
separate the matrix Hamiltonian $H_n(z)$ in Eqn. 3 into its orders
of differential operator:

$$H_n(z) = A(z)d^2/dz^2 + B(z)d/dz + C(z) \qquad (5)$$

where A, B and C depend on z via the material dependence of the
Luttinger parameters etc.. We can now examine three
prescriptions for the interface condition:

(i)   simple continuity of $(f_i(z))$ and $d(f_i(z))/dz$;

(ii) continuity of $(f_i(z))$ and $(A(z)d/dz + B(z))(f_i(z))$,
      derived[9] by integrating the eigenvalue Eqn. 3 across
      the interface;

(iii) continuity of $(f_i(z))$ and $(A(z)d/dz + 1/2B(z))(f_i(z))$,
derived[17] by integrating a symmetrized Hamiltonian,
$d/dz(A(z)d/dz) + 1/2(B(z)d/dz + d/dzB(z)) + C(z)$ across the
interface. This prescription is also equivalent to that of
Altarelli[18].

In fact when γ_1, γ_2 and γ_3 are the same for well and cladding
these three interface conditions coincide, and in the present
case the difference causes a variation of at most 1/4% in the
eigenvalue over the energy range considered. It is interesting
that this equivalence depends only on the equality of the γ_i and
not on that of the Bloch basis.

3. MEASUREMENTS

3.1 Sample Growth and Preparation

Samples were grown along (001) by MOCVD under conditions
optimized for bulk GaSb growth[19]. The material grows
naturally p-type at around 10^{16} cm^{-3} because of a native defect.
X-ray diffraction confirms pseudomorphic growth of the alloy
showing a lattice expansion of the well in the growth direction,
from which we calculate an indium concentration of $x = 0.15$,
corresponding to 1% strain, and a 9.3 nm thick well. Two
samples were studied in detail, 290 and 449 with respective
mobilities of 1,900 cm^2/Vs and 9,300 cm^2/Vs and hole
concentrations of 2.7×10^{11} cm^{-2} and 1.6×10^{11} cm^{-2}.

3.2 Cyclotron Resonance

The selection rule $\Delta n = \pm 1$ for cyclotron resonance holds
irrespective of our simplifying arguments. At low temperatures
and for sufficiently high fields only the lowest levels are
filled. Figure 2 shows a comparison of measured and predicted
cyclotron absorption energy for $n = -2 \rightarrow -1$. Our simple
arguments show that this energy difference depends primarily on
γ_1 and γ_2 for well and barrier, and not on κ, although the
curvature in Fig. 2 indicates the extent to which other
parameters are coupled in by off-diagonal terms.

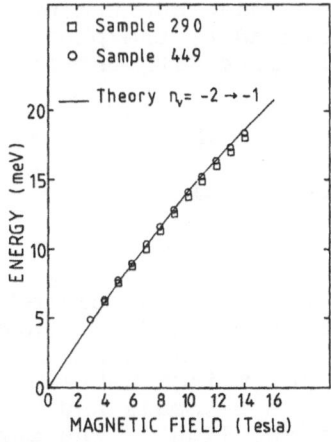

Fig. 2. Cyclotron absorption energy versus
magnetic field.

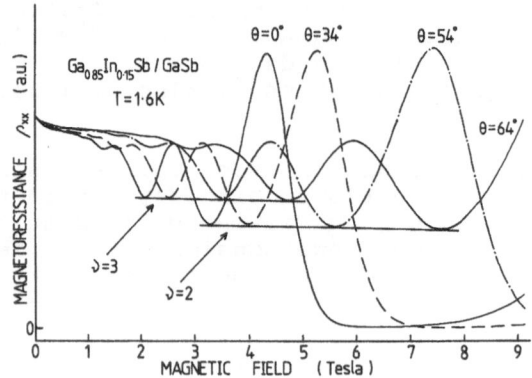

Fig. 3. Shubnikov-de Haas measurements in sample
449 with field tilted at an angle θ to the
normal.

3.3 Shubnikov-de Haas Effect

These measurements involve no selection rules. The change
of Landau level degeneracy with B_z moves the Fermi level between
Landau levels and the height of the maxima in ρ_{xx}, the in-plane
resistance, is an indicator of the lack of overlap and the
separation of adjacent, broadened Landau levels. Figure 3 shows
traces of ρ_{xx} vs B_z for a number of tilt angles for sample 449.
Since $|m_J|\kappa$ is large, around 9, the level separations in Fig. 1
are quite sensitive to κ. Figure 1 represents values calculated
using our best estimate for κ of 5.7 to reproduce the oscillations in ρ_{xx} for $\theta = 0°$ (field parallel to the growth
direction).

3.4 Interband Magnetophotoconductivity

This process involves excitation between Landau-spin levels
across the gap. Comparison with theory requires a calculation
of such levels in the nonparabolic conduction band, together
with an estimate of the exciton binding energy. Details are
described in Ref. 5. The momentum operator driving the
transition operates on the band edge Bloch functions and the
selection rules conserve Landau quantum number and involve a
spin change of unity, carried by the photon. Our simple picture
leads to principle transitions with the following selection
rules:

$$n_v, \ m_J = +\tfrac{3}{2} \rightarrow n_c = n_v - 1, \ m_S = +\tfrac{1}{2}$$

$$n_v, \ m_J = -\tfrac{3}{2} \rightarrow n_c = n_v + 2, \ m_S = -\tfrac{1}{2}$$

where $n_c = 0, 1, 2, \ldots$

Figure 4 shows measured and calculated transition energies.
Within our simple picture these energies are determined by γ_1,
γ_2, m_c and the spin splittings are determined by a cancellation
between κ and $g*$. Spin splittings between the two types of
transition are predicted to be small and are not resolved in our
experiments with unpolarized light. The curvature in Fig. 4 is
due to conduction band nonparabolicity as well as off-diagonal
terms.

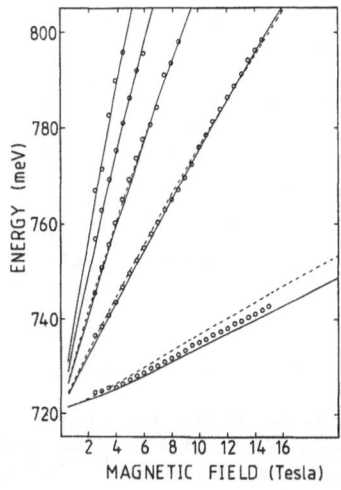

Fig. 4. Interband transition energies. Dotted (solid) lines correspond to spin up (down transitions. Circles are measurements on sample 449.

4.   TILTED FIELD EFFECTS

An in-plane magnetic field is responsible for the terms in $B_\pm$ in Eqn. 1 and also for terms implicit in $a_\pm$. It is still true, however, that $[a_-, a_+] = 1$ so that $a_\pm$ are still Landau ladder operators for cyclotron motion about B_z. Returning to our earlier arguments, symmetry breaking again decouples the pair of 2 x 2 block diagonal matrices and the in-plane field makes no explicit appearance in the $m_J = \pm\frac{3}{2}$ block, whose eigenvalues are still given approximately by Eqn. 4a. This simple picture is confirmed by our measurements of Shubnikov-de Haas effect in a tilted field (Fig. 3, $\theta \neq 0°$) where the shape of the traces is independent of tilt angle but the scale expands with increasing tilt as $\sec\theta$, confirming that the level separation of the $m_J = \pm\frac{3}{2}$ set depends only on B_z. In fact we expect small diamagnetic corrections from the in-plane field terms in the $a_\pm$, analogous to those found in n-type samples [20,21].

To assess the validity of our simple picture we can evaluate the lowest order correction involving in-plane magnetic field from the off-diagonal 2 x 2 blocks in Eqn. 1. The first order perturbation correction is zero since the off-diagonal terms only connect $m_J = \pm\frac{3}{2}$ with $m_J = \pm\frac{1}{2}$, respectively. The second order correction is simplified by the orthogonality of the Landau wavefunctions so that only one term contributes in the sum over Landau levels. The correction for the $m_J = \pm\frac{3}{2}$ levels is at most

$$E^{(2)}_{\pm 3/2} = 3(\gamma_3 \pm \kappa)^2 \mu_B (B_x^2 + B_y^2)/\Delta E$$

with a further strong reduction from the overlap between the z-dependent part of the envelope functions for the $m_J = \pm\frac{3}{2}$ and $\pm\frac{1}{2}$ states, spatially separated owing to the type II behavior of the $m_J = \pm\frac{1}{2}$ levels in the present system. This correction is around 1.3 meV for $m_J = +\frac{3}{2}$ and 0.1 meV for $m_J = -\frac{3}{2}$ for the highest recorded in-plane field of 8 T.

This lack of dependence on tilt angle in Shunbikov-de Haas measurements is also seen in p-type InGaAs/GaAs strained quantum wells[22] and in narrow, unstrained GaAs/AlGaAs quantum wells[23] where the degeneracy removal is entirely due to quantum confinement.

Within the 2 x 2 block describing the $m_J = \pm\frac{1}{2}$ levels the in-plane field contributes to off-diagonal terms which modify the approximate energy levels in Eqn. 4b to

$$E_{n,\pm 1/2} = E_L - \hbar\omega_c (\gamma_1 - \gamma_2)(n + 1/2) \pm \kappa\mu_B (B_z^2 + 4B_\parallel^2)^{1/2} \qquad (4c)$$

showing an enhanced dependence on in-plane field $B_\parallel = (B_x^2 + B_y^2)^{1/2}$. This effect should be observable in Shubnikov-de Haas measurements in quantum wells under sufficient strain that this splitting can more than compensate for the effect of quantum confinement, reversing the order of the levels E_L and E_H.

It is interesting to note that the part of the Hamiltonian Eqn. 1 involving spin-magnetic field interaction and level splitting due to symmetry reduction is entirely equivalent to the spin Hamiltonian of a localized paramagnetic $S = \frac{3}{2}$ ion in a tetragonal environment[24]:

$$H = g\mu_B B.S + D(S_z^2 - S(S + 1))$$

where g and D are replaced by 2κ and $\Delta E/2$.

SUMMARY

Using GaInSb/GaSb strained layer quantum wells we have shown how a combination of different magnetospectroscopic measurements can be interpreted within a strained layer Hamiltonian model to provide electronic structure for opto-electronic device modeling. The symmetry reduction and energy splitting between $m_J = \pm\frac{3}{2}$ and $\pm\frac{1}{2}$ levels due to strain and quantum confinement can greatly simplify electronic structure in a magnetic field leading to simple fan diagrams, selection rules and a 2D spin system for heavy holes with complementary behavior for light holes.

Acknowledgments

The authors wish to thank Drs. E.O'Reilly and J. Owen for helpful discussions and Mr. A.C.G. Wood for a preview of unpublished work. We wish to thank SERC and the Royal Society for support of this work.

REFERENCES

1. M. Asada, A. Kameyama and Y. Suematsu, **IEEE**, OE20:745 (1984)
2. J. Nagle, S. Hersee, M. Krakowski, T. Weil and C. Weisbuch, **Appl. Phys. Lett.**, 49:1325 (1986)
3. D.J. Robbins, Proc. Conf. SPIE, Novel Optoel. Devices

800:34 (1984)

4. A.J. Holden, Proc. 17th ESSDERC, Bologna, Italy, p487 (1987)

5. R.J. Warburton, G.M. Sundaram, R.J. Nicholas, S.K. Haywood, G.J. Rees, N.J. Mason and P.J. Walker, **Surf. Sci.**, 228:270 (1990)

6. R.W. Martin, R.J. Nicholas, G.J. Rees, S.K. Haywood, N.J. Mason and P.J. Walker, to be published.

7. E.P. O'Reilly and G.P. Witchlow, **Phys. Rev. B**, 34:6030 (1986)

8. M.H. Weiler, Semiconductors and Semimetals, Eds. R.K. Willardson and A.C. Beer (Academic, NY 1981) 16:119

9. L.R. Ram-Mohan, K.H. Yoo and R.L. Aggarwal, **Phys. Rev. B**, 38:6151 (1988)

10. F. Ancilotto, A. Fasolino and J.C. Maan, **Phys. Rev. B**, 38:1788 (1988)

11. M. Altarelli, U. Ekenberg and A. Fasolino, **Phys. Rev. B**, 32:5138 (1985)

12. L.C. Andreoni, A. Pasquarello and F. Bassani, **Phys. Rev. B**, 36:887 (1987)

13. N.O. Lipari and M. Altarelli, **Phys. Rev. B**, 15:4883 (1977)

14. C.G. Van de Walle, **Phy. Rev. B**, 39:1871 (1989); C.G. Van de Walle and R.M. Martin, to be published.

15. A.P. Roth and E. Fortin, **Can. J. Phys.**, 56:1486 (1978)

16. M.F.H. Schuurmans and G.N. 't Hooft, **Phys. Rev. B**, 31:8041 (1985), A.C.G. Wood, private communication.

17. R. Eppenga, M.F.H. Schuurmans and S. Colak, **Phys. Rev. B**, 36:1554 (1987)

18. M. Altarelli, **Phys. Rev. B**, 28:842 (1983)

19. S.K. Haywood, A.B. Henriques, N.J. Mason, R.J. Nicholas and P.J. Walker, **Semicond. Sci. Technol.**, 3:315 (1988)

20. F. Stern and W.E. Howard, **Phys. Rev.**, 163:816 (1967)

21. F. Stern, **Phys. Rev. Lett.**, 21:1687 (1968)

22. R.W. Martin, R.J. Warburton, R.J. Nicholas, G.J. Rees, S.K. Haywood, N.J. Mason, P.J. Walker, M. Emeny and L.K. Howard, to be published.

23. Y. Iye, E.E. Mendez, W.I. Wang and L. Esaki, **Phys. Rev. B**, 33:5854 (1986)

24. K.D. Bowers and J. Owen, **Reports on Prog. in Phys.**, 18:304 (1955)

HIGHLY STRAINED InAs/GaAs SHORT PERIOD SUPERLATTICES

Jean-Michel Gerard*

Centre National d'Etudes des Télécommunications
196 Av. Henri Ravera, 92220 Bagneux, France

1.   INTRODUCTION

InAs/GaAs short period superlattices (SPS) are highly
strained heterostructures (7% lattice mismatch between InAs and
GaAs) which can be grown approximately lattice matched to InP.
Such SPS thus constitute an ordered counterpart of the $In_x Ga_{1-x}$
As random alloy. A larger mobility (at low temperature mainly)
can be surmized for this artificially layered structure due to
the suppression of alloy scattering. Due to the change from a
T_d cubic (in InGaAs) to a D_{2d} tetragonal symmetry in InAs/GaAs
SPS), marked **qualitative** differences are also expected between
these two materials' band structures. An anisotropy of the
electron effective mass, and a lifting of the valence band
degeneracy are for example expected in the SPS. Both transport
and optical properties might thus be quite different in these
two materials, and open interesting opportunities for this
ordered alloy in the field of devices. Among others, a larger
spin-orbit splitting would reduce the efficiency of the Auger
nonradiative process when compared to InGaAs.

It is therefore highly desirable to compare experimentally
SPS and random alloy band structures. Until recently, only
photoluminescence (PL) data had been obtained, on thick SPS
layers[1-5], or on multi-quantum well structures (MQW) built
with InAs/GaAs as the well material[5]. Furthermore, MBE grown
InAs/GaAs layers displayed a poor structural and optical quality
(X-ray diffraction satellites larger than $0.5°$ [6,7], 50 to 100
meV PL spectral width[2]), which did not allow a detailed
further investigation. Complete thermal annealing performed at
the same temperature as the MBE growth ($500°C$) showed that these
imperfect SPS were not stable[8].

Some recent developments of the study of highly strained
InAs/GaAs SPS are presented below. We first recall why the
growth of highly strained InAs or GaAs layers tends to be three-
dimensional (3D), and briefly show that an alternate deposition

---

*Member of the Direction des Recherches Etudes et Techniques
(French Ministry of Defense).

Condensed Systems of Low Dimensionality
Edited by J. L. Beeby *et al.*, Plenum Press, New York, 1991

of arsenic and metal atoms by monolayer increments at low
temperature (350°C) can induce a layer by layer growth. An
extended characterization of InAs/GaAs SPS highlights the clear
quality improvement obtained when using this modulated molecular
beam epitaxy (MMBE) technique. Finally, an optical study of
thick SPS layers, and of quantum well structures using InAs/GaAs
as a material for the well, allows us to compare the band
structures of an $In_{0.5}Ga_{0.5}As$ random alloy and of the SPS with
the same average composition, and discuss the potential interest
of this artificially layered material.

2.   GROWTH OF HIGHLY STRAINED LAYERS

Some light is shed on the difficult MBE growth of strained
layers when *in situ* reflection high energy electron diffraction
(RHEED) is performed. A shift from 2D to 3D growth mode is
revealed by a bulk type spotty pattern when about 5 monolayers
(ML) InAs or GaAs have been deposited on a buffer layer lattice
matched to InP under standard growth conditions (500°C, As-
stabilized surface). During the growth of a $(InAs)_n(GaAs)_n$ SPS,
for n smaller than that apparent critical thickness for 3D
growth of the individual layers (2 < n < 5), a continuous shift
to 3D growth is observed by RHEED. Reconstruction streaks
vanish, some intensity modulation of the integral order streaks
gradually appears and a bulk-type spotty pattern is finally
obtained. The SPS thickness necessary for this sequence to be
completed is smaller for large n's, and is typically 500 Å for
n = 4 and 900 Å for n = 3. This degraded growth mode is
responsible for the poor structural and optical quality of MBE
grown SPS.

These results are explained by the close relationship
between the strain state of the film and its growth mode. Large
islands of relaxed material constitute the thermodynamically
stable configuration of heavily mismatched layers. Since MBE
growth does not proceed at equilibrium however, the first stages
of the growth are better understood from a kinematical point of
view. During the growth of unstrained material on a flat
surface, a stationary surface rugosity is reached when the RHEED
oscillations are completely damped. Its lateral scale is of the
order of the diffusion length (during the deposition of one
monolayer) of the metal atoms migrating on the surface. Any
rugosity at a lower scale is smoothed during the subsequent
growth. For highly strained layers on the other hand, no
stationary rugosity can be reached as discussed now for the

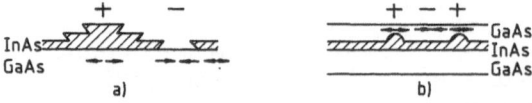

Fig. 1. Intertwining between the MBE growth mode and
        strain state for highly strained epilayers:
        progressive shift to 3D growth for InAs on GaAs
        (a) and correlation of defects in InAs/GaAs
        superlattices (b). → and ⇥ mark zones under
        in-plane tensile or compressive strains. + and
        - indicate respectively areas of preferential
        and inhibited subsequent growth of <u>InAs</u>.

growth of InAs on GaAs. Such defects as steps or terraces allow the strained epilayer to relax partly (Fig. 1a). The surface lattice parameter is larger than that of the substrate on top of InAs terraces, whereas the material is under compression in between. As a result, the growth becomes inhomogeneous: the nucleation of a new terrace is more probable on top of the islands rather than between them; the incorporation at steps is also favored at upper rather than lower monolayers. This preferential incorporation tends to increase the rugosity of the film and possibly leads to 3D growth (regime of full incorporation in islands). For strained superlattices (Fig. 1b), islands in the buried sublayers create long range strain fields which reach the surface. Here again preferential growth occurs, leading to a progressive roughening of the surface. Correlation in the position of the defects in the different sublayers could also be observed for InAs/GaAs superlattices[9].

The growth parameters in standard MBE have a rather small influence on this growth mode shift. The requirement of a good optical quality limits further the useful temperature or arsenic pressure ranges. The growth of highly strained layers thus requires a drastically different approach.

MMBE, which consists in the alternate deposition of As_4 and metal atoms at low temperature (350°C) by monolayer increments, allows a breakthrough in the improvement of highly strained structures[4,10-12]. This technique, known as migration enhanced epitaxy[13] (MEE) or atomic layer MBE[14] (ALMBE), originally allowed the growth of (unstrained) GaAs of good optical quality at very low temperature. To account for this result, quite different growth mechanisms have been proposed. In the ALMBE scheme, an enhancement of the two-dimensional nucleation process is thought to result from the cyclic perturbation of the surface stoichiometry. In the MEE scheme on the other hand, a layer by layer growth mode is favored by the enhancement of the cation migration under very low arsenic pressure. However, the growth study, and measurement of Ga diffusion length (1900 Å at 550°C) have been performed for a standard growth temperature[15]. Since the activation energy of

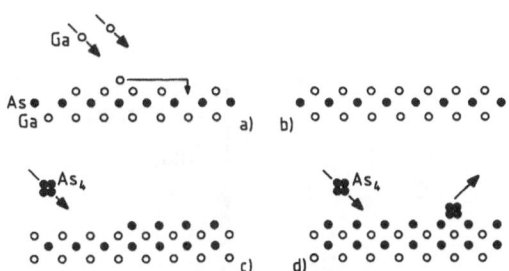

Fig. 2. Schematic model for the MMBE growth of a GaAs monolayer. Ga deposition: large migration of Ga on Ga islands a); smooth and complete Ga atomic layer at the end of a Ga deposition b). As_4 deposition: initial incorporation with a large sticking coefficient c); complete desorption of excess arsenic d). MMBE avoids terrace formation and forces a layer by layer growth mode.

the migration is unknown but certainly rather large (≈ 1 eV) for Ga atoms on a surface covered with arsenic, one may wonder if the related diffusion length is actually large at low temperature.

The successful growth of highly strained structures allows one to discuss more precisely the growth process. A large surface migration helps in reaching a thermodynamically stable configuration, which is possibly 3D. (One can even note that a Weber-Volmer growth process requires a large transport on the surface.) Large surface diffusion length alone thus cannot account for an improved flatness of the epilayer's surface as soon as the grown material is highly strained. We think in fact that the alternate deposition of As_4 and metal atoms forces a layer by layer growth mode.

An original scheme which could be relevant for MMBE growth of one monolayer GaAs on a plane surface covered with arsenic is described on Fig.2. One monolayer Ga is first deposited under As-poor atmosphere. Even if the migration of Ga is small on a surface covered with arsenic, a rapid diffusion of Ga on Ga terraces is expected. Ga-Ga bonds are indeed much weaker than Ga-As bonds. As a result, Ga atoms deposited on top of a Ga terrace will migrate until they reach its edge and establish stable covalent bonds with As atoms (a). A plane and complete Ga monolayer is thus obtained at the end of Ga deposition (b). This surface is highly reactive for As_4 molecules, which are sent to the surface in a second step. The surface is quickly and homogeneously covered with arsenic (c) and excess As_4 molecules fully desorb if the substrate temperature T_s is sufficiently high (e.g. $T_s > 50°C$). On the other hand, a low T_s ($T_s < 400°C$ approximately) avoids arsenic desorption from the deposited monolayer and degradation of the layer.

Although certainly oversimplified, this model suggests that a layer by layer growth mode can be forced if metal atoms tend actually to form a smooth layer rather than gather into droplets. This result, which is valid for unstrained material [13], is also obtained during the MMBE growth of strained InAs or GaAs[4,11]. At 370°C, and for an InP (001) substrate, well

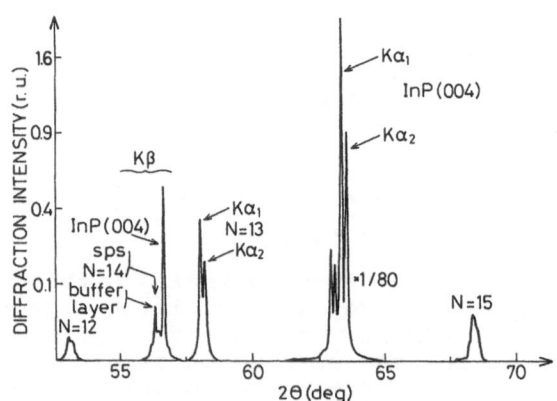

Fig. 3. X-ray diffraction profile of an $(InAs)_4(GaAs)_3$ SPS near InP (004) diffraction order. Cu Kα and Kβ radiations have been used.

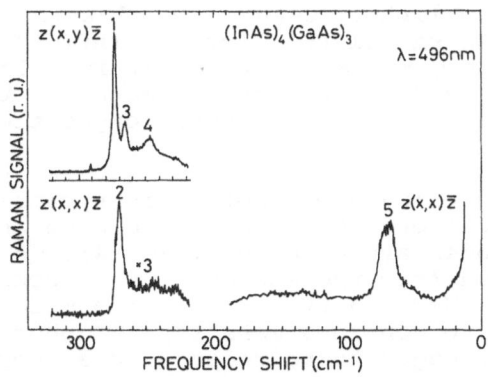

Fig. 4. Raman spectra taken at 77 K (optical modes) and
room temperature (acoustic modes) for an
$(InAs)_4(GaAs)_3$ SPS.

established RHEED reconstructions are observed, namely 2 x 4 (or
2 x 2), 4 x 2 and 3 x 1 at the end of As_4, Ga and In deposition
sequences respectively. This indicates that the surface
stoichiometry is significantly and homogeneously changed at each
step of the deposition cycle.

Finally, this forced layer by layer growth mode allows the
deposition of two-dimensional InAs and GaAs layers on InP up to
the critical thickness for plastic relaxation (and obviously
after). MMBE thus overcomes the tendancy to 3D growth inherent
to highly strained layers in MBE. The suppression of oval
defects for unstrained GaAs in MMBE[14] is also explained by the
proposed picture as well as by the ALMBE model.

3.   CHARACTERIZATION OF THE HIGHLY STRAINED SPS

The structural characterization of a 1200 Å thick
$(InAs)_4(GaAs)_3$ SLS grown on an (InGaAl)As buffer layer lattice-
matched to InP confirms the high quality of this MMBE grown
structure. Scanning transmission electron micrographs reveal
the good in-plane homogeneity and reproducibility from period to
period within the SPS. No misfit dislocations could be detected
within the SPS, nor at its interface with the buffer layer. In
fact, the SPS is approximately lattice matched to InP, and the
individual sublayers are well below the critical thickness for
plastic relaxation.

A θ-2θ diffraction profile of the structure is shown in
Fig.3. Sharp diffraction peaks specific to SPS appear; the
width of the more intense diffraction peaks (150 seconds of arc
for N = 13) is comparable to the experimental resolution
($\Delta\theta$ = 100 seconds of arc near InP(004) diffraction order), which
attests to the high quality of the structure when compared to
the typical values of 0.4° or 1° published for similar MBE
samples[6,7]. The SPS period is very close to its nominal value
of 7 ML, since no other satellites appear than expected ones
[1]. A period of 20.5 Å is determined from the absolute or
relative positions of SPS diffraction peaks, which satisfyingly
agrees with the 20.6 Å theoretical estimate calculated for an

elastically strained $(InAs)_4(GaAs)_3$ superlattice. In particular, the 14th satellite, close to InP(004), is resolved when $K\beta$ monochromatic radiation is used; it allows a direct measurement of the average lattice parameter of the SPS in the growth direction, and thus of its mean composition (56% indium instead of a nominal 57%).

The Raman scattering spectrum[16] is typical of a well-defined superlattice; one folded acoustic mode and several confined optical modes are observed (Fig. 4). A period of 20.5 Å is determined from the energy of the acoustic mode, in good agreement with X-ray diffraction. The modes labeled 1,2 and 3 are assigned to confined GaAs type LO phonon modes, whereas mode 4 is thought to be related to a propagative InAs type LO phonon mode.

A high degree of perfection has thus been demonstrated, as far as the periodicity is concerned, both from X-ray diffraction and Raman scattering experiments. However, the analysis of the X-ray satellite intensities as well as the nearly equal spacing between the confined optical phonons indicate deviations from a perfectly abrupt SLS composition profile at interfaces, though these observations could not be made quantitative up to now. Disorder activated transverse and longitudinal acoustic modes (DATA and DALA) are also observed in the Raman spectra of the $(InAs)_4(GaAs)_3$ superlattice, as in the InGaAs random alloys. These imperfections are presently attributed to the indium segregation in GaAs, which partly intermixes InAs and GaAs sublayers without breaking the periodicity of the structure [17]. This interface effect, whose amplitude may depend somewhat on the growth technique, has been mainly studied by surface techniques and for MBE grown layers. It plays an exacerbated role for monolayer superlattices: the compositional modulation is hardly detectable for MBE or MMBE grown $(InAs)_1(GaAs)_1$ SPS.

Thermal annealing experiments have also been performed so as to test the stability of this highly strained SPS (Fig. 5).

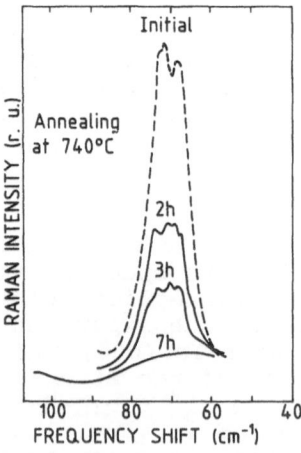

Fig. 5. Evolution of the first folded Raman acoustic doublet of an $(InAs)_4(GaAs)_3$ SPS for successive thermal annealings at 740°C.

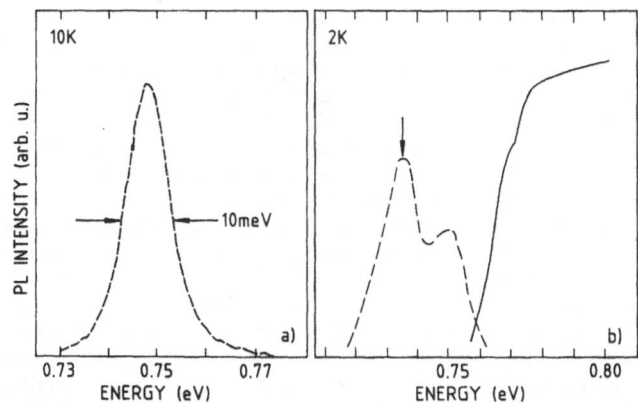

Fig. 6. PL (--) and PL excitation (——) spectra obtained
for an $(InAs)_4(GaAs)_3$ SPS. The excitation power
is in the 1 W/cm^{-2} (a) and 10^{-3} W/cm^{-2} (b) ranges
respectively. The arrow marks the PL collection
energy in PL excitation.

The first Fourier component of the composition profile has been
studied by Raman scattering, since it is simply proportional to
the squared intensity of the first acoustic folded doublet.
Successive annealings were performed at increasing temperatures
under flowing Ar-H$_2$; the sample was covered with a GaInAs
substrate, so as to protect its surface. The temperature has
been calibrated using the melting point of Sb. No significant
change in the Raman spectra could be seen after 3 hours at
640°C. At 750°C however, a progressive and nearly complete
intermixing is obtained after 7 hours of annealing. An
interesting comparison can be undertaken with GaAs/AlAs
unstrained SPS. Complete intermixing after annealing at 700°C
for two hours has been reported for a 10 Å - 10 Å structure[8].
InAs/GaAs SPS are therefore at least as stable as GaAs/AlAs
structures of the same period, despite the very large strains
experienced by the sublayers. The amount of elastic energy
(about 10 meV per metal atom for InAs/GaAs on InP) is in fact
much smaller than the diffusion activation energy. The previous
report of annealing temperatures in the 500°C range for 10 Å -
10 Å InAs/GaAs SPS[8] (observed from a shift of GaAs type
optical phonon) might be explained by a large amount of defects
in layers grown by standard MBE.

Finally, the low temperature photoluminescence (PL)
spectrum obtained for this sample consists of a single intense
line at 0.752 eV under Kr+ ion laser 6764 Å line excitation
(1 W/cm^{-2}) as shown in Fig. 6a). As for unstrained material,
MMBE thus allows one to grow layers of good optical quality at
very low temperatures. The PL peak FWHM (10 meV) compares
favorably to previous reports (e.g. 50 to 100 meV for MBE grown
layers), with the single exception of a 7 meV FWHM observed for
an MOCVD grown $(InAs)_1(GaAs)_1$ monolayer superlattice[1].

4.   OPTICAL INVESTIGATION OF THE SPS's BAND STRUCTURE

We present below the results of an optical study of "bulk"
InAs/GaAs layers and of multi-quantum well structures (MQW)

built with InAs/GaAs as the well material. The optical study of the $(InAs)_4(GaAs)_3$ SPS has first been performed. A source built with a quartz-iodine lamp and a monochromator has been used to perform PL excitation. For such a low excitation level (< 10^{-3} W/cm^{-2}), an additional extrinsic PL peak appears at a 20 meV lower energy (Fig. 6b). The overall shape of the PL excitation spectrum is typical of a direct bulk semiconductor: a sharp edge is observed (10-15 meV), its onset being located at 0.76 eV. The same spectrum is obtained for different PL collection energies (here 0.73 eV). This study gives thus a measure of the actual band gap of $(InAs)_4(GaAs)_3$ ($\approx$ 0.76 eV). Finally, this experiment does not reveal two well-defined excitation edges, as would be the case for a large difference of the light hole and heavy hole related band gaps of the SPS. A small structure near 775 meV is not perfectly resolved, but might be related to the second edge. The SPS valence band splitting would then be in the 15 meV range. We have studied this SPS by modulated reflectivity to measure the spin-orbit splitting Δ_{so} of this material; the split-off to conduction band gap is 1.09 eV. Δ_{so} is subsequently not very different for the SPS ($\approx$ 0.32eV) and $In_{0.53}Ga_{0.47}As$.

The above technique has an obvious limitation which is the need for thick SPS layers. Due to the decreasing quality of highly strained SPS with their thickness, the study of bulk properties may only reveal this degradation, and hide information on the quality of the first few hundred Å. We therefore study thin SPS layers embedded in a barrier material. Beyond a test of the SPS quality, the optical study of the MQW allows one to investigate the band structure of the SPS.

We study three MQW structures built with seven 105 Å quantum wells separated by 400 Å thick InGaAlAs barriers grown by MBE and lattice matched to InP. The well material is either a 9 period $(InAs)_2(GaAs)_2$ SPS grown by standard MBE (sample C) or by MMBE at 350°C (sample B), or the InGaAs random alloy of the same average composition (sample A). For all three samples, we kept the same deposition rates for In and Ga (nominally 0.25 monolayer per second). $(InAs)_n(GaAs)_n$ SPS (obtained for identical In and Ga delivery times) and the alloys obtained for a simultaneous delivery of In and Ga species have the same average In composition (within about 1%) since: i) small flux transients (about 1%) are observed when In or Ga cells are opened and ii) shutter operation delays (about 0.2 s) are similar for In and Ga cells.

High resolution transmission electron microscopy confirms the successful alternate deposition of InAs and GaAs bilayers, and the in-plane homogeneity of this sophisticated well material when grown by MMBE (Fig. 7). On the other hand, a micrograph of the MBE grown MQW C reveals strong in-plane inhomogeneities in this material: alternate bilayers are locally identified for some well-ordered zones, whereas elsewhere marked contrasts reveal fluctuations of the composition (and thus of the strain). A 108 Å overall well thickness is measured for both MQW B and C, in satisfying agreement with their designed value.

PL and PL excitation spectra obtained at 2 K for these samples are shown in Fig. 7. Whereas the staircase-like shape, characteristic of two-dimensional systems, and marked excitonic peaks are seen for the reference MQW A, these structures are

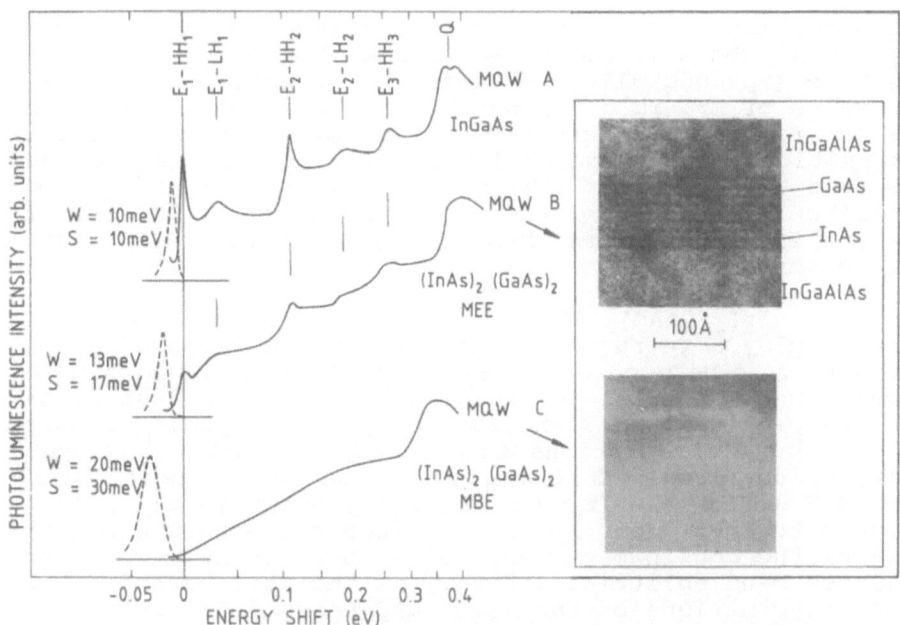

Fig. 7. PL and PL excitation spectra obtained at 2 K for
three MQW samples; different well materials
having the same nominal indium composition have
been used. The spectra are plotted as a
function of the shift from the first structure
seen on the excitation spectra. Q marks the
absorption edge of the InGaAlAs barrier layers.
W and S are the PL spectral width and Stokes
shift. High resolution transmission electron
micrographs are shown in inset for MQW B and C.

completely smeared out for MQW C. This confirms that the MBE
grown SPS is still highly inhomogeneous at the first stages of
its elaboration. Such defects can also account for the PL
Stokes shift previously published for MQW containing InAs/GaAs.
The observation of two-dimensional step-like features and the
recovery of excitonic peaks in the excitation spectrum for MQW B
confirm a remarkable improvement for the MMBE grown
$(InAs)_2(GaAs)_2$ material.

The striking coincidence of the energy spacings of MQW A
and B optical transitions is highlighted in Fig. 7. These
transitions can thus be assigned to E_n-HH_n (n = 1 to 3) and E_n-LH_n (n = 1,2) for MQW B as for MQW A. In an effective mass
framework, a fit of the transition energies for MQW A is
possible for a 108 Å well width; it allows one to check and
correct the indium composition (x = 0.55) common to InGaAs in
MQW A and the SPS in MQW B and C.

Some key electronic properties of InAs/GaAs SPS are
revealed:

i) Since well width and barrier height are kept constant
for both MQW, the spacing between E_n-HH_n optical transitions
strongly depends on m^*_z, the electron effective mass along the

growth direction. $m*_z$ is thus very similar ($\Delta m*_z / m*_z < 7\%$) for $(InAs)_2 (GaAs)_2$ SPS and the InGaAs alloy of same average composition ("pseudo-alloy" behavior). An estimate of this common value of $m*_z$ is given by the Kane electron mass which enters our fit of the transitions (0.04 m_0). Although the E_n-LH_n transitions are not as well resolved, the spacing between the LH_1 and LH_2 levels is nevertheless roughly the same in MQW A and B (since electron confinement energies are). The effective masses for light holes are thus also not markedly different in the two materials.

ii) $(InAs)_2 (GaAs)_2$ SPS as well as biaxially strained InAs or GaAs display a tetragonal symmetry. Large valence band splittings Δ_{h-1} between heavy and light hole bands at the Brillouin zone center are calculated for InAs and GaAs strained on InP (respectively -160 and +360 meV). The spacing between E_1-HH_1 and E_1-LH_1 transitions in MQW samples A and B are very similar, which reveals that Δ_{h-1} is, on the other hand, small for the SPS as for the $In_{0.5}Ga_{0.5}As$ alloy which is slightly mismatched to InP. This spacing is indeed due to the combined effects of the confinement in a 108 Å thick quantum well, and of the valence band splitting (if any) of the well material. When possible fluctuations of the light and heavy hole confinement energies in MQW A and B and experimental errors are taken into account, the values of Δ_{h-1} are found to differ less than 12 meV for these two materials. (The confinement energies for heavy holes should not differ markedly in these samples, since heavy hole effective masses are large and nearly common to all III-V semiconductors). On the other hand, Δ_{h-1} is estimated at 18 meV for the alloy. Consequently, $(InAs)_2 (GaAs)_2$ SPS have (as $In_{0.5}Ga_{0.5}As$ strained on InP) a light hole to conduction fundamental band gap transition and a rather small valence band splitting ($6 < \Delta_{h-1} < 30$ meV).

iii) The first excitonic structure in the PL excitation spectra is observed at 825 and 830 meV respectively for MQW A and B. The light or heavy hole related band gaps of $(InAs)_2 (GaAs)_2$ are quite similar to those of $In_{0.5}Ga_{0.5}As$ strained on InP, since the confinement energies and MQW band gaps are similar. The alloy or SPS band gaps are very sensitive to the average indium composition, which makes difficult a more detailed comparison; a 1% fluctuation of this composition from sample to sample would entail an 8 meV shift of the E_1-HH_1 transition and may thus explain alone the different MQW band gaps. (This sensitivity, often combined with the lack of a precise enough characterization of the samples, may also account for discrepancies between previously published PL data concerning InAs/GaAs[1,3,5]).

This experiment thus highlights a clear pseudo-alloy behavior for $(InAs)_2 GaAs)_2$. Despite its strong ordered character, this artificially layered material displays a band structure which is pretty close to that of a random alloy of the same average composition. Our study is however limited to the Brillouin zone center and the growth direction. A measure of the in-plane conduction effective mass $m*_\parallel$ (0.05m_0) has also recently been performed by cyclotron resonance. Beyond the fact that an anisotropy of the SPS conduction band is expected, the comparison of these experimental data for $m*_z$ (0.04m_0) and $m*_\parallel$ remains quite striking. Indeed, the simpler calculations (in the Ben Daniel - Duke frame for example) ascribe a lighter

542

effective mass to the in-plane electron dispersion curve. A novel determination of m^*_z and $m^*_\parallel$ using a single experimental approach would help to clarify this point.

5.   COMPARISON TO SOME THEORETICAL ESTIMATES

An accurate effective mass calculation of the SPS band structure first requires a satisfying description of the highly strained bulk constituents, over a large energy range. (Since the band gaps of the sublayers are quite different, the SPS states are built with bulk states far from the related band extrema). It is furthermore well known that the validity of the effective mass approach is questionable when very thin layers are involved. We demonstrate below that an adequate formalism, which will be extensively described elsewhere, gives a very satisfying understanding of our experimental results.

A three-band envelope function model, including a Kane description of the host's band structure (suitably modified to include strain effects), is particularly appropriate when weakly strained structures are considered[18] (up to ±1% strain). In the Kane model however, the light hole dispersion curve displays an asymptote ($E = E_a$) between the light hole and spin-split-off band extrema. The description of the light hole dispersion curve is therefore valid over a small energy range, in particular for a material under biaxial compression (about 50 meV below the band maxima for InAs strained on InP). For $E = E_a$, the Kane mass is infinite: as a result, the confinement of the light holes in a quantum well or SPS cannot push their levels beyond that limit.

To analyze the valence states of highly strained structures, we thus use a more refined model. The band structure is calculated for $k_x, k_y = 0$; Luttinger type k_z^2 terms are included in the valence band Hamiltonian of the bulk materials, which account for the $k_z p_z$ interaction of the bands of the Kane set with the remote bands. The calculated band structure in the z direction in these different frames are compared on Fig. 8 for both InAs and GaAs strained on InP. A much more satisfying description of the coupled "light hole" and "split-off" bands is obtained when k^2 terms are included.

The band gap and spin-orbit splittings are strongly modified by the large biaxial strains, which also efficiently lift the valence band degeneracy at the Brillouin zone center. We can note that the experimental pseudo-alloy behavior of $(InAs)_n (GaAs)_n$ SPS appears as essentially nontrivial (mostly when compared to the unstrained case). When one-band effective mass models are used, indeed, and at the limit of very small periods, the SPS's band edge and effective mass are given for each band by an arithmetic average of the constituent's ones, weighted by the sublayer thicknesses. One band models are valid for the conduction and heavy hole bands. For unstrained material, these remain correct for the light hole band if the valence band discontinuity is small when compared to the host's spin-orbit splittings. In such a case, this approach predicts mean values, close to those which are relevant for the random alloy, for the band gap and for the effective masses along the growth direction. In particular, Δ_{H-L} is vanishingly small for the SPS. For highly strained SPS on the other hand, the strong

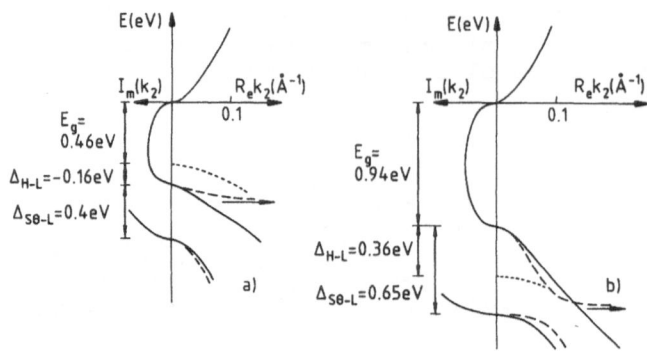

Fig. 8. Light particle dispersion curves in the growth
direction for InAs and GaAs strained on InP at 0
K, as calculated in a three-band Kane model
(---) and corrected using Luttinger-type k^2
terms (— —). Heavy hole dispersion curves
are also plotted (...). One can note the marked
effect of large strains on the band gap,
effective masses, and valence band splittings at
Brillouin zone center, and the good description
of conduction states in the simple Kane model.

mixing of the light valence bands has always to be considered.
Irrelevant one-band descriptions would in fact predict for
$(InAs)_n(GaAs)_n$ large valence band splittings at Γ (Δ_{H-L} =
-0.1 eV, Δ_{so-L} = 0.52 eV).

The band gaps of $(InAs)_2(GaAs)_2$ and $(InAs)_4(GaAs)_3$ are
displayed on Fig. 9, for calculations based on the Kane-type and
Luttinger-type descriptions of the host materials. The
conduction band discontinuity ΔE_C at the InAs/GaAs interface is
used as an adjustable parameter to fit our experimental data.
The unrealistic variation of the light hole to conduction band

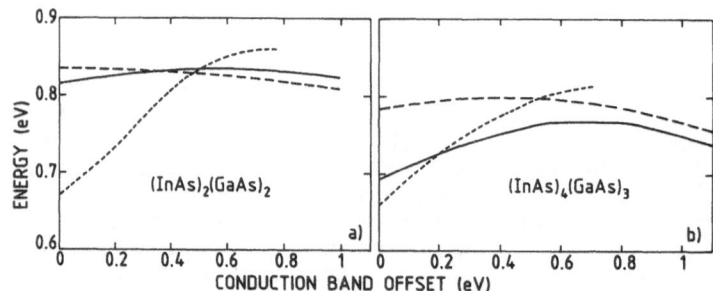

Fig. 9. Calculated low temperature ($\approx$ 0 K) band gaps of
$(InAs)_2(GaAs)_2$ and $(InAs)_4(GaAs)_3$ SPS in an
envelope function framework are plotted as a
function of the conduction band discontinuity.
For E_1-LH_1, the model is either based on a Kane
(---) or a Kane + Luttinger (— —) description of the band structure of the constituents.
Both models give equivalent results for E_1-HH_1
(——).

gap in the first approach is due to the previously mentioned artefact related to the Kane model. When the k^2 terms are included, a very satisfying agreement is found for both SPS with our experimental results for large band offsets (0.6 eV $< E_C <$ 0.8 eV). In particular, the band gap is related to E_1-LH_1 for $(InAs)_2(GaAs)_2$ only if $E_c > 0.4$ eV. When taken into account, the smoothing of the composition profile due to In segregation should not modify significantly these results. For $(InAs)_2(GaAs)_2$ indeed, the theoretical band structure is already quite close to that of the $In_{0.5}Ga_{0.5}As$ random alloy. On the other hand, the influence of this interface effect should become small for SPS of larger period such as $(InAs)_4(GaAs)_3$.

To conclude, high quality InAs/GaAs SPS can be grown by low temperature modulated molecular beam epitaxy. This technique appears therefore as particularly well suited to the growth of highly strained films, or the improvement of less strained layers. A study of InAs/GaAs SPS by Raman scattering and X-ray diffraction reveal clear superlattice effects; as a result, the structural properties of these SPS and of the random alloy of the same average composition should significantly differ. Although this property has not been detailed here, the in-plane equilibrium lattice parameter is smaller for the ordered than for the random alloy. As a result, $(InAs)_n(GaAs)_m$ SPS extend toward the lower energy range the band gaps that can be obtained for (InGaAl)(AsP) based material lattice matched with InP.

Theoretical as well as experimental data highlight a clear pseudoalloy behavior of the SPS's electronic properties. This material could therefore be used as an ordered substitute of similar band structure for the random alloy of the same average composition. On the other hand, no benefit resulting from a change of the band structure should be gained from this substitution (and in particular no reduction of Auger non-radiative processes). Up to now, no significant enhancement of the mobility could be observed for the ordered alloy (typically 20000 to 30000 $cm^2V^{-1}s^{-1}$ for low n type doping for $(InAs)_2(GaAs)_2$ in our case at 77 K). Any improvement in this field will require one to overcome indium segregation in GaAs, which accounts for some residual disorder within InAs/GaAs SPS.

Acknowledgment

The author gratefully acknowledges J.Y. Marzin, B. Jusserand, J. Primot (CNET Bagneux), C. d'Anterroches (CNET Meylan), P. Voisin and B. Soucail (Ecole Normale Supérieure) for their respective contributions to this work.

REFERENCES

1. T. Fukui and H. Saito, **Japan J. Appl. Phys.**, 23:L521 (1984)
2. P. Voisin, M. Voos, J.Y. Marzin, M.C. Tamargo, R.E. Nahory and A.Y. Cho, **Appl. Phys. Lett.**, 48:1476 (1986)
3. B.T. McDermott, N.A. El-Masry, M.A. Tischler and S.M. Bedair, **Appl. Phys. Lett.**, 51:1830 (1987)
4. J.M. Gerard, J.Y. Marzin, B. Jusserand, F. Glas and J. Primot, **Appl. Phys. Lett.**, 54:30 (1989)
5. M. Razeghi, P. Maurel, F. Omnes and J. Nagle, **Appl. Phys. Lett.**, 51:2218 (1987)
6. Y. Matsui, H. Hayashi, M. Takahashi, K. Kikuchi and

K. Yoshida, **J. Cryst. Growth**, 71:280 (1985)

7. H. Ohno, R. Katsumi, T. Takama and H. Hasegawa, **Japan J. Appl. Phys.**, 24:L682 (1985)
8. K. Kakimoto, H. Ohno, R. Katsumi, Y. Abe, H. Hasegawa and T. Katoda, **Int. Phys. Conf. Ser.**, 74:253 (1985)
9. L. Goldstein, F. Glas, J.Y. Marzin, M.N. Charasse and G. LeRoux, **Appl. Phys. Lett.**, 47:1099 (1985)
10. J.M. Gerard and J.Y. Marzin, **Appl. Phys. Lett.**, 53:568 (1988)
11. M. Recio, G. Armelles, A. Ruiz, A. Mazuelas and F. Briones, **Appl. Phys. Lett.**, 54:805 (1989)
12. J.M. Gerard, J.Y. Marzin, C. d'Anterroches, B. Soucail and P. Voisin, **Appl. Phys. Lett.**, 55:559 (1989)
13. Y. Horikoshi, M. Kawashima and H. Yamaguchi, **Japan J. Appl. Phys.**, 27:169 (1988)
14. F. Briones, L. Gonzales, M. Regio and M. Vazquez, **Japan J. Appl. Phys.**, 26:L1125 (1987)
15. S. Nagata and T. Tanaka, **J. Appl. Phys.**, 48:940 (1977)
16. B. Jusserand and J.M. Gerard, p.799 **in**: "Proceedings of the 19th Int. Conf. Phys. Semicon.", Institute of Physics, Polish Academy of Science (1988)
17. C. Guille, F. Houzay, J.M. Moison and F. Barthe, **Surf. Science**, 189/190:1041 (1987)

HETEROJUNCTION BAND DISCONTINUITIES FOR PSEUDOMORPHICALLY

STRAINED $In_xGa_{1-x}As/Al_yGa_{1-y}As$ HETEROINTERFACES

D.J. Arent, C. Van Hoof,* G. Borghs* and H.P. Meier

IBM Research Division, Zurich Research Laboratory
8803 Rüschlikon, Switzerland
*Interuniversity Microelectronics Center (IMEC)
Kapeldreef 75, B-3030 Leuven, Belgium

ABSTRACT

A general description is presented for calculating the strain-induced variations in the band edge discontinuities for pseudomorphically strained III-V heterointerfaces grown in the (100) direction. $In_xGa_{1-x}As/Al_yGa_{1-y}As$ ternary/ternary heterointerfaces are specifically treated within the virtual crystal approximation, accounting for band parabolicity and composition dependent material parameters. In conjunction with the development of an equation describing the strained $In_xGa_{1-x}As$ band gap as a function of In concentration, the conduction band offset ratios, calculated as a function of both In and Al content, are shown to be nonconstant and are in very good agreement with experimental data derived from strained single quantum well samples grown by molecular beam epitaxy and analyzed using room temperature photoreflectance spectroscopy and data from the literature.

1.   INTRODUCTION

The influence of strain on III-V heterojunctions is of continued interest owing to the additional degree of freedom allowed in optimizing their optoelectronic properties[1,2]. Strain-induced band energy variations directly affect the band edge discontinuities at a heterointerface and are of fundamental interest and extremely important for the proper design of strained heterojunction devices with optimized carrier confinement and transport properties. Accurate models describing the band edge movements under strain are thus desirable, particularly as epitaxial tools allow for atomically abrupt interfaces and excellent compositional control.

The properties of pseudomorphic $In_xGa_{1-x}As/Al_yGa_{1-y}As$ quantum structures grown on GaAs are suitable for optical and electronic microdevices. Extensive theoretical[3,4] and experimental[5-11] investigations of InGaAs/GaAs heterointerfaces have reported results for the band energy discontinuities with some, though moderate, agreement[5].

The bulk of results indicates a constant conduction band offset ratio $\Delta Q_{cb} = \Delta E_{cb}/\Delta E_{gap}$ of ~ 0.65 where ΔE_{cb} is the difference in conduction band minima (in eV), and ΔE_{gap} is the difference in band gaps of the strained InGaAs and GaAs. However, recent reports suggest that in fact ΔQ_{cb} may be composition dependent[9-11].

Thus, interesting questions arise as to the nature of the possible composition dependence and, moreover, when AlGaAs is utilized as the barrier material, as to the influence of three binary compounds where two independent bowing factors and strain may influence the band lineup.

In the present paper, band offsets for strained $In_xGa_{1-x}As/Al_yGa_{1-y}As$ interfaces pseudomorphically grown on GaAs are calculated based upon the virtual crystal approximation where strain is introduced as a perturbative effect in a generalized theory of (100) interfaces. Calculated band offsets are then used to determine the transitions for strained single quantum wells (QWs) which were grown by molecular beam epitaxy and show excellent agreement with the transitions exhibited in photoreflectance spectra.

2. EXPERIMENT

Undoped strained single QWs were grown on (001) GaAs substrates by molecular beam epitaxy. For these studies, two $In_xGa_{1-x}As/Al_yGa_{1-y}As$ samples were investigated with the following compositions: sample A with x = 0.10 and y = 0.20, and sample B with x = 0.15 and y = 0.28. Following a 0.2 μm GaAs buffer layer, a 0.5 μm $Al_yGa_{1-y}As$ layer was grown at 705°C. One monolayer of GaAs was then deposited and growth interrupted for 45 sec while the substrate temperature was lowered to 550°C and the 70 Å $In_xGa_{1-x}As$ QW was grown. All QWs were capped with 0.1 μm of $Al_yGa_{1-y}As$.

Photoreflectance spectroscopy was performed at room temperature using an apparatus described in detail in Ref. 5. Transition energies for the strained QWs were calculated according to the equations developed in Section 3 and with a finite square well model which accounts for exciton binding energy[12]. All parameters related to the strain response of the system were linearly interpolated between the most recently published values for the binary constituents which are summarized in Table 1.

3. THEORETICAL FORMALISM

The valence band discontinuities for GaAs/AlAs and unstrained InAs/GaAs have been recently published and are given in Table 1. The value quoted for ΔE_{vb} (GaAs/AlAs) of -0.57 eV represents a constant $\Delta Q_{cb} = 0.58$ applied to the linear composition dependence of the band gap, consistent with Ref. 3. We have adopted the average value ($\Delta E_{vb} = 0.17 \pm 0.07$ eV) found in a large number of publications[5]. For the unstrained InAs/AlAs interface, we have adapted two methods to estimate the valence band offset. First, the transitivity rule[13] was applied to the values quoted in Table 1 and those recently compiled in Ref. 3, resulting in $\Delta E_{vb} = 0.71 \pm 0.1$ eV for the

Table 1. Material parameters used in the calculations.   a) All
        values are listed for 2 K; b) Values are from Ref. 9
        and references therein unless otherwise noted; c) Ref.
        25; d) Ref. 26; e) Ref. 5 and references therein; f)
        Ref. 3; g) Calculated using the transitivity rule from
        values in Refs. 3 and 4 and extrapolating the lattice-
        matched valence band offsets reported in Refs. 14 and
        15 to the binary constituents; h) Ref. 27; i) Ref. 28;
        j) Ref. 29.

Parameter (units)	GaAs	InAs	AlAs
E_g (2 K) (EV)	1.519	0.416	3.099 [c]
Δ_0 (eV)	0.34	0.37	0.28 [d]
a_0 (Å)	5.6536	6.059	5.6611 [d]
$C_{11} \left[10^{11} \dfrac{dyn}{cm^2} \right]$	12.11	8.54	
$C_{12} \left[10^{11} \dfrac{dyn}{cm^2} \right]$	5.48	4.66	
a_c (eV)	-7.1	-5.4	
b_v (eV)	-1.7	-1.8	
$\dfrac{dEg}{dP} \left[10^{-6} eV \dfrac{cm^2}{kg} \right]$	11.3	9.8	
$\dfrac{d(E+\Delta_0)}{dP} \left[10^{-6} eV \dfrac{cm^2}{kg} \right]$	1.27 [d]	1.08 [d]	
ΔE_{vb} relative to GaAs (eV)		0.16 [e]	-0.57 [f]
ΔE_{vb} relative to InAs (eV)			-0.73 [g]
$E_{g,AlGaAs}(y)$ (eV)	$1.519+1.36y+0.22y^2$ [c]		
$\Delta_{0,AlGaAs}(y)$ (eV)	$0.341-0.281y+0.22y^2$ [c]		
$E_{g,InGaAs}(x)$ (eV)	$1.519-1.603x+0.50x^2$ [h]		
$\Delta_{0,InGaAs}(x)$ (eV)	$0.341-0.101x+0.14x^2$ [i]		
m_e (InGaAs)$(x)(m_0)$	$0.067-0.005x$ [j]		
m_{hh} (InGaAs)$(x)(m_0)$	$0.34+0.07x$		
m_{lh} (InGaAs)$(x)(m_0)$	$0.094-0.062x$		
m_e (AlGaAs)$(y)(m_0)$	$0.067+0.0835y$ [d]		
m_{hh} (AlGaAs)$(y)(m_0)$	$0.34+0.042y$ [d]		
m_{lh} (AlGaAs)$(y)(m_0)$	$0.094+0.043y$ [d]		

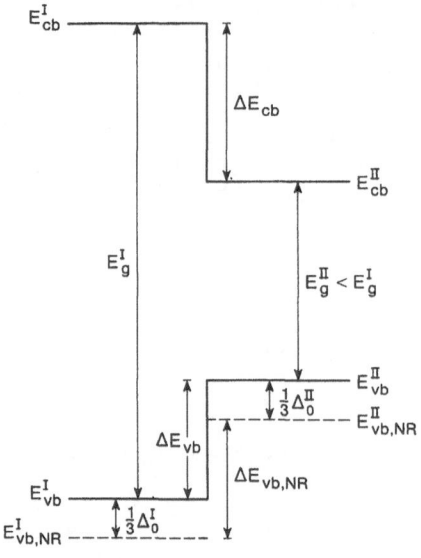

Semiconductor I   Semiconductor II

Fig. 1. Energy band diagram at k = 0 for a generic type
         I unstrained heterointerface between semi-
         conductors I and II where E_{vb} represents the top
         of the degenerate Γ_8 valence bands and Δ_0 is the
         spin-orbit band. ΔE_{vb} is the energy difference
         in the valence band maxima whereas $\Delta E_{vb,NR}$ is the
         difference between the nonrelativistic valence
         band maxima where $E_{vb,NR} = E_{vb} - \Delta_0/3$ for each
         material. The conduction band positions are
         found by adding the band gap energies to the
         respective valence band maxima, and the
         conduction band discontinuity ΔE_{cb} is calculated
         accordingly (from Ref. 16).

unstrained InAs/AlAs interface. Secondly, results from investigations on lattice matched $In_{0.53}Ga_{0.47}As/In_{0.52}Al_{0.48}As$ heterostructures[14,15] were extrapolated to yield a valence band
offset of $\Delta E_{vb} = 0.75 \pm 0.04$ eV for the binary interface, in
general agreement with the value predicted by the model solid
theory ($\Delta E_{vb} = 0.78 \pm 0.2$ eV) in Ref. 4. The average value of
these results is reflected in Table 1 with a relatively high
error of ±0.2 eV.

     The general valence band interface at k = 0 is depicted in
Fig. 1. The quantity of interest ΔE_{vb} (or correspondingly ΔE_{cb},
ΔQ_{cb}, and ΔQ_{vb} where $\Delta E_{cb} + \Delta E_{vb} = \Delta E_{gap}$ and $\Delta Q_{vb} = \Delta E_{vb}/\Delta E_{gap}$) for
a ternary/ternary interface is derived from knowledge of the
unstrained binary interfaces (i.e. InAs/GaAs, InAs/AlAs and
GaAs/AlAs) and the behavior of the band gaps and spin-orbit band
($E_0 + \Delta_0$) with composition and strain.

     The unstrained nonrelativistic valence band offset $\Delta E_{vb,NR}$
for the binary interfaces is then calculated as[4,16]

$$\Delta E_{vb,NR} = \Delta E_{vb} + \frac{1}{3}\,\Delta_0^{I} - \frac{1}{3}\,\Delta_0^{II} \,, \qquad (1)$$

where ΔE_{vb} is given in Table 1 and the Δ_0 values are the corresponding energies of the spin-orbit splitting in semiconductors I and II (see Fig. 1). Considering the mixed ternary interface in the virtual crystal approximation, the unstrained $In_xGa_{1-x}As/Al_yGa_{1-y}As$ valence band offset is then[4,17]

$$\Delta E_{vb,unstrained} = x[y\Delta E_{vb,NR}(InAs/AlAs) + (1 - y)\Delta E_{vb,NR}(InAs/GaAs)]$$

$$+ (1 - x)[y\Delta E_{vb,NR}(AlAs/GaAs)]$$

$$+ \frac{1}{3}\Delta_0(x)(In_xGa_{1-x}As) - \frac{1}{3}\Delta_0(y)(Al_yGa_{1-y}As) \qquad (2)$$

$$+ 3x(1 - x)(\Delta a_v)\Delta a_0/a_0(x) \, ,$$

where the last term is due to strain correction when considering the alloy as a superposition of atoms in the virtual crystal[17], $\Delta a_v = a_v(GaAs) - a_v(InAs)$, $a_v = dE/dP - a_c$, $\Delta a_0 = a_0(InAs) - a_0(GaAs)$, $a_0(x) = xa_0(InAs) + (1 - x)a_0(GaAs)$, a_0 is the lattice constant, and the Δ values contain the appropriate bowing factors.

Strain is calculated according to our recent work[5] and the results are summarized below. For thin InGaAs layers pseudomorphically grown to AlGaAs, we assume that all the strain is incorporated in the InGaAs. For growth in the [001] direction, the InGaAs layer is thus biaxially compressed in [100] and [010] directions and exhibits a corresponding expansion along the [001] plane. The strain ϵ of the sandwiched InGaAs layer is defined as

$$\epsilon = \frac{a_{InGaAs} - a_{AlGaAs}}{a_{AlGaAs}} \, , \qquad (3)$$

where a represents the lattice constant. Pollak and Cardona[18] have developed an accurate description of the valence band manifolds under strain which is given by the Hamiltonian

$$H_{strain} = \begin{bmatrix} & \left|\frac{3}{2}, \frac{1}{2}\right\rangle & \left|\frac{3}{2}, \frac{1}{2}\right\rangle & \left|\frac{1}{2}, \frac{1}{2}\right\rangle \\ -E_H - E_S & 0 & 0 \\ 0 & -E_H + E_S & E_S\sqrt{2} \\ 0 & E_S\sqrt{2} & -E_H - \Delta_0 \end{bmatrix} \qquad (4)$$

with the corresponding energy shifts given by

$$\delta E_0(1) = E_H + E_S \qquad (5a)$$

$$\delta E_0(2) = E_H - \Delta_0/2 - E_S/2 + \frac{1}{2}(\Delta_0^2 + 2\Delta_0 E_S + 9E_S^2)^{1/2} \qquad (5b)$$

$$(5b)\delta E_0 + \Delta_0 = E_H - \Delta_0/2 - E_S/2 - \frac{1}{2}(\Delta_0^2 + 2\Delta_0 E_S + 9E_S^2)^{1/2} \qquad (5c)$$

with $E_H = a(2 - K)\epsilon$, $E_s = b(1 + K)\epsilon$ and $K = -2S_{12}/(S_{11} + S_{12})$ where S_{ij} are the standard elasticity relations. The strain-shifted energy bands labeled $\delta E_0(1)$ and $\delta E_0(2)$ denote the now spin-split nondegenerate Γ_8 valence band components, and $\delta E_0 + \Delta_0$ the spin-split spin-off band. Under compressive stress $\delta E_0(1)$ and $\delta E_0(2)$ correspond to the heavy and light hole valence bands, respectively. The biaxial hydrostatic strain influences both the conduction band (by a relative amount $(1 - \Delta Q_H)E_H$) and the valence bands (by $\Delta Q_H E_H$). The overall effect is an increased band gap and ΔQ_H is given by

$$\Delta Q_H = 1 - \left[a_c(x) \left(\frac{dE_g}{dP}(x) \right)^{-1} \right] . \qquad (6)$$

The corresponding band gap for the strained InGaAs is then $E_{g,strained}(x) = E_g(x) + E_H - E_s$. Using the appropriate composition dependent material parameters, the low temperature (2 K) strained band gap is given by

$$E_g(x) = 1.519 - 0.945x + 0.193x^2 + 0.019x^3 , \qquad (7)$$

where the cubic component results from the composition dependent strain induced variations.

4.   RESULTS

Figure 2 shows the photoreflectance spectra for the two samples. The calculated transition energies for the m-th conduction band - n-th heavy (light) hole (m - nH) (m - nL) are indicated by the arrows at the top of the figure. The overall agreement is quite good. We have not done the extensive computer fitting to the spectra since each transition involves four fitting variables and the usual difference between fitted transition energies and those calculated as the weighted average of the peak values is < 10 meV[19].

Figure 3 shows the variation of the conduction band offset ratio as a function of both In and Al composition. The solid triangles represent the calculated offset ratios for the two samples and the solid dots are data taken from Ref. 20. A few observations are noteworthy. First, the overall agreement with the experimental data is very good, both for the mixed compo-sition cases investigated herein and the $In_xGa_{1-x}As/Al_yGa_{1-y}As$ data of Ref. 20. Secondly, the extreme variation in band offset ratio, as a function of both x and y, primarily arises from the strain-induced band energy changes. For example, the valence band maxima rises ~ 0.15 eV whereas the total shift in the band gap is ~ 0.37 eV. The corresponding conduction band offset ratios ΔQ_{cb} vary from 0.85 to 0.60 for InAs/GaAs and from 0.90 to 0.75 for $InAs/Al_{0.3}Ga_{0.7}As$. The striking behavior of the y = 0.05 curve is somewhat surprising and under current investi-gation. Further, the behavior at higher x reflects the relative difference between the hydrostatic strain and uniaxial strain contributions to the (heavy hole) valance band movement. At higher strain values, the stress related elongation along the growth direction induces nearly equivalent or greater band shifts than the biaxial hydrostatic deformation. This increases

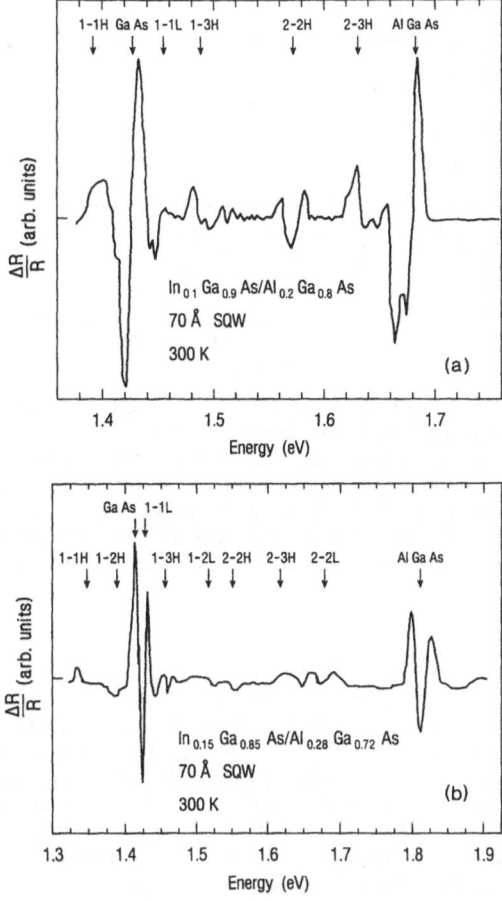

Fig. 2. Room temperature photoreflectance signal $\Delta R/R$ as a function of excitation energy for two 70 Å' strained $In_x Ga_{1-x} As/Al_y Ga_{1-y} As$ single quantum wells with: (a) $x = 0.10$ and $y = 0.20$, and (b) $x = 0.15$ and $y = 0.28$. The arrows mark the calculated transition energies for the m-th conduction band - n-th heavy (light) hole band (m - nH or m - nL) transition.

the effective valence band discontinuity or, equivalently, decreases ΔQ_{cb}.

The data in Ref. 20 is reported to follow a linear dependence on x, though this may be an artifact arising from the limited composition range investigated. Additionally, the extrapolated strained valence band offset calculated with Eqns. 1-6 for InAs/AlAs is found to be 0.77 eV, which is in good agreement with the experimental findings of Moisin et al.[21] who report $\Delta E_{vb} = 0.74 \pm 0.07$ eV and with the value of 0.95 ± 0.2 eV predicted by the model solid theory[4].

In relation to the band offsets of InGaAs/GaAs hetero-interfaces, a recent proposal[10] has suggested that ΔQ_{cb} varies from ~ 0.4 to ~ 0.7 as x increases from 0 to ~ 0.15. Using the calculation scheme outlined above, we find that ΔQ_{cb} *decreases*

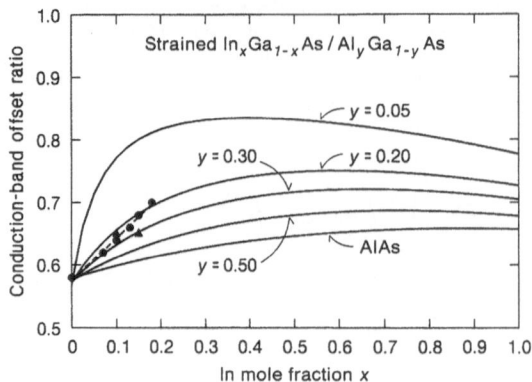

Fig. 3. Strained conduction band offset ratio ΔQ_{cb} as a function of indium mole fraction x for $In_xGa_{1-x}As/Al_yGa_{1-y}As$ interfaces calculated with ΔQ_H and compared to recent experimental data (full circles) for = 0.2 (Ref. 20) and the corresponding least squares fit (dashed line). Full triangles represent data from this work.

only moderately with increasing z. That is, for low x (< 0.05), ΔQ_{cb} ~ 0.70 and decreases to ~ 0.65 at x = 0.4, in general agreement with many other reports[5-9,22,23] which conclude minor, if any, variation of ΔQ_{cb} over this composition range. The large variations in ΔQ_{cb} reported[10,11,24] may arise from poor concentration calibration, or possibly from low sensitivity of the measurements in the low x range where the absolute band edge discontinuities are only ~ 50 meV. Further investigation is required to resolve this issue.

CONCLUSIONS

Within a general framework for strained (001) III-V heterointerfaces, the conduction band offsets for $In_xGa_{1-x}As/Al_yGa_{1-y}As$ heterointerfaces have been calculated and compared to experimental evidence with good agreement found between the two. In general, it is shown that the conduction band offset ratio is nonconstant and extremely sensitive to both In and Al composition. In heterointerfaces where the band discontinuity plays a crucial role in carrier confinement, increasing the Al content gives the added benefit of increased differences in the band gaps of the materials, but only at the expense of a reduction in the conduction band offset ratio. This could be of significant importance in the design of strained layer high electron mobility transistors and quantum well lasers where the band discontinuities directly influence carrier confinement and device performance.

Acknowledgment

Helpful discussions with W. von der Linden, A. Baldereschi and Y. Galeuchet as well as the support of the Laser Science and Technology Department are gratefully acknowledged. CVH acknowledges the support of the Instituut tot Aanmoediging van het Wetenschappelijk Onderwijs in de Nijverheid en de Landbouw.

REFERENCES

1. C.G. Osbourn, **Phys. Rev. B**, 27:5126 (1989)
2. J.Y. Marzin, M.M. Charasse and B. Sermage, **Phys. Rev. B**, 31:8298 (1985)
3. J.M. Langer, C. Delerue, M. Lannoo and H. Heinrich, **Phys. Rev. B**, 38:7723 (1988)
4. C.G. Van De Walle, **Phys. Rev. B**, 39:1871 (1989) and references therein
5. D.J. Arent, K. Deneffe, C. Van Hoof, J. De Boeck and G. Borghs, **J. Appl. Phys.**, 66:1739 (1989)
6. F. Iikawa, F. Cerdeira, C. Vazquez-Lopez, P. Motisuke, M.A. Sacilotti, A.P. Roth and R.A. Masut, **Phys. Rev. B**, 38:8473 (1988)
7. S.H. Pan, H. Shen, Z. Hang, F.H. Pollak, W. Zhuang, Q. Xu, A.P. Roth, R.A. Masut, C. Laselle and D. Morris, **Phys. Rev. B**, 38:3375 (1988)
8. S. Niki, C.L. Lin, W.S.C. Chang and H.H. Wieder, **Appl. Phys. Lett.**, 55:1339 (1989)
9. D., Gershoni, J.M. Vandenberg, S.N.G. Chu, T. Tanbun-Ek and R. A. Logan, **Phys. Rev. B**, 40:10017 (1989)
10. M.J. Joyce, M.J. Johnson, M. Gal and B.F. Usher, **Phys. Rev. B**, 38:10978 (1988)
11. A. Ksendzov, H. Shen, F.H. Pollak and D.P. Bour, **Solid State Commun.**, 73:11 (1990)
12. See for example K. Tai, A. Mysyrowicz, R.J. Fischer, R.D. Slusher and A.Y. Cho, **Phys. Rev. Lett.**, 62:1784 (1989) and references therein
13. N.E. Christensen, **Phys. Rev. B**, 38:12687 (1988) and references therein
14. R. People, K.W. Wecht, K. Alavi and A.Y. Cho, **Appl. Phys. Lett.**, 43:118 (1983)
15. A. Sandhu, Y. Nakata, S. Sas, K. Kodama and S. Hiyamizu, **Jpn. J. Appl. Phys.**, 26:1709 (1987)
16. D.J. Arent, to appear in **Phys. Rev. B**, (March 1990)
17. M. Cardona and N.E. Christensen, **Phys. Rev. B**, 37:1011 (1988)
18. F.H. Pollak and M. Cardona, **Phys. Rev.**, 172:816 (1968)
19. D.E. Aspnes and J.E. Rowe, **Phys. Rev. Lett.**, 27:188 (1971)
20. N. Debbar, D. Biswas and P. Bhattacharya, **Phys. Rev. B**, 40:1058 (1989)
21. J.M. Moisin, C. Guille, M. Van Rompay, F. Barthe, F. Houzay and B. Bensoussan, **Phys. Rev. B**, 39:1772 (1989)
22. T.G. Andersson, Z.G. Chen, V.D. Kulakovskii, A. Uddin and J.T. Vallin, **Phys. Rev. B**, 37:4032 (1988)
23. J.-P. Reithmaier, R. Höger, H. Reichert, A. Heberle, G. Abstreiter and G. Weimann, **Appl. Phys. Lett.**, 46:536 (1990)
24. J. Menéndez, A. Pinczuk, D.J. Werder, S.K. Sputz, R.C. Miller, D.L. Sivco and A.Y. Cho, **Phys. Rev. B**, 36:8165 (1987)
25. C. Basio, J.L. Staehli, M. Guzzi, G. Burri and R.A. Logan, **Phys. Rev. B**, 38:3263 (1988)
26. "Semiconductors", O. Madelung, M. Schulz and H. Weiss, eds., Landolt-Bornstein, New Series, Group 3, Vol. 17a , Springer-Verlag, Berlin (1982)
27. K.H. Goetz, D. Bimberg, H. Jur, J. Selders, A.V. Solomonov, G.F. Glinskii, M. Razeghi and J.J. Robin, **J. Appl. Phys.**, 54:4543 (1983)
28. O. Berolo and J.C. Wooley, p.1420 **in**: "Proceedings of 11th Int. Conf. on the Physics of Semiconductors, Warsaw, 1972", Polish Scientific, Warsaw (1972)

29. C.T. Liu, S.Y. Kin, D.C. Tsui, H. Lee and D. Ackley, **Appl. Phys. Lett.**, 53:2510 (1988)

ELECTRONIC PROPERTIES AND STABILITY OF SEMICONDUCTOR

HETEROSTRUCTURES

Inder P. Batra, S. Ciraci† and A. Baratoff‡

IBM Research Division, Almaden Research Center
K08/282 650 Harry Road, San Jose
₁ Ca. 95120-6099, USA
† Department of Physics, Bilkent University
  06533 Bilkent, Ankara, Turkey
‡ IBM Research Division, Zurich Research Laboratory
  8803 Rüschlikon, Switzerland

1.   INTRODUCTION

Owing to advances in crystal growth techniques, synthetic semiconductor structures, i.e. superlattices or heterostructures, can now be fabricated with high perfection and unusual electronic properties. Such electronic properties undergo dramatic changes because of the reduced dimensionality and the formation of novel quantum states arising from the new periodicity imposed by the superlattice. The novel behavior of carriers discovered in these systems of lower dimensionality have attracted the attention of many prominent research centers around the world. New concepts in solid state electronics have led to the design of devices with very short response times. Not only technological applications, but also the quantum behavior of electrons in such systems has become an active field of fundamental study.

A pseudomorphic semiconductor superlattice[1,2] $(A_x B_{1-x})_n / (C_y B_{1-y})_m$ provides several degrees of freedom for controlling the electronic properties. In such a superlattice the sublattices $(A_x B_{1-x})$ and $(C_y B_{1-y})$ have the same lateral lattice constant, $a_\parallel$, and contain n and m atomic layers (or pairs of layers in compound constituents) perpendicular to the superlattice growth direction z. The letters A, B and C denote elemental (Si, Ge) or compound (AlAs, GaAs, InP etc.) semiconductors.

Each sublattice by itself will have a different electronic band structure and different energy gaps. Upon superlattice formation these bands undergo a relative shift (owing to charge transfer between the sublattices). Consequently, the spatial variation of the band edges along z displays[1] wells and barriers. As a result, the lowest conduction band state of one sublattice decays into the adjacent sublattice if the conduction band edge of the latter occurs at a higher energy and thus

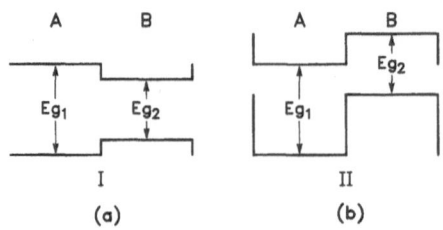

Fig. 1. Alignments of the energy gaps of materials
A and B leading to (a) Type I nonstaggered
and (b) Type II staggered arrangements.

becomes a barrier. This happens because the solution of the
Schrödinger equation in one sublattice can not find a matching
partner (with the same momentum and energy) in the adjacent one.
Then this state is localized in the sublattice where it would
have propagating solutions. If the overlap of the tails in the
barrier is negligible, the localized state becomes confined in
its sublattice, which is denoted as a quantum well. On the
other hand, if the overlap of the tails in the barrier is
significant, the localized states form a subband in the minizone
along the superlattice direction. The localized states are
still propagating in the plane perpendicular to the superlattice
direction.

The two superlattice gaps E_{g1} and E_{g2} involved in a
heterojunction can line up to give nonstaggered (type I) or
staggered band (type II) alignments shown in Fig. 1. For type I
alignment, material A acts as a barrier region for both
electrons and holes. In type II alignment, material A acts as a
barrier region for holes but material B acts as a barrier for
electrons. In this paper we consider two kinds of short period
superlattices from the viewpoint of the first principles self-
consistent field pseudopotential method[3]. In one the
constituents are compound semiconductors and in the other they
are elemental semiconductors. Besides energy band alignments,
we examine stability and electronic structure issues with a view
towards applications.

2.   COMPOUND SUPERLATTICE $(GaAs)_n/(AlAs)_n$

We have investigated $(GaAs)_n/(AlAs)_n$ superstructures with
n = 2,3,4 in the [001] orientation. Total energy calculations[4-7] show that all these superlattices are unstable at
T = 0 relative to disproportionation into the constituent
compounds. These superlattices have only a small lattice
mismatch (< 1%) and hence strain plays a minor role in their
stability.

The electronic states near the band edges exhibit a
confined character even for heterostructures with thin
alternating layers. The charge distributions of the highest
occupied and the lowest unoccupied states obtained[8] from our
electronic structure calculations are shown in Fig. 2 for the
three superlattices (n = 2,3 and 4). By the time we reach n = 4
the plots in Fig. 2 exhibit well defined confinement. At the Γ
point, the lowest conduction band state is localized in AlAs

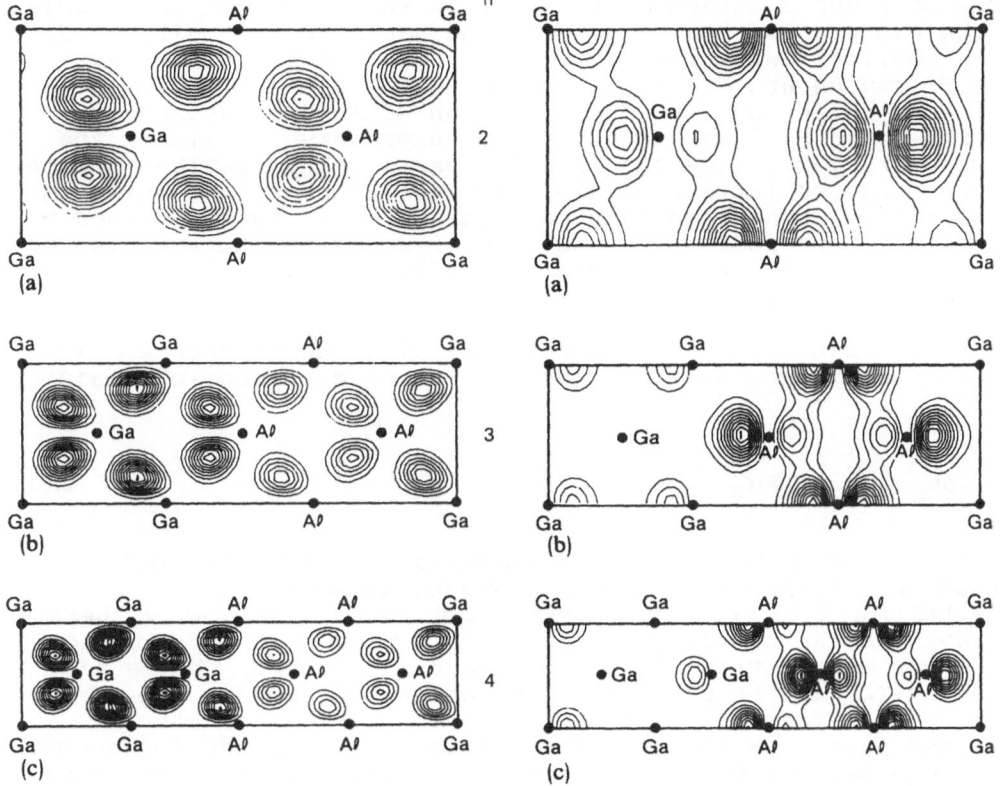

Fig. 2. Charge density for the highest occupied (panels
a, b, c on the left) and the lowest unoccupied
states (panels a, b, c on the right) of
$(GaAs)_n/(AlAs)_n$ at Γ for n = 2,3 and 4 (adapted
from Ref. 8).

whereas the state at the top of the valence band is localized in
GaAs. These states can be identified as confined states, and
their localization suggest[9-14] a staggered band lineup. The
corresponding quantum wells occur in AlAs for electrons, and in
GaAs for holes. This behavior is quite different from that of
common $GaAs/Ga_{1-x}Al_xAs$ heterostructures with large superlattice
periodicity. This result suggests that the band lineup undergoes a change from type II to type I upon increasing the superlattice periodicity. We estimated[8] that this transition
occurs at about n = 10. These results are supported by several
experiments[15-19]. The key finding here is that the superlattice periodicity can be used to tune the band offset and
alignment.

ELEMENTAL SUPERLATTICE Si_n/Ge_m

If the constituents of the superlattice have an appreciable
lattice mismatch, as in Si_n/Ge_m, then the lattice strain
provides an additional degree of freedom in band gap engineering. The effects of the lateral lattice constant $a_\parallel$ and of the
sublattice and superlattice periodicities on the electronic
structure is in general very important for pseudomorphic

(strained) superlattices. Activity in this area has been fueled
by the expectation that a direct band gap material suitable for
laser applications may be fabricated based on Si technology.
Both Si and Ge are indirect band gap materials and their lattice
constants differ by 4%. The formation of a superlattice by
growing Ge on Si leads to a biaxial compressive strain in the
grown overlayers. The superlattice periodicity results in zone
folding and the lattice strain leads to splitting and shifting
of the bands. The resultant band structure is thus governed by
superlattice periodicity and strains imposed by lattice mismatch
and growth conditions. A brief overview of the subject has been
provided by Abstreiter[20].

The grown layers are in registry with the substrate. The
excess energy (per unit volume) due to the macroscopic strain
is[21]

$$\Delta E_s = (\frac{2}{3} - \frac{1}{3}\sigma)\epsilon_0 + \frac{1}{2}C_{11}(\frac{c - c_0}{c_0})^2 , \tag{1}$$

where $\epsilon_0 = 1/2\ B_0(\delta V/V)^2$ is the change in energy under uniform
strain $(a - a_0)/a_0 = 1/3(\delta V/V)$, the bulk modulus,
$B_0 = 1/3\ (C_{11} + 2C_{12})$, $c = c_0 - \sigma(a - a_0)$, and $\sigma = 2C_{12}/C_{11}$. Here
we have used a to represent $a_\parallel$ and c to represent $a_\perp$. Due to
the compressive strain in the overlayers $(a < a_0)$, the lattice
constant (c) in the perpendicular (growth) direction expands
leading to a tetragonal distortion. In the case of unconstrained expansion the last term in the above expression goes to
zero $(c = c_0)$ and thus a part of the strain energy is relieved
by the perpendicular expansive strain.

The growth of up to six Ge layers pseudomorphically
restricted to a Si(001) substrate has been realized by Pearsall
et al.[22]. More importantly, they observed relatively strong
optical transitions in the Si_4/Ge_4 semiconductor superlattice
which are found neither in the constituent crystals, nor in the
$Si_{0.5}Ge_{0.5}$ alloy. This has been viewed as an encouraging step
towards controlling electronic properties relevant for laser
applications.

We have carried out total energy and charge density
calculations[23] on pseudomorphic strained Si_n/Ge_n superlattices
with $1 \leq n \leq 6$. To assure registry with the substrate in the
(001) plane, the lateral lattice constant $a_\parallel$ is taken equal to
a^0_{Si}, the cubic lattice constant of bulk Si. As seen in Fig. 3
the lattice constant perpendicular to the epilayers, R_3 (or $a_\perp$
in terms of the cubic lattice parameter) is determined by three
types of interlayer spacings, i.e. d_1 (Si-Si), d_2 (Si-Ge) and
d_3 (Ge-Ge). The change in d_1 upon superlattice formation on a
Si(001) substrate is, however, negligibly small. · So fixing d_1
equal to $a^0_{Si}/4$, we concentrate on the tetragonally distorted Ge
sublattice.

Equilibrium values for the intelayer spacings we obtained
are $d_1 = 2.56$ a.u., $d_2 = 2.61$ a.u. and $d_3 = 2.70$ a.u. It should
be noted that the calculated d_2 is very close to the average
$(a^0_{Si} + a^0_{Ge})/8$, whereas the value of d_3 implies a tetragonal
distortion, $\epsilon_T = (a_\perp - a_\parallel)/a^0_{Ge}$ of ~ 5% in the Ge sublattice.
The formation energy (or formation enthalpy at T = 0) of the
pseudomorphic Si_n/Ge_n superlattice is calculated from

560

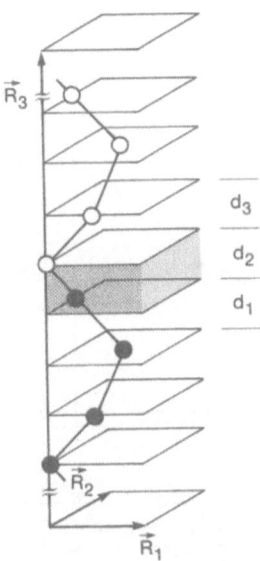

Fig. 3. The tetragonal unit cell of the strained Si_n/Ge_n superlattice with filled and open circles denoting the position of Si and Ge atoms, respectively. $R_1 = R_2 = a^0_{Si}/\sqrt{2}$ and $d_1 = a^0_{Si}/4$. R_3, d_2 and d_3 are obtained by minimization of the total energy (adapted from Ref. 23).

$$\Delta E^{f,s}(Si_n/Ge_n) = E^s_T(Si_n/Ge_n) - [E^0_T(Si_{2n}) + E^0_T(Ge_{2n})]/2,$$

where the total energies of the constituent strain free crystals $E^0_T(Si_{2n})$ and $E^0_T(Ge_{2n})$ are calculated in a unit cell corresponding to that of Si_n/Ge_n, but with the equilibrium lattice constants (a^0_{Si} and a^0_{Ge}) determined for the (strain free) bulk crystals.

The calculated formation energies of the strained super-lattices are summarized in Fig. 4. The following conclusions can be drawn: (i) all superlattice formation energies have positive values. By definition $\Delta E^{f,s} > 0$ indicates instability. Consequently, decomposition into constituent crystals (i.e. segregation) is possible, as long as it is permitted by the kinetics of the reaction. Alternatively, the strain energy accumulated in the Ge sublattice can be relieved by the creation of misfit dislocation; (ii) The value of $\Delta E^{f,s}$ increases with increasing n because the Ge sublattice has more strained layers. Moreover for large n, $\Delta E^{f,s}/n$, saturates to a constant value.

The above conclusions suggest that the formation energy of the superlattices with n > 2 has two major components, the interfacial energy and the strain energy. If the transfer of charge from one sublattice to the other is confined in a narrow region near the interface (and thus the resulting redistribution of charge in the sublattice is insignificant), these two components of the formation energy, $\Delta E^{f,s}$, can be dealt with separately[23,24]. The superlattice formation energy can then be written as,

561

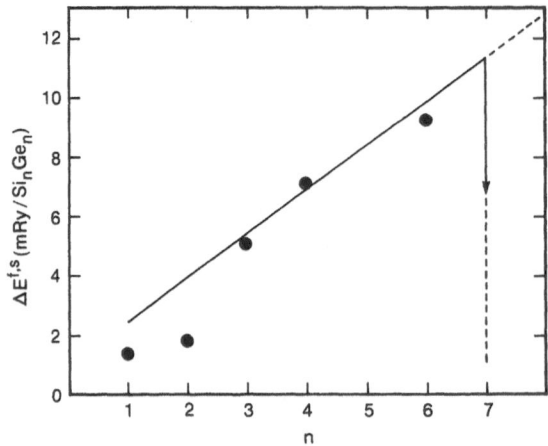

Fig. 4. Comparison of the calculated formation
energies of Si_n/Ge_n superlattices (filled
circles) with those estimated from a
simple linear relation (reproduced from
Ref. 23).

$$\Delta E^{f,s}(Si_n/Ge_n) \simeq 2\ I + n\zeta,$$

where I is the interfacial energy (about 0.5 mRy per atom) and ζ
is about 1.5 mRy. The value of I obtained by Martins and
Zunger[25] is about 0.6 mRy.

The formation energies estimated from this simple model are
compared with our *ab initio* results in Fig. 4. A good match is
obtained for n ≥ 3. Small period superlattices Si_1/Ge_1 and
Si_2/Ge_2 deviate from this because the interface term has a
residual n dependence which becomes dominant at small n values.
Within this simple approach, the formation energy of Si_7/Ge_7 is
predicted to be ~ 11.3 mRy. As revealed by experimental
work[22], only six epitaxial Ge layers can grow on Si(001), and
beyond that thickness misfit dislocations are generated. The
above model thus yields an estimate for the energy barrier to
form a misfit dislocation of 9.8 < ΔQ < 11.3 mRy/cell. All
structures we studied[23] are found to be indirect band gap
semiconductors. An interesting feature we noted was that the
energy difference between the direct and indirect gap, δE_g,
continued to decrease with increasing n, and reached a value of
0.07 eV for n = 6.

Charge density plots of the conduction band states at Γ
(and of the lowest indirect gap state Δ_{c1}) are presented in Fig.
5 for Si_4/Ge_4. The lowest two states at Γ (above Δ_{c1}) are
strongly confined in the Si sublattice. This can be understood
as follows. The minimum of the conduction band of Si occurs
along six equivalent Δ directions of the strain free bulk.
Under the tetragonal strain they split into two degenerate $\Delta_\perp$
states, along the superlattice direction, and four degenerate $\Delta_\parallel$
states in the (001) plane of the superlattice. The states at
$X_\perp$ fold into Γ due to superlattice periodicity. The states Γ_{c1}
and Γ_{c2} are such folded Si-confined states. Similarly, the
states Γ_{c3} and Γ_{c4} are folded Ge-confined states. The state Γ_{c5}
is the lowest unfolded state analogous to the $\Gamma_{2'}$ bulk state.
These results are in accord with other calculations[26,27].

The Si_4/Ge_4 superlattice has[27] a type II band offset with $\Delta E_v = 0.8$ eV. If the minimum $\Delta_\perp$ of the band along the super-lattice direction were to occur below the minimum $\Delta_\parallel$ of the band parallel to epilayers then Si_n/Ge_m would become a direct band gap semiconductor. This does in fact happen[28,29] for certain specific superlattice and sublattice periodicities with $n + m = 10$. The minimum of the conduction band is then folded to the center of the superlattice Brillouin zone. Based on this fact it was hoped that pseudomorphic Si/Ge superlattices would compensate for the shortcomings of silicon and open new horizons in application of Si based devices in photonics. However, the oscillator strength of the lowest transition found so far has not been too encouraging.

On the other hand, electroreflectance data[22] on Si_4/Ge_4 yielded optical transitions at 0.76, 1.25 and 2.3 eV. The lowest transition at 0.76 eV raised some hopes because it is neither present in the parent constituents nor in their alloys. It is still uncertain whether this transition is direct or indirect. Most theoretical studies[26,30,31] attribute the roughly ten times weaker transitions observed at 0.76 and 1.25 eV to an *indirect* and to a *direct* band gap originating from $\Delta_\parallel$ and $X_\perp$ states with $\vec{k}$ parallel and perpendicular to the layers, respectively. These studies also agree that the $X_\perp$ states are folded into the center, $\Gamma(\vec{k} = 0)$, of the superlattice Brillouin zone and are strongly confined to the Si layers.

Wong et al.[32] suggested that the lowest transition is direct, based on its relative intensity and on empirical

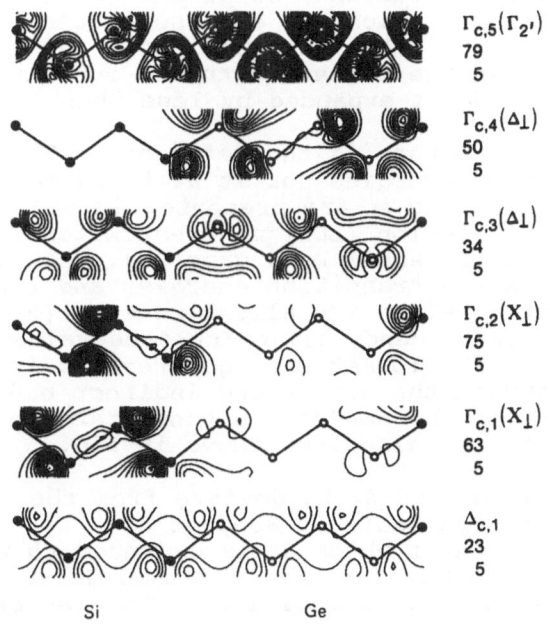

	$\Gamma_{c,5}(\Gamma_{2'})$ 79 5
	$\Gamma_{c,4}(\Delta_\perp)$ 50 5
	$\Gamma_{c,3}(\Delta_\perp)$ 34 5
	$\Gamma_{c,2}(X_\perp)$ 75 5
	$\Gamma_{c,1}(X_\perp)$ 63 5
	$\Delta_{c,1}$ 23 5

Si   Ge

Fig. 5. Charge density contour plots of the first six conduction band states of Si_4/Ge_4. Upper and lower numerals at each panel are the values of maximum charge density and the contour spacing in 10^{-4} a.u., respectively (adapted from Ref. 23).

pseudopotential calculations. Like others[23,27], they assume a larger Ge interlayer spacing along [001] so as to maintain registry with the Si substrate in accordance with macroscopic strain considerations. Assuming in addition a Ge form factor as in bulk cubic Ge, they find a *direct* band gap of 0.9 eV attributed to $X_\perp$-derived states. The transition observed at 1.24 eV is attributed to a higher $\Delta_\perp$-derived state also folded into Γ which they find at 1.4 eV. Lastly, Wong et al.[32] claim that an analogous picture results if the Ge-Ge bonds in the superlattice are stretched to their values in bulk Ge, a situation which they presume is favored by a few defects at the Si-Ge interfaces. This would imply a tetragonal distortion in excess of that dictated by the macroscopic strain or by other first principles total energy calculations. Several objections to these arguments have recently been raised by Van de Walle[33] and Froyen et al.[34].

It is clear both from first principles calculations as well as from simple strain energy considerations that stretching the Ge-Ge bond length to achieve bulk values is energetically unfavorable in an ideal superlattice. However, we find the suggestion by Wong et al.[32] sufficiently intriguing and are willing to accept for a moment that somehow the growth process circumvents those objections. Is it possible that such a stretched superlattice produces the direct transition proposed by Wong et al.?

To this end we first obtained[35] a fully relaxed structure of the strained Si_4/Ge_4 superlattice pseudomorphic with Si(001). Previously calculated interlayer spacings were $d_1 = 2.56$, $d_2 = 2.61$ and $d_3 = 2.70$ a.u. Residual atomic forces calculated for that geometry indicated that Si-Si and Ge-Ge spacings must be subject to small variations in each sublattice. The fully relaxed structure showed that the spacings adjacent to the interface and that at the center of the Ge sublattice are contracted, whereas d_2 is expanded by less than 1%.

We take this structure as the reference. The calculated electronic transition energies for it at Γ and to the minimum of the conduction band along $\Delta_\parallel$ are presented in Fig. 6a. The local density approximation (LDA) underestimates conduction band energies and yields too small band gaps[30]. However, differences between LDA transition energies and the corresponding experimental values are almost constant[27] at symmetry points (~ 0.5-0.6 eV). Hence all calculated values are shifted upwards by 0.6 eV. The lowest band gap remains indirect, and the difference between the direct and indirect band gaps is 0.2 eV. The parentage of the conduction band states did not change significantly from that discussed above in Fig. 5.

We next permit d_2 and d_3 to deviate from the optimized values to test the hypothesis of Wong et al. We consider two structures in both of which d_3 is expanded to make the length b_{Ge-Ge} of Ge-Ge bonds in the strained Ge sublattice equal to b^0_{Ge-Ge}. In the second, in addition d_2 is also expanded so that $b_{Si-Ge} = (b^0_{Si-Si} + b^0_{Ge-Ge})/2$. Total energies relative to that of the fully relaxed structure are higher for both of these structures. This is also obvious from Eqn. 1 because for these structures $c \neq c_0$. The transition energies relative to the top of the valence band are presented in Fig. 6b-c. Different states are affected differently by these small but significant

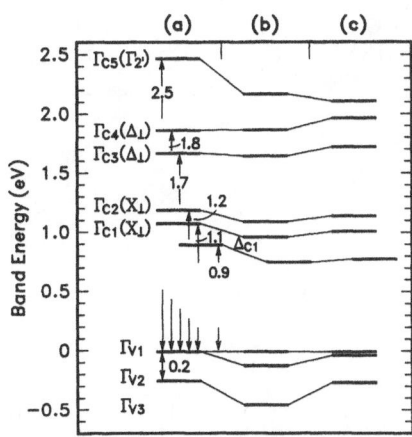

Fig. 6. Transition energies in Si_4/Ge_4 at Γ for (a)
fully relaxed reference structure (b) d_3
expanded to let the Ge-Ge distance in the
strained superlattice achieve its bulk
unstrained value, b^0_{Ge-Ge} and (c) d_3 expanded
as in (b) but in addition d_2 also expanded so
that $b_{Si-Ge} = (b^0_{Si-Si} + b^0_{Ge-Ge})/2$ (adapted from
Ref. 35).

changes from the optimum structure. For example, the minimum of
the conduction band, $E(\Delta_{c1})$, and the first and second conduction
band states at Γ are only slightly lowered, and the indirect
character of the lowest band gap (i.e. the difference between
the lowest direct and indirect band gaps) is increased. Most
importantly, the energy of the lowest optically allowed direct
transition to Γ_{c2} stays above 1.1 eV. It thus appears risky to
argue that all transition energies should be lowered by the same
amount. The lowest unfolded state at k = 0, Γ_{c5}, is lowered
quite appreciably, i.e. by 0.3 to 0.4 eV. These changes are
consistent with those described by Van der Walle[33] and by
Froyen et al.[34]. On the basis of our results we find the
hypothesis of Wong et al.[32] is not supported. It is also
unlikely that the required *additional* increase in the Ge-Ge
spacing can be accommodated without disrupting lateral registry.
Froyen et al.[34] find that to lower the direct band gap one
needs to reduce the Ge-Ge spacing.

We conclude that a proper explanation of the new
transitions (observed to be 10 to 100 times stronger than
predicted but still disappointing from the viewpoint of laser
applications) must be sought elsewhere. The energy of the
$\Gamma_{c2}(X\perp) - \Gamma_{v1}$ of the lowest allowed direct transition is only
slightly lowered to 1.1 eV and remains well above the observed
lowest transition energy (0.76 eV). On the other hand, the
energy $\Gamma_{c5} - \Gamma_{v1}$ of the strong direct transition would be
significantly lowered, in contrast to observations[22].

3.1 Order-Disorder Transition

Finally, we should touch upon the neostructural order-
disorder transition reported in $Si_{1-x}Ge_x$ alloys grown on a
Si(001) substrate. It is generally believed that SiGe bulk

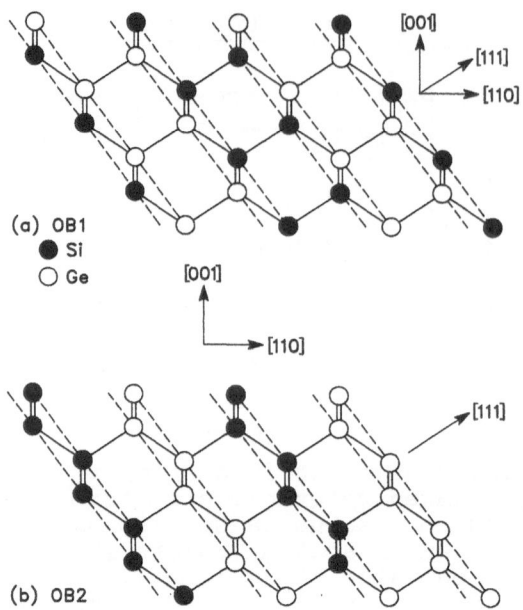

Fig. 7. Two types of long range order observed in
SiGe superlattices grown on Si(001). The
ordering consists of alternating double
layers of Si and Ge along the [111]
direction, (a) in OB1 the double layers
are mixed and (b) in OB2 the double layers
are pure Ge or Si.

alloys are totally random. The observations of two types[36,37]
of long range ordering in $Si_{0.5}Ge_{0.5}$ alloys are really
fascinating and deserve fundamental considerations. The
observed ordered phases have two equipolar rhombohedral (Si_2Ge_2
unit cell) variants, shown in Fig. 7, both allowing a bilayer
segregation normal to the [111] direction. The first one
reported by Ourmazd and Bean[36] (OB1), by growing a thin alloy
film pseudomorphically on an (001) Si substrate, showed the
bilayer sequence (SiGe)(GeSi)..., along the [111] direction. In
the OB1 structure each Si (Ge) has three heteropolar bonds, and
pairs of widely separated planes are occupied by atoms of the
same kind. The second one reported by LeGoues et al.[37] (OB2),
in a thick bulklike alloy on an (001) Si substrate, showed the
bilayer sequence (SiSi)(GeGe)..., along the [111] direction. In
this OB2 structure each Si(Ge) has three homopolar bonds, and
pairs of widely separated planes are occupied by atoms of the
opposed kind. The structure is microscopically strained because
bond lengths deviate from the ideal tetrahedral bond length.

Various first principles total energy calculations[6,23]
showed that the epitaxial OB1 structure is more stable than the
coherently strained constituents. Furthermore, the OB1
structure was shown to be more stable than the OB2 structure.
Nevertheless disproportionation into bulk constituents had the
lowest energy. The observation of the OB2 structure has led to
the speculation[37] that it might have a lower chemical energy
than the OB1 structure. This was found not to be the case in

our calculations reported[6] in 1987, as confirmed recently by Zunger et al.[38]. Thus the origin of the stability of the OB2 structure has to be sought elsewhere, possibily in surface effects[39]. Furthermore, the ordered phase is only obtained if the growth is on a (001) Si substrate. The (001) surface shows a strong reconstruction in the form of Si dimers which are energetically strongly bound. Dimer breaking can release energy in the lattice and may be accommodated by inducing long range order.

CONCLUDING REMARKS

In summary, we have reviewed the results of self-consistent field pseudopotential calculations which provide valuable insights into the stability and the electronic properties of short-period pseudomorphic AlAs/GaAs and Si/Ge superlattices. In particular, strained Si/Ge heterostructures have attracted much attention because of the promise of emitting light across a direct band gap significantly lower than those of the elemental constituents or alloys thereof. A direct band gap superlattice can be fabricated by growing Si on Ge and by carefully choosing the n and m values in Si_nGe_m. In this case the oscillator strength to the zone folded conduction band state is disappointingly small, however. Alternatively, by growing Si on Ge substrate a direct band gap material can be obtained. This, however, is only of academic interest because the intent is to use Si as the substrate material in practical devices. At the present time Si/Ge superlattices do not appear to be suitable for laser applications.

Acknowledgment

One of us (IPB) wishes to thank the organizers of the workshop for the opportunity to present this work.

REFERENCES

1.   L. Esaki, **IEEE Journal of Quantum Electronics**, Vol. QE-22, No. 9 (1986)
2.   N. Narayanamurti, **Science**, 235:1024 (1987)
3.   See, for the method and related references, I.P. Batra, S. Ciraci, G.P. Sirivastava, J.S. Nelson and C.Y. Fong, **Phys. Rev. B**, 34:8246 (1986)
4.   D.M. Bylander and L. Kleinman, **Phys. Rev. B**, 34:5280 (1986); 36:3229 (1987)
5.   D.M. Wood, S.H. Wei and A. Zunger, **Phys. Rev. Lett.**, 58:1123 (1987)
6.   S. Ciraci and I.P. Batra, **Phys. Rev. Lett.**, 58:2114 (1987); in: "Properties and Applications of Impurity States in Semiconductor Superlattices", C.Y. Fong, I.P. Batra and S. Ciraci, eds., Plenum, New York, (1988)
7.   A. Oshiyama and M. Saito, **Phys. Rev. B**, 36:6156 (1987)
8.   J.S. Nelson, C.Y. Fong and I.P. Batra, **Appl. Phys. Lett.**, 50:1595 (1987)
9.   D. Ninno, K.B. Wong, M.A. Gell and M. Jaros, **Phys. Rev. B**, 32:2700 (1985)
10.   M.A. Gell, D. Ninno and J. Jaros, **Phys. Rev. B**, 34:2416 (1986);   M.A. Gell and D.C. Herbert, **Phys. Rev. B**, 35:9591 (1987)

11. J. Ihm, **Appl. Phys. Lett.**, 50:1068 (1987)
12. G. Duggan and H.I. Ralph, **Phys. Rev. B**, 35:4152 (1987)
13. S. Ciraci and I.P. Batra, **Phys. Rev. B**, 36:1225 (1987)
14. I.P. Batra, S. Ciraci and J.S. Nelson, **J. Vac. Sci. Technol.**, B5:1300 (1987)
15. A. Ishibashi, Y. Mori, M. Itabashi and N. Watanabe, **J. Appl. Phys. Lett.**, 58:2691 (1985)
16. E. Finkman, M.D. Sturge and M.C. Tamargo, **Appl. Phys. Lett.**, 49:1299 (1986)
17. P. Dawson, B.A. Wilson, C.W. Tu and R.C. Miller, **Appl. Phys. Lett.**, 48:541 (1987)
18. G. Danan, B. Etienne, F. Mollot and R. Planel, **Phys. Rev. B**, 35:7784 (1987)
19. G. Duggan, H.I. Ralph, P. Dawson, K.J. Moore, C.T.B. Foxon, R.J. Nicholas, J. Singleton and D.C. Rogers, **Phys. Rev. B**, 35:7784 (1987)
20. G. Abstreiter, **Thin Solid Films**, 183:1 (1989)
21. D.M. Wood and A. Zunger, **Phys. Rev. B**, 40:4062 (1989)
22. T.P. Pearsall, J. Bevk, L.C. Feldman, J.M. Bonar, J.P. Manaerts and A. Ourmazd, **Phys. Rev. Lett.**, 58:729 (1987)
23. S. Ciraci and I.P. Batra, **Phys. Rev. B**, 38:1835 (1988)
24. R.G. Dandrea, J.E. Bernard, S.-H. Wei and A. Zunger, **Phys. Rev. Lett.**, 64:36 (1990)
25. J.L. Martins and A. Zunger, **J. Mater. Res.**, 1:523 (1986)
26. S. Froyen, D.M. Wood and Z. Zunger, **Phys. Rev. B**, 36:4547 (1987)
27. S. Satpathy, R.M. Martin and C.G. Van de Walle, **Phys. Rev. B**, 38:13237 (1988)
28. R. Zachai, K. Eberl, G. Abstreiter, E. Kasper and H. Kibbel, **Phys. Rev. Lett.**, 64:1055 (1989)
29. O. Gulseren and S. Ciraci, to be published.
30. M.S. Hybertsen and M. Schlüter, **Phys. Rev. B**, 36:19683 (1987)
31. M.S. Hybertsen, M. Schlüter, R. People, S.A. Jackson, D.V. Lang, T.P. Pearsall, J.C. Bean, J.M. Vanderberg and J. Beyk, **Phys. Rev. B**, 37:10195 (1988)
32. K.B. Wong, M. Jaros, I. Morrison and J.P. Hagon, **Phys. Rev. Lett.**, 60:2221 (1988)
33. C.G. Van de Walle, **Phys. Rev.**, 62:974 (1989)
34. S. Froyen, D.M. Wood and A. Zunger, **Phys. Rev. Lett.**, 62:975 (1989)
35. S. Ciraci, A. Baratoff and I.P. Batra, **Phys. Rev. B.**, 41:6069 (1990)
36. A. Ourmzd and J.C. Bean, **Phys. Rev. Lett.**, 55:765 (1985)
37. F.K. LeGoues, V.P. Kesan and S.S. Iyer, **Phys. Rev. Lett.**, 64:40 (1990)
38. A. Zunger, R.G. Dandrea and J.E. Bernard, **Phys. Rev. Lett.**, to be published.
39. P.C. Kelires and J. Tersoff, **Phys. Rev. Lett.**, 63:1164 (1989)

MICROSCOPIC THEORY OF SEMICONDUCTOR SUPERLATTICES

M. Jaros, L.D.L. Brown, A.W. Beavis, J.P. Hagon,
P. Harrison, I. Morrison, R.J. Turton and K.B. Wong

Physics Department, The University
Newcastle upon Tyne NE1 7RU, UK

ABSTRACT

We have carried out the first microscopic calculations of
the electronic structure and optical properties of semiconductor
superlattices (e.g. Si-Ge, HgTe-CdTe) with real interfaces, i.e.
structures in which the perfect translational symmetry in the
interface plane is broken due to interdiffusion, defects and
interface roughness. The modeling of such structures lies
outside the scope of simple envelope function schemes. It is
also inaccessible to a priori calculations since microscopic
potentials of clusters of hundreds of atoms must be considered.
We find new states, some of which lie in the forbidden gap (e.g.
in HgTe-CdTe), and significant changes in optical transitions
(e.g. in Si-Ge). Our results provide quantitative guidelines
for establishing a microscopic signature of semiconductor
interfaces.

1.   INTRODUCTION

The physics and applications of semiconductor superlattices, including strained layer materials such as Si-Ge
heterostructures have been studied without exception in terms of
a supercell model. In this picture the interface is formed
between layers of alternating species (e.g. Si and Ge) matched
to a chosen substrate whose lattice constant determines the
atomic separations in the interface plane. In all theoretical
schemes used to describe these systems, periodic boundary
conditions are applied along the growth axis, implying that the
structure is infinite. In the local density and quasi-particle
(the so called first principles) calculations, the width of the
alternating layers is typically six atomic monolayers (i.e.
approximately $1\frac{1}{2}$ bulk lattice constant). The unit cell in the
interface plane is that of a perfect bulk material. Such cells
are large enough to ensure separation of the total surface
electron charge density per atom from the interface in the
adjacent period but far from adequate for modeling individual
quantum states peculiar to the interface and located near the
band edges. In fact practically all theoretical studies of

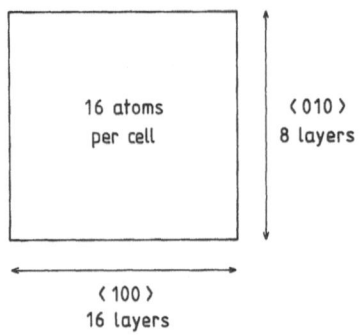

16 atoms
per cell

⟨010⟩
8 layers

⟨100⟩
16 layers

Fig. 1. The enlarged unit cell in the
interface plane.

semiconductor surfaces and interfaces have been implemented as
calculations of electronic states of a suitably chosen ultra-
thin superlattice[1-6].

The purpose of our research project is to make the first
step beyond the above idealizations. In order to establish a
link between theory and experiment, it is necessary to consider
the effect upon observables of interdiffusion across the
interface and presence of impurities and defects. All these
imperfections lead to the breakdown of the translational
symmetry invoked in idealized models and to deviations from the
corresponding selection rules for electronic transitions.
Furthermore, in most cases of practical interest we have to deal
with a finite structure along the growth direction, for instance
with a single interface, double barrier structure, or a superlattice with a few periods only such as when the structure is
grown on Si or Ge buffers[7]. We find that the consequence of
these deviations from ideal condition is a change in localization of confined states and in some cases new levels.

2.   Si-Ge SUPERLATTICES

We begin by choosing a suitable bulk crystal Hamiltonian H_0
(e.g. a pseudopotential for a Si crystal under appropriate
strain) such that the eigenfunction $\varphi(n,k)$ and the band
structure energies $E(n,k)$ can be generated in the relevant range
of energies in good agreement with experiment. We then expand
the superlattice wavefunction ψ in terms of the complete set of
$\varphi(n,k)$. The Schrödinger equation for the heterostructure is
$\{H_0 - E + V\} \sum_{n,k} A(n,k)\varphi(n,k) = 0$ where V is the microscopic
potential representing the difference between the atoms forming
the superlattice layer and that of the starting material.

In modeling the structures grown on Si or Ge, the starting
Hamiltonian we use has tetragonal symmetry (point group D_γ).
In order to model deviations from the bulk translational
symmetry in the interface plane, we enlarge the unit cell in the
interface plane into a rectangular block whose dimensions are
sixteen and eight atomic layers along the <100> and <010>
directions, respectively (Fig. 1).

In the case of an ideal infinite Si-Ge superlattice
consisting of four monolayers of Si and four monolayers of Ge

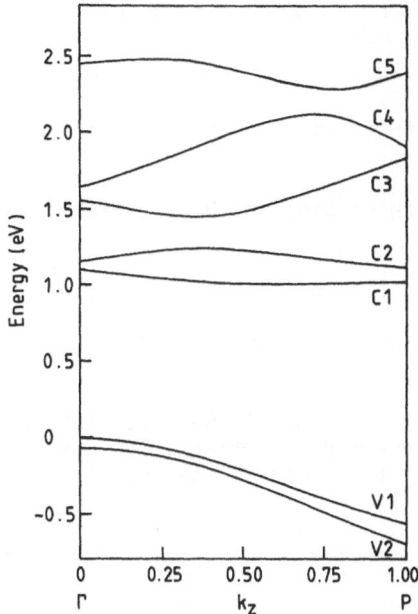

Fig. 2. The electronic band structure of an ideal
Si_4Ge_4 superlattice. The lattice
separation in the interface plane is that
of bulk Si. The state V1 is the top of the
valence band and state C1 is the bottom of
the conduction band. The first finite
optical transition at k = 0 (Γ) is between
V1 and C2.

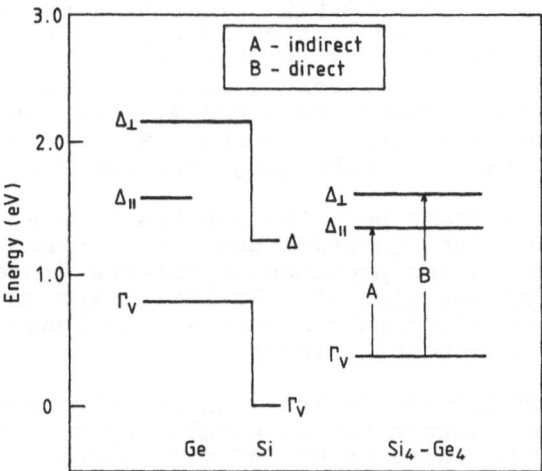

Fig. 3. The band offset diagram for Si–Ge heterostructures on Si. The lowest direct and
indirect transitions are indicated.

Table 1. The modulus squared of the optical matrix
element (in atomic units) for the lowest
direct and indirect transitions in the Si_4Ge_4
superlattice.

1.   Every sixteenth Si atom in the interface layer replaced
      with Ge.

2.   Every sixteenth Ge atom in the interface layer replaced
      with Si.

3.   Protruding Ge atom in (1) displaced by 0.05 Å.

4.   Every sixteenth Ge atom in the interface layer of Ge
      replaced with Sn.

5.   The result for a perfect Si_4Ge_4 superlattice. The
      magnitude of the indirect transition quoted here is the
      effect zero in the numerical calculations. A and B are
      shown in Fig. 3.

Transition	1	2	3	4	5
Indirect (A)	0.97×10^{-5}	0.77×10^{-5}	0.54×10^{-5}	0.30×10^{-3}	0.11×10^{-12}
Direct (B)	0.10×10^{-2}	0.15×10^{-2}	0.10×10^{-2}	0.60×10^{-3}	0.14×10^{-2}

our calculations show, in agreement with other authors, that the
fundamental gap is indirect at about 0.9 eV with the minimum
lying at the nonfolded $\Delta_\parallel$ conduction band minimum. The first
direct transition at the zone centre is found at about 1.2 eV.
This zone folded state (C2 in Fig. 2) is derived from the
longitudinal valleys and is located mainly in Si. It lies at a
higher energy because of the large barrier potential (Fig. 3)
separating the conduction states of Si and Ge layers in the
idealized defect free superlattice model. The same procedure
has been used to generate the electronic structures of related
systems discussed in the forthcoming paragraphs.

    We shall not address here the important question[6] as to
whether the nuclear configuration used in this and other models
is a realistic one in the presence of defects since there is at
present no means of settling it. We shall also confine
ourselves to purely electronic processes and leave out the
question of electron-phonon interactions.

    The effect of interdiffusion upon the electronic structure
of a Si-Ge heterostructure can be modeled by replacing Si atoms
from the Si layer by those of Ge and vice versa. The best
studied structure is that of Si_4Ge_4 described in the above
paragraph. In the ideal structure the purely electronic
transition across the fundamental gap (A in Fig. 3) is forbidden
(i.e. it can occur only via phonons). The existence of foreign
atoms in the layer breaks the translational symmetry in the

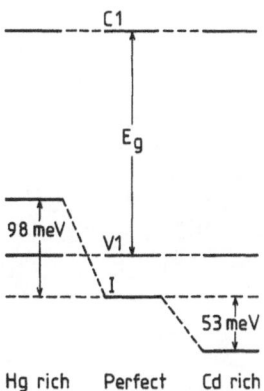

Fig. 4. Middle section: the position of the lowest
conduction band state (C1), the uppermost
valence state (V1) and the state localized at
the interfaces (1) in an ideal symmetric HgTe-
CdTe superlattice of period 25.8 Å, at the
center of the Brillouin zone. The magnitude of
the forbidden gap E_g = 228 meV is indicated.
Left and right sections (the dashed lines are
added to guide the eye only) show that the
interface state is raised in energy in
structures with Hg-rich interfaces and lowered
in those with Cd-rich interfaces.

interface plane and relaxes the selection rule. The magnitude
of the optical matrix element is a good measure of the strength
of this effect. In particular, we can compare the strength of
transition A with the first direct transition (B in Fig. 3) at
the centre of the Brillouin zone involving the zone folded
state. Five different nuclear configurations have been
considered. The results are summarized in Table 1. We can see
that when every sixteenth atom is replaced in the interface
layer by the species from the adjacent material the optical
matrix element for the indirect electronic transition is no
longer zero. However, the effect is not strong enough to make A
and B comparable. In fact, we found that when we replaced two
Si atoms with Ge ones and vice versa there was no significant
increase in the matrix element for the indirect transition A
compared to the case when only one atom is replaced! This is
because the difference between the Si and Ge atoms is small and
the corresponding difference potential is shallow. It means
that the interdiffusion would have to involve more than one out
of eight atoms in the interface layers to make A and B
comparable.

The difference between Si (or Ge) and Sn is larger than
that between Si and Ge. When Sn is used to replace one of the
interface atoms, the corresponding enhancement of transition
probability for A is also larger.

3.   HgTe-CdTe SUPERLATTICES

The electronic properties of the HgTe-CdTe superlattice
reported here have been calculated in the manner described in

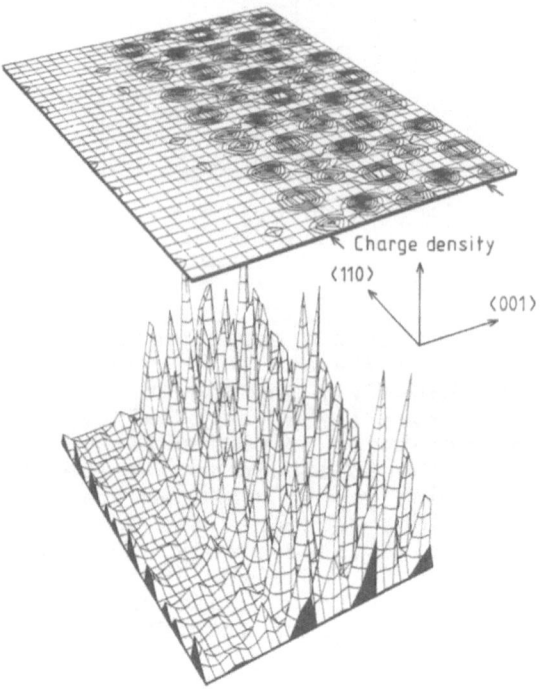

Ideal Interface
Uppermost valence state

Charge density

⟨110⟩

⟨001⟩

Fig. 5. The contour plots of the charge density
associated with state 1 and V1 of the ideal
structure. The corresponding energies are shown
in the middle part of Fig. 4. The position of
the interface plane is indicated by an arrow.
The area of the plot has the dimensions of one
superlattice period (25.8 Å) in the growth ⟨001⟩
direction and 32.3 Å in the ⟨110⟩ direction
parallel to the interface plane. Both two- and
three-dimensional plots are shown to indicate
clearly the degree of localization resulting
from the breakdown of the bulk translational
symmetry along the interface.

Ref. 8. Our predictions based on semi-empirical relativistic
pseudopotentials are in good agreement with the results of more
recent first principle calculations. In particular, the deep
localization predicted in Ref. 8 of the interface state in this
superlattice - associated with the large difference in the
relativistic correction at Hg and Cd atoms - has also been
recovered[9].

In all existing calculations[10] it has been assumed that
the translational symmetry in the interface plane is preserved.
In such a calculation the unit cell in this plane therefore
remains the same as that in a bulk crystal. We shall model
deviations from this ideal interface geometry by assuming that
one of the interface monolayers is either Cd or Hg rich. This
means that we replace, say, one Cd atom in the interface layer
of CdTe by a Hg atom. We place this site in a larger unit cell

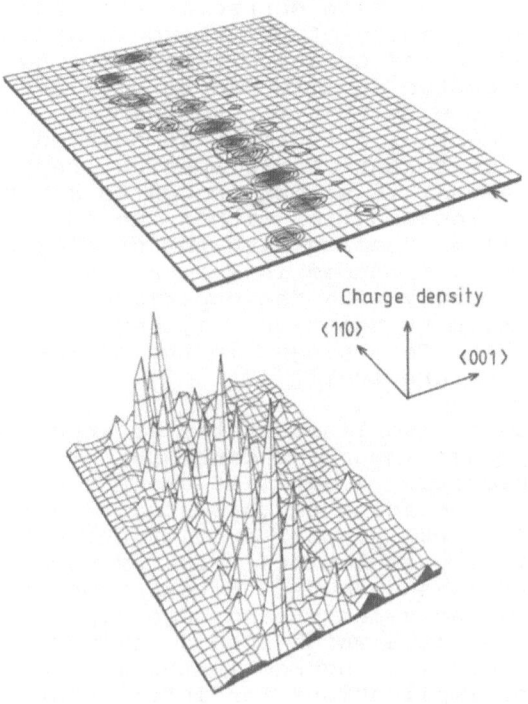

HgTe Rich Interface
Interface state

Charge density

⟨110⟩   ⟨001⟩

Fig. 6. The contour plots of the electron charge
densities of state 1 and V1 of the nonideal
superlattice with Hg-rich interfaces. The
corresponding energies are shown in Fig. 4.
The arrangement is the same as in Fig. 5.

in the interface (x,y) plane whose length in both <100> and
<010> directions equals two bulk lattice constants. This is
equivalent to assuming that, for example, on average every
eighth Cd atom in the interface monolayer is replaced with Hg.
In the growth direction <001>, the superlattice consists of
alternating HgTe and CdTe layers of equal width of 12.9 Å.

In order to appreciate the changes in the electronic
structure brought about by the breakdown of the perfect crystal
translational symmetry in the interface plane, we present our
results for both ideal and nonideal structures. These results
are summarized in Figs. 4-6.

The effect of creating either a Cd or Hg-rich interface
layer upon the energy levels at the center of the Brillouin zone
is shown in Fig. 4. The interface state energy moves either
down or up, respectively, compared to the ideal interface level.
The changes in the magnitude of the forbidden gap between the
"usual" superlattice band edge states (V1 and C1 in Fig. 4)
obtained in these calculations are within the error of the
numerical procedure (of order 5 meV). The dispersion of the V1,
C1 and interface state (I) bands along the growth direction is
not significantly different from their counterparts in the ideal
structure.

The form of the electron charge density of the interface state I in the ideal structure is shown in Fig. 5. We can see that the computer calculation correctly accounts for the bulk-like translational symmetry of the charge distribution along the interface. The effect of creating the Cd and Hg-rich structures upon the electron charge density can be obtained by comparing the result in Fig. 5 with that for the nonideal structure shown in Fig. 6. The potential difference associated with the introduction of a different atomic species (replacement of Cd with Hg and vice versa) is strong enough to bring about significant charge localization within the new (enlarged) unit cell. This is well in keeping with the concept of anomalous "deep" localization introduced in Ref. 8. Since the interface level interacts strongly with the uppermost valence level, the breakdown of the bulk translational symmetry is also clearly visible in state V1. The changes in the lowest conduction band state (C1 in Fig. 4) are small.

The formation of localized states at imperfect interfaces demonstrated above has important consequences for interpretation of experimental results. A detailed account of these implications lies outside the scope of this study. However, it is possible to make several general predictions without recourse to additional numerical calculations. For example, in transport (e.g. Hall effect) measurements, the existence of localized levels lying above the valence band edge reported in Fig. 4 will give rise to rich spectra which should dominate over extrinsic impurity effects normally invoked in the literature[11]. There are also important implications for interpretation of optical data. If the uppermost occupied levels are highly localized at the interface, their optical spectra (line width, oscillator strength) depend weakly on the well width. This is in contrast to the properties of confined envelope functions whose optical properties scale in a simple fashion with the well width and superlattice period. Finally, the pronounced difference between Cd and Hg rich interfaces may be used as a means of characterizing the interface quality.

Acknowledgment

This work has been supported in part by the S.E.R.C. (U.K.) and O.N.R. (U.S.A.).

REFERENCES

1. C.G. Van de Walle and R.M. Martin, **Phys. Rev. B**, 34:5621 (1986)
2. I. Morrison, M. Jaros and K.B. Wong, **Phys. Rev. B**, 35:9693 (1987).
3. S. Froyen, D.M. Wood and A. Zunger, **Phys. Rev. B**, 36:4547 (1987); 37:6893 (1988).
4. M.S. Hybertson and M. Schluter, **Phys. Rev. B**, 36:9683 (1987).
5. I. Morrison and M. Jaros, **Phys. Rev. B**, 37:916 (1988).
6. K.B. Wong, M. Jaros, I. Morrison and J.P. Hagon, **Phys. Rev. Lett.**, 60:2221 (1988).
7. K.B. Wong and M. Jaros, **Appl. Phys. Lett.**, 53:657 (1988).
8. M. Jaros, A. Zoryk and D. Ninno, **Phys. Rev. B**, 35:8277 (1987).
9. N.E. Christensen, **Phys. Rev. B**, 38:12687 (1988).

10. An excellent summary of the relevant references can be found in J.N. Schulman and Y.C. Chang, **Phys. Rev. B**, 33: 2594 (1986).

11. C.A. Hoffman, J.R. Meyer, E.R. Youngdale, J.R. Lindle, F.J. Bartoli, K.A. Harris, J.W. Cook, Jr., and J.F. Schetzina, **Phys. Rev. B**, 37:6933 (1988) and references therein.

STRAINED LAYER LASERS IN THE InGaAs/GaAs/AlGaAs HETEROSTRUCTURE

SYSTEM

H. Morkoç

University of Illinois, Coordinated Science Laboratory
and Materials Research Laboratory
1101 West Springfield Avenue, Urbana, Ill. 61801, USA

ABSTRACT

Recent developments in the technology and fundamentals of
strained layer epitaxial systems have generated overwhelming
interest in the exploitation of such heterostructures for
optical and electronic device applications. This is in part due
to the additional degrees of freedom provided for device
structures to be tailored for the particular application and in
many cases improved performance over what is possible with the
lattice matched systems alone. For example, quantum well lasers
with strained InGaAs active layers have achieved threshold
currents comparable to those with GaAs channels but with much
less edge losses due to the smaller surface recombination
velocity in InGaAs and very stable power outputs. Reduced
threshold currents and increased differential gains are expected
to lead to modulation at higher frequencies.

1.   INTRODUCTION

With the recently introduced epitaxial growth schemes such
as Molecular Beam Epitaxy and Organometallic Vapor Deposition,
it has become possible to grow multi-component epitaxial
heterostructures with excellent compositional and thickness
control. This precise control, along with UHV in the case of
MBE, paved the way for the investigation of growth mechanisms in
a previously impossible amount of detail. Armed with an
improved understanding of crystal growth, scientists were able
to begin investigating strained layer heteroepitaxy in detail.
For SL structures, the constraint of having a lattice matched
layer on a suitable substrate is relaxed, up to a critical
thickness. Material scientists and the device designers were
able to choose a stack of various layers on a substrate without
worrying about the immediate lattice matching problem. In a
sense, an additional degree of freedom is provided for the
optimization of the experiment and/or the device structure. The
electronic and optical properties such as the band gap can be
designed beyond what is already made available by revolutionary
quantum well structures.

Following the early work of Matthews and Blakeslee[1] who primarily concentrated on the materials aspects of the lattice mismatched systems, Osbourne[2] diligently was able to revive the field by pointing out the interesting optical and electrical properties of these systems. The Matthews and Blakeslee work had shown that dislocation free material could be grown up to a critical thickness below which the strain is taken coherently by the crystal. Osbourn and colleagues, initially with GaAsP and later on with InGaAs and InAsSb, were able to show the advantages to be gained by the exploitation of these strained systems for both devices and scientific merit.

In this paper we will review the progress and the issues involved in three different device structures, all of which rely on the use of strained InGaAs layers, mostly on GaAs substrates. First, MODFETs based on strained layer epitaxy will be discussed, including their microwave properties. This will be followed by the strained InGaAs quantum well laser which is expected to yield lower threshold current densities due to the associated smaller density of states in the valence band. The last topic to be discussed is the Heterojunction Bipolar Phototransistor designed with a strained InGaAs absorber in the collector region operating out to 1 micron and utilizing quarter wave stack reflectors in order to enhance the quantum efficiency.

The entire discussion will begin with a simple theoretical treatment of the band structure of strained layers and quantum wells near the zone center. A highly simplified theory will then follow to treat the effective mass and density of states particularly in the valence band. This will be complemented again with a simple theory of laser threshold current in an effort to gauge the likely advantages of these systems in low threshold and high speed lasers.

2.   THEORY

In an effort to determine the basic advantages of strained layer systems, the background theory is simplified so that a clear understanding of the important parameters can be gained. A discussion of how strained systems are able to enhance the performance of the devices discussed will also be given. After the band energies of strained InGaAs are calculated, quantization is added to find the conduction and valence band quantum states. This is followed by the treatment of the effective masses in bulk and quantum wells (in plane and out of plane). The effective masses so derived are extrapolated to the InAs mole fraction desired by linear extension between the end binary points. The effective mass treatment is bridged to the density of states in three and two-dimensional systems and is later used for the treatment of strained lasers in Section 5.

2.1 Quantized States in Strained InGaAs Wells

Eigenstates of strained layer superlattices require the consideration of both strain and quantum size effects. In the envelope function formalism, two approaches have been reported for the calculation of the strain effect in quantum wells. The first is Bastard's original method[3], where the strain effect was not considered. In this method, one has to calculate the strain dependent band gap. The result is then used in Bastard's formula to extract the transition energies in the $In_xGa_{1-x}As$/GaAs quantum well. The second method[4] is a complement of Bastard's original formula, which includes the strain dependent terms in the Hamiltonian. The unstrained band gap of bulk $In_xGa_{1-x}As$ is used as an input parameter to calculate the InGaAs/GaAs subband energies. If the spin-orbit splitting Δ and the band gap E_g are large compared with the strain effect, the first method will be a good approximation to the more exact Marzin's formula.

When growing on a GaAs buffer, the $In_xGa_{1-x}As$ layers sustain a biaxial in-plane compression and a corresponding extension along the [001] growth direction. The in-plane lattice constant of such strained layers, $a_\perp$ is simply given by:

$$a_\perp = \frac{a_1 d_1 + a_2 d_2}{d_1 + d_2} \tag{1a}$$

where a, d (1,2) denote the lattice constants and thicknesses of the InGaAs (1) and GaAs (2) layers, respectively. In general, the GaAs buffer layer is much thicker than the InGaAs layers ($d_2 \gg d_1$), so we obtain $a_\perp = a_2 = a_{GaAs}$. If one defines $z \parallel [001]$ along the growth direction, the strain tensor ϵ_{ij} has only diagonal components, i.e.:

$$\epsilon_{xx}^{(1,2)} = e_{yy}^{(1,2)} = \frac{a_\perp - a^{(1,2)}}{a^{(1,2)}} \tag{1b}$$

$$\epsilon_{ij} = 0 \quad i \neq j \tag{1c}$$

$$\epsilon_{zz}^{(1,2)} = - \frac{2C_{12}^{(1,2)}}{C_{11}^{(1,2)}} \epsilon_{xx} \tag{1d}$$

where $C_{11}^{(1,2)}$, $C_{12}^{(1,2)}$ are the elastic constants of $In_xGa_{1-x}As$ and GaAs, respectively. Eqns. 1a-1d completely define the alloy strain tensor.

Since the spin-orbit splitting of $In_xGa_{1-x}As$, $\Delta = 350$ meV, and the Kane matrix element, P, is much larger than the strain effect, we can still consider the total angular momentum J and the magnetic quantum number M_j to be good quantum numbers as long as $k_\perp = (k_x, k_y) = 0$. The strain effect will split the degenerate Γ_8 valence band into two twofold subbands. In the case of InGaAs/GaAs MQWs on a GaAs buffer, the heavy hole $|\frac{3}{2}, \frac{3}{2}>$ will move up and the light holes $|\frac{3}{2}, \frac{1}{2}>$ move down.

A 6 x 6 strain Hamiltonian is an accurate description of the valence band splittings under high stress, as given by Pollak[5]. This strain Hamiltonian can be easily diagonalized and yields three eigenvalues at k = 0 (each doubly degenerate). We obtain[6]:

$$\overline{\lambda} = \delta(x) \quad \text{(heavy hole)} \tag{2a}$$

581

$$\tilde{\lambda} = -1/2[\Delta(x) + \delta(x)] \pm 1/2[9\delta(x)^2 - 2\delta(x)\Delta(x) + \Delta^2(x)]^{1/2} \qquad (2b)$$

$$(+ \text{ for light hole, } - \text{ for split-off})$$

$$\delta(x) = -1/2\Delta\epsilon_g = -b(\epsilon_{zz} - \epsilon_{xx}) = b\frac{(C_{11} + 2C_{12})}{C_{11}}\epsilon_{xx} \qquad (2c)$$

$$\tilde{\lambda} = \lambda + E_g^0(x) + \Delta E_H \qquad (2d)$$

where $\tilde{\lambda}$ is the shift of the valence band edge due to the shear strain, λ is the strain dependent valence band edge with conduction band edge as the origin, ΔE_g is the shear term in the strain Hamiltonian, b is the valence band deformation potential associated with [001] distortion, and ΔE_H is the hydrostatic term in the strain Hamiltonian. ΔE_H is given by:

$$\Delta E_H = a''(\epsilon_{xx} + \epsilon_{yy} + \epsilon_{zz}) = 2a''\frac{C_{11} - C_{12}}{C_{11}}\epsilon_{xx} \qquad (2e)$$

where a" is the hydrostatic deformation potential in the valence band. Similarly, the shift in conduction band edge due to the hydrostatic deformation is:

$$\Delta E_c = 2a'\frac{C_{11} - C_{12}}{C_{11}}\epsilon_{xx} . \qquad (2f)$$

Here, a = a' + a" is the total hydrostatic contribution, which can be obtained from the pressure dependence of the band gap, dE_g^0/dP:

$$a = a' + a'' = -\frac{(C_{11} + 2C_{12})}{3}\frac{dE_g^0}{dP} . \qquad (2g)$$

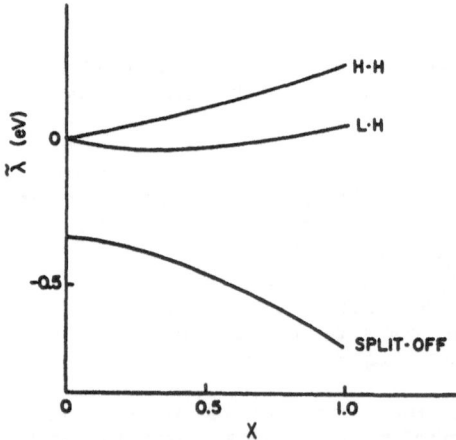

Fig. 1. The shift of the valence band edge from shear deformation, $\tilde{\lambda}$ vs the InAs mole fraction on GaAs buffer. The three curves correspond to heavy hole, light hole and split-off valence band edge, respectively[6].

Table 1. The material parameters used in calculating the band gap of $In_xGa_{1-x}As$ with $0 < x < 1$ at 300 K.

	a_0 (A)	E_g^0	$\dfrac{dE_g^0}{dp}$	a	b	Δ
			$(10^{-6}\,eV/Kgcm^2)$	(eV)	(eV)	(eV)
GaAs	5.6544	1.424	11.5	-8.45	-1.7	0.341
InAs	6.0583		10.2	-6.0	-1.8	0.381
AlAs	5.6611		10.2		-1.5	0.275

	C_{11}	C_{12}	ϵ	$\dfrac{m_{hh}^*}{m_0}$	$\dfrac{m_e^*}{m_0}$
	$(10^{11}\dfrac{dyne}{cm^2})$	$(10^{11}\dfrac{dyne}{cm^2})$			
GaAs	11.88	5.38	13.18	0.5	0.067
InAs	8.329	4.526	14.6	0.41	0.023
AlAs	12.02	5.70	10.06	0.76	0.15

$E_g^0(x)$ is the composition dependent band gap of unstrained $In_xGa_{1-x}As$ bulk material. The composition dependence of $E_g^0(x)$ at different temperatures is given by:

$$E_g^0 = 1.43 + 1.53x + 0.45x^2 \,(eV) \qquad 300\ K \qquad [7]$$

$$E_g^0 = 1.508 + 1.47x + 0.375x^2 \,(eV) \qquad 77\ K \qquad [8] \qquad\qquad (3)$$

$$E_g^0 = 1.5192 + 1.5837x + 0.475x^2 \,(eV) \qquad 2\ K \qquad [9] \ .$$

The shift of the valence band edge from the shear deformation, $\bar{\lambda}$, is given in Fig. 1. The heavy hole state moves up almost linearly with x, and the light hole state moves down first, reaching a broad minimum at about x = 0.3 and then increases monotonically at large x. The split-off band decreases with x.

The parameters of GaAs and InAs are summarized in Table 1. A linear interpolation is used for $In_xGa_{1-x}As$ from 0 < x < 1. The temperature dependence of all the parameters is neglected. Finally, the strain dependent energy gap is calculated as:

$$E_g = E_c - \lambda = E_c^0 + \Delta E_c + \Delta E_H - \bar{\lambda} + E_g^0(x) \ . \qquad\qquad (4)$$

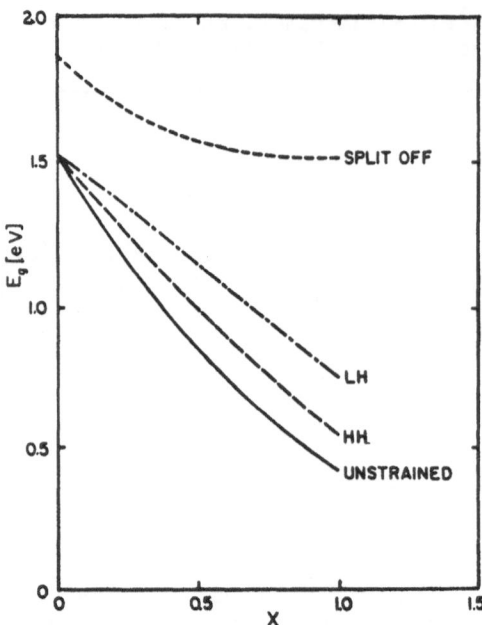

Fig. 2. The band gap (E_g x) of bulk $In_x Ga_{1-x}As$ on GaAs
vs the mole fraction at 2 K. The compressive
strain in $In_x Ga_{1-x}As$ increases the band gap for
the heavy hole, light hole and split-off valence
band, respectively[6].

We chose the bottom of the conduction band as the energy
origin, so $E_c^0 = 0$. ΔE_c and ΔE_H are the band edge shifts arising
from the hydrostatic strain for the conduction and valence
bands, respectively. Only the sum $\Delta E_c + \Delta E_H$ is observable
experimentally. Sometimes, two-thirds of the change is
arbitrarily assigned to the conduction band and one-third to the
valence band[10] (a' = 2a/3, a" = a/3). The choice of a' and a"
will not influence the final result in Eqn. 4.

The strain dependent band gap E_g of the bulk $In_x Ga_{1-x}As$ on
a GaAs substrate is shown in Fig. 2 as a function of x at
T = 2 K. The compressive strain in InGaAs increases the band
gap enormously.

After obtaining the band gap for heavy and light holes, the
subband energies are calculated in InGaAs/GaAs MQWs using the
envelope function method[3]. Only the Γ point is considered
with $k_\perp = 0$. Kane's three band model is used because Δ is not
large compared to the band gap which is about 1500 meV at 2 K.

In Marzin's revision of Bastard's method, an 8 x 8 Kane
Hamiltonian matrix is used for an accurate description of a
strained MQW. When $k_\perp = 0$, there is no coupling between the up
and down spin states and the Hamiltonian is reduced to two
identical 4 x 4 matrices.

The dispersion relation of the conduction subbands is
determined by the following equations:

$$\cos[q(L_A+L_B)] = \cos(K_A L_A)\cos(K_B L_B) - \tfrac{1}{2}(\zeta+1/\zeta)\sin(K_A L_A)\sin(K_B L_B) \qquad (5a)$$

$$\zeta = \frac{K_B \overline{M}_A(E)}{K_A \overline{M}_B(E)} \qquad (5b)$$

$$\overline{M}_A(E) = 2P^2 \frac{E + E_{gA}^0 + 2/3\Delta_A - 5/6\Delta E_s^A}{(E + E_{gA}^0 - 1/2\Delta E_s^A)(E + E_{gA}^0 + \Delta_A) - 1/2(\Delta E_s^A)^2} \qquad (5c)$$

$$\overline{M}_B(E) = 2P^2 \frac{E - V_s + E_{gB}^0 + 2/3\Delta_A - 5/6\Delta E_s^B}{(E - V_s + E_{gB}^0 - 1/2\Delta E_s^B)(E - V_s + E_{gB}^0 + \Delta_B) - 1/2(\Delta E_s^A)^2} \qquad (5d)$$

$$(E - \Delta E_H^A)[(E + E_{gA}^0 - 1/2\Delta E_s^A)(E + E_{gA}^0 + \Delta A) - 1/2(\Delta E_s^A)^2]$$
$$= \hbar^2 K_A^2 P^2 [E + E_{gA}^0 + 2/3\Delta_A - 5/6\Delta E_s^A] \qquad (5e)$$

$$(E - V_s - \Delta E_H^B)[(E - V_s + E_{gB}^0 - 1/2\Delta E_s^B)(E - V_s + E_{gB}^0) - 1/2(\Delta E_s^B)^2]$$
$$= \hbar^2 K_B^2 P^2 [E - V_s + E_{gB}^0 + 2/3\Delta_B - 5/6\Delta E_s^B]. \qquad (5f)$$

In Eqns. 5a – 5f, the unstrained conduction band edge is chosen as the energy origin. The relations between V_s, V_p, V_δ and E_{gA}^0, E_{gB}^0, Δ_A, Δ_B are shown in Fig. 3.

The heavy hole subband is treated separately and described with a standard single band Kronig-Penney model since it is decoupled from other states. The final formulae are similar to those associated with the unstrained case, but the strain dependent band edge shift must be included. As a first order approximation, we used the unstrained effective masses for the heavy holes which are given in Table 1.

2.2 Minimalist's View of Effective Masses Near k = 0

A simplified treatment of the bulk layer will first be described as a way of introduction followed by the discussion of the unstrained and then the strain quantum well case.

The conduction band is generally very simple to treat since this band is nondegenerate and parabolic even under strain. As a result only the compositionally dependent change in the effective masses is needed for the problem. The bulk effective mass shown in Fig. 4 for InGaAs throughout the entire range between the two binary end points has been used for quantum wells with the appropriate in-plane and out-of-plane considerations.

The valence bands of GaAs and InGaAs, on the other hand, are characterized by four fold degenerate states, two of which are termed heavy holes and the other two light holes. These states come about from the solution of Schrödinger's equation and are associated with total angular momentum and magnetic quantum numbers $|\tfrac{3}{2},\pm\tfrac{3}{2}>$ and $|\tfrac{3}{2},\pm\tfrac{1}{2}>$. We will neglect the spin orbit split-off band described by quantum numbers of $|\tfrac{1}{2},\pm\tfrac{1}{2}>$

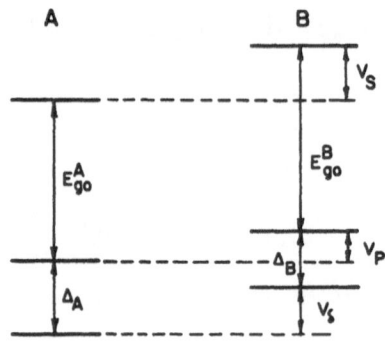

Fig. 3. The relation between V_s, V_p, V_s, Δ_A, Δ_B, $E_{gA}{}^0$ and $E_{gB}{}^0$ for a hetero-junction A-B system[6].

because it is well separated from the heavy and light hole states along with band mixing which is beyond the scope of this treatment.

In bulk, the effective mass associated with the $|\frac{1}{2},\pm\frac{3}{2}>$ state is given by:

$$m^*_{hh} = \frac{m_0}{\gamma_1 - 2\gamma_2} = 0.353m_0 \text{ for GaAs,} \tag{6}$$

$$= 0.341m_0 \text{ for InAs}$$

$$= 0.342m_0 \text{ for In}_{0.3}\text{Ga}_{0.7}\text{As}$$

where m_0, γ_1 and γ_2 are the free electron mass and Luttinger parameters, respectively.

The mass associated with the $|\frac{3}{2},\pm\frac{1}{2}>$ state is given by:

$$m^*_{lh} = \frac{m_0}{\gamma_1 + 2\gamma_2} = 0.082m_0 \text{ for GaAs,} \tag{7}$$

$$= 0.0274m_0 \text{ for InAs}$$

$$= 0.051m_0 \text{ for In}_{0.3}\text{Ga}_{0.7}\text{As} \ .$$

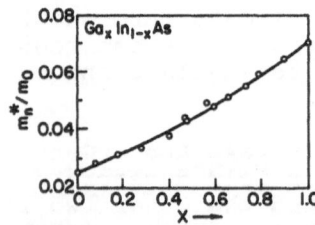

Fig. 4. Bulk effective mass of electrons in $\text{In}_{1-y}\text{Ga}_y\text{As}$ vs GaAs composition

2.3 Three-dimensional Density of States

In the parabolic band approximation:

$$E = \frac{\hbar^2 k^2}{2m^*} = \frac{p^2}{2m^*} \tag{8}$$

or:

$$p = \sqrt{2m^* E} \tag{9}$$

where $p = \hbar k$, momentum and $p\,dp = m^* dE$.

The differential volume in p space between E and E + dE is:

$$dV_p = 4\pi p^2 dp = 4\pi m^{*3/2}\sqrt{2}\sqrt{E}\,dE . \tag{10}$$

The volume in p space is $V_p = p_x p_y p_z = h^3/V$ where V is the volume in real space. Taking spin into consideration:

$$V_p = \frac{1}{2}\frac{h^3}{V} . \tag{11}$$

The total number of states in a volume V is:

$$\frac{dV_p}{V_p} = \frac{8\pi\sqrt{2}m^{*3/2}V}{h^3}\sqrt{E}\,dE . \tag{12}$$

The density of states per unit volume in space per unit energy is:

$$g(E) = \frac{dV_p}{V_p}\frac{1}{VdE} = \frac{8\pi\sqrt{2}}{h^3}m^{*3/2}\sqrt{E} . \tag{13}$$

The hole concentration in a single band is given by:

$$p_0 = \int [1 - f_v(E)]g_v(E)dE \cong \int \frac{8\pi\sqrt{2}}{h^3}m_p^{*3/2}\frac{\sqrt{E_v - E}}{1 + \exp\{(E - E_F)/kT\}}dE. \tag{14}$$

For $E - E_F$ much smaller than kT, the Boltzmann approximation is valid and:

$$p_0 = \int_{-\infty}^{E_v} dp_0 = \frac{8\pi\sqrt{2}}{h^3}m^{*3/2}e^{E_F/kT}\int_{-\infty}^{E_v}\sqrt{E_v - E}\;e^{E/kT}dE \tag{15}$$

$$p_0 = 2(2\pi m_p^* kT/h^2)^{3/2}\exp\{(E_v - E_F)/kT\} \tag{16}$$

$$p_0 = U_v \exp\{(E_v - E_F)/kT\} \text{ with } U_v = 2(2\pi m_p^* kT/h^2)^{3/2} . \tag{17}$$

Table 2. Luttinger Parameters from[33].

	γ_1	γ_2	γ_3
GaAs	7.65	2.41	3.28
InAs	19.67	˙8.37	9.29
AlAs	4.04	0.78	1.57

In the case of heavy and light hole $|\frac{3}{2},\pm\frac{3}{2}>$ and $|\frac{3}{2},\pm\frac{1}{2}>$ degenerate states:

$$p_0 = p_{hh} + p_{1h} = U_{v1}\exp\{(E_v - E_F)/kT\} + U_{vh}\exp\{(E_v - E_F)/kT\} \qquad (18)$$

where U_{v1} and U_{vh} represent the density of states for light and heavy hole bands. The other terms have their usual meanings. For degenerate bulk layers,:

$$U_{v1} + U_{vh} = 2(2mkT/h^2)^{3/2}(m_{hh}^{*3/2} + m_{1h}^{*3/2}) \qquad (19)$$

where h, k and T represent Planck's constant, Boltzmann's constant and the absolute temperature.

$$p_0 = 2(2\pi kT/h^2)^{3/2}(m_{1h}^{*3/2} + m_{hh}^{*3/2})\exp\{(E_v - E_F)/kT\}$$
$$= 2(2\pi m^*_{ds}kT/h^2)^{3/2}\exp\{(E_v - E_F)/kT\} \qquad (20)$$

where the density of state effective mass $m_{ds}^{*3/2} = m_{1h}^{*3/2} + m_{hh}^{*3/2}$ is introduced. Using $\gamma_1 = 7.65/19.67$, $\gamma_2 = 2.41/8.32$ and $\gamma_3 = 3.28/9.29$ for GaAs and InAs, respectively (Table 2), and a linear extrapolation for $In_xGa_{1-x}As$ m_{ds} is found to be equal to $0.3779m_0$, $0.3459m_0$ and $0.3548m_0$ for GaAs, InAs and $In_{0.3}Ga_{0.7}As$, respectively.

2.4 Effective Masses in Two-dimensional Systems

The picture changes somewhat in a quantum well, because of the confinement in the z direction with the result that the degeneracy at k = 0 is lifted and the dispersion relation becomes orientation dependent (anisotropic). It is therefore necessary to investigate the applicable Hamiltonian with the specific task of determining the effective masses.

As discussed earlier, the 8 x 8 Kane Hamiltonian can be reduced to a 4 x 4 coupled Hamiltonian by discarding the coupling between spin up and down states and assuming parabolic bands[11]:

$$H = (1/2m_0)(\gamma_1 + 5/2\gamma_2)p^2 - (\gamma_2/m_0)^2[p_x^2J_x^2 + p_y^2J_y^2 + p_z^2J_z^2]$$
$$- (2\gamma_3/m_0)[\{p_xp_y\}\{J_xJ_y\} + \{p_yp_z\}\{J_yJ_z\} + \{p_zp_x\}\{J_zJ_x\}] \qquad (21)$$

where $\{a,b\} = (ab + ba)/2$, m_0 is the free electron mass, P the linear momentum, J the angular momentum operator for $\frac{3}{2}$ spin hole and γ_i are the Luttinger parameters.

The kinetic Energy term in k space is

$$
\begin{array}{cccc}
|\frac{3}{2},\frac{3}{2}> & |\frac{3}{2},\frac{1}{2}> & |\frac{3}{2},-\frac{1}{2}> & |\frac{3}{2},-\frac{3}{2}>
\end{array}
$$

$$
\begin{bmatrix}
A_- & B & C & 0 \\
B* & A_+ & 0 & C \\
C* & 0 & A_+ & -B \\
0 & C* & -B* & A_-
\end{bmatrix}
\tag{22}
$$

where:

$$
A_\pm = \frac{\hbar^2 \gamma_1}{2m_o}[(1 \pm \frac{\gamma_2}{\gamma_1})(k_x^2 + k_y^2) + (1 \mp \frac{2\gamma_2}{\gamma_1})k_z^2]
\tag{23a}
$$

$$
B = \frac{\hbar^2 \gamma_1}{2m_o}\{-2\sqrt{3}\frac{\gamma_3}{\gamma_1}[ik_y k_z - k_z k_x]\}
\tag{23b}
$$

$$
C = \frac{\hbar^2 \gamma_1}{2m_o}[\sqrt{3}\frac{\gamma_2}{\gamma_1}(k_y^2 - k_x^2) + 2\sqrt{3}\frac{\gamma_3}{\gamma_1}ik_x k_y]
\tag{23c}
$$

$k_x^2 + k_y^2 = k_\perp^2$ (in plane) and quantization is in the z direction.

This Hamiltonian must be modified to include the effect of strain. Assuming that the $\frac{3}{2}$ state is decoupled from the $\frac{1}{2}$ state (justified due to the large spin orbit splitting and small k values), we can obtain the applicable Hamiltonian as the sum of the strain Hamiltonian and the quantum well Hamiltonian. This can be accomplished by adding the $-\Delta E_H - \Delta E_s/2$ term to the 1st, 2nd and 4th diagonal terms and the $-\Delta E_H + \Delta E_s/2$ term to the 3rd diagonal term.

In k_z direction ($k_\parallel$ or [001]) and for $k_\perp = 0$ ($k_x = k_y = 0$), (assume $\gamma_2 = \gamma_3$ since we neglect band mixing), the E(k) relation reduces to:

$$
E = \frac{\hbar^2 k_z^2}{2m_o}(\gamma_1 - 2\gamma_2) \text{ for the } \pm\frac{3}{2} \text{ heavy hole}
\tag{24}
$$

and:

$$
E = \frac{\hbar^2 k_z^2}{2m_o}(\gamma_1 + 2\gamma_2) \text{ for the } \pm\frac{1}{2} \text{ light hole.}
\tag{25}
$$

Thus:

$$
m_{\parallel h}^* = \frac{m_o}{\gamma_1 - 2\gamma_2} \text{ for the } \pm\frac{3}{2} \text{ hole}
\tag{26}
$$

and:

Table 3. Effective Masses

	Bulk			Quantum Well$_{hh}$			Bulk or QW
	$\dfrac{m_{hh}}{m_0}$	$\dfrac{m_{lh}}{m_0}$	$\dfrac{m_{ds}^{*}}{m_0}$	$\dfrac{m_{\perp}}{m_0} = \dfrac{m_{ds}^{*}}{m_0}$		$\dfrac{m_{\parallel}}{m_0}$	$\dfrac{m_e}{m_0}$
GaAS	0.353	0.082	0.3779	0.0994		0.353	0.067
InAs	0.341	0.0274	0.3459	0.0354		0.341	0.023
In$_{0.3}$Ga$_{0.7}$As	0.342	0.051	0.3548	0.0648		0.342	0.538

$$m_{\parallel 1}^{*} = \frac{m_o}{\gamma_1 + 2\gamma_2} \quad \text{for the } \pm\tfrac{1}{2} \text{ hole} \tag{27}$$

For $k_x = k_\perp$ and $k_y = k_z = 0$, $k_\perp$ or [100] direction ($\perp$ to z):

$$E = \frac{\hbar^2 k_\perp^2}{2m_o}(\gamma_1 + \gamma_2) \quad \text{for the } \pm\tfrac{3}{2} \text{ heavy hole} \tag{28}$$

$$E = \frac{\hbar^2 k_\perp^2}{2m_o}(\gamma_1 - \gamma_2) \quad \text{for the } \pm\tfrac{1}{2} \text{ light hole.} \tag{29}$$

It follows that:

$$m_{\perp h}^{*} = \frac{m_o}{\gamma_1 + \gamma_2} \quad \text{for the } \pm\tfrac{3}{2} \text{ state} \tag{30}$$

and:

$$m_{\perp 1}^{*} = \frac{m_o}{\gamma_1 - \gamma_2} \quad \text{for the } \pm\tfrac{1}{2} \text{ state.} \tag{31}$$

Note that the in plane heavy hole mass is actually smaller than the light hole mass which results in anisotropic band structure even before any band mixing, see Table 3 and Fig. 5.

In compressive strain layer quantum wells, the heavy and light hole bands are well separated (by ΔE_s). For low hole concentrations, one can assume that only the hh$_1$ state is occupied. One can then go about defining a density of states effective mass in the following manner:

$$m_{ds}^{*} = (m_x^{*} m_y^{*})^{1/2} \tag{32}$$

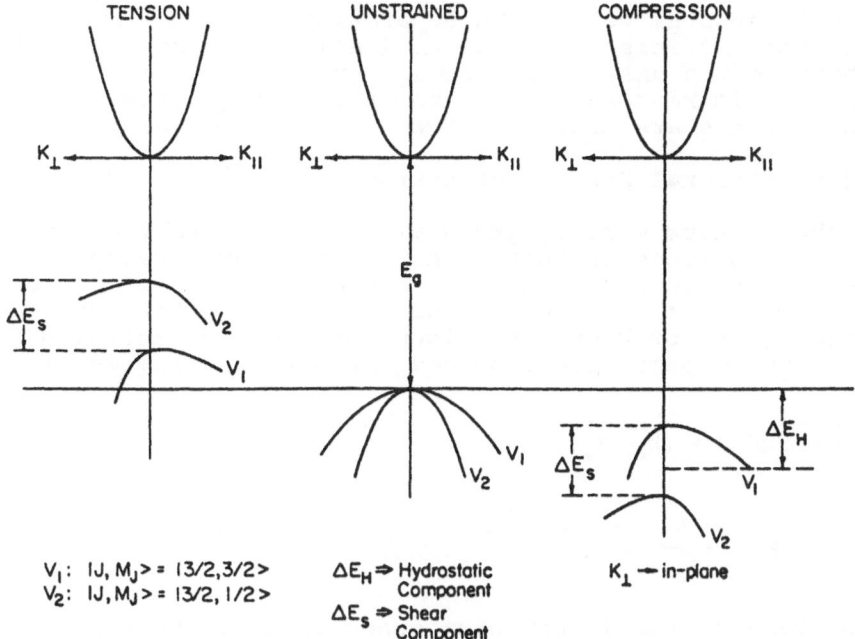

V_1: $|J, M_J> = |3/2, 3/2>$ $\Delta E_H \Rightarrow$ Hydrostatic Component
V_2: $|J, M_J> = |3/2, 1/2>$ $\Delta E_s \Rightarrow$ Shear Component $K_\perp \rightarrow$ in-plane

Fig. 5. Schematic band representation in strained layers under tensile and compressive strain along with the unstrained case as the reference.

$$m^*_{ds}\bigg|_h = m_o \left(\frac{1}{\gamma_1 + \gamma_2} \frac{1}{\gamma_1 + \gamma_2} \right)^{1/2} \tag{33}$$

$$= m_o \left(\frac{1}{\gamma_1 + \gamma_2} \right) \tag{34}$$

$$m^*_{ds}\bigg|_l = m_o \left(\frac{1}{\gamma_1 - \gamma_2} \right) \tag{35}$$

and if only the heavy hole band is filled:

$$\frac{m^*_{ds}}{m_o} = 0.0994 \text{ for GaAs well} \tag{36}$$

$$\frac{m^*_{ds}}{m_o} = 0.0648 \text{ for } In_{0.3}Ga_{0.7}As \text{ well} \tag{37}$$

$$\frac{m^*_{ds}}{m_o} = 0.0356 \text{ for InAs well.} \tag{38}$$

In GaAs quantum wells, the separation between $\frac{3}{2}$ and $\frac{1}{2}$ states is not sufficiently large to justify the assumption that the $\frac{3}{2}$ state is the only filled band. As indicated above, this approximation is reasonable, for the strained quantum well case where these bands are separated further due to strain.

2.5 Two-dimensional Density of States

If the quantum well is grown along the z direction, the elctrons are confined in that direction and move freely in the direction perpendicular to z. The k-space density of states can be transformed to a real space density of states through the usual calculation of k states allowed between the energies E and E + dE (assuming parabolic band and infinite quantum well):

$$k_z = \frac{n\pi}{L}, \quad p^2 = \hbar^2(k_x^2 + k_y^2) \tag{39}$$

$$E = \frac{\hbar^2}{2m*}(k_\perp)^2 = \frac{\hbar^2}{2m*}(k_x^2 + k_y^2) \ . \tag{40}$$

The surface area in momentum space is given by $A'_p = h^2/Area$ where Area is the surface in real space. Considering spin up and down, the area unit in momentum space becomes $A_p = h^2/2 \times Area$.

The area in momentum space contained within E and E + dE circles is:

$$dA_p = 2\pi p\,dp \tag{41}$$

$$p^2 = 2m*E \quad \text{and} \quad p\,dp = m*dE \tag{42}$$

then:

$$dA_p = 2\pi m*dE \ . \tag{43}$$

The 2D density of states in real space:

$$g_{2D}(E) = \frac{dA_p}{A_p\,dEArea} = \frac{2\pi m*dE}{\frac{1}{2}h^2\,dE} \tag{44}$$

$$= \frac{4\pi m*}{h^2} = \frac{m*}{\pi\hbar^2} \ . \tag{45}$$

Any reduction in the effective mass leads to a linear reduction in the density of states and since:

$$p_0 = \int_{E_{hh1}}^{\infty} g_{2D}(E)f(E)dE = \frac{kTm_{hh}^*}{\pi\hbar^2}\ln[1 + \exp\frac{E_F - E_{hh1}}{kT}] \tag{46}$$

the required hole concentration to move the hole Fermi level by a given amount is also reduced. The picture becomes fairly complicated if more than one subband is occupied. In that case,

the sum of all bands with appropriate effective masses must be
considered. In fact, the hole mass is energy dependent because
of nonparabolicity and band mixing which complicate the picture
further. For multiple states, both heavy and light holes, we
have:

$$g_{2D}(E) = \frac{1}{\pi\hbar^2}\left[\sum_i m^*_{hhi}u(E - E_{hhi}) + \sum_i m^*_{lhi}u(E - E_{lhi})\right] \qquad (47)$$

where u is a step function.

If the separation between the light and heavy hole bands is
equal to or greater than 3kT, which is reasonable for many of
the structures investigated, the contribution to the hole
concentration of Eqn. 46 from the light hole band can be
neglected even for equal masses.

The situation in the conduction band is much simpler as
compared to the valence band. First of all, the dispersion
relation is isotropic, that is constant energy surface is
spherical in the bulk and the band is parabolic. It suffices
then to extrapolate the effective mass from the binary points to
arrive at the applicable effective mass. Due to the isotropy
and nondegeneracy, the density of states effective mass and
conduction mass are identical and are given by:

$$m^*_c = m^*_c\Big|_{GaAs} - (x)[m^*_c\Big|_{GaAs} - m^*_c\Big|_{InAs}]$$

$$= m^*_c\Big|_{GaAs}(1 - x) + xm^*_c\Big|_{InAs} = m_0[0.067(1 - x) + 0.023x] \qquad (48)$$

where x is the mole fraction of InAs in InGaAs. For x = 0.3,
(using $0.067m_0$ and $0.023m_0$ for GaAs and InAs effective masses,
respectively), one finds an effective mass of $0.0538m_0$ for
$In_{0.3}Ga_{0.7}As$ bulk (Table 3). It is generally a good
approximation to assume that the electron mass in bulk is about
the same as that in quantum wells with strain as well.

3.   MEASURED EFFECTIVE ELECTRON AND HOLE MASSES

It should be pointed out again that the n-type transport in
strained quantum wells is not noticeably affected by strain.
However, the effective mass changes somewhat in accordance with
the channel composition. It is customary to assume that the
conduction band is parabolic for Fermi energies up to about
100 meV into the band. This leads to an effective mass being
independent of the doping level. The effective mass of
electrons in $Al_{0.15}Ga_{0.85}As/In_{0.10}Ga_{0.90}As$ modulation doped
structures has been measured directly by cyclotron resonance
measurements[12]. The same group also made Shubnikov-de Haas
measurements to determine the electron concentration as well.
Samples with InAs mole fractions in the range of 10-20% with 2D
electron concentrations on the order of 7×10^{11} cm^{-2} exhibited
an m_{CR} of about 0.06 m_0.

Far Infrared (FIR) cyclotron resonance (CR) measurements were made in p-type modulation doped GaAs/$In_{0.2}Ga_{0.8}As$ samples with a sheet hole concentration of 8.5×10^{11} cm^{-2} to determine the hole mass. The mass determined by CR is found to be $m_{CR} = (0.191 \pm 0.08)m_0$[13] when the best fit is achieved to a Ge bolometer detected transmittance versus the magnetic field. The tight binding calculations of Osbourn[14] indicate a zone center mass of $0.09m_0$. This discrepancy is attributed to band nonparabolicity which is present in the valence band. The CR mass at the Fermi level can be expressed as:

$$m_{CR}(E_F) = m_v^o[1 + 2E_F(C/\Delta E_s)] \quad E_F < \Delta E_s/2 \tag{49}$$

using the calculated value of $0.09m_0$ for m_v^0 and measured m_{CR},[13] finds a $\Delta E_s/C$ nonparabolicity value of 26 meV for the particular sample investigated. Here C denotes the band non-parabolicity factor.

Valence band dispersion was also studied by Jones et al.[15], using selection rule breaking magnetoluminescence along with conduction band dispersion in modulation doped samples with mole fractions of 15, 20 and 25% and electron concentration of about 8×10^{11} cm^{-2}. The investigation was carried out with the application of a 5.0 T magnetic field and the transitions between the valence band and the conduction band were observed.

From the Landau levels at 4 K, the conduction band and valence band edges were determined as a function of the reduced k vector. It should be pointed out that, with the application of the magnetic field, diagonal (direct subband with n = n') and selection rule breaking off-diagonal (n ≠ n') transitions are observed.

The luminescence energy is given by:

$$E(n',n) = E_g + E_c(n') + E_v(n) , \tag{50}$$

here n and n' correspond to Landau level quantum numbers. In a parabolic conduction band:

$$E_c(n') = (n' + \tfrac{1}{2})e\hbar B/m_c^* c \tag{51}$$

where B and c denote magnetic field and velocity of light.

Manipulation of Eqns. 50 and 51 leads to $E(n',n) = E(n,n) + E_c(n' - n - 1/2)$. For $n' \geq n$ and at low temperature holes reside in the n = 0 ground state. Thus, $E_c(n' - 1/2)$ can be determined by the difference between E(0,0) (direct) and E(n',0) with n' = 1, 2, 3 symmetry breaking transitions. Once $E_c(n')$ is determined, then $E_v(n)$ can be found from symmetry retaining n' = n transitions.

Despite the nonparabolic valence band, the crystal momentum can be found from:

$$\frac{\hbar^2 k^2}{2m^*} = (n + \frac{1}{2})\frac{e\hbar B}{m^* c} \tag{52}$$

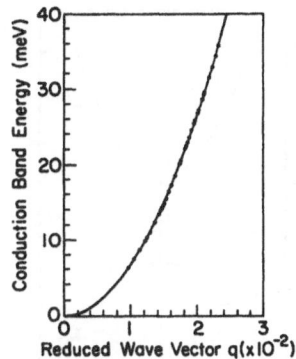

Fig. 6. Conduction band edge E-k diagram
with k replaced by the reduced wave
vector q = ak/2π. The solid line is
the parabolic approximation with
$m_n^* = 0.07m_0$ [15].

with

$$k = \sqrt{(2n+1)} \ L^{-1} \tag{53}$$

where

$$L = \sqrt{\frac{\hbar c}{eB}} \tag{54}$$

is the classical magnetic length.

Using q = ak/2π where a is the lattice constant, $E_c(n')$ and $E_v(n)$ versus k can be plotted as shown in Figs. 6 and 7, for a 15% InAs channel. The conduction band is parabolic as seen from

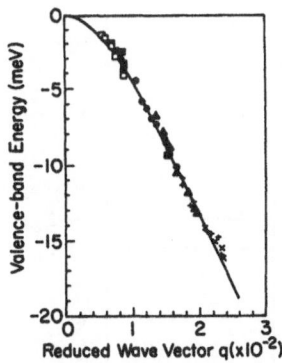

Fig. 7. Valence band edge E-k diagram where
k is replaced by the reduced k
defined as ak/2π. The solid line is
the best fit obtained from Eqns. 50
and 51[15].

595

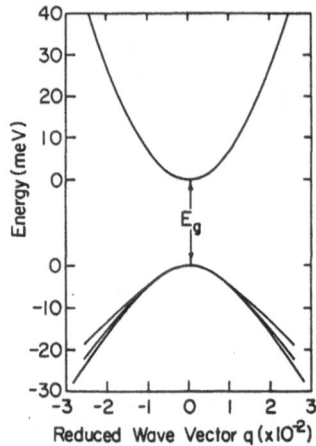

Fig. 8. The conduction and valence band
edges for strained InGaAs with InAs
mole fractions of 15, 20 and
25%[15].

the close agreement between the parabolic curve (solid line and
data points) and leads to an effective conduction mass of
$0.071m_0$ which is in excellent agreement with $(0.07 - 0.076)m_0$
deduced from cyclotron resonance measurements of Liu et al.[16].
For samples with InAs mole fractions of 20 and 25%, the conduction band effective mass is $0.071m_0$ and $0.072m_0$, respectively.
Although small and often neglected, there is still a degree of
nonparabolicity in the conduction band, and CR simply measures
the mass at the Fermi level which is doping dependent and may
explain the discrepancy with that of Szydlik et al.[12].

As discussed earlier, the valence band is not parabolic and
the solid lines in Fig. 7 are the best fit to the data. Tight
binding calculations of Osbourn[2], suggest a nonparabolicity of
the form:

$$E[1 + \frac{C}{\Delta E_s} E] = \frac{\hbar^2 k^2}{2m^o_v} \quad . \tag{55}$$

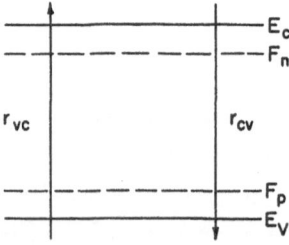

Fig. 9. Upward and downward transitions
between conduction and valence
bands.

596

Using the data of Fig. 8 and values of ΔE_s, for which the calculations were shown earlier, one finds, C and m_v^0/m_0 values of, in order, 2.4/0.084, 1.7/0.086 and 1.3/0.094 for 15, 20 and 25% InAs containing channels, respectively.

The average effective mass obtained by fitting a parabola to the data of the valence band dispersion results is $0.15m_0$ as compared to $0.1m_0$ obtained by magnetotransport measurements[17]. We should point out that as the strain is increased, ΔE_s get larger, the band mixing is diminished and the band becomes more parabolic near k = 0.

4.   A FEW FUNDAMENTALS OF LASERS

Let us now consider the requirements for laser oscillations in a very simple manner and point out how a strained InGaAs active layer could possibly reduce the laser threshold current density. As a first step we must treat the process of absorption and emission of photons that simultaneously take place in a semiconductor laser[18].

For the absorption process between valence and conduction band states, the upward transition rate (see Fig. 9) is:

$$r_{vc} = B_{vc}f_v(1 - f_c)P(E) \tag{56}$$

where B_{vc} is the probability that a transfer occurs, f_v the probability that the valence band contains an electron, $1 - f_c$ probability that the conduction band has an empty state, P(E) the density of photons with energy $E \approx (E_c - E_v)$ (in a quantum well E_c and E_v represent the applicable lowest confined states, i.e., E_{1c}, E_{1hh} and E_{11h}, respectively) and:

$$f_v = \frac{1}{1 + \exp\{(E_v - F_p)/kT\}} \quad \text{and} \quad f_c = \frac{1}{1 + \exp\{(E_v - F_n)/kT\}} \tag{57}$$

where F_p and F_n denote the quasi-Fermi levels for holes and electrons, respectively.

The downward transition rate is:

$$r_{cv}(\text{stim}) = B_{cv}f_c(1 - f_v)P(E) \tag{58}$$

where B_{cv} is the probability that such transition occurs. The downward transition rate without interacting with the optical field is:

$$r_{cv}(\text{spon}) = A_{cv}f_c(1 - f_v) \tag{59}$$

where A_{cv} is the probability that such transition occurs.

At equilibrium:

$$r_{vc} = r_{cv}(\text{stim}) + r_{cv}(\text{spon}) \tag{60}$$

or:

$$B_{vc}f_v(1 - f_c)P(E) = B_{cv}f_c(1 - f_v)P(E) + A_{cv}f_c(1 - f_v) \tag{61}$$

then:

$$P(E) = \frac{A_{cv} f_c (1 - f_v)}{B_{vc} f_v (1 - f_c) - B_{cv} f_c (1 - f_v)} \tag{62}$$

The equilibrium expression above must hold even for the case of a semiconductor under nearly equilibrium conditions, that is, $F_n = F_p$. Substituting $h\nu = E$ and utilizing[18]

$$P(E) = \frac{8\pi \bar{n}^3 (h\nu)^2}{h^3 c^3 [\exp(h\nu/kT) - 1]} \tag{63}$$

where $\bar{n}$ is the refractive index, (the other parameters have their usual meaning), one obtains:

$$\frac{8\pi \bar{n}^3 (h\nu)^2}{h^3 c^3 [\exp(h\nu/kT) - 1]} = \frac{A_{cv}}{B_{vc}\{\exp(h\nu/kT)\} - B_{cv}} \tag{64}$$

or:

$$\frac{8\pi \bar{n}^3 (h\nu)^2}{h^3 c^3}[B_{vc} \exp(h\nu/kT) - B_{cv}] = A_{cv}[\exp(h\nu/kT) - 1] . \tag{65}$$

Equating the temperature dependent parts leads to the spontaneous emission probability:

$$A_{cv} = \frac{8\pi \bar{n}^3 (h\nu)^2}{h^3 c^3} B_{cv} . \tag{66}$$

Equating temperature independent parts leads to:

$$B_{vc} = B_{cv} . \tag{67}$$

Equations 66 and 67 show that the spontaneous emission probability is related to the absorption and stimulated emission probabilities.

The absorption coefficient is:

$$r_{vc}(abs) = r_{vc} - r_{cv} = [B_{vc} f_v (1 - f_c) - B_{cv} f_c (1 - f_v)]P(h\nu)$$
$$= B_{vc}(f_v - f_c)P(h\nu) . \tag{68}$$

The absorption coefficient $\alpha(h\nu)$ is given by:

$$\alpha(h\nu) = \frac{r_{vc}(abs)}{F(h\nu)} = \frac{r_{vc}(abs)}{v_g P(h\nu)} = \frac{B_{cv}(f_v - f_c)}{c/\bar{n}}$$

$$= -\left[\frac{\bar{n}}{c}\right]\left[\frac{B_{cv}}{A_{cv}}\right]A_{cv}(f_c - f_v) \tag{69}$$

where v_g and $F(h\nu)$ are the group velocity and photon flux densities, respectively.

Equation 66, together with Eqn. 69 leads to the stimulated emission rate:

$$r_{stim}(h\nu) = A_{cv}(f_c - f_v) \quad \text{and}$$

$$\alpha(h\nu) = -\frac{h^3 c^2}{8\pi \bar{n}^2 (h\nu)^2} r_{stim}(h\nu) \tag{70}$$

Similarly:

$$r_{cv}(spon) = r_{spon}(h\nu) = \frac{8\pi \bar{n}^2 (h\nu)^2}{h^3 c^2} \alpha(h\nu) \frac{f_c(1 - f_v)}{f_v - f_c} \tag{71}$$

$$r_{spon}(h\nu) = \frac{8\pi \bar{n}^2 (h\nu)^2}{h^3 c^3} \alpha(h\nu) \left\{ \exp[\{h\nu - (F_n - F_p)\}/kT] - 1 \right\}^{-1} . \tag{72}$$

From Eqns. 70 and 72:

$$r_{stim}(h\nu) = r_{spon}(h\nu)[1 - \exp\{(h\nu - F_n + F_p)/kT\}] . \tag{73}$$

Thus, $r_{stim}(h\nu)$, $r_{spon}(h\nu)$ and $\alpha(h\nu)$ are all interrelated.

The absorption coefficient is determined by the difference between upward and downward transition rates with the inclusion of the join density of states (g_{cv}). Substitution of:

$$m_r^* = (1/m_n^* + 1/m_p^*)^{-1} \tag{74}$$

for the effective mass in Eqns. 13 and 45 would lead to the joint density of states, $g_{cv}(h\nu)$, without the k selection rule:

$$\alpha(h\nu) = \int_{-\infty}^{\infty} \frac{B_{vc}}{c/\bar{n}}(f_v - f_c) g_{cv}(h\nu)\delta(E_c - E_v - h\nu)dE . \tag{75}$$

Here the integral is taken over the product of the conduction and valence band density of states for various E_c and E_v separated by $h\nu$ as dictated by the $\delta(E_c - E_v - h\nu)$ term. In quantum wells or in particular, strained quantum wells, B_{vc} must be calculated for the structure in use and the density of states must be replaced by the $g_{2D}(E)$, and E_c and E_v are replaced by E_{lh} and E_{hh} quantum states. In quantum well lasers, the active layer is undoped and k-selection rule applies. However, the salient features of the absorption coefficient would be similar to that in heavily doped bulk lasers which break the k-selection rule.

If we select $E' = E - E_c$ (meaning if $E' = 0$, E is at E_c), then $E'' = E' - h\nu$, and use the matrix element in the calculation of α, we get:

$$\alpha(h\nu) = \text{constant} \int_{-\infty}^{\infty} g_{cv}(h\nu)|M|^2[f(E'') - f(E')]dE . \tag{76}$$

α in bulk layers is proportional to:

$$\text{constant}(E - E_g)^{1/2} \tag{77}$$

defined for $E > E_g$, and has the form of the density of states.

For the quantum well case, the actual picture is more complex, but we will assume that only the HH1 and C1 transition is involved, in which case:

$$\alpha_{HH1-C1} = \frac{\text{constant}}{E} \int g_{cv}(h\nu)|<HH1|P^2|C1>|^2 [f(E'') - f(E')]dE \tag{78}$$

$$= \text{constant}\; g_{cv}(h\nu)|M|^2$$

where $g_{cv} = m_r^*/\pi\hbar^2$ with $m_r^* = (1/m_{hh\perp}^* + 1/m_n^*)^{-1}$ representing the joint density of states. The matrix element for this transition is about unity.

For the light hole transition (if this band is also involved):

$$\alpha_{LH1-C1} = \frac{\text{constant}}{E} \int g_{cv}(h\nu)|<LH1|P|^2C1>|^2 [f(E'') - f(E')]dE \tag{79}$$

$$= \text{constant}\; g_{cv}(h\nu)|M|^2$$

Here, $|M|^2 = \frac{1}{3}$. Based on these expressions for bulk and quantum wells, we can construct the absorption coefficient versus energy for bulk and quantum wells. In the quantum well case, for illustrative purposes only, we show the HH1 - C1 related transitions only. The expressions are more valid for nearer k = 0. For $k \neq 0$, higher injection levels, these simplified expressions are not applicable and the bulk like masses become applicable.

Likewise (without the k-selection rule):

$$r_{spon}(h\nu) = \int_{-\infty}^{\infty} A_{cv} g_{cv}(h\nu) f(E')[1 - f(E'')]dE' \tag{80}$$

(where $h\nu = E' - E''$, photon emission energy) and:

$$r_{stim}(h\nu) = \int_{-\infty}^{\infty} A_{cv} g_{cv}(h\nu)[f(E') - f(E'')]dE' . \tag{81}$$

The terms A_{cv} and B_{cv} must be evaluated for the particular structure. Evaluation of transition probabilities requires the calculation of matrix elements. The presentation so far has indicated that the valence band is not parabolic and that the mass is dependent on the hole concentration. In addition, at high injection levels, even the conduction band is not parabolic. However, for gauging the threshold current and

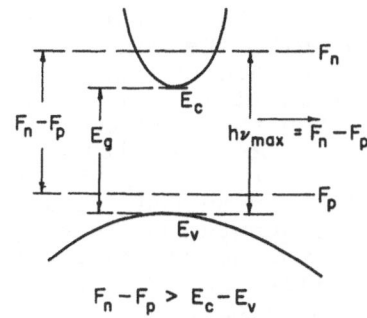

$$F_n - F_p > E_c - E_v$$

Fig. 10. Energy vs k diagram of conduction and
valence band edges with the quasi-Fermi
levels shown under the lasing conditions.
Note that large valence band density of
states (at least in bulk) prevents the
hole quasi-Fermi level from moving into
the valence band.

comparing various structures, the approximations made so far are
reasonable. Additionally, the Fermi level clamping in lasers
would help keep the quasi-Fermi levels close to the band edges.

The necessary condition for stimulated emission (Bernard
and Daraffourg[18]) is that a photon be more likely to cause an
electron transition to the valence band with the emission of a
photon than an upward transition with the absorption of a
photon. In other words:

$$r_{cv} > r_{vc} \tag{82}$$

or:

$$B_{cv} f_c (1 - f_v) P(E) > B_{vc} f_v (1 - f_c) P(E) \tag{83}$$

or:

$$f_c (1 - f_v) > f_v (1 - f_c) \Rightarrow \exp[(F_n - F_p)/kT] > \exp[(E_c - E_v)/kT] \tag{84}$$

which reduces to:

$$F_n - F_p > E_c - E_v = h\nu \quad \text{or} \quad np > U_c U_v . \tag{85}$$

This simply states that the separation of the quasi-Fermi levels
must exceed the band gap energy, or the product of the excess
carrier concentrations must exceed the product of the total
density of states.

In order for $F_n - F_p > E_c - E_v$, both n and p must be
substantially increased by injection. In which case, the
expressions above are not absolutely correct but could be used
to make a qualitative argument. In quantum wells, E_c and E_v
must be replaced with the first electron and hole quantum
subbands.

Due to the disparity in electron and hole masses,
particularly in bulk III-V, g_c is much smaller than g_v causing
the hole quasi-Fermi level to be far less sensitive to

injection. As a result the electron quasi-Fermi level does move up into conduction band while the hole quasi-Fermi level remains in the energy gap near the valence band when the lasing condition is satisfied (Fig. 10). In bulk GaAs, the lasing condition is satisfied when the injected electron concentration is 1.5×10^{18} cm^{-3}. If the hole density of states were small, much smaller electron concentrations, thus much lower threshold current for lasing would be required. This is the reason why the use of strained layer quantum wells becomes attractive. The large separation between the heavy and light hole states in strained quantum wells allows for the population of the heavy hole state only with reduced effective hole mass.

The current density for 1 μm active region needed to sustain the spontaneous emission is:

$$J_{nom} = er_{spon}(\text{total}) \tag{86}$$

$$J_{nom} = -e \int_{0}^{\infty} \frac{8\pi\bar{n}^2 E^2 \alpha(E)}{h^3 c^2 [\exp\{(E - F_n + F_p)/kT\} - 1]} dE \tag{87}$$

where e is the electronic charge. The lower limit is determined by $\alpha(E)$ and the upper limit by the exponential term in the denominator.

The absorption coefficient follows the shape of the joint density of states in both bulk and quantum well structures. It is once again clear that reduced joint density of states through the reduction of hole mass, is desirable. In Fig. 11, we show the absorption coefficient and emission spectrum for bulk and quantum well cases.

In semiconductors when pumped beyond transparency, the absorption coefficient changes its sign and leads to gain. The propagating power can be expressed as:

$$F(z) = F_0 \exp(\bar{g}z) \tag{88}$$

where F(z) is the photon density at z and $\bar{g}$ is the gain per unit length. The gain, from Eqn. 87:

$$\bar{g} = \frac{\Delta F(z)/\Delta z}{F(z)} = \frac{r_{stim}}{F(z)} \tag{89}$$

with no external losses, the gain is given by:

$$\bar{g}(h\nu) = \alpha_i(h\nu) = \frac{h^3 c^2}{8\pi\bar{n}^2(h\nu)^2} r_{stim}(h\nu) . \tag{90}$$

It is, therefore, also true that the smaller the $\alpha(E)$, the smaller the gain coefficient. In fact, in quantum wells, α is step-like and, unless the upper quantum states are occupied, saturates causing a gain saturation as well.

In terms of the gain spectrum, J_{nom}, which is normalized to 1 μm active area, becomes:

$$J_{nom} = e \int_0^\infty \frac{8\pi\bar{n}^2 E^2 \bar{g}(E)}{h^3 c^2 [\exp\{(E - F_n + F_p)/kT\} - 1]} dE \ .$$ (91)

When we include the losses through the facets and the confinement factor, Γ, the model gain becomes:

$$\Gamma\bar{g} = \alpha_i + (1/L) \ln(1/R)$$ (92)

where L and R are the cavity length and facet reflectivity, respectively. This equation is reasonably accurate even 10-15% above threshold.

Generally, the maximum gain-current relationship is linear, that is:

$$\bar{g}_{max} = \beta(J_{nom} - J_0)$$ (93)

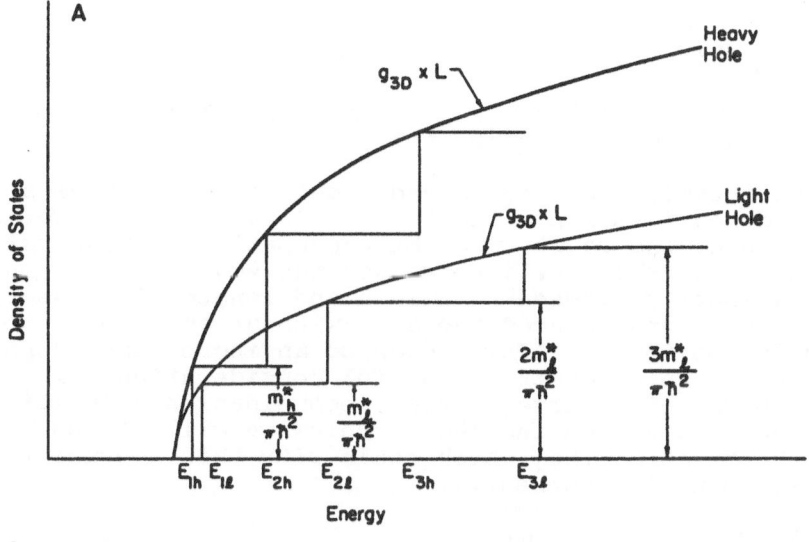

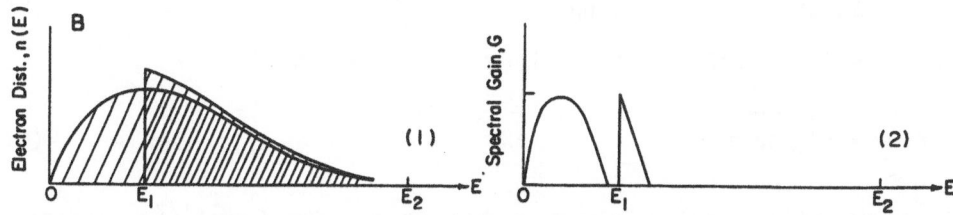

Fig. 11. (a) Valence band density of states in a quantum well (schematic) shown for heavy and light hole hands in parabolic approximation. Three-dimensional density of states multiplied with the quantum well thickness is also shown for both bands. (b) Schematic carrier distribution (upper) and spectral gain vs energy (e.g., conduction band) in bulk and quantum well lasers. Note that in a quantum well, maximum gain identical to that in bulk can be obtained at slightly reduced carrier concentration[27].

where β is the gain coefficient and J_0 is the current density for zero g_{max}. Thus:

$$J_{nom} = J_0 + \bar{g}_{max}/\beta = J_0 + (1/\Gamma\beta)\{\alpha_i + (1/L)\ln(1/R)\}. \tag{94}$$

J_{nom} is the valence current density required to turn the semiconductor transparent and to overcome the internal and external losses. With the quantum efficiency taken into account, this expression reduces to:

$$J_{nom} = (1/\eta)[J_0 + (1/\Gamma\beta)\{\alpha_i + (1/L)\ln(1/R)\}] . \tag{95}$$

In GaAs bulk lasers this J_0 figure is about 4500 A/cm^{-2} per μm at the lasing threshold which is also consistent with the experimentally observed threshold current densities in quantum well lasers as well. The threshold current density for a laser with a pumped region thickness of d becomes:

$$J_{th} = J_{nom}d = J_0 d/\eta + (d/\Gamma\beta\eta)\{\alpha_i + (1/L)\ln(1/R)\} \tag{96}$$

for the GRINSCH lasers under discussion the optical field is Gaussian[19]:

$$\Gamma \cong (2/\pi)^{1/2} d/W_0 \tag{97}$$

with W_0 being the Gaussian beam radius, about 2300 Å for GRINSCH lasers. For d = 100 Å quantum wells, Γ = 0.034.

Lasers intended for high speed modulation must have a small area for reduced junction capacitance. Substantial reductions in the length of the cavity must be accompanied by high reflectivity coating of the facets to reduce the edge losses. Due to the large disparity between the width and length of a laser, any scaling in the width reduces the area without any appreciable change in its periphery. This leads to an increased periphery over area ratio which deserves careful consideration. In material systems with large surface recombination velocities, nonradiative edge losses increase. Below we will present the theory of Henry et al.[20] which deals with this nonradiative surface recombination current.

$$\text{Recombination Rate} = s \sqrt{np} \tag{98}$$

where s is the surface recombination velocity,

$$\text{Flux} = s \int_l \sqrt{np} \, dl . \tag{99}$$

Surface current = e(Periphery = P/Area = A)$\{sL'(np)^{1/2}\}$, which, by assuming Boltzmann distribution, reduces to:

$$e \frac{P}{A} sL' \sqrt{U_c U_v} \exp\{(F_n - F_p)/2kT\} . \tag{100}$$

Here U_c and U_v represent the total density of states in conduction and valence bands and L' is the surface diffusion length.

This surface recombination current must be added to the threshold current as an additional size dependent loss term.

604

As can be seen, a reduction of about a factor of 100 in the recombination velocity, a feasible amount for $In_xGa_{1-x}As$, reduces the contribution to the threshold current in small geometry stripe lasers quite substantially.

The threshold current density in bulk and quantum well lasers reduces linearly with the thickness of the area that is pumped. The rate of this drop is characterized by $J_0 = 4500$ A/cm^2 per μm in bulk GaAs lasers. Experimental results obtained in GaAs quantum wells show the same dependence with the same slope down to about 150 Å or so, below which the J_0 is independent of thickness[21]. However, the threshold current density increases rapidly for quantum wells that are extremely thin (< 35 Å) due to the very small confinement factor. A model based on a two-dimensional recombination coefficient, as will be described below, can qualitatively explain the observations.

Band to band recombination is bimolecular and in 3D given by:

$$R^{3D} = B^{3D}np \tag{101}$$

where R^{3D} and B^{3D} are the 3D recombination rate and recombination coefficient, respectively, with $B^{3D} = 10^{-10}$ cm^3/s.

In the two-dimensional case, $R^{2D} = B^{2D}n_sp_s$ with B^{2D} being about 1.9×10^{-4} cm^2/s at 300 K for $n_s = 10^{12}$ cm^{-2}[22]. At threshold the sheet carrier concentration is about 1.2×10^{12} cm^{-2} and is independent of well thickness. The current density at transparency is:

$$J_0 = eB^{2D}n_s^{transp}p_s^{transp} \cong 50 \text{ A/cm}^2 \tag{102}$$

and is independent of the quantum well thickness[23]. As the quantum well thickness is reduced below about 50 Å, the 2nd term in Eqn. 93 which is applicable to all lasers increases due to the decrease in Γ which leads to an increase in the threshold current density.

The relaxation frequency of a laser is given by:

$$f_r = (2\pi)^{-1}(cg'P_0/\bar{n}\tau_p)^{1/2} \tag{103}$$

where $g' = d\bar{g}/dn$, P_0 and τ_p represent differential gain, photon density in the mode of cavity where the lasing occurs, and the photon lifetime, respectively[24]. The term g' is the parameter that is affected by the material properties. Laser geometry can be changed by reducing the cavity length, which decreases τ_p as well. By the use of quantum wells, doping the active region p type and using strained quantum wells, the differential gain, g' can be increased. The maximum gain, however, saturates[25] in quantum well lasers as shown in Fig. 12.

In Eqn. 103, the photon density is limited by the facet damage and/or other device failure mechanisms and can not be increased indefinitely. We are, therefore, left with the term g' for the additional optimization of f_r. Stated in words, the material system and the laser structure must be chosen in such a way as to lead to a sharp rise in gain for incremental injection measured by current or injected carrier concentration. A small hole effective mass would lead to rapid reduction of α (below

605

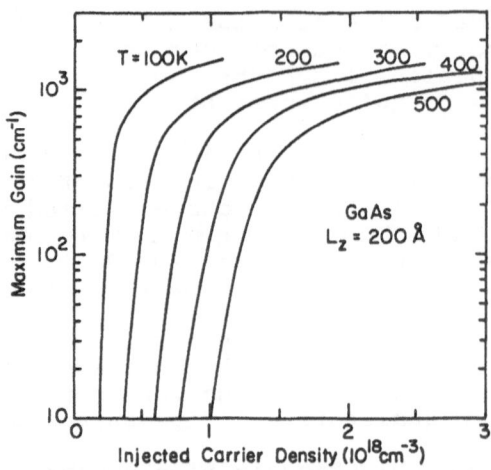

Fig. 12. Calculated maximum gain vs injection level in a
GaAs quantum well (200 Å) laser where only the
heavy hole band is assumed occupied, in the
temperature range of 100-500 K. Note the fast
initial rise (large differential gain) and
saturation later on[33].

threshold) with eventual reversal of its sign (gain). In other
words, both the electron and hole quasi-Fermi levels move
rapidly with injection and the lasing condition is satisfied at
relatively low injection levels. At the same time, the slope of
the gain curve with respect to injection, (current, carrier
density or energy above the band gap) would be higher which is
needed for large f_r (Fig. 13). However, the constant density of
states, if only the lowest subbands are occupied, would lead to
gain saturation which implies a limitation in P_0 in quantum well
lasers if Γ is very small. However, if the waveguide is
designed correctly, the clamping of the quasi-Fermi levels would
insure operation near the high g' region. In other words,
quantum well lasers, particularly with p doping in the pumped
region, short cavity lengths and a small hole effective masses,
would lead to larger f_r than possible in bulk lasers.

As can be seen in Table 3, the effective masses become
smaller as the InAs mole fraction is increased. In addition the
quantum wells can be doped p-type to place the Fermi level near
the valence band so that only the electron quasi-Fermi level
needs to be moved to satisfy the lasing condition. Using 90%
InAs (on InP substrates) and p doping of 5×10^{18} cm^{-3}, Suemene
et al.[26] predict f_r up to 90 GHz.

5.   PERFORMANCE OF STRAINED QUANTUM WELL LASERS

The majority of laser development with strained quantum
wells has so far been on the InGaAs pumped region with AlGaAs
waveguide grown on GaAs substrates. Using organometallic vapor
deposition for growth, Fischer et al.[27] reported low threshold
lasers with 37% InAs mole fraction in the well (40 Å) exhibiting
17 mA threshold currents when fabricated in the ridge laser
configuration with 2.5 or 3.75 μm wide ridges. Power levels of

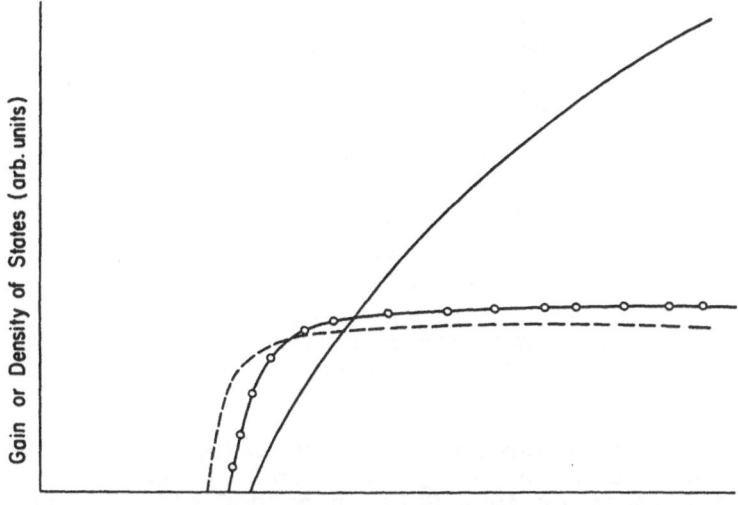

Fig. 13. Schematic representation of maximum gain vs injected carrier density in bulk (—) quantum well (-o-) and strained (- - -) quantum well laser structures. In this highly simplified picture, it is apparent that the threshold current is lower and the differential gain is higher in quantum well lasers as compared to bulk. Strained InGaAs quantum well laser offers lower threshold and higher differential gain.

24 μm facets (100 mA) were obtained in CW conditions without bonding. Bonded devices exhibited lifetimes of about 144 hours under CW conditions at room temperature.

More recently, Eng et al.[28] reported on a similar structure with the exception that the layers were grown by molecular beam epitaxy. Additionally, following the stripe formation, a blocking AlGaAs p/n junction was grown by LPE for

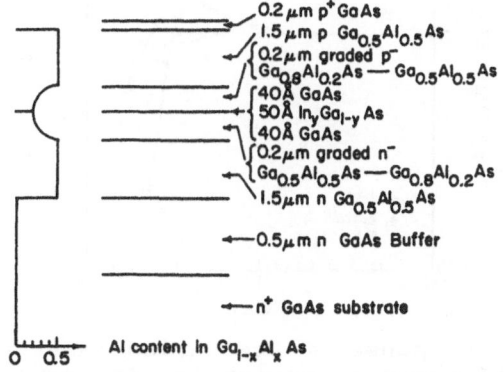

Fig. 14. Layer structure for the strained laser used in Ref. 36.

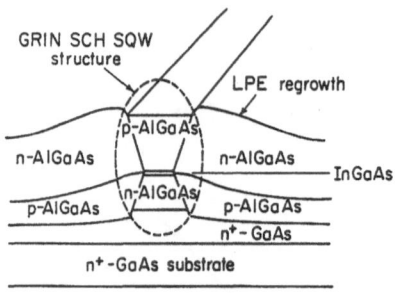

Fig. 15. Cross-sectional diagram of the
        MBE/LPE InGaAs strained well
        GRINSCH laser[36].

lateral confinement to ensure low threshold current. The layer
structure used is schematically shown in Fig. 14, where the 50 Å
$In_yGa_{1-y}As$ is flanked by 40 Å GaAs on either side. A roughly
parabolically graded 0.4 μm waveguide with AlAs mole fractions
from 20% near the well to 50% near the 1.5 μm $Al_{0.5}Ga_{0.5}As$
cladding layers. The substrate was cut 4% off (001) towards
(011).

    Broad area lasers fabricated from the MBE grown samples
exhibited threshold current densities of about 114 A/cm^2 at
990 nm for 100 μm x 1540 μm lasers. Following the buried laser
process, (see Fig. 15 for details), a threshold current of 2.4
mA at 950 nm for a 2 μm x 425 μm laser cavity was achieved.
When the facets were coated with high reflectivity films (R =
0.85), 2 x 225 μm devices have shown threshold currents of as
low as 0.9 mA. In 1540 μm long broad area lasers, an internal
quantum efficiency of 60% and internal distributed loss of
9 cm^{-1} were typical despite the long cavity length.

    The buried lasers were capable of delivery 5 mW of optical
power per facet with a slope efficiency of 22% mW/mA, which is
indicative of low internal losses. These lasers were also
tested for high speed modulation at power level of 1, 2.5 and

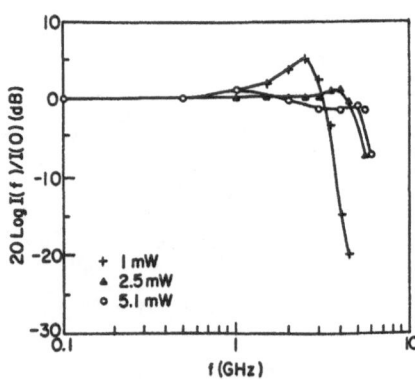

Fig. 16. RF response of the buried laser under
        different injection levels corresponding
        to 1, 2.5 and 5.1 mW of optical
        power[36].

5 mW with a maximum band width of 5.5 GHz obtained at injection corresponding to 5 mW optical power as shown in Fig. 16. The 5.5 GHz data may be limited by the detector response. The 5.5 GHz figure is comparable to GaAs BH laser. The effects of the reduced hole effective mass which would lead to high differential gains and, in turn, to higher corner frequencies might have been observed if not for detector response limit. One should, however, be cautious as the reduced mass is applicable only near k = 0.

Similar lasers were also explored, in terms of their power output which in itself is remarkable. Bour et al.[29] have used OMCVD grown strained quantum well $In_{0.2}Ga_{0.8}$ (70 Å) lasers inserted in the active region of otherwise GaAs/AlGaAs graded refractive index separate confinement heterojunction lasers for CW power output characteristics. For 90 μm (wide) x 600 μm (cavity length) lasers, the measured CW threshold current was 110 mA (200 A/cm^2), quoted to be about the same as those obtained from GaAs/AlGaAs lasers of similar structures grown in that particular reactor. The laser spectrum centered around 930 nm was found to be wider than that in GaAs active layer devices and was attributed to the strain induced broadening of the gain spectrum. A differential quantum efficiency, η of 23% was maintained up to 1 A of injection. Single facet output power exceeding 1 W was measured. In short cavity lasers, the end losses are increased which require high gain for maintaining laser oscillations. Due to the gain saturation observed in quantum wells (caused by the saturating density of states), the threshold current increases dramatically (Fig. 17). In 250 μm devices, the lasing wavelength shifts to about 916 nm from about 930 nm associated with much longer cavity devices. This marked shift, however, was found to be smaller than that can be explained by lasing on n = 1 lh - e (80 meV) and 130 meV to n = 2 hh - e transition.

The temperature dependence of the lasing threshold (J_{th} = J_0 exp(T/T_0) indicates T_0 = 150 K which falls between those associated with GaAs $In_{0.53}Ga_{0.47}$As channels. In $In_{0.53}Ga_{0.47}$As channels Auger recombination has been shown to be linked to reduced T_0. This would imply that the $In_{0.2}Ga_{0.8}$As strained well lasers also show the size of limitation by Auger recombination.

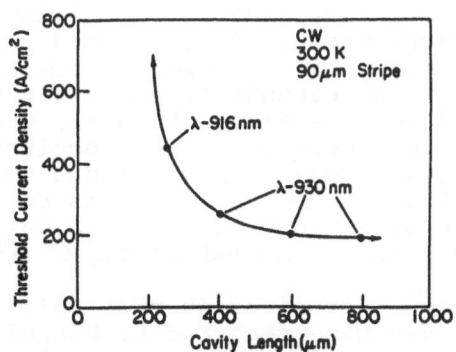

Fig. 17. Threshold current density vs cavity length in a strained laser family along with the fundamental lasing wavelength[37].

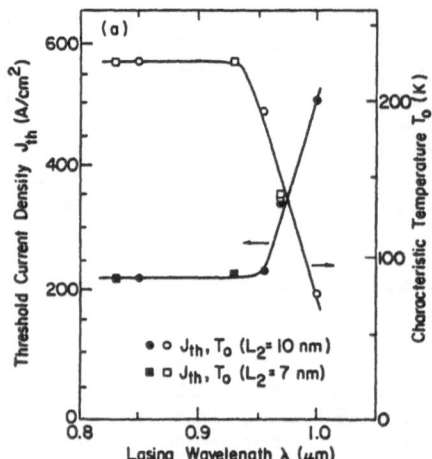 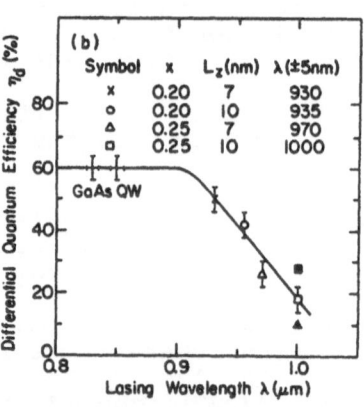

Fig. 18. (a) Threshold current and characteristic
temperature for strained $In_xGa_{1-x}As$ quantum
well lasers with x varied to cover the
wavelength range shown in the horizontal
scale[37]. (b) The differential quantum
efficiency for the same set of lasers. The
stripe width and cavity length of the lasers
tested are 90 and 400 μm, respectively[39].

However, Suemune et al.[30] argue that Auger rates in the strain
quantum wells are lower and that lower threshold current should
result. T_0 values of 100 K for $In_{0.3}Ga_{0.7}As$ (strained) and 67 K
for $In_{0.3}Ga_{0.7}As$ (matched to InP) lasers have been measured.
Since the rate of Auger recombination is proportional to the
third power of the carrier density, any laser showing gain
saturation, such as quantum well lasers, will be affected by the
Auger recombination process. This and other nonradiative
recombination centers, may be responsible for the small quantum
efficiency observed.

Since the quantum efficiencies in strained quantum well
lasers are inferior to those in GaAs ones, other nonradiative
processes may be at play. One of the most likely candidates is
the interfacial nonradiative process as alluded to by Bour et
al.[31], where a sharp increase in threshold current accompanied
by a sharp decrease in T_0 was observed as the InAs mole fraction
was increased to what corresponds to about 960 nm wavelength
(Fig. 18a). In addition, as Fig. 18b shows, the differential
quantum efficiency also decreases rather noticeably. The author
attributes these degradations in performance to interfacial nonradiative recombination as the strain is increased. The
distribution loss obtained in these devices was about 7 cm^{-1}
which compares with 9 cm^{-1} obtained by Eng et al.[28].

Reliability data on strained quantum well lasers with 25%
InAs mole fraction have been reported by Beernick et al.[32]
also. While those with 100 Å of quantum well showed no
degradation during a 1000 hour test, devices with $\approx$ 150 Å well
degraded rather quickly. The 150 Å is above the single kink
dislocation model and below the double kink dislocation model,
touched upon in the FET section, which may explain the shorter

lifetime. It is shown that 150 Å devices may be well within the
metastable region with smaller barrier to dislocation generation
under high density of optical fields such as those present in
semiconductor lasers.

CONCLUSIONS

We have discussed the impact of strained layers on lasers.
The reduced in-plane mass and the resultant density of states
attainable with quantum wells are certainly attractive for
lasers. Strain induced enhanced separation of heavy and light
hole bands allows the population of the heavy hole states,
hopefully only the first one, which actually have light masses.
These properties lead to a smaller joint density of states or
absorption coefficient, which rises initially rather steeply
with respect to energy, carrier concentration, or the injection
current for lasing. The reduced density of states implies
smaller threshold current, while the sharp rise leads to
increased differential gain. The relaxation oscillation
frequency is proportional to the square root of the differential
gain which can thus be increased. However, the experimental
threshold currents obtained in all quantum well lasers can be
explained simply by the reduced volume being pumped without any
quantum well physics. Quasi-Fermi level clamping assures
operation near the largest g', leading to large f_r as compared
to that attainable in bulk lasers. With strained InGaAs layers,
however, avoidance of the light hole (actually with a heavy in-
plane mass) and reduced surface recombination velocity may make
it possible to reduce the stripe width without invoking
additional complicating technology to minimize the edge losses.

Acknowledgment

This work has been supported by Office of Naval Research
under the contract N00014-88-K-0724 and Air Force Office of
Scientific Research under grant #AFOSR-89-0239. The author
would like to thank Profs. Y.C. Chang and K. Kishino for
discussions, L. Arsenault, S. Strite and J. Reed for assistance
and Ms. S. White for manuscript preparation.

REFERENCES

1.   J.W. Matthews and A.E. Blakeslee, **J. Crystal Growth**, 32:265
     (1976)
2.   G.C. Osbourn, **J. Vac. Sci. and Technol.**, A3:826 (1985)
3.   G. Bastard, **Phys. Rev. B**, 25:7584 (1982)
4.   J.Y. Marzin, **in**: "Heterojunction and Semiconductor
     Superlattices", G.Allen, G. Bastard, N. Boceara, M. Lannoo
     and M. Voss, Eds., Springer Verlag, Berlin, Heidelberg,
     (1982)
5.   F.H. Pollak and M. Cardona, **Phys. Rev.**, 172:816 (1968)
6.   G. Ji, D. Huang, U.K. Reddy, T.S. Henderson, R. Houdré and
     H. Morkoç, **J. Appl. Phys.**, 62:3366 (1987)
7.   S. Adachi, **J. Appl. Phys.**, 53:8775 (1982)
8.   Y.T. Leu, F.A. Thiel, H. Scheiber, B.I. Miller and J.
     Bachmann, **J. Electron. Mater.**, 8:663 (1979)
9.   K.H. Goetz, D. Bimberg, Jür, J. Selders, A.V. Solononov,
     G.F. Glinskii, M. Razeghi and J.J. Robin, **J. Appl. Phys.**,
     54:4543 (1983)

10. K. Nishi, K. Hirose and T. Mizutani, **Appl. Phys. Lett.**, 49:794 (1986)
11. W.T. Masselink, Ph.D. Thesis, University of Illinois (1986)
12. P.P. Szydlik, S. Alterovitz, E.J. Haugland, B. Segall, T.S. Henderson, J. Klem and H. Morkoç, **Superlattices and Microstructures**, 4:4 (1988)
13. Y.S. Lin, C.T. Liu, D.C. Tsui, E.D. Jones and L.R. Dawson, **Appl. Phys. Lett.**, 55:666 (1988)
14. G.C. Osbourn, J.E. Schirber, T.J. Drummond, L.R. Dawson, B.L. Doyle and I.J. Fritz, **Appl. Phys. Lett.**, 49:731 (1986)
15. E.D. Jones, S.K. Lyo, I.J. Fritz, J. Klem, J.E. Schirber, C.P. Tigges and T.J. Drummond, **Appl. Phys. Lett.**, 54:2227 (1989)
16. C.T. Liu, S.Y. Lin, D.C. Tsui, H. Lee and D. Ackley, **Appl. Phys. Lett.**, 53:2510 (1988)
17. I.J. Fritz, J.E. Schirber, E.D. Jones, T.J. Drummond and G.C. Osbourn, **Inst. Phys. Conf. Series**, 83:233 (1986)
18. H.C. Casey, Jr. and M.B. Panish, "Heterostructure Lasers", Academic Press, (1978)
19. W.T. Tsang, **in**: "Applications of Multiquantum Wells, Selective Doping and Superlattices", R. Dingle, ed.; Vol. 24 of "Semiconductors and Semi Metals", Willardson and Beer, eds., Academic Press (1987)
20. C.H. Henry, R.A. Logan and F.R. Merritt, **J. Appl. Phys.**, 49:3530 (1978)
21. H.Z. Chen, H. Wang, A. Ghaffari, H. Morkoç and A. Yariv, **Appl. Phys. Lett.**, 51:990 (1987)
22. T. Matsusue and H. Sakaki, **Appl. Phys. Lett.**, 50:1429 (1987)
23. H.Z. Chen, private communication.
24. Y. Arakowa and A. Yariv, **IEEE J. Quantum Electronics**, QE22:1887 (1986)
25. N.K. Dutta, **J. Appl. Phys.**, 53:7211 (1982)
26. I. Suemene, L.A. Coldren, M. Yamanashi and Y. Kan, **Appl. Phys. Lett.**, 53:1378 (1988)
27. S.E. Fischer, D. Fekete, G.B. Feak and J.M. Ballantyne, **Appl. Phys. Lett.**, 50:714 (1987)
28. L. Eng. T.R. Chen, S. Sanders, Y.H. Zhurng, B. Zhao, A. Yariv and H. Morkoç, **Appl. Phys. Lett.**, 55:1378 (1989)
29. D.P. Bour, D.B. Gilbert, L. Elbaun and M.G. Harvey, **Appl. Phys. Lett.**, 53:2371 (1988)
30. I. Suemene, **Appl. Phys. Lett.**, 55:2579 (1989)
31. D.P. Bour, R.U. Martinelli, D.B. Gilbert, L. Elbaun and M.G. Harvey, **Appl. Phys. Lett.**, 55:1501 (1989)
32. K.J. Beernink, P.V. York, J.J. Coleman, R.G. Waters, J. Kim and C.M. Wayman, **Appl. Phys. Lett.**, 55:2167 (1989)
33. From Table II.

GaAs ON InP BASED OPTOELECTRONIC INTEGRATED CIRCUITS FOR OPTICAL SWITCHING NETWORKS

A. Ackaert, I. Pollentier, P. Demeester,
P. Van Daele, D. Rondi,† G. Glastre,† A. Enard,†
R. Blondeau,† P. Jarry,‡ J. Cavaillès,‡ M. Renaud‡ and
H. Angenent‡

University of Gent (LEA) - IMEC
Sint Pietersnieuwstraat 41, B-9000 Gent, Belgium
† Thomson CSF/LCR
Domaine de Corbeville, BP 10,91401 Orsay, France
‡ LEP, Avenue des Cartes 3
94451 Limeil Brevannes Cx, France

1. INTRODUCTION

For the fabrication of long wavelength OptoElectronic Integrated Circuits (OEICs) there is a strong tendency towards the integration of InP/InGaAsP based optical and electronic components. The problems encountered due to the immature InP-related technology for the fabrication of electronic devices are mainly due to the low Schottky barrier on InP materials. This makes it impossible to use the classical Metal Semiconductor Field Effect Transistor (MESFET) structures, which are well developed in the GaAs technology. Alternatively, more complex structures, such as Metal Insulator Semiconductor FETs (MISFET), Junction FETs (JFET) and Heterostructure Bipolar Transistors (HBT) are used.

In this work two alternative integration schemes of a GaAs MESFET/InP active waveguide integration have been studied. First, the heteroepitaxial growth of GaAs directly on the InP substrate and secondly, the transfer of a fully processed GaAs MESFET to the InP substrate by epitaxial lift off (ELO). Results on light modulation in active waveguides using the MESFETs as driving sources will be discussed. Both proposed integration techniques seem to be valid and a comparison will be made between process related characteristics of both techniques.

2. GaAs MESFETs ON InP USING HETEROEPITAXIAL GROWTH

2.1 Literature Review

The use of GaAs MESFET structures monolithically integrated with long wavelength optoelectronic devices has been illustrated by the work of Suzuki et al.[1,2]. Although there exists a large lattice mismatch between the GaAs and the InP material

Condensed Systems of Low Dimensionality
Edited by J. L. Beeby *et al.*, Plenum Press, New York, 1991

(±3.8%), resulting in relatively high dislocation densities
(±10^7 cm^{-2}) during epitaxial growth, no important degradation of
the majority carrier MESFET devices is noticed compared to
similar test devices in GaAs on GaAs material. Some influence
on the GaAs on InP MESFET performance of the use of thick GaAs
buffer layers[3] and of the implementation of GaAs/AlGaAs
superlattices in this GaAs buffer layer[4] has been reported.
Making use of two low temperature deposited nucleation layers,
Lo[4] obtained X-ray FWHM values of 315 arcsec in 2.5 µm GaAs on
InP layers. The corresponding MESFET devices show a transconductance of 220 mS/mm for a gate length of 1 µm. Propagation
losses of waveguides in GaAs on InP seem to be slightly higher
for GaAs on exactly (100) oriented InP compared to the values
measured in GaAs on off-oriented (100) InP wafers[5]. Material
quality improvement through the use of off-oriented InP wafers
is also reported by Harbison[6]. The first laser devices
operating under pulsed conditions at room temperature, have been
reported by Van Ackere[7] and by Chang-Hasnain[8]. These
minority carrier devices consist of GRINSCH (Graded Index
Separate Confinement Heterstructure) QW laser structures fabricated on both exact[7,8] and off-oriented[8] InP substrates. On
off-oriented substrates Van Ackere[7] obtained on 450 µm x 10 µm
stripes threshold currents of l_{th} = 56 mA and differential
quantum efficiencies (η) up to 20%. The use of off-oriented
wafers seems to have an effect on both the threshold current and
on the quantum efficiencies. Chang-Hasnain[8] reports for 500
µm x 50 µm stripes (η, l_{th}) values of (36%, 340 mA) on (100)
exact substrates, and of (20%, 275 mA) on 3° off-oriented InP
material.

2.2 Experimental Results

Epitaxial growth is carried out in a commercial atmospheric
pressure MOVPE system with a small horizontal reactor and an
infra-red heated Si susceptor. Trimethylgallium (TMG), Arsine
(AsH$_3$) and Silane (SiH$_4$) are used as source materials with
typical molefractions for AsH$_3$: 2 10^{-3} and for TMG : 10^{-4},
resulting in a growth velocity of 100 nm/min. Heteroepitaxial

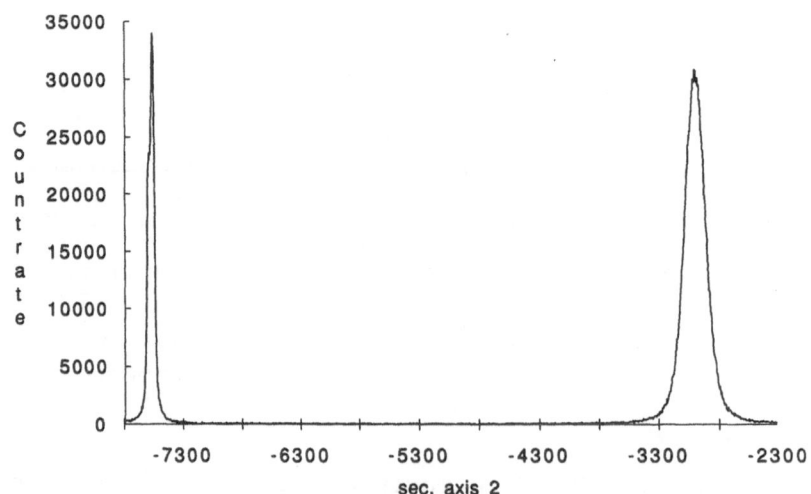

Fig. 1. Double crystal X-ray diffraction spectrum of a
        3 µm GaAs on InP layer.

GaAs on InP layers were grown using a thin GaAs nucleation layer deposited at low temperature (450°C). There was no PH_3 available in the growth system. This procedure results in good heteroepitaxial material quality with FWHM X-ray values of 200 arcsec for 3 μm layers (see Fig. 1). This value indicates a residual dislocation density of $\pm$ 1.5 10^7 cm^{-2}, indicating a significant improvement compared to typical values obtained in very thin GaAs on InP layers.

The MESFET layer structure consisted of a 3 μm GaAs undoped buffer layer (n-type, low 10^{14} cm^{-3}) and an active 200 nm n-type doped GaAs layer (1.5 10^7 cm^{-3}). This active layer showed at room temperature a mobility of 3700 cm^2/Vs, which is comparable to values obtained in similar test structures in homoepitaxial GaAs on GaAs. The variation of the nucleation layer growth parameters concerning both thickness ($\pm$ 5 to 20 nm) and deposition temperature (420 to 475°C) seems to have little influence on the resulting optical and electrical layer quality. Transistors were fabricated using mesa etching for isolation, AuGe/Ni for ohmic source and drain contact and TiW/Au for the Schottky gate contact. A transconductance of 100 mS/mm and a cut-off frequency of 7.1 GHz was obtained for MESFET configurations with a 1.5 μm gate length. The breakdown voltage was higher than 10 V. These results were independent of the substrate material (GaAs or InP). High spatial resolution PL-measurements performed at LEP, show the feasibility of the GaAs on InP growth for the fabrication of larger circuits. On micro-scale a variation of 20 mV over a typical area of one chip (300 x 300 μm^2) in threshold value is estimated[9]. Compared to a variation of 10 mV for high quality implanted GaAs wafers this result is very acceptable. The fabrication feasibility of more complex structures in the GaAs on InP material is illustrated through the realization of an 11-stage ring oscillator using about 100 Schottky gates (gate length : 2 μm). The oscillator frequency was 30 MHz at a power consumption of 44 mW. The rather large delay time of 3 ns per stage is only due to the circuit layout and not due to the intrinsic MESFET properties.

3.   GaAs MESFETs ON InP USING EPITAXIAL LIFT OFF (ELO)

3.1 **Literature Review**

The ELO technique was introduced by Yablanovitch in 1987[10]. It consists essentially of a lift off procedure of GaAs layers homoepitaxially grown on GaAs substrates. This is carried out by means of a very selective etching procedure of an underlying AlAs layer. The thin epitaxial layers can then be transferred and attached to arbitrary substrates, e.g. Si, glass, LiNbO$_3$, InP etc., without significant degradation of the epilayer quality. Different research groups have used the ELO technique to transport processed MESFET devices[11,12]. This resulted in high quality devices on glass, on Si and on InP substrates and no degradation was observed when comparing with device characteristics before lift off. The major advantage of this technique lies in the fact that the epitaxial layers are grown lattice matched and therefore consist of better material quality.

Other results obtained recently, using the ELO technique, are: the fabrication of GaAs LEDs on Si[13], laser diodes on glass[14] and MIS photodetectors on glass waveguides[15].

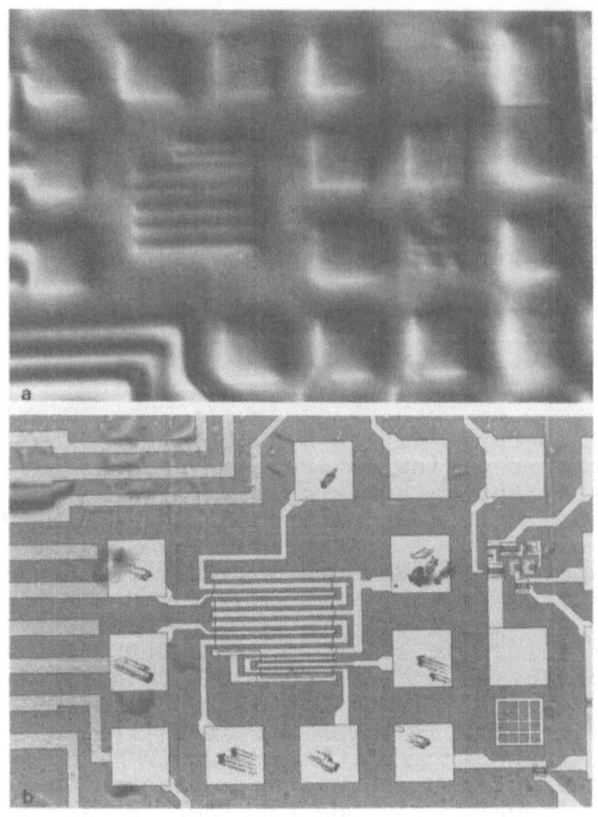

Fig. 2. ELO MESFET structure a) GaAs buffer layer up, b) metallization pattern up.

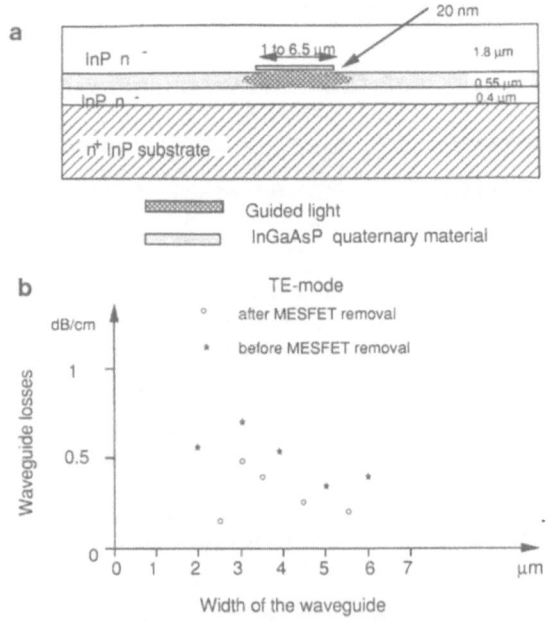

Fig. 3. a) Waveguide structure and, b) loss measurements

3.2 Experimental Results

To realize quasi-monolithic integration two approaches can be used: a) devices can be fully processed before ELO and then transferred to an arbitrary substrate; or b) as grown layer structures can be removed through ELO from the GaAs substrate and attached on another "host" substrate, where the further processing steps are carried out. Although attention is being paid to both integration schemes[13], the results reported here were obtained using the first technique.

After the growth of the MESFET structure on a GaAs substrate, involving an extra 500 nm AlAs lift-off layer, followed by a 4 nm undoped GaAs buffer layer and a 200 nm GaAs active layer, the same processing technique as described under Literature Review is carried out. The fully fabricated MESFET structure - buffer layer + active layer - were then lifted off from the original GaAs substrate using a wax carrier and a HF/H_2O (1:5) solution at 0°C. A dilution of HF is used for reducing the formation of H_2 gas bubbles, which can induce cracks in the thin GaAs structure. When the lift off etching technique is completed, the released film of only a few microns thick, and typical dimensions of a few mm^2, is very fragile and needs to be supported in further manipulations. Figure 2a shows a picture taken off the back side of the thin ELO structures. A deformation of the ELO layers, due to tensile forces induced by the metal patterns on the other side of the structure, is clearly visible. In Fig. 2b the same ELO structure is shown in an upward configuration, now attached to an InP substrate using Van der Waals bonding. MESFET characteristics before and after the transplantation were compared and showed no significant operation difference. The transconductance of the large MESFET shown in Fig. 2 was about 15 mS/mm.

4.   INTEGRATION OF GaAs MESFETs AND InP WAVEGUIDES

4.1 GaAs Growth on InP Waveguides

An important research topic in the development of an integration scheme is the investigation of the influence of the GaAs MESFET growth and processing on the underlying waveguide structure. A cross-section of the Thomson waveguides used in this study is visible in Fig. 3a. The lateral confinement in those structures is ensured by a thin quaternary layer (10 to 20 nm thick) embedded in a thick InP cladding layer[16]. Using the procedure described in the previous section, GaAs MESFETs were fabricated on top of these buried waveguides. Before growth typical losses were 0.5 dBcm. After MESFET growth and processing there was a slight increase in losses, but this effect disappears again after the removal of the metallization and the GaAs layers. The experimental values obtained for the TE mode are presented in Fig. 3b, but similar results hold for the TM polarization.

The etched channels visible in Fig. 5 gave rise to an extra parasitic waveguiding effect in the InP substrate. During cooling down from the growth temperature the difference in thermal expansion behavior between the GaAs and the InP material gives rise to a tensile stress in the GaAs layer and to equal, but opposite compressive stress in the InP substrate. Etching

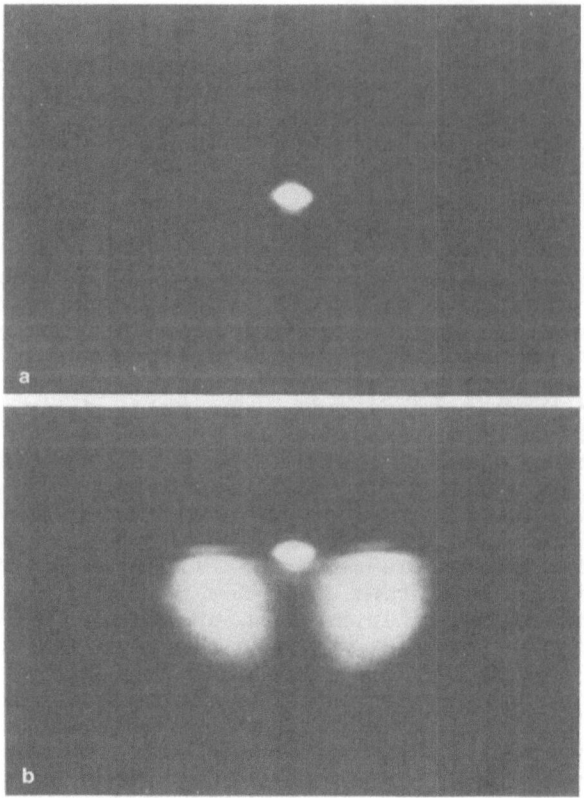

Fig. 4. Near field pattern a) all light coupled into the
waveguide, b) light partially coupled into the
substrate.

away part of the GaAs structure induces – due to partial stress
relief – a locally photoelastic effect near the borders of the
etched channel in the underlying InP substrate. The near field
output of the waveguides depends strongly on the input coupling
of the light: when all the light is coupled into the waveguide,
a single spot is observed at the output (Fig. 4a); coupling part
of the light into the substrate, however, results in parasitic
guiding near the edges of the etched channel.

4.2 Monolithic Integration of GaAs MESFETs and InP Waveguide Modulators

A schematic overview of the integration of GaAs MESFETs and
InP active waveguides is shown in Fig. 5. The waveguide growth

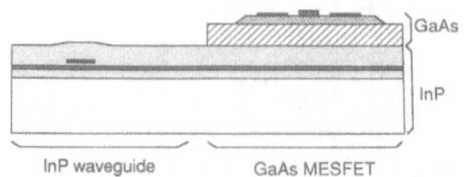

Fig. 5. Schematic overview OEIC.

(Thomson CSF) was followed by the heteroepitaxial GaAs on InP growth (IMEC-RUG). After processing of the active waveguides (Thomson CSF) and of the MESFET devices (IMEC) both components were connected via external means. A special mask design is currently developed for interconnecting the devices by means of lithographically defined metal paths.

In Fig. 6a we observe the modulation characteristic of the active waveguide. Optical modulation can be established using a voltage modulation in the range of 30 to 40 V, reverse bias (see lower curve Fig. 6a). The corresponding current-voltage characteristic is shown in the upper curve of the same figure. Since the breakdown voltage of the MESFETs is 10 to 15V, an external DC bias of -40 V was applied in the measurement set-up. The AC coupling for light modulation was established via the GaAs MESFETs fabricated on the InP substrates. The upper curve in Fig. 6b shows the AC gate source voltage signal applied to the MESFET, while the lower curve illustrates on the same time-scale the light output response of the waveguide. When we calculate the modulation efficiency we obtain a value of 26°/Vmm, which should be compared with a value of 4°/Vmm without the use of a MESFET.

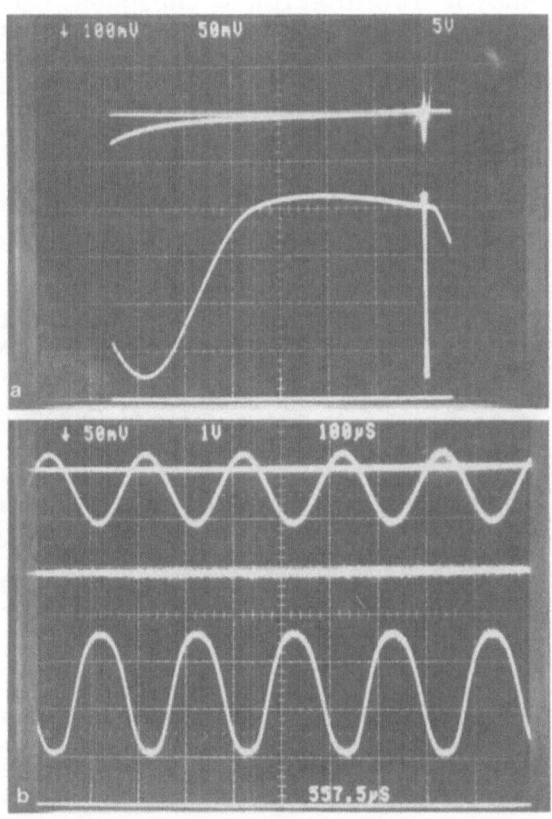

Fig. 6. Modulation characteristics; a) single
waveguides, b) OEIC configuration.

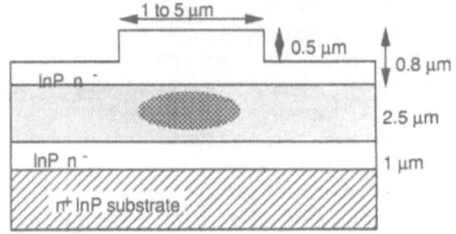

Fig. 7. Cross-section of LEP ridged waveguide.

4.3 ELO Integration of GaAs MESFETs and InP Optical Switches

The feasibility of the ELO technique for the development of an alternative integration scheme, has been illustrated using a directional coupler consisting of two InP/InGaAsP ridge waveguides. A cross-sectional view of such a waveguide fabricated at LEP is shown in Fig. 7. Typical losses obtained are 15 dBcm.

Using the ELO technique described in Section 3, fully processed GaAs MESFET structures were transferred to InP directional coupler chips. A main advantage of this technique is that it combines two components, grown and processed under ideal conditions. There is no cross influence possible due to combined growth and/or fabrication techniques. The OEIC configuration is again obtained via external interconnection means. In this measurement set-up a 0.5 V applied V_{GS}-voltage applied to the MESFET, resulted in a 25% optical power modulation. A result which is acceptable for the first integration realization. On chip interconnection between the GaAs ELO-MESFET and the InP optical switch is under further study.

CONCLUSION

In this paper two alternative integration schemes of a GaAs MESFET integrated with an InP active waveguide have been illustrated. First, through the use of the heteroepitaxial GaAs on InP growth technique; and secondly through the ELO transplantation of a fully processed GaAs MESFET to the InP substrate. Both techniques provided high quality GaAs MESFETs on InP substrates. No marked influence of the integration schemes on the waveguide losses and/or MESFET performances is being noticed. As a result we obtained the monolithic integration of GaAs MESFETs and InP active waveguides and the first ELO integration of similar GaAs components with InP optical switches.

Acknowledgment

This work was supported by the European Community RACE project No. 1033 . A. Ackaert and I. Pollentier wish to thank the IWONL Instituut ter bevordering van het Wetenschappelijk Onderzoek in Nijverheid en Landbouw) for the financial support.

REFERENCES

1. A. Suzuki, T. Itoh, T. Terakado, Y. Inomoto, K, Kasahara, K. Asano, S. Fujita, T. Torikai, **J. Lightwave Techn.**, 5:1479 (1987)
2. A. Suzuki, T. Itoh, T. Terakado, K. Kasahara, K. Asano, Y. Inomoto, H. Ishihara, T. Torikai, S. Fujita, **Electronic Letters**, 23:954 (1988)
3. K. Kasahara, K. Asano, T. Itoh, **Inst. Phys.**, Ser.No.91:195 (1987)
4. Y. Lo, R. Bhat, T. Lee, **Electronics Letters**, 24:865 (1988)
5. Y. Lo, J. Harbison, J. Abeles, T. Lee, R. Nahory, **IEEE Electron Device Lett.**, 9:383 (1988)
6. J. Harbison, Y. Lo, J. Abeles, R. Deri, B. Skromme, D. Hwang, L. Florez, M. Seto, L. Nazar, T. Lee, R. Nahory, **J. Vac. Sci.**, B7:345 (1989)
7. M. Van Ackere, A. Ackaert, I. Moerman, D. Lootens, P. Demeester, P. Van Daele, R. Baets, P. Lagasse, **Electronics Letters**, 25:47 (1989)
8. C. Chang-Hasnain, Y. Lo, R. Bhat, N. Stoffel, T. Lee, **Appl. Phys. Lett.**, 54:354 (1989)
9. M. Erman, G. Gillardin, J. LeBris, M. Renaud, **Journ. of Crst. Growth**, 96:469 (1989)
10. E. Yablanovitch, T. Gmitter, J. Harbison, R. Bhat, **Appl. Phys. Lett.**, 51:2222 (1987)
11. Van Hoof, W. De Raedt, M. Van Rossum, G. Borghs, **Electronics Letters**, 25:137 (1989)
12. P. Demeester, P. Van Daele, I. Pollentier, W. Temmerman, P. Lagasse, D. Rondi, G. Glastre, A. Enard, R. Blondeau, P. Jarry, J. LeBris, M. Renaud, H. Angement, Proc. ECOC Conf. (1989) p365.
13. I. Pellentier, P. Demeester, A. Ackaert, L. Buydens, P. Van Daele, R. Baets, **Electronic Letters**, 26:194 (1990)
14. E. Yablanovitch, E. Kapon, T. Gmitter, C. Yun, R. Bhat, **IEEE Photonics Techn. Lett.**, 1:41 (1989)
15. A. Yi-Yan, W. Chan, T. Gmitter, L. Florez, J. Jackel, E. Yablanovitch, R. Bhat, J. Harbison, **IEEE Photonics Techn. Lett.**, 1:379 (1989)
16. Y. Bourbin, A. Enard, R. Blondeau, M. Razeghi, D. Rondi, M. Papuchon, B. de Cremoux, **Electronics Letters**, 24:221 (1988)

STRAINED LAYER LASERS AND AVALANCHE PHOTODETECTORS

A.R. Adams, J. Allam, I.K. Czajkowski, A. Ghiti,
E.P. O'Reilly and W.S. Ring

Physics Department, University of Surrey
Guildford, Surrey, GU2 5XH, UK

1.   INTRODUCTION

The ability to grow high quality epitaxial layers of group III-V and group IV semiconductors in a state of permanent strain has two important implications for electronic and optoelectronic devices. First, it relaxes the constraints imposed by the requirement of lattice-matching. Thus, for example, by an appropriate choice of well width and composition $In_xGa_{1-x}As$/GaAs quantum well lasers can be made to operate at a convenient wavelength for the selective pumping of erbium doped fibre lasers and amplifiers. Also $In_xGa_{1-x}As$ on GaAs can provide a high mobility channel for n-type FETs or the base for n-p-n bipolar transistors, improving their emitter injection efficiency. Finally, if such devices can be grown on silicon substrates, further considerable advantages would arise for the integration of optoelectronic circuits. For these applications strain is a problem that must be overcome in order to obtain the desired effects.

The second aspect of strained layers refers to the advantages that arise due to the improved band structure that strain introduces and this is the subject of the present paper. The most important effect of shear strain is that it removes the cubic symmetry of group III-V and group IV semiconductors. This has two effects which we will consider. First, it removes the degeneracy at the top of the valence band and permits for the first time valence band structure engineering. This has a number of potential device applications; we emphasise here the advantages for lasers. Secondly, if the strained layer is grown on an (001) substrate it removes the degeneracy of the X or Δ minima, reducing the indirect band gap. We discuss here how this conduction band engineering can improve the band structure for low noise SiGe avalanche photodiodes.

2.   LASERS

Lattice-matched $In_xGa_{1-x}As$ quantum well lasers operating at 1.55 μm, where optical fibres have their minimum loss, are as

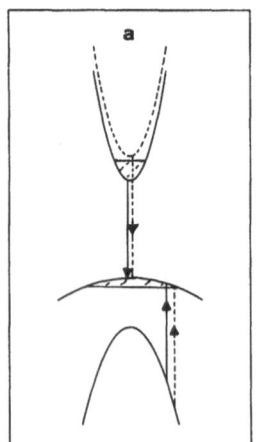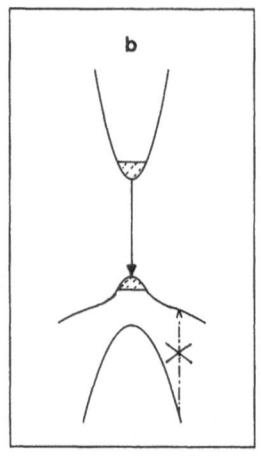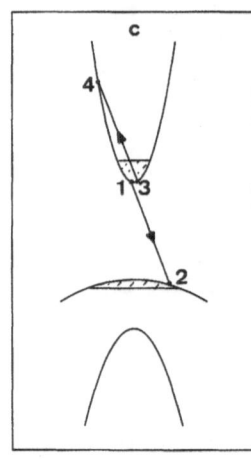

Fig. 1. Illustrates the loss mechanisms in 1.55 μm
        lasers. 1a shows how intervalence band
        absorption is reduced by the application of
        hydrostatic pressure (dotted curves) while
        1b shows that it can be eliminated in strained
        structures. 1c shows Auger recombination.

sensitive to changes in temperature as bulk devices. This is
believed to be due to the two processes of intervalence band
absorption (IVBA) and Auger recombination (AR) which are
illustrated in Figs. 1a and 1b respectively. IVBA occurs when
an emitted photon is reabsorbed by lifting an electron from the
spin-split-off band into an injected hole. This loss reduces

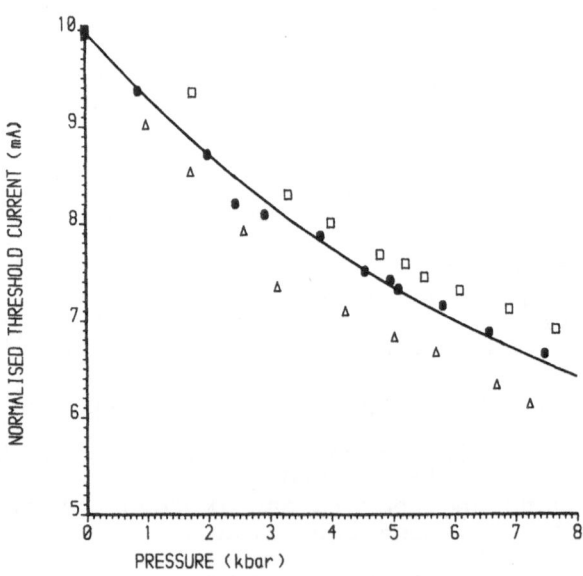

Fig. 2. Measured pressure dependence of the
        normalized threshold current in 1.55 μm
        quantum well lasers together with
        theoretical fit assuming IVBA and AR.

the quantum efficiency of the laser and hence increases the threshold current, I_{th}[1]. The effect is temperature sensitive because the number of holes available at the large k value required increases exponentially with temperature. Only one of the many AR mechanisms[2] can be illustrated in Fig. 1c. Here the energy and momentum of a recombining electron-hole pair(1) and(2) are transmitted to a second electron(3) lifting it further into the conduction band(4). The interpretation that these processes are occurring is now supported by a number of studies, including recent high pressure measurements of the threshold current of 1.55 μm quantum well lasers. The results are shown in Fig. 2. As can be seen, I_{th} decreases with increasing pressure. This is due to the fact that the direct band gap and hence the photon energy increases with increasing pressure at about 10 meV/kbar while leaving the valence band relatively unchanged (see Fig. 7). Thus at high pressure, the intervalence band absorption occurs at larger k values as shown by the dotted arrows in Fig. 1a. Here there are fewer holes and therefore the loss is decreased. Auger recombination processes also decrease with increasing pressure because as the energy exchanged increases so the momentum exchange must increase and holes with a larger k are again required. There is a complicated interplay between IVBA and AR[3] since IVBA increases the threshold carrier density N_{th} and hence the amount of AR which depends on N_{th}^3. The theoretical curve through the points in Fig. 2 was obtained with an IVBA coefficient of 40 cm^{-1} per 10^{18} cm^{-3} of holes and an Auger coefficient of 1.4 x 10^{-28} cm^6/s. Since the loss mechanisms decrease as the band gap increases IVBA and AR do not present a problem in shorter wavelength GaAs lasers. We will now show that efficient operation can occur if, instead of going to shorter wavelengths, we stay at 1.55 μm and restrict the holes in k-space[4]. This can be achieved in strained layer lasers.

Consider a layer under biaxial compression forming a type-I quantum well. Strain splits the valence band degeneracy at the Γ point by about 60 meV for each 1% lattice mismatch[5]. When the degeneracy is lifted, the valence bands become anisotropic and for biaxial compression the highest valence band has a large effective mass $m*_\perp$ (heavy-hole-like) in the growth direction and a comparatively low effective mass $m*_\parallel$ (light-hole-like) in the plane of the layer. The quantum well confinement energies are determined by $m*_\perp$ while the subband dispersion and quantum well density of states at the valence band edge are determined by the low in-plane mass, $m*_\parallel$. The highest valence subband is expected to remain light hole-like until it interacts and mixes with lower lying bands with a larger in-plane mass. Because of the strain-induced splitting and confinement effects, the in-plane mass can remain light hole-like over a wide energy range (greater than kT at room temperature). The effect in IVBA of reducing the in-plane hole mass is illustrated in Fig. 1b. First, the reduced density of states means that the requirement for lasing, that the quasi-Fermi levels are separated by more than the band gap, can be achieved with lower injected carrier density. Secondly, the holes are restrained in k-space. Both effects taken together are predicted to practically eliminate IVBA and AR in 1.55 μm lasers.

We have undertaken band structure calculations using the Luttinger-Kohn Hamiltonian in the axial approximation. Figure 3 shows the valence band structure calculated for two laser

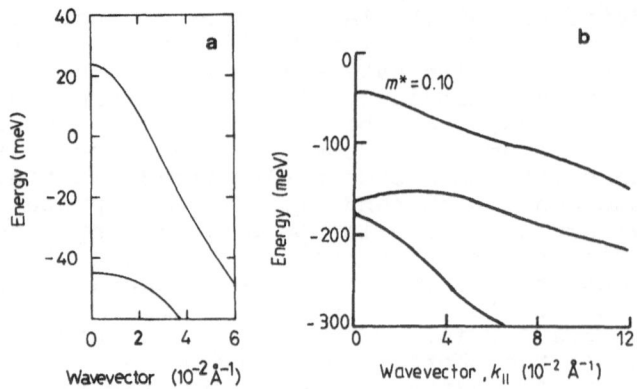

Fig. 3. Calculated valence band structures of a) 40 Å
$Ga_{0.33}In_{0.67}As$ wells between $Ga_{0.71}In_{0.29}As$
barriers and b) 35 Å $Ga_{0.17}In_{0.83}As$ wells between
$Ga_{0.17}In_{0.83}As_{0.3}P_{0.7}$ barriers. Both give 1.5 μm
laser operation.

structures designed to operate at 1.55 μm. Figure 3a is for
40 Å quantum wells of strained $In_{0.67}Ga_{0.33}As$ between unstrained
$In_{0.3}Ga_{0.7}As$ barriers[6]. The effective mass is found to be
approximately 0.087 over 50 meV. Figure 3b refers to 35 Å wide
strained $In_{0.83}Ga_{0.17}As$ wells between $In_{0.83}Ga_{0.17}As_{0.3}P_{0.7}$
barriers lattice matched to InP which provides the optical
confinement[7]. The effective mass in this structure is
predicted to be 0.1 over about 50 meV. Using the band structure
of Fig. 3a we have undertaken calculations of the optical gain

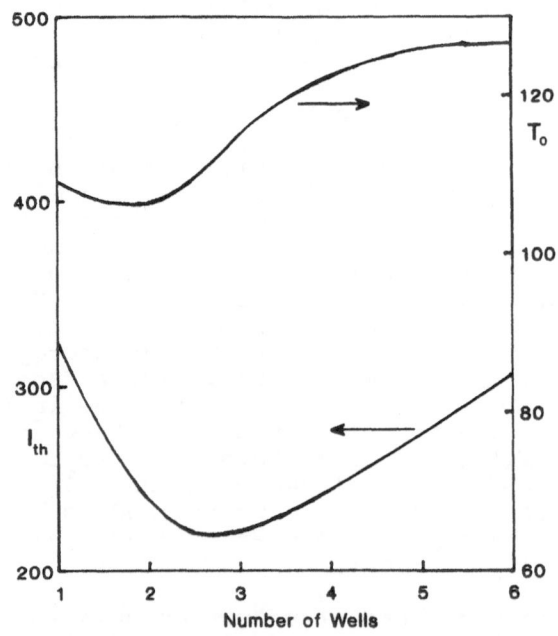

Fig. 4. Calculated I_{th} in A/cm^2 and T_0 in ^0K using the
band structure of Fig. 3b.

spectrum following Asada[8] and deduced the threshold current for a 300 μm long laser assuming a facet reflectivity of 0.3. Fig. 4 shows the predicted threshold current density at 300 K as a function of the number of wells. Also shown is the temperature sensitivity parameter, T_o, obtained assuming an exponential dependence on temperature of the form $J_{th}(T) = J(300)\exp\{(T - 300)/T_o\}$. These results predict that J_{th} is reduced by almost an order of magnitude and T_o doubled compared to 1.55 μm bulk lasers.

Because of the reduced density of states in the valence bands of strained layer lasers, they exhibit a very sharp increase in gain with increasing carrier density[9] and less gain saturation than normal quantum well lasers. Fluctuations in the required gain due to either modulation or inherent instabilities therefore require relatively small changes in carrier density and therefore only small changes in refractive index ensue. This means that the chirp and the line width should be greatly reduced. Also, the increased rate of change of gain with carrier concentration increases the relaxation oscillation frequency f_r. Figure 5 compares the rate of change of gain with carrier density calculated for an unstrained QW laser assuming the hole mass $m_h = 0.5$ with that calculated for a strained laser where it is assumed $m_h = 0.1$. As can be seen dg/dn is increased by about a factor of four at a gain of 3×10^3 at which lasing occurs, leading to a doubling of f_r while the line width enhancement factor α is reduced by a factor of two-thirds. The most recent experimental work on InP-based strained layer lasers known to the authors is by Thijs et al.[10] who used LP-OMVPE to grow four 30 Å wide $In_{0.8}Ga_{0.2}As$ quantum wells in InGaAsP (1.3 μm) lattice-matched to InP to form DCPBH lasers operating at 1.5 μm. They observed a differential external efficiency of 82% and a characteristic temperature T_0 of 97 K. CW powers as high as 200 mW were observed from 1500 μm long devices and the CW threshold current at 20°C was as low as 29 mA for devices with only two wells. These very good results can be interpreted as support for the reduction in the internal loss mechanisms of IVBA and AR proposed above. What is more, lifetime tests performed at a heatsink temperature of 60°C and a CW output power of 5 mW showed almost no degradation after 4000 hours.

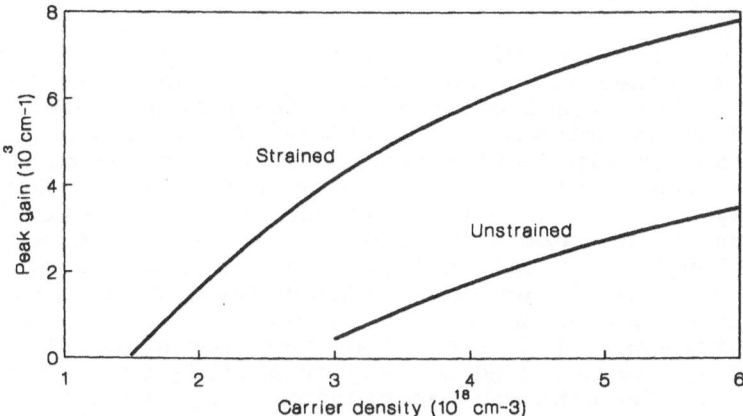

Fig. 5. Variation of peak gain with carrier concentration using the band structure of Fig. 3b.

InGaAs strained layers on GaAs grown by MBE as part of a collaborative program between the University of Surrey and RSRE have been studied in some detail. A single 90 Å wide $In_{0.19}Ga_{0.81}As$/GaAs quantum well modulation p-doped to 3.3×10^{11} cm^{-2} was found by cyclotron resonance to have a hole effective mass of 0.155 at the Fermi level[11]. A hole mobility μ_h of about 17,000 cm^2/Vs was observed at liquid helium temperatures. μ_h was found to be pressure dependent with $1/\mu$ dμ/dp varying with pressure at -0.7%/kbar[11]. This is intermediate between the observed rate of change of mobility of electrons and of heavy holes. These observations support the light-hole properties illustrated in Fig. 3.

Strained quantum well lasers based on the $In_xGa_{1-x}As$/GaAs/Al_yGa_{1-y} system were shown to operate CW at 300 K as early as 1983[12] and since then their technology has proceeded rapidly. In the wavelength range out to about 1.1 μm, IVBA and AR are small. However the lasers do benefit from a reduced density of states at the valence band and recently Offsey et al.[13] have reported consistently lower threshold current densities and a significant increase in band width of $In_{0.3}Ga_{0.7}As$ strained layer lasers over comparable GaAs devices. Very significantly, InGaAs/GaAs devices also show good operating stability even at 50°C with high output powers (200 mW)[14]. Since such short wavelength lasers with high output powers are subjected to some of the most extreme working conditions of any semiconductor electronic devices, these results are most encouraging for future applications of strained layers. To the knowledge of the authors no long wavelength lasers based on GaAs substrates have yet been reported and this awaits the development of suitable relaxed buffer layers.

3. AVALANCHE PHOTODIODES

The noise performance and stability of operation of an avalanche photodiode (APD) is largely determined by the ratio of the impact ionization rate for electrons (α) to that for holes (β). For a low noise APD, it is necessary that this ratio should be far from unity[15]. Unfortunately for most III-V semiconductors and for Ge the ionization rate ratio is of order unity. However, Si is an exception with $\alpha/\beta \approx 10^1 - 10^2$ depending on the electric field[16] and it makes excellent APDs for wavelengths less that $\approx$ 1 μm.

In this section we will discuss APDs fabricated from Ge_xSi_{1-x}/Si strained layer heterostructures, where the band gap is indirect. The effect of strain on the band structure of Ge_xSi_{1-x} alloys is potentially beneficial for both the absorption and multiplication characteristics of APDs. As mentioned in the introduction, the Δ minima will be split in a strained layer grown on an (001) substrate. The indirect band gap is therefore decreased and the absorption edge moves to longer wavelengths. Strained layers may be produced by growing $Ge_{1-x}Si_x$/Si strained layer superlattices (SLS) on Si substrates, where the thickness of the Ge_xSi_{1-x} layers is kept below the critical thickness. The absorption characteristics of such superlattices were measured by Lang et al.[17] as a function of Ge content x. The band gap was significantly reduced below that for the corresponding unstrained alloy. Subsequently, $Ge_{0.6}Si_{0.4}$/Si_{1-x} SLS waveguide photodectors and APDs were

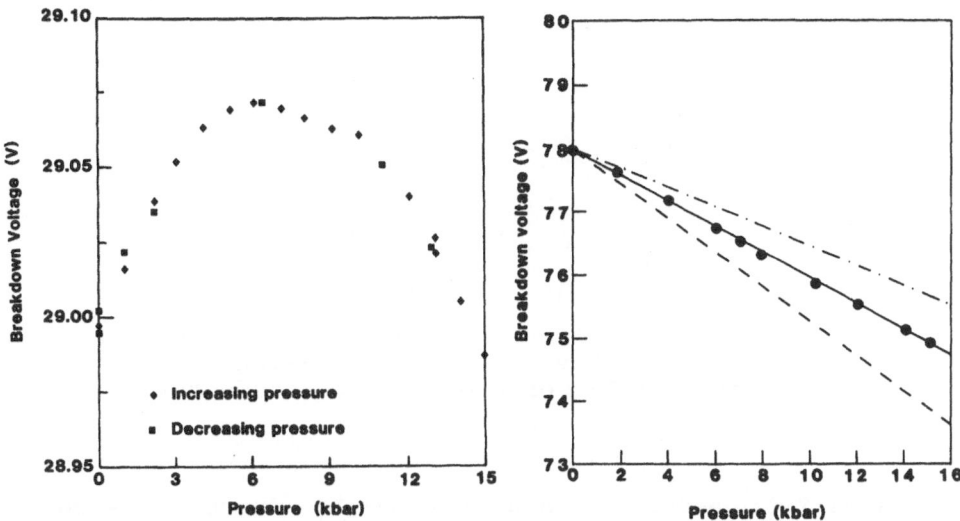

Fig. 6. Pressure dependence of the breakdown voltage
a) in Ge APDs; b) Si APDs.   Theoretical curves
--- P = ∞, -·- P = 0.02, —— P = 0.15.

demonstrated with high efficiency at 1.3 μm.   The reader is
referred to a detailed review by People[18].

     Strain may also be expected to influence the ionization
rates in Ge_xSi_{1-x} alloys.   Impact ionization is the reverse of
the Auger recombination described in the previous section,
involving the creation of an electron-hole pair (e.g. (1) and
(2) in Fig. 2) by a high energy carrier (e.g. (4)) relaxing to a
lower state (3).   The threshold energy for ionization can be
determined from the band structure by minimizing the energy of
the initiating electron or hole.   Because of the complexity of
the bands at high energies, and because experimental determination of threshold energies from ionization rate measurements is
difficult, we have applied hydrostatic pressure to Si and Ge
APDs and have measured the effect of perturbation of the band
structure on the avalanche multiplication.   This has given
insight into the ionization processes in Si and Ge and allowed
us to make predictions about strained Ge_xSi_{1-x} systems.

     Figure 6a shows the variation of the breakdown voltage with
pressure for a Ge APD in the range 0-15 kbar[19].   The sharp
increase of current and its positive temperature coefficient
showed that the breakdown was due to impact ionization rather
than tunneling.   The data is representative of that obtained for
a number of devices of different areas.   To understand this
behavior we have calculated threshold energies at various
pressures from empirical pseudopotential band structures
(including spin-orbit interactions) using the method of Anderson
and Crowell[20].

     A large number of different ionization processes were found
for both electrons and holes, each associated with one of the
conduction band minima at the Γ, X or L points in the Brillouin
zone.   The pressure variation of the threshold energy for each
process was approximately equal to the variation of the
associated minima.   The effect of pressure on the band structure

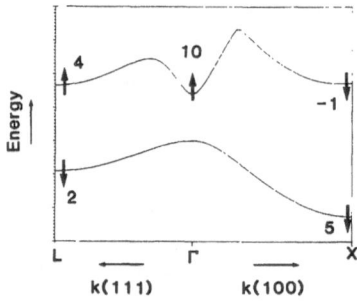

Fig. 7. Pressure coefficients in meV/kbar with
respect to the valence band maximum for a
typical III-V or group IV semiconductor.

of a typical tetrahedrally-coordinated semiconductor is shown in
Fig. 7, and the positions of the conduction band minima relative
to the top of the valence band in Ge are shown in Fig. 8, taken
from the data of Ahmed et al.[21]. It would appear that the
peak in the avalance breakdown voltage at ≈ 6 kbar is due to the
crossing of the Γ and Δ minima. If this interpretation is
correct then these results represent direct experimental
evidence for multiple ionization processes and "soft" ionization
thresholds. Since the lowest energy thresholds are associated
with the L minima over the pressure range studied, the ionization cross-section must be finite since carriers are also
ionizing via Γ and Δ processes at higher energies. This
"softness" results from the anisotropy of the threshold energy
since at a given energy only carriers from a restricted region
of the Brillouin zone can undergo ionization.

Measurements were also performed on Si APDs[19] and the
observed variation of breakdown voltage with pressure is shown
in Fig. 6b. The behavior is qualitatively different from that

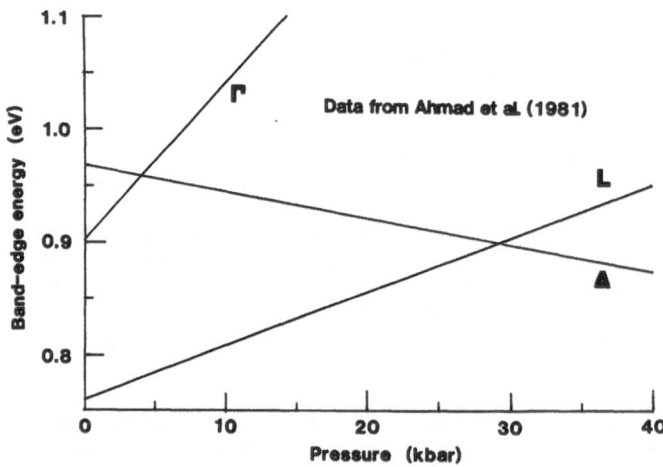

Fig. 8. Pressure dependence with respect to the valence
band maximum of the conduction band minima in
germanium.

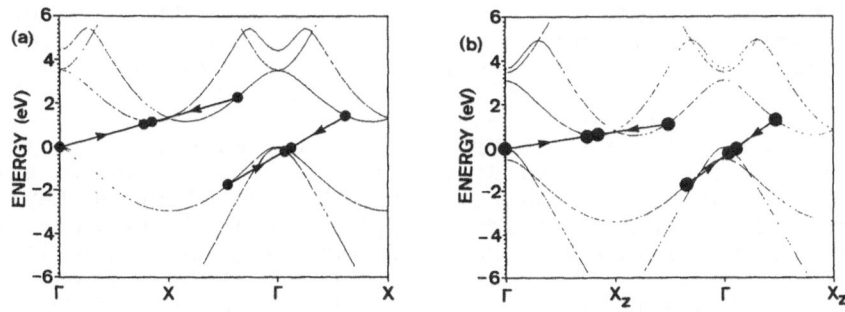

Fig. 9. Illustrations of the minimum thresholds for
impaction ionization by electrons and holes in
a) cubic silicon b) strained silicon.

of Ge, showing a linear decrease of breakdown voltage with
pressure. Unlike Ge, the conduction band minima in Si are well
separated in energy and multiplication is dominated by
ionization in the Δ valley. The lowest electron and hole
threshold we calculated for Si are shown in Fig. 9a. The
threshold energy for electrons (1.18 eV) is almost equal to the
indirect band gap, E_g, while for holes the threshold is
considerably larger (1.71 eV ≈ $1.5E_g$). This behavior can be
directly correlated to simple features of the band structure,
viz the heavy longitudinal electron mass and the ratio of the
valence band width to the indirect band gap. The ratio of
thresholds is 1.5 which is significantly larger than that in
most materials and contributes to the large α/β ratio in Si.
The experimental results for the pressure variation were fitted
with a lucky-drift model with soft thresholds. A good fit is
obtained with a Keldysh softness parameter P = 0.15[22]. The
prediction of a hard threshold model, P = ∞ and a much softer
model, P = 0.02, are shown for comparison.

Having observed that the threshold energy for electrons
decreases for decreasing band gap under hydrostatic pressure we

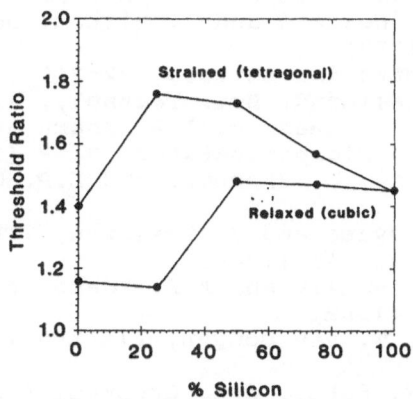

Fig. 10. Ratio of the threshold energy for hole impact
ionization to that for electron impact ionization as a function of composition in strained
and unstrained SiGe alloys on (001) Si
substrates.

consider the advantages that might be obtained with strained layer systems. For growth on an (001) substrate, the Δ minima will be split and thus the electron threshold, which remains approximately equal to E_g, is decreased. However, we find that the hole threshold is relatively unchanged. Figure 9b illustrates the hole and electron thresholds we calculate for the extreme case of strained Si on cubic Ge while Fig. 10 shows our calculated values of the ratio of hole and electron thresholds for the range of Si alloys on (001) cubic Si substrates. Also shown is the ratio of thresholds assuming the alloy is relaxed (i.e. cubic). The increase in the threshold ratio suggests that the strain may increase the α/β ratio. This effect, added to the high quantum efficiency at 1.3 μm discussed above, implies that strained layer Ge_xSi_{1-x} alloys have potential importance for low-noise detectors for use in optical fibre communications.

CONCLUSION

In conclusion, we have briefly reviewed some of the benefits that might be gained by fabricating optoelectronic devices from strained layer systems. Experimental evidence becoming available not only supports the theoretical proposals but also indicates that the strained layer devices are stable even under quite extreme operating conditions.

Acknowledgment

The authors gratefully acknowledge the support of SERC, BTRL and STC. They wish to thank M.A. Gell for providing band structures of strained Ge_xSi_{1-x} alloys and colleagues in the Strained-layer Structures Research Group at Surrey for many stimulating discussions.

REFERENCES

1. A.R. Adams, M. Asada, Y. Suematsu and S. Arai, **Jpn. J. Appl. Phys.**, 19:L621 (1980)
2. N.K. Dutta and R.J. Nelson, **J. Appl. Phys.**, 53:74 (1982)
3. A.R. Adams, K.C. Heasman and J. Hilton, **Semicond. Sci. Technol.**, 2:7 (1987)
4. A.R. Adams, **Electronics Letters**, 22:249 (1989)
5. E.P. O'Reilly, **Semicond. Sci. Technol.**, 4:121 (1989)
6. E.P. O'Reilly, K.C. Heasman, A.R. Adams and G.P. Witchlow, **Superlattices and Microstructures**, 3:99 (1987)
7. A. Ghiti, W. Batty, U. Ekenburg and E.P. O'Reilly, **SPIE**, 861:96 (1988)
8. M. Asada, A. Kameyama and Y. Suematsu, **IEEE J. Quantum Electronics**, QE-20:745 (1984)
9. A. Ghiti, E.P. O'Reilly and A.R. Adams, **Electronics Letters**, 25:821 (1989)
10. P.J.A. Thijs and T. van Dongen, **Electronics Letters**, 25:1735 (1989)
11. D. Lancefield, W. Batty, C.G. Crookes, E.P. O'Reilly, A.R. Adams, K.P. Homewood, G. Sundaram, R.J. Nicholas, M. Emeny and C.R. Whitehouse, **Proc. EP2DS8**, Grenoble, (1989)
12. M.J. Ludowise, W.T. Dietze, C.R. Lewis, M.D. Camras, N. Holonyak, Jr., B.K. Fuller and M.A. Nixon, **Appl. Phys. Lett.**, 42:487 (1983)

13. S.D. Offsey, W.J. Schaff, P.J. Tasker and L.F. Eastman, **IEEE Photonics Technol. Letts.**, 2:9 (1990)
14. J.S. Major, Jr., W.E. Plano, A.R. Sugg, D.C. Hall, N. Holonyak, Jr. and K.C. Hsieh, **Appl. Phys. Lett.**, 56:105 (1990)
15. R.J. McIntyre, **IEEE Trans Electron Devices**, ED-13:164 (1966)
16. C.A. Lee, R.A. Logan, R.L. Batdorf, J.J. Kleimack and W. Wiegmann, **Phys. Rev.**, 134:A761 (1964)
17. D.V. Lang, R. People, J.C. Bean and A.M. Sergent, **Appl. Phys. Lett.**, 47:1333 (1985)
18. R. People, **IEEE J Quantum Electron**, QE22:1969 (1986)
19. I.K. Czajkowski, J. Allam, M. Silver, A.R. Adams and M.A. Gell, **IEE Proc.**, 137:79 (1990)
20. C.L. Anderson and C.R. Crowell, **Phys. Rev.**, B5:2267 (1972)
21. C.N. Ahmad, A.R. Adams, B.J. Sealy, R.J. Nicholas and G.D. Pitt, **High Pressure in Research and Industry**, 2:464 (1981)
22. L.V. Keldysh, **Sov. Phys. JETP**, 10:509 (1960)

THE InGaAlAs SYSTEM FOR STRAINED AND UNSTRAINED HETEROSTRUCTURES ON InP SUBSTRATE

M. Quillec, M. Allovon and J.M. Gerard

Centre National d'Etudes des Télécommunications
196 Avenue Henri Ravera
92220 Bagneux, France

1. INTRODUCTION

In recent years, the quaternary system InGaAlAs grown on InP has been the subject of increasing interest, especially in the microelectronic field. Recently, good results on lasers have extended that field to optoelectronics. On the other hand, whereas epitaxy was for a long time restricted to lattice-matched material, pseudomorphic epitaxy has now proved extremely fruitful for broadening the possibilities of a given material combination. In this paper, we give an overview of the use of the microstructures based on the InGaAlAs/InGaAs system for devices, and their pseudomorphic extensions.

2. InGaAlAs VERSUS InGaAsP

The InGaAlAs system provides a large band gap range with a single group V element , on an InP substrate. Compared to the InGaAsP system, several features listed below make it more attractive for many applications. Besides, the development of growth of pseudomorphic epilayers increases some of these advantages, as will be developed in the following.

* Based on a compilation of several independent measurements found in the literature, we believe that the conduction band discontinuity at the $In_{0.52}Al_{0.48}As/In_{0.53}Ga_{0.47}As$ interface is about 0.53 eV (75% of the band gap difference), whereas it seems to be only 0.236 eV (40% of the band gap difference) at the $InP/In_{0.53}Ga_{0.47}As$ interface (Fig. 1).

* High Schottky barriers are obtained on $In_{0.52}Al_{0.48}As$ (0.62 eV[1]).

* Semi-insulating $In_{0.52}Al_{0.48}As$ can be obtained easily, for instance by low temperature MBE, as demonstrated recently[2].

* More abrupt interfaces are expected between materials with the same group V element.

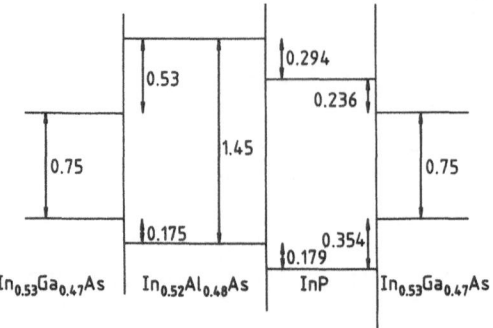

Fig. 1. Conduction and valence bands discontinuities
between InP and lattice-matched InGaAs and
InAlAs.

* Controlling gradual composition variations is easier with a
  single group V element.

* Efficient mixing was demonstrated on superlattices with
  common group V element.

3. MBE GROWTH OF InGaAlAs ON InP

As we have shown earlier[3], two indium sources are
required for an easy and flexible control of MBE growth of
InGaAlAs. The two ternary compounds, InGaAs and InAlAs, can
then be independently adjusted, for instance lattice matched to
the substrate. Then, the quaternary alloy obtained by
superimposing them is also lattice matched to the substrate, and
can be written as:

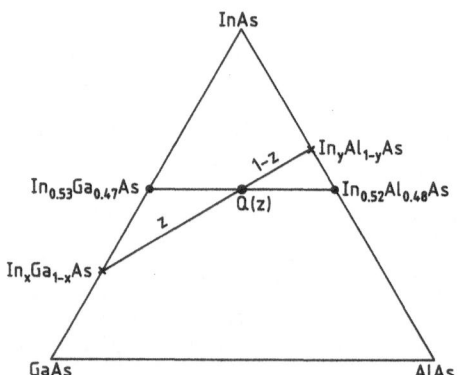

$$Q(z) = (In_xGa_{1-x}As)_{1-z} \ (In_yAl_{1-y}As)_z$$

Fig. 2. Representation of the InGaAlAs compositions.
Simple geometrical considerations show that a
given quaternary can result from superimposition
of two related ternaries, InGaAs = G(x) and
InAlAs = A(y), with growth rates in the same
ratio as the segments G(x)Q(z) and A(y)Q(z).

636

$$Q(z) = (In_{0.53}Ga_{0.47}As)_{1-z}(In_{0.52}Al_{0.48}As)_z$$

the composition z is related to the growth rates R1 and R2 of the two ternary alloys: $z/(1 - z) = R2/R1$.

3.1 Strained Material and Heterostructures

In Fig. 2, the locus of InP lattice-matched quaternary alloys is a line between $In_{0.53}Ga_{0.47}As$ and $In_{0.52}Al_{0.48}As$. The composition of Q(z) can also be obtained from two lattice-mismatched ternaries: $G(x) = In_xGa_{1-x}As$ and $A(y) = In_yAl_{1-y}As$, as shown in Fig. 2, provided that the following relation is fulfilled between the segments G(x)Q(z) and A(y)Q(z) and the growth rates R1 and R2:

$$G(x)Q(z)/A(y)Q(z) = z/(1-z) = R1/R2.$$

It is thus possible with a single set of In1, In2, Ga and Al atom fluxes, to grow a lattice-matched quaternary alloy and lattice-mismatched InGaAs and InAlAs ternaries. Strained structures in the InGaAlAs family on InP substrate are of particular interest because they allow compensation of one strained layer by the adjacent ones, so that the strained structure remains lattice matched as a whole to the substrate. This is the case (the elastic parameters being close are taken equal for simplicity) for a multiquantum well involving L_b thick A(y) barriers and L_z thick G(x) wells (as shown in Fig. 2); lattice-matching to InP requires:

$$L_z/L_b = (1-z)/z = (y-0.52)/(x-0.53)$$

Figure 3 presents the calculated band gaps related to the valence subbands of the two strained ternary bulk materials; unstrained gaps are plotted for comparison[4]. Strained layer superlattices lattice matched as a whole to the InP substrate have been grown by MBE. Figure 4 shows an X-ray double diffraction profile obtained on such a SLS, near the (004) reflection. The parameters are: number of periods: 10, A(0.64), L_b = 14.2 nm, G(0.41), L_z = 15.8 nm. The two sets of satellites, separated by the substrate contribution, refer to each material, as demonstrated in[5], and the position of the SLS "zero order" shows that the structure is well lattice matched to the substrate.

Figure 5[6] shows the two possible configurations; in case 1 of Fig. 5a, presented in Fig. 5b, the smaller band gap material, InGaAs, is under biaxial compression. The highest valence band contains $|J = \frac{3}{2}, m_J = \pm\frac{3}{2}>$ states, which are the heavy holes (HH) in the bulk material. Due to the biaxial compression, however, this band is relatively light hole-like in the directions parallel to the interfaces, and heavy-like only in the direction perpendicular to the interfaces. Thus, in the quantum well, only the confinement is determined by this latter heavy mass, whereas the dispersion is determined by the in-plane light mass.

4. INTEREST IN STRAINS IN InGaAlAs BASED DEVICES

In recent years, many papers have shown interest in strained layers for device improvement. In the following, we

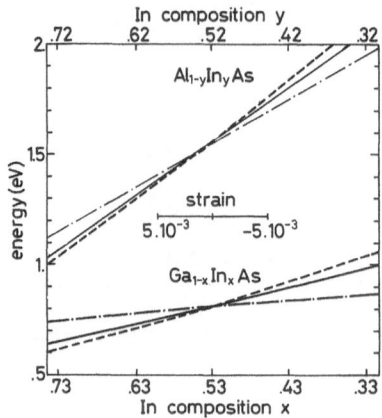

Fig. 3. Calculated light-hole (-·-) and heavy-hole (——)
related band gaps for bulk InGaAs and InAlAs
strained to InP. The unstrained band gap is
also shown (---). The strain scale is represented in terms of relative lattice mismatch
with InP.

gather the improvements obtained or expected due to strains in
InGaAlAs based opto or microelectronic devices. We will
emphasize recent results obtained in the optoelectronic field
since this family of materials is often believed to be badly
suited to such applications.

4.1 Lasers

Until recently, only the InGaAsP/InP system provided good
lasers in the 1.3-1.6 μm range of interest for optical fiber
communications. In that system, following the suggestion of
Adams et al.[7], Thijs and Van Dongen[8] have shown that introducing biaxial compression in the quantum wells does improve the
device, especially in terms of temperature sensitivity

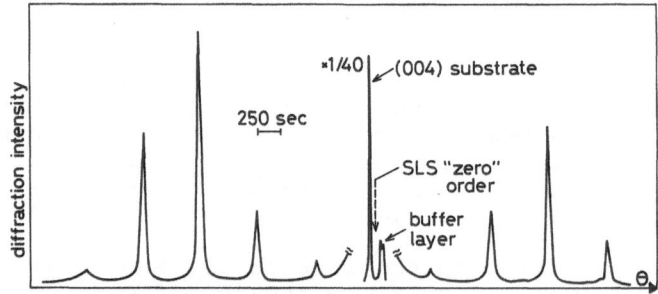

Fig. 4. X-ray double diffraction profile obtained on a
10 period $In_{0.41}Ga_{0.59}As(L_z = 15.8$ nm)/
$In_{0.64}Al_{0.36}As(L_b = 14.2$ nm) strained layer
superlattice near 004 reflection. The position
of the "zero order" diffraction peak of the SLS
shows that the whole structure is lattice
matched to InP, since the strains are
compensated.

638

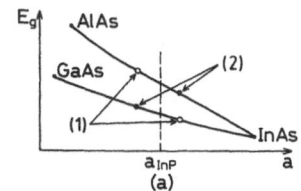

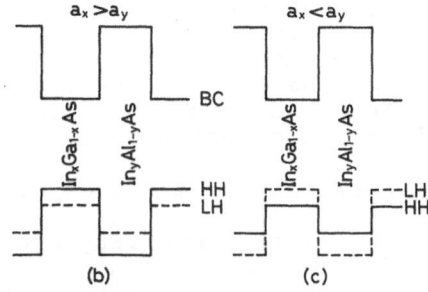

Fig. 5. The two possible strain configurations (a) and
the resulting band extrema line-ups (b), (c) for
InGaAs/InAlAs MQW lattice matched to InP. Cases
(1) and (2) in (a) refer to (b) and (c) respectively.

(high T_0). The idea is that, since the top of the valence band
is then light hole-like in the in-plane directions, the population inversion is obtained with lower carrier density (lower
threshold current); besides, the latter effect, plus the
increased splitting of the valence subbands are responsible for
the reduction of CHCC Auger and IVBA losses (see Ref. 7), thus
resulting in improved T_0.

Very low threshold current density has recently been
obtained in the InGaAlAs/InGaAs system, both by MOVPE with InP
cladding layers[9], and by MBE, with InAlAs cladding layers,
i.e. a phosphorus free structure[10]. In this latter case, we
have recently obtained threshold current density as low as 820
A/cm^2, and achieved CW operation on a buried ridge structure
involving MOCVD regrowth with current as low as 17 mA[11].

Strained MQW in the InGaAlAs/InGaAs system are expected to
improve the device more than in the InGaAsP/InP one since:

*   Compression in the wells can easily be compensated for by
        tension in the barriers, as shown above.

*   In order to keep a given laser emission wavelength, InGaAs
        QW's under biaxial compression, due to the indium rich
        composition, must be narrower than unstrained ones (see
        Fig. 3). Due to the higher conduction offset at the
        InGaAlAs/InGaAs interface, however, relatively wider wells
        are required in that system; for instance, taking barriers
        with equivalent band gaps $\lambda = 1.1$ μm, 1.5 μm emission is
        obtained with 8 nm and 9.5 nm with InGaAsP and InGaAlAs
        barriers respectively. This is favorable for two reasons
        (Ref. 7): i) the inversion population is inversely propor-

tional to L_z, ii) the optical confinement factor decreases like L_z^2.

*   Better confinement in the conduction band decreases electron coupling between wells, thus resulting in negligible gain broadening.

4.2 High Electron Mobility Transistors

Since the extremely high transconductance obtained by Hirose et al.[12] on InAlAs/InGaAs HEMT's grown on InP by MBE, numerous publications have shown the superiority of the InAlAs/InGaAs system over the GaAlAs/GaAs combination. Recently, very low noise W-band HEMTs have been demonstrated[13], using lattice-matched InGaAs/InAlAs/InP.

These remarkable performances are due to:

*   The relatively high electron mobility and peak drift velocity in the $In_{0.53}Ga_{0.47}As$.

*   The high conduction band discontinuity at the interface (the sheet density is roughly proportional to the square root of this discontinuity).

*   The high doping efficiency of Si in InAlAs.

*   The good quality of Schottky barriers on InAlAs; the barrier height is 0.62 ± 0.05[14].

Ng et al.[15] have analysed both theoretically and experimentally the effect of an increase of In concentration in the InGaAs channel. They have shown that by varying the In from 53% to 65%, the electron effective mass is reduced by 6% and the conduction band discontinuity is increased by 14%; the occupation of the lowest subband level is increased by about 30%, whereas the total sheet concentration is little affected. The measured mobility enhancement is about 13% at room temperature. Maximum intrinsic transconductances measured on 1.5 μm gate HEMTs[16] show improvement by 47% (191 to 359 mS/mm) when In concentration is varied from 53 to 65%.

4.3 Resonant Tunneling Structures (RTS)

The important criteria for this device are the current peak-to-valley ratio and the peak current density. By far the best results have been obtained, especially at room temperature, with the InAlAs/InGaAs system on InP substrate. The reasons for this are[17]:

*   Lower effective mass in InAlAs ($0.075m_0$) compared to $Al_{0.3}Ga_{0.7}As$ ($0.092m_0$); the incidence of this parameter was shown by Inata et al.[18].

*   Increased barrier height with a direct band gap material (enhancement of peak-to-valley ratio).

*   Larger mobility in the InGaAs well (for high frequency operation).

*   Easy low resistance nonalloyed contacts on InGaAs.

        Good results have been reported with lattice-matched
structures. Peak-to-valley ratios of 7 at room temperature and
21 at 77 K, with a current density of 10 kA/cm2 have been
reported in Ref. 17. We have recently obtained a peak-to-valley
ratio as high as 14 at room temperature on similar devices,
which is the best reported data for a lattice-matched system.

        Impressive improvement of RTS performances has been
obtained with the same system using highly strained material in
the barriers. In 1986, Inata et al.[18] improved the peak-to-
valley ratio to 14 at 300 K, due to the enhancement of the
conduction band gap discontinuity. This result was recently
increased again by Mehdi et al.[17] to 24, for a peak density as
high as 15 kA/cm2. An even higher peak-to-valley ratio (30 at
300 K, 63 at 77 K, 5 kA/cm2) was obtained by Broekaert et
al.[19] who used in addition a strained InAs well inside the
$In_{0.53}Ga_{0.47}As$ well region, thus allowing a decrease of its
thickness while keeping the same energy level in it.

4.4 Modulators

        Modulators based on the quantum confined Stark effect[20]
have been demonstrated in the 1.5 μm wavelength range using
either ternary and quaternary wells in both InAlAs/InGaAs[21]
and InP/InGaAs systems, lattice matched to InP. Variational
calculations show that for a given electric field the energy
shift of an exciton resonance is proportional to the fourth
power of the well width[22], so that efficient modulation needs
relatively thick wells. This favors the InAlAs/InGaAs
combination compared to InP/InGaAs, since the same operating
wavelength requires narrower wells in that latter system (see
the section above on lasers). In order to increase even more
the well's thickness, for operation in the 1.5 μm range, Bigan
et al.[23] used strained $In_{0.49}Ga_{0.51}As$ material in 10 nm wide
wells; opposite strain was applied in the barriers ($In_{0.6}Al_{0.4}As$,
L_b = 5 nm), so that the 40 periods MQW was lattice matched as a
whole to the InP substrate. An excitonic shift as large as 60
nm was observed for 10 V reverse bias. Under waveguiding
configuration, an 18 dB extinction has been measured for 6 V
reverse bias at 1.55 μm operating wavelength, with a 0.16 mm
device. With unstrained material, operation at the same
wavelength, needs 8 nm wells (instead of 10 nm), leading to an
exciton resonance shift three times smaller (20 nm instead of 60
nm).

SUMMARY

        During the past years, several teams have shown the
interest of the InP lattice-matched InGaAlAs system for
microelectronics applications. More recently, the field of
optoelectronics was also successfully covered by this phosphorus
free combination. It has been proven that pseudomorphic
material was able to enhance the ultimate performances obtained
with lattice-matched material, without drawback due to strains,
as long as the layers are thin enough not to generate misfit
dislocations.

REFERENCES

1. P. Chu, C.L. Lin and H.H. Weider, **Appl. Phys. Lett.**, 53:2423 (1988)
2. A.S. Brown, U.K. Mishra, C.S. Chou, C.E. Hooper, M.A. Melendes, M. Thompson, L.E. Larson, S.E. Rosenbaum and M.J. Delaney, **IEEE Electron. Dev. Lett.**, 10:565 (1989)
3. J.P. Praseuth, M.C. Joncour, J.M. Gérard, P. Hénoc and M. Quillec, **J. Appl. Phys.**, 63:400 (1988)
4. J.M. Gérard, J.M. Marzin and J. Primot, **Journ. de Physique**, C5:48 (1987); C5:169 (1987)
5. M. Quillec, L. Goldstein, G.LeRoux, J. Burgeat and J. Primot, **J. Appl. Phys.**, 55:2904 (1984)
6. J.Y. Marzin, J.M. Gérard, P. Voisin and J.A. Brum, Optical Studies of Strained III-V Heterolayers, to be published in **Semiconductor and Semimental**, 32:55, T.P. Pearsall, ed., Acad. Press.
7. A.R. Adams, K.C. Heasman and E.P. O'Reilly, **in**: "Band Structure Engineering in Semiconductor Microstructures", R.A. Abram and M.Jaros, eds., published in cooperation with NATO SAD by Plenum Press, series B, vol. 189.
8. P.J.A. Thijs and T. Van Dongen, **Electron. Lett.**, 25:1735 (1989)
9. R.W. Glew, B. Garret and P.D. Greene, **Electron. Lett.**, 25:1103 (1989)
10. M. Quillec, M. Allovon, F. Brillouet, A. Gloukhian, J.P. Praseuth and B. Sermage, **Electron. Lett.**, 25:1731 (1989)
11. K. Kazmierski, M. Blez, M. Quillec, M. Allovon and B. Sermage, **Electron. Lett.**, 26:889 (1990)
12. K. Hirose, K. Ohata, T. Mizatani, T. Itoh and M. Ogawa, Chap.10:529 **in**: "Institute of Physics Conference Series No. 79", Adam-Hilger, Bristol, UK (1986)
13. P.C. Chao, A.J. Tessmer, K-H. G. Duh, P. Ho, M-Y. Kao, P.M. Smith, J.M. Ballingall, S-M.J. Liu and A.A. Jabra, **IEEE Electron. Device Lett.**, 11:59 (1990)
14. P. Chu, C.L. Lin and H.H. Wieder, **Appl. Phys. Lett.**, 53:2423 (1988)
15. G.I. Ng, D. Pavlidis, M. Quillec, Y.J. Chan, M.D. Jaffa and J. Singh, **Appl. Phys. Lett.**, 52:728 (1988)
16. Y.J. Chan, D. Pavlivis, G.I. Ng, M.Jaffe, J. Singh and M. Quillec, proceedings of IEDM (1987)
17. I. Mehdi and G. Haddad, **J. Appl. Phys.**, 67:2643 (1990)
18. T. Inata, S. Muto, Y. Nakata, T. Fujii, H. Ohniski and S. Hiyamitsu, **Jpn. J. Appl. Phys.**, 25:L983 (1987)
19. T.P.E. Broekaert, W. Lee and C.G. Fonstad, **Appl. Phys. Lett.**, 53:1545 (1988)
20. D.A.B. Miller, D.S. Chemla, T.C. Damen, A.C. Gossard, W. Wiegmann, T.H. Wood and C.A. Burrus, **Phys. Rev. B**, 32:1043 (1985)
21. K. Wakita, I. Kotaka, O. Mitomi, Y. Kawamura and O. Mikami, 7th IOOC 89, Technical Digest, Paper 19C2-2.
22. G. Bastard, E.E. Mendez, L.L. Chang and L. Esaki, **Phys. Rev.B**, 28:3241 (1983)
23. E. Bigan, M. Allovon, M. Carre and A. Carenco, **Electron. Lett.**, 26:355 (1990)

MATERIALS, DEVICE AND CIRCUIT PROPERTIES OF AlInAs-GaInAs HEMTs

Umesh K. Mishra* and April S. Brown

Hughes Research Laboratories
Malibu, Ca., USA

1. INTRODUCTION

GaInAs based HEMTs are superior for high frequency applications compared to GaAs based devices because of the better electronic transport properties in GaInAs. The electron mobility is 13,000 $cm^2V^{-1}s^{-1}$, the peak velocity is 2.7 x $10^7 cms^{-1}$ and the conduction band discontinuity between $Al_{.48}In_{.52}As$ and $Ga_{.47}In_{.53}As$ is 0.51 eV, all higher than the GaAs system. The large conduction band discontinuity coupled with the high doping possible in $Al_{.48}In_{.52}As$ ($N_D > 1$ x $10^{19}cm^{-3}$) provides for large two-dimensional electron gas concentrations in modulation doped structures. Furthermore, as in the GaAs based system, the electronic properties can be enhanced by incorporating a strained In rich channel in device design. These materials properties have been utilized to produce discrete HEMTs with $f_T > 200$ GHz, ring oscillators with room temperature gate delay of 4ps and frequency dividers operating at 26.7 GHz.

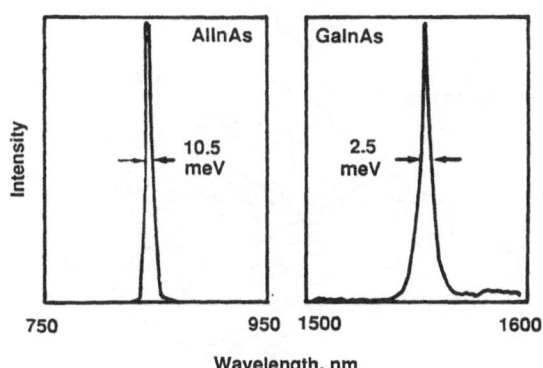

Fig. 1. Photoluminescence spectra of $Ga_{.47}In_{.53}As$ and $Al_{.48}In_{.52}As$ grown at the Hughes Research Laboratory.

---

*Now at University of California, Santa Barbara, CA 93106, USA.

2.   MATERIALS GROWTH AND CHARACTERIZATION

High quality $Ga_{.47}In_{.53}As$ in the channel and $Al_{.48}In_{.52}As$ in the cladding layers is necessary for well behaved, high performance devices.   Low temperature photoluminescence and Hall measurements were used to determine the optical and electronic properties of both materials.   Fig. 1 shows the photo-luminescence obtained from GaInAs and AlInAs grown at 500°C at a growth rate of 0.75 μm/hr.   The linewidth of 10.5 meV obtained from AlInAs is the best reported but is still far from the theoretical prediction of 4 meV.   The quality of the AlInAs is primarily determined by the miscibility gap in the AlInAs phase diagram[1] and the volatility of temperature (< 550°C) and a high V/III beam equivalent pressure ratio (P_{be}(V/III) ~ 100)[2]. These growth conditions limit the cation surface mobility which degrades the quality of the bulk AlInAs and the AlInAs/GaInAs heterointerface.   A way to reduce interface roughness and improve surface morphology without increasing surface temperature is to use vicinal substrates.   We studied the effect of growth on vicinal substrates on the quality of the component materials and the interfaces for both lattice matched and pseudomorphic cases[3].   This was done by analyzing AlInAs-GaInAs quantum well photoluminescence (PL) and relating the narrower full widths at half maximum of the excitonic transitions to smoother heterojunction interfaces.   For quantum wells 2.0-8.0 nm thick, the FWHM for the PL peak is determined

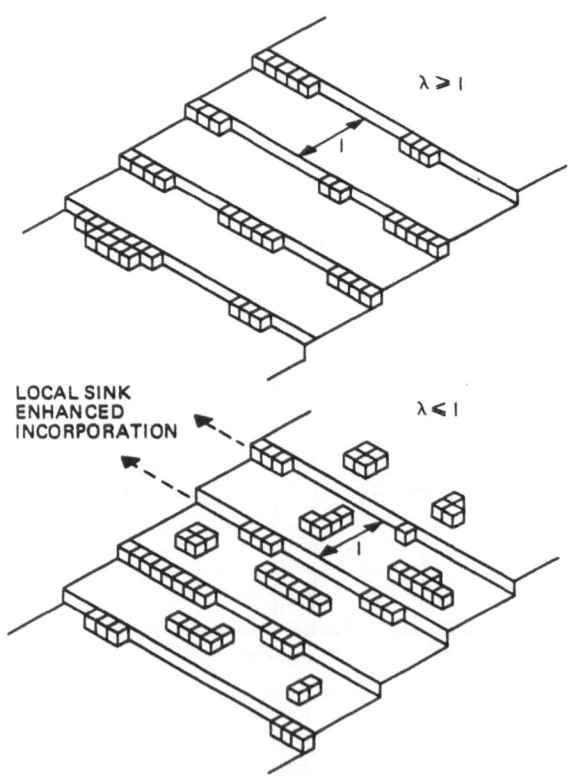

Fig. 2. Growth dependence on step dimension on vicinal
        substrates.

primarily by the heights of the islands at the interface, assuming that the islands are greater than or equal to 20 nm in lateral extent. The study showed that by using InP substrates misoriented 2 degrees off the (100) InP surface, increased disorder is induced in the AlInAs alloy, which will result in a rough heterojunction interface. TEM micrographs verify that significant disordering occurs in the AlInAs with Al-rich and In-rich regions propagating in regions of approximately 30 nm periodicity when growth occurs on the 2°-off substrates[4]. It is believed that it results from the disparity in diffusion lengths of Al and In atoms. For 2°-off substrates, the misorientation produces terraces of 8.0 nm length on the surface. If the In atoms have a diffusion length of 8.0 nm or more they will diffuse to the step edge and incorporate. The Al atoms, on the other hand, probably do not have a large enough diffusion length to diffuse to the step edge. Thus, In-rich regions can be induced at the step edges and the strain associated with these regions can induce clustering throughout the epitaxial layer. For a 4°-off substrate the terrace length is 4.0 nm and it is probable that both the Al and the In atoms will be mobile enough to diffuse to the step ledges and incorporate. This uniform array of good incorporation sites will aid in promoting layer growth that produces high quality interfaces. The degree to which these effects will arise also depends on the differences in the bond configurations and strains associated with different atomic configurations at the step edges. The most dramatic clustering occurs for the 2°-off substrate misoriented towards the (111)-P face and the smoothest interface occurs for the 4° towards the (111)-In face. A schematic description of the hypothesis and the resulting differences in optical material quality are presented in Figs. 2 and 3.

3.   HEMT DEVICE AND CIRCUIT PERFORMANCE

On the basis of these materials studies, the epitaxial layer design shown in Fig. 4 was chosen for fabricating AlInAs-GaInAs HEMTs. In a previous study[5], it was determined that the poor structural and optical properties obtained for material grown on substrates oriented 2°-off the (100) were reflected in the poor D.C. and R.F. performance of HEMTs. The substrate chosen was therefore either (100) or 4°-off Fe-doped InP. The growth temperature was 500°C. First, a 250 nm thick AlInAs buffer was grown, followed by a 40 nm thick AlInAs-GaInAs superlattice. The AlInAs layer served the dual purpose of separating the active layer from the InP substrate and also providing a high band gap layer which confines the electrons to the active GaInAs channel, which reduces the HEMT output conductance. The superlattice also serves two other important purposes[6]. It smooths the AlInAs growth front, minimizing interface roughness scattering resulting in higher electron mobilities in the channel. Also, the superlattice can getter outdiffusing impurities from the substrate which further improves the transport properties of the active channel. Next, a 40 nm thick GaInAs channel was grown, followed by a 2 nm undoped AlInAs spacer. The AlInAs layer was then grown, 12.5 nm thick, and doped with Si at 4×10^{18} cm^{-3}. This was followed by a 20 nm thick undoped AlInAs layer which serves as a Schottky barrier enhancing layer and the structure was capped by a 5 nm thick n^+ GaInAs contact layer. Both lattice matched $Ga_{47}In_{53}As$ and pseudomorphic $Ga_{.38}In_{.62}As$ channels were used. The major

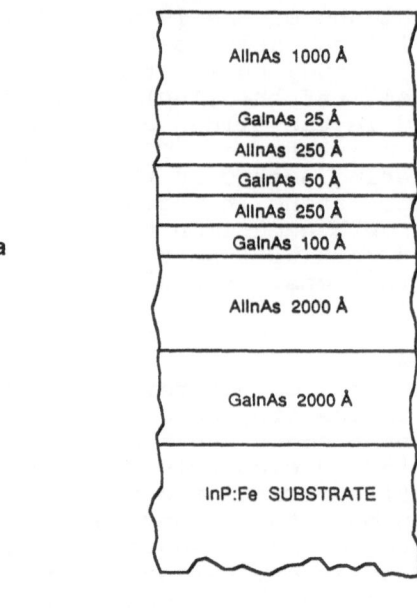

a

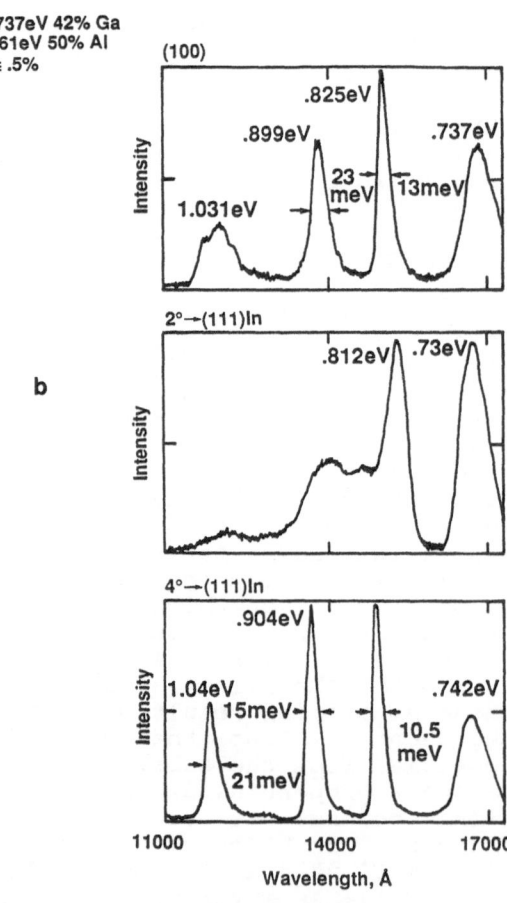

b

Fig. 3. Schematic of (a) quantum well structure
studied and (b) the PL spectra.

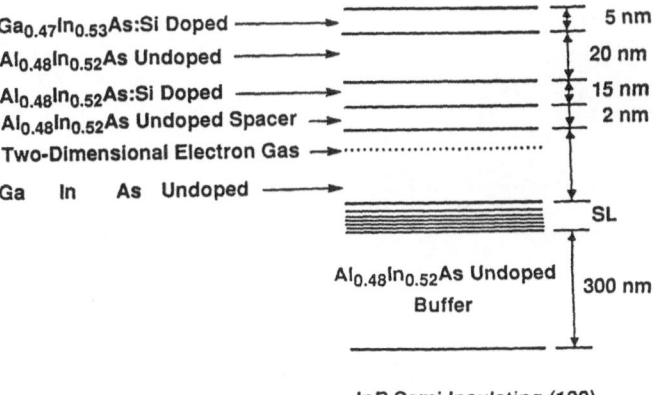

Fig. 4. Schematic cross-section of AlInAs-GaInAs
structure.

differences in the epitaxial layer design between the lattice
matched wafer (L) and the pseudomorphic wafer (P) were:

(i)   the indium mole fraction was ~ 53% in the lattice matched
      case and 62% in the pseudomorphic wafer and,

(ii) the thickness of the GaInAs channel; 40 nm in wafer L and
      27 nm in wafer P.

    The reduced thickness in wafer P ensured that the critical
thickness for the formation of dislocations was not exceeded.
The properties of the two-dimensional gas (2DEG) are a strong
function of the nature and magnitude of the Si doping in the
AlInAs, the width and composition of the $Ga_{.47-u}In_{.53+u}As$
conductive channel and the spacer layer thickness. As the
doping in the AlInAs donor layer is increased, the number of
electrons available for transfer into the GaInAs to form the
2DEG is increased. The available electrons are shared between
the AlInAs-GaInAs heterojunction at the channel-donor interface

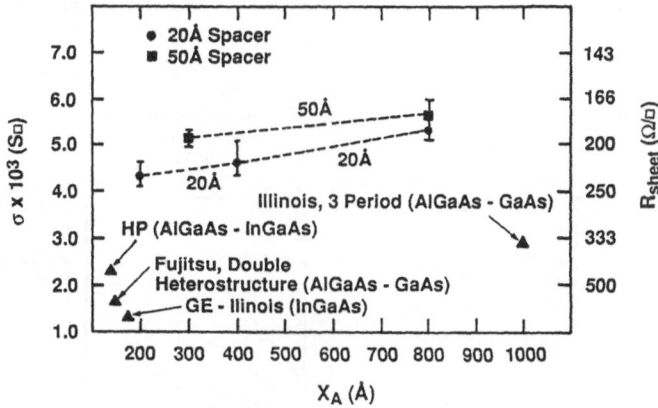

Fig. 5. Effect of epitaxial layer design on the
conductivity of the two-dimensional gas.

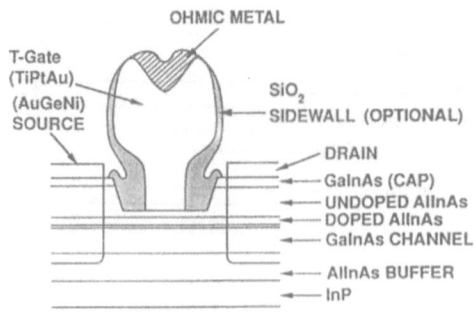

OHMIC METAL

T-Gate
(TiPtAu)
(AuGeNi)
SOURCE

SiO₂
SIDEWALL (OPTIONAL)

DRAIN
GaInAs (CAP)
UNDOPED AlInAs
DOPED AlInAs
GaInAs CHANNEL

AlInAs BUFFER
InP

Features
• L$_g$ = 0.15 μm
• L$_{SD}$ = 0.45 μm
• Passivated Surface (with Sidewall)
• Lowered Gate Resistance

Fig. 6. Schematic cross-section of a self-aligned
AlInAs-GaInAs HEMT.

and the donor-contact layer interface. The relative distribution is determined by; (i) the distance of the doped region
from the two interfaces and (ii) the nature of the doping, i.e.
uniform or planar. The width of the channel affects the 2DEG
mobility through the interface roughness scattering at the
inverted AlInAs-GaInAs interface. Lastly, as has been stated
before, increasing the indium mole fraction in the channel
increases both the electron affinity in the channel (and hence
ns) and the electron mobility. The most important and easily
measurable parameter which characterizes the channel is the
conductivity of the 2DEG, which governs the parasitic
resistances of the device. Figure 5 summarizes the effect of
the various parameters on the conductivity (or resistivity) of
the two-dimensional electron gas and illustrates the superiority
of the AlInAs-GaInAs system over the AlGaAs-GaAs and the AlGaAs-
InGaAs systems[7].

    The second major factor that determines device performance
other than the quality of the epitaxial material is the nature
of the device fabrication. The two approaches are (i) the non-

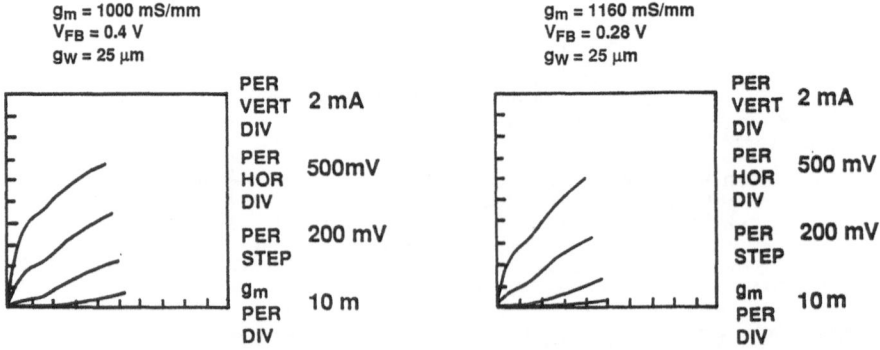

Fig. 7. I-V characteristics of 0.1 μm x 25 μm devices.

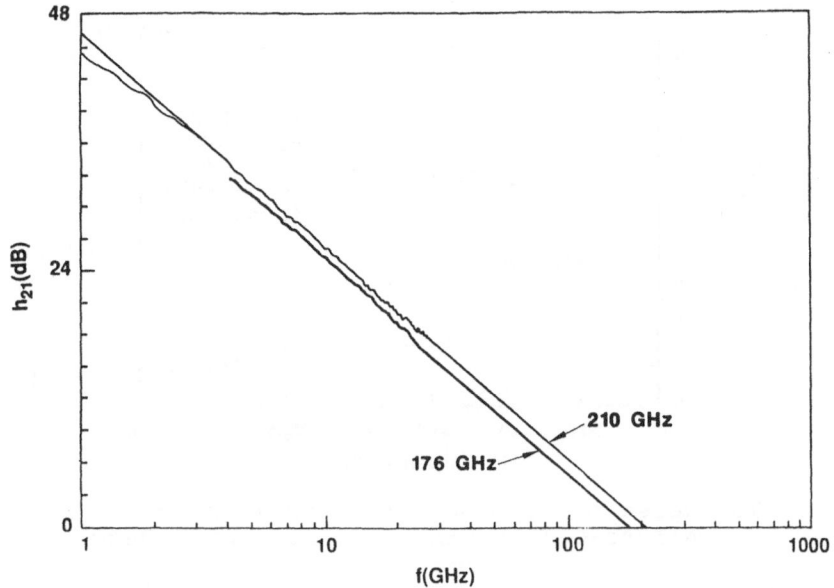

Fig. 8. Current gain versus frequency of self-aligned
HEMTs with 0.15 μm gate length.

self-aligned and (ii) the self-aligned approaches. In the
former the gate is placed between the source and the drain of
the device within the realignment tolerances of the lithographic
tool that is used. In the self-aligned approach the gate is
defined first and the source-drain electrodes are evaporated in
a self-aligned fashion to the gate. Figure 6 shows the

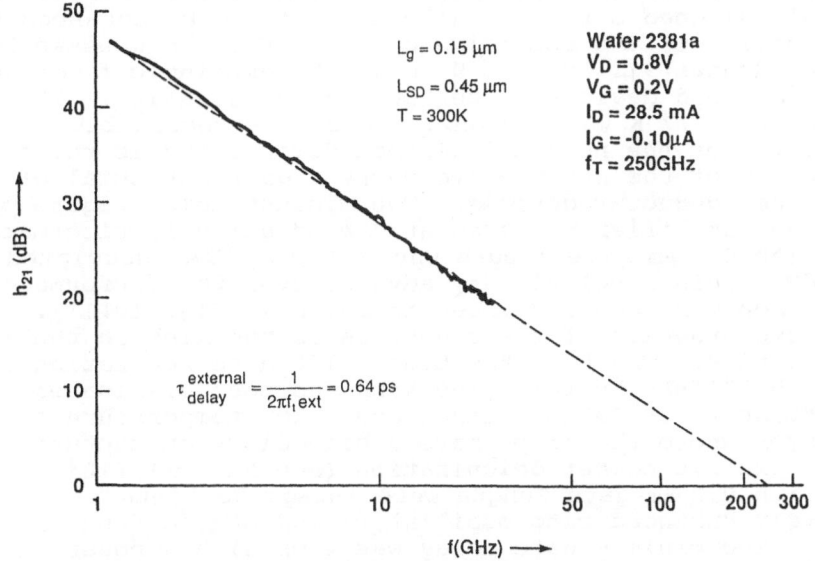

Fig. 9. Current gain versus frequency of self-aligned
HEMTs with 0.15 μm gate length.

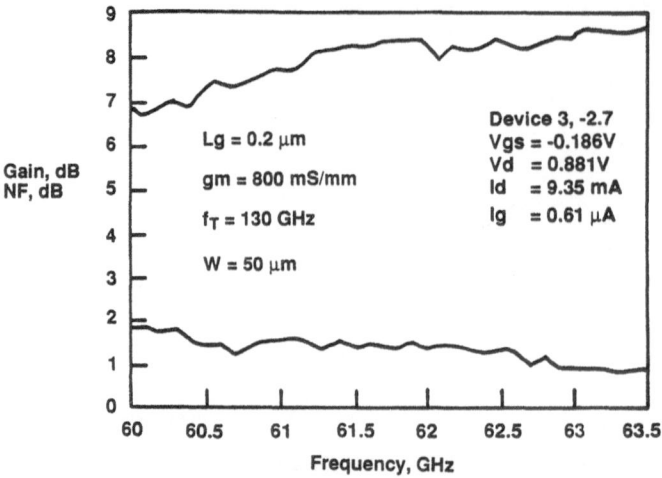

Fig. 10. Noise performance of a single stage
AlInAs-GaInAs HEMT amplifier at V Band.

schematic cross-section of such a device. The I-V characteristics of typical 0.1 μm x 25 μm wide lattice matched and
pseudomorphic HEMTs employing a non-self-aligned technology is
shown in Fig. 7. The extrinsic transconductance of the devices
were 1080 mS/mm and 1160 mS/mm respectively. Figure 8 shows the
current gain versus frequency of 200 μm wide devices from wafers
L and P. The extrapolated f_T were 176 GHz and 210 GHz
respectively[8]. The f_T measured on devices with 50 μm gate
length was ~ 150 GHz. The reason for the reduced f_T in 50 μm
wide devices is that the parasitic pad capacitance is a larger
fraction of the total input capacitance and hence reduces the
extrinsic f_T of the device. The reduced parasitic resistances
in the self-aligned structure allows the f_T to be enhanced by
the elimination of the resistive divider effect discussed by
Tasker and Hughes[9]. Figure 9 shows the extrinsic f_T of 50 μm
wide self-aligned devices to be 250 GHz at 300 K[10]. The
extrinsic f_T at 77 K was 292 GHz. This is a remarkable
improvement over the non-self-aligned devices and is the first
demonstration of the sub mm-wave current gain potential of three
terminal semiconductor devices. The minimum noise figure of a
single stage amplifier measured at V band using lattice-matched
devices with 0.2 μm gate length was 0.8 dB. The associated gain
was 8.7 dB. This revolutionary advance over the performance
obtained from GaAs based devices is shown in Fig. 10[11]. The
most serious drawback of these devices is the kink in the I-V
characteristics. The kink has been related to ionization of
traps in the AlInAs buffer. The kink has been eliminated by
using a "trap free" GaInAs buffer and a low temperature AlInAs
buffer layer where the traps have a high electron capture rate
and hence exhibit no net deionization (see Fig. 11)[12].
Devices with 0.2 μm gate length were integrated into
capacitively enhanced ring oscillators and static frequency
dividers. The minimum gate delay was 4 ps with a power
dissipation of 1 mW per gate[13]. The highest operating
frequency for the dividers was 26.7 GHz[4].

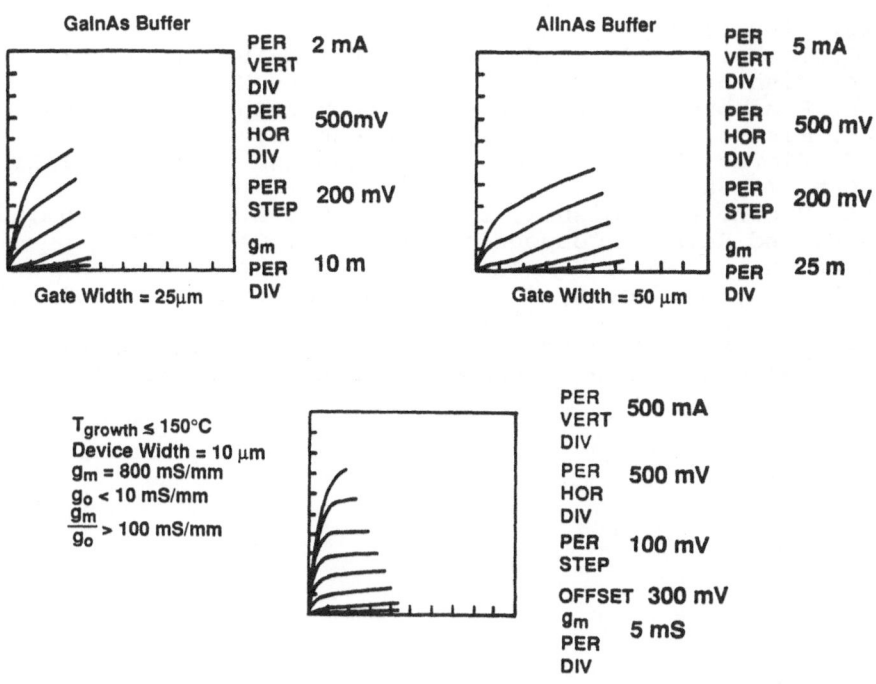

Fig. 11. Dependence of kink in the I-V characteristics
on the traps in the buffer.

CONCLUSIONS

AlInAs-GaInAs HEMTs fabricated with epitaxial material
grown either lattice matched or strained on InP substrates have
shown remarkable analog and digital performance. The full
realization of the incredible potential that these devices offer
will be realized with the improvement of the AlInAs material
quality, device isolation technologies and general maturing of
the device and circuit processing. We believe that the needs
which these devices will serve are primarily in the ultra low
noise and high speed areas.

REFERENCES

1.   J. Singh, S. Dudley, B. Davies and K.K. Bajaj, **J. Appl.
     Phys.** 60:3167 (1986)
2.   S. Chika, H. Kato, M. Makayama and N. Sano, **JJAP** 25:1441
     (1986)
3.   A.S. Brown, U.K. Mishra, J.A. Henige and M.J. Delaney, **J.
     Appl. Phys.**, 64:3476 (1988)
4.   Y.P. Hu, P.M. Petroff, X. Qian and A.S. Brown, **Appl. Phys.
     Lett.**, 53:2194 (1988)
5.   A.S. Brown, U.K. Mishra and S.E. Rosenbaum, **IEEE Trans. on
     Electron Dev.**, 36:641 (1989)
6.   A.S. Brown, J.A. Henige and M.J. Delaney, **Appl. Phys.
     Lett.**, 52:1142 (1988)
7.   A.S. Brown, U.K. Mishra, J. Henige and M.J. Delaney, **J.
     Vac. Sci. Technol.**, B6:678 (1988)
8.   U.K. Mishra, A.S. Brown and S.E. Rosenbaum, **IEDM Tech.**

Dig., 180 (1988)

9.   P.J. Tasker and Brian Hughes, **IEEE EDL-10** 291 (1989)
10. U.K. Mishra, A.S. Brown, L.M. Jelloian, M. Thompson,
     S.E. Rosenbaum and L.D. Nguyen, **IEDM Tech. Dig.**, (1989)
11. U.K. Mishra, A.S. Brown, S.E. Rosenbaum, C.E. Hooper,
     M.W. Pierce, M.J. Delaney, S. Vaughn and K. White, **IEEE
     EDL-9**, (1988)
12. A.S. Brown, U.K. Mishra, L. Larson and S.E. Rosenbaum, **in**:
     "GaAs and Related Compounds", Atlanta, GA. (1988)
13. A. Brown, C. Chou, M. Delaney, C. Hooper, J. Jensen,
     L. Larson, L. Nguyen, M. Thompson and U. Mishra, **GaAs IC
     Symposium Tech. Dig.**, (1989)
14. U.K. Mishra, J.F. Jensen, A.S. Brown, M.A. Thompson,
     L.M. Jelloian and R.S. Beaubien, **IEEE EDL-9**, 482 (1988)

STRAIN INDUCED BAND STRUCTURE MODIFICATIONS IN SEMICONDUCTOR
HETEROSTRUCTURES AND CONSEQUENCES FOR ELECTRONIC AND OPTICAL
DEVICES

Jasprit Singh

Center for High Frequency Microelectronics
Department of Electrical Engineering
and Computer Science
The University of Michigan, Ann Arbor
MI 48109-2212, USA

ABSTRACT

Coherent strain produced in lattice mismatched epitaxy can
be exploited to alter the band structure of semiconductor
heterostructures to a significant extent. The dominant effect
of strain is to lift certain degeneracies in the Brillouin zone,
alter the density of states at band edges and the band gaps.
These effects can have significant effects on transport and
optical properties of devices. This will be illustrated by
examining the role of strain in electrical properties of SiGe
strained alloys and on the performance of strained quantum well
lasers.

1.   INTRODUCTION

The effect of hydrostatic and uniaxial strain on the
electronic band structure of semiconductors has been an area of
active study for several decades. The information yielded by
these studies is valuable in developing a better theoretical
understanding of both the band structure as well as deformation
potential scattering of electrons and holes due to phonons.
However, until comparatively recently, there has been little
interest in studying device related issues such as optical
absorption/emission and charge carrier transport in the presence
of strain. This was because strain had to be introduced by some
external apparatus such as a high pressure diamond anvil cell.
This rendered the phenomena difficult to study, with the issue
remaining of little technological importance.

The development of lattice mismatched heteroepitaxy has
resulted in an increase in the study of strain related changes
in important physical processes. Theoretical and experimental
studies show that if a material with a bulk lattice constant a_L
is grown as a film on a thick substrate with lattice constant
a_S, it will grow epitaxially, with an in-plane lattice constant
of a_S and an adjustment, via the Poisson effect, in the

perpendicular lattice constant[1-4]. This pseudomorphic growth
continues up to a critical thickness determined by a balance
between strain and chemical energy. Beyond this thickness, the
overlayer relaxes, producing dislocations, and the in-plane
lattice constant of the film reverts to its bulk value, a_L[1-4].
For film thicknesses less than the critical thickness a large
strain can be produced in the film, which can greatly change its
band structure, both by changing effective masses and lifting
degeneracies[5-8]. Since the pseudomorphic layer is
thermodynamically stable, it is possible to fabricate semiconductor devices with strained layer components. The strain
induced band structure changes may lead to increased charged
carrier mobility within the pseudomorphic layers. This, in
turn, becomes a useful way to increase the speed of semiconductor device operation[9-12]. Strain may also play an
important role in optical devices such as lasers since the laser
properties are strongly coupled to the band structure[13-16].

In this paper we will examine the effect of the strain
achievable in strained epitaxy on band structures of direct
(e.g. GaAs) and indirect (e.g. Si) band gap semiconductors.
More importantly, we will examine the consequences of the strain
on transport and optical properties as well as on performance of
devices. In the next section we will examine the band structure
related changes. In Section 3 we will examine the effect of
strain on shallow impurity levels. In Section 4 we will study
the consequences of strain on transport of carriers at low and
high fields. The effect of strain on quantum well lasers will
be examined in Section 5. Discussions and conclusions will be
presented in Section 6.

This paper will focus only on the theoretical aspects of
the problem. Many of the results calculated and shown here have
been verified, but some, like the effect of strain on shallow
levels and transport in strained layers, have not yet been fully
explored.

2.   STRAIN INDUCED BAND STRUCTURE CHANGES

The crystalline structure of essentially all important
semiconductors consists of an FCC lattice with a two atom basis.
This leads to either a diamond crystal structure (when the two
basis atoms are identical) or zinc blende structure. This
underlying structure has important consequences for the band
structure of the semiconductors. We will briefly recall some
important symmetry issues resulting from the crystalline
structure since an understanding of the symmetries is crucial to
understanding the role of strain.

2.1 Conduction Band Edge States

In direct band gap semiconductors such as GaAs, InAs, etc.
the conduction band minimum is at the zone center and the
central cell part of the Bloch function (i.e. the u_{nk} in $\psi_k \sim$
$u_{nk} \exp(i\mathbf{k.r})$) has spherical symmetry since it is made up of s-
type functions as shown in Fig. 1. This state has only spin
degeneracy. As one moves away from the conduction band edge,
small amounts of p-type states are mixed with s-type states
leading to nonparabolicity effects. As we shall note below,
due to the s-type nature of the conduction band edge, strain
only shifts the edge without any splitting of states.

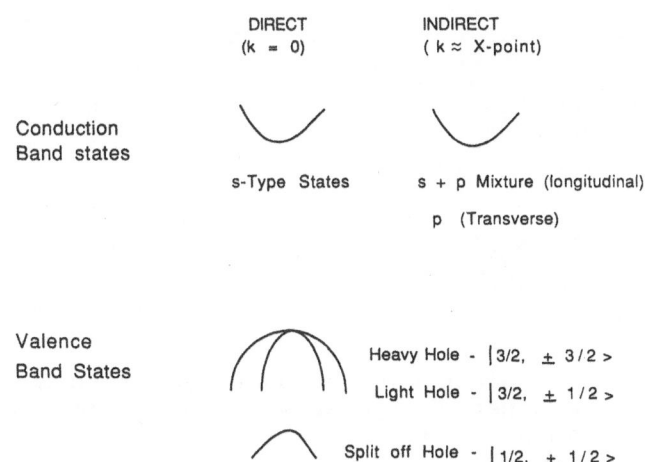

Fig. 1. A schematic showing the symmetry of the
states describing the important states in
direct and indirect band gap materials.

For indirect band gap materials such as Si (with conduction
band edge near the X-point with 6-fold degeneracy) and Ge (L-
point conduction band edge with 4-fold degeneracy), the eigenstates have strong directional anisotropies. These anisotropies
by themselves do not lift the degeneracies of the unstrained
material, but may cause degeneracy liftings in presence of the
strain tensor produced in strained epitaxy.

2.2 Valence Band Edge States

The valence band edge states of all semiconductors are
essentially similar. At the zone center one has a doubly
degenerate state corresponding to the heavy hole (HH) and light
hole (LH) band while the split-off (SO) state is present at an
energy given by the spin-orbit coupling. The character of the
HH, LH, SO states is given in the total angular momentum basis
by ($|\ \frac{3}{2},\ \pm\frac{3}{2} >$, $|\ \frac{3}{2},\ \pm\frac{1}{2} >$ and $|\ \frac{1}{2},\ \pm\frac{1}{2} >$) states respectively. The
spatial dependence of these states is given by p-type functions.
Figure 1 gives an illustrative example of the cases discussed by
us.

2.3 Strain Tensor in Lattice Mismatched Epitaxy

In order to study the effect of strain on electronic
properties of semiconductors, it is first essential to establish
the strain tensor produced by epitaxy. The work of Frank and
van der Merwe[1,2] and of Matthews and Blakeslee[3] demonstrated
that careful growth of an epitaxial layer whose lattice constant
is close, but not equal, to the lattice constant of the
substrate can result in a coherent strain, as opposed to
polycrystalline or amorphous incoherent growth. If the strain
is incorporated into the epitaxial crystal coherently, the
lattice constant of the epitaxial layer in the directions
parallel to the interface is forced to be equal to the lattice
constant of the substrate. The lattice constant of the
epitaxial layer perpendicular to the substrate will be changed
by the Poisson effect. If the parallel lattice constant is
forced to shrink, or a compressive strain is applied, the

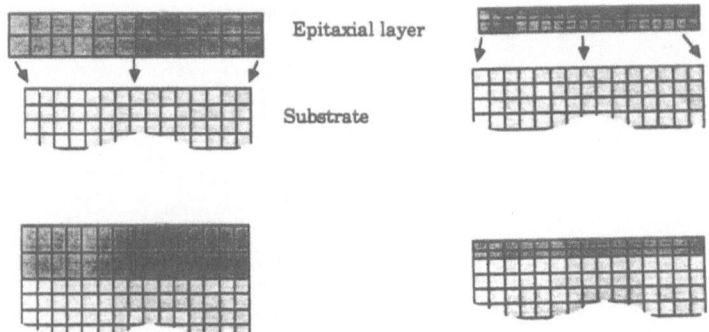

Fig. 2. Pseudomorphic growth on overlayer with a
lattice constant larger (a) and smaller (b)
than the substrate.

perpendicular lattice constant will grow. Conversely, if the
parallel lattice constant of the epitaxial layer is forced to
expand under a tensile strain, the perpendicular lattice
constant will shrink. These two cases are depicted in Fig. 2.
This type of coherently strained crystal is called
pseudomorphic.

For systems of interest in the present work, the epitaxial
semiconductor layer is biaxially strained in the plane of the
substrate, by an amount $\epsilon_\parallel$, and uniaxially strained in the
perpendicular direction, by an amount $\epsilon_\perp$. For a thick
substrate, the in-plane strain of the layer is determined from
the bulk lattice constants of the substrate material, a_S, and
the layer material, a_L:

$$\epsilon_\parallel = a_S/a_L - 1 = \epsilon . \tag{1}$$

Since the layer is subjected to no stress in the perpendicular
direction, the perpendicular strain $\epsilon_\perp$, is simply proportional
to $\epsilon_\parallel$:

$$\epsilon_\perp = -\epsilon_\parallel/\sigma \tag{2}$$

where the constant σ is known as Poisson's ratio.

Noting that there is no stress in the direction of growth
it can be simply shown that for the strained layer grown on a
(001) substrate[18]:

$$\sigma = c_{11}/2c_{12}$$

$$\epsilon_{xx} = \epsilon_\parallel, \ \epsilon_{yy} = \epsilon_{xx}, \ \epsilon_{zz} = (-2c_{12}/c_{11})\epsilon_\parallel \tag{3}$$

$$\epsilon_{xy} = 0, \ \epsilon_{yz} = 0, \ \epsilon_{zx} = 0$$

while in the case of a strained layer grown on a (111)
substrate[18]:

$$\sigma = \frac{c_{11} + 2c_{12} + 4c_{44}}{2c_{11} + 4c_{12} - 4c_{44}}$$

$$\epsilon_{xx} = \left[\frac{2}{3} - \frac{1}{3} \left(\frac{2c_{11} + 4c_{12} - 4c_{44}}{c_{11} + 2c_{12} + 4c_{44}} \right) \right] \epsilon_\parallel, \quad \epsilon_{yy} = \epsilon_{xx} = \epsilon_{zz},$$

$$\epsilon_{xy} = \left[\frac{-1}{3} - \frac{1}{3} \left(\frac{2c_{11} + 4c_{12} - 4c_{44}}{c_{11} + 2c_{12} + 4c_{44}} \right) \right] \epsilon_\parallel, \quad \epsilon_{yz} = \epsilon_{xy} = \epsilon_{zx}. \tag{4}$$

2.4 Consequences of Strain on Band Structure

Once the strain tensor is known, one can apply the deformation potential theory to calculate the effects of strain on various eigenstates in the Brillouin zone. The strain perturbation Hamiltonian is defined and its effects are calculated in the simple first order perturbation theory. In general we have[18-20]

$$H_\epsilon^{\alpha\beta} = \sum_{ij} D_{ij}^{\alpha\beta} \epsilon_{ij}$$

where D_{ij} is the deformation potential operator which transforms under symmetry operations as a second rank tensor. $D_{ij}^{\alpha\beta}$ are the matrix elements of D_{ij}.

We will summarize the effect of strain on band structure by examining various high symmetry points in the Brillouin zone as depicted in Fig. 1.

2.4.1 Conduction band edge for direct band materials: The effect of the strain is to cause a shift in the energy given by

$$\delta E^{(000)} = H^\epsilon = D_{xx}(\epsilon_{xx} + \epsilon_{yy} + \epsilon_{zz}) \tag{5}$$

where D_{xx} is conventionally denoted by $\Xi_d^{(000)}$.

2.4.2 Conduction band edge for X-point-like states: The effect of strain on the six valleys is given by

$$\delta E^{(100)} = \delta E^{(\bar{1}00)} = D_{xx} \epsilon_{xx} + D_{yy}(\epsilon_{yy} + \epsilon_{zz}) \tag{6}$$

Writing $D_{yy} = \Xi_d^{(100)} = D_{zz}$ and $D_{xx} = \Xi_d^{(100)} + \Xi_u^{(100)}$

$$\delta E^{(100)} = \delta E^{(\bar{1}00)} = \delta E' + \Xi_u^{(100)} \epsilon_{xx}$$

$$\delta E^{(010)} = \delta E^{(0\bar{1}0)} = \delta E' + \Xi_u^{(100)} \epsilon_{yy}$$

$$\delta E^{(001)} = \delta E^{(00\bar{1})} = \delta E' + \Xi_u^{(100)} \epsilon_{zz}$$

where $\delta E' = \Xi_d^{(100)}(\epsilon_{xx} + \epsilon_{yy} + \epsilon_{zz})$.

It is clear from the expression above that if a strained epitaxy occurs along the (001) direction (so that $\epsilon_{xx} = \epsilon_{yy} = \epsilon_\parallel$ and $\epsilon_{zz} = -\epsilon_\parallel/\sigma$), a splitting will occur between the four X-valleys (110, 010, $\bar{1}$00, 0$\bar{1}$0) in the x-y plane and the X valleys in the z-direction. This has important consequences for the Si-Ge strained alloy as discussed later.

2.4.3 Valence band structure: The effect of the strain on the valence band structure is somewhat more complicated since the eigenstates at the valence band edge are degenerate in k-space and E-space. A strain perturbation Hamiltonian for diamond lattices in the $|x\rangle$, $|y\rangle$, $|z\rangle$ basis has been developed by Bir and Picus[20].

Restricting ourselves to the HH and LH states, the strain Hamiltonian can be written as

$$
H_\epsilon = \begin{bmatrix}
H_{hh}^\epsilon & H_{12}^\epsilon & H_{13}^\epsilon & 0 \\
H_{12}^{\epsilon\,*} & H_{1h}^\epsilon & 0 & -H_{13}^\epsilon \\
H_{13}^{\epsilon\,*} & 0 & H_{1h}^\epsilon & H_{12}^\epsilon \\
0 & -H_{13}^{\epsilon\,*} & H_{12}^{\epsilon\,*} & H_{hh}^\epsilon
\end{bmatrix}
\tag{7}
$$

where the matrix elements are given by

$$
H_{hh}^\epsilon = a(\epsilon_{xx} + \epsilon_{yy} + \epsilon_{zz}) - b\left[\epsilon_{zz} - \frac{1}{2}(\epsilon_{xx} + \epsilon_{xx})\right]
$$

$$
H_{1h}^\epsilon = a(\epsilon_{xx} + \epsilon_{yy} + \epsilon_{zz}) - b\left[\epsilon_{zz} - \frac{1}{2}(\epsilon_{xx} + \epsilon_{xx})\right]
$$

$$
H_{12}^\epsilon = -\frac{\sqrt{3}}{2}b(\epsilon_{yy} - \epsilon_{xx}) - id\epsilon_{xy}
$$

$$
H_{13}^\epsilon = -d(\epsilon_{yz} + i\epsilon_{zx}).
$$

Here, the quantities a, b, and d are valence band deformation potentials. As discussed earlier, strains achieved by lattice mismatched epitaxial growth along the (001) direction can be characterized by $\epsilon_{xx} = \epsilon_{yy} = \epsilon$, and $\epsilon_{zz} = -(2c_{12}/c_{11})\epsilon$. All of the off diagonal strain terms are zero. Using this information, we get

$$
H_{hh}^\epsilon = \left[2a\left(\frac{c_{11}-c_{12}}{c_{11}}\right) + b\left(\frac{c_{11}+2c_{12}}{c_{11}}\right)\right]\epsilon
\tag{8}
$$

$$
H_{1h}^\epsilon = \left[2a\left(\frac{c_{11}-c_{12}}{c_{11}}\right) - b\left(\frac{c_{11}+2c_{12}}{c_{11}}\right)\right]\epsilon
\tag{9}
$$

with all other terms zero. If the hole dispersion is to be described in a quantum well, the hole states $|m, k\rangle$ can be written as[21,22],

$$\langle r_h | m, k \rangle = \frac{\exp(ik \cdot \rho_h)}{2\pi} \sum_\nu g_m^\nu(k, z_h) U_o^\nu(r_h),$$ (10)

where k is the in-plane two-dimensional wave vector, ρ_h is the in-plane radial coordinate, z_h is the coordinate in the growth direction, the U_o^ν are the zone-center Bloch functions having spin symmetry ν, and m is a subband index. The envelope functions $g_m^\nu(k, z_h)$ and subband energies $E_m(k)$ satisfy the Kohn Luttinger equation[21] along with the strain effect[22,23],

$$
\begin{bmatrix}
H_{hh} + \frac{1}{2}\delta & b & c & 0 \\
b^\dagger & H_{1h} - \frac{1}{2}\delta & 0 & c \\
c^\dagger & 0 & H_{1h} - \frac{1}{2}\delta & -b \\
0 & c^\dagger & -b^\dagger & H_{hh} + \frac{1}{2}\delta
\end{bmatrix}
\begin{bmatrix}
g_m^{3/2,3/2}(k, z_h) \\
g_m^{3/2,1/2}(k, z_h) \\
g_m^{3/2,-1/2}(k, z_h) \\
g_m^{3/2,-3/2}(k, z_h)
\end{bmatrix}
= E_m(k)
\begin{bmatrix}
g_m^{3/2,3/2}(k, z_h) \\
g_m^{3/2,1/2}(k, z_h) \\
g_m^{3/2,-1/2}(k, z_h) \\
g_m^{3/2,-3/2}(k, z_h)
\end{bmatrix}.
$$ (11)

The δ is the separation of the hh and lh states in a bulk material due to the strain and is given by Eqns. 8 and 9. For the $In_y Ga_{1-y} As$ system it is given by $\delta = -5.966\epsilon$ eV. Note that the function $g_m^\nu(k, z_h)$ depend on k as well as z_h and that the energy bands are not, in general, parabolic. The matrix entries in Eqn. 11 are given by [20-23]

$$H_{hh} = -\frac{\hbar^2}{2m_0}\left[(k_x^2 + k_y^2)(\gamma_1 + \gamma_2) - (\gamma_1 - 2\gamma_2)\frac{\partial^2}{\partial z_h^2}\right] + V^h(z_h),$$

$$H_{1h} = -\frac{\hbar^2}{2m_0}\left[(k_x^2 + k_y^2)(\gamma_1 - \gamma_2) - (\gamma_1 + 2\gamma_2)\frac{\partial^2}{\partial z_h^2}\right] + V^h(z_h),$$

$$c = -\frac{\sqrt{3}\hbar^2}{2m_0}\left[\gamma_2(k_x^2 - k_y^2) - 2i\gamma_3 k_x k_y\right],$$

$$b = \frac{i\sqrt{3}\hbar^2}{m_0}(-k_y - ik_x)\gamma_3\frac{\partial}{\partial z_h},$$

where m_0 is the free-electron mass, V_h is the potential profile for the hole, and for GaAs $\gamma_1 = 6.85$, $\gamma_2 = 2.1$, and $\gamma_3 = 2.9$.

Equation 11 can be solved by writing it in finite difference form and diagonalizing the resulting matrix to obtain the in-plane band structure[22].

The effect of strain on band structure for both conduction band and valence band states is illustrated by examining the direct band gap material $In_x Ga_{1-x} As$ grown on GaAs and the indirect band gap material $Ge_x Si_{1-x}$ alloy grown on Si.

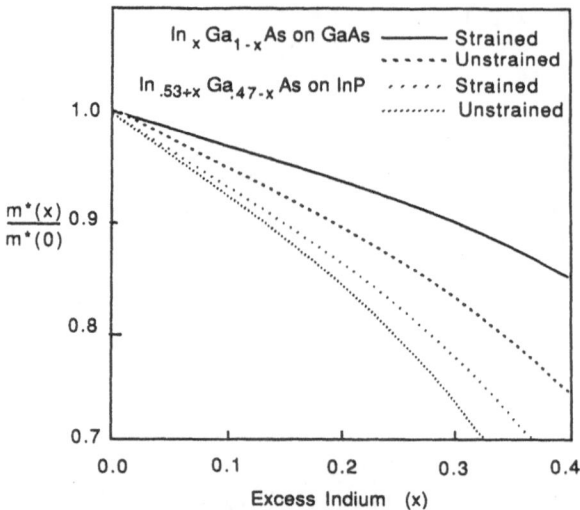

Fig. 3. Change in the in-plane electron mass in
the $In_x Ga_{1-x}As$ system lattice matched to
GaAs, and the $In_{0.53+x}Ga_{0.47-x}As$ on InP.
The results are shown for both unstrained
and strained semiconductors (from Ref.24).

For direct band gap materials conduction bands, the strain
tensor only moves the position of the band edge and has a rather
small effect on the carrier mass. The results shown in Fig. 3
are from calculations done for the unstrained and strained
system using the tight binding method adapted to include the
deformation potential theory[24] and show that the mass decrease
occurs due to addition of indium. However, in the valence bands
the strain has a dramatic effect as shown in Fig. 4 by the hole
dispersion curves for a $GaAs/Al_{0.3}Ga_{0.7}As$ and an $In_{0.1}Ga_{0.8}As/$
$Al_{0.3}Ga_{0.7}As$ quantum wells. As can be seen quite clearly, for
the strained structure the hole dispersion curves represent a
much lower effective mass near the band edge. The hole masses
can be decreased by up to a factor of 3 by increasing the In
content[25].

In the case of the indirect band gap $Si_{1-z}Ge_z$ alloy grown
on Si, the hole masses can again be reduced with strain. The
conduction band also is significantly affected according to Eqn.
6. For (001) growth there is splitting in 6 equivalent valleys.
The results on the band edge states is shown in Fig. 5.

It is quite evident from the discussions in this section
that strain can significantly affect the band structure of semiconductors. The level of strain we have considered can be
incorporated in the semiconductor reasonably easily through
strained epitaxy. It is now important to identify whether or
not the changes produced by the strain have any impact on the
physical properties of the semiconductors. This is explored
next.

3.   CONSEQUENCES OF STRAIN ON SHALLOW LEVELS

The effect of the changes in hole dispersion curves
manifests itself in acceptor levels[26]. In Fig. 6 we show

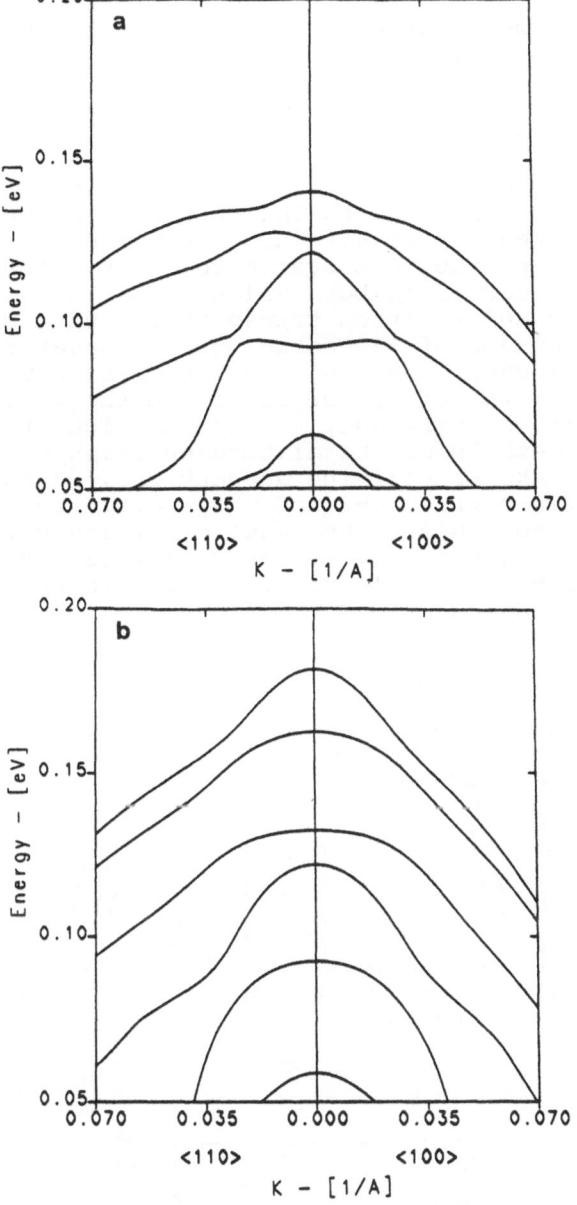

Fig. 4. Hole dispersion in a 100 Å (a)
GaAs/$Al_{0.3}Ga_{0.7}$As and (b) $In_{0.1}Ga_{0.8}$As/
$Al_{0.3}Ga_{0.7}$As quantum well.

results of calculations on the acceptor level energies in a
strained quantum well. As can be seen, the strain has a very
dramatic effect on the acceptor binding energy. In fact the
acceptor binding energy could be used as a spectroscopic tool to
identify the amount of strain in the system.

The effect of strain on direct band gap conduction band
edges is very small and is expected to alter the donor energies
only marginally as can be estimated from the results shown in

Fig. 3. However, for indirect band gap materials (e.g. SiGe alloys) one can expect large changes similar to the acceptor level changes discussed above.

4. CONSEQUENCES OF STRAIN ON TRANSPORT

Since carrier dispersion curves and density of states are affected by strain, one expects the transport properties to be considerably altered due to strain. The Si-Ge alloy grown on Si substrate offers an excellent system to study these effects. Due to high degeneracies in both valence and conduction band edges in Si, hole and electron transport is quite poor. Of course, the robustness of Si technology compensates for this. However, as we discuss below, strain can improve the transport properties. Since the effect of strain on the valence band structure is quite significant, transport calculations must retain the full details of the dispersion relations and it is not possible to make simplifying assumptions. Such calculations have recently been carried out and they retain the full band structure of the hole states and explicitly evaluate the matrix element for scattering as a function of initial and final wave vectors separately[8,18]. This, of course, increases the complexity of the calculations.

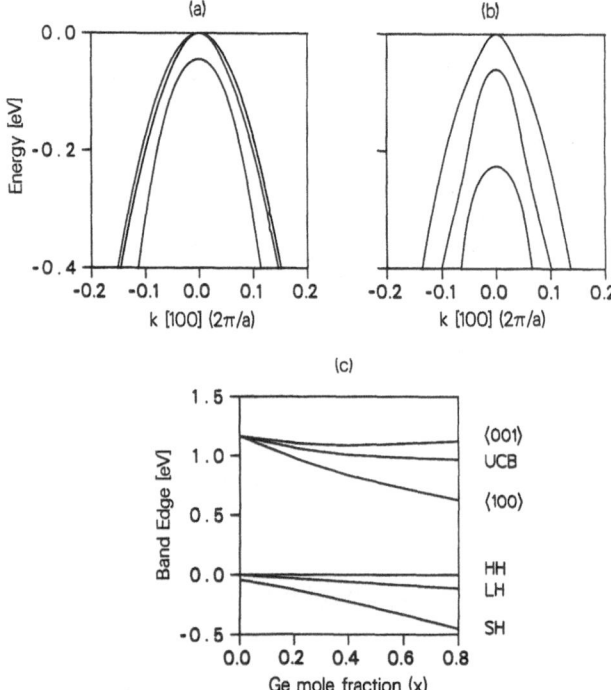

Fig. 5. Hole dispersion curves for (a) Si and (b) $Si_{0.6}Ge_{0.4}$; (c) splittings of the conduction band and valence band are shown as a function of alloy composition. UCB: unstrained conduction band, <001>: out-of-plane X valleys, <100>: in-plane X valleys, HH: heavy hole band, LH: light hole band, SH: split-off band.

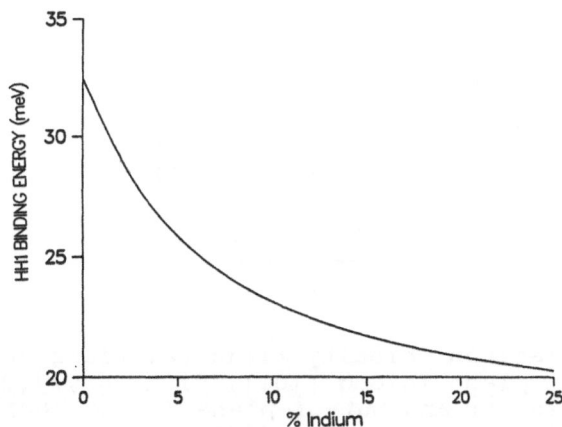

Fig. 6. Effect of strain on hydrogenic acceptor
levels. The effect is illustrated by
examining the HH1 acceptor level energy as
a function of In content in a 100 Å
$In_x Ga_{1-x} As/Al_{0.3} Ga_{0.7} As$.

In Fig. 7a the calculated in-plane (along the (001) plane
if growth is along (001) direction) hole velocity-field
relationships for Si and Si-Ge alloys is shown. The results are
for 10% step increases in Ge content from Si(x = 0) to $Si_{0.6} Ge_{0.4}$
(x = 0.4). Also shown are experimental points reported in the
literature for Si and Ge. Agreement with the published data
provides confidence in the formalism. There is a steady
improvement in the carrier transport, both at low fields and
high fields, as the strain increases. However, it must be noted
that there is currently no good estimate for the effect of the
alloy scattering potential in this system. Figure 7b shows the
effect of the choice of alloy scattering potential on the
velocity-field curve of $Si_{0.8} Ge_{0.2}$. For Fig. 7a the alloy
potential is 0.2 eV (i.e. half of the band gap difference
between Si and Ge). From Fig. 7b it can be seen that it is
important to quantify the alloy potential. The out-of-plane

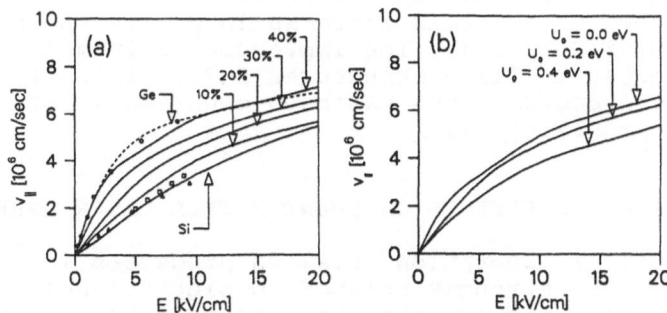

Fig. 7. (a) Calculated in-plane (along [100]) hole
velocity field results for Si and $Si_{1-x} Ge_x$
alloys (with X increasing in steps of 0.1),
experimental points are shown for Si and Ge;
(b) effect of alloy scattering potential on
transport properties.

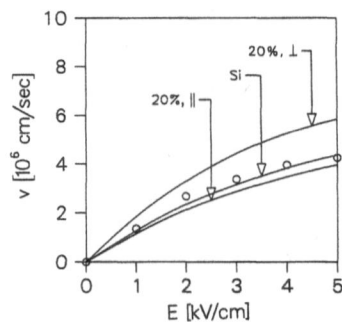

Fig. 8. Electron velocity-field relations for Si,
        in-plane (along [100]) $Si_{0.8}Ge_{0.2}//(001)Si$
        (20% ‖) and out-of-plane (along [001])
        $Si_{0.8}Ge_{0.2}//(001)Si$ (20% ⊥). Experimental
        points are shown (circles) for Si.

transport for holes is also improved, but to a lesser extent
than for in-plane.

     In Fig. 8 the simulation results for electron velocity-
field relations for $Si_{0.8}Ge_{0.2}$ and Si, along with experimental
points for Si are shown. For the alloy, both in-plane and out-
of-plane transport are shown. The in-plane transport is
slightly degraded while the out-of-plane transport is improved.
Two aspects contribute to the improvement in the perpendicular
transport. First, the band edge splitting allows a high
fraction of carriers to be in the lowered in-plane X valleys
which have a lower transverse (in the field direction) effective
mass. Second, the intervalley scattering is suppressed due to
the splitting of the in-plane and out-of-plane X valleys. On
the other hand, for in-plane electron transport, the longitudinal mass of the in-plane valleys dominates the transport and
since the longitudinal mass is higher than the conduction band
edge density-of-states effective mass in Si, the transport
deteriorates somewhat.

     The transport improvements expected from the calculations
predict a three to four fold increase in device speeds for
n-Si/p-SiGe/n-Si HBTs[27]. The improvements in hole transport
are also expected in III-V structures[5-7]. In fact, a dramatic
increase in transconductance has been observed in strained
InGaAs/AlGaAs MODFETs[9-11].

5.   STRAIN INDUCED EFFECTS ON QUANTUM WELL LASER PERFORMANCE

     Since optical absorption/emission processes in
semiconductors are strongly related to electron-hole joint
density of states, one expects the band structure changes to
influence performance of optical devices such as quantum well
lasers. We have already discussed the influence of strain on
quantum well band structures. We will now focus on a typical
quantum well laser to see what role strain plays.

     The subband-to-subband optical gain can be calculated using
the Fermi golden rule, which gives[16]

$$g_{nm}(\hbar\omega) = \frac{4\pi^2 e^2 \hbar}{n_0 c m_0^2 V \hbar\omega} \sum_{k,\sigma} |\hat{\epsilon}\cdot P^{\sigma}_{nm}(k)|^2 \delta(E^e_n(k) - E^h_m(k) - \hbar\omega).$$

(12)

$$\left[f^e\left[E^e_n(k)\right] - f^h\left[E^h_m(k)\right]\right]$$

where n_0 is the index of refraction, c is the speed of light, V is the crystal volume, $\hat{\epsilon}$ is the polarization of the light, $\hbar\omega$ is the photon energy, and f^e and f^h are the distribution functions for electrons and holes, respectively. The optical matrix element is $P^{\sigma}_{nm}(k)$.

To get the total gain g one must sum over all subbands n, m and integrate against a broadening function. The result is

$$g(E) = \int dE' \sum_{nm} g_{nm}(E')\Delta(E - E')$$

(13)

We take $\Delta(E)$ to be a Lorentzian function of half width $k_b T/2$ where T is the temperature and k_b is Boltzmann's constant. Since the density of states is significantly affected by strain, the gain curves show a large dependence on strain.

The TE mode ($\hat{x}$-polarization) couples ~ 3 times as strongly to the HH ($|\frac{3}{2}, \pm\frac{3}{2}\rangle$) as to the LH ($|\frac{3}{2}, \pm\frac{1}{2}\rangle$) states. The TM mode (z-polarization), on the other hand, couples exclusively to the LH states. Since the strain strongly affects the HH-LH

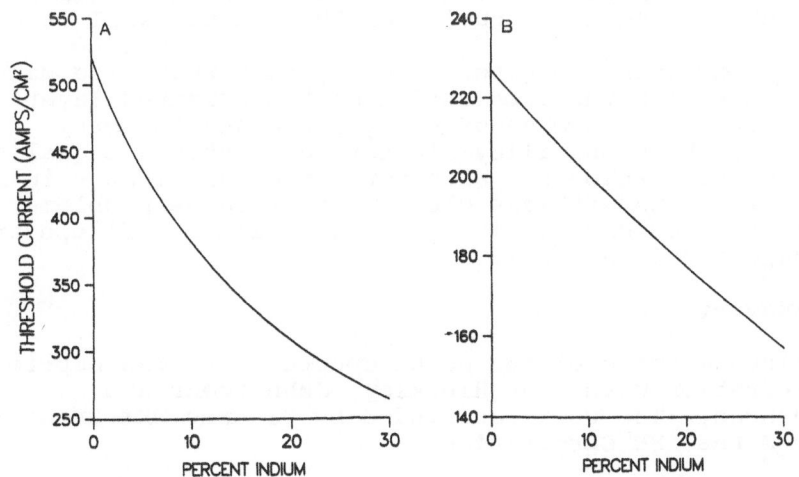

Fig. 9. Plot of the calculated threshold current at 300 K as a function of excess indium content for $In_x Ga_{1-x}As/Al_{0.3}Ga_{0.7}As$ 50 Å well on a GaAs substrate (A), and $In_{0.53+x}Ga_{0.47-x}As/In_{0.53}Al_{0.47}As$ 80 Å well on InP substrate (B). Cavity loss values of 46 cm^{-1} and 48 cm^{-1} was assumed. The well sizes were chosen for lowest threshold current.

separation, one expects the polarization dependence of the gain to be a strong function of strain. In compressive strain, the strain raises HH states so that most of the holes are in HH states and the gain is primarily in the TE mode. Also, the reduced hole mass allows the gain to be positive at smaller injection values. Thus a significant decrease in threshold is expected for the compressively-strained laser. The improvement results from the higher value of the function $[f^e - f^h]$ in Eqn. 12 due to the lighter hole mass.

The calculated threshold current values are plotted in Fig. 9 as a function of strain (or In content) for both GaAs based and InP based strained quantum well lasers[28]. The effect of strain clearly manifests itself in the improved device performance[12-17] of lasers.

CONCLUSIONS

We have seen from theoretical calculations that strain can play a major role in altering the bandstructure of hetero-structures. The band structure changes are particularly dramatic for band edges which are degenerate (such as valence band edge and X and L point conduction valleys). The band structure modifications play an important role in a variety of physical phenomenon. The shallow impurity levels are considerably altered and are found to provide a good signature of the amount of strain in the system. The transport properties can see substantial improvements and in this paper we examined the SiGe alloy system grown on Si(001) substrate. Both the hole transport and the out-of-plane electron transport are found to improve. This is a direct result of the effects of strain on degeneracy lifting. Finally the strain has a profound influence on laser threshold properties as well. As shown by calculations, compressive strain can significantly improve the laser threshold and modal purity of the quantum well laser.

It is important to note, however, that there are many real life problems with the fabrication of the strained layer heterostructures. In case of alloys, one has to worry additionally about the alloy clustering. Interface quality and dislocation generation are also very important issues in real systems. It is neverthless clear that if these problems are overcome, one can expect rich payoffs in almost all electronic and optoelectronic devices.

Acknowledgment

Various aspects of the work reported here was carried out in collaboration with John Hinckley, John Loehr and Vasu Sankaran. The work was funded by US Army URI Program and a grant from the IBM Corporation.

REFERENCES

1. F.C. Frank and J.H. van der Merwe, **Proc. Roy. Soc. London Ser.** 198: 205,216 (1949)
2. J.W. Matthews, **Epitaxial Growth**, 8:559 (1975)
3. W.A. Jesser and J.H. van der Merwe, **Dislocations in Solids,** 8 41:421 (1989)

4. B.W. Dobson and J.Y. Tsao, **Appl. Phys. Lett.**, 51:1325 (1987)

5. T.E. Zipperian, L.R. Dawson, T.J. Drummond, J.E. Shirber and J.J. Fritz, **Appl. Phys. Lett.**, 53:975 (1988)

6. S. Smith and A.D. Welbourn, Proc. of the 1987 Bipolar Circuits and Tech. Meeting, J. Jopke, ed., IEEE Electron Device Soc., Minneapolis, p.57 (1987)

7. J. Hinckley and J. Singh, **Appl. Phys. Lett.**, 53:784 (1988)

8. J. Hinckley, V. Sankaran and J. Singh, **Appl. Phys. Lett.**, 55:2010 (1989)

9. C.P. Lee, W.T. Wang, G.J. Sullivan, N.H. Sheng and D.L. Miller, **IEEE Electron Device Lett.**, 8:85 (1987)

10. S. Tiwari and W. Wang, **IEEE Electron Device Lett.**, EDL-5:333 (1984)

11. Y.J. Chan, D. Pavlidis, R. Razeghi, M. Jaffe and J. Singh, **15th International GaAs and Related Compounds Conference**, Atlanta, 1988.

12. C.A. King, J.L. Hoyt and G.F. Gibbons, **IEEE Trans. Elec. Dev.**, 36:2093 (1989)

13. I. Suemune, L.A. Coldren, M. Yamanishi and Y. Kan, **Appl. Phys. Lett.**, 53:1378 (1988)

14. A. Ghiti, E.P. O'Reilley and A.R. Adams, **Electron. Lett.**, 25:821 (1989)

15. T. Ohtoshi and N. Chinone, **IEEE Photon. Tech. Lett.**, 1:117 (1989)

16. G. Feak, D. Nichols, J. Singh, J. Loehr, J. Pamulapati, B. Bhattacharya and D. Biswas, p.362 **in**: "Proceedings of the IEEE Cornell Conference on Advanced Concepts in High Speed Semiconductor Devices and Circuits", (1989)

17. W. Rideout, B. Yu, J.LaCourse, P.K. York, K.J. Beernick, J.J. Coleman, **Appl. Phys. Lett.**, 56:706 (1990)

18. J. Hinckley and J. Singh, **Phys. Rev.B**, 41:2912 (1990)

19. B.K. Ridley, "Quantum Processes in Semiconductors", Oxford

20. G.L. Bir and G.E. Picus, "Symmetry and Strain Induced Effects in Semiconductors", John Wiley & Sons, New York. (1974)

21. J.M. Luttinger, **Phys. Rev.**, 102:1030 (1956)

22. S. Hong, M. Jaffe and J. Singh, **IEEE J. Quant. Elect.**, 23:2181 (1987)

23. E.P. O'Reilly, **Semiconductor Sci. Technol.**, 4:121 (1989)

24. M. Jaffe and J. Singh, **J. Appl. Phys.**, 65:329 (1988)

25. M. Jaffe, J.E. Oh, J. Pamulapati, J. Singh and P.K. Bhattacharya, **Appl. Phys. Lett.**, 54:2345 (1989)

26. J. Loehr and J. Singh, **Phys. Rev. B**, 41:3695 (1990)

27. J. Hinckley, V. Sankaran, J. Singh and S. Tiwari, p.141 **in**: "Proceedings of the IEEE Cornell Conference on Advanced Concepts in High Speed Semiconductor Devices and Circuits", (1990)

28. J. Loehr, P. Bhattacharya and J. Singh, Integrated Photonic Research Conference, Hilton Head, S. Carolina, March 1990

ELECTRICAL CHARACTERIZATION OF SiGe HETEROJUNCTION BIPOLAR TRANSISTORS (HBTs) FORMED BY MOLECULAR BEAM EPITAXY (MBE)

D.J. Godfrey

British Telecom Research Laboratories
Martlesham Heath, Ipswich, Suffolk, UK

1.   INTRODUCTION

The use of silicon germanium (SiGe) as a narrow band gap base material for heterojunction bipolar transistors (HBTs) offers many important performance benefits in comparison with more conventional silicon (Si) base devices. It has been shown[1] that the reduced band gap of SiGe alloys is seen primarily as a valence band offset. As a result, the barrier height for electron injection from the emitter into the SiGe base is reduced and the collector current is enhanced in comparison with an equivalent silicon transistor for a given emitter/base bias. This modification in the fundamental operating characteristics of the device may be used to provide transistors with significantly increased frequency performance. Indeed, SiGe heterojunction devices with predicted cut-off frequencies (F_T) up to 75 GHz have been reported[2].

In this work, molecular beam epitaxy (MBE) has been used to produce npn HBTs with germanium concentrations between 0% and 30% and base widths between 25 nm and 150 nm. The electrical behavior of the resultant devices is described. In particular, the temperature dependence of the reverse leakage in both Si and SiGe base devices is detailed, as is the use of temperature dependent collector characteristics to determine the level of valence band offset produced by the germanium concentration in the alloy base. Finally, the requirements of optimized SiGe HBTs are considered.

2.   MBE GROWTH AND MATERIALS ISSUES

The HBT layers used in this work were grown on Si(001) substrates by MBE using a VG Semicon V80 system fitted with an Inficon Sentinel 3 flux controller. The arsenic doping was introduced by low energy ion implantation during the MBE growth and boron doping was achieved using a boric oxide source.

A key feature in the growth of strained layers is the critical thickness at which relaxation begins to occur. A

Direct base contact device

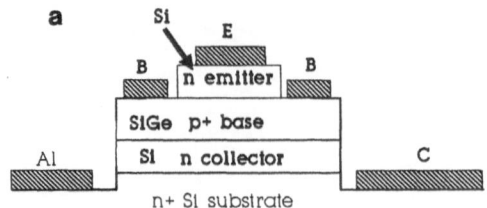

Implanted base contact device

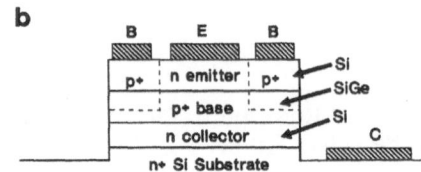

Passivated direct contact structure

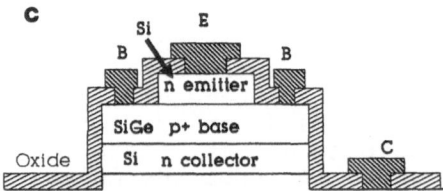

Fig. 1. Schematic cross-section of mesa HBTs. a) Direct base contact device; b) implanted base contact device; c) passivated device for direct base contact approach.

detailed study of the stability of the Si/SiGe system is given in Ref. 3. For the majority of devices fabricated in this study, the SiGe layers were at thicknesses greater than the critical thickness but less than the value where the layer will relax during MBE growth. For layers in this metastable region, complete relaxation requires an elevated temperature to cause both nucleation of dislocations and their glide through the material. From the description of the device processes given in the next section, it will be seen that transistors have been produced using fabrication sequences which either involve only limited heat-treatments or require both ion implantation (which may provide additional nucleation centers) and rapid thermal heating (which will then cause further strain relaxation).

3.   DEVICE FABRICATION

In this work, HBTs have been produced from planar MBE layers using simple mesa processes as illustrated in Fig. 1. The direct contact process requires the use of a Si/SiGe selective[4] etch to enable the emitter layer to be removed without affecting the thickness of the SiGe material used to

provide contact to the base. Alternatively, a masked p$^+$ implant
may be used to overdope the emitter layer in the base contact
region. In this process, a high temperature rapid thermal
anneal (typically 825°C for 10 s) was used to activate the
implanted dopant. Both these processes can be used with or
without the incorporation of a dielectric passivating layer.

Using these processes, devices were produced with a Ge
content in the base of 5% and 10%. As expected, increasing the
Ge concentration in the base of the device resulted in an
enhanced collector current. However, some of this increase was
offset by the increase in the transistor base current.

4. DIODE MEASUREMENTS

Collector/base (C/B) and emitter/base (E/B) diodes formed
in the transistor structure have been studied. In addition, C/B
diodes have been produced by using the Si/SiGe selective etch to
remove the emitter layer over the complete area of the device.
Measurements of devices with differing area:perimeter ratios
were performed and demonstrated that, for both diode ideality
factors and reverse leakage levels, edge effects do not have a
significant influence on the electrical properties of the
devices.

Typical C/B diode characteristics obtained over a
temperature range of 240 K to 300 K are given in Fig. 2. The
SiGe concentration in this sample was 15% and the base thickness
was 150 nm. By plotting the logarithm of the leakage current
against 1/absolute temperature, an activation energy for the
reverse leakage of 680 meV was obtained. This figure is
consistent with a trap at a mid-band gap level. Equivalent
levels of leakage ($\sim 10^{-7}$ Acm^{-2} for 1 V reverse bias at 300 K)
and a trap level (606 meV) were found for diodes produced with
Si bases by MBE. This indicates that the leakage performance of
the devices studied in this work is influenced more by the

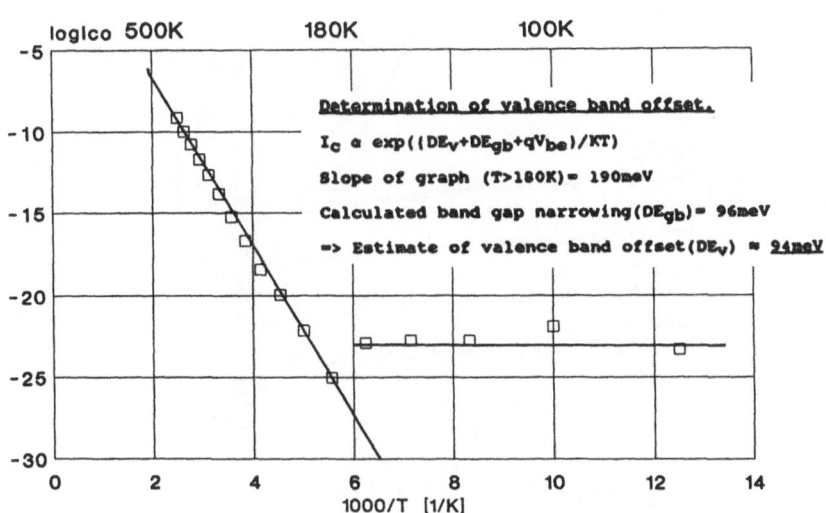

Fig. 2. Temperature dependent diode characteristics.

low processing and deposition temperatures used rather than by
the strain or defects associated with relaxation in the SiGe
material.

5.   TEMPERATURE DEPENDENT COLLECTOR CHARACTERISTICS

As indicated earlier, the collector current (I_C) in a SiGe
base device is enhanced with respect to an equivalent Si homojunction transistor by the degree of valence band offset
produced in the SiGe base. Experimentally, this effect may be
studied by measuring the temperature dependence of I_C (see Fig.
3 where results from a 150 nm 15% Ge base transistor are given).
The expression for I_C as a function of temperature from simple
drift-diffusion considerations is given in the inset of Fig. 3.
The collector current used in the figure has been obtained by
extrapolating the experimental I_C to a V_{BE} of 0 V. The remaining temperature dependent terms in the collector current expression are the valence band offset and the band gap narrowing in
the base. It can be seen that over the temperature range 180 K
to 450 K, the experimental data follow the expected activation
energy type behavior. By assuming the Del Alamo model for band
gap narrowing, the valence band offset for the 15% Ge base has
been estimated as being 94 meV. This value is in good agreement
with equivalent measurements of valence band offsets reported by
other groups producing SiGe HBTs (e.g. Ref. 1).

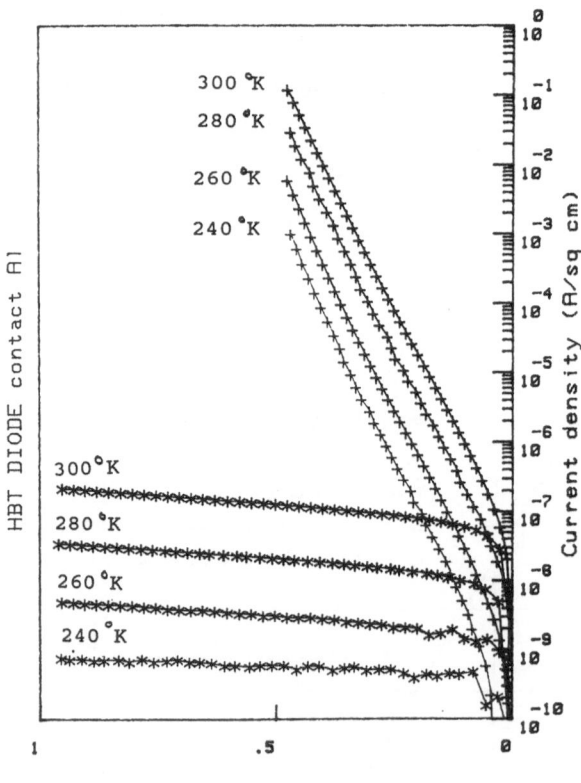

Fig. 3. Temperature dependent collector current.

6. OPTIMIZED SiGe HBTs

In order to optimize the cut-off frequency of bipolar transistors, it is important to minimize the components of the forward transit time of the device. These include the neutral emitter delay, the emitter/base depletion region delay (usually negligible compared with other delay terms), the neutral base delay and the collector/base depletion region delay. The base delay can be minimized by the use of a narrow base width. This requirement is compatible both with the capabilities of MBE growth and the importance of maintaining the SiGe thickness near or below the critical thickness (see Section 2). In this work, the thickness for narrow bases was chosen to be 25 nm. For such base widths, the base doping needs to be made high ($> 10^{19}$ cm^{-3}) to prevent punchthrough conduction and to provide an acceptable value for the base resistance of the device. The neutral emitter delay is reduced for SiGe HBTs compared with Si homojunction devices due to the relative increase in the collector current. This improvement may be 'traded' for a reduction in the emitter doping compared with Si base transistors. For the devices produced in this work, an emitter doping of 10^{18} cm^{-3} and a Ge content of 30% was chosen as representing the current 'best-guess' for an optimized HBT.

Devices have been produced using the above layer specification by the implanted contact process described earlier. The process does not include an n$^+$ 'top-up' implant into the emitter. The heat treatment used to activate the boron implant was 925°C for 10 s. The resulting transistors showed a collector current which was enhanced by a factor of ~ 26 compared with the value expected from a Si base device. However, the I_C calculated on the assumption that the full valence band offset for 30% Ge alloy (~176 meV) was present gave an enhancement of ~900 times. This degradation in the experimentally measured I_C may be related both to relaxation of the SiGe and to the diffusion of boron from the SiGe region into the Si emitter. The latter effect would result in the loss of the emitter/base heterojunction with the consequent reduction in collector current.

CONCLUSIONS

It has been shown that SiGe heterojunction bipolar transistors formed by MBE provide enhanced collector currents compared with Si homojunction devices. Some of this improved performance is offset by an increase in the transistor base current with increasing Ge concentration.

The temperature dependence of reverse leakage in homo and heterojunction diodes has been studied and it is found that the electrical performance of Si and SiGe devices is comparable. This implies that the characteristics of the diodes are controlled more by the low processing and deposition temperatures used than by the presence of strain or relaxation in the SiGe material.

For optimized SiGe HBTs, it is necessary to produce devices with narrow base widths (~25 nm) and high Ge concentrations (up to 30%). First devices produced in such layers show a degradation in the expected collector current which is believed to be

673

caused by diffusion of boron from the SiGe material into the Si emitter, and possible relaxation of the SiGe layer.

Acknowledgment

The research work on SiGe HBTs at BTRL involves primarily Dr. M. Jones, Dr. M. Gell, A.S. Martin, Dr. C. Tuppen, Dr. C. Gibbings and Dr. A. Reeder. The MBE layers used in this work were grown by Dr. C. Gibbings and the electrical measurements were performed by A.S. Martin and Dr. A. Reeder.

REFERENCES

1. C. King, J. Hoyt and J Gibbons, **IEEE Electron Device Lett.** 10:52 (1989)
2. S. Iyer, G. Paton, J. Stork, B. Meyerson and D. Harame, **IEEE Electron Device Lett.**, 36:2943 (1989)
3. C. Tuppen and C. Gibbings, **J. Appl. Phys.**, to be published.
4. P. Narozny, H. Dambkes, H. Kibbel and E. Kasper, **IEEE Trans Elect. Dev.**, 36:2363 (1989)

LAYERED COMPOUNDS

PHYSICS AND CHEMISTRY OF LAYERED COMPOUNDS

W.Y. Liang

Cavendish Laboratory
Madingley Road
Cambridge CB3 OHE, UK

1. INTRODUCTION

A layered solid bears a strong anisotropy in all its
physical properties. Such a solid usually displays an axial
symmetry and its properties can be distinguished between those
belonging to the basal plane and those perpendicular to it. It
is therefore often loosely called a 'two-dimensional' solid.
The simplest of them is graphite but there is also a large class
of compounds with similarly anisotropic structural, electrical,
magnetic and optical properties. Due to the weak interlayer
bonding and the highly anisotropic electronic band structure and
Fermi surface, these properties show up, for example, in the
ease of cleaving and in compressibility, in polarization
dependent reflectivity, in conductivity and 'two-dimensional'
superconductivity, etc. In the present chapter, a layered solid
should be structurally as well as electrically anisotropic;
therefore an artificial superlattice heterostructure
semiconductor which exhibits strong electrical anisotropy only
will be excluded.

In addition to graphite and BN in which each layer consists
of a simple sheet of atoms, the next layered type solids have
layers each consisting of three atomic planes. These include
compounds of transition metal dichalcogenides such as ZrS_2, NbS_2
and MoS_2, various metal dihalides such as $NiCl_2$, PbI_2 and CdI_2,
nontransition metal dichalcogenides such as SnS_2 and $SnSe_2$. As
shown in Fig. 1a, a layer of one of these compounds is made up
from three atomic sheets strongly bound by either a covalent or
ionic bond, but the layers are only weakly bound with each other
by what is commonly called the van der Waals interaction.
Instead of a metal atom sheet being sandwiched by two nonmetal
atom sheets, it is also possible to form the so-called anti-
structure layered compounds such as Tl_2S and Ag_2F, as well as
numerous other variants including the trihalide BiI_3, the first
row transition metal phosphorus trichalcogenides such as VPS_3
and $NiPSe_3$, the oxychalcogenides and oxyhalides such as TiOS,
TiOCl, the group III chalcogenides such as GaSe, InSe and In_2Se_3
and other A_2B_3 such as the semimetallic Bi_2Te_3 and the semiconducting As_2Se_3. Some layered compounds have complicated

Condensed Systems of Low Dimensionality
Edited by J. L. Beeby *et al.*, Plenum Press, New York, 1991

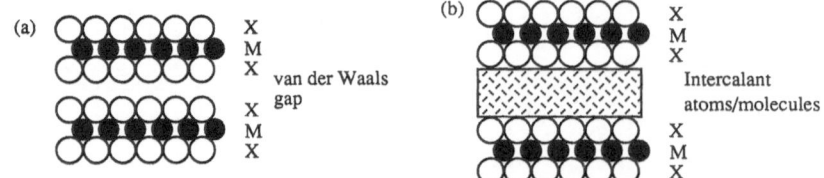

Fig. 1. Schematic representation of (a) transition metal
dichalcogenide MX_2 sandwiches and the interlayer
van der Waals gap, (b) an intercalation complex
formed with intercalant species inserted to the
van der Waals gap; M = transition metal and X =
S, Se or Te.

multi-sandwich structure. The most well known of them is mica
and some of the recently discovered cuprate oxide
superconductors may also be regarded as of this class of
materials.

The weak interlayer forces offer the possibility of
introducing foreign atoms or molecules to the interstitial sites
between the layers and this is the process of intercalation. A
wide range of new compounds can be generated through this
process with physical properties contrasting those of the parent
compounds[1,2]. Figure 1b shows a typical intercalation complex
of the layered compound shown in 1a. Intercalation process is
the basis on which certain layered compounds are used as the
cathode of solid state batteries. Its practical applications
are likely to increase, particularly in microelectronics. At
the same time, intercalation is also now being recognized as an
important practical way of controlling oxygen content and
introducing cation dopant to oxide superconductors.

The most widely known and the largest group of layered
compounds are the transition metal dichalcogenide MX_2 where M =
transition metal and X = S, Se or Te[3]. These materials will,
therefore, be dealt with in some detail. Their two-dimensional
properties, however, should be seen as examples for the wider
family of layered compounds.

2.   CRYSTAL STRUCTURE

With the exception of VS_2 and chromium compounds which are
difficult or impossible to grow, all other transition metals
from group IV, V, VI and some from group VII and VIII of the
periodic table, can form layered structure dichalcogenides. The
dichalgogenide family MX_2 is made up from any of these
transition metals M and one of the chalcogen X from Table 1.

Each atomic plane is a two-dimensional close packed
structure (Fig. 2a) and the planes are stacked along the c-axis
as X-M-X-(vdw)-X-M-X-(vdw)-, where (vdw) represents the van der
Waals gap created by removing alternative metal M layers from an
otherwise regular three-dimensional structure similar to the
rocksalt viewed along the (111) axis. In the ideal structure,
the atomic distances for the remaining atoms are unchanged,
giving a c/a ratio of 1.633. In practice, only compounds having

Table 1. Elements in the dichalcogenide family.

Group	Transition Metal M					Chalcogen X
	IV	V	VI	VII	VIII	VI
						S
	Ti	(V)	(Cr)		Ni	Se
	Zr	Nb	Mo		Pd	Te
	Hf	Ta	W	Re	Pt	

octahedrally coordinated structure can have such an ideal c/a ratio. The octahedrally coordinated structure is shown in Fig. 2b in which the sandwich layer is made up of atomic planes arranged in AbC, and the layers may be stacked in a variety of ways resulting in a large number of polytypes. The simplest polytype has the form AbC-(vdw)-AbC-, called the 1T structure (lower case letters are used to denote metal layers and upper case letters the chalcogen layers). This structure is adopted mainly by the group IV, VII and VIII metal dichalcogenides. The group VI metal compounds, on the other hand, mainly adopt the trigonal prismatic coordination in which the sandwich layer is made up of atomic planes arranged in AbA (Fig. 2c), and the simplest polytype has the form AbA-(vdw)-BaB-(vdw)-AbA, called $2H_b$ or AbA-(vdw)-CbC-(vdw)-AbA-, called $2H_a$ sequence. Many other polytypes can also be formed containing different mixtures of octahedral and trigonal prismatic layers.

As already noted, a true layered solid must be structurally anisotropic as characterized by its easy cleaving property. Phonon spectroscopy offers an excellent way of quantifying the degree of anisotropy not only by distinguishing inter- and intralayer phonon modes but also through these frequencies the shear moduli in different directions can be deduced. The Raman spectrum of MoS_2 at room temperature is shown in Fig. 3 in which the lowest frequency mode E_{2g}^2 at 33.5 cm^{-1} has been identified as the rigid layer mode while all higher frequency modes involve both intralayer as well as interlayer forces. A shear force constant of 152 Nm^{-1} corresponding to intralayer forces and 2.67 Nm^{-1} corresponding to interlayer forces have been deduced. A ratio of 57 between these two shear force constants is a very good indicator of the high degree of anisotropy in MoS_2.

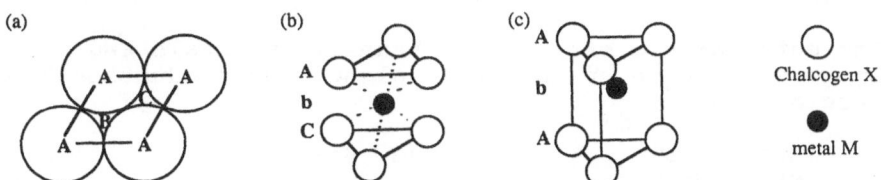

Fig. 2. (a) basal plane view showing two-dimensional close-packed structure, (b) octahedrally coordinated unit and (c) trigonal prismatically coordinated unit of MX_2.

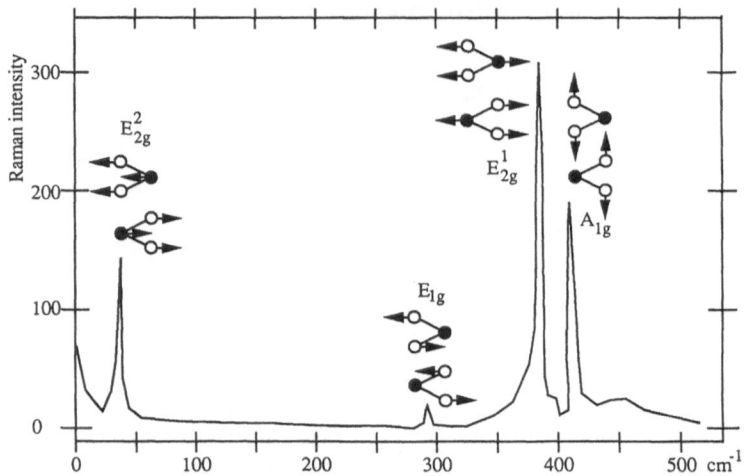

Fig. 3. Raman spectrum of MoS_2 at room temperature;
individual modes are indicated.

3. ELECTRONIC PROPERTIES

3.1 Transition Metal Dichalcogenides

The dichalcogenides of transition metals from group IV, V and VI are characterized by distinct electronic properties. These are, in the sequence of the groups, semiconductor with a large band gap ($TiSe_2$ is a semimetal due to overlapping valence and conduction bands), metal and superconductor, and semiconductor with a small band gap respectively. This is the result of increasing the number of valence electrons by one on each move to a higher group. The same can also be achieved by adding electrons to the host compound by intercalation. Thus $Na_x ZrS_2$ with $x = 1$ behaves in many respects like NbS_2 and is also a superconductor. These transport properties are dominated by the occupation of the lowest d conduction band often referred to as the $d_z{}^2$ band. An additional feature is the tendency of the d electrons to become tightly bound due to the narrow d-band, with the consequence of Anderson localization or the formation of charge density waves. In both cases, electrons are immobilized at low temperatures and the solid behaves like a semiconductor. In the case of transition metal phosphorus trichalcogenides such as $MnPS_3$, the d electrons are believed to be completely bound to the transition metal ions. They have a strong influence on magnetic properties but do not contribute to conduction.

The outer atomic orbitals of transition metals and chalcogens which made up the main valence and conduction band structure are

$$
\begin{array}{ll}
\text{Zr:} & d^2 s^2 \\
\text{Nb:} & d^3 s^2 \\
\text{Mo:} & d^4 s^2
\end{array}
\qquad
\text{S, Se:} \quad s^2 p^4
$$

All the sixteen valence electrons in ZrS_2 are responsible for the strong "skeleton" chemical bonds, and are shown to occupy the valence bands in the energy level density diagram in

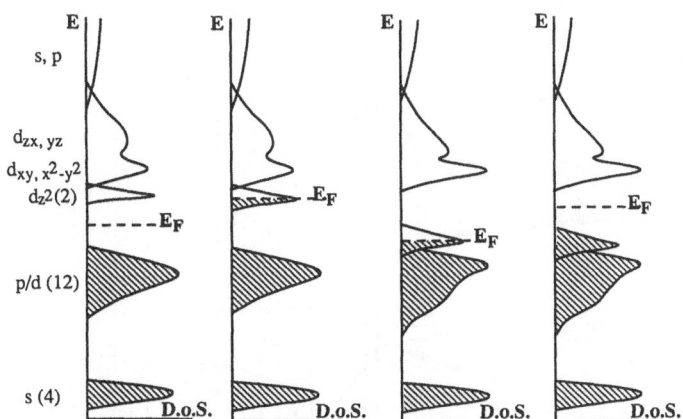

Fig. 4. Schematic density of energy level diagrams for
groups IV, V and VI transition metal
dichalcogenides. NbS_2 does not crystallize in
octahedral form and the second diagram
represents an extrapolated hypothetical
situation. Number in parentheses next to the
orbital notations gives the number of states in
a given band.

Fig. 4. In fact these strong chemical bonds serve to underpin a
common structure for all members of the dichalcogenide family,
and the addition of further electrons to the d_z^2 conduction band
produces relatively small perturbations to this part of the band
structure. This is the basis of the 'rigid band approximation'
often used as a first approximation to describe effects on the
electronic structure resulting from intercalation. The main
difference between the three main groups of compounds is, of
course, the occupation of the d_z^2 band. Figure 4 shows that the
d_z^2 band does shift in relation to other bands with increasing
electron occupation and this is accompanied by a trigonal
distortion as measured in terms of c/a ratios. The main driving
force for such a distortion is to minimize the total energy by
lowering electron occupied conduction band. In the trigonal
prismatic structure, the disadvantage of ion-ion Coulomb
repulsion may therefore be offset by the lowering of the d_z^2
band with increasing electron filling and trigonal distortion,
making it the more stable form. Group V metal compounds, on the
other hand, have only one electron in the d_z^2 band and adopt
both coordination structures.

Results of detailed band structure calculations support the
band scheme shown in Fig. 4. One of these calculations based on
muffin-tin orbitals and atomic sphere approximation which also
include exchange correlation corrections has been applied to 2H-
TaS_2 and the result is shown in Fig. 5[4]. 2H-TaS_2 is a metal
having trigonal prismatic coordination whose density of states
function is expected to be similar to that of 2H-NbS_2 as shown
in Fig. 4. At the same time, the triple peak structure of the
conduction band of the 1T structure group IV and group V
compounds can be clearly seen in Fig. 6 from the optical joint
density of states functions determined by reflectivity
measurements[5].

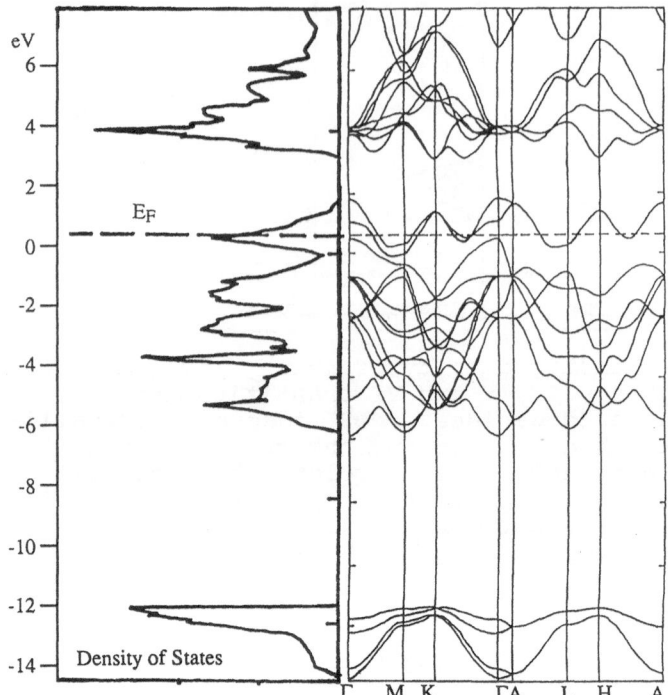

Fig. 5. Band structure and calculated density of states function of $2H\text{-}TaS_2$.

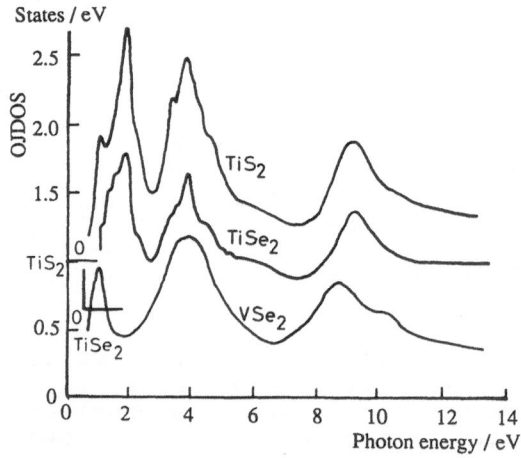

Fig. 6. Optical joint density of states functions for TiS_2, $TiSe_2$ and VSe_2, determined by reflectivity measurements; transitions to the d_z^2 band are responsible for the lowest peak in the spectra.

3.2 Group III Nontransition Metal Chalcogenides

This family of layered compounds is rich in polytypes not only because of the different possible stacking sequence of layer sandwich, but also the basic sandwich can vary greatly in composition. Typical examples are In_2Se_2, In_2Se_3 and Ga_2Se_2. There are no d orbitals involved and the states near the gap are of p_z character having contributions from both Ga (or In) and Se. These orbitals overlap appreciably in the c direction and their energies are strongly influenced by interlayer interactions. Both pressure and intercalation will therefore change the minimum band gap radically as shown in Fig. 7. On the other hand, it appears that these p_z orbitals have little overlap in the basal plane and most of the band width is the consequence of the orbital overlap in the c-direction. This provides a possible explanation for the observations of exciton features and sharp luminescence line spectra being retained in samples of Li_xInSe with $x \sim 0.3$; and even with $x = 1.5$, the carrier concentration remaining as low as 4×10^{17} cm^{-3}[6]. Although electron carriers are introduced into the InSe host material following Li intercalation, they appear to be trapped by Anderson localization. The rigid band approximation is clearly not valid in this system.

3.3 Transition Metal Phosphorus Trichalcogenides MPX₃ (M = V, Mn, Fe, Co, Ni, Zn, Cd X = S, Se)

This family of compounds may be regarded as having the analogous structure[7] to the dichalcogenide family if viewed as $[M_{2/3}(P_2)_{1/3}]X_2$. The detailed crystal structure is a complicated monoclinic derived from a distorted $CdCl_2$ structure[8]. This gives rise to an equally complicated electronic structure and magnetic properties[9]. By analogy to transition metal dichalcogenides, every third metal atom in the basal plane is replaced by a diphosphorus pair. While the metal atoms are octahedrally coordinated similar to ZrS_2, each phosphorus atom is tetrahedrally coordinated with another phosphorus atom and three chalcogen atoms. The increased metal atom distance gives rise to an interesting situation for the d orbital states. It is known that all MPX_3 compounds are semiconductors. Except for the vanadium compounds in which mixed valent states are observed and full stoichiometry can not be obtained[8], other transition metal ions are in the high spin

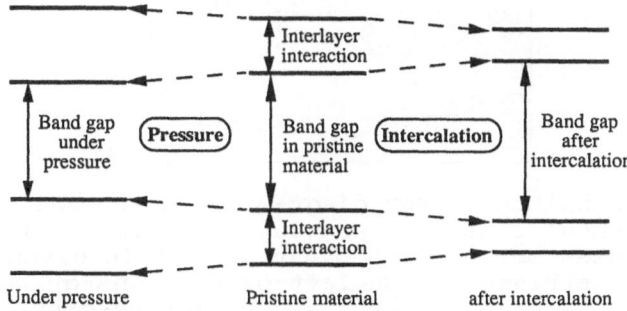

Fig. 7. Effects of pressure and intercalation on the band gap of Ga(or In)Se.

state and have charge 2+. This implies a negligible overlap and tightly bound d-orbitals. Consequently electrons in these d-levels would have no mobility. It is interesting to note that the diphosphorus pairs (P_2) in the lattice not only occupy the position of the metal ions, they also behave as cations to the chalcogen atoms. By labeling individual bands with their dominant orbital characters a possible band scheme can be deduced and this is shown in Fig. 8 for $MnPS_3$. This band scheme is similar to that published recently by Glasso et al.[10]. An appropriate description of the bands must, of course, take into account the effects of hybridization and include in particular mixing of s and p orbitals. Figure 8 also shows numbers of electron states contained in a given band. These numbers correspond to the formula unit $Mn_2(P_2)S_6$ within a unit cell. The Mn d-electrons are tightly bound and there is an antiferromagnetic order which is the appropriate state at low temperatures. It is believed that the states near the large band gap are of p_z character, resulting from the tightly bound phosphorus atoms. The s, p band in the conduction band in fact contains appreciable antibonding character of two sulphur atoms. However, by analogy to the transition metal dichalcogenides, the main bonding and antibonding chalcogen p- and s-states lie in the valence band. It is clear that as long as the d-orbitals remain tightly bound and nonoverlapping, the materials will behave like a semiconductor irrespective of the number of d-electrons present, and there will be a strong magnetic ordering.

4. EFFECTS OF INTERCALATION

Intercalation is made possible because atoms within a layer are strongly bound by covalent or ionic forces while the layers

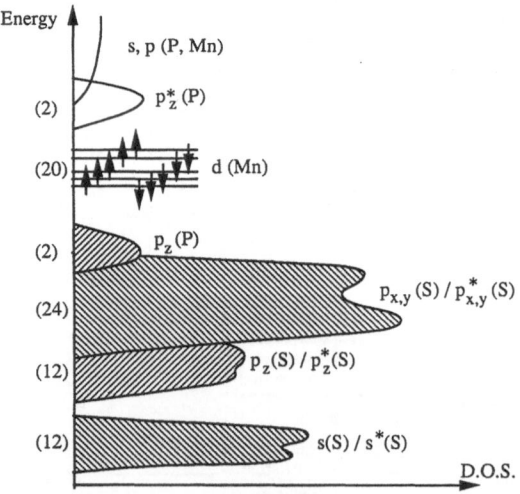

Fig. 8. A possible energy diagram for $MnPS_3$ and similar compounds at low temperatures. The number of states per $M_2(P_2)X_6$ formula unit is given in parentheses to the left of each energy band. The labeling of orbital characters gives an indication of dominant contributions to the band.

are held together by a relatively much weaker force - the van der Waals interaction. Intercalant species are inserted into the van der Waals gap, and in most cases they occupy the octahedral or the tetrahedral interstitial sites, and the process is generally accompanied by a charge transfer (electrons) to the host layers and an expansion of the layer-layer distance.

It is believed that the tendency for the charge transfer is the driving force for the intercalation reaction. Transition metal dichalcogenides are known only to form intercalation complexes with electron donor species, although other layered materials such as graphite may be intercalated with either electron donating or electron accepting species. The charge transfer process may be used to "fine tune" the electronic properties of the host material. Control of band filling in this way is a feature unique to low dimensional structures of this type, and provides an extra variable of enormous value. It is thus possible to achieve semiconductor to metal and metal to semiconductor transitions with intercalation. Changes in band filling also allow us to investigate the Fermi-surface driven structural instabilities caused by the formation of charge density waves. In the case of MPX_3 electrons are tightly bound to the transition metal ions and intercalation does not result in any appreciable changes in conductivity - the material remains an insulator - but can produce magnetic phase transition of the first order.

There are broadly four categories of intercalation complexes for the transition metal dichalcogenides, some of them also apply to other systems. They are classified according to the intercalant species and the main effects they produce.

(a)   Simple Metals

Li, Na, K, Rb and Ag are highly electro-positive ions which makes a very good electron donor and the effectiveness of these intercalant species is to increase electron filling with a minimum perturbation to the host band structures. Other poly-valent metals such as Ca, In, Sn, and Pb also intercalate well but they tend to modify the lattice through a strong covalent interaction between the intercalant species and the host lattice.

Changes in electrical properties as a result of potassium intercalation are illustrated as follows:

(Group IV) HfS_2 + xK   → $K_x HfS_2$
Wide gap;   Metal
semiconductor

(Group V) $NbSe_2$ + xK   → $K_x NbSe_2$
Metal and   Poor metal and for x = 1
superconductor, T_c = 7.4 K   expect semiconductor

(Group VI) MoS_2 + xK   → $K_x MoS_2$
Diamagnetic semiconductor   Paramagnetic metal; for
                                          x ~ 0.5, superconductor
                                          with T_c ~ 6.5 K

Trigonal prismatic
coordination   Octahedral coordination

The last example clearly demonstrates that changing electron occupation can lead to structural changes and therefore a complete breakdown of the rigid band approximation in this case.

(b) Organic Molecules

These are mainly nitrogen containing molecules such as amines, amides, pyridine, ammonia, hydrazine. The main effect of organic molecule intercalation is the possibility of greatly increasing layer separation with a minimum charge transfer. For example, Gamble and Geballe[11] and their colleagues found that when intercalating $2H-TaS_2$ with n-octadecylamine, the layers can be separated by a bilayer of the intercalant molecules resulting in an expansion of ~ 60 Å from a separation of ~ 3 Å before intercalation. Since the interactions between the layers are now considerably weakened, this makes it possible to study the physical properties of nearly separated layers, such as superconductivity as well as normal conduction.

(c) "3d" Transition Metals

These include transition metal V, Cr, Mn, Fe, Co, and Ni. "3d" transition metals form insertion compounds with transition metal dichalcogenides. They are classed as intercalation compounds because the foreign atoms occupy interstitial sites between the layers. On the other hand, these compounds can only be prepared from synthesizing the appropriate stoichiometric mixture of the elements and the process is, of course, not reversible. Electrochemical method has also been used but this process is very slow and difficult to achieve homogeneity. The main interest of these compounds is their magnetic properties. Depending on the local moment of the transition metal ions and the precise Fermi surface topology, either ferromagnetic or antiferromagnetic ordering has been observed. The anisotropy of the magnetic exchange interaction adds further interests to the general magnetic and transport properties of these materials as represented by a recent study on $Fe_{1/3}TaS_2$ and $Mn_{1/3}TaS_2$ by Dijkstra et al.[12].

(d) "Misfit" Layer Compounds $(MS)_n TS_2$

The symbols used are M = Sn, Pb, Bi, rare earth elements, T = Nb or Ta, and n = 1.08 - 1.19. These are relatively new additions to the family of intercalation compounds[13,14]. They are built of alternating double layers of MS with M in distorted square pyramidal coordination by sulphur and the sandwich TS_2 has T in distorted trigonal prismatic coordination.

5. DIMENSIONALITY OF SUPERCONDUCTIVITY AND OXIDE
   SUPERCONDUCTORS

Many metallic members of the transition metal dichalcogenides are superconductors which exhibit considerable anisotropy particularly in their normal state conductivity and upper critical fields H_{c2}, see Fig. 9a[15]. As we have shown, superconducivity is observed in materials which are intrinsically metallic (e.g. $NbSe_2$) or having acquired the metallicity via intercalation (e.g. $K_{.5}MoS_2$). An interesting question is whether superconductivity can persist in individual layers and

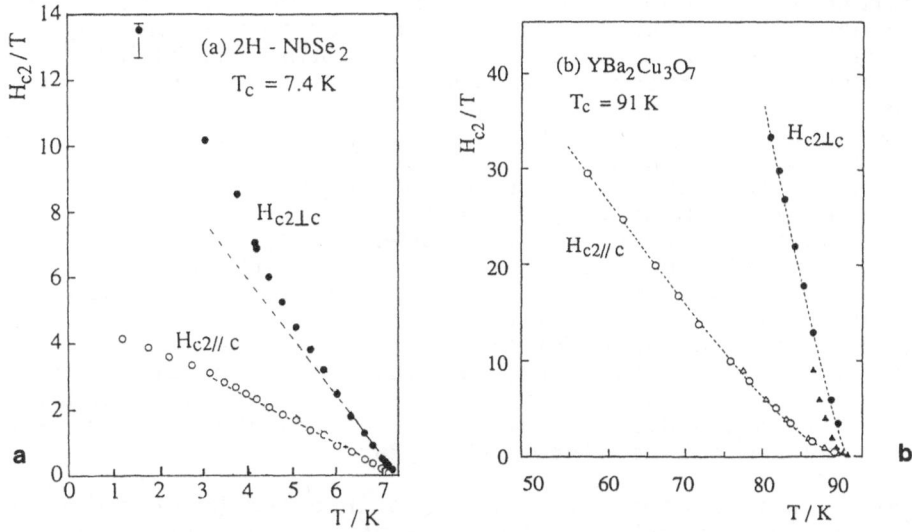

Fig. 9. Extreme anisotropy in H_{c2} with respect to field
applied parallel and perpendicular to the c-
axis, (a) 2H-NbSe$_2$ and (b) YBa$_2$Cu$_3$O$_7$.

whether this may be regarded as examples of two-dimensional
superconductivity. Evidence for believing individual layer
superconductivity comes from a number of observations. (1) The
transition temperature T_c appears to be insensitive to the
changing separation between layers. Coleman et al.[16] showed
that T_c for 2H-TaS$_2$ remained at around 4 K irrespective of the
size of the organic molecules used as intercalants which gave a
range of layer separation from ~ 6 Å to 12 Å. (2) Intercalation
with chromocene which possesses strong local moments does not
affect T_c[17]. (3) NbSe$_2$ crystals of unit cell thickness (two
layers) remains superconducting with a T_c of ~ 5 K compared with
7.4 K in the thick specimen[18]. Unfortunately, the fact that
each layer consists of at least three sub-atomic layers makes it
impossible to conclude that superconductivity phenomenon had
been demonstrated in a truly two-dimensional system.

Many recently discovered oxide superconductors have layered
structure and it is therefore not surprising that they share
many of the behaviors described above. The anisotropy for
normal conductivity with current path parallel and perpendicular
to the c-axis generally fall in the range 30 to 100 at room
temperature which is similar to NbSe$_2$. The anisotropy in H_{c2} is
also similar as shown in Fig. 9b[19]. Furthermore, because of
the short coherence length in the c-direction ($\leq$ unit cell
thickness), the superconducting properties tend to be unaffected
by changing unit cell thickness and by the presence of high
local moment ions such as lanthanide substitution of Y in
YBa$_2$Cu$_3$O$_7$ (except Ce, Pr and Tb).

The basic structure of high temperature oxide
superconductors is the perovskite containing Cu. The perovskite
may also be "split" into two opposing pyramids and further
layers of CuO$_2$ and Y (or lanthanides or Ca) may be added to the
space between. It is well known that the majority of the
electronic states at the Fermi level in the oxide system

concentrate in the CuO_2 layers. It is also believed that their spin exchange interactions and the specific corner sharing structure of CuO_2 are essential in determining the high T_c in these materials. A very large group of oxide superconductors has the basic building block containing the split pyramids sandwiching an atomic layer of Y (or Ca), and the full formula for the block is $YBa_2Cu_2O_6$ as shown in Fig. 10(a and b). Variations of this unit structure are possible by a full or partial substitution of Y by Ca, Fig. 10(c and d), or substitution of Ba by Sr (as in Bi-Sr-Ca-Cu-O). However, any attempt to substitute Cu ions in the CuO_2 layers (as distinct from the CuO chains) so far has led to a rapid degradation of superconductivity. The structures illustrated in Fig. 10 are representative of a number of this group of oxide super-conductors. The fact that they are made up of the above basic building blocks with a variety of chemical layers inserted is emphasized in these diagrams in order to facilitate comparison with the layered solids discussed so far.

Two further factors are important in order to optimize the superconducting properties. One relates to the carrier density and the other to ionic sizes. Both factors, however, have implications on the chemical stability of the compounds. In all the compounds shown here, it has been possible to raise T_c by a suitable substitution of cations within the basic building block or by an appropriate choice of the insertion layers. By this means the highest T_cs for the representative structures in Fig. 10 have been narrowed to a range of values within 90-110 K. These T_cs are, respectively for the four structures, 96 K for $GdBa_2Cu_3O_7$, 91 K for $(Y_{.9}Ca_{.1})Ba_2Cu_4O_8$[21] and rising to 108 K under pressure[22], 85 K for $Tl_{1.1}Ca_{.9}Ba_2Cu_{2.1}O_{7.1}$, and 108 K for $Tl_{1.7}Ca_{.9}Ba_2Cu_{2.3}O_{8.1}$[23]. It is possible that many more higher T_c materials can be found by applying these chemical rules. A recent discovery by Liu et al.[24] of a 108 K superconductor $(Y_{.2}Ca_{.8})Sr_2Cu_2(Pb_{.5}Tl_{.5})O_7$ is a result of such a systematic search for new materials. Although chemically complex this material is structurally akin to that of Fig. 10c with the substitution $(Y_{.2}Ca_{.8})$ for Ca, Sr for Ba, and $(Pb_{.5}Tl_{.5})$ for Tl. It can be seen also that Fig. 10c is the tetragonal form of Fig. 10a which is orthohormbic (the same may be said of Figs. 10d and 10b). The interesting study of metal-superconductor-insulator transmutation induced by cation substitutions as applied to this system is discussed elsewhere[25]. Given the basic unit of $YBa_2Cu_2O_6$ plus a variety of insertion layers, it is indeed quite remarkable that a whole range of superconductors can be fabricated whose optimized T_cs fall within such a narrow range of temperature. There is evidence also which shows that as the CuO_2 layers are further separated by the insertion layers, or the interaction between CuO_2 layers is weakened by disorder in the c-direction, two-dimensionality is enhanced and fluctuation phenomenon near T_c prevails[26]. Other important systems not mentioned here include those whose basic building block contains one and three CuO_2 layers.

6.   PRACTICAL APPLICATIONS

We return to the main materials of this chapter which are the transition metal dichalcogenides. MoS_2 has long been used in lubrication and in heterogeneous catalysis, particularly in the oil and synthetic chemical industry[27]. Molybdenum and

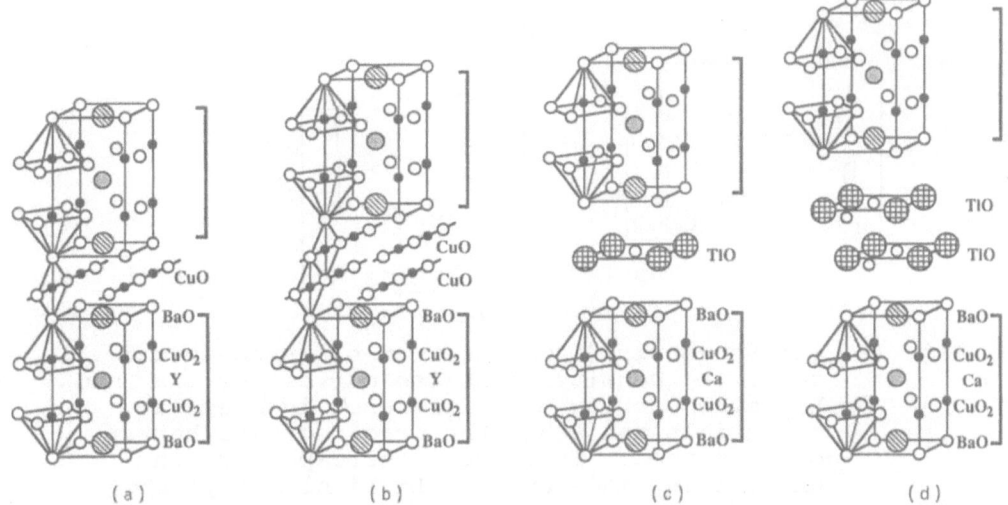

Fig. 10. Structure of oxide superconductors which can be built up from basic building block $YBa_2Cu_2O_6$ or $CaBa_2Cu_2O_6$ together with a variety of insertion layers. The insertion layers and the final products shown are: (a) CuO giving $YBa_2Cu_3O_7$, c/n = 11.65 Å and T_c = 93 K, (b) 2 CuO giving $YBa_2Cu_4O_8$, c/n = 13.63 Å and T_c = 81 K, (c) TlO giving $TlCaBa_2Cu_2O_7$, c/n = 12.7 Å and T_c = 80 K*, (d) 2 TlO giving $Tl_2CaBa_2Cu_2O_8$, c/n = 14.65 Å and T_c = 100 K*, where c/n represents one formula unit sandwich thickness in the c-direction (*Parkin et al. [20]).

tungsten dichalcogenides provide efficient photoelectric conversion and have therefore also been used in photocells. Because of the layer nature, crystals of $NbSe_2$ may be cleaved into a shape consisting of a thin section connecting two thick sections, thereby forming a weak link exhibiting Josephson effect. This is the basis of applying $NbSe_2$ superconductor in making SQUIDs. Single crystal SQUIDs made from $NbSe_2$ have the superior property of extra mechanical as well as thermal robustness and reliability over other SQUIDs made up of alloys. The application that has attracted most attention in recent years, however, has been in secondary solid state batteries, particularly with the Li/MoS_2 and Li/TiS_2 systems. Other layered materials have also been tried as cathodes, including for example, the recent success of the Paris group in demonstrating solid state batteries in a thin film package of Li/B_2O_3-Li_2O-Li_2SO_4/In_xSe_{1-x} where In_xSe_{1-x} is acting as the cathode[28].

An ideal electrochemical cell should provide a constant potential, a large charge capacity but no internal resistance. By means of intercalation process a cathode made up of a layered solid can, in principle, provide a very large specific volume capacity for positive ions such as Li, and it is in this role that layered type solids are of great importance. The simple operation (Fig. 11a) involves electrons being driven from the Li_3Al anode which has a high electrochemical potential to a cathode with a lower electrochemical potential (Fig. 11b). At the same time Li^+ ions are released to the cell and attracted to

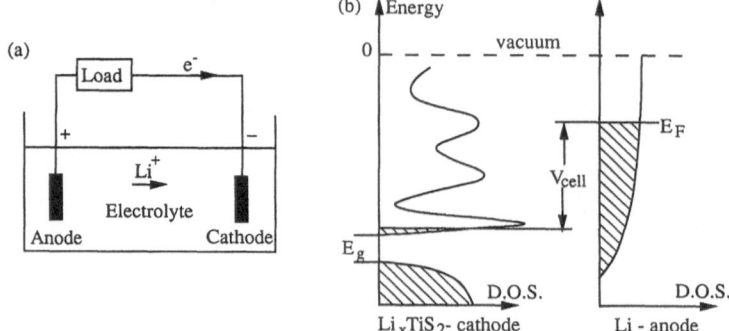

Fig. 11. (a) Schematic diagram for an electrochemical
cell; (b) Density of states curves for TiS_2 and
Li metal showing the difference in chemical
potentials and hence the origin of the battery
cell voltage. This would correspond to the
anode being made up of a Li/Li_3Al alloy, the
cathode being of TiS_2 and the electrolyte being
an ionic conductor which could be made up of
$LiClO_4$ in B_2O_3-SiO_2 glass or in an organic
solvent such as polyethylene oxide polymer.

the cathode which has become negatively charged. Intercalation
process then allows the Li^+ ions to enter and populate the
internal crystal structure of the layered type cathode until
saturation, which occurs when there is approximately one Li ion
per formula unit (i.e. $x = 1$ in Li_xTiS_2). If the process is
allowed to continue, there will generally be a sudden drop in
the cell voltage and possibly a degradation of the cathode
material (see Fig. 12). The discharge must therefore be stopped
and the cell recharged to its original potential. While Li ion
has been accepted as the easy mobile ion, the understanding and
choice of both electrolyte and cathode are still a subject
requiring considerable research.

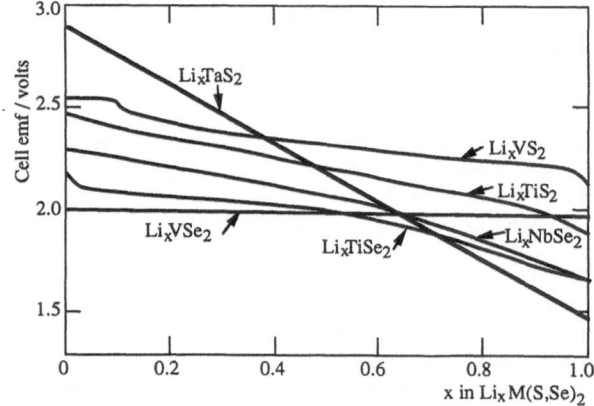

Fig. 12. Variation of open-circuit voltages at room
temperature for six Li/Li_xMX_2 cells as a
function of Li concentration x in Li_xMX_2
compound electrode.

6.1 Layered Cathode

Considerations for a good cathode include (a) a high mobility for positive ions and electrons; (b) a high positive ions uptake; and (c) a high structural stability independent of degree of intercalation. The cathode should be both a good ionic and an electronic conductor in order to achieve low internal resistance and to ensure uniform intercalation. Not only its structure must not change as a result of repeated cycling, but its volume should ideally also remain constant throughout the charging and discharging periods. These requirements can not be easily met in practice. However, experiments show that some materials are good in one respect but fail in another. For example MoS_2 and TiS_2 provide high mobility and high positive ion uptake, but they do not generally provide sufficient structural stability. On the other hand, MPX_3 cathodes offer excellent structural stability but have relatively poor ionic and electronic mobility.

6.2 Cell Emf

As electrons begin to enter the layered cathode, its energy bands and chemical bonds usually undergo a renormalization in order to lower the total energy of the system. This will inevitably involve a considerable shift of the Fermi level and change in crystal structure, both are undesirable feature for a battery particularly in terms of maintaining a constant cell emf. It is important therefore that the energy bands remain rigid throughout the whole range of intercalation concentrations. It turns out that this "rigid-band" picture is a reasonable first approximation for many transition metal dichalcogenides when intercalated with alkaline metal atoms because the basic skeleton structure is made up of sixteen electrons and the addition of an extra electron represents a relatively small perturbation to the total energy. However, this approximation clearly breaks down in the case of GaSe and InSe layered solids.

In addition to the requirement of the rigid band approximation, the energy band in the host material receiving intercalation electrons should be narrow and have a high density of states at the Fermi level. A d-band or a nonbonding band will be most suited for this purpose. Figure 12 shows a series of open circuit voltages for lithium intercalation into different transition metal dichalcogenide hosts[29]. The d_z^2 band in VSe_2 is very narrow (0.6 eV)[30], and this shows up as giving the most stable emf almost independent of Li concentration.

6.3 Electrolytes

An essential property of an electrolyte is its high ionic conductivity but zero electronic conductivity. In recent years solid electrolyte has increasingly become a viable alternative to its liquid counterpart. The main difficulty has been the low ionic conductivity at room temperature. This problem appears to have been overcome by the use of $LiClO_4$ - solvates of organic solvents immobilized in a polymer network of polyacrylonitrile (PAN), poly-tetraethylene glycol diacrylate (PEGDA) or poly-vinyl pyrrolidone (PVP)[31]. A typical electrolyte might contain 38 mole % ethylene carbonate (EC) and 33 mole % propylene carbonate (PC) with 8 mole % $LiClO_4$ immobilized in

21 mole % PAN. A conductivity of 1.7×10^{-3} ohm^{-1}cm^{-1} at 20°C and 1.1×10^{-3} ohm^{-1}cm^{-1} at -10°C has been achieved. These values compare favorably with those of liquid electrolytes and are the right magnitude which will make solid state batteries operating at ambient temperature a practical reality. The above authors also claimed that the films were free standing and dimensionally stable. As solid electrolyte can be used in thin film form the actual resistance across the electrodes this conductivity represents is very small. These electrolytes have not been tested for their long term stability nor have the combined characteristics between these electrolytes and layered type cathodes been fully studied.

CONCLUSION

Layered type solids represent a large and important group of materials ranging from graphite with a simple structure to oxide superconductors with multiple atomic planes within a unit cell. The richness in structure is accompanied by equally varied physical properties and their possible applications. Intercalation process adds a further dimension to this already interesting system. New materials are being discovered, for example, by inserting a square lattice on a hexagonal close-packed matrix which hitherto was thought not feasible. More efforts are needed to understand the nature and ways of exploiting these materials particularly in connection with their two-dimensionality.

REFERENCES

1. W.Y. Liang, p.31 **in**: "Intercalation in Layered Materials", NATO ASI Series, M.S. Dresselhaus, ed., Plenum Press, New York (1986)
2. R.H. Friend and A.D. Yoffe, **Advances in Phys.**, 36:1 (1987)
3. W.Y. Liang, p.459 **in**: "Physics and Chemistry of Electrons and Ions in Condensed Matter", NATO ASI Series, J.V. Acrivos, N.F. Mott and A.D. Yoffe, eds., Reidel Publishing Co. (1984)
4. G.Y. Guo and W.Y. Liang, **J. Phys. C**, 19:995 (1986)
5. S.C. Bayliss, Ph.D. Thesis, University of Cambridge (1984)
6. C. Julien, M. Jouanne, P.A. Burret and M. Balkanski, **Mat. Sci. and Eng.**, B3:39 (1989)
7. J.W. Johnson, p.267 **in**: "Intercalation Chemistry", M.S. Whittingham and A.J. Jacobson, eds., Academic Press (1982)
8. F. Hulliger, **in**: "Structural Chemistry of Layer-type Phases", F. Lévy, ed., (1976)
9. R. Brec, p.93 **in**: "Intercalation in Layered Materials", NATO ASI Series, M.S. Dresselhaus, ed., Plenum Press, New York (1986)
10. V. Grasso, F. Neri, L. Silipigni and M. Piacentini, **Phys. Rev. B**, 40:5529 (1989)
11. F.R. Gamble and T.H. Geballe, Vol.3:Ch.3 **in**: "Treatise on Solid State Chemistry, Inclusion Compounds", Plenum, New York (1976)
12. J. Dijkstra, P.J. Zijlema, C.F. van Bruggen, C. Haas and R.A. de Groot, **J. Phys. Condens. Matter**, 1:6363 (1989)
13. G.A. Wiegers, A. Meetsma, R.J. Haange and J.L. de Boer, **Mat. Res. Bull.**, 23:1551 (1988)
14. G.A. Wiegers, A. Meetsma, S. van Smaalen, R.J. Haange,

J. Wulff, T. Zeinstra, J.L. de Boer, S. Kuypers, G. Van Tendeloo, J. Van Landuyt, S. Amelinckx, A. Meerschaut, P. Rabu and J. Rouxel, **Solid State Commun.**, 70:409 (1989)

15. N. Toyota, H. Nakatsuji, K. Noto, A. Hoshi, N. Kobayashi, Y. Muto and Y. Onodera, **J. Low Temp. Phys.**, 25:485 (1976)

16. R.V. Coleman, G.K. Eiserman, S.J. Hillenius, A.T. Mitchell and J.L. Vicent, **Phys. Rev. B**, 125:139 (1983)

17. M.B. Dines, **Science**, 188:1210 (1975)

18. F.R. Frindt, **Phys. Rev. Lett.**, 28:299 (1972)

19. Y. Iye, T. Tamegai, T. Sakakibara, T. Goto, N. Miura, H. Takeya and H. Takei, **Physica**, C153-155:26 (1988)

20. S.S.P. Parkin, V.Y. Lee, A.I. Nazzal, R. Savoy, T.C. Huang, G. Gorman and R. Beyers, **Phys. Rev. B**, 38:6531 (1988)

21. T. Miyatake, S. Gotoh, N. Koshizuka and S. Tanaka, **Nature**, 341:41 (1989)

22. E.N. van Eenige, R. Griessen, R.J. Wijngaarden, J. Karpinski, E. Kaldis, S. Rusiecki and E. Jilek, **Physica C**, 168:482 (1990)

23. S.S.P. Parkin, V.Y. Lee, A.I. Nazzal, R. Savoy, R. Beyers and S.J. La Placa, **Phys. Rev. Lett.**, 61:750 (1988)

24. R.S. Liu, P.P. Edwards, Y.T. Huang, S.F. Wu and P.T. Wu, **J. Sol. State Chem.**, 86:334 (1990)

25. W.Y. Liang, submitted to **Mat. Sci. and Eng. B**, (1990)

26. J.W. Loram, K.A. Mirza and P.F. Freeman, (1990) to be published

27. M.C. Zonneyvylle, R. Hoffman and S. Harris, **Surface Sci.**, 199:320 (1988)

28. C. Julien, I. Samaras, M. Tsakiri, P. Dzwonkowski and M. Balkanski, **Mat. Sci. and Eng. B**, 3:25 (1989)

29. M.S. Whittingham, **Prog. in Solid State Chem.**, 12:41 (1978)

30. H.P. Hughes C. Webb and P.M. William, **J. Phys. C**, 13:1125 (1980)

31. K.M. Abraham and M. Alamgir, **J. Electrochem.**, 137:1657 (1990)

THE GRAPHITE INTERCALATION COMPOUNDS AND THEIR APPLICATIONS

Daniel Guérard and Hervé Fuzellier

Laboratoire de Chimie du Solide Minéral (URA CNRS 158)
Université de Nancy I, BP 239
54506 Vandoeuvre lès Nancy Cédex, France

1.   INTRODUCTION

Graphite is one material whose properties are quite remarkable: high electrical conductivity, in spite of a low carrier concentration, high electrical anisotropy, high melting and boiling points. Its lamellar structure is characterized by a heterodesmic behavior which leads the planes to be easily separable under mechanical or chemical interactions. On the other hand, the strength of the in-plane bonds allows the planes to remain coherent.

The graphene planes can be separated either by mechanical action or by reaction with chemical species to lead to Graphite Intercalation Compounds (GICs) in which the intercalated reactant (whose name becomes intercalate) occupies the intervals between the graphite (graphene) planes.

By its electronic structure, graphite is able to react either with compounds which give electrons to graphite (electron donors) or accept electrons from graphite (electron acceptors). On the donor side, there are, first of all, the alkali metals[1-4], the alkaline and some rare earth metals[5,6], and also metallic alloys containing one alkali metal[7-9], the alkali metal hydrides[10,11]. For the electron acceptors, one has the halogens[12,13], transition metal halides[14,15], some nitrates[16], acids[17,18], oxides[19,20] and oxihalides[21].

A GIC is characterized by its:

-   formula, e.g.: KC_8, C_5HNO_3...which is independent of the respective valences of carbon and the intercalate;

-   stage s, which is equal to the number of carbon layers between two successive intercalated sheets, according to the occurrence of intercalate in each gallery between the carbon planes, in one gallery in two,..., in s, s being an integer (Fig. 1). This phenomenon of staging is character- istic of the GICs, compared to other lamellar host materials like: transition metals dichalcogenides, clays...;

Condensed Systems of Low Dimensionality
Edited by J. L. Beeby *et al.*, Plenum Press, New York, 1991

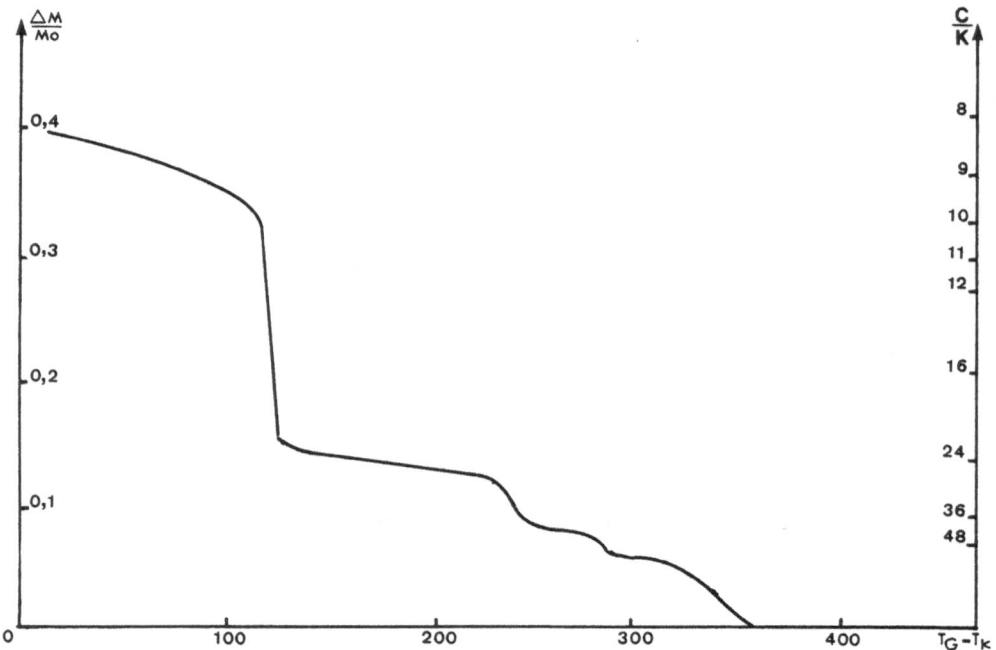

Fig. 1. Intercalation of potassium into graphite after
Ref. 29.

-    interplanar distance d_i which corresponds to the distance
between two successive carbon layers surrounding the
intercalate. This distance depends on the intercalate and
varies from 3.70 Å (with lithium) up to around 13 Å
($RbTl_{1.5}C_4$);

-    identity period (or repeat distance) along the $\vec{c}$ axis: I_c
which corresponds to the distance between two successive
intercalated planes.

Those three last numbers: s, d_i and I_c are related by:

$$I_c \sim d_i + (s - 1)3.35 \qquad (\text{Å}) \quad .$$

The value of 3.35 Å means that the graphene planes keep, in
a GIC, their distance as in pristine graphite.

The names of the GICs are derived from that of the
classical compounds: for instance, the compound $KHgC_4$ is named
first stage potassium mercurographitide whereas the compound
$C_{24}^+HSO_4^-,2.2H_2SO_4$ becomes first stage graphite sulfate.

Altogether, whatever its combination, almost each element
of the periodic table can be intercalated into graphite. For
instance, hydrogen can be intercalated as proton with acids[17],
neutral H_2 molecules by physisorption in the MC_{24} compounds[22]
or anion H^- as alkali metal hydride; europium intercalates as
metal[6], ammonia complexes[23] or chloride[24]. Even rare
gases can be intercalated as fluorides or oxyfluorides[25].

By multiplying the number of possible intercalates by the
number of possible stages for a given intercalate, one obtains

about 3 to 400 difference GICs. Each of them is characterized
by its chemical composition, stage, structure, chemical and
physical properties.

The applications of the GICs are as numerous as their
properties; however, there is one important limitation to those
applications: their fragility towards air and water.

The field of these compounds is still very actively studied
as shown by the seven International Conferences devoted to them
since 1977[26] and many special Colloquia, satellites of
Materials and Carbon Conferences.

2.   SYNTHESIS

Since there are many kinds of reactants, there are also
many ways of synthesis. When possible, the best way is to
realize the direct intercalation of the reactant (in vapor,
liquid or solid phase) on graphite[27-29].

The **reaction of the vapor** of the reactant on graphite is
often used. It is done generally in glass tubes sealed under
vacuum, heated in a two zone furnace. The temperature of
graphite T_G is higher than that of the reactant T_R (whose value
governs the vapor pressure of the reactant) in order to avoid a
simple distillation of the reactant on graphite. According to
the difference of temperature $\Delta T = T_G - T_R$, one obtains
different stages, as shown in Fig. 1.

By **compressing** a mixture of lithium and graphite **powders**
(Li + 6C) under inert atmosphere for two days at 10kbars, one
obtains the gold-like colored first stage compound LiC_6(4).

With donor compounds, one uses **solvents** like ammonia or THF
when starting from ammonium-metals in an excess of ammonia or
complexes like, for instance, naphtalen alkali metals in
solution in THF. There is a cointercalation of the solvent[30].
This kind of ternary containing solvated atoms, ions or
molecules can also be obtained (with a better control of the
reaction) by **addition of the solvant molecules** on binary
compounds[31,32]. In this case, there is a complete
reorganization of the intercalated layers with a change of the
stage whose number decreases in most cases:

$$KC_{24} + x\ THF \rightleftharpoons KC_{24}(THF)_x \quad (with\ x = 1,2\ or\ 3)$$

Stage 2   Stage 1

but increases when the addition reaction is done by
hydrogen[33,34]:

$$KC_8 + 1/3H_2 \rightleftharpoons KH_{2/3}C_8$$

Stage 1   Stage 2

The intercalation of Lewis acids can be realized by:

-      Chemical reaction, directly if the intercalant is oxidizing
       enough: intercalation of HNO_3 or $SbCl_5$ [18,19]) or by means

of an oxidant: intercalation of H_2SO_4 in presence of HNO_3; halides with chlorine...

- Electrochemical reaction with an anodic oxidation of graphite[35,36], according to the following equations:

$$24C \rightleftharpoons C_{24}^+ + e^-$$

$$3.2H_2SO_4 + e^- \rightleftharpoons HSO_4^- + 2.2H_2SO_4 + 1/2H_2 \uparrow$$

$$24C + 3.2H_2SO_4 \rightleftharpoons C_{24} + HSO_4^-2.2H_2SO_4 + 1/2H_2 \uparrow$$

GICs can be considered as formed by a lamellar carbon macrocation and an intercalate totally or partly ionized with ions solvated by either the neutral intercalated reactant itself or by molecules of the solvent such as, for instance, nitromethane, dimethylsulfate,... which are often used as electrolyte media.

On the other hand, the advantage in using solvents can be done in order to decrease either the intercalation temperature (CrO_3 in acetic acid[37]) or the activity of the reactant (bromine in carbon tetrachloride[38]).

By **reduction in situ of an intercalated halide**, one obtains a lower oxidation degree of the metal (e.g. ferric chloride is reduced to $FeCl_2$ which is unable to intercalate into graphite), even free metallic iron. However, the question of a true intercalation compound or inclusions of the free metal as small lentils whose diameter is around 100 Å is still open and controversial[39,40].

The **bi-intercalation compounds** (also called heterostructures) are prepared by two successive intercalations: one uses, as host structure for the second intercalant, a compound already intercalated. This host structure must be of a stage ≥ 2 in order to present empty galleries for the second intercalation.

The first bi-intercalation compounds were obtained by reaction on the second stage $KH_{2/3}C_8$ of calculated quantities of a heavy alkali metal heated just above its melting point[41].

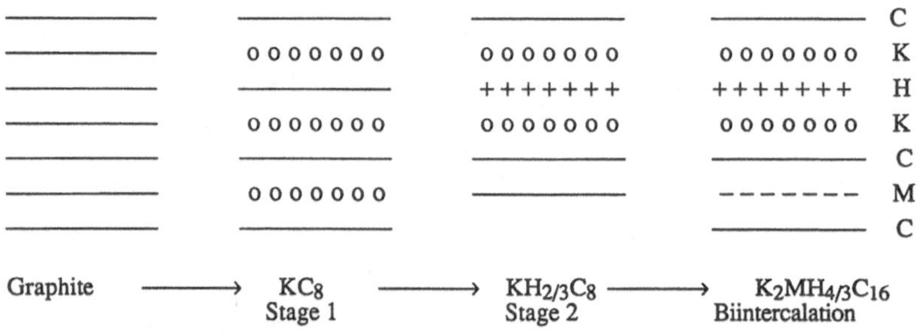

Fig. 2. Successive reactions for the preparation of a bi-intercalation compound.

Starting from the elements, there are successive reactions to perform:

-    Preparation of the binary first stage

-    Ternarization by hydrogen, with increasing state to 2

-    Second intercalation.

These steps are shown in Fig. 2.

Other preparation methods are to be noted: **Ion implantation**, which disturbs the graphite lattice and allows the intercalation of sodium[42], **compression** which decreases the stage whereas the in-plane density increases[43].

On the whole, the synthesis of GICs can be done by numerous methods and, according to the reaction conditions (temperature, duration, method...), one obtains different compounds as illustrated by the intercalations of potassium vapor into graphite: according to the temperature gradient, there appears the 1st, 2nd, ..., nth stage as shown in Fig. 1.

In most cases, the intercalation can be considered as completely reversible; however, there are some irreversible reactions - which are in fact side reactions when:

-    the "solvant" becomes too reactive as, for instance, oxygen,

-    the intercalate is oxidizing enough to form covalent bonds with graphite, as it appears with fluorine at temperatures above 500°C which gives graphite fluoride,

-    the temperature of the reaction is too high and leads to the formation of metal carbides, e.g. with lithium above 500°C, one obtains Li_2C_2 instead of the intercalation compound LiC_6,

-    the compound is steeply heated at a temperature above the boiling point of the intercalate. This involves the exfoliation of the graphite lattice, which is, of course, an irreversible reaction.

3.   STRUCTURE

The structure of the GICs depends doubly on the **graphite network**:

-    along the $\vec{c}$ axis, the graphene planes are almost unchanged by intercalation, they remain parallel and the intercalated layer lies in the middle of the gap between the graphene planes.

-    in-plane, the hexagonal network of the carbon atoms creates potential wells and the intercalated species tends to lie on top of those potential wells. This involves a change in the graphene plane stacking: A/A (where the slash represents the intercalate) instead of ABA in the hexagonal graphite.

The nature of **the intercalate** plays a major role in the structure: along the $\vec{c}$ axis, it increases the interplanar distance d_i, according to its size and structure (single or multiple layers). The d_i value varies from 3.70 Å (LiC_6) up to more than 13 Å ($RbTl_{1.5}C_4$). The bond strength also has an effect on the stacking of the intercalated species: in the case of the alkali metal, there is only one intercalated layer, but in most cases the intercalate presents several layers, up to five different metal sheets in the case of the graphite-cesium-antimony compounds[44].

In-plane, according to the bonding strength of the intercalate, its structure can accommodate or not with the graphite lattice and the compounds are commensurate, partly commensurate or totally incommensurate with respect to the graphite lattice.

For intercalates whose melting point is close to room temperature, the intercalated species are randomly dispersed in the layer and behaves as a "liquid" or a "gas" in between the graphene planes[17]. Many structural transitions as a function of the temperature occur in the GICs and are related to variations of the physical properties[45].

When the intercalate is in epitaxy with respect to the graphene layers, it occupies only a part of the potential wells (which would correspond to a formula MC_2 since one hexagon contains two carbon atoms). In the first stage compounds of the metals whose formulae are either MC_6 (M = Li; Ca, Sr, Ba; Sm, Eu, Yb) or MC_8 (M = K, Rb, Cs), one hexagon in three or four respectively is occupied by the metal atoms. The possible sites are called α, β, γ and α, β, γ, δ respectively. In that case, various structures were established:

   $-A\alpha A\alpha$ ... in LiC_6, Space Group P6/mmm[4],

   $-A\alpha A\beta A\alpha$ ... in the other MC_6 compounds, SG: $P6_3$/mmm[5,6]

   $-A\alpha A\beta A\gamma A\alpha$ ... in CsC_8, SG: $P6_2 22$ or $P6_4 22$[46]

   $-A\alpha A\beta A\gamma A\delta A\alpha$ ... in KC_8 and RbC_8, SG: Fddd[46].

In fact, even in these compounds, the structure remains "ideal" compared to the actual behavior and there are many defects due to translation, stacking faults, grain boundaries, polysynthetic character: due to the three crystallographic equivalent directions, the crystals are systematically twin. Then, the classical crystallographic study methods as, for instance, the automatic diffractometer are inadequate for the GICs and one had to set up methods especially devoted for these studies:

   - double rotation of single crystals for the hk0
     reflexions[47],

   - triple rotation for the hkl reflexions[48],

   - scanning along hk rods[49],

   - monochromatic Laue photographs[50].

In summary, the structure along the $\vec{c}$ axis is easily determined, the in-plane structure is often obtained, but the 3D

structure is more difficult to determine and remains only
"ideal".

4.   PROPERTIES

     The chemical properties are roughly those of the free
reactant which involves a wide range of possible applications;
on the other hand, there are two effects due to the
intercalation:

-     dispersion of the intercalate, which involves a high
reactivity, for instance the first stage compounds of the heavy
alkali metals burn spontaneously in the air. However, this
behavior does not apply to the whole of the GICs. For instance,
some compounds containing even an alkali metal (as metallic
alloys or metal hydrides) are poorly reactive towards air and
even water: the graphite-CsH compounds can be stored for weeks
in water at room temperature without any change of stage and
physical properties[11],

-     role of the carbon matrix, which involves an orientation
for the reactions, especially in catalysis or for the addition
of atoms (e.g. chlorine) on organic molecules.

     Those effects are turned to account for the applications.

     The intercalation phenomenon involves a **charge transfer** to
(or from) the graphite lattice. Since the chemical properties
of the intercalate are roughly those of the intercalate, this
charge transfer cannot be total; on the other hand, since the
GICs are stable, the charge transfer is definitely not zero.

     It varies, according to the nature of the intercalate: in a
given series, for instance with the alkali metals, it increases
as the electronegativity decreases (from lithium to cesium). It
depends also on the amount of electrons given to the system by
electrochemical intercalation: the charge for a first stage
graphite-sulfuric acid varies from C_{28}^+ to C_{21}^+. This
overcharging process is accompanied by a weak decreasing of the
interplanar distance: from 7.98 Å with C_{28}^+ to 7.96 Å with C_{24}^+.
The change is also characterized by an organization of the
intercalated layer from a liquid-like intercalate to a 3D
compound, where the c parameter becomes three times the identity
period along the $\vec{c}$ axis: c = 3 x 7.96 = 23.88 Å with a graphite
stacking A/B/C/A.

     The charge transfer is demonstrated by several factors:

-     the **chemical formula** of the GICs: the intercalated species
in the compound graphite-sulfuric acid is HSO_4^-, $2.2H_2SO_4$; with
halides, the intercalate is $NiCl_{2.13}$ instead of being
stoichiometric $NiCl_2$[51]. The excess of chlorine has been
related to the charge transfer,

-     the **in-plane carbon-carbon distance** is modified by the
intercalation: it increases of around 1% in the first stages
graphite-donor compounds but decreases by about 0.1% with
electron acceptors of first stage. These variations were
related to the charge transfer[52] and a relation between the
variation u (Å) of the in-plane d_{c-c} due to the intercalation

and the charge transfer f_C on the carbon atoms was proposed a few years ago by Pietronero and Strässler[53]:

$$u = .157f_C + .146 \left| f_C \right|^{3/2} + .236f_C^2 \quad ,$$

- the study by **spectroscopic methods** (NMR, Raman...) of the carbon shows a shift of the carbon peak with, in addition, several peaks which correspond to the 1st, 2nd, ...nth layer with respect to the intercalate[53,54],

- the **color** of the GICs is quite different from that of graphite: with metals, the first stage compounds are gold-like colored, the second stages are steel-blue, as are the first stage compounds with acceptors. The colors are as shiny as those of metals and this is one of the reasons for which Ubbelohde called them "Synthetic Metals", the other one being their high electrical conductivity[55],

- graphite itself presents a fairly high electrical **in-plane conductivity**, in spite of a low carrier concentration (10^{-4} per carbon atom). Its resistivity - about 40 $\mu\Omega$cm - is due to a very high in-plane mobility of the carriers. On the other hand, the resistivity perpendicular to the planes is 10^3 to 10^4 times higher.

The GICs present a metal-like in-plane conductivity, with values close to those of copper or aluminum. The conductivity perpendicular to the planes becomes higher than that of pristine graphite when the intercalate is an electron donor and the anisotropy decreases down to 10 in LiC_6 which is the most 3D GIC. The $\vec{c}$ axis conductivity decreases when the intercalate is an acceptor species, then the anisotropy reaches values as high as 10^6 in the case of graphite-AsF_5 compounds[45].

The in-plane conductivity increase is due to the higher number of the carriers, but is limited by their mobility which decreases drastically because of the defects created during the intercalation. A spread maximum of the in-plane conductivity occurs at stages 4 to 6 with donors, 2 to 5 with acceptors but even for a stage 10, the electrical conductivity remains higher than that of graphite.

Between 1975 and 1985, the critical temperature of the **superconducting behavior** increased from 170 mK (KC_8) up to 4.05 K ($CsBiC_4$)[8]. This was very exciting before the discovery of the High T_c superconductors, however, one interesting fact remains: the anisotropy of the critical fields[56].

5.   APPLICATIONS

The presence of the carbon matrix allows the reactant to remain "solid" even at relatively high temperature (largely above its melting point). On the other hand, it induces an orientation in addition reactions: chlorination of organic molecules[57] as well as copolymerization of styrene and isoprene. In this case, the graphene layers involves a good regularity in the sequences sty-iso-sty...[58].

The intercalate is dispersed and one admits generally that this dispersion effect is the main factor for the high reactivity of the GICs. In fact this assumption is not true: the main factor is the high mobility of some intercalates in the galleries, as shown, for instance, by the oxidation of the binary alkali metal compounds: the oxide is formed at the edge (out of the carbon planes) of the sample. When the intercalate is either as inclusions between the carbon layers or very far from its melting point, the mobility remains low and this explains why such compounds cannot be used as catalysts: most of the reactants do not penetrate into the galleries unless their size is small enough in comparison with that of the intercalate like hydrogen for instance.

Before describing some applications of the GICs, one has to take into account the negative effects of the intercalation phenomenon, which occurs in several industrial processes and, then, are of importance. For instance, in the atmosphere of the blast-furnaces there is always potassium vapor. The coke reacts, as well as any carbon material, with potassium, becomes fragile, breaks and reduces the porosity to the gases of the mixture of ore and coke.

Many patents are related to the applications of the GICs[59]. Actually there are only two industrial processes based on GICs.

The first one in the manufacturing of **carbon cardboards**: French Papyex[R] and American Graphoil[R]. The way for the production is as follows:

- preparation of a GIC (generally a graphite sulfate),

- rapid heating at high temperature (largely above the boiling point of the intercalate) which involves the exfoliation of graphite and leads to expanded graphite. This material can be used as prepared or compacted in very light bricks whose apparent density is around 0.1, as heat insulator,

- rolling under pressure of the expanded graphite to form the graphite cardboard which can be used as substitute for asbestos for high temperature gaskets.

The present production of this material is in terms of hundreds of tons per year around the world.

The second actual application is the **extinguishing of sodium fires** by complexes containing ferric chloride and ammonia intercalated into graphite[60]. The effects of these compounds are multiple:

- desorption and vaporization of the intercalate,

- formation of sodium chloride which melts on the surface of sodium,

- formation of a graphite blanket which covers the whole and prevents a further contact of oxygen with sodium.

By using this compound instead of a mixture of sodium chloride, sodium carbonate and graphite, one reduces the amount of the extinguisher by a factor of 10[61].

An important interest of the binary GICs containing an alkali metal consists of their **selective reaction with hydrogen**. The first attempt in this direction was carried out in 1960[33] by reacting hydrogen on the first stage compound KC_8. This reaction leads to a new intercalation phase: a ternary compound graphite-potassium-hydrogen, according to the equation:

$$KC_8 + 1/3H_2 \rightleftharpoons KH_{2/3}C_8 \qquad (1)$$

This addition, totally reversible, involves a complete reorganization of the intercalated layer of potassium: in gathers in one half of the surface between the carbon sheets, according to the Daumas-Hérold model[62] and splits in two layers surrounding a medium plane of hydrogen[34].

The reactions of hydrogen and deuterium with KC_8 are quite different: a mixture of $H_2 + D_2$ is enriched in deuterium in the gaseous phase. This reaction could be used for the isotopic protium-deuterium separation; unfortunately, there is also formation of HD molecules. On the other hand, the physisorption at low temperature on a second stage MC_{24} (with M = heavy alkali metal), which presents a higher separation coefficient, leads to an enrichment in deuterium of the solid phase, without any formation of HD molecules and can be applied for **isotopic separation**[22]. The physisorption occurs at 77 K and disappears at 120 K:

$$MC_{24} + 2H_2 \rightleftharpoons MC_{24},2H_2 \qquad (2)$$

This equilibrium can be also used for the **storage of hydrogen**. The amount of stored hydrogen is 1.2% in weight when the metal is potassium, a little less than the value of 1.6% considered as the minimum content for practical uses. However, the reaction is totally reversible, unlike to the other hydrogen storage materials.

The high electrical conductivity of some GICs makes them possible **electrical wires**. There are two limitations to this application: the fragility of the compounds towards air and humidity and, up to now, one does not know how to prepare graphite fibers which possess both good mechanical properties and a low resistivity (the conductivity of the final compound is tightly correlated to that of the precursor). On the other hand, the GICs present two advantages compared to the classical wire materials like copper or duralumin: they are light and have a very small (even negative) thermal expansion coefficient along the $\vec{a}$ axis, which is parallel to the fiber axis. In the case of use of GICs based on fibers, one could multiply by a factor 2 or 3 the distance between two power pylons, distance which is limited by the thermal expansion of the materials.

There is research on **heat storage** through the intercalation-deintercalation of ammonia[63], e.g.:

$$MC_{12}(NH_3)\beta + (\alpha - \beta)NH_3 \rightleftharpoons MC_{12}(NH_3)_\alpha$$

The main application as **electrodes for batteries** are due to compounds derived from graphite but which are not actually GICs: for instance graphite fluoride[64] or graphite oxide[65], however, there are some applications based on electrodes made of GICs containing Lewis acids[66].

The high reactivity with oxygen and water can be used for the **purification of gases**. For instance, a gas like nitrogen, rare gases, CO, CH$_4$... passing through a powder of CsC$_{24}$ has a content of oxygen as low as 10^{-2} ppm[67].

At the scale of the laboratory, one can prepared **high purity hydrogen** using reactions (1) or (2) since those reactions are specific to hydrogen.

It is also possible to do isotopic separation by intercalation-deintercalation: as a rule, when two isotopes can be intercalated into graphite, the heavier intercalates preferentially[68].

REFERENCES

1. K. Fredenhagen and G. Cadenbach, **Z. Anorg. Allgem. Chem.**, 3:158 (1926)
2. A. Hérold, **Bull. Soc. Chim. Fr.**, 999 (1955)
3. R.C. Asher and S.A. Wilson, **Nature**, 181:409 (1958), A. Métrot and A. Hérold, **C.R. Acad. Sc. Paris**, 2C:883 (1967)
4. R. Juza and V. Wehle, **Natw.**, 52:560 (1965), D. Guérard and A. Hérold, **C.R. Acad. Sc. Paris**, 262C:557 (1966) and **Carbon**, 13:337 (1975)
5. D. Guérard and A. Hérold, **C.R. Acad. Sc. Paris**, 279C:455 (1974) and 280C:729 (1975)
6. D. Guérard and A. Hérold, **C.R. Acad. Sc. Paris**, 281C:929 (1975); M. El Makrini, D. Guérard, P. Lagrange and A. Hérold, **Physica**, 99B:257 (1980)
7. P. Lagrange, M. El Makrini, D. Guérard and A. Hérold, **Physica**, 99B:473 (1980); M.G. Alexander, D.P. Goshorn, D. Guérard, P. Lagrange, M. El Makrini and D.G. Onn, **Sol. St. Comm.**, 38:103 (1981)
8. E. McRae, J.F. Marêché, A. Bendriss-Rerhrhaye, P. Lagrange and M. Lelaurain, **Ann. Phys.**, 11:12 (1986)
9. P. Lagrange, A. Essaddek and J. Assouik, **Synth. Met.**, 34:9 (1989)
10. D. Guérard, C. Takoudjou and F. Rousseaux, **Synth. Met.**, 7:43 (1983)
11. D. Guérard, L. Elansari, N.E. Elalem, J.F. Marêché and E. McRae, **Synth. Met.**, 34:27 (1989)
12. A. Frenzel, Dissertation Berlin 1933; W. Rüdorff, **Z. Anorg. Allgem. Chem.**, 245:383 (1941)
13. R. Juza and A. Schmeckenbecher, **Chem. Ber.**, 291:46 (1957); G. Colin and A. Hérold, **C.R. Acad. Sc. Paris**, 244C:46 (1957); G.R. Hennig, **J. Chem. Phys.**, 20:1443 (1952); B. Bach and A. Hérold, **C.R. Acad. Sc. Paris**, 257C:1706 (1963); G. Furdin, B. Bach and A. Hérold, **C.R. Acad. Sc. Paris**, 271C:683 (1970)
14. H. Thiele, **Z. Anorg. Allgem. Chem.**, 207:340 (1932)
15. J. Mélin and A. Hérold, **C.R. Acad. Sci. Paris**, 269C:877 (1969)
16. P. Scharff, E. Stumpp and C. Eberhart, **Synth. Met.**, 34:121 (1989)
17. A.R. Ubbelohde, **Proc. Roy. Soc.**, A304:25 (1968); H. Fuzellier, Thèse Nancy (1974)
18. W. Rüdorff, **Z. Phys. Chem.**, 45:42 (1940)
19. G.R. Hennig, **J. Chem. Phys.**, 19:922 (1951); H. Fuzellier and A. Hérold, **C.R. Acad. Sc. Paris**, 267C:607 (1968)
20. H. Fuzellier, A. Hérold and M. Bagouin, **C.R. Acad. Sc. Paris**, 262C:1074 (1966)

21. R. Vasse, G. Furdin, J. Mélin and A. Hérold, **Carbon**, 4:248 (1981)
22. P. Lagrange, D. Guérard, J.F. Marêché and A. Hérold, **J. Less Com. Met.**, 131:371 (1987)
23. H. Schäfer-Stahl and G. von Eynatten, **Synth Met.**, 7:73 (1983)
24. E. Stumpp, **Mat. Sc. Eng.**, 31:53 (1977)
25. H. Selig and O. Gani, **Inorg. Nucl. Chem. Lett.**, 11:75 (1975)
26. Conferences devoted to the GICs: a) La Napoule, France, May 1977, **Mat. Sc. Eng.**, 31: (1977), b) Provincetown, USA May 1980, **Synth. Met.**, 2: (1980), c) Pont-à-Mousson, France, May 1983, **Synth. Met.**, 7-8: (1983), d) Tsukuba, Japan, May 1985, **Synth. Met.**, 12: (1985), e) Jerusalem, Israel, May 1987, **Synth. Met.**, 23: (1988), f) Pont-à-Mousson, France, March 1988, Proc. of Int. Coll. on Layered Compounds, D. Guérard and P. Lagrange, eds., g) Berlin, FRG, May 1989, **Synth. Met.**, 34: (1989)
27. D. Guérard and A. Hérold, **C.R. Acad. Sc. Paris**, 275C:571 (1972)
28. S. Basu, C. Zeller, P. Flanders, C.J. Fuerst, W.D. Johnson and J.E. Fischer, **Mat. Sc. Eng.**, 38:275 (1979); D. Billaud, E. McRae, J.F. Marêché and A. Hérold, **Synth. Met.**, 3:21 (1981)
29. B. Carton, Thèse Nancy 1971
30. L. Bonnetain, P. Touzain and A. Hamwi, **Mat. Sc. Eng.**, 31:45 (1977); F. Béguin, R. Setton and L. Facchini, **Synth. Met.**, 7:263 (1983)
31. F. Béguin, R. Setton, A. Hamwi and P. Touzain, **Mat. Sc. Eng.**, 40:167 (1979)
32. S.A. Hark, B.R. York and S.A. Solin, **Synth. Met.**, 7:257 (1983)
33. D. Saehr and A. Hérold, **Bull. Soc. Chim. Fr.**, 3:130 (1965)
34. D. Guérard, P. Lagrange and A. Hérold, **Mat. Sc. Eng.**, 31:29 (1977)
35. W. Rüdorff and R. Zeller, **Z. Anorg. Allgem. Chem.**, 279:182 (1955)
36. B. Bouayad, A. Marrouche, M. Tilhi, H. Fuzellier and A. Metrot, **Synth. Met.**, 7:159 (1983)
37. N. Platzer, **Bull. Soc. Chim. Fr.**,:1177 (1961)
38. J.G. Hooley, **Can. J.Chem.**, 35:374 (1957); **Synth. Met.**, 7:159 (1983)
39. R. Gross, Thèse Nancy (1962)
40. M.E. Vol'pin, Y.N. Novikov, N.D. Lapkina, V.I. Kastochkin, Y.T. Struchkov, M.E. Kazakov, R.A. Stukan, V.A. Povitsktj and A.V. Zvarikina, **J. Amer.Chem. Soc.**, 97:3366 (1975)
41. A. Métrot, R. Vangélisti, P. Willmann and A. Hérold, **Eletroch. Acta.**, 24:685 (1979)
42. M.Z. Tahar, L. McNeil, M.S. Dresselhaus and M. Endo, **Proc. Carbone 1984**, Bordeaux, France, :176 (1984)
43. N. Wada, R. Clarke and S.A. Solin, **Synth. Met.**, 3:27 (1980) and **Phys. Rev. Lett.**, 44:1616 (1980); J.E. Fischer, C.D. Fuerst and K.C. Woo, **Synth. Met.**, 7:1 (1983)
44. A. Essaddek, M. Lelaurain, J.F. Marêché, E. McRae and P. Lagrange, **Synth. Met.**, 34:365 (1989)
45. M. Lelaurain, J.F. Marêché, E. McRae, G. Furdin and A. Hérold, **J. Mat. Res.**, 3:87 (1988)
46. P. Lagrange, D. Guérard and A. Hérold, **Ann. Chim. Fr.**, 3-2:143 (1978); D. Guérard, P. Lagrange, M. El Makrini and A. Hérold, **Carbon**, 16:285 (1978) and **C.R. Acad. Sc. Paris**, 285C:405 (1977)

47. D. Guérard, M. Lelaurain and A. Aubry, **Bull. Soc. Fr. Minér. Crist.**, 98:43 (1975)
48. D. Guérard and P. Lagrange, **Carbon**, 22:579 (1984)
49. J. de Courville, D. Tchoubar and C. Tchoubar, **J. Appl. Cryst.**, 12:332 (1979)
50. J. Clément, P. Lagrange, D. Guérard and F. Rousseaux, **Analusis**, 18:18 (1990)
51. J. Gaultier, C. Hauw, J.M. Masson, J.C. Rouillon and S. Flandrois, **C.R. Acad. Sc. Paris**, 289C:45 (1979)
52. L. Pietronero and S. Strässler, **Phys. Rev. Lett.**, 47:593 (1981)
53. R.J. Nemanich, S.A. Solin and D. Guérard, **Phys. Rev.B**, 16:2965 (1977)
54. P. Lauginie, H. Estrade-Szwarckopf, B. Rousseau, J. Conard, D. Guérard, N.E. Elalem and L. Ansari, **Synth. Met.**, 34:563 (1989)
55. A.R. Ubbelohde, **Proc. Roy. Soc.**, 309:297 (1963)
56. L.A. Pendrys, R. Wachnik, F.L. Vogel, P. Lagrange, G. Furdin, M.El Makrini and A. Hérold, **Sol. St. Comm.**, 38:677 (1981)
57. K. Laali and J. Sommer, **Nouv. J. Chim.**, 5:469 (1981)
58. G. Merle, J.P. Pascault, Quang Tho Pham, C. Pillot, R. Salle, J. Golé, I. Rashkov, I.M. Panayotov, D. Guérard and A. Hérold, **J. Polym. Sc.**, 15-9:2067 (1977)
59. R. Setton, **Synth. Met.**, 23:511 and 519 (1988)
60. GB Pat.1 588 876 (1981)
61. D. Berger and J. Maire, **Mat. Sc. Eng.**, 31:331 (1977)
62. N. Daumas and A. Hérold, **C.R. Acad. Sc. Paris**, 268C:373 (1969)
63. P. Touzain, J. Michel and P. Blum, **Synth Met.**, 8:313 (1983)
64. JPN Pat. 28 246 (1980)
65. FRG Pat.2 741007 (1984)
66. Eur Pat. 126 701 (1984); JPN Pat. 20 466 (1985)
67. A. Métrot, M. Hanna and A. Hérold, **Ann. Chim.**, 3:361 (1978)
68. D. Billaud, A. Hérold and F. Leutwein, **C.R. Acad. Sc. Paris**, 268C:373 (1969)

MODIFICATION OF ELECTRONIC TRANSPORT PROPERTIES BY INTERCALATION

REACTIONS

A. Lerf and W. Biberacher

Walther-Meissner-Institut
8046 Garching, FRG

ABSTRACT

The possibilities and limitations of intercalation reactions in tuning electronic transport properties are demonstrated for the host lattice $2H-TaS_2$. The two-phase situation at low intercalate content prevents the study of the charge density wave/superconductivity interplay in the intercalation compounds. The transport properties in the normal conductivity state of the intercalation compounds seems to follow the expectation from the rigid band model. In contrast to this, the superconductivity properties are determined not only by the change of the density of states at varying charge transfer, but also by the intercalate properties, mainly by the intercalate arrangement in the van der Waals gap. It is shown that the very low T_c values of $2H-TaS_2$ and $2H-TaSe_2$ cannot be explained solely by the decrease of the density of states due to charge density wave formation.

1.   INTRODUCTION

There are many electronically conducting compounds (mainly of the transition metals) which can serve as host lattices for intercalation reactions[1-3]. (For further information on intercalation processes see the paper of Liang in this volume). Although these reactions can lead to a large expansion of the lattice, the essential structural elements remain nearly unchanged. Hence, the electronic structure should be preserved at least in first approximation. The uptake of atoms or molecules is accompanied by a reduction or an oxidation of the host[4]. These redox processes are connected with an electron transfer to or from the electronic system of the host lattice. It is assumed that this electron transfer leads to a change in the Fermi energy and as a consequence in the density of states (DOS) within an otherwise "rigid band"[5].

Consequently the intercalation process can be applied, like doping, to "tune" the electron budget of a compound and hence its electronic transport properties[5]. In this paper the limitations and possibilities of intercalation reactions for modifying physical properties are demonstrated for the special host lattice 2H-TaS$_2$. This is a very versatile host, which forms intercalation compounds of very different chemical species[6]; in this material the effect on different properties (conductivity, superconductivity, charge density wave (CDW) phenomena) can be studied simultaneously.

2. HOMOGENEITY RANGES AND PREPARATION PROBLEMS

In 2H-TaS$_2$ the S-Ta-S layers are built up from two densely packed sulphur monolayers lying one on top of the other. Half of the resulting trigonal holes are occupied by Ta atoms. These S-Ta-S layers are stacked in a hexagonal close packing scheme forming octahedral and tetrahedral sites in the van der Waals gap[6].

As in the case of doping, one needs information on the homogeneity ranges and their size. Only within these single phase regions is it reasonable to look at the effect on the physical properties. From the investigation of intercalation processes it is known that all intercalation reactions that have been studied start with a two-phase region[7-9]. Sometimes there is a sequence of two or more two-phase regions (due to staging). There is no single-phase intercalation compound below a charge transfer n = 0.1 e$^-$/TaS$_2$. The extent of these two-phase regions depends on the type of the intercalate, but there is no simple relationship to intercalate properties. This two-phase situation can be explained if the intercalation proceeds within islands with definite intercalate content[10]. The homogeneity ranges in the single-phase compounds depend also on the type of the intercalate:

i)   no variation of the intercalate content is possible in most of the intercalation compounds of molecules[6]. The same effect is observed for the electrochemical intercalation of metal complexes (La(K222)$_1$), organic cations (S(CH$_3$)$_3$) or solvated cations[11];

ii) the largest homogeneity range is observed for Li$_x$TaS$_2$[12,13]. It amounts to 0.3 < x = n < 0.9. However, the preparation of homogeneous samples is difficult, since the lithiation tends to nonequilibrium states[13];

iii) in the case of the hydrated metal intercalation compounds M$_{n/v}$(H$_2$O)$_y$TaS$_2$ (M = Li, Na, K, Rb, Cs, Mg, Ca, Sr, Ba, Mn, Fe, Co, Ni, Zn, all rare earths ions; v = valency) there is a homogeneity range too, at least within the limits 0.27 < n < 0.4. This range is nearly independent of the type of the intercalated ion. Only the upper limit varies slightly with the intercalated ion[14]; sometimes it is determined by the onset of the water decomposition. With this series of compounds the influence of different ions and their hydration state in the same charge transfer range can be investigated.

In addition there is a particular problem which is common to all intercalation reactions. Since this solid state reaction can lead to a considerable crystal expansion (10 - 200%), the danger of crack formation or complete disintegration of the crystals is very large. As a consequence all intercalation reactions are connected with a decrease in crystal quality, an unfortunate effect for the study of physical properties. Therefore, it is essential to find such intercalation procedures as will minimize this damage. Until now this task has been solved only for a few intercalation compounds of 2H-TaS$_2$. The best samples - with respect to the residual resistance ratio - are obtained for the hydrated metal intercalation compounds via electrointercalation. This method has the further advantage that the desired charge transfer to the host lattice can be adjusted very easily[14,15].

3. SUPPRESSION OF THE CHARGE DENSITY WAVE

Before describing the influence of intercalation on the CDW, we will review the essential features of this phase transition. 2H-TaS$_2$ shows anomalies in the electronic transport properties at about 80 K[5,6,16]. Below this temperature a new phase is observed in electron diffraction by the appearance of superlattice spots at a*/3[17]. By Time Differential Perturbed Angular Correlation (TDPAC) studies three inequivalent Ta sites with populations 3:3:3 as in the case of 2H-TaSe$_2$ are observed[18].

By applying pressure to the sample the transition temperature to the CDW state (T$_{CDW}$) decreases continuously[19]. This is accompanied by a simultaneous increase of the transition temperature to the superconducting state (T$_c$). The CDW transition is suppressed completely at a pressure of about 50 kbar, and T$_c$ increases to about 4.5 K. This result is explained in the following way: the CDW formation leads to a small band gap at the Fermi energy. As a consequence the DOS decreases and, hence, the T$_c$ value. Applying pressure the bandgap closes, the DOS increases and therefore T$_c$ also. The suppression of the CDW can be achieved also by intercalation. This effect can be seen e.g. in the resistivity data. In Fig. 1 the R(T) curves for the empty host lattice and two intercalation compounds are compared. In 2H-TaS$_2$ a small jump in the resistivity versus temperature curve occurs at 78 K. It is removed completely in both intercalation compounds. In order to see whether this depression of the CDW by intercalation is also seen in the DOS, the specific heat data of 2H-TaS$_2$ and some of its intercalation compounds are considered. From these data the electronic part γ can be deduced, which is proportional to the DOS. In the intercalation compounds the γ values are about 10% higher than in the empty host lattice[21]. These authors consider this as evidence that intercalation works in a similar way to the application of pressure. This interpretation needs some further comments. In Fig. 2 T$_c$ values of all isomorphous 2H-compounds and of some intercalation compounds of NbS$_2$ and TaS$_2$ are plotted versus the γ term of the specific heat. There is only a small difference in the γ and T$_c$ values for NbS$_2$ and NbSe$_2$. The scatter of the data for NbS$_2$ (4 samples) is caused by the difficulties in preparing stoichiometric samples. In the case of the NbS$_2$ intercalation compounds a decrease in T$_c$ and γ with increasing charge transfer is observed (charge transfer n increases from n = 0 at $\gamma \simeq$ 18m J/moleK^2 to n = 0.5 at $\gamma \simeq$ 5mJ/moleK^2; the same region of n,γ,T$_c$ values is covered in Li$_x$NbS$_2$[24]). This trend is in agreement with the band structure: the Fermi energy in e.g. NbS$_2$ lies near a maximum of

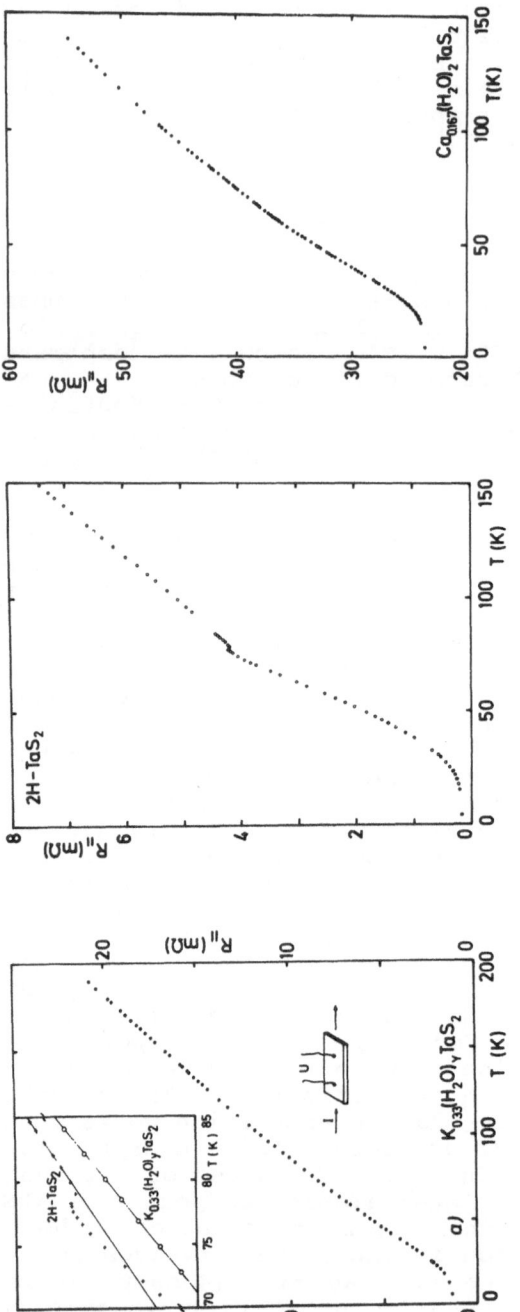

Fig. 1. Temperature dependence of the resistivity parallel to the S-Ta-S-layers for the empty 2H-TaS$_2$ (middle) and two intercalation compounds: K$_{0.33}$(H$_2$O)$_y$TaS$_2$ (left) and Ca$_{0.16}$(H$_2$O)$_y$TaS$_2$ (right). The insert in the left figure shows the region near the phase transition in expansion. According to Refs. 14 and 20.

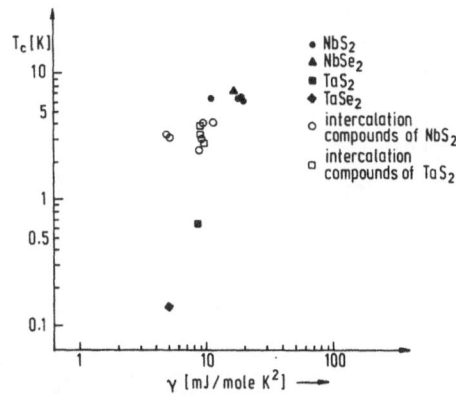

Fig. 2. Double log plot of T_c values versus the γ
term of the specific heat for the
isomorphous 2H layered dichalcogenides and
intercalation compounds of 2H-TaS_2 and 2H-
NbS_2. Data taken from Refs. 6, 21-24.

the DOS[25]; the charge transfer to the host accompanied with
intercalation can lead to a drop of the DOS only and therefore
of γ and T_c. It is not possible to differentiate between the γ
values of the intercalation compounds of TaS_2 and NbS_2. This
behavior is expected from the similar band structure of all
isomorphous dichalcogenides. Clearly separated from this area
of γ-T_c values are the corresponding values of the empty Ta-
dichalcogenides. Whereas the γ values of TaS_2 and $TaSe_2$ are of
the same order as those of the intercalation compounds, the T_c
values are essentially lower. This means, in our opinion, that
the very low T_c values of the empty materials cannot be
explained by the band gap formation in the CDW state and the
accompanying decrease in DOS alone.

From Fig. 2 one can see that intercalation shifts the T_c
values to temperatures above 2 K, in the γ-T_c region of those
compounds with slight or no CDW. Measurements of the resistance, the Hall effect (see also below), and the magnetic
susceptibility confirm the suppression of the CDW[5,6,20,26,27].
Hence the increase in T_c serves as an indication for the
suppression of the CDW[27]. Since superconducting intercalation
compounds, which are single phase and homogeneous, show a charge
transfer higher than 0.1 e^- TaS_2, this value seems to be high
enough to suppress the CDW completely. The same effect is
obtained if 5% of Ta are replaced by Nb atoms[28]. These
observations lead to the conclusion that the coexistence of
superconductivity and CDW in intercalation compounds can only be
observed at very low charge transfer to the host (a few %).
This range is not accessible by normal intercalation reactions
because of the two-phase region at low intercalation contents.
Therefore, it is not possible to tune the CDW-superconductivity
interrelation by intercalation in the manner which is possible
by the application of pressure.

There are a few intercalation compounds which are
considered as examples of the coexistence of CDW and superconductivity. But the interpretation given in the literature is
not unambiguous. In H_xTaS_2, the whole volume of the sample is

713

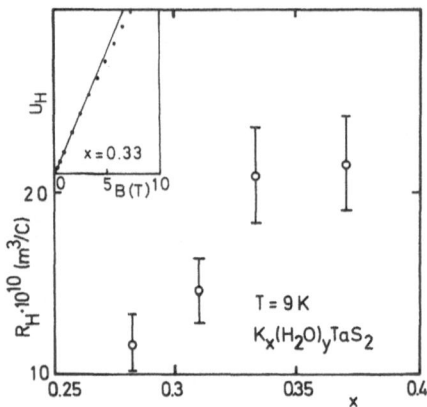

Fig. 3. Zero-field Hall constant R_H as function of the charge transfer n for $K_x(H_2O)_yTaS_2$ according to Ref. 20. In the insert the field dependence of the Hall-voltage for the compound x = 0.33 is shown. Consider that x = n for monovalent cations.

superconducting (T_c = 4 K) only at n = 0.1[27]. At concentrations below and above n = 0.1 the content of the superconducting phase decreases, indicating two-phase situations. In agreement with TDPAC studies[9] it is assumed that empty TaS_2 and $H_{0.1}TaS_2$ coexist below n = 0.1. Above n = 0.1 the protons switch to the trigonal prismatic holes within the layers[29] leading to a decrease in superconducting volume, because this compound is also not superconducting. In Fe_xTaS_2 (x < 0.1)[30], from a comparison with Fe_xNbS_2[31], the question arises whether the Fe is distributed homogeneously in the van der Waals gap.

4. INFLUENCE OF CHARGE TRANSFER ON THE ELECTRONIC TRANSPORT PROPERTIES

The electronic properties have been studied only for a few intercalation compounds of $2H-TaS_2$[5,6]. Here we report on the data for $K_x(H_2O)_yTaS_2$. Within experimental error the resistivity at room temperature does not change with increasing charge transfer[20]. In contrast to this, the Hall constant increases by a factor of 2 when the charge transfer increases from n = 0.27 to n = 0.35, as shown in Fig. 3[20]. The increase in R_H is expected from the band filling as outlined above. The different trends in resistivity and the Hall-constant can be an indication of a change of the charge carrier mobility.

More dramatic effects of n can be observed in the super-conducting transition temperatures. In Fig. 4 T_c is shown as a function of n for different intercalation compounds[15]. There is a strong change in T_c with n for the alkali metal intercalates; a smaller change is observed in the calcium compound and no change in the lanthanum compound. The decrease of T_c with increasing n seems at first sight to be in agreement with the decreasing DOS. However, the different behavior for different intercalates is in conflict with the rigid band concept.

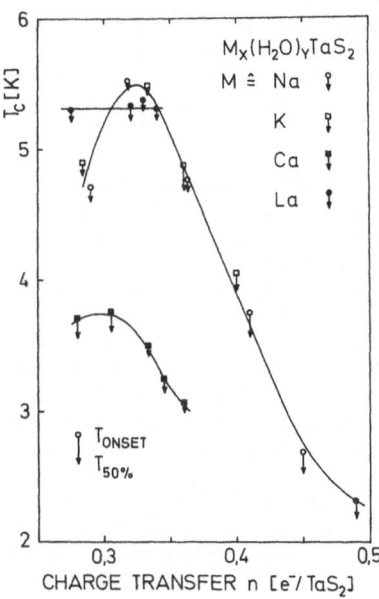

Fig. 4. T_c values as function of charge transfer n for
various hydrated metal intercalation
compounds[15].

This can be explained by the following alternatives:

i)   the rigid band concept is violated. Hall constant[20],
      specific heat (see discussion above) and magnetic suscept-
      ibility data[32] seem to confirm the rigid band concept.
      However, the dHvA-oscillations in $K_x(H_2O)_yTaS_2$ [33] and the
      change of the nuclear quadrupole interaction frequency with
      the charge transfer[7,34] cannot be understood in this
      model;

ii)   the T_c values of intercalation compounds of $2H-TaS_2$ are not
      determined solely by the charge transfer and the DOS.

A detailed look at the T_c data of the hydrated metal
intercalation compounds shows that the second possibility is not
unrealistic. The T_c value at the same charge transfer
(n = 0.33) and at the same layer distance is changed by a factor
of 2 if two sodium ions are replaced by one calcium ion. The
compounds of Na and K show the same charge transfer dependence
of T_c, although the layer separations in the two compounds
differ a factor of 2: 50% for K and nearly 100% for Na. If the
layer distance of the Na compound is reduced to that of the
potassium compound by dehydration the T_c decreases also by a
factor of 2 and is comparable to that of the Ca compound with a
layer separation of 6 Å. For an explanation of these results
one has to consider the following problem: these results show
for the first time for complex intercalates (layer separation in
the order of that of the molecular intercalation compounds;
Δd = 3-6 Å corresponding to one or two water layers surrounding
the ions) the influence of charge transfer on T_c. Although the
type of intercalate is not changed in order to obtain different
charge transfer values (as in the case of the molecular

715

intercalation compounds), it is not possible to avoid a change in the ratio of water to ion, and hence the intercalate arrangement. That the intercalate arrangement is of importance for the T_c values is shown most obviously in the following two examples.

Sodium ions can be intercalated also by a reaction of 2H-TaS$_2$ with NaOH solutions[35]. At low concentrations the compound Na$_{0.33}$(H$_2$O)$_y$TaS$_2$ is formed. The T_c value of this compound is identical with that of the electrochemically prepared sample. However, if the concentration of the NaOH solution is increased, the T_c values of the resulting compounds drops more dramatically than those of the electrochemically prepared compounds with the same charge transfer (Fig. 5). The difference in the two types of compounds is an additional uptake of NaOH in the van der Waals gap. This leads probably to a destruction of the usual water/ion arrangement.

It is possible to intercalate all rare earth ions from their aqueous solutions[36]. In all compounds with a charge transfer of n = 0.33 the layer distance is nearly the same. But the compounds with RE = Gd - Lu change the layer distance to 11.5 Å within a few minutes when kept in a normal laboratory atmosphere. In addition, by intercalation of these ions to charge transfer values higher than n = 0.35 the crystals shrink to the same layer distance of about 11.5 Å. The compounds intercalated with the rare earth ions which do not show these effects (RE La - Lu) have T_c values of 5.5 K in the 12.8 Å phase. If they are transferred to the 11.5 Å phase within a few hours, the T_c values drops to 2-3 K. For the intercalation compounds with the other ions, which are less stable in the state with the higher distance, T_c values of about 2 K are observed in this state; but T_c increases to 4-5 K, if the layer distance is reduced to 11.5 Å. Crystal field splitting obtained from magnetic susceptibility measurements allows the conclusion that e.g. Nd has a different coordination from that of the Tb compound. According to the coordination behavior of the rare earth ions in aqueous solutions[37] we assume that the RE ions La - Eu are ninefold coordinated and Gd - Lu have the coordination number eight. As a consequnce the packing of the

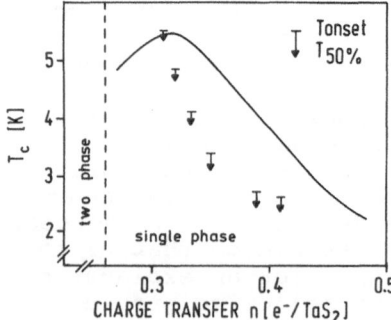

Fig. 5. T_c values as function of the charge transfer n for Na$_x$(H$_2$O)$_y$TaS$_2$ prepared by the treatment of 2H-TaS$_2$ with NaOH solutions (symbols) and by electrointercalation (full line). The data of the first series are taken from Ref. 35, those of the second series from Fig. 4.

rare earths/aquo complexes in the interlayer gap should be different. This may be the reason for the different hydration behavior and the different T_c values. The example of the rare earth intercalates shows most obviously how, e.g., the coordination of the intercalate ions can control the structure of the interlayer electrolyte film and thus indirectly the physical properties.

OUTLOOK

The existence of narrow homogeneity ranges and especially the two-phase situation at low intercalate content prevent the continuous tuning of physical properties. This phenomenon is not restricted to 2H-TaS$_2$. It seems to be very general in intercalation chemistry as is shown by a look at 15 different host lattices and most of their intercalation derivatives. At the moment it is not clear whether the two-phase regions at low intercalate content are at least in part of kinetic origin (due to the nature of the intercalation process). Besides this problem there are two-phase regions and very small homogeneity ranges, which are definitely of thermodynamic origin (cf. the phase diagram of LiVS$_2$). In these cases a well defined stoichiometry (accompanied with a superstructure of the intercalate) is energetically favoured over a situation with randomly distributed intercalate of variable composition (effects on the band structure?). Since the size of the homogeneity ranges depends on the intercalate, the accompanying breakdown of continuous tuning of physical properties can be considered as the strongest influence of the intercalate on the electronic transport properties.

Besides this dramatic influence of the intercalate there are more subtle effects, which cannot be observed in all electronic transport properties. These can be caused by the details of the intercalate arrangements (as in the case of the T_c values of the hydrated metal intercalation compounds), by polarization effects or by the charge distribution between host and lattice (it is not generally true that the charge separation is nearly as complete as in the 2H-TaS$_2$ intercalation compounds discussed above). In our opinion the intercalate-specific effects on the electronic transport properties should be considered in more detail in further investigations.

REFERENCES

1. R. Schöllhorn, in: "Inclusion Compounds", J.L. Atwood, J.E. Davies, D.D. MacNicol, eds., Academic Press 1:249 (1984)
2. "Intercalation in Layered Materials", NATO ASI Series B vol. 148, M.S. Dresselhaus, ed., Plenum Press, New York (1986)
3. "Chemical Physics of Intercalation", NATO ASI Series B vol. 172, A.P. Legrand, S. Flandois, eds., Plenum Publishing Corp., New York (1987)
4. R. Schöllhorn, Angew. Chemie. Int. Ed. Engl., 19:983 (1980)
5. R.H. Friend, A.D. Yoffe, Adv. Phys., 36:1 (1987)
6. G.V. Subba Rao, M.W. Shafer in: "Intercalated Materials", Physics and Chemistry of Materials with Layered Structures, F. Levy, ed., Reidel, Dordrecht 6:99 (1979)
7. T. Butz, A. Lerf, Rev. Chim. Min., 19:496 (1982)

8. A. Lerf, T. Butz in: "Reactivity of Solids", P. Barret, L-C Dufour, eds., Elsevier, Amsterdam Part A:473 (1985)

9. T. Butz, A. Lerf, **Ber. Bunsenges. Phys. Chem.**, 90:638 (1986)

10. S.A. Safran, D. Hamann, **Phys. Rev. Lett.**, 42:1410 (1979)

11. C. Ramos, Diploma Thesis, TU München (1986)

12. A.H. Thompson, **Physica**, 99B:100 (1980)

13. T. Butz, A. Lerf, J.O. Besenhard, **Rev. Chim. Min.**, 21:556 (1984)

14. W. Biberacher, Thesis, TU München (1984)

15. W. Biberacher, A. Lerf, J.O. Besenhard, H. Möhwald, T. Butz, S. Saibene, **Il Nuovo Cim.**, 2D:1706 (1983)

16. J.A. Wilson, F.J. DiSalvo, S. Mahajan, **Adv. Phys.**, 24:117 (1975)

17. J.P. Tidmann, O. Singh, A.E. Curzon, R.F. Frindt, **Phil. Mag.**, 30:1191 (1974)

18. T. Butz, K-H. Ebeling, E. Hagn, S. Saibene, E. Zech, A. Lerf, **Phys. Rev. Lett.**, 56:639 (1986)

19. D. Jerome, C. Berthier, P. Molinie, J. Rouxel, **J. Phys. (Paris)**, 37:C4 (1976)

20. W. Biberacher, A. Lerf, **Mol. Cryst. Liq. Cryst.**, 121:149 (1985)

21. R.E. Schwall, G.R. Stewart, T.H. Geballe, **J. Low Temp. Phys.**, 22:557 (1976)

22. Y. Hamaue, R. Aoki, **J. Phys. Soc. Japan**, 55:1327 (1986)

23. H. Schwenk, Diploma Thesis, TU München (1977)

24. D.C. Dahn, J.F. Carolan, R.R. Haering, **Phys. Rev. B.**, 33:5214 (1986)

25. N.J. Doran, B. Ricco, D.J. Titterington, G. Wexler, **J. Phys.**, C11:685 (1978)

26. F.J. DiSalvo, **Ferroelectrics**, 17:361 (1977)

27. D.W. Murphy, F.J. DiSalvo, G.W. Hull, J.V. Waszczak, S.F. Meyer, G.R. Stewart, S. Early, J.V. Acrivos, T.H. Geballe, **J. Chem. Phys.**, 62:967 (1975)

28. M. Ikebe, K. Katagiri, Y. Watanabe, Y. Muto, **Physica**, 105B:453 (1981)

29. C. Riekel, H.G. Reznik, R. Schöllhorn, C.J. Wright, **J. Chem. Phys.**, 70:5203 (1979)

30. S.J. Hillenius, R.V. Coleman, E.R. Domb, D.J. Sellmeyer, **Phys. Rev. B**, 19:4711 (1979)

31. F.W. Boswell, A. Prodan, W.R. Vaughan, J.M. Corbett, **Phys. Stat. Sol. (A)**, 45:469 (1978)

32. D.C. Johnston, **Sol. State Comm.**, 43:533 (1982)

33. W. Biberacher, W. Joss, J.M. v Ruitenbeek, A. Lerf, **Phys. Rev.B**, 40:115 (1989)

34. P. Blaha, in preparation

35. W. Biberacher, A. Lerf, F. Buheitl, T. Butz, A. Hübler, **Mat. Res. Bull.**, 17:633 (1982)

36. C. von Wesendonk, W. Biberacher, A. Lerf, **Sol. State Comm.**, 74:183 (1990)

37. A. Habenschuss, F.H. Spedding, **J. Chem. Phys.**, 70:2797,3758 (1979)

MOLECULAR DYNAMICS SIMULATION OF ION TRANSPORT IN BETA"-ALUMINA

C. Lane and G.C. Farrington

Department of Materials Science and Engineering
University of Pennsylvania
3231 Walnut Street, Philadelphia, PA 19104, USA

1.   INTRODUCTION

Na(I)-β"-alumina, ($Na_{1.67}Mg_{0.67}Al_{10.33}O_{17}$), is a layered oxide well known for its ability to undergo ion exchange in which the sodium ion content is replaced by a variety of mono, di, and trivalent cations[1]. It consists of alternating layers of Al, O, and Mg, commonly referred to as 'spinel' blocks, and conduction layers. The spinel blocks comprise four close-packed oxygen layers with aluminum ions at interstitial sites. The conduction layers contain both the so-called 'column' oxygens, which are approximately 5.6 Å apart and which link the spinel blocks together through Al-O-Al bonds, as well as the entire sodium ion content. The sodium ions diffuse rapidly along hexagonal pathways around the column oxygens via what is widely accepted to be a vacancy mechanism.

The general chemical formula for Na(I)-β"-alumina is $Na_{1+x}Mg_xAl_{11-x}O_{17}$, where x is typically 0.67. This nonstoichiometry results in excess sodium in the conduction layers, compared to the sodium content of its ideal formula, $NaAl_{11}O_{17}$, and the excess charge is compensated by Mg(II) substitution for Al(III) in the spinel blocks. The excess sodium can be easily accommodated in the structure because the conduction layers have a large population of vacant cation sites. Sodium ions are primarily found in the tetrahedrally-coordinated Beevers-Ross (BR) site, which is most favorable, or the octahedrally-coordinated mid-oxygen (mO) site.

It has been widely conjectured that order/disorder reactions might occur among the mobile ions and vacancies in the β"-alumina conduction layers which would influence both ion arrangement and mobility. Diffuse X-ray scattering has shown evidence for the formation of a $\sqrt{3}$a by $\sqrt{3}$a superlattice of sodium and vacancies at BR sites at room temperature [2]. A schematic diagram of this superlattice is shown in Fig. 1. Furthermore, the sodium ion conductivity is nonlinear with respect to temperature in an Arrhenius-type plot, also shown in Fig. 1. It has been suggested that this nonlinearity is the result of a subtle order/disorder transition in which a low

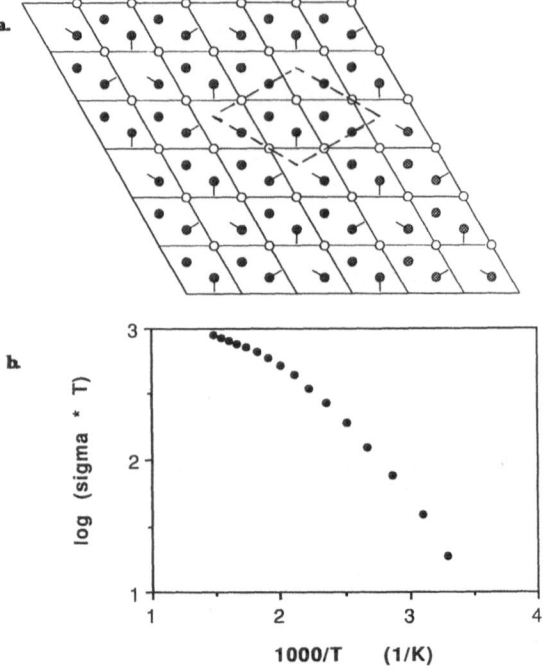

Fig. 1. (a) Schematic representation of the $\sqrt{3}a$ by $\sqrt{3}a$
cation-vacancy superlattice observed by Boilot
et al.[2] in Na(I)-beta"-alumina. Short lines
drawn from cations that border vacancies
indicate the direction of relaxation. (b)
Arrhenius plot of the conductivity as a function
of temperature for Na(I)-beta"-alumina.

temperature ordered configuration becomes disordered at
temperatures approximately greater than 300°C[3].

The possible order/disorder reactions differ greatly among
the various compositions of β"-alumina. One of the most
intriguing variations of composition can be created by changing
the stoichiometry of the conduction layer by altering the ratio
of mobile ions to vacancies. This is accomplished by exchanging
multivalent cations for the Na(I). For example, when divalent
cations, such as Ba(II), are introduced into Na(I)-β"-alumina,
one Ba(II) ion replaces two Na(I) ions. The result is an
increase in the number of vacancies. It turns out that at 80%
exchange of Ba(II) for Na(I), a stoichiometric number of mobile
cations in the β"-alumina chemical formula is reached; i.e.
there are one vacancy and one mobile ion present per unit cell.
This is precisely the composition at which ion-vacancy ordering
would seem most likely. In fact, evidence for ordering at this
composition has been observed for Na(I)-Eu(II)-β"-alumina by X-
ray diffraction[4] and for Na(I)-Pb(II)- and Na(I)-Ba(II)-β"-
alumina by conductivity measurements[5,6].

This paper describes the results of molecular dynamics
simulations of ion-vacancy ordering in Na(I)- and mixed Na(I)-
Ba(II)-β"-alumina.

Table 1. Interatomic potential parameters: 1. Fit to
bulk properties of Na(I)-beta-alumina[9]; 2.
Fit to bulk properties of MgO and BaO[10]; 3.
Calculated by H. Hummel (private communication)
using the electron gas approximation[11].

	A	ρ	C	Source
O-O	22764.0	0.14900	27.88	1
O-Al	1460.3	0.29912	0.0	1
O-Mg	1428.5	0.29450	0.0	2
O-Na	1226.8	0.30650	0.0	1
O-Ba	905.7	0.39760	0.0	2
Al-Al	0.0	0.1	0.0	
Al-Mg	0.0	0.1	0.0	
Al-Na	15240.7	0.14620	0.0	1
Al-Ba	9997.6	0.22600	0.0	3
Mg-Mg	0.0	0.1	0.0	
Mg-Na	0.0	0.1	0.0	
Mg-Ba	0.0	0.1	0.0	
Na-Na	9597.4	0.16789	0.0	1
Na-Ba	5626.7	0.25300	0.0	3
Ba-Ba	12654.0	0.27500	0.0	3

2. MD SIMULATION

The MD code used in this work is loosely based on a
constant volume, number, and energy code written by Walker[7].
Simulations were carried out on the Pittsburgh Supercomputing
Center Y-MP and a Stardent Titan located in our laboratory.
Since the MD technique has been described in great detail
elsewhere[8], only those aspects of the simulation particular to
this study are presented here.

2.1 Interionic Potentials

All ions were treated as rigid spheres in order to minimize
computation time. The Coulombic contribution to the interionic
potentials was calculated by an Ewald summation. Short range
potentials were in the Born-Mayer-Huggins form. The values of
the parameters A_{ij}, ρ_{ij}, and C_{ij} and their sources are listed in
Table 1.

2.2 Equilibration

For both studies of pure Na(I)- and mixed Na(I)-Ba(II)-β''-
alumina, the MD box included nine full unit cells, or
approximately 800 atoms. Initially, the cations were placed
randomly in the conduction layers. In each case, the system was
allowed to equilibrate for 3000 timesteps, each 5×10^{-15}
seconds long. During the equilibration period, the velocities
of all of the atoms were scaled periodically to achieve a
desired temperature. The lattice parameters were also scaled
periodically to assure that the internal stresses were equal to
zero. At the end of the equilibration time, the temperature and
the lattice parameters were found to oscillate around average
values. The volume of the box was then held fixed for the rest
of the simulation run by incorporating the average lattice
parameters.

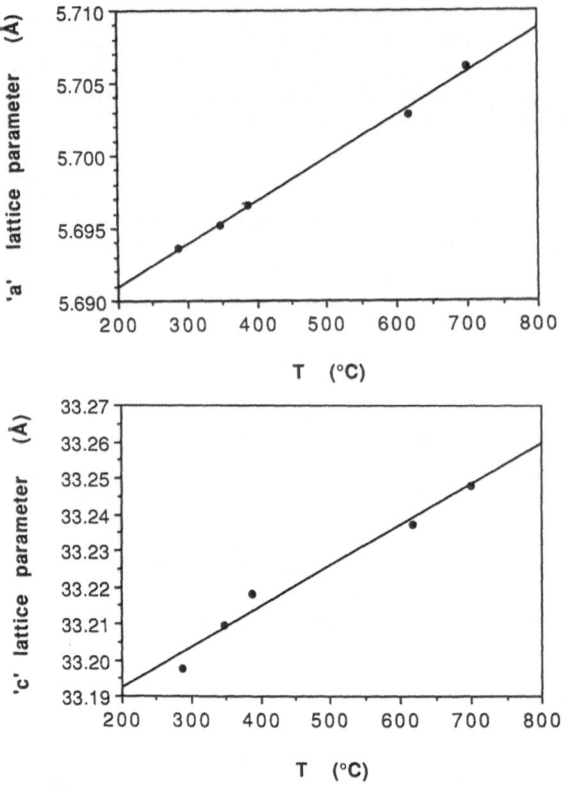

Fig. 2. Simulated temperature dependence of the lattice parameters in the 'a' and 'c' directions in Na(I)-β"-alumina.

2.3 Diffusion

The tracer diffusion coefficient was obtained from the random walk equation:

$$D_T = 4<r^2(t)>t.$$

The mean square distance traveled, $<r^2(t)>$, was calculated by averaging over the number of mobile cations and over time origins[12]. D_T is related to the experimental conductivity by the Nernst-Einstein relation. For simplicity, f_c, the charge correlation factor, was taken as unity, in accordance with a vacancy mechanism for ionic conduction[13]. Values of f_t, the tracer correlation factor, as a function of composition were taken from the work of Murch and Thorn[14].

3.   SIMULATION RESULTS

3.1 Na$^+$-β"-alumina

Simulations were carried out over a temperature range of 290°C to 700°C. The simulated temperature dependence of the lattice parameters is indicated in Fig. 2. The slopes of the lines are 2.99 x 10^{-5} Å/°C and 1.12 x10^{-4} Å/°C for the a and c

722

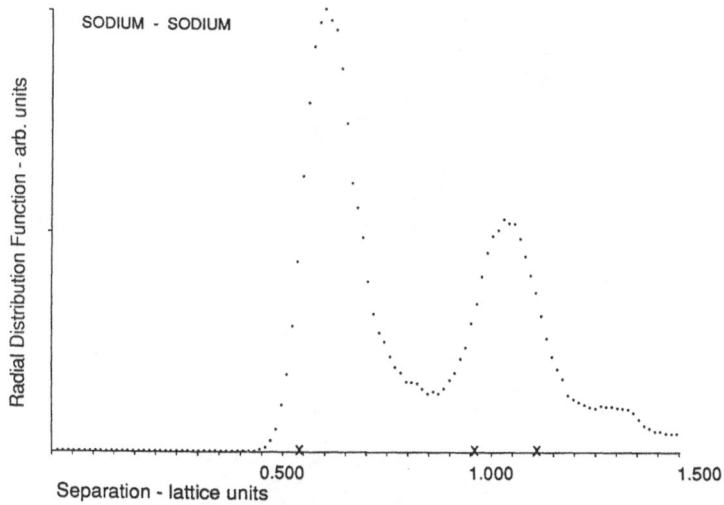

SODIUM - SODIUM

Radial Distribution Function - arb. units

Separation - lattice units

0.500   1.000   1.500

Fig. 3. Simulated radial distributuion function at 290°C
for Na(I)-β"-alumina. Small x's along the
horizontal axis indicate the crystallographic
first, second and third nearest neighbor
distances.

directions respectively. These values correspond reasonably
well with the experimental values 4.56 x 10^{-5} Å/°C and
2.34 x 10^{-4} Å/°C measured by May and Henderson[15].

   At all temperatures, Na(I)-Na(I) radial distribution
functions obtained from our simulations indicate that the Na(I)-
Na(I) nearest neighbor interatomic separation is shifted by
approximately 0.4 Å from its stoichiometric value, thus showing
the distance between Na(I) in unrelaxed BR sites and Na(I) in
relaxed sites bordering vacancies. At moderately low
temperatures, a very small hump is observed between the first
and second nearest neighbor peaks, indicative of the distance
between the relaxed Na(I) ions themselves. An example of such a
distribution at 290°C is shown in Fig. 3. These results agree
generally with previous MD simulations of pure Na(I)-β"-alumina
by Wolf et al.[16].

   The simulations presented in [16] utilized a 2 by 2 unit
cell which, although sufficient to predict local structural
properties, is, as they acknowledge, not large enough to
accommodate the formation of the $\sqrt{3}a$ by $\sqrt{3}a$ superlattice which
has been experimentally observed at low temperatures. Zendejas
et al.[17] extended this work by using a simulation box that is
large enough to accommodate the formation of such a superlattice
and they did, in fact, observe it.

   Our simulations both confirm the formation of this
superlattice in the conduction layers and show its nonstatic
nature. Figure 4 shows sequential 'snapshots', each lasting
7.5 ps, of one conduction layer at 290°C. The figure shows that
the conduction layers in the simulation box alternate between
periods of relative quiescence interrupted by periods of rapid
ion motion. During the quiet periods, the sodium ions are
relatively immobile and the expected vacancy superlattice and

a.

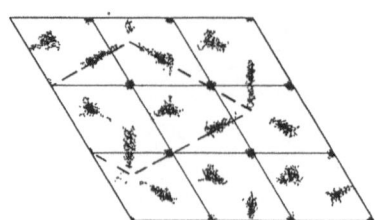

b.

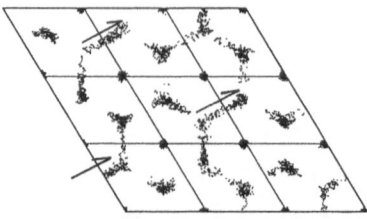

c.

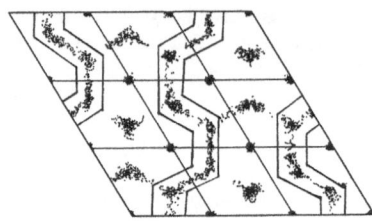

d.

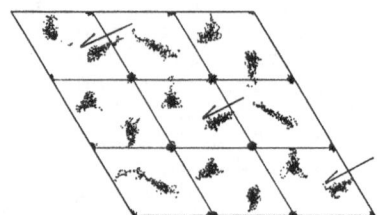

Fig. 4. Sequential 7.5 ps 'snapshots' of a simulation
run at 290°C for Na(I)-β"-alumina. 4a Initial
quiet period showing evidence of the $\sqrt{3}a$ by $\sqrt{3}a$
superlattice; 4b transition involving correlated
motion of cations; 4c indication of 'diffusion
channels' around static pairs of occupied BR
sites; 4d correlated motion allowing for the
reformation of the superlattice.

sodium ion relaxation toward adjacent vacancies are observed, as
can be seen in Fig. 4a. A transition period then occurs
involving the correlated jumps of cations into a static
configuration involving pairs of occupied BR sites with the $\sqrt{3}a$
by $\sqrt{3}a$ periodicity. The result is the formation of temporary
'channels' through which cations can move freely until the
static configuration dissolves by means of another set of
correlated jumps. At this point, the vacancy superlattice
appears again, as seen in Fig. 4d. This process has been

observed to repeat itself with other cations forming the static
configuration and the channels proceeding in different
directions.

The implication of these observations is that sodium ion
motion at low temperatures in β''-alumina does not occur by a
simple vacancy mechanism, as has long been thought, but appears
to be the result of strongly cooperative ion motion that occurs
only when just the right ion-vacancy arrangement is established.
It has been conjectured that the ion-vacancy superlattice does
not form at higher temperatures. If so, then ion transport
under these conditions can be expected to occur by the expected
vacancy mechanism. Our preliminary simulations at higher
temperatures reveal some evidence that the channel forming
process may still be occurring, but more rapidly. Further
analysis is underway.

3.2   Na(I)-Ba(II)-β''-alumina

Simulations of mixed Na(I)-Ba(II)-β''-alumina compositions
have also had some success at reproducing expected microscopic
ordering processes and conductivity trends. Thomas et al. have
already reported preliminary MD simulations of a single
composition of Na(I)-Ba(II)-β''-alumina. Their work was aimed

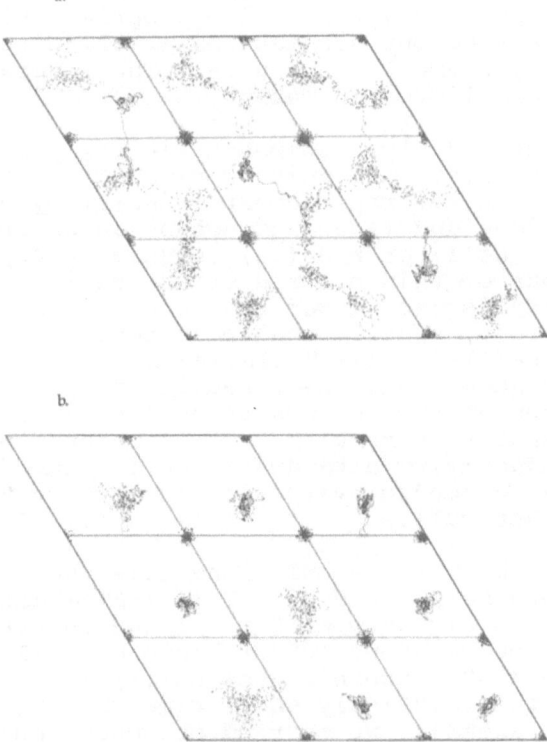

Fig. 5. (a) 7.5 ps 'snapshot' of one conduction layer in
        40% exchanged Na(I)-Ba(II)-β''-alumina (3Ba(II),
        9 Na(I), and 6 vacancies); (b) 80% exchanged
        Na(I)-Ba(II)-β''-alumina (6 Ba(II), 3Na(I), and 9
        vacancies).

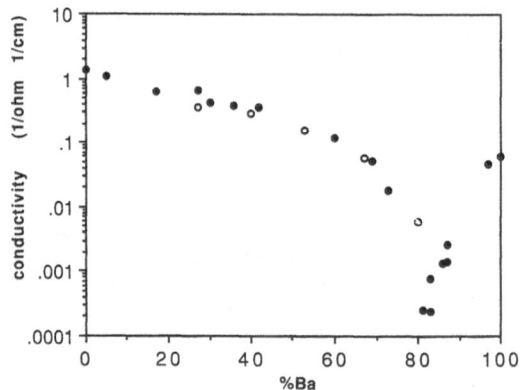

Fig. 6   Simulated and experimental conductivity as a
function of composition in mixed Na(I)-Ba(II)-
β"-alumina.   Closed circles represent
experimental data measured at 673 K.   Open
circles represent conductivity values calculated from MD simulation results at 1200 K.

principally at examining the influence of Ba(II) on the local
structure, and they chose to ignore the short-range repulsive
potentials associated with Ba(II) for the sake of
simplicity[17].   We have introduced appropriate Ba(II)
potentials and carried out simulations at 927°C for $Na_{1.67-2x}$
$Ba_xMg_{0.67}Al_{10.33}O_{17}$, where $x = 0.0-0.78$.   The interatomic
potentials used are listed in Table 1.

As shown in the 7.5 ps 'snapshot' in Fig. 5a, the cations
in the conduction layers are quite mobile at low Ba(II)/Na(I)
ratios.   However, at 80% replacement of Na(I) by Ba(II) shown in
Fig. 5b, the cations and vacancies assume an ordered arrangement
in which either a Na(I) or a Ba(II) occupies every other BR
site.   Simulations of this ordered structure indicate that its
ionic conductivity is much lower than compositions containing
more Na(I).   In fact, experimental measurements of the ionic
conductivity of Na(I)-Ba(II)-β"-alumina as a function of
composition have shown that the conductivity passes through a
pronounced minimum at this composition, just as would be
expected from simulation results.   The simulated and
experimental conductivity data are shown in Fig. 6.   The general
trend of decreasing conductivity with increasing Ba(II) content
is predicted rather well.

Difficulties with the simulations arise near the end
compositions, pure Na(I)- and pure Ba(II)-β"-alumina.   The
simulations at this high temperature become unstable at low
Ba(II) contents, below about 20% exchange of Ba(II) for Na(I).
In addition, above 80% exchange, the Ba(II) ions are essentially
immobile during the relatively short time span of the
simulation.   For example, we carried out one simulation of 93%
exchanged Na(I)-Ba(II)-β"-alumina which was equilibrated with
the cations placed initially at their ordered sites.   After the
equilibration period, 6000 timesteps passed before one Ba(II)
jumped into an adjacent vacancy.   During 9000 subsequent
timesteps, only 35 jumps were recorded.   It is essentially
impossible to translate such meager statistics into any

meaningful value of a tracer diffusion coefficient, with the results that the simulations do not predict the experimentally observed increase in conductivity above 80% exchange.

4.   DISCUSSION

The goal of this work has been to predict with some accuracy the properties - structure, lattice parameters, ion-vacancy arrangement, and ionic conductivity - of a set of known β"-aluminas, with a clear eye on the possibility of then extending the technique to the exploration of the simulated properties of unknown compositions. Our investigations illustrate the opportunities and pitfalls presented by MD simulation. Clearly, MD is an exceptional structural probe, as our results and those of many others have shown. However, using MD to simulate time-dependent properties is far more challenging. The ultimate dilemma is that of the contrast between the short time scale of each simulation and the relatively long times over which macroscopic phenomena, such as ion transport, actually occur. This point is made very clear by the data obtained for 93% exchanged Na(I)-Ba(II)-β"-alumina, in which only 35 Ba(II) jumps were recorded over 15000 timesteps or 75 ps (5 hours of cpu time on the PSC Y-MP or 45 hours on the Stardent Titan). Ultimately, MD results are limited by the need to run simulations over long periods of time to reproduce phenomena that can only be observed on relatively long timescales. It is actually quite impressive that the calculated magnitudes of the tracer diffusion coefficients, averaged over simulation runs that last only 30 ps, are so close to the bulk experimental values measured over a timescale of microseconds.

Of course, the major challenge in MD is in obtaining accurate interatomic potentials. The fact that the Ba(II) potentials used in this work produced results that are consistent with experimental observations is gratifying and encouraging, but not any cause for believing that the simulation of other compositions will be as successful. As is so often the case, many of the most interesting β"-aluminas contain ions such as Pb(II), Hg(II), and Ag(I) whose more complex electronic structure and polarizability present daunting challenges in obtaining useful interatomic potentials.

Finally, achieving realistic simulation results depends somewhat on prior knowledge of the results expected, or at least a good hunch. For example, finding ordering is tricky with MD. Simply by defining a box size and shape, periodicity is imposed on the simulated system because it is the box that is periodically replicated over all space to form a 'bulk' material. Thus, if one is trying to reproduce an ordered structure with a specific periodicity, the box must also share this periodicity. Otherwise, the periodicity of the box will override the desired ordered structure and conceivably produce results that are not physical.

CONCLUSIONS

MD simulation and experimental measurements have been used to study the relationship between ion-vacancy ordering and aspects of the structure and ionic diffusion in pure Na(I)- and

mixed Na(I)-Ba(II)-β"-alumina. In the former case, the results
indicate the formation of a mobile $\sqrt{3}$a by $\sqrt{3}$a superlattice and
suggest that, while high temperature ionic conductivity may be a
result of a vacancy mechanism, low temperature conductivity
appears to be dominated by highly correlated cation motion
through short-lived 'diffusion channels' that periodically form
and disappear.

In Na(I)-Ba(II)-β"-alumina, order has a strong influence on
ionic conductivity and ion arrangement. The composition at
which 80% of the Na(I) has been replaced by Ba(II) shows clear
evidence of long-range ordering - at least over the length scale
of the simulation box - which greatly inhibits ionic motion.
The simulated results are confirmed by experimental measurements
of ionic conductivity as a function of composition.

Acknowledgment

This work was supported by the Office of Naval Research.
Additional support from the National Science Foundation, MRL
program, under Grant No. DMR-8819885 is gratefully acknowledged.
C. Lane would also like to express her gratitude to H. Hummel
for the Ba(II) related interatomic potential parameters.

REFERENCES

1.   Y.-F.Y. Yao and J.T. Kummer, **J. Inorg. Nucl. Chem.**, 29:2453
     (1967); G.C. Farrington and J.L. Briant, **Science**, 204:1371
     (1979); G.C. Farrington and B. Dunn, **Solid State Ionics**,
     7:267 (1982); G.C. Farrington, B. Dunn and J.O. Thomas,
     Appl. Phys.A, 32:159 (1983)
2.   J.P. Boilot, G. Collin, Ph. Colomban and R. Comes, **Phys.
     Rev.B**, 22:5912 (1980)
3.   J. Briant and G.C. Farrington, **J. Solid State Chem.**, 33:385
     (1980)
4.   M.A. Saltzberg, J.O. Thomas and G.C. Farrington, **Chem. of
     Materials**, 1:19 (1989)
5.   G.S. Rohrer and G.C. Farrington, **J. Solid State Chem.**,
     85:4628 (1990)
6.   C. Lane, G.C. Farrington, J.O. Thomas and M.A. Zendejas,
     Solid State Ionics, in press.
7.   J.R. Walker, **in**: "Computer Simulation of Solids", C.R.A.
     Catlow and W.C. Mackrodt, eds., Springer, Berlin (1982)
8.   M.J.L. Sangster and M. Dixon, **Advances in Physics**, 25:247
     (1976)
9.   J.R. Walker and C.R.A. Catlow, **J. Phys.C**, 15:6151 (1982)
10. G.V. Lewis, **Physica**, 131B:114 (1985)
11. R.G. Gordon and Y.S. Kim, **J. Chem. Phys.**, 56:3122 (1972)
12. M.J. Gillan, **Physica**, 131B:157 (1985)
13. D. Wolf, p.341 **in**: "Fast Ion Transport in Solids",
     Vashishta, Mundy and Shenoy, eds., Elsevier North Holland
     Inc. (1979)
14. G.E. Murch and R.J. Thorn, **Phil. Mag.**, 36:529 (1977)
15. G.J. May and C.M.B. Henderson, **J. Mat. Sci.**, 14:1229 (1979)
16. M.L. Wolf, J.R. Walker and C.R.A. Catlow, **Solid State
     Ionics**, 13:33 (1984)
17. M.A. Zendejas and J.O. Thomas, **Solid State Ionics**, 28-30:46
     (1988); J.O. Thomas and M.A. Zendejas, submitted for
     publication.

LAYERED CUPRATES WITH DOUBLE COPPER LAYERS: RELATIONSHIPS

BETWEEN STRUCTURE, NONSTOICHIOMETRY AND SUPERCONDUCTIVITY

B. Raveau, C. Michel, J. Provost, M. Hervieu
and D. Groult

Laboratoire de Cristallographie et Sciences des
Matériaux, CRISMAT, ISMRA, Université de Caen
Boulevard du Maréchal Juin, 14050 Caen Cedex-France

ABSTRACT

Layered cuprates with double copper layers form a rather
large family which appears as very promising for superconductivity at high temperature. The structure of those
phases, the nonstoichiometry phenomena, especially extended
defects observed by high resolution electron microscopy and
oxygen nonstoichiometry are presented here. The relationships
between those structural features and the superconducting
properties of those materials are discussed.

Many superconductive layered cuprates have been synthesized
since the beginning of 1987. In spite of the variety of their
chemical compositions, all those materials belong to the same
structural family. They consist of an intergrowth of multiple
oxygen deficient perovskite layers (p) with multiple rock salt-
type layers (r.s) according to the general formula
$(ACuO_{3-x})_m^p (AO)_n^{r.s}$ where m and n are the number of Cu-O and AO
planes respectively. They can also be represented by the symbol
[m,n]. The rock salt slabs are, in fact, built up from n + 1
planes: (n - 1) AO planes sandwiched by 2 $A^{II}O$ planes (A^{II} = Ba,
Sr) which ensure the junction between the two structures. Thus,
the structural principles which allow these phases to be
synthesized appear as very simple, taking into consideration the
fact that besides the two-dimensional character of the
structure, the mixed valency of copper Cu(II)-Cu(III), should be
introduced in order to create hole carriers on the copper-oxygen
framework. In reality, the physics and chemistry of those
materials are more complex than expected. We present here some
recent issues dealing with oxygen nonstoichiometry, extended
defects, in connection with superconductivity and new structures
in the m = 2-members, involving single rock salt layers.

1. THE IDEAL STRUCTURE AND FORMULA OF THE [2,n] CUPRATES

Among the different [m,n] layered cuprates, the [2,n]
cuprates form a rather important series, since about twenty

Table 1. The different layered cuprates $(A\,Cu\,O_{3-x})_2$ $(A'O)_n$ with double copper layers.

[2,3] (n = 3)		[2,2] (n = 2)	
$Tl_2Ba_2CaCu_2O_8$	S.C.	$TlBa_2CaCu_2O_7$	S.C
$Tl_2Ba_{2-x}Sr_xCaCu_2(O_8$	S.C.	$TlSr_2CaCu_2O_7$	S.C.
$Bi_2Sr_2CaCu_2O_8$ (55-63)	S.C.	$Tl_{0.5}Pb_{0.5}Sr_2CaCu_2O_7$	S.C.
$Bi_{2-x}Pb_xSr_2Ca_{1-x}Ln_xCu_2O_8$	S.C.	$TlBa_2LnCu_2O_7$	N.S.C.
$Tl_{2-x}Ba_{1-x}Tl_{1-x}LnCu_2O_8$	N.S.C.	(Ln = Y, Nd)	
		$TlBa_2Ca_{1-x}Ln_xCu_2O_7$	S.C.
		$(Tl,Bi)Sr_2Ca_{1-x}Y_xCu_2O_7$	S.C.
		$Pb_{0.5}Sr_{2.50}Y_{1-x}Ca_xCu_2O_7$	S.C.
		$Pb_{0.5}Ca_{0.5}Sr_2Y_{1-x}Ca_xCu_2O_7$	S.C.

[2,1] (n = 1)	
$La_{2-x}Sr_{1+x}Cu_2O_6$	N.S.C.
$La_{2-x}Ca_{1+x}Cu_2O_6$	N.S.C.
$La_{2-x}Sr_xCaCu_2O_6$	S.C
$A_{2-x}Ln_{1+x}Cu_2O_{6-y}$	N.S.C.
$Sr_2NdCu_2O_{5.76}$	N.S.C.
$Sr_{1-9}La_{1.1}Cu_2O_{5.83}$	N.S.C.
$Sr_6Nd_3Cu_6O_{17}$	N.S.C.

compounds[1-37] are known up to now (Table 1). The ideal structure of all those compounds has in common the existence of pyramidal copper layers interleaved with calcium ions in pseudo-cubic coordination (Fig. 1).

The different members of this series differ by the nature of the rock salt-type slabs. The [2,3] phases, also denoted "2212" (see Note:), are characterized by triple rock salt layers

Note: Two types of nomenclature are usually used for the high Tc superconductors. As an example, [2,3] or "2212" can be used for $Tl_2Ba_2CaCu_2O_8$. The first one indicates the number of perovskite and rock salt layers, respectively. The second one describes either the nominal composition, i.e. the coefficient of Tl-Ba-Ca-Cu respectively, or the layers stacking; in the latter case, the coefficients can be correlated with the n and m values: [2,2,1,2] corresponds to [n-1,2,m-2,m].

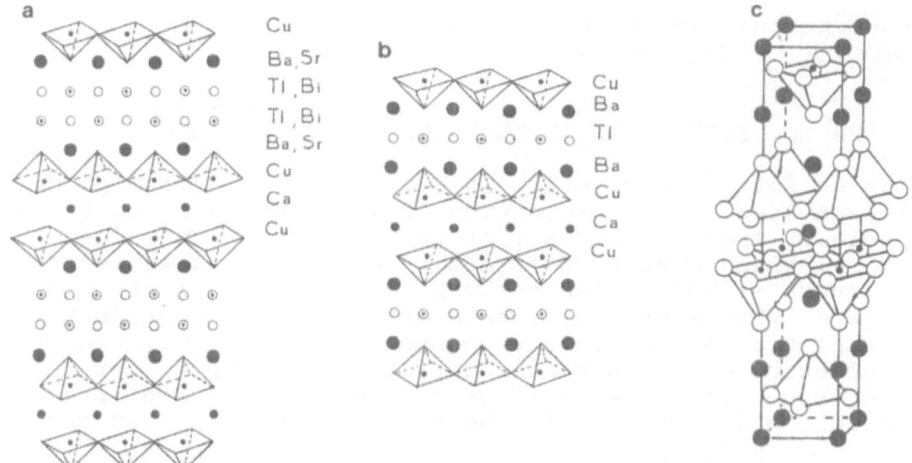

Fig. 1. Idealized structures of the [2,n] cuprates:
(a) $Bi_2Sr_2CaCu_2O_8$ or $Tl_2Ba_2CaCu_2O_8$ (n = 3),
(b) $TlBa_2CaCu_2O_7$ (n = 2), (c) $La_{2-x}A_{1+x}Cu_2O_6$,
A = Ca,Sr (n = 1).

(Fig. 1a) and can be formulated $A^{III}_2A^{II}_2CaCu_2O_8$ with A^{III} = Tl or Bi and A^{II} = Sr or Ba[18-24,28-37]. The [2,2]-oxides, also denoted "1212", exhibit double rock salt layers (Fig. 1b) and can be represented by the formula $A'A^{II}_2CaCu_2O_7$ with A' = Tl or $Tl_{0.5}Pb_{0.5}$ or $Pb_{0.5}Sr_{0.5}$ and A^{II} = Ba or Sr[9-17, 25-27]. The [2,1] or "0212" compounds, which are characterized by single rock salt layers (Fig. 1c) are only observed in the case of

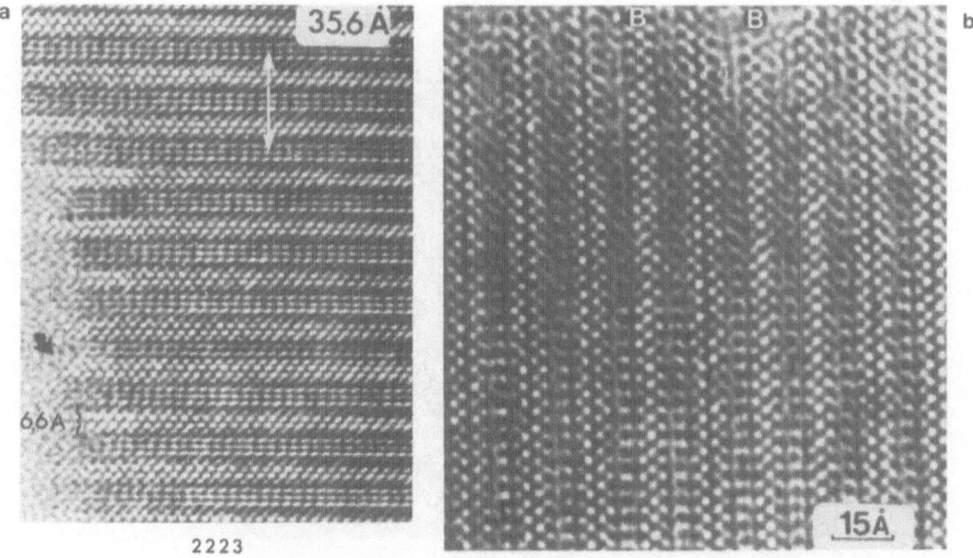

Fig. 2. a) m' = 2 defective layer (arrowed) is observed in a $Tl_2Ba_2Ca_2Cu_3O_{10}$ matrix [3,3]; b) n' = 2 defective layers are observed in a $Tl_2Ba_2CaCu_2O_8$ matrix [2,3].

lanthanum and neodymium, according to the formula $Ln_{2-x}A^{II}_{1+x}$ Cu_2O_6 with A^{II} = Ca, Sr[1-8].

2. EXTENDED DEFECTS AND INCOMMENSURABILITY EFFECTS

The ability of perovskite layers to accommodate rock salt-type layers leads to the formation of intergrowth defects. This is especially true in the case of thallium cuprates which can exhibit, although pure RX diffraction patterns, variations in the thickness either of the perovskite layers or of the rock salt layers, the existence of such defects being dependent on the thermal treatment. This phenomenon is easily observed by high resolution electron microscopy[38,39] as shown, for instance, from Fig. 2 which shows two sorts of intergrowth defects currently obtained in the matrix of the Tl-2223 and Tl-2212 phases.

The first one (Fig. 2a) corresponds to the appearance of an $m' = 2$-defect, i.e. of $Tl_2Ba_2CaCu_2O_8$ layers in a 2223 matrix, involving double copper layers instead of triple copper layers. The second defect (Fig. 2b) corresponds to the appearance of an $n' = 2$ defect in the 2212 matrix, i.e. to the replacement of a triple rock salt-type layer by a double rock salt-type layer, so that locally one gets the "1212" $TlBa_2CaCu_2O_7$ structure instead of the $Tl_2Ba_2CaCu_2O_8$ composition.

The introduction of additional "CaO" layers in the rock salt layers (Fig. 3) is also an intergrowth defect which appears in the thallium cuprates[39].

Contrary to the thallium cuprates, the bismuth cuprates do not exhibit many intergrowth defects. This is due to the fact that one only observes bismuth bilayers so that only triple rock salt layers are possible and also that there exists only one

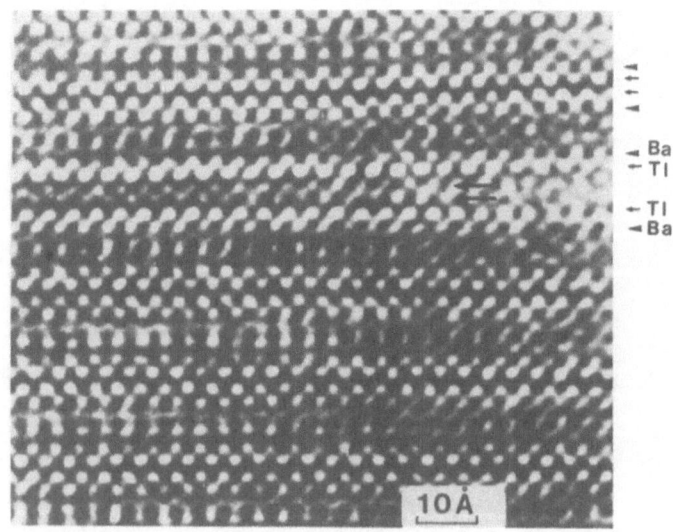

Fig. 3. The introduction of additional "CaO" layers, between the two "TlO" layers is observed in the $Tl_2Ba_2CaCu_2O_8$ oxides.

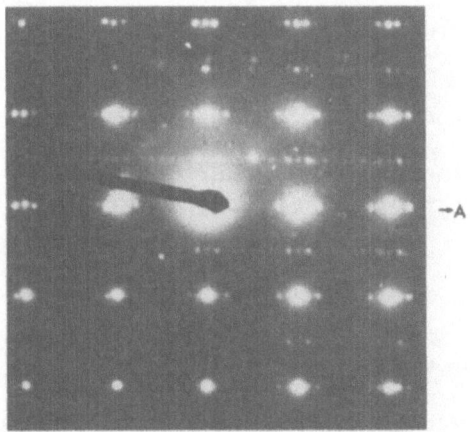

Fig. 4. [001] electron diffraction pattern of
$Bi_2Sr_2CaCu_2O_8$: satellites in incommensurate
positions (q~4.7) are observed along A.

member with m ≥ 3, $Bi_2Sr_2Ca_2Cu_3O_{10}$, which, moreover, is difficult
to prepare. However, in those materials another feature must be
pointed out which deals with the existence of incommensurability. One, indeed, observes on the electron diffraction
patterns of $Bi_2Sr_2CaCu_2O_8$ (Fig. 4) satellites corresponding to a
nonintegral q vector with respect to the fundamental lattice
[33, 40-43]. This behavior can be explained by the stereo-
activity of the $6s^2$ lone pair of Bi^{3+} which allows a great
distortion of the rock salt layers and consequently leads to a
possible introduction of additional oxygen in those layers. It
results in a waving of the bismuth oxygen layers (Fig. 5). It
is clear that this phenomenon is not particular to the superconductive cuprates but is also observed for isotypical iron
oxides[44,45]; nevertheless, it is not known so far how it can
influence the superconducting properties of those oxides.

3.   A DRAMATIC EFFECT OF THE OXYGEN NON STOICHIOMETRY UPON T_c:
      CASE OF THE "1212" AND "2212" THALLIUM CUPRATES

      The critical temperatures of the two superconductors,
$TlBa_2CaCu_2O_7$ and $Tl_2Ba_2CaCu_2O_8$ were found to be variable
according to the authors ranging from 50 K[10] to 100-112 K
[9,46] for the first one and from 95 K to 108 K[20] for the
second one.

      These results suggested to us very early that the oxygen
nonstoichiometry could drastically influence their critical
temperatures so that their formula should be written
$TlBa_2CaCu_2O_{7\pm\delta}$ and $Tl_2Ba_2CaCu_2O_{8\pm\delta}$ respectively. This dramatic
influence of oxygen nonstoichiometry upon T_cs was indeed shown
first for the oxides $TlBa_2CaCu_2O_7$ and $Tl_2Ba_2CuO_6$[10]. These
latter compounds prepared under oxygen pressure of several
atmospheres in sealed ampoules were found to be superconducting
(T_c = 50 K) and nonsuperconducting respectively; annealing these
samples at 450°C-500°C under an argon flow allowed T_c to be
increased up to 65 K for the first one and superconductivity at
30 K to be obtained for the second one. The fact that the

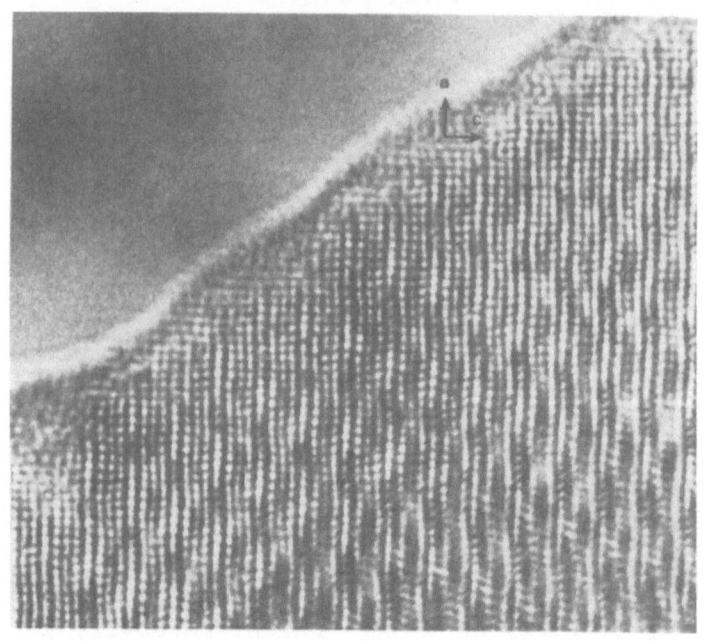

Fig. 5. [010] high resolution electron microscopy image
showing the waving of the bismuth-oxygen layers.

superconducting transition can be increased by decreasing the
oxygen content was then confirmed later for $Tl_2Ba_2CuO_6$ [47,48]
whereas the opposite was found for $Tl_2Ba_2CaCu_2O_8$ [49] and
$Tl_2Ba_2CaCu_3O_{10}$ [50]. However, a study of the annealing of
$Tl_2Ba_2CaCu_2O_8$ crystals under argon and vacuum[46] showed a more
complex behavior: for one, the crystal's T_c decreased by vacuum
annealing, whereas for the others T_c increased significantly.

A recent study of $Tl_2Ba_2CaCu_2O_8$ clearly shows that in those
cuprates the critical temperature is mainly governed by the
oxygen stoichiometry[51]. Starting from the as-synthesized
sample under an oxygen pressure of about 4-7 bars in sealed
tubes, three sorts of annealings were performed in a first step
corresponding to samples A, B and C respectively. For sample A,
annealing was performed at 400°C in an oxygen flow. For sample
B, annealing was carried out in an argon flow at 400°C. For
sample C, the sealed silica ampoule containing the "as-
synthesized" sample was cooled down to 400°C and kept at this
temperature for 12 hours and then slowly cooled down to room
temperature. Figure 6a shows the real part χ' of the AC-
susceptibility of A (or B) "as-synthesized" sample (A and B were
synthesized in the same evacuated ampoule), of oxygen annealed
sample A and of argon annealed sample B. The data are *not
corrected* for the demagnetizing effect as shown from the χ'
values smaller than -1. The first important result deals with
the fact that the critical temperature is increased as well by
argon annealing as by oxygen annealing. One, indeed, observes
that T_c is about 97 K for the as-synthesized samples, whereas it
is increased up to 104 K by oxygen annealing (sample A) and up
to about 112 K by argon annealing (sample B). The AC-magnetic
field amplitude (5 Oe) was the same for the three curves

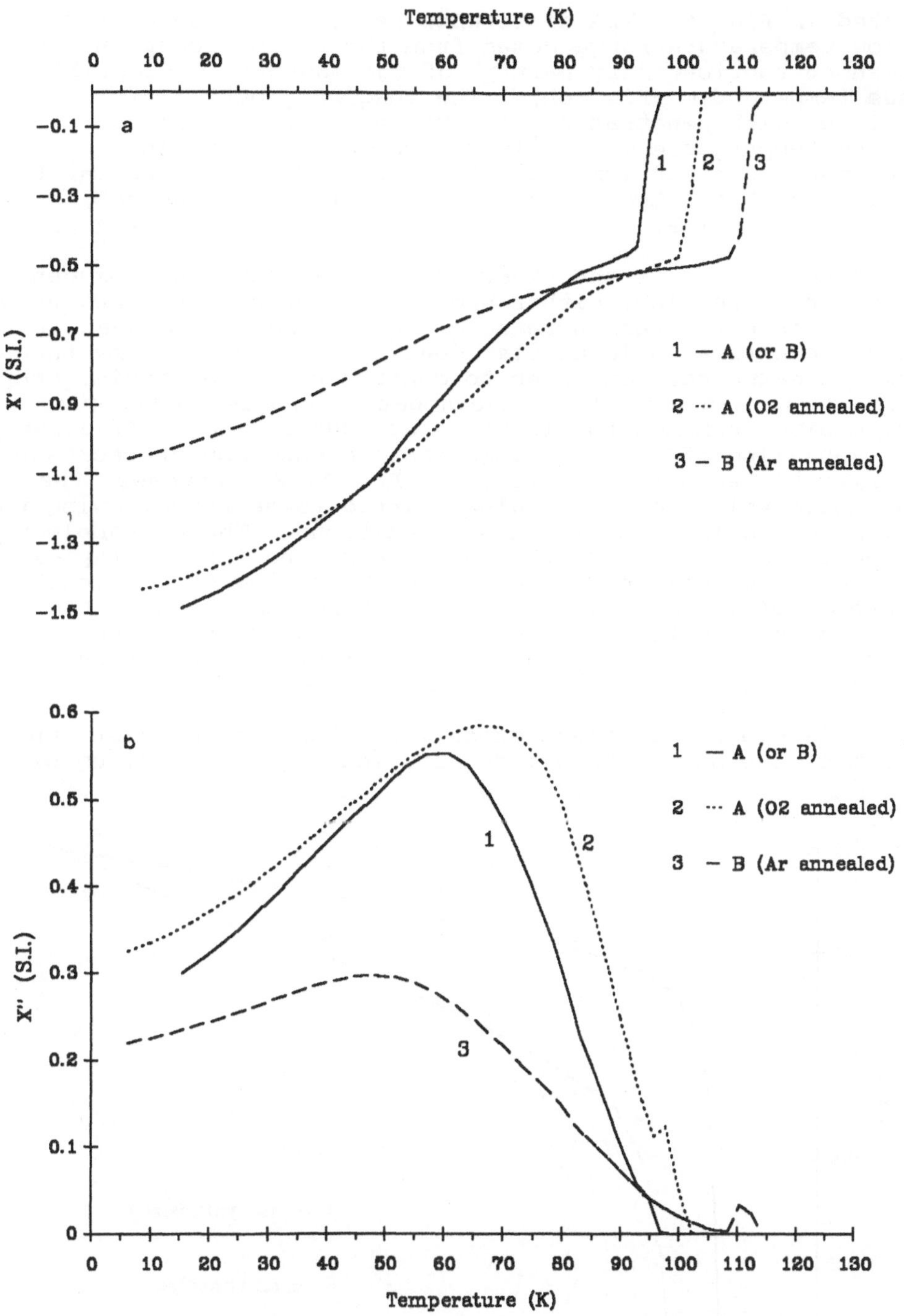

Fig. 6   (a) Real part χ′(T) of the magnetic AC
          susceptibility of thallium cuprates
          Tl$_2$Ba$_2$CaCu$_2$O$_8$: (1) as synthesized sample (T$_c$ ~ 96
          K), (2) sample annealed under oxygen flow at
          400°C, (3) sample annealed under argon flow at
          400°C.   (b) Imaginary part χ″(T) of the magnetic
          AC susceptibility corresponding to the curves in
          (a).

reported in Fig. 6. The difference between the magnetization in the low temperature range comes from the grain decoupling as can be seen on the imaginary part χ'' of the magnetic susceptibility versus temperature (Fig. 6b). The imaginary part χ'' is associated with penetration of vortices in the grain boundaries when the temperature T is raised. These magnetization measurements are confirmed by electrical resistance as can be seen in Fig. 7 for the argon annealing effect on the as-synthesized samples.

At this stage of the experiment, it appears that oxygen loss during annealing, either under argon or oxygen flow, should be responsible for the increase of T_c's , suggesting that it allows the optimal hole carrier density to be reached by those thermal treatments. In order to confirm this hypothesis, sample C ($T_c \sim 98$ K) was heated in an evacuated ampoule in the presence of zirconium turnings for 12 hours at 400°C. The $\chi'(T)$ curve after annealing of the C' sample shows a spectacular increase of the critical temperature (Fig. 8): $\Delta T_c \approx 17$ K. However, the diamagnetic volume of the grains started to decrease, owing to the long annealing time in these conditions. The C' sample was then heated at 800°C for 1/2 hour in air, and finally rapidly cooled down to room temperature. The resulting C" sample shows a decrease of T_c down to 105 K (Fig. 8) but the diamagnetic volume in the low temperature range is largely increased compared to C', showing good evidence of grain boundaries "reconstruction".

It is clear from these experiments that there exists an optimum of the hole carriers density leading to a maximum of T_c.

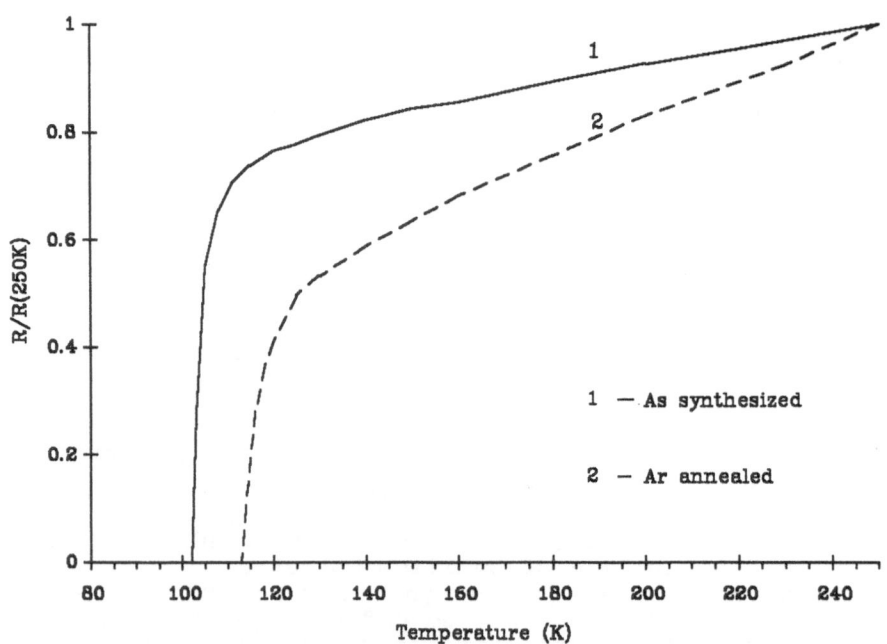

Fig. 7. Electrical resistances versus T for a $Tl_2Ba_2CaCu_2O_8$ sample: (1) as synthesized sample ($T_c \sim 100$ K), (2) the same sample annealed under argon flow at 400°C.

Temperature (K)

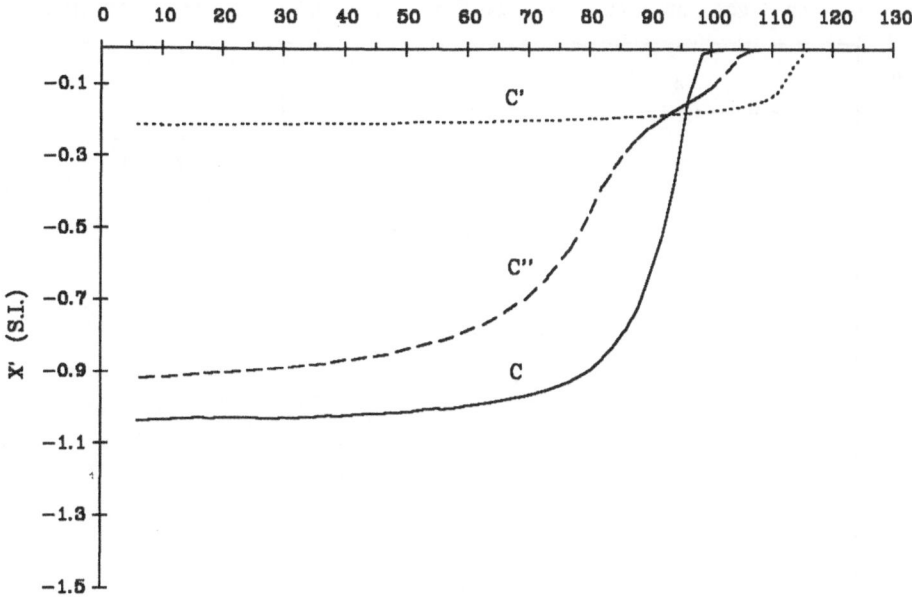

Fig. 8. Real part χ′(T) of the AC susceptibility of a
$Tl_2Ba_2CaCu_2O_8$ sample prepared with the same
synthesis conditions as samples A and B (Fig.
6a): (C) sample annealed at 400°C under oxygen
pressure in sealed silica ampoule, (C′) the same
sample annealed in evacuated silica ampoule with
Zr turnings at 400°C for 12 hours, (C") the C′
sample heated during 1/2 hours at 800°C in
evacuated sealed ampoule.

Nevertheless, although the oxygen nonstoichiometry seems to play
an important role in this phenomenon, the weight loss may also
result from the thallium oxide volatilization. The volatility
of thallium oxide at 400°C was, indeed, demonstrated for this
compound by chemical analysis. In order to avoid this thallium
loss, it was decided to work at lower temperature, 300°C, in a
reducing atmosphere (argon-hydrogen); in these conditions no
thallium loss was observed. A spectacular effect of this method
of annealing upon T_cs was observed as shown on Fig. 9. T_c can,
indeed, be increased from 96 K to 118 K for a very short
annealing time of 15 minutes. It can also be seen that the
variation of the annealing time does not drastically influence
the T_cs but that a prolonged annealing tends to decrease the
diamagnetic volume. The weight losses determined by TGA show
that they are very small, and that the smallest weight loss of
0.07% leads to a dramatic increase of T_c's: $\Delta T_c \approx 22$ K.

Similar effects were observed for the "1212" phase
$TlBa_2CaCu_2O_7$. It must also be emphasized, from the HREM
observations, that the treatment under hydrogen at 300°C does
not alter the materials, contrary to the annealings at 400°C-
500°C under different atmospheres.

The oxygen nonstoichiometry is the key for the optimization
of the superconducting properties and especially the critical

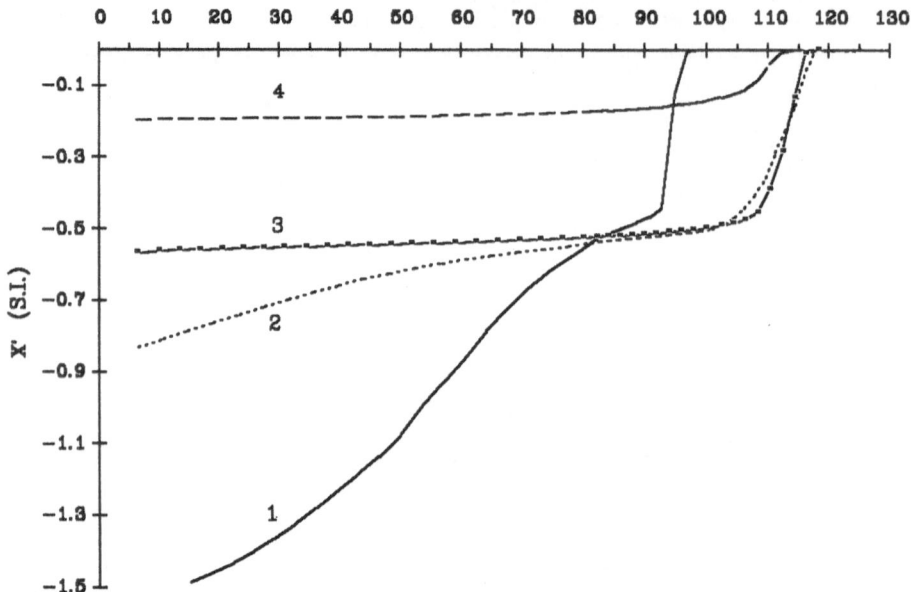

Fig. 9. AC susceptibilities χ'(T) of the same sample
Tl"2212" for different annealing times in a gas
mixture (Ar 90% + H_2 10%): (1)-as synthesized,
(2)-15 mn, (3)-120 mn (4)-720 mn.

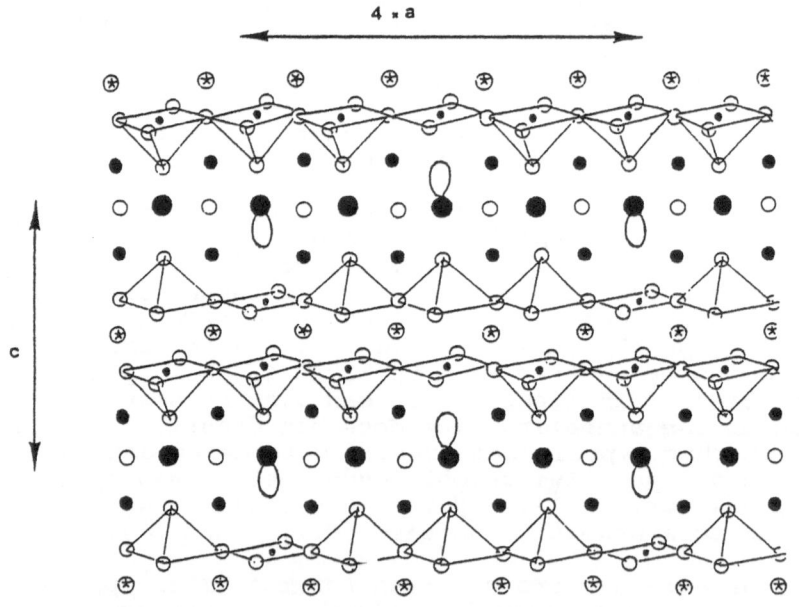

Fig. 10. $Pb_{0.5}Sr_{2.5}Y_{1-x}Ca_xCu_2O_{7-\delta}$: model of a perfect
ordering of the Pb(II) and Sr in the
intermediate rock salt layers. This idealized
drawing shows that 4a x c superstructure can be
built up. Other models can be obtained by
translation of the adjacent layers.

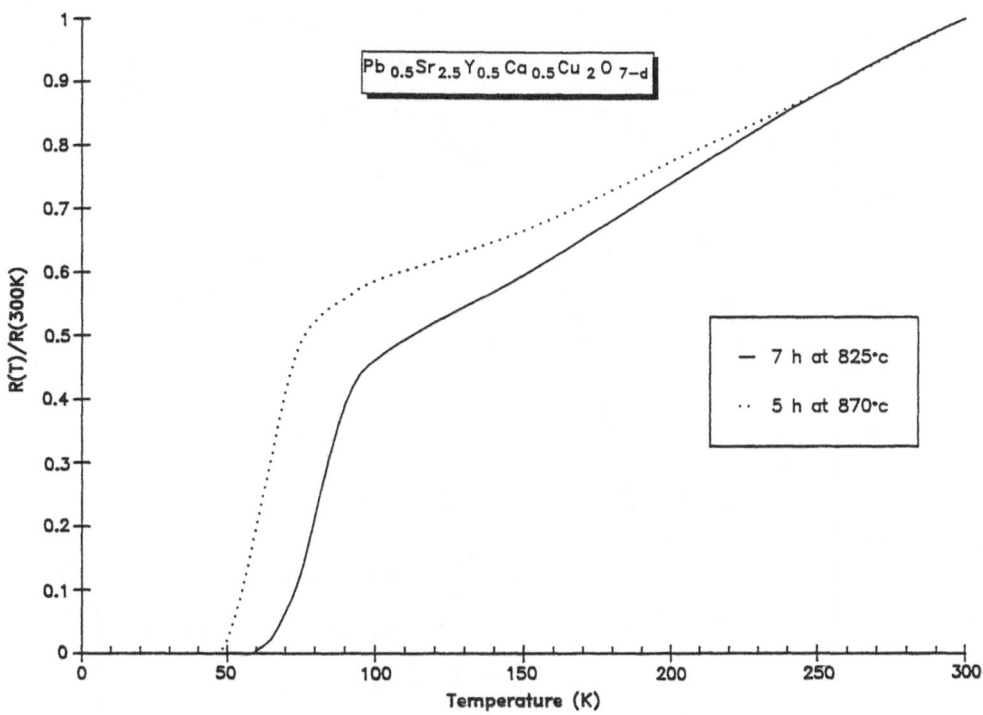

Fig. 11. Temperature dependence of the ratio R (T)/R
(300 K) for different synthesis conditions
(unannealed samples) for $Pb_{0.5}Sr_{2.5}Y_{1-x}Ca_xCu_2O_{7-\delta}$
(x = 0.50).

temperature of the thallium cuprates, owing to its great
influence upon the hole carrier density. This explains why the
values of T_c observed for the "2212" phase could be rather
different from one author to the other, depending upon the
oxygen pressure, temperature and time used for its
preparation[46,51-53].

It is worth pointing out that the effect of oxygen nonstoichiometry upon T_cs is all the more important since the
number of copper layers forming the perovskite slabs is smaller,
i.e. since m is smaller. One, indeed, observes ΔT_c variations
of 90 K for $Tl_2Ba_2CuO_6$ (m = 1), of more than 20 K for
$TlBa_2CaCu_2O_7$ and $Tl_2Ba_2CaCu_2O_8$ (m = 2), and of 0-5 K for
$TlBa_2Ca_2Cu_3O_9$ and $Tl_2Ba_2Ca_2Cu_3O_{10}$ (m = 3). Thus, the important
factor which governs the T_cs in thallium cuprates is the hole
carrier density per mole of copper in the structure (equivalent
to the number of Cu(III) per mole Cu) and this factor seems to
be predominant with respect to the effect of the number m of
copper layers in the perovskite slabs.

The main issue which remains to be solved deals with the
knowledge of oxygen content and distribution in the structure,
and consequently the actual carrier density. Nevertheless, the
"annealing at 300°C in a reducing atmosphere" appears as a very
powerful method to control the hole carrier density and
consequently, the T_cs of those cuprates without altering the
structure.

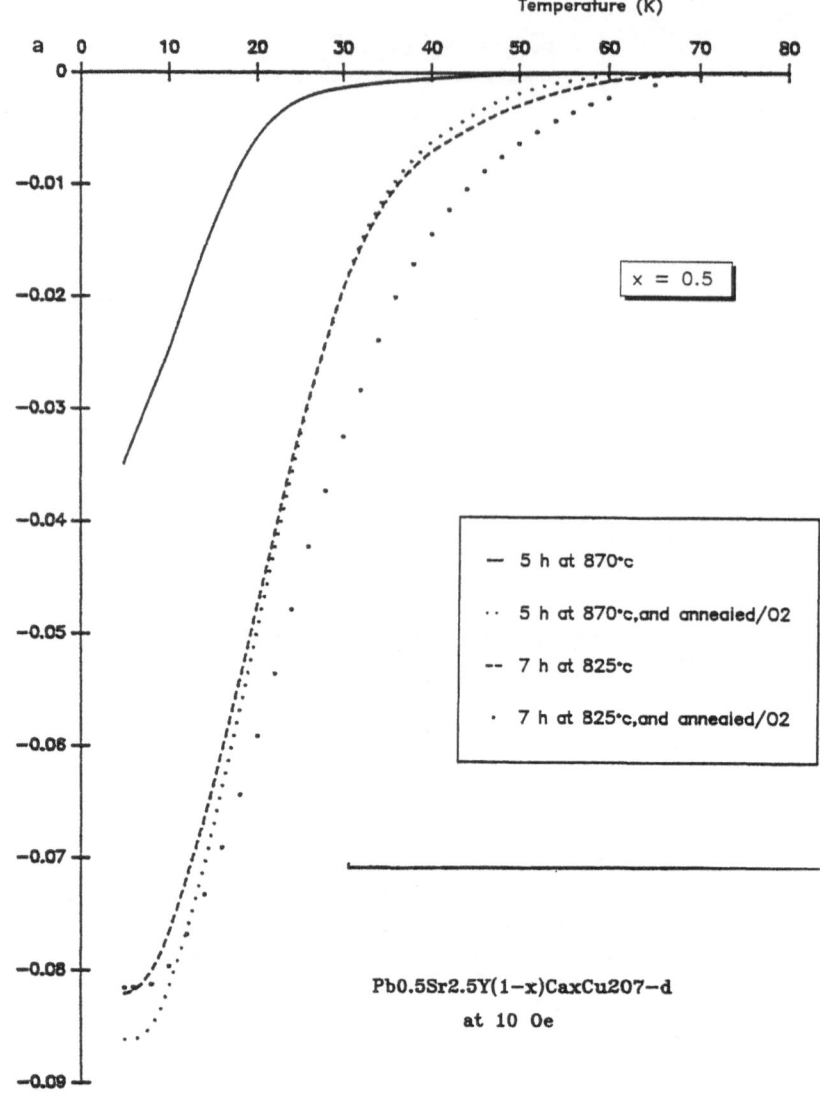

Fig. 12. (a) Magnetization vs T for $Pb_{0.5}Sr_{2.5}Y_{1-x}$
$Ca_x Cu_2 O_{7-\delta}$ (x = 0.5) at 10 Oe for different
thermal treatments.

4. OXYGEN NONSTOICHIOMETRY AT THE BOUNDARY BETWEEN ROCK SALT AND PEROVSKITE LAYERS: THE "1212" LEAD CUPRATE

The oxide $Pb_{0.5}Sr_{2.5}Y_{1-x}Ca_x Cu_2 O_{7-\delta}$ [25,26] is the only pure lead cuprate which exhibits double copper layers. Its structure (Fig. 10) belongs indeed to the "1212"-type already described for $TlA_2 CaCu_2 O_7$ (A = Ba,Sr) (Fig. 1b). Mixed lead-strontium monolayers $[Pb_{0.5}Sr_{0.5}O]_\infty$ replace the $[TlO]_\infty$ monolayers in the double rock salt slabs. Nevertheless, an important difference with the thallium cuprates deals with the existence of oxygen vacancies at the boundary between the perovskite and rock salt-type layers. It results in the existence of several super-

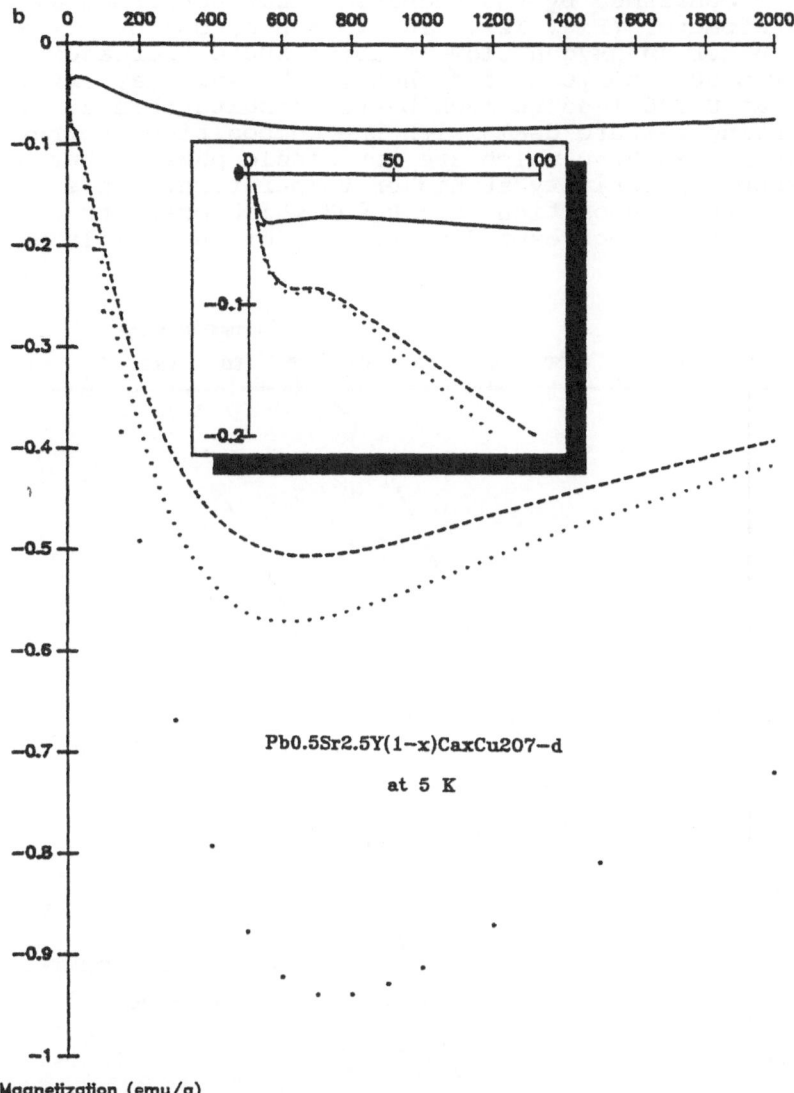

Fig. 12. (b) Magnetization vs H(H < 2000 Oe) at 5 K for
the sample of Fig. 12a. Insert shows the low
field part of the curves.

structures which can be interpreted in terms of ordering of
lead, strontium and anionic vacancies. An isostructural solid
solution has been isolated by replacing completely strontium by
calcium in the mixed monolayers leading to the formulation
$Pb_{0.5}Ca_{0.5}Sr_2Y_{1-x}Ca_xCu_2O_{7-\delta}$ [27]. Both oxides exhibit a wide
homogeneity range, $0 \leq x \leq 0.60$, but curiously superconductivity
is only observed for $0.50 \leq x \leq 0.60$.

In all cases, one again observes very broad resistive tran-
sitions as shown, for instance, for $Pb_{0.5}Sr_{2.5}Y_{0.5}Ca_{0.5}Cu_2O_{7-\delta}$
(Fig. 11). Moreover, it can be seen that the experimental
conditions, especially the temperature, influence dramatically

741

the superconducting properties of those materials. This
behavior is confirmed by the magnetic study performed with a
SQUID magnetometer (Fig. 12). For instance, one observes that
annealing under an oxygen flow at about 500°C increases T_c
(onset) from 60 K to 70 K, and that in the same way, synthesis
performed at 825°C lead to much better results than at 870°C.
An interesting feature deals with the compositions corresponding
to x greater than 0.60 which are not single phased, but which
present superconductivity at higher temperature. This is indeed
the case of the composition "x = 0.90" which presents a
diamagnetic signal corresponding to 3-4% of the sample volume

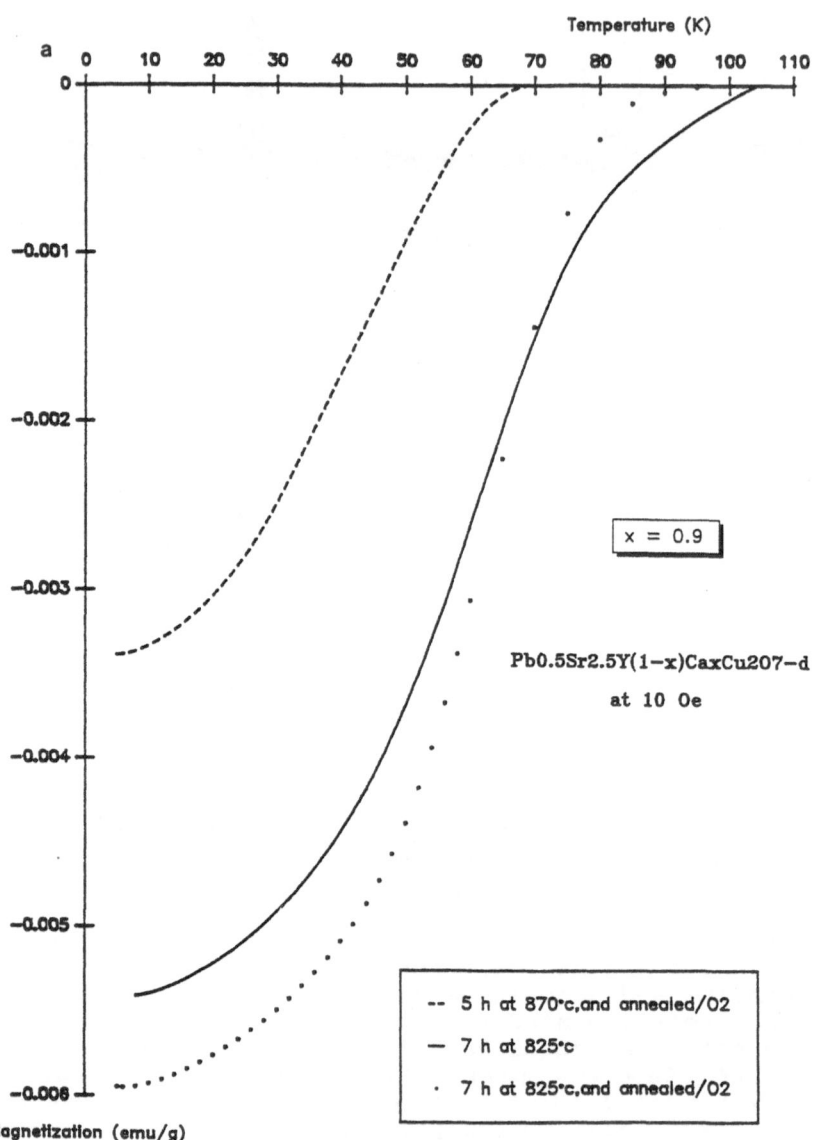

Fig. 13. Multiphase sample of nominal composition
$Pb_{0.5}Sr_{2.5}Y_{1-x}Ca_xCu_2O_{7-\delta}$ (x = 0.9):
(a) magnetization versus T at 10 Oe for
different thermal treatments.

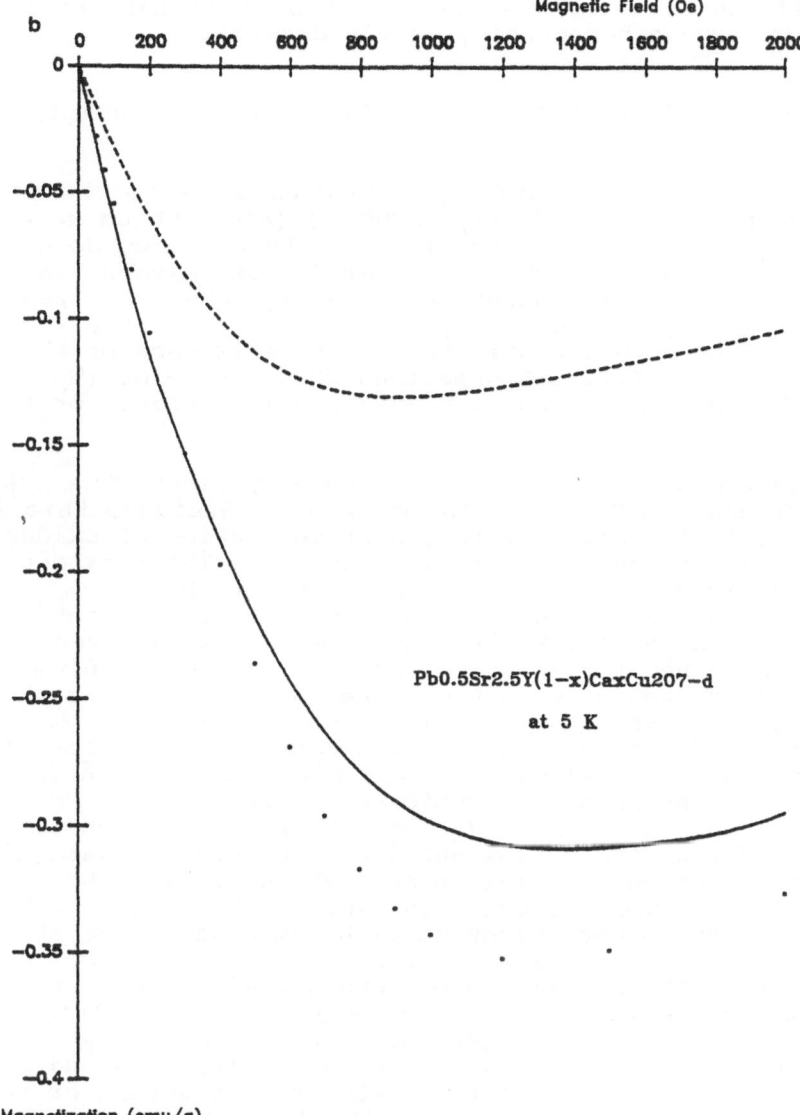

Fig. 13. (b) Magnetization vs H(H < 2000 Oe) at 5 K.

and a T_c of 104 K (Fig. 13); of course, for this latter composition, zero resistance cannot be reached owing to the large volume of the insulating phase belonging to the Sr_2PbO_4-type which prevents the percolation. Nevertheless, this confirms the possibility of superconductivity up to 100 K in those systems, and suggests the possible existence of another superconductive lead cuprate.

Thus, the layered cuprates form a potential family of high T_c superconductors. However, their crystal chemistry is complicated by the two possible states Pb(II)/Pb(IV), which are also important species for the creation of new superconductors. A systematic study of the oxygen nonstoichiometry in these oxides will be necessary to control their synthesis and to improve the oxygen homogeneity in the crystal.

5. A NEW SERIES OF [2,1] OXIDES; THE SYSTEM "$Ln_2SrCu_2O_6$ - $LnSr_2Cu_2O_{6-y}$"

The oxides $La_{2-x}A_{1+x}Cu_2O_{6-x/2+\delta}$ (A = Ca,Sr) were the first layered cuprates of the family $(ACuO_{3-x})_m(A'O)_n$ which were synthesized[1]. It is remarkable that these oxides do not exhibit superconductivity when prepared under oxygen flow in spite of their lamellar structure (Fig. 1c) characterized by double copper layers absolutely similar to those observed in thallium and bismuth cuprates and of the existence of the mixed valency Cu(II)-Cu(III). A superconducting behavior (T_c = 60 K) can only be obtained using high oxygen pressure[55] for the composition $La_{1.6}Sr_{0.4}CaCu_2O_6$.

Trying to change the number of hole carriers, the system "$Ln_2SrCu_2O_6$-$LnSr_2Cu_2O_{6-y}$" was investigated. Besides, this first class of [2,1] layered cuprates, a second series of oxides, with richer strontium contents were isolated[5]. These oxides, which correspond always to the general formula $Ln_{2-x}Sr_{1+x}Cu_2O_{6-x/2+\delta}$ differ from the lanthanide rich oxides by a significant deficiency with respect to the "O_6" ideal formula given above. They were obtained for samarium, europium and gadolinium with x values ranging from 0.70 to 0.90. Recently, a similar phase was synthesized for lanthanum[5]. Those oxides differ from the preceeding oxides by a tripling of one of the parameters leading to an orthorhombic cell with a ≈ a_p, b ≈ $3a_p$, c ≈ 20 Å (a_p ≈ 3.8 Å, the parameter of the cubic perovskite cell). Recently, a new layered cuprate $Sr_6Nd_3Cu_6O_{17}$ was synthesized by solid state reaction under an argon flow and its structure was determined by neutron diffraction[6]. This phase, which can also be formulated $Sr_6NdCu_2O_{5.66}$, exhibits similar parameters to those observed for the other strontium rich compounds. The structure of this oxide (Fig. 14) is characterized by a similar arrangement of the metallic atoms and a great part of the oxygen atoms with respect to the $La_2SrCu_2O_6$-type structure (Fig. 1c). However, several of the oxygen atoms and anionic vacancies are distributed in a different way. Consequently, it appears that this structure is intermediate between the structure of the "123" phase $YBa_2Cu_3O_7$[54] and the "0212" structure of $La_2SrCu_2O_6$. One can describe it as a regular intergrowth of the "123" type ribbons $[Sr_{7/6}Nd_{5/6}Cu_2O_{4.66}]_\infty$ running along $\vec{b}$ and from $[Sr_{5/6}Nd_{1/6}O]_\infty$ rock salt-type layers parallel to (100). Thus $Sr_6Nd_3Cu_6O_{17}$, can be considered as the member m = 2, n = 1 of a series of intergrowths of general formula $[ACuO_{2.33}]_m^{"123"}[AO]_n^{RS}$ in which A will correspond to an adequate content of alkaline earth cations (Sr, Ba) and of lanthanides. The "123" $[Sr_{7/6}Nd_{5/6}Cu_2O_{4.66}]_\infty$ ribbons, which are three polyhedra wide along $\vec{b}$ ($2CuO_5 + 1CuO_4$), exhibit a significant difference with the classical $YBa_2Cu_3O_7$ structure[54]. The Cu-O interatomic distances show indeed that the CuO_4 groups and the CuO_5 pyramids are strongly distorted with Cu-O distances ranging from 1.85 Å to 2.22 Å. The apical Cu-O distance of the CuO_5 pyramid (2.22 Å) is shorter than that observed for $YBa_2Cu_3O_7$ (2.29 Å), whereas a large distortion is observed in the basal plane (1.88 to 2.02 Å) compared to $YBa_2Cu_3O_7$ (1.93 to 1.96 Å). The CuO_4 groups are no more planar; Cu(1), two O(1) atoms and O(5) remain approximately in the same plane whereas O(8'), which ensures the

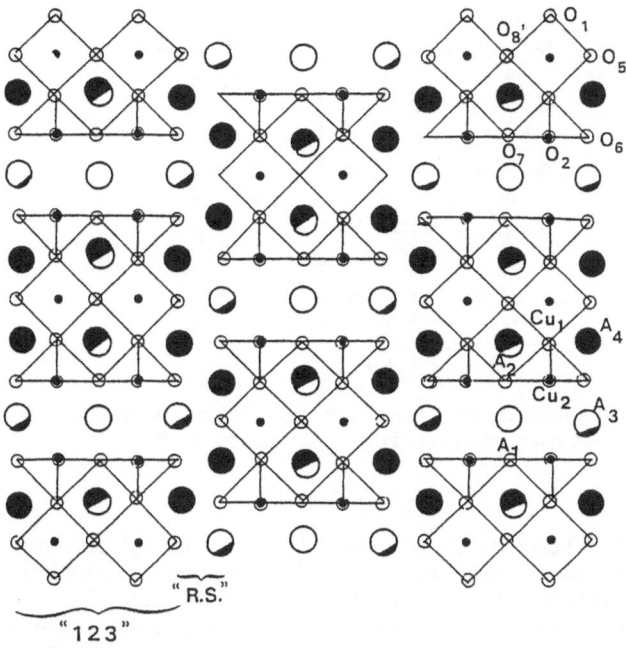

Fig. 14. Schematic drawing of the structure of the oxide $Sr_6Nd_3Cu_6O_{17}$.

junction between two CuO_4 groups, is significantly out of this plane leading to a Cu-O-Cu angle of 110° instead of 180°. These features are easily understandable if one considers the sequence of "123" ribbons along $\vec{c}$ (Fig. 14). Two successive ribbons are indeed shifted by about 3.8 Å along $\vec{b}$, so that at a same level along this latter direction, a neodymium plane is replaced by a copper plane. Such an arrangement induces strains in the structure which can only be decreased by a distortion of the CuO_4 and CuO_5 polyhedra and by their tilting.

The distribution of the strontium and neodymium cations is remarkable and is certainly an important factor for the stabilization of such a structure. One, indeed, observes that in the middle of the "123" type ribbons the smaller cations, neodymium, take the place of yttrium in eightfold coordination (A_1 sites) whereas, in the rock salt-type layers, the sites which exhibit an eightfold coordination (A_4 sites) are only occupied by strontium. On the other hand, the cationic sites which are located at the intersection of the rock salt layer and of the basal planes of the CuO_5 pyramids (A_3 sites) are occupied half by strontium and half by neodymium. In the same way, it is worth pointing out that the barium sites, in the "123" ribbons, are occupied by much smaller cations, half neodymium and half strontium (A_2 sites). This latter occupancy may be at the origin of the distortion of the CuO_4 square groups and may allow an adjustment of the "123" and rock salt layers.

The structure of the other strontium rich compounds, with oxygen contents intermediate between "$O_{5.66}$" and "O_6" is up to now not completely elucidated. Nevertheless, the neutron diffraction studies of the oxides $La_{1.1}Sr_{1.9}Cu_2O_{5.83}$[7] and

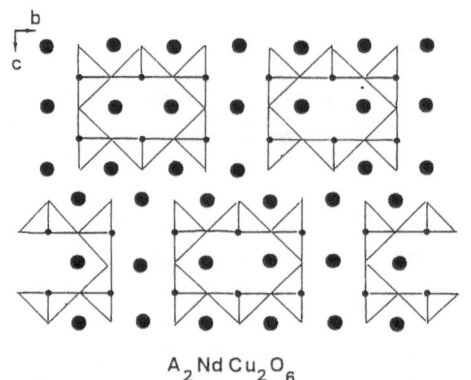

$A_2NdCu_2O_6$

Fig. 15. Schematic drawing of the structure of the
hypothetical oxide "$Sr_2NdCu_2O_6$".

$NdSr_2Cu_2O_{5.76}$[8] clearly show that their structure is closely
related to that of $Sr_6Nd_3Cu_6O_{17}$. The oxygen distribution in the
structure can indeed be interpreted as the result of the
intimate intergrowth of two extreme structures, the first one
corresponding to $Sr_6Nd_3Cu_6O_{17}$ (Fig. 14) and the second one
corresponding to the hypothetical compound $Sr_2NdCu_2O_6$ whose
ideal structure (Fig. 15) would consist of six-sided tunnels
built up from rings of six corner-sharing CuO_5 pyramids. Such
tubes would be arranged in such a way that they form fluorite-
type cages, and rock salt-type layers where the Nd and Sr
cations are located.

From all these results it is clear that the oxygen nonstoichiometry in the "0212" oxides is so far not completely
understood and that it should be possible to synthesize new
phases by playing with ordering of oxygen and vacancies, i.e. by
changing the thermal treatments (temperature, time) and also the
nature of the gaseous atmosphere. Although they do not
superconduct, the knowledge of their magnetic and electron
transport properties will be of capital importance in order to
understand superconductivity in layered cuprates.

REFERENCES

1.   N. Nguyen, L. Er-Rakho, C. Michel, J. Choisnet and B.
      Raveau, **Mat. Res. Bull.**, 15:891 (1980)
2.   F. Izumi, E. Takayama-Muromachi, Y. Nakai and H. Asano,
      Physica C., 157:89 (1989)
3.   V. Caignaert, N. Nguyen and B. Raveau, **Mat. Res. Bull.**,
      25:199 (1990)
4.   N. Nguyen, J. Choisnet and B. Raveau, **Mat. Res. Bull.**,
      17:567 (1982)
5.   D.M. de Leeuw, C.A.H.A. Mutsaers, G.P.J. Geelen and C.
      Langereis, **J. Solid State Chem.**, 80:276 (1989)
6.   V. Caignaert, R. Retoux, M. Hervieu, C. Michel and B.
      Raveau, **J. Solid State Chem.**, in press
7.   R.A. Steeman, D.M. de Leeuw, G.P.J. Geelen and E. Frikkee,
      Physica C., 162-164:542 (1989)
8.   V. Caignaert, R. Retoux, C. Michel, M. Hervieu and B.
      Raveau, **Physica C.**, 167:483 (1990)

9.   S.S.P. Parkin, V.F. Lee, A.I. Nazzal, R. Savoy, T.C. Huang, G. Gorman and R. Beyers, **Phys. Rev. B**, 38:6531 (1988)

10. M. Hervieu, A. Maignan, C. Martin, C. Michel, J. Provost and B. Raveau, **J. Solid State Chem.**, 75:213 (1988)

11. M.A. Subramanian, C.C. Torardi, J. Gopalakrishnan, P.L. Gai, J.C. Calabrese, T.R. Askew, R.B. Flippen and A.W. Sleight, **Science**, 242:249 (1988)

12. C. Martin, J. Provost, D. Bourgault, B. Domenges, C. Michel, M. Hervieu and B. Raveau, **Physica C**, 157:460 (1989)

13. C. Martin, D. Bourgault, C. Michel, M. Hervieu and B. Raveau, **Modern Phys. Lett. B**, 3:93 (1989)

14. A.K. Ganguli, R. Nagarajan, K.S. Nanjun Daswamy and C.N.R. Rao, **Mat. Res. Bull.**, 24:103 (1989)

15. S. Nakajima, M.Kikuchi, N. Kobayashi, H. Iwasaki, D. Shindo, Y. Syono and Y. Mato, 2nd Int. Symp. on Superconductivity, Nov. 14-17, Tsukuba, Ibaraki, Japan (1989)

16. C.N.R. Rao, A.K. Ganguli and R. Vijayaraghavan, **Phys. Rev. B**, 40:2565 (1989)

17. R. Vijayaraghavan, A.K. Ganguli, N.Y. Vasanthacharya, M.K. Rajumon, G.V. Kulkarni, G. Sankar, D. Sarma, A.K. Sood, N. Chandrabhas and C.N.R. Rao, **Supercond. Sci. Technol.**, 2:195 (1989)

18. R.M. Hazen, D.W. Finger, R.J. Angel, C.T. Prewitt, N.L. Ross, C.G. Adidiacos, P.J. Heaney, D.R. Veblen, Z.Z. Sheng, A. El Ali and A.M. Hermann, **Phys. Rev. Lett.**, 60:1657 (1988)

19. M.A. Subramanian, J.C. Calabrese, C.C. Torardi, J. Gopalakrishnan, T.R. Askew, R.B. Flippen, K.J. Morrissey, U. Chowdhry and A.W. Sleight, **Nature**, 332:420 (1988)

20. S.S.P. Parkin, V.Y. Lee, E.M. Engler, A.I. Nazzal, T.C. Huang, G. Gorman, R. Savoy and R. Beyers, **Phys. Rev. Lett.**, 60:1539 (1988)

21. C. Politis and H. Luo, **Modern Physics Lett. B**, 2:793 (1988)

22. A. Maignan, C. Michel, M. Hervieu, C. Martin, D. Groult and B. Raveau, **Modern Physics Lett. B**, 2:681 (1988)

23. E.A. Hayri and M. Greenblatt, **Physica C.**, 156:775 (1988)

24. D. Bourgault, C. Martin, C. Michel, M. Hervieu and B. Raveau, **Physica C.**, 158:511 (1989)

25. T. Rouillon, J. Provost, M. Hervieu, D. Groult, C. Michel and B. Raveau, **Physica C**, 159:201 (1989)

26. T. Rouillon, J. Provost, M. Hervieu, D. Groult, C. Michel and B. Raveau, **J. Solid State Chem.**, 84:375 (1990)

27. T. Rouillon, A. Maignan, M. Hervieu, C. Michel and B. Raveau, **Physica C.**, in press

28. M. Maeda, Y. Tanaka, M. Fukutoni and T. Asano, **Jpn. J. Appl. Phys. Lett.**, 27:L209, L548 (1988)

29. R.M. Hazen, C.T. Prewitt, R.J. Angel, N.L. Ross, L.W. Finger, C.G. Hadidiacos, D.R. Veblen, P.J. Heaney, P.H. Hor, R.L. Meng, Y.Y. Sun, Y.Q. Wang, Y.Y. Sue, Z.J. Huang, L. Gao, J. Bechtold and C.W. Chu, **Phys. Rev. Lett.**, 60:1174 (1988)

30. M.A. Subramanian, C.C. Torardi, J.C. Calabrese, J. Gopalakrishnan, T.R. Askew, R.B. Flippen, K.J. Morrissey, U. Chowdhry and A.W. Sleight, **Science**, 239:1015 (1988)

31. S.A. Sunshine, T. Siegrist, L.F. Schneemeyer, D.W. Murphy, R.J. Cava, B. Batlogg, R.B. Van Dover, R.M. Fleming, S.H. Glarum, S. Nakara, R. Farrow, J.J. Krajewski, S.M. Zahurak, J.V. Vaszczak, J.H. Marshall, P. Marsh, L.W. Rupp and W.F. Peck, **Phys. Rev. B**, 38:893 (1988)

32. J.M. Tarascon, Y. Lepage, P. Bardoux, B.G. Bagley, L.H.

Greene, W.R. McKinnon, G.W. Hull, M. Giroud and D.M. Hwang, **Phys. Rev. B**, 37:9382 (1988)

33. M. Hervieu, C. Michel, B. Domenges, Y. Laligant, A. Lebail, G. Ferey and B. Raveau, **Modern Phys. Lett. B**, 2:491 (1988)

34. P. Bordet, J.J. Capponi, C. Chaillout, J. Chenavas, A.W. Hewat, E.A. Hewat, J.L. Hodeau, M. Marezio, J.L. Tholence and D. Tranqui, **Physica C**, 153-155:623 (1988)

35. T. Kajitani, K. Kusaba, M. Kibuchi, N. Kobayashi, Y. Syono, T.B. Williams and M. Hirabayashi, **Jpn. J. Appl. Phys. Lett.**, 27:L587 (1988)

36. H.G. von Schnering, L. Walz, M. Schwartz, W. Beker, M. Hartweg, T. Popp, B. Hettlich, P. Müller and G. Kämpf. **Angew. Chem.**, 27:574 (1988)

37. R. Retoux, V. Caignaert, J. Provost, C. Michel, M. Hervieu and B. Raveau, **J. Solid State Chem.**, 75:157 (1989)

38. M. Hervieu, C. Michel, A. Maignan, C. Martin and B. Raveau, **J. Solid State Chem.**, 74:428 (1988)

39. M. Hervieu, C. Martin, J. Provost and B. Raveau, **J. Solid State Chem.**, 76:419 (1988)

40. M. Hervieu, B. Domenges, C. Michel and B. Raveau, **Modern Phys. Lett. B**, 2:835 (1988)

41. G. van Tandeloo, H.W. Zandbergen, J. van Landuyt and S. Amelinckx, **Appl. Phys. A**, 46:233 (1988)

42. T. Kijima, J. Tamaka, Y. Bando, M. Onoda and F. Izuni, **Jpn. J. Appl. Phys.**, 27:L369 (1988)

43. E.A. Hewat, P. Bordet, J.J. Capponi, C. Chaillout, J.L. Hodeau and M. Marezio, **Physica C.**, 153-155:619 (1988)

44. M. Hervieu, C. Michel, N. Nguyen, R. Retoux and B. Raveau, **Europ. J. Solid State Inorg. Chem.**, 25:375 (1988)

45. R. Retoux, C. Michel, M. Hervieu, N. Nguyen and B. Raveau, **Solid State Comm.**, 69:599 (1989)

46. B. Morosin, R.J. Banghman, D.S. Ginley, J.E. Schirberand and E.L. Venturini, **Physica C**, 161:115 (1990)

47. Y. Kubo, Y. Shimikawa, T. Manako, T. Sato, S. Ijima, T. Ichihashi and H. Igarashi, **Physica C**, 162-164:991 (1989)

48. K.V. Ramanujachary, S. Li and M. Greenblatt, **Physica C.**, 165:377 (1990)

49. I.K. Gopalakrishnan, P. Sastry, H. Rajagpal, A. Saqueira, J.V. Yakümi and R.M. Iyer, **Physica C**, 159:811 (1989)

50. A. Schilling, H.R. Ott and F. Hulliger, **Physica C**, 157:144 (1989)

51. C. Martin, A. Maignan, J. Provost, C. Michel, M. Hervieu, R. Tournier and B. Raveau, **Physica C**, 168:8 (1990)

52. Y. Shimikawa, Y. Kubo, T. Manako, T. Satoh, S. Ijima, T. Ichiashi and H. Igarashi, **Pysica C**, 157:279 (1989)

53. A. Sulpice, B. Giordamengo, R. Tournier, M. Hervieu, A. Maignan, C. Martin, C. Michel and J. Provost, **Physica C.**, 156:243 (1988)

54. J.J. Capponi, C. Chaillout, A.W. Hewat, P. Lejay, M. Marezio, N.Nguyen, B. Raveau, J.L. Tholence and R. Tournier, **Europhysics Lett.**, 3:1301 (1987)

55. R.J. Cava, B. Batlogg, R.B. van Dover, J.J. Krakewski, J.V. Waszczak, R.M. Fleming, W.F. Peck Jr., L.W. Ruppo Jr., P. Marsh, A.C.W.P. James, L.F. Schneemeyer, **Nature**, 345:602 (1990)

IS THERE SPIN ON THE QUASI-PARTICLES IN THE COPPER OXYGEN PLANES OF HIGH TEMPERATURE SUPERCONDUCTORS?

George Reiter

Physics Department and Texas Center for High
Temperature Superconductors
University of Houston, Texas, USA

ABSTRACT

We show that the three band model for the copper-oxygen
planes is not equivalent to the t-J model in its low energy
physics by calculating the oxygen-oxygen spin correlation
function exactly in a ferromagnetic background, and comparing it
with the result given by the Zhang and Rice approximation, i.e.
zero The effect of direct oxygen-oxygen hopping on the spin of
the hole on the oxygen is calculated exactly for a ferromagnetic
background, and an argument is given that there will be spin
associated with the charge on the quasi-particle in the
antiferromagnet.

1.   INTRODUCTION

The three band Hubbard model proposed by Emery[1] is
generally regarded as being a sufficiently accurate representation of the copper-oxygen planes (with perhaps the addition of
direct oxygen-oxygen hopping) to capture the physics of the
situation. Shortly after it was proposed, Zhang and Rice[2]
argued that a simplified version of the model was actually
equivalent, in its low energy physics, to the t-J model. The
mapping they propose maps singlet combinations of an oxygen
Wannier state with the copper hole at the same site onto
vacancies in the t-J model. As such, the particles which appear
in the Hamiltonian have no spin. Emery and the author[3],
making use of an exact solution of the three band model in a
ferromagnetic background, have shown that the quasi-particles of
the lowest energy band do, in fact, have a nonzero average spin
on the oxygen sites, and, in fact, that there is an oxygen-
oxygen spin-spin correlation function which is nonzero in the
three band model, with a frequency spectrum which extends to
zero. This correlation function must be zero for the t-J model.
In a strict sense, this demonstrates that the two models are not
equivalent. I will reproduce part of that demonstration here.

There have been arguments made subsequently that one must
include a direct oxygen-oxygen hopping in the tight binding

model to make it consistent with band structure calculations, and the question arises as to how this would affect the spin. I present here an exact solution for the ferromagnetic background of the three band model with direct oxygen-oxygen hopping included. In the limit that this hopping, t_p, dominates the hopping through the copper, it is clear that the carriers will carry the full spin of the hole, and hybridization with the copper spin will be irrelevant. It is therefore of some interest to determine the spin value for the oxygen at the values of the parameters which are believed to hold in the actual materials, and we will find that it is substantial. We can also show by a plausible continuity argument, that there will be a net spin on the quasi-particles in the physically more relevant antiferromagnetic background.

2. DISCUSSION

The Hamiltonian we will begin with is the strong coupling limit of Emery's original model

$$H_2 = 2(t_1 + t_2)/N \sum_{\vec{k}\vec{k}'\sigma\sigma'} \alpha_k \alpha_{k'} \cdot \vec{S}_{\vec{k}-\vec{k}'} \cdot \vec{s}_{\sigma\sigma'} \cdot b^{\dagger}_{\vec{k}',\sigma} \cdot b_{\vec{k}\sigma}$$

$$+ \frac{1}{2}(t_1 - t_2) \sum_{\vec{k}\sigma} \alpha_k^2 b^{\dagger}_{\vec{k}\sigma} b_{\vec{k}\sigma} \tag{1}$$

where

$$\vec{S}_{\vec{k}} = \sum_{\vec{m}} \exp(i\vec{k}\cdot\vec{m}) \vec{S}_{\vec{m}}, \quad b_{\vec{k}\sigma} = (\alpha_k N^{1/2})^{-1} \sum_{\vec{m},\vec{\delta}} a^{\dagger}_{\vec{m}+\vec{\delta},\sigma} \exp(-i\vec{k}\cdot\vec{m})$$

and

$$\alpha_k^2 = 2(2 + \cos k_x + \cos k_y).$$

$a^{\dagger}_{\vec{m}+\vec{\delta},\sigma}$ creates a hole of spin σ in an $O(2_{px})$ or $O(2_{py})$ state at $\vec{m} + \vec{\delta}$, where $\vec{\delta}$ is $(1/2,0)$ or $(0,1/2)$.

When $t_1 = t_2$, this model is identical with what has been called the Kondo-Heisenberg model. The model is exactly solvable for a single oxygen hole in a ferromagnetic background[3], and for $t_1 = 0$, the solution is obtainable without having to do any quadrature as

$$|\psi_{\vec{k}}\rangle = \left[\alpha_k b^{\dagger}_{\vec{k}\uparrow} - 1/N \sum_q \alpha_q b^{\dagger}_{\vec{q}\downarrow} S^{+}_{\vec{k}-\vec{q}} \right] |FM\downarrow\rangle \tag{2}$$

with an energy

$$\lambda_k = -t_2(\alpha_k^2 + 4) . \tag{3}$$

To bring out the nature of Zhang and Rice's approximation, we introduce the Fourier transform of in-cell triplets and singlets

$$|\phi_k^\pm> = \frac{1}{\sqrt{2N}} \sum_m \exp(i\vec{k}.\vec{m})(b_{m\uparrow}^\dagger \pm b_{m\downarrow}^\dagger S_m^+)|FM\downarrow> \qquad (4)$$

where $b_{\vec{m},\sigma}^\dagger$ is the Wannier state in the cell $\vec{m}$ associated with the low energy band. We find that

$$|\psi_{\vec{k}}> \cong \left[\alpha_k^2 + \alpha(0)^2\right]^{1/2} \left[\cos\theta_k |\phi_{\vec{k}}^-> + \sin\theta_k |\phi_{\vec{k}}^+>\right] \qquad (5)$$

where $\alpha(0) = N^{-1}\Sigma_{\vec{k}}\alpha_k$ with the singlet-triplet mixing angle given by

$$\tan\theta_k = \{\alpha_k - \alpha(0)\}/\{\alpha_k + \alpha(0)\}. \qquad (6)$$

The ZR approximation sets $\theta_k = 0$, but it is clear that $\tan\theta_k$ ranges from essentially zero to -1. To illustrate the physical significance, we calculate the z-component of the oxygen-oxygen spin-spin correlation function, in the state $|\psi_{\vec{k}}>$ using the approximation (5)

$$S_k^z(q,\omega) = (1/4N) \sin^2(\theta_k + \theta_{k+q})\delta(\omega-\lambda_k+\lambda_{k+q}) . \qquad (7)$$

This is the result for a single oxygen hole of wavevector k. For many oxygen holes, the result is intensive, because (7) should be summed over k values in a Fermi sea (assuming independent particles). This is, in principle, a measurable quantity, by NMR for instance, (in fact Berthier et al.[4] have concluded that, by comparing oxygen, copper and yttrium relaxation rate data, there is indeed a contribution from the spin on the oxygen) has spectrum near $\omega = 0$, is clearly part of the "low energy physics", and is zero within the approximation of ZR. ZR's argument is that there is a large gap between the singlet and the triplet states, so that the triplets may be neglected. This is true, the gap is $2t_2\alpha_k\alpha(0)$. But there is also a significant mixing matrix element $t_2\{\alpha(0)^2 - \alpha_k^2\}$, with the same energy scale.

There are not two energy scales in the problem to make ZR's approximation exact in any limit.

When we include direct oxygen-oxygen hopping, for which some approximation could also be included in the t-J model, giving the t-t'-J model, the model remains solvable, in a ferromagnetic background. The Hamiltonian now becomes

$$H = H_2 - 4t_p \sum_{k\sigma}\cos(k_x/2)\cos(k_y/2)\left[a_{+k\sigma}^\dagger a_{+k\sigma} - a_{-k\sigma}^\dagger a_{-k\sigma}\right] \qquad (8)$$

where

$$a_{\pm k\sigma}^\dagger = (1/\sqrt{2})\left[\exp(ik_x/2)a_{k\sigma}^{\dagger x} \pm \exp(ik_y/2)a_{k\sigma}^{\dagger y}\right] \qquad (9)$$

are the oxygen band creation operators. It is convenient to

represent H_2 in terms of these modes, rather than the modes introduced earlier(1), and we find

$$b_{k\sigma} = \frac{1}{\sqrt{2}} \frac{[\{\cos(k_x/2) + \cos(k_y/2)\}a_{+\vec{k}\sigma} + \{\cos(k_x/2) - \cos(k_y/2)\}a_{-\vec{k}\sigma}]}{[\cos^2(k_x/2) + \cos^2(k_y/2)]^{1/2}} \qquad (10)$$

It should be noticed that when $k_x = k_y$, or at the bottom of the band, where both are small, the antibonding states are close to the states created by $a_{+\vec{k}\sigma}^{\dagger}$. If the sign of t_p is such that these latter states are the lower band, i.e. positive with our conventions, then the two sets of states will hybridize well. We believe that the physical sign is positive, but we will present results for both choices of sign.

We represent the solution as

$$|\psi_{\vec{k}}\rangle = \mu_{\vec{k}}a_{\vec{k}\uparrow}^{\dagger} + (1/N)\sum_{\vec{k}'} W(\vec{k},\vec{k}')S_{\vec{k}-\vec{k}'}^{+} a_{\vec{k}',\downarrow}^{\dagger} |FM\downarrow\rangle \qquad (11)$$

where all quantities are actually two-dimensional vectors, i.e.

$$a_{\vec{k}\sigma}^{\dagger} = \begin{bmatrix} a_{+\vec{k}\sigma}^{\dagger} \\ a_{-\vec{k}\sigma}^{\dagger} \end{bmatrix}, \quad \mu_{\vec{k}} = \begin{bmatrix} \mu_{\vec{k}}^{+}, \mu_{\vec{k}}^{-} \end{bmatrix}, \quad W(\vec{k},\vec{k}') = \begin{bmatrix} W^{+}(\vec{k},\vec{k}'), W^{-}(\vec{k},\vec{k}') \end{bmatrix}. \qquad (12)$$

The solution is obtainable by a straightfoward extension of the methods of Ref. 3 and is

$$W(\vec{k},\vec{k}') = 2(t_1+t_2)\mu_{\vec{k}}\left[C(\vec{k},\vec{k}') - \{\overline{\lambda}_{\vec{k}} - \overline{\epsilon}_{\vec{k}} + 2t_2 C(\vec{k},\vec{k})\}A(\vec{k},\vec{k}')/2(t_1+t_2)\right] \qquad (13)$$

$$\times \left[\overline{\lambda}_{\vec{k}} + \overline{\epsilon}_{\vec{k}} - 2t_1 C(\vec{k}',\vec{k}')\right]^{-1} \equiv \mu_{\vec{k}} G(\vec{k},\vec{k}')$$

where the eigenvalue $\overline{\lambda}_{\vec{k}}$ and $\mu_{\vec{k}}$ are defined by

$$\mu_{\vec{k}}\left[\overline{\lambda}_{\vec{k}} - \overline{\epsilon}_{\vec{k}} + 2t_2 C(\vec{k},\vec{k}') - 2(t_1 + t_2)(1/N)\sum G(\vec{k},\vec{k}')C(\vec{k}',\vec{k})\right] = 0 \qquad (14)$$

with

$$A(\vec{k},\vec{k}') = \begin{bmatrix} c_{\vec{k}'}^{+}/c_{\vec{k}}^{+} & 0 \\ 0 & c_{\vec{k}'}^{-}/c_{\vec{k}}^{-} \end{bmatrix}, \quad C(\vec{k},\vec{k}') = \begin{bmatrix} c_{\vec{k}'}^{+} \\ c_{\vec{k}'}^{-} \end{bmatrix}\begin{bmatrix} c_{\vec{k}}^{+}, c_{\vec{k}}^{-} \end{bmatrix},$$

$$\qquad (15)$$

$$\overline{\epsilon}_{\vec{k}} = \begin{bmatrix} \epsilon_{\vec{k}} & 0 \\ 0 & -\epsilon_{\vec{k}} \end{bmatrix}, \quad \overline{\lambda}_{\vec{k}} = \lambda_{\vec{k}}\begin{bmatrix} 1 & 0 \\ 0 & 1 \end{bmatrix}.$$

It may be checked that when $t_1 = t_p = 0$, the solution reduces to that already obtained. For the general case, the eigenvalue and the eigenvectors may be obtained numerically by an iterative procedure. I show in Fig. 1 the results of the numerical solution for the average spin on an oxygen site as a function of the wavevector of the quasi-particle for several values of t_p and $t_1 = t_2 = .5$. The value of $t_p = .75$ is approximately the physical[5] value. The value of the spin is insensitive to the ratio of t_1/t_2. Several things should be noticed. With the physical choce of sign, there is a significant enhancement of the spin over most of the zone, to about half the value for a bare hole. With the opposite choice of sign, the spin is nearly cancelled over much of the zone. Also, the spin in the center of the zone, which is where the band minimum of the quasi-particles in the antiferromagnetic t-J model would occur, is strongly affected at these values of t_p. It is clear that there will be a spin on the oxygen for a general set of parameters for the three band model, and this will be poorly approximately by the t-J model, or the t-t'-J model. It might be argued that, since the spin up and spin down bands for the oxygen hole are well separated in energy, the residual spin on the oxygen is unlikely to play any role in the low energy interaction between quasi-particles, since the spin cannot be excited, and the point would be well taken. This separation does not occur, however, in the antiferromagnet.

The problem of the antiferromagnet is not exactly solvable, but we can make some progress, and keep track of the spin, by beginning with the large t_p limit, and treating the hopping through the copper as a perturbation. We can also make use of the fact that the ground state and its excitations in the absence of any holes are given by spin wave theory, to very good approximation[6]. We will simplify the problem by neglecting the coupling to the upper oxygen band (created by $a_{-\vec{k}\sigma}$) in the perturbation, which we take to be H_2. t_p is assumed positive. We will treat the operator $S^z_{\vec{k}',-\vec{k}'}$ in H_2 by mean field theory, replacing it by its average $\lambda\delta_{\vec{k}-\vec{k}'+\vec{k}_0}$, where $\vec{k}_0$ is the antiferromagnetic ordering wavevector, (π, π) and explicitly diagonalize the Hamiltonian which results from neglecting spin-flip hopping, for the moment.

$$H = (t_1 + t_2)\lambda N^{-1} \sum_{\vec{k}} c^+_{\vec{k}+\vec{k}_0} c^+_{\vec{k}} \left[a^\dagger_{+(\vec{k}+\vec{k}_0)\uparrow} a_{+\vec{k}\uparrow} - a^\dagger_{+(\vec{k}+\vec{k}_0)\downarrow} a_{+\vec{k}\downarrow} \right]$$

(16)

$$+ \sum_{\vec{k}\sigma} \epsilon_{\vec{k}} a^\dagger_{+\vec{k}\sigma} a_{+\vec{k}\sigma} \ .$$

The result is that the energy eigenvalues, for either spin, are

$$\vec{\epsilon}_{\vec{k}} = (1/2)(\epsilon_{\vec{k}} + \epsilon_{\vec{k}+\vec{k}_0}) \pm \left[(\epsilon_{\vec{k}} - \epsilon_{\vec{k}+\vec{k}_0})^2 + [(t_1 + t_2)\lambda c^+_{\vec{k}+\vec{k}_0} c^+_{\vec{k}}]^{1/2} \right. .$$

(17)

The spin is, however, correlated with the relative phase difference for holes on the two distinct sublattices of the antiferromagnet, since the creation operators for the eigenstates are

$$\alpha^{\dagger}_{\vec{k}\sigma} = M_1 a^{\dagger}_{+\vec{k}\sigma} + \sigma M_2 a^{\dagger}_{+\vec{k}+\vec{k}_0\sigma}$$

(18)

where the coefficients M_1, M_2 are wavevector dependent, and $M_1^2 + M_2^2 = 1$. This may be readily seen by constructing the Wannier states which follow from (16).

$$\alpha^{\dagger}_{i\sigma} = N^{-1}\sum_{\vec{k}}\exp(ik.r_i)[M_1 + (-1)^n\sigma M_2]a^{\dagger}_{+k\sigma} \ .$$

(19)

The index n is 0 if i is on one sublattice, 1 if it is on the other. When the spin-flip term is added, the eigenstates will no longer be eigenstates of the oxygen spin, and there will be flipped spins on the copper. The two spin bands will remain degenerate, however. This may be shown by explicit calculation, but it is clear from the symmetry of the problem. If the total number of spins is odd, the ground state of the Heisenberg model will be one of the two degenerate spin wave-like ground states, with total spin zero. The only physical difference between the

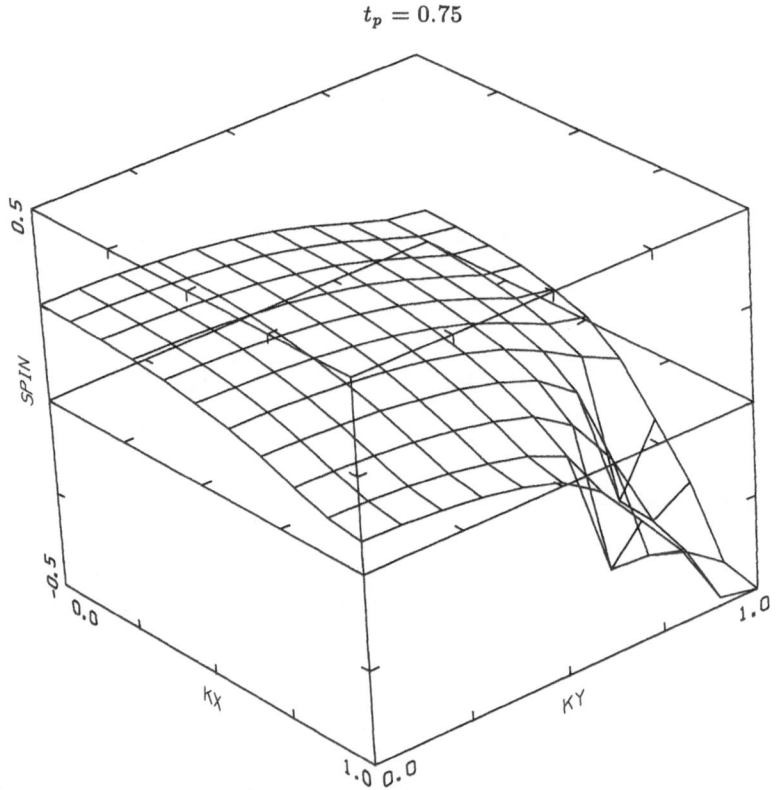

$t_p = 0.75$

Fig. 1. The average spin on the oxygen for $t_p = 0.75$ as a function of the wavevector of the eigenstate in the ferromagnetic background. $t_1 = t_2 = .5$.

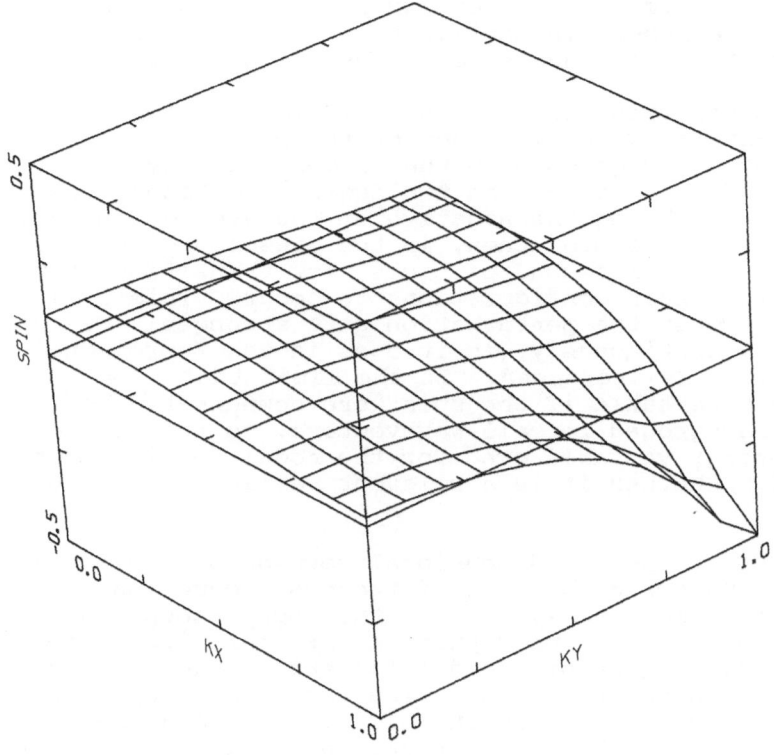

$t_p = -0.75$

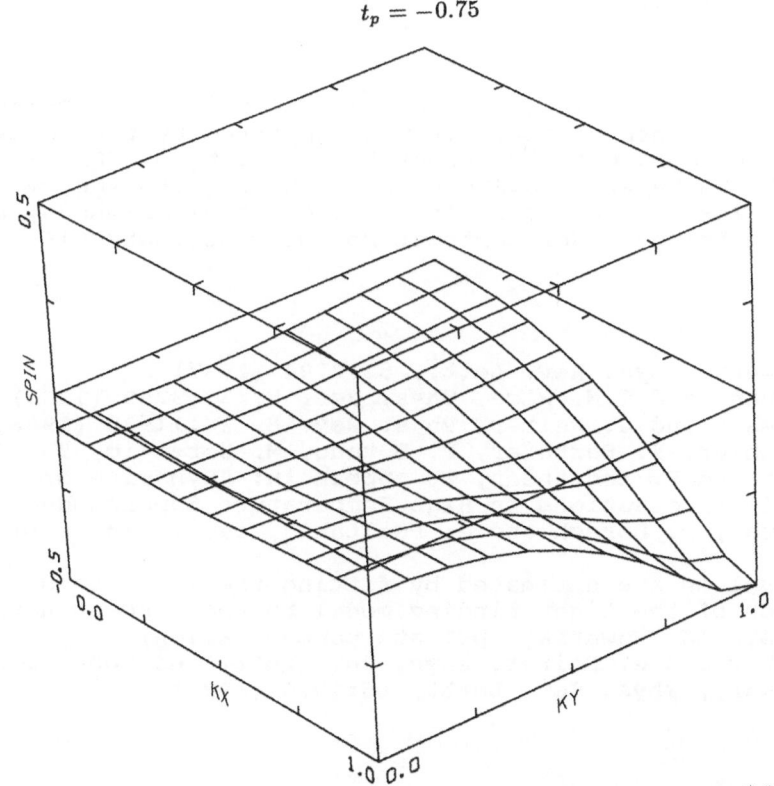

Fig. 1. (continued) b) t_p = 0.0; c) t_p = -0.75

states generated by adding an oxygen spin up or down, will be the sublattice they are associated with. The two sublattices are equivalent, however, so the energy of the up and down spin states must be the same. In the limit of infinite system size, the restriction to an odd number of spins doesn't matter.

It seems clear from the previous calculations for the ferromagnet, that for general values of the parameters there will be a nonzero average spin on the oxygen, for any wavevector, and for either of the two degenerate bands, the average in one band being the negative of the average in the other. It will be possible, then, in the presence of additional interactions, such as a second particle, for transitions between the two degenerate spin configurations to occur. This will remain true even when the perturbation gets stronger. We expect the spin to vary continuously, as it does in the ferromagnet, as the strength of H_2 is increased. In the limit that $t_p = 0$, it would be quite remarkable if the spin were compensated perfectly by the copper background for all wavevectors, so that it would remain a low energy variable even for the model considered by Zhang and Rice. Whether it is a relevant variable is an open question.

In the case of the one-dimensional version of the model, where the spin and charge degrees of freedom become independent, Emery[7] has discussed the limit that the copper-copper and oxygen-oxygen on site Coulomb repulsions are infinite. There is an additional spin degree of freedom for each oxygen hole added, not one less, as in the t-J model. The oxygen and copper spins participate on an equal footing in this limit, and are described by a Heisenberg model. There is, then, an a priori case to be made that the spin of the quasi-particle remains a relevant low energy dynamical variable.

Acknowledgment

This work was initiated in collaboration with Vic Emery and owes much to the interaction. It was supported by the Texas Center for Superconductivity, under Prime Grant No. MDA 972-8-G-0002 from the US Defense Advanced Research Projects Agency and the State of Texas, and the Division of Material Sciences, US Department of Energy under Contract No. DE-AC02-76ch00016

REFERENCES

1.   V.J. Emery, **Phys. Rev. Lett.**, 58:2794 (1987)
2.   F.C. Zhang and T.M. Rice, **Phys. Rev. B**, 37:3759 (1988)
3.   V.J. Emery and G. Reiter, **Phys. Rev. B**, 38:11938 (1988)
4.   C. Berthier, Y. Berthier, P. Butaud, M. Horvatic, Y. Kitaoka and P. Segransan, **to appear in**: "Dynamics of Magnetic Fluctuations in High Temperature Superconductors", G. Reiter, P. Horsch and G. Psaltakis, eds., Plenum Press (1989)
5.   These values are estimated by fitting the mean field solution of the tight binding model to the band structure results. (G. Sawatzky, private communication)
6.   T. Becher and G. Reiter, **Phys. Rev. Lett.**, 63:1004 (1989)
7.   V.J. Emery, **Phys. Rev. Lett.**, 65:1076 (1990)

LOW DIMENSIONAL MOLECULAR STRUCTURES

LOW DIMENSIONAL MOLECULAR CONDUCTORS:

VARIOUS ASPECTS OF THEIR PHYSICAL PROPERTIES

D. Jérome

Laboratoire de Physique des Solides (associé au CNRS)
Université Paris-Sud, 91405
Orsay, France

1.   INTRODUCTION

A search for superconductivity in organic conductors was
proposed by Little in 1964[1]. This proposal has boosted the
research of molecular solids showing metal-like conduction.
Materials of this kind first appeared in the early seventies.
However, it took some time before superconductivity was
discovered in certain classes of organic conductors[2]. The
actual superconducting materials discovered so far, Fig. 1, do
not resemble those proposed by Little, and it is far from
obvious that the nature of the electron pairing has anything to
do with Little's excitonic mechanism. However, these new
materials are fascinating compounds, in which several new
properties have been particularly investigated: low dimensional
conductivity, electron-electron and electron-phonon
interactions, collective conduction of an electron-hole
condensate and the possibility for a metal-insulator transition

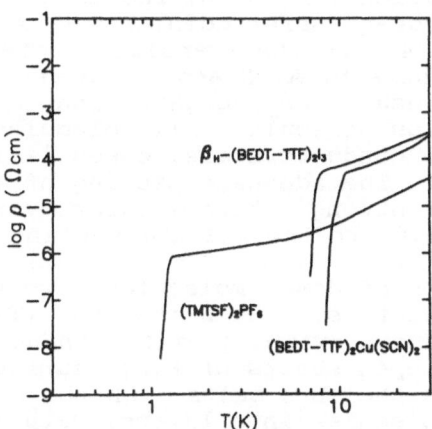

Fig. 1. Superconducting transition in various
        organic conductors.

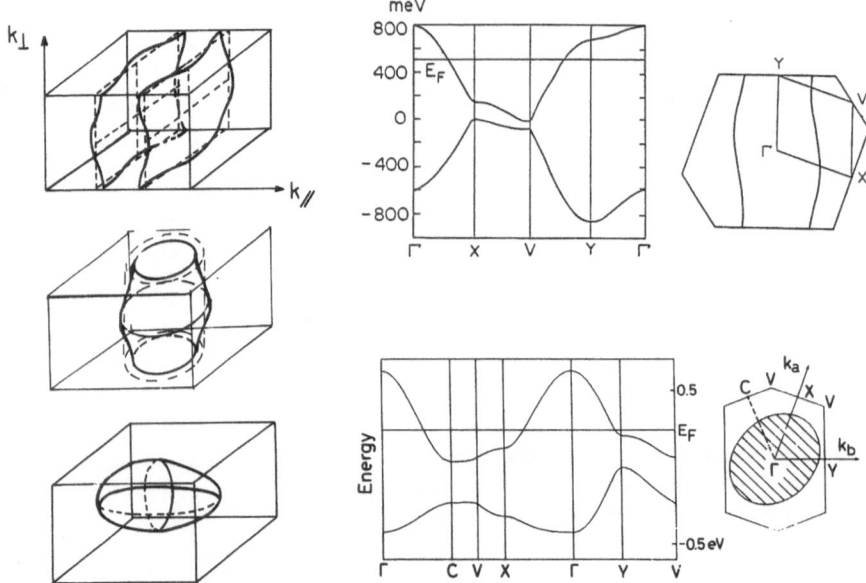

Fig. 2. Model Fermi surface for 1, 2 and 3D conductors
(left) and band structures of 1 and 2D
conductors (right).

to be driven by disorder-order phenomena. These concepts are
briefly reviewed in this article.

2. LOW DIMENSIONALITY

The vast majority of molecular crystals are known to be
insulators or semiconductors. Even if electrons can be
delocalized over a conjugated ring in a given molecule there
remains a finite energy gap to overcome before some
intermolecular conduction of any reasonable size is obtained.
Intermolecular conduction has been achieved, so far, only in
organic salts with cation (or anion) radicals. Some examples
are given by the charge-transfer complexes $D^{+?}A^{-?}$[3], the
organic salts $D_x^{+1}X^{-1}$[4] and the transition metal coordination
complexes $D^{+?}M_x^{-?}$[5] where D, A, X and M are respectively a π-
donor or acceptor molecule, an inorganic ion, and a transition
metal coordinated anion molecule. All molecules involved in the
building of molecular conductors are, essentially, flat
conjugated molecules. The adequate packing of these molecules
in a crystal allows a strong intermolecular overlap of the
π-orbitals perpendicular to the molecular plane.

Segregated chains of donor molecules give rise to one-
dimensional (1D) conductors. The historical $(TMTSF)_2X$, X = PF_6,
ClO_4, etc. are well known examples of 1D conductors,
characterized by two open sheets of Fermi surface, Fig. 2, and
known collectively as the TM_2X salts. On the other hand, the
packing of organic molecules into layers, with strong
intermolecular interactions between layers, leads to more 2D
conductors, characterized by a tube-like Fermi surface, for
example β-$(ET)_2X$[6] or κ-$(ET)_2Cu(SCN)_2$[7], Fig. 2. As we shall

see in what follows, there exist drastically different behaviors between conductors exhibiting open (1D) and tube-like (2D) Fermi surfaces. This difference is already observed in their superconducting properties, since T_c is usually higher, $\approx$ 8 to 10 K in 2D superconductors, as compared to the 1D ones ($T_c \approx$ 1-3 K)[8]. Furthermore, besides superconductivity (SC), the physics of the (TMTTF-TMTSF)$_2$X series is governed by the one-dimensionality of the Fermi surface (FS). Since the FS of TM$_2$X compounds is open with the possibility of a slight warping, resulting from a nonzero transverse coupling $t_\perp$, the important concepts of 1D physics should apply to these materials, namely the absence (or at least the severe suppression) of long range order at finite temperature, and the existence of equally divergent electron-electron (SC) and electron-hole (charge density wave (CDW), spin density wave (SDW)) pairing correlations in the 1D regime[9-11]. This mixture of diverging channels is, however, limited at low temperature by a cross-over temperature T_{x1}. The experimental data and the theory suggest that T_{x1} is indeed smaller than the value $t_\perp/\pi$, which is expected for a quasi-1D (Q-1D) noninteracting electron gas[12]. It has been argued that the cross-over can be renormalized by the existence of Coulombic interactions in the 1D electron gas. In the (TMTSF)$_2$X series T_{x1} is about 10 K[13]. This is ten times smaller than the bare value, which can be inferred from the transverse coupling $t_\perp \approx$ 20-30 meV in band structure calculations.

As long as the transverse coupling remains small enough to enable a nesting of the FS by the vector Q over a large fraction of its area, the electron-hole instability prevails at low temperature and a 3D long range magnetic order SDW of wave vector Q occurs, with the concomitant opening of a gap over the entire FS (insulating state). The nesting vector Q is equal to the transverse vector $Q_t = (2k_F, _2\pi/b)$, provided the transverse coupling is small, i.e. $t'_\perp = t_\perp^2/t_\parallel \ll T_0$, where T_0 is the transition temperature for the onset of a SDW state, calculated within the mean field approximation[14a,b]. If the material deviates from one-dimensionality, applying pressure for example, $t'_\perp$ can be increased. Hence the nesting vector shifts from

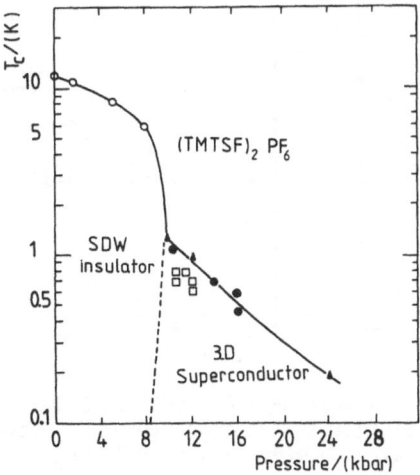

Fig. 3. T-P phase diagram of (TMTSF)$_2$PF$_6$.

$Q_t = (2k_F, \pi/b)$ to $Q_o = (2k_F + q_\parallel, q_\perp)$, where $q_\parallel$ is a component of small amplitude and $q_\perp$ is usually of order $(0.5-0.4)\pi/b$ [15]. Q_o is, thus, the vector which optimizes the nesting of the non-sinusoidal FS for an actual Q-1D conductor.

Figure 3 shows that the stability of the SDW ground state of $(TMTSF)_2PF_6$ is suppressed under pressure and allows the onset of superconductivity above 9 kbar[2]. This feature is a direct consequence of the increased transverse coupling, induced under pressure. The restoration of 1D features in Q-1D compounds is, however, possible, as shown by the unexpected effect of a large magnetic field on the ground state of a Q-1D conductor. Crudely speaking, the field suppresses the deviations from nested Fermi surfaces[14a,b]. The quantity $t'_\perp$ is effectively decreased by the field, and 1D instabilities are recovered. However, the experimental data of $(TMTSF)_2PF_6$, Fig. 4 (left), show that a sequence of phases must be crossed before the so-called "N = 0" phase (corresponding to the insulating SDW ground state, which is stable at ambient pressure) can be reached above 18 T[16]. The capital N labels the phase containing N fully occupied Landau levels. N = 0 means no free carriers, namely a gap opened over the whole Fermi surface. Phases with N = 1,2,3 etc., which are stable below 18 T, correspond to semimetallic states with a very low density of unnested carriers ($\approx 10^{-2}$/unit cell) and 1, 2 or 3 completely filled Landau levels. The situation, where the Fermi level falls between completely filled and completely empty levels minimizes the diamagnetic energy of the 2D carriers. This situation prevails in a finite range of magnetic fields provided the wave vector of the magnetic modulation within a given subphase has some flexibility and can vary linearly with the field[14a,b,c]. First order phase transitions are expected between various subphases, in agreement with the hysteresis shown in Fig. 4 (left). A direct consequence of the field dependent Q vector is the quantization

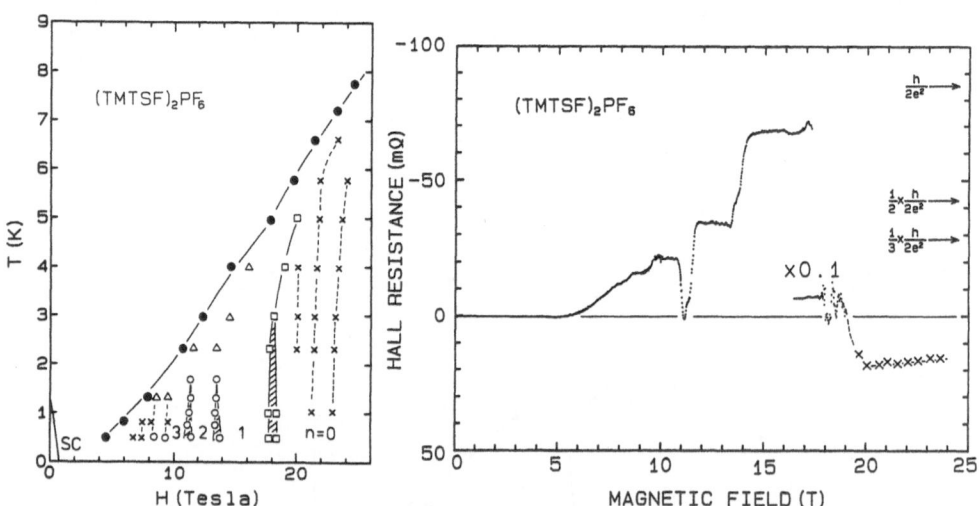

Fig. 4. T-H phase diagram of $(TMTSF)2PF6$ at a pressure of 8 kbar (left), the magnetic field is along the axis of lowest conductivity. Quantized Hall voltage in the field induced spin density wave states (right).

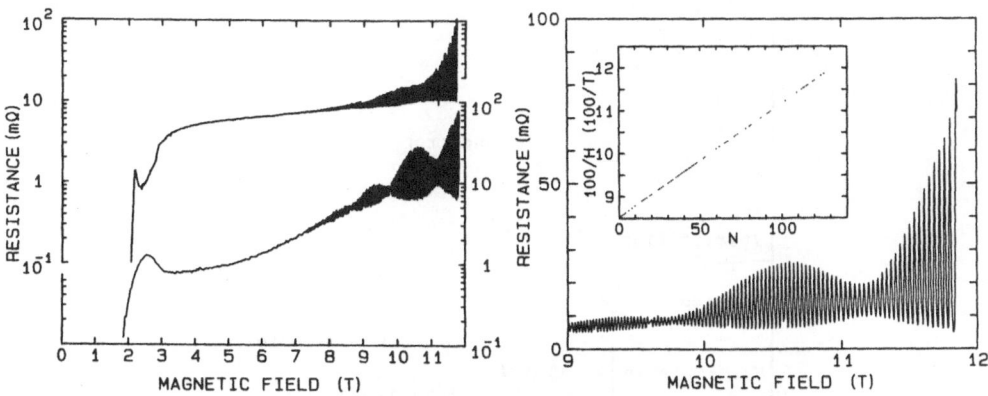

Fig. 5. Transverse magnetorestistance of β_H-(ET)$_2$I$_3$
(left) and magnification of the data between 9
and 12 T at T = 0.38 K (right). The insert
shows the plot of the oscillations versus
integral numbers, N.

of the Hall resistance $V_H/I = R_H$, since $R_H = H/ne$ and the
density of unnested carriers becomes n = NeH/h. Thus, the Hall
resistance reads $R_H = h/2Ne^2$ (where the factor 2 in the
denominator comes from the spin degeneracy). The (TMTSF)$_2$PF$_6$
Hall data, Fig. 4 (right), agree quantitatively with the
predicted quantized values of R_H.

Another spectacular effect of the low dimensionality is
observed in the series of (ET)$_2$X conductors with a 2D tube-like
FS. In Fig. 5, we present some magnetoresistance data, obtained
in the β-phase of (ET)$_2$I$_3$ cooled down to very low temperature,
using a special procedure[17] in order to avoid the formation of
an incommensurate lattice modulation at low temperature. The
so-called β_H(T$_c$ = 8.1 K) is, thus, stabilized at low
temperature[18]. The behavior of the magnetoresistance of the
β_H phase at T = 0.38 K is shown in Fig. 5 (left) up to 12 T, and
for the field range 9-12 T in Fig. 5 (right). The magnetic
field is parallel to the axis of the tubular FS. A "fast"
oscillation of extremely large amplitude is visible above 9 T.
It is periodic versus 1/H, with a fundamental field H$_0$ = 3730 T,
Fig. 5 (right). In addition, a low frequency oscillation
(beating) is clearly observed, with the frequency H$_1$ = 36.8 T.
Both the fast oscillations and the slow beating are the
signature of a 2D Fermi surface, resembling the schematic
drawing in Fig. 2. The frequency H$_0$ measures the density of
carriers $\approx$ 1 per formula unit, in agreement with the degree of
oxidation, and the beating frequency H$_1$, which is related to the
FS warping, provides an estimate for the interlayer coupling,
namely $t_\perp \approx 0.5$ meV (if $t_\parallel \approx 70$ meV).

3.   INTERACTIONS

Superconductivity is only one among the various
instabilities observed at low temperature in low dimensional
organic conductors. The variety of ground states is a direct
consequence of the interplay between electron-electron and
electron-phonon interactions in a low dimensional gas. This is

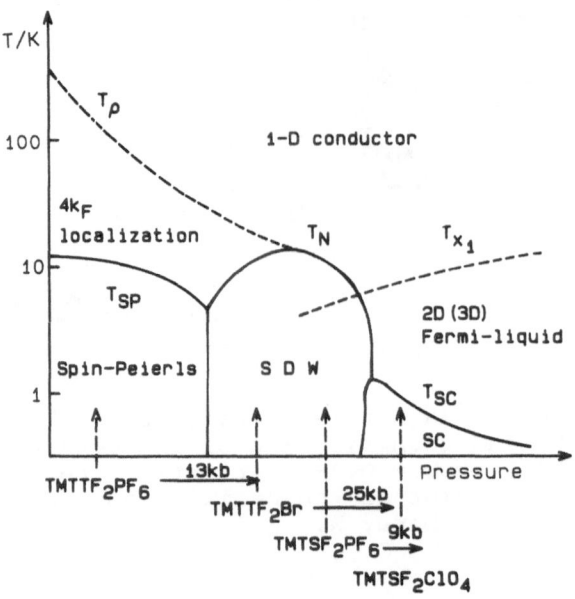

Fig. 6. General phase diagram of the (TMTTF-
TMTSF)₂X series.

illustrated by the generalized phase diagram of the (TMTTF-
TMTSF)$_2$X series, Fig. 6. The all-sulphur compound (TMTTF)$_2$PF$_6$
displays a charge localization below T_ρ ≈ 200 K (a 4k$_F$
localization, accompanied by a loss of the charge degrees of
freedom but no change in the spin susceptibility, which follows
the law of the 1D quantum antiferromagnet down to 20 K). The
charge localization has been attributed to the role of strong on
site Coulomb repulsions enhanced by commensurability effects
(Umklapp scattering)[19]. The electron-phonon interaction gives
rise to 1D lattice modulation at the wave vector 2k$_F$, as seen by
X-ray diffuse scattering experiments. Below 19 K the 1D quantum
antiferromagnet undergoes a phase transition towards a 3D spin
singlet state (Spin-Peierls (SP) state with a 2k$_F$ lattice
dimerization). The ground state of this compound changes over
from SP to antiferromagnetism above 10 Kbar. A similar
antiferromagnetic ground state (SDW) is observed for (TMTTF)$_2$Br
under atmospheric pressure. In this latter system, the role of
Coulombic forces is decreased, as indicated by the 4k$_F$
localization, which is suppressed down to 100 K. Furthermore,
the electrons do not couple to the lattice until SDW long range
order is established below 15 K. One can go one step further
with (TMTSF)$_2$PF$_6$, since no 4k$_F$ charge localization is observed
before the onset of a SDW order at 12 K. Finally, for
(TMTSF)$_2$ClO$_4$, the SDW ground state is destabilized due to the
enhanced transverse coupling.

Additional evidence for the role of e-e interactions is
given by the study of the magnetic properties. First, it is
well known that the spin susceptibility of χ_s of all Q-1D
conductors presents a pronounced growth in temperature above
30 K which cannot be explained by a picture of noninteracting
electrons. Secondly, the temperature dependence of the nuclear
spin-lattice relaxation rate, induced by the modulation of the
hyperfine field, does not follow the behavior expected for non-

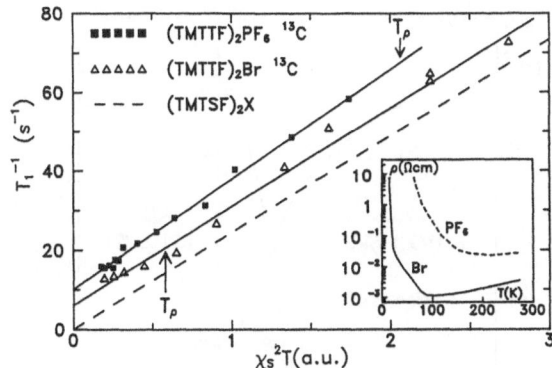

Fig. 7. T_1^{-1} versus $\chi_S^2 T$ in various organic
conductors. The insert shows the
corresponding T dependence of the
resistivity.

interacting 1D particles. T_1^{-1} of all members of the (TMTTF-TMTSF)$_2$X series follows the experimental temperature dependence:

$$T_1^{-1} = C_0 T \chi_s^2(T) + C_1 \tag{1}$$

which can be fairly well understood within the 1D theory[13].

The first term in Eqn. 1 is related to the uniform (q = 0) spin correlations of the 1D electron gas, whereas the second term probes the 1D (q = 2k$_F$) part. Figure 7 shows that 2k$_F$ correlations are well developed, up to room temperature, in compounds exhibiting a Coulomb-assisted localization (TMTTF$_2$Br and TMTTF$_2$PF$_6$) and behaving like a 1D quantum antiferromagnet (AF) with a temperature independent C$_1$. The 2k$_F$ contribution is no longer visible for selenium salts in Fig. 7. However, a different presentation of the data, namely T_1^{-1} versus T, reveals the existence of active AF correlations in (TMTSF)$_2$ClO$_4$, up to 30 K, despite the presence of a superconducting ground state. The analysis of the NMR data leads to the intermediate coupling regime (g$_1$/2πv$_F$) ≈ 0.6 and 0.5 for selenium and sulphur compounds, respectively.

4. ANION ORDERING

In the TM$_2$X series the anions occupy an inversion center in the structure. Therefore a problem arises if the anion itself is not centrosymmetric. The anions can be statistically disordered with equal occupancy of two symmetry related configurations. However, this state should transform into an ordered state at low temperature, for entropy minimization[20]. The order-disorder transition is accompanied by a modification of the lattice periodicity. In some cases the periodicity along the stacking axis is unaffected by the ordering process, for example (TMTSF)$_2$ClO$_4$ with ordering (0, 1/2, 0) at 24 K, but for other systems the ordering may open gaps at the Fermi level, as is the situation which prevails in (TMTSF)$_2$ReO$_4$ with ordering (1/2, 1/2, 1/2) at 180 K. The anion ordered state can be changed under pressure into another ordered state, (0, 1/2, 1/2)

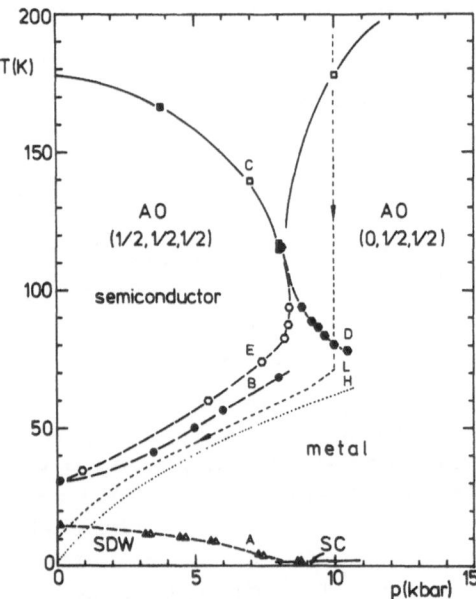

Fig. 8. Phase diagram of $(TMTSF)_2ReO_4$. The SDW
phase can be reached following the dashed
line with arrows.

which is stable at low temperature and preserves the conducting
properties.

The latter configuration can be reached in a metastable
state even under atmospheric pressure, see Fig. 8, and a new
insulating state (presumably SDW) is revealed below 12 K[21].
Once the anions can be frozen at low temperature in a
configuration which does not create new gaps at the Fermi level,
the SDW/SC competition, as observed in $(TMTSF)_2PF_6$, is
recovered. This experiment shows that the competition between
ground states is a property of organic conductors, deeply rooted
in the physics of 1D conductors.

5.   CONDUCTING COLLECTIVE MODELS

The peculiarity of the 1D electron gas is the nesting of
the Fermi surface, with the well known consequence: the metallic
phase becomes unstable at low temperature against the formation
of an insulating state. Electron-phonon and electron-electron
interactions lead to CDW-periodic lattice distortion and SDW
ground states, respectively, which are characterized by a
modulation of the condensate with the wave vector $2k_F$. In
addition to the single particle dispersion, which opens a gap at
the Fermi level, new modes appear. They correspond to the
excitations of the condensate. The phase mode is gapless, with
the dispersion law:

$$\omega_{ph} = m_o/m^* \, v_F k \qquad (2)$$

where m^* is the effective mass of the condensate. As long as
$2k_F$ is not commensurate with the underlying lattice a rigid

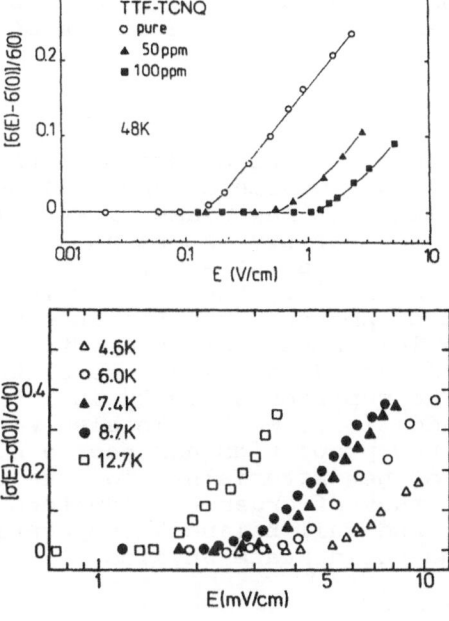

Fig. 9. Nonlinear conductivity in TTF-TCNQ with
different concentrations of irradiation induced
defects (top). Nonlinear conduction from
sliding SDW in the antiferromagnetic state of
$(TMTSF)_2PF_6$.

motion of the density wave can be achieved without any cost in
energy. This translational invariance gives rise to a
collective contribution to the conductivity:

$$\sigma(\omega) = ne^2/m^*\delta(\omega_0) \tag{3}$$

where n is the density of condensed carriers and ω_0 is either
zero, in an ideally pure material, or a finite frequency when
the collective mode is pinned by commensurability or impurities.
Therefore, a threshold electric field must be reached before the
condensate can contribute to the conduction collectively, in
addition to the single particle part. The sliding of a CDW
condensate has been extensively investigated in inorganic 1D
conductors[22]. However, the evidence for a sliding of both
kinds (CDW or SDW) has been given only recently in low
dimensional molecular conductors. Figure 9 (top) shows the
onset of nonlinear conduction, above 100 meV/cm, in a "pure"
sample of TTF-TCNQ[23]. The data also show that the threshold
field is increased when charged impurities are created by
irradiation. A high-pressure study has shown that the pinning
can be drastically increased (by at least three orders of
magnitude) when the CDW wavelength is made commensurate with the
underlying lattice (Λ_{CDW} = 3b; under 18 kbar)[24]. Quite
similarly, the sliding of a SDW condensate also contributes to
collective conduction, as shown in the SDW state of $(TMTSF)_2PF_6$,
Fig. 9 (bottom)[25]. However, the pinning of a SDW is about ten
times smaller than that of a CDW condensate. The SDW can be
visualized as the superposition of two out-of-phase spin-

polarized CDW. Thus, in first order (sinusoidal approximation) the SDW is not pinned by charged impurities. Only deviations from sinusoidality can give rise to a source of pinning.

CONCLUDING REMARKS

This review of some investigations performed in organic conductors has presented a wealth of physical phenomena which have been discovered in this new class of materials. It is needless to say that the variety can be attributed to the relevance, in these materials, of several important concepts, such as the low dimensionality, the interaction of electrons with other electrons or phonons, and the incommensurability. The knowledge of the low temperature properties is now under control, although understanding the nature of the driving interaction leading to superconductivity in 1 or 2D materials, will still require additional work. We also wish to emphasize that the finding of new properties has been a strong stimulus for the development of new materials. To make this story short we may say that the study of organic conductors has also initiated a fruitful and well balanced cooperation between Organic Chemistry and Solid State Physics.

Acknowledgment

I wish to acknowledge the close cooperation of my colleagues, who have been very active in various domains, K. Bechgaard, P. Batail and their groups for the chemistry, C. Bourbonnais and G. Montambaux for their contributions to the theory. A lot of the experimental work has been performed by P. Auban, J.R. Cooper, F. Creuzet, W. Kang, S. Tomic and P. Wzietek. I have also enjoyed several stimulating discussions with J. Friedel. This work has largely benefited from the technical help of J.C. Ameline. The activity in organic conductors was partly supported by the Stimulation Contract DG XII, ST 2000315, the ESPRIT BRA contract 3121 and the DRET contract 88/198.

REFERENCES

1.   W.A. Little, **Phys. Rev.**, 134A:1416 (1964)
2.   (a) D. Jérome, A. Mazaud, M. Ribault and K. Bechgaard, **J. Phys. Lett. Paris**, 41:L-95 (1980) (b) K. Bechgaard, K. Carneiro, M. Olsen and F.B. Rasmussen, **Phys. Rev. Lett.**, 46:852 (1981)
3.   J. Ferraris, D.O. Cowan, V. Walatka and J.H. Perlstein, **J. Am. Chem. Soc.**, 95:948 (1973) and L.B. Coleman, M.J. Cohen, D.J. Sandman, F.G. Yamagishi, A.F. Garito and A.J. Heeger, **Solid State Comm.**, 12:1125 (1973)
4.   K. Bechgaard, C.S. Jacobsen, K. Mortensen, H.J. Pedersen and N. Thorup, **Solid State Comm.**, 33:1119 (1980)
5.   M. Bousseau, L. Valade, M-F. Bruniquel, P. Cassoux, M. Garbauskaus, L. Interrante, and J. Kasper, **Nouveau Journal de Chimie**, 8:3 (1983)
6.   V.N. Laukhin, E.E. Kostyuchenko, Y.V. Susko, I.F. Schegolev and E.B. Yagubskii, **JETP Lett.**, 41:81 (1985) and K. Murata, M. Tokumoto, H. Anzai, H. Bando, G. Saito, K. Kajimura and T. Ishiguro, **J. Phys. Soc. Japan**, 54:1236 (1985)
7.   H. Urayama, H. Yamochi, G. Saito, K. Nozawa, M. Kinoshita,

S. Sato, K. Oshima, A. Kawamoto and J. Tanaka, **Chem. Lett.**, 55 (1988)

8. F. Creuzet, G. Creuzet, D. Jérome, D. Schweitzer and H.J. Keller, **J. Physique Lett.**, 46:L-1079 (1985)

9. J. Friedel and D. Jérome, **Contemp. Phys.**, 23:583 (1982)

10. L.D. Landau and E.M. Lifshitz, p.482 **in** "Statistical Physics", Pergamon, London (1959)

11. Y. Byschkov, L.P. Gorkov and I.E. Dzyaloshinskii, **Sov. Phys. JETP**, 23:489 (1966), J. Sólyóm, **Adv. in Physics**, 28:201 (1979) and D. Jérome and H.J. Schulz, **Adv. in Physics**, 31:299 (1982)

12. V.J. Emery, **J. Physique**, C-3, 44:977 (1983)

13. C. Bourbonnais, p.155 **in**: "Low Dimensional Conductors and Superconductors", D. Jérome and L.G. Caron, eds., NATO ASI Series, Plenum (1987) and C. Bourbonnais, **Mol. Cryst. Liq. Cryst.**, 119:11 (1985)

14. (a) G. Montambaux, p.233 **in**: "Low Dimensional Conductors and Superconductors", D. Jérome and L.G. Caron,eds., NATO ASI Series, Plenum (1985) and (b) K. Yamaji, **Mol. Cryst. Liq. Cryst.**, 119:105 (1985)

15. T. Takahashi, Y. Maniwa, H. Kawamura and G. Saito, **J. Phys. Soc. Japan**, 55:1364 (1986)

16. J.R. Cooper, W. Kang, P. Auban, G. Montambaux, D. Jérome and K. Bechgaard, **Phys. Rev. Lett.**, 63:1984 (1989)

17. W. Kang, G. Creuzet, D. Jérome and C. Lenoir, **J. Physique**, 48:1035 (1987)

18. W. Kang, G. Montambaux, J.R. Cooper, D. Jérome, P. Batail and C. Lenoir, **Phys. Rev. Lett.**, 62:2559 (1989

19. V. Emery, R. Bruinsma, S. Barisic, **Phys. Rev. Lett.**, 48:1039 (1982)

20. J.P. Pouget, p.17 **in**: "Low Dimensional Conductors and Superconductors", loc cit.

21. S. Tomic and D. Jérome, **J. Phys. Cond. Matter**, 1:4451 (1989)

22. P. Monceau, p.369 **in**: "Low Dimensional Conductors and Superconductors", loc cit.

23. R.C. Lacoe, H.J. Schulz, D. Jérome, K. Bechgaard and I. Johansen, **Phys. Rev. Lett.**, 55:2351 (1985)

24. R.C. Lacoe, J.R. Cooper, D. Jérome, F. Creuzet, K. Bechgaard and I. Johansen, **Phys. Rev. B**, 58:262 (1987)

25. S. Tomic, J.R. Cooper and K. Bechgaard, **Phys. Rev. Lett.** 62:462 (1989)

CONDUCTION IN LANGMUIR-BLODGETT FILMS

André Barraud and Michel Vandevyver

C.E.N. Saclay
DPhG/SCM
91191 Gif Sur Yvette, France

1.   INTRODUCTION

Since the discovery, in 1984, of a strategy leading to
conducting films one or a few molecules thick[1,2], many
electronic conducting Langmuir-Blodgett (LB) films have been
fabricated[3-12], mainly based on tetracyanoquinodimethane
(TCNQ), tetrathiafulvalene (TTF), di-mercapto-dithiol-thione
(dmit) and their derivatives. A great deal of interest for
potential applications, especially for molecular electronics,
has arisen from some of their unique features:

-     thin, Angström thick conducting plane;
-     self insulation and decoupling of each conducting plane
      from the next one;
-     optically quasi-transparent electrode;
-     very well defined interelectrode spacing.

However, two series of puzzling observations are to be
noted concerning these thin conducting films:

1a)   the better the quality of the LB film, the poorer their
      conduction, and conversely;
1b)   in TCNQ salts series, unexpectedly, it is the most
      conjugated cations, which should favor conduction by
      smoothing the lattice potentials, which give rise to the
      worst conductors;
1c)   no metal-like conduction (and a fortiori super-conduction)
      has been obtained in LB films of long chain compounds to
      date even when non LB parent compounds exhibit a noteworthy
      metallic conduction;
2a)   dc conductivity decreases as the number of superimposed LB
      layers is decreased, going to zero for a single conducting
      plane;
2b)   conducting Langmuir-Blodgettable mixed valence compounds
      are conducting when spread in the form of a single layer at
      the surface of a glycerol subphase[13,14], while they are
      insulating when transferred from a water subphase.

In this paper these unanswered questions will be discussed
in the light of the large experience now accumulated on

Condensed Systems of Low Dimensionality
Edited by J. L. Beeby *et al.*, Plenum Press, New York, 1991

molecular Langmuir-Blodgett conductors. Some (series 1) involve the intrinsic structure of LB conducting films; others (series 2) involve extrinsic structural effects.

2.   INTRINSIC STRUCTURAL EFFECTS

Observation 1a is quite general. It can be exemplified by a series of semi-amphiphilic* salts of TCNQ. The TCNQ compound is spread and transferred in the form of an insulating[1], purely ionic salt of $TCNQ^-$. $TCNQ^-$ is then oxidized by exposure to iodine vapors[15]. The lattice of the hydrophobic chains of the amphiphilic cation controls the solid state reaction and stops it when all the sites available for iodine in the lattice are filled with I_3^-. As a result, TCNQ is half oxidized and left as a mixed valence anion ($TCNQ^- TCNQ^O$). Since it bears no impeding chain, it can reorganize into the energetically most favorable state. If allowed by the cation lattice, one-dimensional conducting stacks of mixed valence TCNQ are the lowest energetic state, because of the gain in electronic energy provided by band creation. Therefore the TCNQ molecules, which originally lay flat in the polar plane, tend to organize in in-plane stacks, with their long axis normal to the film plane[3].

However, these stacks with their in-stack intermolecular distance of 3.2 Å impose a stress on the lattice of the chains, which counterbalances the gain in electronic energy. If the lattice is soft, we expect conduction to win, but the cohesion of the LB film will then be lost. If the lattice is stiff, conduction is not favored and the stable energy dip corresponding to the conducting mixed valence structure does not exist, so that iodine is either rejected (no reaction), or leads to completely oxidized TCNQ.

For example, cation $C_{18}N^+(CH_3)_3$ automatically gives rise to conducting LB films upon iodination, whatever the conditions[6]. But the lattice is so loose that inert molecules must be added to transfer the film properly. The same holds for sulfonium[4] and phosphonium[5] cations, which are known to give loose LB films.

Cation C_{18} or C_{22} pyridinium[15,16] is an intermediate case: iodination is delicate, the successful concentration range is narrow, and often infrared spectra reveal that oxidation has proceeded beyond the mixed valence state in some regions of the films[15]. Nevertheless, once obtained, the conducting structure is stable.

Cation C_{18} quinolinium[17] never gives rise to conducting films upon iodine doping. This is because the inter-ring π-π interactions are very strong (two rings) and render the cation lattice very stiff. Building up LB films indeed requires no additional molecule and the films are quite stable.

The extreme case is that of $(C_{18})_3$ phenantrolinium[17] (3 rings) which not only gives no conducting phase with iodine, but gives rise to ionic mixed stacks from the origin. In this case no reorganization is possible upon any further treatment. On

----

* Semi-amphiphilic compounds are compounds made of two partners, only one of which is amphiphilic.

Table 1. Conductivity of LB films of metal(dmit)$_2$ complexes (from Ref.12).

Materials	Bulk conductivity (Scm^{-1})	
	Bromine oxidation	Electrochemical oxidation
2C10-Au	0.12	
2C14-Au	0.15	
2C18-Au	0.005	
2C22-Au	Below the limit of detection	
3C10-Au	15	33
3C14-Au	5.4	19
3C16-Au	2.6	0.46
3C18-Au	1.4	0.12

the other hand the corresponding films are remarkably stable and easy to transfer.

A quite similar trend is observed in the series of Ni- and Au-dmit with various chain lengths[12]. Table 1, taken from Ref. 12, compares the conductivities of Au-dmit with chains from C_{10} to C_{22}, rendered conducting by bromine or electrochemical oxidation. When the chain length is increased, the film conductivity decreases markedly, and when the chain is long enough to induce lattice organization (C_{22}), conductivity falls below a measureable limit.

This has been found to be a general trend with LB conductors. High conductivities and a metallic behavior near room temperature have been observed only in short chain compounds which have not been proved to transfer in the form of lamellar organized multilayers[12,18]. Long chain compounds with high film cohesion have not exhibited high conductivity so

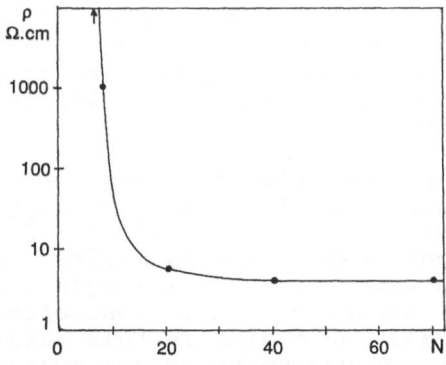

Fig. 1. d.c. resistivity ρ of multilayers of C_{18} TCNQ-TMTTF vs number of layers N. A quite similar curve (asymptotic value 10 ohm.cm instead of 4.2) is obtained with multilayers of C_{18} Pyridinium-TCNQ.

far. This is because there is no reason why the lattice of the chains should adopt the same spacing and directions as the ones of the conducting stacks. A big effort should be made for the determination of the in-plane structure of several simple examples in order to try to design couples of partner molecules with compatible lattices. This field of supermolecular engineering has many common features with that of solid state topochemical polymerization, in which LB films proved to be a handsome tool for investigation[19].

3.   EXTRINISIC STRUCTURAL EFFECTS

Most conductivity measurements are performed on multilayers, most of the times 20, 40 or more layers. Very interesting are the dc conductivity measurements which compare the conductivity of films made up of a different number of layers. Figure 1 gives a typical example: above 20 layers the dc conductivity reaches its asymptotic value and becomes independent of film thickness. Below ten layers, the conductivity decrease becomes significant. For four layers (two conducting polar planes) the film is dc insulating. However, the infrared spectrum (Fig. 2), although noisy and unprecise because of the small amount of material involved, seems to keep all the main features of the conducting TCNQ salts: broad band to band electron transition, absorption band around 3000 cm^{-1}, Fano effect on the C≡N stretching band at 220 cm^{-1}, and broadening and red shift of the dimer-activated a_g modes at 1150 and 1350 cm^{-1}. This means that most of the material is still conducting, although dc current does not pass.

This can be understood if one assumes that layers are made of adjoining small two-dimensional crystallites, and impurities or defects segregate and gather at grain boundaries. Then a small amount of segregated material is enough to make thin insulating walls which hinder dc conductivity. This is supported by the UV-visible absorption spectra of all the LB TCNQ conducting films, which exhibit two slight "poison" peaks, one between 420 and 440 nm arising presumably from $TCNQ_0$ clusters, another near 500 nm arising presumably from dicyano-p-toluoyl-cyanide (DCTC), an oxidized derivative of TCNQ. The latter species, which is formed during film fabrication, may be expelled from the mixed valence TCNQ stacks because it tends to increase electron energy by shortening the conducting stacks. Such self-purification mechanisms are quite common and take place every time cooperation or ordering lowers the total energy of a molecular assembly.

Insulating grain boundaries in conducting LB films have been visualized by scanning electron microscopy (SEM) by the method of the voltage induced contrast[20]. Even if each wall is not fully efficient, many walls in series are able to cut down conductivity to almost zero in a single conducting plane. As far as the author knows, no single conducting LB layer has ever been obtained; all the claims in this field either were not true LB films or were proved to be actual multilayers.

These insulating species are insoluble in water and hence remain in the film. A possible difference between glycerol and water is their dissolution power. The insulating species may be slightly soluble in glycerol. As a consequence, glycerol

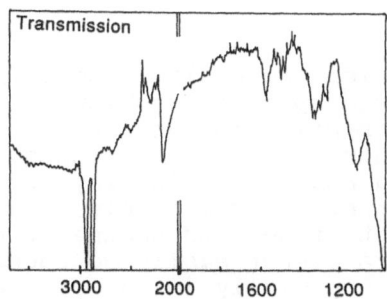

Fig. 2. IR absorption spectrum of 4 layers (two
conducting planes) of C_{18} Pyridinium-TCNQ
"conducting" films.

behaves as a cleaning subphase toward a conducting film spread
on its surface. This may explain why a single conducting layer
exhibits high dc conductivity on glycerol while it becomes dc
insulating on water. Unfortunately the high viscosity of
glycerol does not allow film transfer.

The onset of dc conduction in multilayers then requires
out-of-plane percolation, via defects such as "dry" micelles or
roof-like structures (presumably of nanometric size). This
allows conduction paths to go round the in-plane barriers. The
more paths, or equivalently the more layers, the higher the
conductivity.

This out-of-plane percolation explains why no care has to
be taken when putting down electrodes to measure the in-plane
conductivity of multilayers: any local electrical contact
between an electrode and one of the conducting planes gives rise
to an arborescent spreading of the current which supplies all
the conducting planes with electrons, whatever film thickness
and contact location, above or underneath the film.

The out-of-plane percolation is also expected to severely
affect the in-plane to out-of-plane conductivity anisotropy.
Because of the very structure of LB films, their intrinsic
anisotropy is expected to be very high, due to the interplane
insulation by the aliphatic chains. Any out-of-plane
percolation breaks this insulation and shorts the film locally.
This is why the few published data on the conductivity
anisotropy are derived from measurements with aluminum
electrodes[21], the oxide of which prevents the electrodes from
being short-circuited. This causes the measured values to be
orders of magnitude higher than the actual anisotropy, but
orders of magnitude lower than the intrinsic one.

This low actual anisotropy prohibits so far the use of LB
conductors as electrodes with molecular size interelectrode
spacing, one of the basic materials needs for the development of
molecular electronics. Hence a careful study of the nature,
density and properties of the defects responsible for in-plane
dc insulation and those responsible for out-of-plane percolation
is needed for any application of the films to molecular
electronics to be considered.

CONCLUSION

Compared to classical molecular conductors, LB molecular conducting films undergo specific problems closely related to their structure. Their main distinctive feature is their very low dimensionality, which is a direct consequence of their alternately conducting and insulating layered structure: the secondary 3D character of classical molecular conductors is lost. Hence the effect of defects becomes tremendously high because charge carriers cannot easily go round obstacles. This is why the situation, especially in the case of dc conduction, is dominated by defects.

A great deal of experience is now gathered on several families and series of LB conductors; it converges to point out two distinct classes of defects which have two different solutions:

1. Intrinsic structural problems arising from a mismatch between the hydrophobic natural lattice and the period of the conducting stack. Unless lucky in a trial shot, only extensive study of in-plane structure followed by careful molecular and supermolecular engineering will bring agreement between the two lattices. The increase in stability due to the cooperation of the two lattices may unexpectedly stabilize compounds whose nonamphiphilic parent compounds are poor conductors and hence unexplored.

2. Extrinsic defects due to the multicrystallinity and the preparation method of the films. Controlling these defects is a prerequisite to applying these conducting layers to molecular size devices. Atomic resolution microscopy seems one of the most suitable tools for getting information on the nature and the working process of these defects.

Procedures must also be developed to try and get rid of grain boundaries and out-of-plane percolation, for instance fabrication methods involving new film cleaning subphases or single crystalline film growth or conversion. This is the price to be paid for LB conducting films to enter successfully the field of molecular electronics.

REFERENCES

1. A. Barraud, A. Ruaudel-Teixier, M. Vandevyver and P. Lesieur, **Nouveau Journal de Chimie**, 9:365 (1985)
2. A. Ruaudel-Teixier, M. Vandevyver and A. Barraud, **Mol. Cryst. Liq. Cryst.**, 120:319 (1985)
3. A. Barraud, P. Lesieur, A. Ruaudel-Teixier and M. Vandevyver, **Thin Solid Films**, 134:195 (1985)
4. A. Barraud, M. Lequan, R.M. Lequan, P. Lesieur, J. Richard, A. Ruaudel-Teixier and M. Vandevyver, **J. Chem. Soc., Chem. Comm.**, 11:797 (1987)
5. A. Barraud, P. Lesieur, J. Richard, A. Ruaudel-Teixier, M. Vandevyver, M. Lequan and R.M. Lequan, **Thin Solid Films**, 160:81 (1988)
6. M. Vandevyver, J. Richard, A. Barraud, A. Ruaudel-Teixier, M. Lequan and R.M. Lequan, **J. of Chem. Phys.**, 87:6754 (1987)
7. M. Vandevyver, A. Barraud, P. Lesieur, J. Richard and A.

Ruaudel-Teixier, **Journ. de Chimie Physique**, 83:599 (1986)

8. J. Richard, M. Vandevyver, A. Barraud, J.P. Morand and P. Delhaès, **J. Colloid Interface Sci.**, 129:254 (1989)

9. A. Otsuka, G. Saïto, T. Nakamura, M. Matsumoto, Y. Kawabata, K. Honda, M. Goto and M. Kurahashi, **Synthetic Metals**, 27:B575 (1988)

10. A.S. Dhindsa, C. Pearson, M.R. Bryce and M.C. Petty, **J. Phys. D : Appl. Phys.**, 22:1586 (1989)

11. Y. Kawabata, T. Nakamura, M. Matsumoto, M. Tanaka, T. Sekiguchi, H. Komizu, E. Manda and G. Saïto, **Synthetic Metals**, 19:663 (1987)

12. T. Nakamura, H. Tanaka, K. Kojima, M. Matsumoto, H. Tachibana, M. Tanaka and Y. Kawabata, **Thin Solid Films**, 179:183 (1989)

13. A. Barraud, J. Leloup and P. Lesieur, **Thin Solid Films**, 133:113 (1985)

14. M. Matsumoto, T. Nakamura, F. Takei, M. Tanaka, T. Sekiguchi, M. Mizuno, E. Manda and Y. Kawabata, **Synthetic Metals**, 19:675 (1987)

15. J. Richard, M. Vandevyver, P. Lesieur, A. Barraud and K. Holczer, **J. Phys. D : Appl. Phys.**, 19:2421 (1986)

16. L. Henrion, G. Derost, A. Barraud and A. Ruaudel-Teixier, **Sensors and Actuators**, 17:493 (1989)

17. A. Barraud, **in**: Annual Report on EURAM research contract MA 1E/0034/C : "Advanced Materials for Molecular Electronics, Ultrathin Conducting Organic Films", (1988)

18. A. Ruaudel-Teixier, unpublished results.

19. A. Barraud, C. Rosilio and A. Ruaudel-Teixier, **Thin Solid Films**, 68:91 (1980)

20. A. Barraud, J. Richard, A. Ruaudel-Teixier, M. Vandevyver, **Thin Solid Films**, 159:413 (1988)

21. K. Ikegami, S. Kuroda, M. Sugi, M. Saïto, S. Iizima, T. Nakamura, M. Matsumoto, Y. Kawabata and G. Saïto, **Synthetic Metals**, 19:669 (1987)

LANGMUIR-BLODGETT FILMS OF POTENTIAL ORGANIC RECTIFIERS#

Robert M. Metzger and Charles A. Panetta*

Department of Chemistry, University of Alabama
Tuscaloosa, AL 35487-0336 USA
*Department of Chemistry,University of Mississippi
University, MS 38677 USA

1.   INTRODUCTION

We review here the progress made towards the goal of a truly unimolecular device, the organic rectifier. A unimolecular device, by definition, is an electronic or other device whose active principle is based on the manipulation of the molecular geometry, or the molecular conformation, of a single organic molecule or a small cluster. This goal is at the forefront of a new field of research, called "molecular electronics" (ME), which aims to fill a future need: conventional "inorganic" microelectronic devices, based on silicon, gallium arsenide or germanium, must become ever smaller because of the need for faster electronic circuits, but the fabrication difficulties will rise rapidly and, at the nanometer level, organic molecules, with the tunability of their molecular orbitals, should offer significant advantages.

Of course, at present one cannot routinely address a group of several isolated single molecules: as yet there are no "molecular wires". However, one can use the tip of a scanning tunneling microscope to address a single molecule, and thus several candidates for molecular electronic devices can finally be studied.

This review focusses mainly on the efforts to realize the Aviram-Ratner unimolecular rectifier[1-3], a proposal advanced in 1973 by Ari Aviram, Mark A. Ratner and co-workers that a single organic molecule of the type D-σ-A could be a rectifier of electrical current. This molecule D-σ-A would do so, because the D end is a good organic one-electron donor, σ is a covalent saturated ("sigma") bridge, and A is a good organic one-electron acceptor. The main advantage of the organic unimolecular rectifier, which may be one of the key experiments in molecular electronics, is that the working thickness promised by such a D-σ-A device is of the order of one or two molecular lengths,

Supported by the National Science Foundation, Grant DMR-88-01924.

i.e. about 5 nm: such a small size is predicted to be unattainable, even by the rosiest forecasts, for silicon or gallium arsenide technology.

We chronicle here the progress[4-24] of our own group, the "Organic Rectifier Project" (ORP) which has been reviewed before [5,8,10,16,19-21,23,25-27]. As is discussed below, the rectification by a single organic molecule, or by an organized Langmuir-Blodgett (LB)[28-34] monolayer of such molecules, has not yet been demonstrated by our group. The revolutionary scanning tunneling microscope (STM)[35] has an obvious potential for confirming the Aviram ansatz; the early and disappointing efforts to detect rectification by an *ad hoc* modification of the STM[20,21,23] will be reviewed; our own continuing efforts will be mentioned briefly. We mention also the results of Sambles' group at Exeter and Ashwell's group at Cranfield, who may have achieved rectification by an LB film[36,37].

2. LARGER-SCALE ORGANIC DEVICES

Before we discuss the unimolecular rectifier, we should rapidly mention some larger-scale organic devices which are well established.

2.1 pn Organic Junction Rectifiers

After the discovery of the junction diode and the transistor, it was of some academic interest to see whether organic molecules could function in bulk pn rectifiers (diodes) or as npn transistors. This would occur if a film or crystal of an organic electron donor were brought in contact with that of an organic electron acceptor. This was indeed verified in the 1960's[38]

2.2 Multilayer LB Organic Rectifiers

Kuhn and co-workers[39] showed that one can obtain a pn (or DA) rectifier in a **LB multilayer** sandwich Al|(CA-D)$_q$|(CA)$_r$|(CA-A)$_s$| Al, where | denotes an interface boundary, Al denotes a bulk Al contact, (CA-D)$_q$ denotes the electron donor system D [=q LB monolayers of cadmium arachidate (CA) randomly doped in the ratio 5:1 with suitable organic π electron donors D] (CA)$_r$ denotes a spacer layer of r undoped monolayers of CA, and (CA-A)$_s$ denotes the electron acceptor system A [=s LB monolayers of CA randomly doped with suitable organic π electron acceptors A]. This work was repeated and confirmed by Sugi and co-workers[40], who observed rectification properties, but **only** if q ≥ 3, r ≥ 1, and s ≥ 3. In these LB films, the nearest-neighbor distance between D and A in successive layers cannot be controlled.

2.3 The CuTCNQ Phase-Change Switch

In 1979 a fast switch was discovered by Potember in CuTCNQ [41]. This was due to the thermodynamic metastability (in crystals or amorphous powders) of the violet, low-conductivity ionic state (IS) Cu$^+$TCNQ$^-$(c), relative to the yellow, low-conductivity, neutral state (NS) Cu^{0}TCNQ^0 with, presumably, an intermediate, mixed-valent, higher-conductivity state (CS); one could switch between the two states IS <--> CS either with an applied voltage over a certain threshold value, or by a moderate laser beam, while heat will restore the IS. This is found also in AgTCNQ, and in a few other related systems[42]. The switching rate is quite fast, and can even be used for optical data storage[43], but has not been incorporated in commercial devices.

2.4 A Conducting-Polymer-based Transistor

Wrighton and co-workers developed a "molecule-based transistor" which uses conducting polymers: either chemically doped polyaniline layers deposited atop Au interdigitated electrodes[44] or on a 50-100 nm "gate" polyaniline polymer between two Au electrodes shadowed with SiO_2; this device still has a rather slow switching rate (10kHz), due to the slowness of ionic conduction, and a gain of almost 1,000[45]. Recently Stubb and co-workers showed that a single Langmuir-Blodgett monolayer can be used in a "molecule-based" transistor[46].

3. THE AVIRAM-RATNER ANSATZ

The 1973 proposal of Aviram and Ratner[1-3] originates from the discovery of highly conducting lower dimensional organic charge-transfer systems based on good one-electron donors (D) such as tetrathiafulvalene (TTF, **1**) (bold numbers refer to molecular configurations illustrated in the figures) and good organic one-electron acceptors (A) such as 7,7,8,8-tetracyanoquinodimethan (TCNQ, **2**). Good donor molecules (i.e. molecules with relatively low gas-phase first ionization potentials I_D) are, at the same time, poor acceptors (they have low electron affinity A_D); good acceptors (i.e. molecules with a relatively high first electron affinity A_A) are, at the same time, rather poor donors (have high I_A) thus the gas-phase energy required for reaction (**3**) (both components at infinite separation) is about 4 eV, while reaction (**4**) would need over 9 eV:

$$TTF(g) + TCNQ(g) \rightarrow TTF^+(g) + TCNQ^-(g) \quad I_D\text{-}A_A = 4.0 \text{ eV} \qquad (3)$$

$$TTF(g) + TCNQ(g) \rightarrow TTF^-(g) + TCNQ^+(g) \quad I_D\text{-}A_A = 9.6 \text{ eV(est)} \qquad (4)$$

Thus, if one makes a D-σ-A molecule like (**5**), and assembles it somehow between two metal electrodes M_1 and M_2, as in (**6**) (discussed hereinafter as $M_1|D\text{-}\sigma\text{-}A|M_2$), then the direction of easy electron flow is from M_1 to M_2, because it utilizes the zwitterionic state $D^+\text{-}\sigma\text{-}A^-$ (while the electron flow from M_1 to M_2 would be inefficient, because the barrier to forming the zwitterion $D^-\text{-}\sigma\text{-}A^+$ would be several eV higher). Using terms

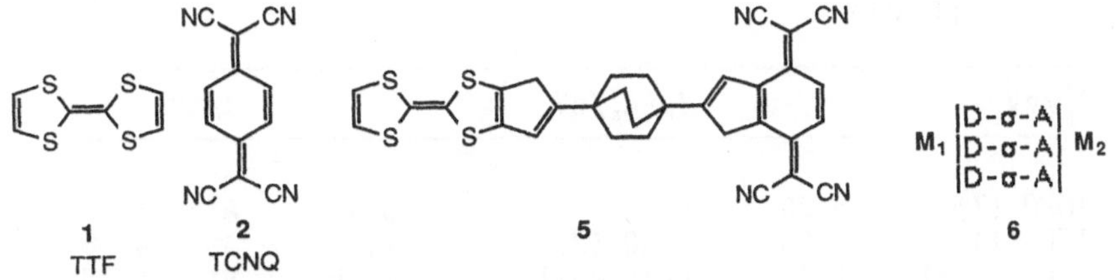

Fig. 1. Molecular configurations **1**, **2**, **5** and **6**.

popularized by Hoffmann[47], the Aviram-Ratner device can work if the tunneling of electrons from A to D is through the bond system, and will fail if the tunneling between the metal electrodes M_1 and M_2 is predominantly through space. Molecule (5) was never synthesized, and the idea languished until the ORP started in earnest.

4.   STRATEGY FOR ASSEMBLY OF A D-σ-A DEVICE

To address a single molecule electrically, one would love to have a "molecular wire" (e.g. a polyacetylene strand) or a "molecular antenna" (e.g. the conjugated portion of β-carotene); at present, neither of these can be easily connected to an external potential source. Until the recent advent of the STM[35], one could not connect a single molecule to an external circuit. Therefore, when the ORP was initiated, in about 1981, to realize the Aviram-Ratner rectifier, one had to content oneself with assemblies of molecules. The three techniques that showed promise were (i) the LB technique[28-34], and the technique of covalently bonding molecules to electrode surfaces either by (ii) silanizing a hydroxyl-coated electrode then attaching molecules covalently[48], or (iii) by silanizing the molecule and attaching it directly by spin-coating by the oleophobic method[49], to a hydroxyl-coated electrode[50]. Good monolayer coverage is claimed for the latter method[50] but not for the former method[48].

The ORP was committed to the LB technique. This implies the need for LB films that are defect-free in the lateral dimension, at least to within the area probed electrically, while it is well known that LB monolayers have many microscopic and macroscopic defects[51].

4.1 Electronic and Synthetic Criteria

There are several interlocking criteria that must be satisfied for the rational synthesis of suitable D-σ-A systems:

(i) I_D for the donor end D must be as small, and as close as possible to the work function ϕ of the metal layer M_1. Typical values are listed in Table 1.

(ii) A_A for the acceptor end (Table 1) must be as large as possible, and as close as possible to the work function ϕ of the

Table 1. Experimental ionization potentials I_D for selected good donors D, experimental electron affinities A_A for selected good acceptors A, and work functions ϕ for selected metals M_1, M_2 ([20]).

I_D/eV		I_D-A_A/eV		ϕ/eV	
TMPD (7)	6.25	DDQ (9)	3.13	Al	3.74
TTF (1)	6.83	TCNQ (2)	2.8	Au	4.58
Pyrene (8)	7.41	Chloranil (10)	2.76	Pt	5.29

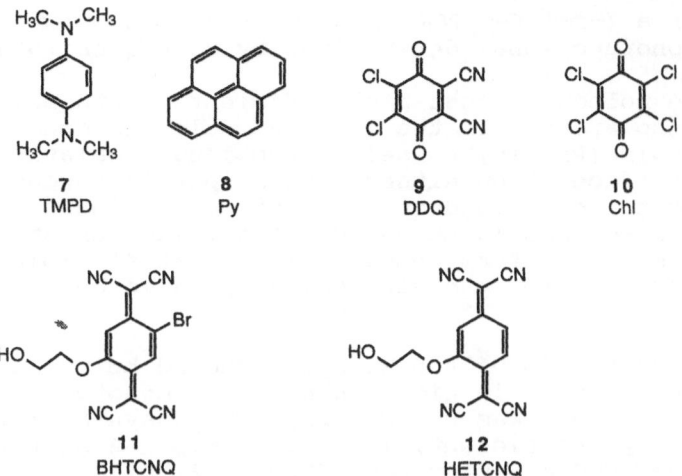

Fig. 2. Molecular configurations **7-12**.

metal layer M_2. It is clear from Table 1 that requirements (i)
and (ii) can be met only approximately.

(iii) In a molecule in which the "sigma" bridge has already
been built, it is extremely difficult to convert, *in situ*, by
chemical synthesis, a weak donor into a strong donor or a weak
acceptor into a strong acceptor. Therefore, instead, one must
synthesize a series of mono-substituted strong donors and mono-
substituted strong acceptors, which can be joined by some
coupling reaction which can avoid the usual, and undesirable,
formation of ionic charge-transfer complexes: such a coupling
reaction is the urethane, or carbamate coupling reaction,

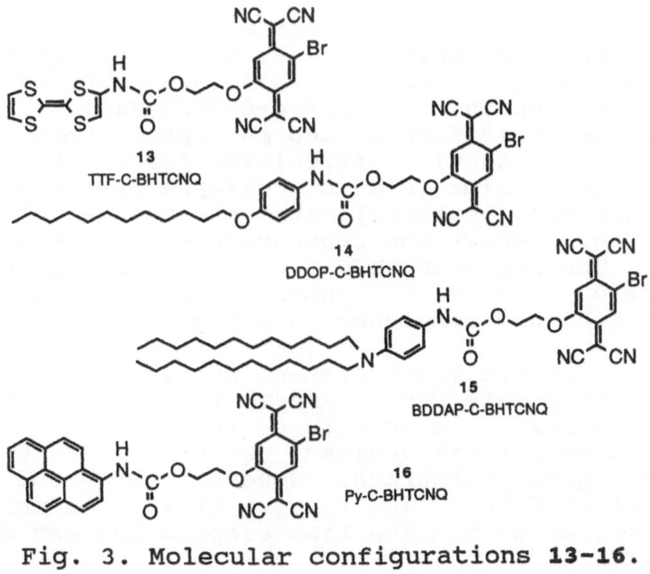

Fig. 3. Molecular configurations **13-16**.

pioneered for a $(-\sigma\text{-TTF-}\sigma\text{-TCNQ-})_x$ copolymer by Hertler[52] and adapted to monofunctional derivatives by Baghdadchi[53].

(iv) The molecules must pack efficiently into self-assembling monolayers. If the designed molecule does not form Pockels-Langmuir (PL)[15,24] self-assembling monolayers at the air-water interface, then either long aliphatic "greasy tails" must be added to form a hydrophobic tail or an ionic, hydrophilic "head" should be added. The molecules should be fairly flat, so as to form compact films, yet flexible enough so as to transfer well (by the vertical dipping method) as LB films.

(v) The synthesis of the acceptor should be relatively easy, and should occur in high yield. The acceptor used in the early work of the ORP was BHTCNQ (11), whose synthesis[52,53] contained a very inefficient, low-yield step. A better mono-functionalized TCNQ acceptor was HETCNQ (12)[17].

(vi) The electron transfer through the D-σ-A molecule, and through its hydrophobic or hydrophilic tails must be fast: as said above, **a molecular device that is small but slow is not predicted to be useful.** That electron transfer is fast through properly designed molecules, e.g. the photosynthetic reaction center, is well known. Miller, Closs, and co-workers[54,55] showed that the intramolecular electron transfer rate through a molecule D-σ-A at first increases steadily with an increasing I_D-A_A. Then, if I_D-A_A is increased further, the electron transfer rate decreases because of an increase in the Franck-Condon reorganization (because now the geometries of D$^+$ and A$^-$ are quite different from the geometry of D and A respectively) [54,55] (Marcus "inverted region"[56]). Thus, a balance must be struck between the available I_D-A_A values and the limitations imposed by the Franck-Condon factor.

(vii) The device (6) must have a finite tolerance for high voltages or heating.

4.2 Langmuir-Blodgett Films

We present in Table 2 an updated catalog of molecules (13-24) prepared by the ORP, which form PL monolayers at the air-water interface and which transfer well onto Al or glass or other slides as LB monolayers. TTF-C-BHTCNQ (13) was difficult to purify; the "neutral" form seemed to deposit "pancake-style" onto the water and synthetic difficulties forced its abondon-ment. The strongest films (highest collapse pressure, most vertical pressure-area isotherm) were obtained with BDDAP-C-BHTCNQ (15). Fig. 5 shows the pressure-area isotherm for BDDAP-C-HETCNQ (17). The acceptor HMTCAQ used in (19) and (20) was easy to prepare[9], but is well known to be a weak two-electron acceptor, with a highly nonplanar geometry.

Molecules 21-24 are model systems for a related project, which aims to incorporate D-σ-A systems into LB-film-forming diacetylenes, polymerizable in situ on the film balance, as potential nonlinear optical devices. 21-24 form Z-type multilayers on a glass substrate. However, no second harmonic signals were detected for Z multilayers of 23: the multilayers may have reorganized within the time elapsed between deposition and measurement[58].

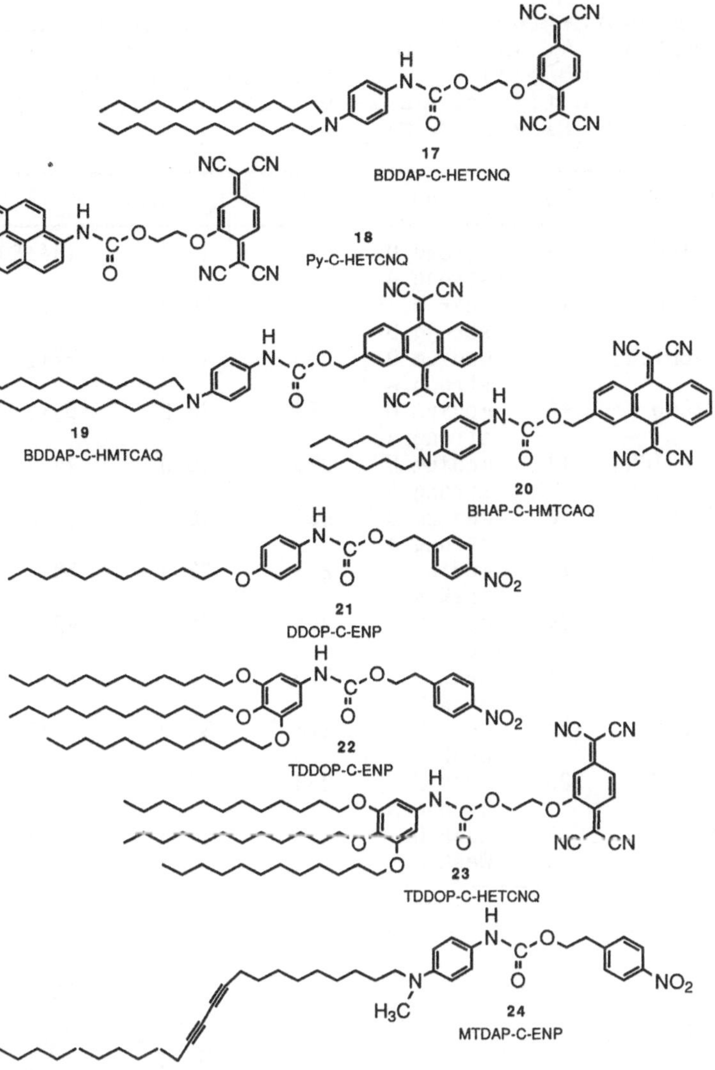

Fig. 4. Molecular configurations **17-24**.

4.3 Cyclic Voltammetry

One donor, several acceptors and several D-σ-A molecules have been characterized by cyclic voltammetry (CV). The results, given in Table 3, confirm that the carbamate linkage preserves the oxidation (reduction) potentials of the D(A) ends of the D-σ-A molecules.

4.4 Crystal Structures of Model Donor, Acceptor, and D-σ-A Molecules

A few crystal structures have been solved: for the donor DMAP-C-Me, **25**[12], for the acceptor BHTCNQ,**11**[13] and for the methyl ester (AETCNQ,**27**) of the acceptor HETCNQ,**12**[14]. The small difference in conformation between AETCNQ and BHTCNQ can be attributed to crystal packing forces, rather than to intramolecular effects. Amphiphilic molecules that form LB

Table 2. Pressure-Area isotherm data for Pockels-Langmuir films. Π_C and A_C are the pressure and molecular area, respectively, at the collapse point. $ indicates that the film makes Z-type LB multilayers (substrate at 22 C, film at 5 C).

Molecule	No.	Type	T/K	Π_c/mN/m	A_c/Å^2	Ref
TTF-C-BHTCNQ	13	strong D strong A	292	12.7	134±50	[5]
DDOP-C-BHTCNQ	14	weak D strong D	292	20.2	50±1	[7]
BDDAP-C-BHTCNQ	15	medium D strong A	293	47.3	57±1	[15]
Py-C-BHTCNQ	16	medium D strong A	283	28.2	53±1	[7]
BDDAP-C-HETCNQ	17	medium D strong A	293	40.0	44±1	[57]
Py-C-HETCNQ	18	medium D strong A	283	46	--	[57]
BDDAP-C-HMTCAQ	19	medium D weak A	293	22.3	58±1	[15]
BHAP-C-HMTCAQ	20	medium D weak A	293	35.8	42±1	[16]
DDOP-C-ENP	21	weak D weak A	278	23.7	38±1	[24]
TDDOP-C-ENP	$22	weak D weak A	278	34.0	76±1	[24]
TDDOP-HETCNQ	$23	weak D strong A	283	47.5	54±1	[24]
MTDAP-C-ENP	$24	weak D weak A	278	16.5	63±1	[23]

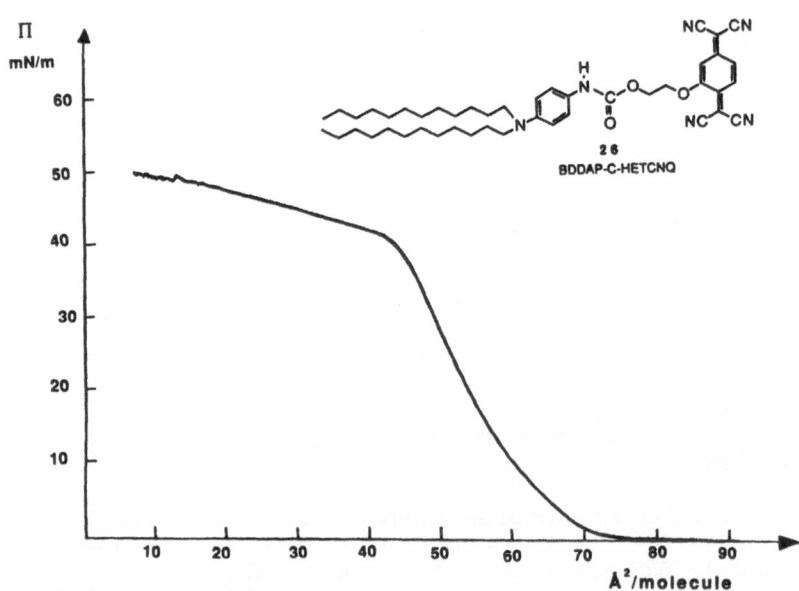

Fig. 5. Pressure-area isotherm of BDDAP-C-HETCNQ, 17, at 293 K [57].

Table 3. Solution cyclic voltammetric potentials[26,27]. All data were obtained at a Pt electrode, and are given in Volts vs SCE.
(a)   Solvent: CH_3CN. Reference electrode: SCE. A peak at 0.37 V (return scan) grows with successive cycles: dimer or polymer formation[12].
(b)   Solvent: CH_3CN. Reference electrode: SCE.
(c)   Solvent: CH_2ClCH_2Cl. Reference electrode: Ag|AgCl. An offset correction of 0.15 V has been applied to convert the values to V vs SCE.
(d)   Solvent: CH_3CN. Reference electrode: Ag|AgNO_3. An offset correction of 0.320 V has been applied to convert the values to V vs SCE.
(e)   Solvent: CH_2ClCH_2Cl. Reference electrode: Ag|AgCl. An offset correction of 0.19 V has been applied to convert the values to V vs SCE.

Molecule	No.	Oxid.(1) $D \rightarrow D^+$		Oxid.(2) $D^+ \rightarrow D^{++}$		Red.(1) $A \rightarrow A^-$		Red.(2) $A^- \rightarrow A^{--}$		Ref.
		E_p	$E_{1/2}$	E_p	$E_{1/2}$	E_p	$E_{1/2}$	E_p	$E_{1/2}$	
Donor:										
DMAP-C-Me	25 a	0.58	0.55	–	–	–	–	–	–	[12]
Acceptors:										
TCNQ	2 b	–	–	–	–	–	0.19	–	-0.35	[59]
TCNQ	2 c	–	–	–	–	0.11	0.13	-0.46	-0.43	[60]
BHTCNQ	11 b	–	–	–	–	–	0.305	–	-0.170	[7]
HETCNQ	12 b	–	–	–	–	–	0.107	–	-0.398	[17]
HMTCAQ	26 b	–	–	–	–	–	–	-0.372	-0.333	[9]
D-σ-A:										
DDOP-C-BHTCNQ	14 d	–	1.21	–	–	–	0.25	–	-0.07	[7]
BDDAP-C-HETCNQ	17 e	0.66i	–	1.10	–	0.02	–	-0.49i	–	[60]
Py-C-HETCNQ	18 e	1.04	1.01	1.18	1.15	0.11	0.08	-0.32	-0.35	[60]
BHAP-C-HMTCAQ	20 c	0.63	0.60	–	–	–	–	-0.39	-0.36	[60]
TDDOP-C-HETCNQ	23 c	1.20i	–	–	–	0.10	0.07	-0.50	-0.47	[24]
DDOP-C-ENP	21 c	1.42	1.39	–	–	-1.16	-1.13	–	–	[24]
TDDOP-C-ENP	22 c	1.17i	–	–	–	-1.12	-1.15	–	–	[24]
MTDAP-C-ENP	24 c	0.57	0.54	–	–	-1.09	-1.06	–	–	[60]

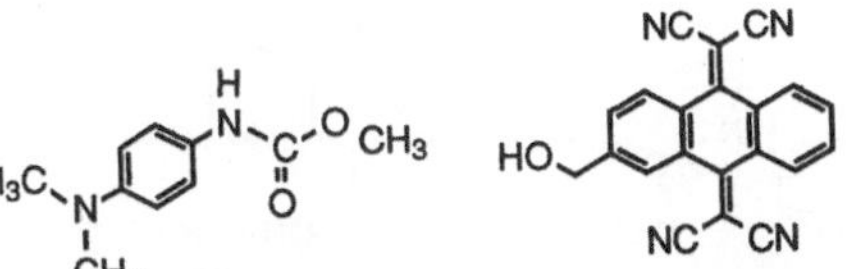

25
DMAP-C-Me

26
HMTCAQ

Fig. 6. Molecular configurations **25** and **26**.

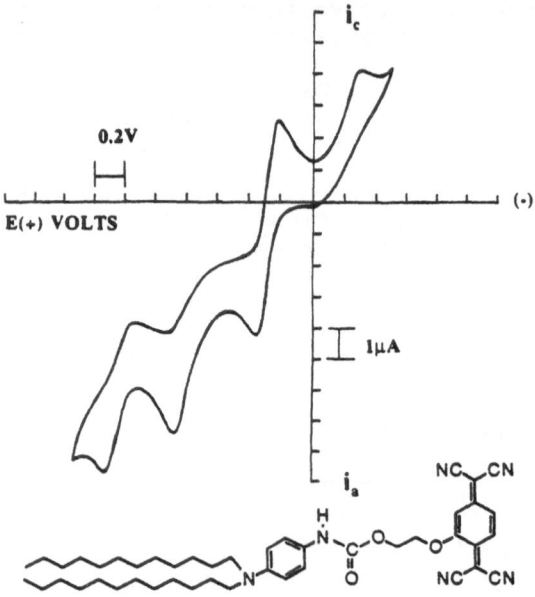

Fig. 7. Cyclic voltammogram of BDDAP-C-HETCNQ, **17**.

films will not usually crystallize, because of the usual
aliphatic "tails" added to them. However, we have solved the
structures of two D-σ-A molecules which do not form PL or LB
films: Ph-C-BHTCNQ,**28**[6], and DMAP-C-HMTCAQ,**29**[22]. Both
structures show an extended carbamate linkage; in Ph-C-
BHTCNQ,the dihedral angle between the phenyl ring and the six-
membered central ring of BHTCNQ is only 8°[7]. This gives hope
that in LB films of related D-σ-A molecules the carbamate
linkage will also be extended.

4.5 Molecular Orbital Calculations

As reported previously [20] semi-empirical molecular
orbital (MO) calculations, using the MNDO algorithm, with full

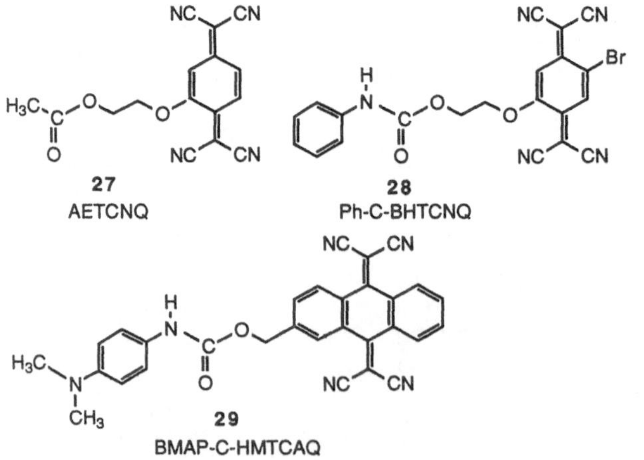

27
AETCNQ

28
Ph-C-BHTCNQ

29
BMAP-C-HMTCAQ

Fig. 8. Molecular configurations **27-29**.

geometry optimization in program MOPAC, have been performed on D-σ-A molecules to predict their geometry and also their HOMO and LUMO energies. Overall, the MNDO structures are extended, as expected, but there are some small surprises[20]. The MNDO ionization energies I_D values are not as low as hoped[20].

4.6   Fourier Transform Infrared Spectra of D-σ-A Monolayers

Grazing-angle Fourier transform infrared (FTIR) spectra of monolayers of BDDAP-C-HMTCAQ, BDDAP-C-BHTCNQ, and TDDOP-C-HETCNQ, have been measured and reported previously[19-21,23]. The C-H stretch bands are well resolved, even for a single monolayer, and a broad structure at about 3500 cm^{-1} is seen for "fresh" samples, but disappears for samples older than about 60 days; this may be water trapped between the LB film layer and the aluminum layer, but the identification is not certain.

5.   RECTIFICATION ATTEMPTS

We' review here seven attempts, by the ORP and by others, to detect unimolecular rectification.

(1)   A 2 mm diameter droplet of Hg was used[7] to probe the conductivity across a single monolayer LB film deposited on either Pt or conducting tin oxide glass: the sandwiches: (i) Pt|DDOP-C-BHTCNQ|Hg, (ii) Pt|Py-BHTCNQ|Hg, and (iii) conducting SnO$_2$ glass|DDOP-C-BHTCNQ|Hg were thus tested: in all cases the background conductivity of the solid support was measured, presumably because of microscopic pinholes in the LB film[7].

(2)   In the hope[19] that defects may be avoided, statistically, in a domain of the order of about 0.5 mm x 0.5 mm, if one searched through enough samples, the left-hand halves of fifteen glass microscope slides were coated (using a mask) with five parallel fingers of Al at least 500 nm thick, 3.5 mm long, and 1.6 mm wide. Then the fifteen slides were coated with a single LB monolayer of BDDAP-C-BHTCNQ, **15**, at room temperature. Then the slides were coated again with five fingers of Al, but this time on the right hand side of the slide, so that the vertical overlap Al|BDDAP-C-BHTCNQ|Al would be in an area of only about 0.5 mm x 0.5 mm. Of the 75 junctions thus prepared, many were open circuits; the rest were short circuits. Thus, a defect-free domain of BDDAP-C-BHTCNQ was not found[19].

(3)   Aviram, Joachim, and Pomerantz reported in 1988[61] that rectification had been observed in a modified STM, for the molecule **30** (which was originally proposed as an intramolecular hydrogen atom transfer switch, based on H bonding in ortho-quinone-cathecol systems[1,2]). It was confirmed that there is indeed intramolecular H atom transfer in **30**, and in addition it was claimed that rectification was also observed for current passing through it[83]. The claim of rectification was later retracted[62].

(4)   Dr. Pomerantz very kindly studied with his modified STM, an LB monolayer of BDDAP-C-BHTCNQ, **15**, deposited on a Au|Ag|mica substrate, using an atomically sharp W tip, as the couple W|BDDAP-C-BHTCNQ|Au. Large "rectification" currents were observed[20,21], but later control experiments by Dr. Pomerantz

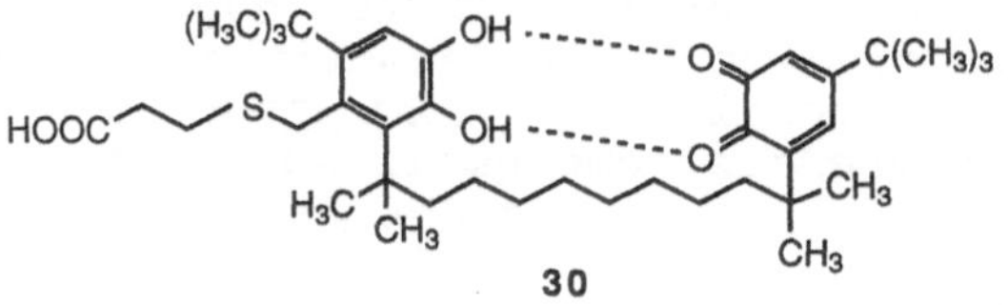

Fig. 9. Molecular configuration **30**.

showed that this "rectification effect", with very large currents, could occur **in the absence of any molecule.** Disclaimers for the preliminary results were issued[20,21].

(5)   The synthesis of BDDOP-C-BHTCNQ was repeated by Sandman.   Recently Sambles and co-workers at the University of Exeter found that a monolayer of BDDOP-C-BHTCNQ, sandwiched between Pt and Mg electrodes, behaved as a rectifying LB film[36].

(6)   Sambles, Ashwell, and co-workers found that also an LB film of Z-β-(1-hexadecyl-4-quinolinium)-α-cyano-4-styryldicyanometanide, a D-π-A molecule, similarly sandwiched between Pt and Mg electrodes (the latter shadowed with Ag) showed either macroscopic rectification behavior, or else some effect of Schottky barriers or dipolar coupling between substrate and Mg[37].

(7)   We have recently acquired an STM (Digital Instruments Nanoscope II), and have studied monolayer LB films of BDDAP-C-HETCNQ and Py-C-HECTNQ, transferred onto highly oriented pyrolytic graphite (Union Carbide grade ZYA) by a horizontal lifting technique[63].   Incontrovertible molecular images of either monolayer have not been found.   At very low set-point currents, some asymmetries in the current-voltage plots can be seen[63].

6.   FUJIHIRA'S LB PHOTODIODE D-σ1-A-σ2-S

We should mention in closing that Fujihira and co-workers have demonstrated that a single LB monolayer can function as a photodiode[64]: **this is probably the first truly unimolecular device.**   They synthesized a D-σ1-A-σ2-S molecule, where D = electron donor = ferrocene, σ1 = $(CH_2)_{11}$ chain, A = final electron acceptor = viologen, σ2 = $(CH_2)_6$ chain, S = sensitizer = pyrene.   This molecule was transferred as an LB monolayer onto a semitransparent Au electrode (with the viologen, or A, part of molecule closest to Au); this electrode was used as a window of an electrochemical cell, which also contained a 0.1 M KCl solution and a Pt counter electrode.   Under bias, an electron is transferred from solution to the ferrocene end of the LB film, and then to the ground state of the pyrene molecule.   Light at 330 nm excited the pyrene radical cation from the ground state to an excited state, from which the electron is transferred to the viologen, thus completing the circuit.   A photocurrent of 2 nA at 0.0 V vs SCE was observed only when the light was turned on[64].

CONCLUSION

We have outlined here several topics in ME, and hope to have convinced the reader that exciting prospects exist in this infant field. In particular, it seems likely that rectification by a single molecule will be demonstrated incontrovertibly in the very near future.

REFERENCES

1. A. Aviram, M.J. Freiser, P.E. Seiden and W.R. Young, **U.S.Patent** US-3,953:874 (27 April 1976)
2. A. Aviram and M.A. Ratner, **Chem Phys. Lett.** 29:277 (1974)
3. A. Aviram, P.E. Seiden and M.A. Ratner, p.5 **in** "Molecular Electronic Devices", F.L. Carter, ed. Dekker, New York, (1982)
4. R.M. Metzger and C.A. Panetta, **J. Phys. (Les Ulis, Fr.) Colloque** 44:C3-1605 (1983)
5. R.M. Metzger and C.A. Panetta, p.1 **in**: "Molecular Electronic Devices, Vol. II", F.L. Carter, ed. Dekker, New York (1987)
6. C.A. Panetta, J. Baghdadchi and R.M. Metzger, **Mol. Cryst. Liq. Cryst.** 107:103 (1984)
7. R.M. Matzger, C.A. Panetta, N.E. Heimer, A.M. Bhatti, E. Torres, G.F. Blackburn, S.K. Tripathy and L.A. Samuelson, **J. Molec. Electronics** 2:119 (1986)
8. R.M. Metzger, C.A. Panetta, Y. Miura and E. Torres, **Synth. Metals** 18:797 (1987)
9. E. Torres, C.A. Panetta and R.M. Metzger, **J. Org. Chem.** 52: 2944 (1987)
10. R.M. Metzger and C.A. Panetta, p.81 **in**: "Proc. of the Eighth Winter Conference on Low Temperature Physics, Cuernavaca, Mexico", (1987)
11. R.K. Laidlaw, Y. Miura, C.A. Panetta and R.M. Metzger, **Acta Cryst.** C44:2009 (1988)
12. R.K. Laidlaw, Y. Miura, J.L. Grant, L. Cooray, M. Clark, L.D. Kispert and R.M. Metzger, **J. Chem. Phys.** 87:4967 (1987)
13. R.K. Laidlaw, J. Baghdadchi, C.A. Panetta, Y. Miura, E. Torres and R.M. Metzger, **Acta Cryst.** B44:645 (1988)
14. Y. Miura, R.K. Laidlaw, C.A. Panetta and R.M. Metzger, **Acta Cryst.** C44:2007 (1988)
15. R.M. Metzger, R.R. Schumaker, M.P. Cava, R.K. Laidlaw, C.A. Panetta and E. Torres, **Langmuir**, 4:298 (1988)
16. R.M. Metzger and C.A. Panetta, p.271 **in**: "Organic and Inorganic Lower Dimensional Materials", NATO ASI Series Vol. B168, P. Delhaès and M. Drillon, eds., Plenum, New York, (1988)
17. Y. Miura, E. Torres, C.A. Panetta and R.M. Metzger, **J. Org. Chem.** 53:439 (1988)
18. Y. Miura, C.A. Panetta and R.M. Metzger, **J. Liquid Chrom.** 11:245 (1988)
19. R.M. Metzger and C.A. Panetta, **J. Mol. Electronics** 5:1 (1989)
20. R.M. Metzger and C.A. Panetta, **J. Chim. Phys.** 85:1125 (1988)
21. R.M. Metzger and C.A. Panetta, **Synth. Met.** 28:C807 (1989)
22. R.M. Metzger, R.K. Laidlaw, E. Torres and C.A. Panetta, **J Cryst. Spectr. Res.** 19:475 (1989)
23. R.M. Metzger and C.A. Panetta, p.293 **in**: "Molecular

Electronics - Science and Technology", A. Aviram and
A. Bross, eds., New York Engineering Foundation, (1990),

24. R.M. Metzger, D.C. Wiser, R.K. Laidlaw, M.A. Takassi,
    D.L. Mattern and C.A. Panetta, **Langmuir**, 6:1515 (1990)
25. R.M. Metzger and C.A. Panetta, **in**: "Lower Dimensional
    Systems and Molecular Electronics", R.M. Metzger,
    P. Day and G.C. Papavassiliou, eds., NATO ASI Series,
    Plenum Press, (in press)
26. R.M. Metzger and C.A. Panetta, **in** "Advanced Organic Solid
    State Materials", L.Y. Chiang, D.O. Cowan and
    P. Chaikin,eds., **Materials Research Society Symposium
    Proceedings Series**, 173:531 (1990)
27. R.M. Metzger and C.A. Panetta, **Nouv. J. Chim.**, in press
28. See e.g. G.L. Gaines, Jr. "Insoluble Monolayers at Liquid -
    Gas Interfaces", Interscience, New York, (1966)
29. K.B. Blodgett **J. Am. Chem. Soc.** 57:1007 (1935)
30. K.B. Blodgett and I. Langmiur, **Phys. Rev.** 51:964 (1937)
31. H. Kuhn, D. Möbius and H. Bücher, p.577 **in**: "Techniques of
    Chemistry, Vol. I-Physical Methods of Chemistry - Part V -
    Determination of Thermodynamic and Surface Properties",
    A. Weissberger and B.W. Rossiter, eds., Wiley -
    Interscience, New York, (1972)
32. H. Kuhn, **Pure Appl. Chem.** 51:341 (1979)
33. H. Kuhn, **Pure Appl. Chem.** 53:2105 (1981)
34. See e.g. **Thin Solid Films** Vols. 68 (1989), 99 (1983), 132-
    134 (1985), 159-160 (1987)
35. G. Binnig, H. Rohrer, Ch. Gerber and E. Weibel, **Phys. Rev.
    Lett.** 49:57 (1982)
36. N.J. Geddes, J.R. Sambles, D.J. Jarvis, W.G. Parker and
    D.J. Sandman, **Appl. Phys. Lett.**, in press
37. G.J. Ashwell, J.R. Sambles, A.S. Martin, W.G. Parker and
    M. Szablewski, **Nature**, submitted
38. J.E. Meinhard, **Appl. Phys. Lett.** 35:3059 (1964)
39. E.E. Polymeropoulos, D. Möbius and H. Kuhn, **Thin Solid
    Films** 68: 173 (1980)
40. M. Sugi, K. Sakai, M. Saito, Y. Kawabata and S. Iizima,
    Thin Solid Films 132:69 (1985)
41. R.S. Potember, T.O. Poehler and D.O. Cowan, **Appl. Phys.
    Lett.** 34:405 (1979)
42. R.S. Potember, R.C. Hoffmann, H.S. Hu, J.E. Cocchiaro,
    C.A. Viands, R.A. Murphy and T.O. Poehler, **Polymer** 28:574
    (1987)
43. R.S. Potember, R.C. Hoffman, R.C. Benson and T.O. Poehler,
    J. Phys. (Les Ulis) 44:C3-1597 (1983)
44. H.S. White, G.P. Kittleson and M.S. Wrighton, **J. Am. Chem.
    Soc.** 106:5375 (1984)
45. E.T. Turner Jones, O.M. Chyan and M.S. Wrighton, **J. Am.
    Chem. Soc.** 109:5526 (1987)
46. J. Paloheimo, P. Kuivalainer, H. Stubb, E. Vuorimaa and P.
    Yei-dahti, **Phys. Scripta**, submitted
47. R. Hoffmann, **Acc. Chem. Res.** 4:1 (1971)
48. R.W. Murray, **Acc. Chem. Res.** 13:135 (1980)
49. W.C. Bigelow, D.L. Pickett and W.A. Zisman, **J. Colloid Sci.**
    1:513 (1946)
50. R. Maoz, L. Netzer, J. Gun and J. Sagiv, **J. Chim. Phys.**,
    85:1059 (1988)
51. I.R. Peterson, **J. Chim. Phys.**, 85:997 (1988)
52. W.R. Hertler, **J. Org. Chem.**, 41:1412 (1976)
53. J. Baghdadchi, Ph. D. dissertation, Univ. of Mississippi,
    Dec. 1982
54. L.T. Calcaterra, G.L. Closs and J.R. Miller, **J. Am. Chem.
    Soc.**, 105:670 (1983)

55. J.R. Miller, L.T. Calcaterra and G.L. Closs, **J. Am. Chem. Soc.**, 106:3045 (1984)
56. R.A. Marcus, **Disc. Faraday Soc.**, 29: 21 (1960)
57. X.-L. Wu, J.L. Parakka, and R.M. Metzger, unpublished results.
58. A.C. Cephalas, private communication.
59. J.R. Anderson and O. Jorgensen, **J. Chem. Soc. Perkin Trans. I.**, 3095 (1979)
60. M.A. Takassi, Ph.D. Dissertation, University of Mississippi, Aug. 1989.
61. A. Aviram, C. Joachim and M. Pomerantz, **Chem. Phys. Lett.**, 146:490 (1988)
62. A. Aviram, C. Joachim and M. Pomerantz, **Chem. Phys. Lett.**, 162:416 (1989)
63. X.-L. Wu and R.M. Metzger, unpublished results.
64. M. Fujihira, K. Nishiyama and H. Yamada, **Thin Solid Films**, 132:77 (1985)

TRANSPORT IN LOW DIMENSIONAL ORGANIC STRUCTURES

B. Movaghar

QMW Polymer Physics Group
QMW Mile End Road, London E41NS, UK

ABSTRACT

Organic materials very frequently constitute systems of low
dimensionality. Conjugated polymers and Langmuir-Blodgett
multilayers are two examples. A survey will be given of novel
phenomena associated with the electron transport along polymer
chains and across LB films. An attempt will be made to compare
and contrast organic and inorganic low dimensionality with
emphasis on potential applications to the subject of Molecular
Electronics.

1. INTRODUCTION

There has been growing interest recently in the physics of
organic materials with a view to developing novel materials for
use in electronic devices. The QMW group (Queen Mary College
polymer physics group) has devoted its research effort in two
directions: Langmuir-Blodgett films, and conjugated polymers.

Langmuir-Blodgett films now constitute a vast domain of
research and to some extent even development[1] and I shall
briefly outline our interests in this particular area. The work
at QMW has focused mainly on electronic conduction perpendicular
to the layers. This investigation is in part motivated by the
idea of eventually fabricating a molecular register or 'memory'
which uses the transport of electronic charge across the
multilayer assembly[2]. The control of delay and arrival times
of charges in the layers is achieved by using electric fields.
Arrival can for example be signaled by a fluorescing molecule or
recombination with an injected hole. The major difficulty one
encounters in this technology is the fabrication of controllable
'hopping' layers. Hopping, because transport perpendicular to
multilayers is usually stochastic. The problem of 'order', or
absence of order, manifests itself by the presence of defects
and in particular of 'conducting filaments' which short-circuit
the transport paths. Unlike semiconductor superlattices where a
high degree of order can be achieved using the technique of
Molecular Beam Epitaxy, Langmuir-Blodgett multilayers cannot as

yet be grown with the same degree of perfection. To some extent this is a material problem, but to some extent this is also intrinisic to the technique: room temperature deposition, weak bonding between the molecules...etc. Doping the material is difficult and the formation of barrier layers is also a nontrivial task. The reader should also note that the formation of conduction bands in molecular solids is quite unlike those of inorganic solids. The excited bands of molecular solids are formed by coupling together molecular exciton-polaron states so that quantum coherence is difficult to achieve over a multilayer. Band formation perpendicular to layers is unlikely; most of the time, the electrons undergo stochastic motion in this direction. Coherent motion parallel is, however, achievable[3], see the work on the metallic $TMTTF-C_nTCNQ$ layers.

Considerable effort has been devoted to trying to produce electrically well defined multilayer structures, however, with varying degrees of success. Insulating structures with 4 layers of ω-tricosenoic acid and magnesium electrodes can now be routinely produced, (see the work of Geddes and Sambles[4]). In the QMW structures, the insulating property of the multilayers is achieved by forming 'blocking layers' (see Fig. 1). In this way it has been possible for Severn, Sudiwala and Wilson[5], and more recently Burrows, Donovan and Wilson[6] to measure a characteristic 'hop time' τ across the layers, typically $\tau \sim 15$ ns. The theory for this kind of transport was worked some years ago in a series of papers [7,8,9]. The time dependent photocurrent can be evaluated rigorously for low and high fields, in the 'drift' regime when $eFa > kT$, a is the interlayer distance, we have

$$I_p(t) = n\, n_1(t) e v_d \tag{1}$$

where

$$n_1(t) = 1 - v_d t/L_x + \sum_{m=1}^{\infty} \sum_{n=m+N_x}^{\infty} (v_d t/a)^{n+1} \exp(-v_d t/a)/N_x\ (n+1)! \tag{2}$$

e is the electronic charge, v_d is the drift velocity, n the carrier density and $N_x a = L_x$ is the length of the system. The double sum is negligible when $(v_d t/L_x) < 1$ so that the decay is a linear function of t almost all the way.

Using an essentially similar analysis, Wilson and co-workers were able to extract 'hop times' from their data (see Fig. 1b). The data agree well with the form given by Eqn. 1. Here it is assumed that the transit time is fast compared to the recombination in the system. Recombination can easily be taken into account when necessary.

Future work envisaged by the group is to use novel materials which improve LB film order and to fabricate novel field effect transistors using LB films as the active material. Potential applications of conduction effects in LB materials are in the area of sensors. Here, the sensitivity of the material to defect creation may be just the right asset to detect molecules! Improving material quality and structural order is clearly the key point of the electrical work. There are,

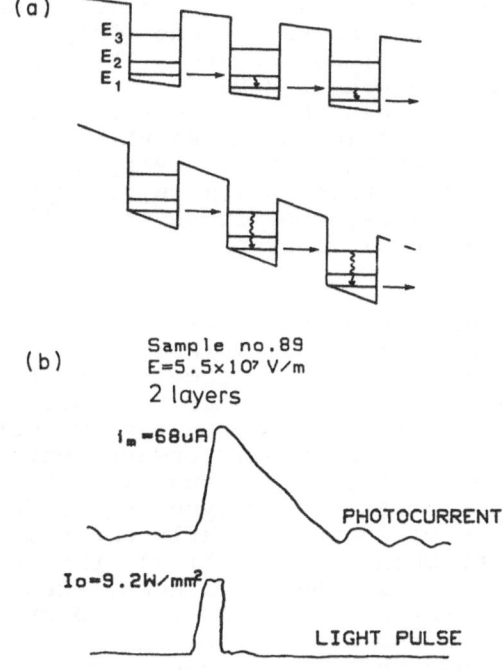

(a)

E_3
E_2
E_1

(b)

Sample no.89
E=5.5×10⁷ V/m
2 layers

i_m =68uA

PHOTOCURRENT

Io=9.2W/mm²

LIGHT PULSE

Fig. 1. a) A multilayer system represented by
quantum wells with blocking contacts.
b) Time decay of the photocurrent for a
two-layer protoporphyrin structure.

however, a large number of possible applications of LB
technology which do not require the same degree of perfection.
This includes for example fluorescence, energy transfer and
nonlinear optics[10]. Let me now turn to the field of
electrical conduction in conjugated polymers.

2. POLYMERS

Electrical properties of polymers can be subdivided into
three broad areas: 1) the work on the disordered doped
materials, for example Bisphenol-A-Polycarbonate, here there is
no anisotropy in the conductivity and it cannot therefore be
termed a low dimensional conductor as such[11]. The transport
is almost exclusively hopping transport between dopant molecules
and defects. These materials are useful in xerography; most of
the experimental work has been carried out in industrial laboratories such as Xerox[11] and IBM[12]. Observed transit signals
have puzzled the experimentalists by exhibiting Poole-Frenkel-
like high field behavior without the charged centers. The
recent transport theory of Movaghar, Meunier and Yelon[13]
identifies this unusual behavior with a Coulomb gap type density
of localized states caused by disorder and/or possibly electron-
electron interaction effects. Disorder effects are also invoked
in the model proposed by Pautmeier et al.[14]. The $\exp(F/F_0)^{1/2}$
mobility law observed in these materials is very common in insulators and does not appear to signal a new kind of transport
mechanism.

Now consider another class of materials, namely the conjugated polymers of the Polyacetylene (PA), Polydiacetylene (PDA) and Polypyrrole (PPy) categories. These materials exhibit some degree of order and can be termed quasi-one-dimensional. The question is: what is new and interesting about this class of materials? In what way could the novel physics be incorporated or used in a novel electronic device? Very briefly the answer to the second question is this: polymers, in contrast to van der Waals bonded organic solids, constitute thermally stable materials with good mechanical properties and can be produced relatively cheaply.

Conjugated polymers have novel electronic properties because of their low dimensionality (anisotropic absorption, transport etc.). I will deal with some of these features later in the text. Let me remind the reader that here, too, structure is a major problem. There are now roughly two categories of conjugated materials. The partially ordered materials exhibit some anisotropy or none at all, for example here we have the Shirakawa PA and most of the other types of conducting polymers[16]. In the category of nearly ordered and crystalline polymers, we have the Durham route PA[17], the Naarman PA[18] and, of course, the crystalline polydiacetylenes PDA-TS and PDA-DCH. The latter are the only genuine ordered crystalline polymers. PA, PPy and the other types of 'conducting' materials are interesting because they can be doped and can nowadays reach conductivities of 10^5 Siemens[18], and anisotropies of the order of 10^2. Polyacetylene can be n or p doped and can therefore, in principle, be manipulated like silicon or germanium. Degradation under ambient conditions, however, often makes these materials unsuitable for practical applications. This is why less conductive but more stable alternatives (PPy, for example) have a role to play in this technology.

Coming back to the question of why these materials are interesting, one can, I think, say that it is not because they can replace usual semiconductor device functions. As metals, however, they do have some novel applications as metallic paints for example, and electromagnetic absorbers. Because they can be electrochemically doped, they might lead to light-weight batteries and possibly to batteries which can take on unusual shapes and therefore be integrable in novel kinds of geometries.

The polydiacetylenes are well known for their excellent crystallinity, their high nonlinear optical coefficients, and the fact that they can be grown as Langmuir-Blodgett films. The PDAs, unfortunately, cannot be doped to metallic limits without damaging the crystal structure. The technological applications have therefore focused on the nonlinear optics. Though an important amount of work has been done in this area, no commercial PDA device has yet emerged[19]. The soluble polydiacetylenes of the BCMU type can be grown into waveguide structures even though they are relatively lossy guides (1 dB/cm); they do combine flexibility of growth with high nonlinear coefficients. This might make them useful for some forms of optical network technology[19].

The enormous field of electronic properties of conjugated polymers, and in particular optical properties, has been covered in many review articles. Here I shall only consider some selected aspects related to electrical transport.

3. ELECTRONIC STRUCTURE OF ORGANIC SOLIDS AND IN PARTICULAR CONJUGATED POLYMERS

Ordinary solids can usually be described by one electron Hartree-Fock type band structure concepts. The disordered systems can be described by invoking extensions of these concepts such as 'localization', Coherent Potential Approximation, etc. This is not so for conjugated polymers. Here it is important to include more rigorously the coupling between atomic displacements and electronic states. The most popular and widely accepted model of electronic and vibrational properties is the SU-SCHRIEFFER-HEEGER Hamiltonian abbreviated as SSH. This can be written in the form[20]

$$H = \sum_{i j \sigma} t_{i j} c^+_{i \sigma} c_{j \sigma} + \sum_n (1/2) K_n (u_n - u_{n+1})^2 + \sum_n (1/2) M_n (\dot{u}_n)^2 . \tag{3}$$

The first term is the kinetic energy with t_{ij} representing the overlap integral, the second is the elastic energy of the lattice with u_n denoting the atomic displacement from the n_{th} equilibrium position and K_n the elastic constant. The last term is the kinetic energy of the lattice. The coupling between electron and lattice is contained in t_{ij} with

$$t_{ij} = t_0 \exp[-\alpha |R_0 + u_i - u_j|] \tag{4}$$

where R_0 is the equilibrium separation and α is the inverse decay length of the wavefunction. This model leads to the well known Peierls distortion, and all the interesting consequences for the electronic excitation spectrum. These are summarized in [20,21] and can be briefly stated as follows: according to Eqn. 3, the ground state of PA is a Peierls semiconductor, excited states are solitons, polarons and bipolarons. Semiconductors which have an intrinsic gap due to their structure acquire an additional Peierls gap so that the total gap is due to a combination of band structure effect and atomic relaxation. The magnitude of the final gap is determined by the trade-off between the energy gained by lowering the valence band and energy lost by increasing the elastic energy[21]. Most of the conjugated polymers have an intrinsic gap and thus electronic excitations are polarons and bipolarons. Polyacetylene is an exception; the gap is exclusively Peierls generated and it therefore also has the famous soliton excitations[20].

This summarizes the conventionally accepted picture of the electronic structure of the conjugated polymers. Considerable effort has gone into trying to verify or check this model. I shall not go into the details of the ingenious and extensive work that has been undertaken to observe solitons and polarons in PA. The interested reader can follow up the subject in detailed reviews and papers[22]. Here I shall focus on the completeness of the theoretical model and suggest that Eqn. 4 does not describe all the important phenomena. I shall also briefly review the work done at QMW on this topic.

Let me first consider the inadequacies of the simple SSH Hamiltonian. It is clear and also generally accepted that to Eqn. 4 one should also add the Hubbard term

$$H_U = U \sum_{i\sigma} n_{i\sigma} n_{i-\sigma} \qquad (5)$$

giving rise to on-site repulsion. The Hubbard term gives rise to antiferromagnetic order in the half filled band limit and is not in contradiction with the chemical bond picture of the polymers. The Hubbard term, when treated in mean field theory, will not destroy the Peierls gap picture, apparently it simply generates a Peierls-Hubbard gap instead[23].

The Hubbard term is therefore not expected to change the qualitative picture we have of the ground state of these materials. It will, however, introduce magnetic correlations and therefore considerably complicate the excited state spectrum. Polarons and solitons are still possible as long as there is dimerization and the right kind of symmetry in the system. Mobilities will, of course, be affected by spin scattering. Bipolarons are strongly affected (prevented) by Coulomb repulsion, as one would expect.

A similar argument applies to the long range Coulomb coupling

$$H_{lr} = \sum_{ij\sigma} (e^2/r_{ij}) \, n_{i\sigma} n_{j-\sigma} \; . \qquad (6)$$

In the ground state Eqn. 6 will not have a great effect on the picture we have of the solution; it can be incorporated in the definitions of t_{ij} and U. In a highly doped system, Eqn. 6 will be screened out by mobile charges and can be treated by mean field theory. For undoped polymers, however, or materials under strong illumination conditions, H_{lr} can represent a serious contribution. This is easily proven by looking at the free carrier generation mechanism in materials such as PDA and, indeed, most organic semiconductors[24]. Free charges can only be produced by breaking up the strongly bound excitons formed by light excitation or, indeed, thermal excitation! The energy cost of an excitation is therefore very much dependent on the distance at which the electron is placed from the positive charge. Weak screening with $\epsilon \sim 2$ in PDA also implies that the extra charge will itself produce a dielectric polarization and, depending on the medium, (surrounding molecules or side groups) will induce a more or less important electronic polaron. Both lattice and electronic contributions must therefore be included in polaron formation. These will be characterized by parameters λ_1 and λ_2 respectively. An electron excited by light from the ground state will therefore obey the Schrödinger equation

$$\{-\hbar^2/2m_\mu \Delta_\mu{}^2 - (\lambda_\epsilon + \lambda_1) \, \psi^* \psi + Ze^2/r - e\mathbf{F}.\mathbf{r}\} \, \psi = E\psi. \qquad (7)$$

The system is in general anisotropic so the effective mass is dependent on the directions in space. Equation 7 applies on the time scale τ_p after the deformation has formed. The polaron is expected to form on the time scale of an optic phonon frequency, i.e. $\tau_p \sim 0.1$ ps. An external field F applied in the x-direction has been included in Eqn.7; Z is the effective nuclear charge seen by the electron. The energy E is measured from the bottom of the conduction band; care has to be taken in going from the discrete to the continuum limit. The parameter Z can be chosen to produce the correct experimental exciton binding energy. In the presence of a strong field, excited states of Eqn. 7 can dissociate and give rise to charged polarons traveling in the field direction. At this stage the quantum and stochastic aspects come into the problem in a very subtle way: quantum mechanically the field can produce tunneling out of the polaron into the continuum[25]; it can, however, also cause the polaron to shrink[26]. Energy relaxation to the lattice can also produce stochastic polaron motion along the bottom of the conduction band. The bound polaron has to drag along its deformation and its velocity will therefore saturate roughly at the speed of response of the deformation, i.e. the speed of sound. The diffusivity of the relaxing or hopping polaron will also be limited by a typical optic phonon frequency. The complete analysis of this problem is rather subtle and involved and will not be pursued in this paper. There are interesting analogies to the problem of motion in a superlattice under the influence of a strong electric field, as one encounters for example in the novel infrared devices[27].

Realistically speaking, we must observe that in organic solids, quantum effects are only relevant to the very short time domain of light excitation. Rapid relaxation and thermal scattering will quickly take us into the stochastic or semi-classical regime. This is indeed the usual way one describes the charge carrier generation process in these materials, namely using the Onsager picture. The excited electron simply diffuses in the combined field of the positive hole and external field. The total time dependent photocurrent can therefore be represented by a superposition of a short time polarization current $I_p(t)$ which is due to the actual creation process, and geminate recombination of charge carriers, and a conduction current $I_c(t)$. The conduction current is due to the liberated carriers, i.e. those which have escaped geminate recombination, see Fig. 2. The polarization current leads to zero net charge transfer provided the carriers are not trapped on the way back; it should in principle go negative on geminate recombination. The

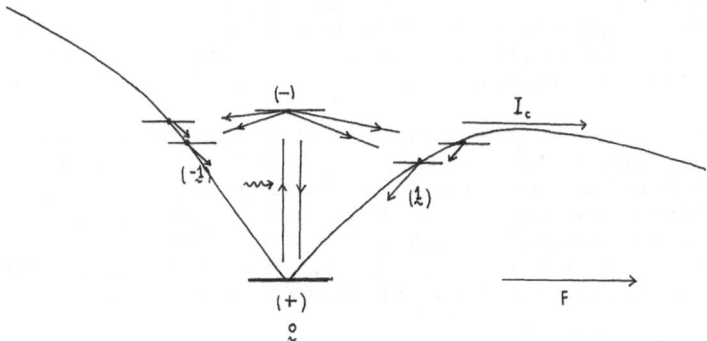

Fig. 2. The charge carrier generation process in a highly excited molecular exciton. The relaxing hot electron has effectively a larger number of channels to relax into in the field direction than against the field. I_c denotes the conduction and F is the external electric field.

generation being, however, much faster than the recombination of the self-trapped carriers to the hole, it could mean that the negative current is hidden by the conduction current. Usually one neglects $I_p(t)$ altogether; it might, however, be of some relevance at low temperatures and when there is strong self-trapping of charge. A quantitative analysis is in progress using simulation techniques [D. Willock QMW].

Returning now to the computation of $I_c(t)$, we have to deal with the well known Onsager problem. In other words we have to evaluate ϕ, the probability of generating free carriers while diffusing in a combined Coulomb and electric field. The problem has been analysed in various dimensions and numerically for anisotropic lattices[28]. For polarons we have to use the polaron diffusivity $D_f(F_{int})$ rather than the bare particle diffusivity D. Note that F_{int} is the true internal field including the hole. The polaron diffusivity has been analysed by Davydov[26], and by Wilson[29]. In one dimension ϕ can be evaluated analytically[28] to give

$$\phi_1 = \frac{\int_0^1 dt \ \exp(-s/t - r_o t)}{\int_0^\infty dt \ \exp(-s/t - r_o t)} \ . \tag{8}$$

where $s = r_k/b$, $r_o = eFb/kT$, $r_k = e^2/4\pi\epsilon_o\epsilon kT$, and b is the initial thermalization length of the problem. The diffusional aspect in this formula is actually only contained in b. To first order in the field, we obtain from Eqn. 8

$$\phi_1 = eFb/kT \ (b/r_k) \ \exp[-r_k/b]. \tag{9}$$

The initial thermalization length b as it appears in Eqn. 9 might seem a simple quantity. In reality, however, it is quite a subtle quantity to evaluate properly in the framework of a first principle model. Remember that it contains information on the initial eigenstate produced by the light, the way the particle subsequently relaxes its excess energy to the lattice in the presence of an external field and the self-trapping process, finally giving rise to a net displacement b which depends on the wavelength λ, F and T. A rigorous analysis is in progress at QMW [D. Willock].

Naturally none of the conjugated polymers is truly one-dimensional in the solid state. In general the field dependence of ϕ is therefore not as strong as in Eqn. 8[30]. The polydiacetylenes are the closest to truly one-dimensional systems. The chain-chain separation is ~ 7 Å, the mobility anisotropy $\mu_\parallel/\mu_\perp$ is 10^3 at least[31]. The long time photo-current I_c can therefore be well described by the equation [32],

$$I_c(t) = eN_e\phi_1 v_d(t) \tag{10}$$

where N_e is the number of excitons/volume.

The drift velocity will be trap limited for $t > \tau_t$ and polaron-like for free sections of the chain. In principle one should therefore be able to see the free polaron dynamics in the short time interval $\tau_p < t < \tau_t$. Indeed, this has been attempted by Donovan and Wilson using 25 ps pulses in PDA-TS[32]. These authors claim to have observed intrinsic polaron motion on this time scale, characterized by a saturated drift velocity v_s of the order of the sound velocity 2000m/s, and a one-dimensional Onsager generation function. In the high excitation limit, and/or long time limit, recombination sets in, so that one can have both deep traps (chain ends, etc.) or bimolecular recombination. In contrast to Moses et al.[33], Donovan and Wilson interpret their field dependent photocurrent decays as due to enhanced bimolecular recombination. Field dependent decays on short samples were interpreted by Moses et al. as carrier sweepout. Donovan and Wilson have shown that similar effects can be generated using the light intensity[34]. Samples with large electrode spacings exhibit no transit or sweepout behavior at all! This suggests that short electrode separations sample over a different distribution of obstacles than large gaps. The latter imply deep traps and chain end type defects on the way; short chains might, however, allow genuine transit to the exit electrode[9]. Given the fact that the decay depends on temperature, (see Fig. 3), field in short samples, and light intensity, and that the field plays a different role at low and high temperatures, a detailed quantitative summary is still in order and will be published soon[34].

The understanding of the high excitation regime may not be as simple as it may seem. In the presence of something like 10^{18} carriers/cm^3, generated by a pulse, one has an important carrier-carrier interaction so that a typical site energy on a π-orbital of the backbone at time t will be given by

$$\epsilon_i(t) = \epsilon_o - e^2 \sum_j^h n_j(t)/\epsilon R_{ij} + e^2 \sum_j^e n_j(t)/\epsilon R_{ij} - eF.R_i, \qquad (11)$$

where e, h, denote electron and hole respectively and $n_j(t)$ is the carrier density at site j at time t. A nonlinear wave or polaron will be only weakly affected by the internal field. A hopping conductor, however, will be strongly perturbed by the Coulomb interaction, giving rise to disorder potentials and a large capture cross section for e-h recombination.

4.   THE STATISTICAL MECHANICS OF EXCITED STATES

In a wide class of organic materials, including the low doped polymers, we must take seriously the excitonic nature of

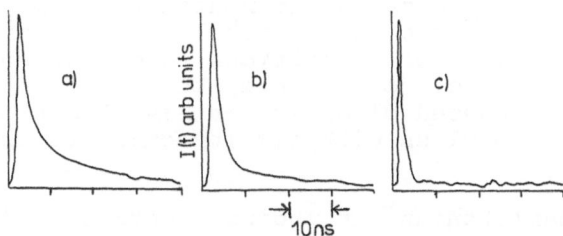

Fig. 3. The photocurrent of PDA-TS for a 10 ns pulse at three different temperatures, T = 300 K left, T = 60 K middle, T = 13 K right curve.

the excited states. This is what charge generation experiments
tell us. If we do, then we discover that we have a new class of
problems to deal with, not unlike spin glasses. These would be
Coulomb glasses. Naïvely we might, for example, expect the
distribution of excited electrons in the conduction band in the
vicinity of the holes to look like

$$n_i = \exp\{-\beta(\epsilon_c - Ze^2/\epsilon R_{io})\} \tag{12}$$

where ϵ_c is the bottom of the conduction band and R_{io} is the
distance to the positive hole. More correctly we should, of
course, proceed to evaluate the true eigenstates and, in the
presence of many excitons, we should include exciton-exciton
coupling and screening effects. Light pulses can routinely
generate 10^{20} excitons/cm^3 in PDA for example. Given the fact
that we have low dimensional materials, often with narrow bands
and low dielectric constants, we should expect localization by
disorder to set in quite early so that our crystal turns into a
kind of Coulomb glass. Remember that for $\epsilon \sim 2$, the Coulomb
interaction is of order 0.06 eV at 100 Å! Gláss formation can
happen even in thermal equilibrium. The presence of electric
fields and illumination will add free carriers and change the
exciton density. This will give rise to nonlinear effects in
the distribution functions which have to be evaluated
selfconsistently using Eqn. 11 for example.

In terms of a model Hamiltonian it implies that the
electrons and holes obey Schrödinger equations of the type

$$H_e = \sum_{ij\sigma} t^c{}_{ij} c^+{}_{i\sigma} c_{j\sigma} + \sum_{i\sigma} \epsilon_{i\sigma} n_{i\sigma} \tag{13}$$

where

$$\epsilon_{i\sigma} = \epsilon^o{}_{i\sigma} - \sum_{j\sigma'}{}^e (e^2/\epsilon R_{ij}) n_{j\sigma'} + \sum_{j\sigma'}{}^h (e^2/\epsilon R_{ij}) n_{j\sigma'} \tag{14}$$

and similarly for holes. The interesting point here is that
particles and holes are generated and annihilated continuously
so that only averages are fixed. When created locally, a pair
will produce a serious disturbance in its vicinity and therefore
affect the creation of other pairs by producing a 'local gap'
for other excitations. Excitations will therefore tend to
cluster in space rather than be randomly distributed giving rise
to order and self-organization. The analysis of such models
which combine stochastic and quantum effects promise to yield
important new insight into the possibility of dynamic glass
formation, metal insulator transitions, and even superconductivity. Assuming, for example, that all the states are
localized by self-induced disorder, we have the problem of
hopping in a Coulomb glass with all its characteristic
peculiarities[35].

The above Hamiltonian, including electric fields and light,
is the kind of problem we have to face up to in order to make
new progress in the field of organic physics. Polaronic self-
trapping can be easily included in Eqn. 13 by adding the term

804

$$H_p = -\lambda \sum_{i\sigma} n_{i\sigma} \, |a_{i\sigma}|^2 \qquad (15)$$

with a_i denoting the amplitude of the wave function at site i. Self-trapping enhances, of course, glass formation and leads also to local variations in dielectric properties.

5.   CONNECTION TO MOLECULAR ELECTRONICS

Can the physical phenomena just described and anticipated in the last section be of technical significance? The answer in my opinion is: solitons and polarons, though fascinating from the physics point of view, do not appear at present to contain, apart from self-localization, any technologically useful novelty. Excitons, on the other hand, do. This is already known and exploited to some extent in the field of fluorescence, nonlinear optics, and energy transfer. Excitonic physics becomes ,particularly powerful when combined with LB technology. Coulomb induced glasses or self-organization effects can become interesting in combination with techniques such as tunneling microscopy which can pick up local variations on molecular scales. A better understanding of excitonic or Coulomb interactions in solids might lead to the design of new materials with which we can generate metal-insulator transitions, light generated superstructures, optical switches and possibly even superconductivity.

Acknowledgment

The author is deeply indebted to his colleagues in the QMW polymer group for helpful advice and new unpublished data; these include Dr. E.G. Wilson, Dr. K. Donovan, Dr. D. Batchelder, Dr. R. Sudiwala and D. Willock.

REFERENCES

1.   G. Roberts, B. Holcroft and A. Barraud and J. Richard, **Thin Solid Films**, 160:53 (1988)
2.   E.G. Wilson, **Electr. Letters**, 19:237 (1988)
3.   M. Matsumoto, T. Nakamura, E. Manda, Y. Kawabata, K. Ikegami, S. Kuroda, M. Sugi and G. Saito, **Thin Solid Films**, 160:61 (1988)
4.   N. Geddes and R. Sambles, 4th Int. Conf. on LB Films, Tsukuba, Japan. (1989)
5.   J. Severn, R. Sudiwala and E.G. Wilson, **Thin Solid Films**, 160:171 (1988)
6.   P.E. Burrows, K. Donovan and E.G. Wilson, 4th Int. Conf. on LB Films, Tsukuba, Japan. (1989)
7.   B. Movaghar, B. Pohlmann, D. Murray and D. Wuertz, **J. Phys. C**, 17:1677 (1984)
8.   B. Movaghar, D. Murray, K. Donovan and E.G. Wilson, **J. Phys. C**, 17:1247 (1984)
9.   B. Movaghar, B. Pohlmann and D. Wuertz, **Z Phys. B**, 60:523 (1987)
10.   See for example **Thin Solid Films**, Vols.1 & 2:160: (1988)
11.   M.A. Abkowitz and M. Stolka, **Phil. Mag. Lett.**, in press (1989)

12. L. Schein, D. Glatz and A Peled, **J. Appl. Phys.**, 66:686 (1989)
13. B. Movaghar, A. Yelon and M. Meunier, **Chem. Phys.**, to be published (1990)
14. R. Richert, L. Pautmeier and H. Baessler, Preprint, Univ. of Marburg, (1989)
15. **J. Molecular Electronics**, 4:79 (1988)
16. A. McDiarmid and A.J. Heeger, **Synthetic Metals**, 1:101 (1980)
17. D.D.C. Bradley, R.H. Friend, H. Lindenberger and S. Roth, **Polymer**, vol. 27 (1986)
18. H. Naarman, **Synthetic Metals**, 17:223 (1987)
19. D. Bloor **in**: "Polydiacetylenes", D. Bloor and R.R. Chance, eds., Martinus Nijhoff, Dordrecht (1985) see also D. Batchelder p. 187 same volume.
20. W.P. Su, A.J. Heeger abnd J.R. Schrieffer, **Phys. Rev. B**, 22:2099 (1980)
21. N.A. Cade and B. Movaghar, **J. Phys. C**, 16:539 (1981)
22. D. Baeriswyl, Springer Series in Modern Science, Springer, Berlin, 63:85 (1985)
23. W.P. Su, S. Kivelson and J.R. Schrieffer, p. 201 **in**: "Physics in One Dimension", J. Bernasconi and T. Schneider, eds., Springer, Berlin, (1981)
24. M. Pope and S. Swenberg, Electronic Processes in Organic Crystals, Clarendon Press, Oxford. (1982)
25. B. Movaghar and I. Howard, **Synthetic Metals**, 27:A61 (1988)
26. A.S. Davydov, **Sov. Phys. JETP**, 51:397 (1980)
27. B.F. Levine, K.K. Choi, C.G. Bethea, J. Walker and R.J. Malik, **Appl. Phys. Lett.**, 50:1092 (1987)
28. R. Haberkorn and Michel-Beyerle, **Chem. Phys. Lett.**, 23:128 (1973)
29. E.G. Wilson, **J. Phys. C, Solid State**, 16:6739 (1983)
30. B. Ries, PhD Thesis, University of Marburg (1984)
31. E.L. Frankevitch, I.A. Sokolik and A.A. Lymarev, **Mol. Cryst. Liq. Cryst.**, 175:41 (1989)
32. K. Donovan and E.G. Wilson, **J. Phys. C**, 19:L357 (1986)
33. D. Moses, M. Sinclair and A.J. Heeger, **Phys. Rev. Lett.**, 58:2710 (1987)
34. K. Donovan and E.G. Wilson to be published
35. B.I. Shklovskii and A.L. Efroes, Electronic Properties of Doped Semiconductors, Springer, Berlin (1984)

ORGANIZED MONOLAYER ASSEMBLIES IN THE SEARCH FOR BIOMIMETIC

MACHINES

Hans Kuhn

Max-Planck-Institute for Biophysical Chemistry
P.O. Box 2841, 3400 Göttingen, West Germany

1.   INTRODUCTION

A biomimetic machine is defined here as a supramolecular device, designed to assist the understanding of basic biological mechanisms and to study nonnatural processes inspired from biosystems, i.e. it is an assembly where purposely constructed molecules have their purposely planned places and cooperate, thus forming a supramolecular functional unit. A simple biomimetic machine can be made by incorporating the molecules of interest in monolayers and superimposing such monolayers in a planned sequence. They are useful in discussing design principles and future prospects. Some examples are considered to illustrate the interplay between theoretical modelling and monolayer assembly, e.g. a photoinduced electron pump, solitons in molecular wires and natural and artificial creativity.

Biosystems are entities that are highly organized on every hierarchic level, and this organization is caused by their organization on the molecular level. To mimic biomachinery on the molecular scale, by constructing functional units composed of single molecules that interact in a planned manner, is a very challenging aim.

It is of interest to search for ways to construct biomimetic machines and to reflect on such machines. Assembling monomolecular layers is useful to study possibilities and to demonstrate feasibilities. Molecules of interest are incorporated in monolayers. The layers are deposited on a solid on top of each other, in planned sequence, and thus used as pincers to pick up single molecules and to arrange different kinds of molecules in a planned manner[1-9].

Biosystems are information-processing devices that get their input data from sensors and lead their output data to activators. The process is driven by an influx of energy rich compounds and an outflow of low energy compounds. The systems posses a copying mechanism with flaws and are therefore subject to evolution by continuous interaction of the copies with the environment, resulting in the survival of the fittest.

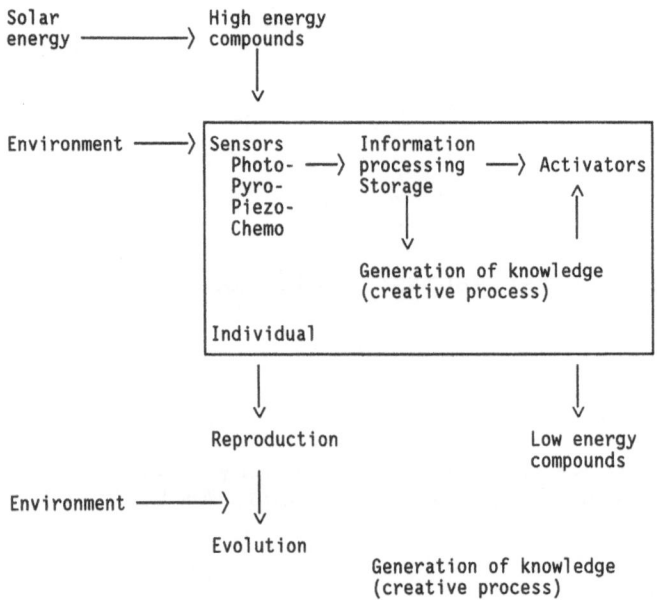

Fig. 1. Bioprocesses stimulating thought in terms of
        biomimetic machineries.

A machine is present to produce energy rich compounds by
converting solar energy, driving all bioprocesses (Fig. 1).

On the one hand, attempts to construct biomimetic machines
are inspired by considering bioprocesses; on the other hand,
reflecting on biomimetic machines helps us in understanding
bioprocesses and seeing these processes from the viewpoint of
supramolecular engineering. We restrict our discussion to a few
examples.

2.   MACHINES TO PRODUCE ENERGY RICH COMPOUNDS

For converting solar energy, photons are collected by an
antenna system and a photoinduced charge separation then takes
place, giving rise to reduction and oxidation processes (Fig.
2).

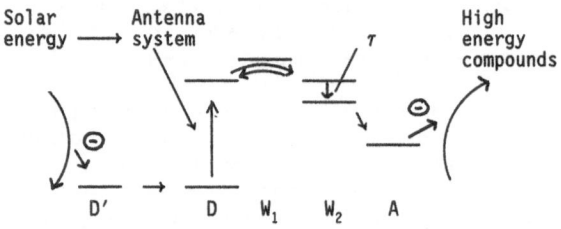

Fig. 2. Photo-induced electron pump.

2.1 Photon Funnel

Extremely efficient antenna systems can be made, by spreading, on a water surface, an oxacyanine dye with long hydrocarbon chain substituents and a long-chain hydrocarbon in 1:1 ratio, together with the corresponding thiacyanine in high dilution (ratio 10^{-4}). The hydrocarbons fill the gaps between the hydrocarbon substituents, and a densely packed monolayer with close-packed chromophores is thus obtained. Light is absorbed by the oxacyanine, the exciton moves over a large distance, and is then trapped by the thiacyanine[10] (Fig. 3a).

2.2 Electron Transfer in Monolayer Assemblies

Constructing a biomimetic machine for efficient energy conversion is more difficult. The presently available systems[11-17] are, by far, less efficient than the photosynthetic reaction centers in plants and bacteria. This is because the components in an efficient system must be extremely well adjusted to each other.

Electron transfer in monolayer assemblies can be easily studied[18]. The results are important for finding design principles of photoinduced electron pumps and for understanding the mechanism of electron transfer in reaction centers. The dye molecule is separated from the electron acceptor molecule by a fatty acid spacer layer (thickness d, Fig. 3b). The electron transfer rate (given by the amount of fluorescence quenching of the dye) decreases exponentially with d, indicating electron tunneling through the spacer. The result can be applied to estimate the arrangement of the component molecules in a photoinduced electron pump[19]. The electron acceptor A must be at a sufficient distance from the donating dye D (Fig. 2), to avoid back-transfer of the electron from A^- to D^+. Donor D' must be separated from D by a barrier, which must be sufficiently narrow, however, to allow tunneling of the electron from D' to D^+.

A basic problem is how the electron, after excitation, can be transferred sufficiently fast from D* (photoexcited D) to A, i.e. fast compared to the de-excitation of D*. A π-electron system, acting as molecular wire W carrying the electron from D* to A, is required. Such arrangements were made by assembling monolayers[20,21] (Fig. 3c), and an increase of the electron transfer rate by a factor of 10, as compared with the transfer through a fatty acid layer, was measured. A more efficient device for removing the electron from the excited dye is needed for an efficient electron pump.

2.3 Primary Processes in Bacterial Reaction Center to Elucidate Design Principles of Photoinduced Electron Pump

In attempting to construct such a device it is useful to consider the situation in the bacterial reaction center, where such a molecular wire is indeed present and is represented by two π-electron systems W_1 (bacteriochlorophyll) and W_2 (bacteriopheophytin)[22] (Fig. 2). We shall discuss the primary step in the electron transfer in the bacterial reaction center in more detail, in order to illustrate the importance of the interplay of molecular biology and theoretical modeling in attempting to construct biomimetic devices.

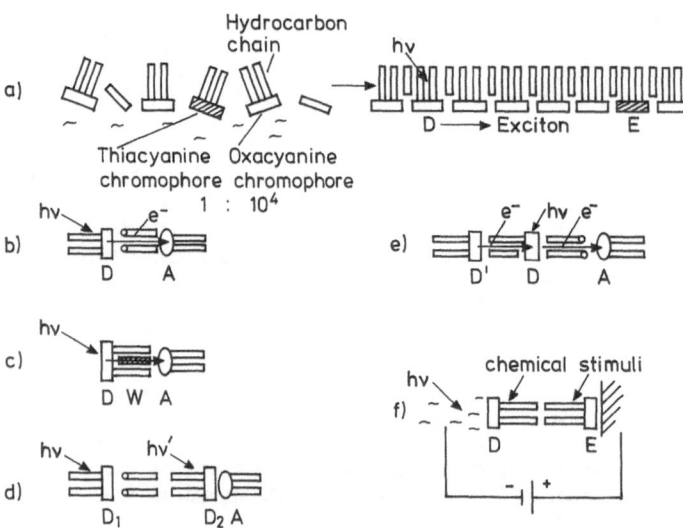

Fig. 3. Examples of organized monolayer assemblies:
a) antenna system. Donor D and energy acceptor
E. b) Arrangement for photoinduced electron
transfer. Donor D and electron acceptor A.
c) Electron transfer through conducting chain W.
d) Molecular switching device: Dye D_1 is excited
and emits fluorescence. This fluorescence is
quenched if D_2 is simultaneously excited: An
electron is transferred to A and A acts as
energy acceptor of D_1[25]. e) Molecular
switching device: Dye D is excited, electron is
transferred to A and an electron is supplied by
D' which is oxidized thereby[25]. f) Molecular
switching device: Dye D is excited and transfers
energy to E. Excited E injects an electron into
the semiconductor, thus producing a photocurrent. Both molecules D and E can be switched
(by a specific chemical change) from the active
into an inactive form, and the photocurrent is
accordingly stimulated[26].

The electron is transferred from D* (special pair of
bacteriochlorophylls) to W_2 within 3 ps, and from there to A
(quinone) in 230 ps[23]. A first absorption change takes place
in 0.5 ps, indicating a short-live intermediate[23].

The result can be understood by a simple model[24], in
which the interaction of the chromophore system with the
environment is assumed to be negligible until vibronic de-
excitation takes place, in contrast to what is assumed in the
conventional theory of electron transfer (which has not offered
an explanation for these experimental findings). The levels of
D* and the vibronically excited level of W_2 coincide and the
electron, after excitation, is spread over D, W_1 and W_2 and is
finally trapped in the ground state of W_2^- by vibronic
relaxation (time constant τ). The time constants of the first
and second process are 0.5 ps and 3 ps respectively, and this
explains the experimental results mentioned above. The
transient absorption change corresponds to 16% W_1 and therefore

W_1 shares 16% of the delocalized electron in this intermediate. It can be concluded that the level of W_1^- is above the level of D* by an amount b = 0.06 eV[24].

The rate of trapping the electron in W_2 from D^*, given by:

$$k = \frac{1}{2\tau} \cdot \frac{(2\epsilon)^2}{a^2 + (2\epsilon)^2} \qquad (1)$$

depends strongly on a, where a is the amount by which the level of vibronically excited W_2^- is higher (or lower) than the level of D* and $\epsilon = 2\epsilon_{DW1}\epsilon_{W1W2}/b$, where ϵ_{DW1} and ϵ_{W1W2} are the energy matrix elements for the electron transfer from D* to W_1 and from W_1 to W_2. The levels of D* and of vibronically excited W_2^- are assumed to coincide at the temperature T = 0, but to differ on the average by $a = \pm(1/2)k_BT$ at temperature T due to thermal noise (k_B = Boltzmann constant). Then the rate k (Eqn. 1) decreases with increasing temperature; this is indeed observed in the reaction center and is found to be in quantitative agreement with theory[24].

The model can be further tested by distinct environmental changes (e.g. exchange of a polar by a nonpolar amino acid residue) that impair coincidence of energy levels and thus reduce the rate. The model explains the unidirectionality of the electron flow in the reaction center by assuming the proposed coincidence of energy levels as a product of evolutionary constraint[24].

This example shows that the fast separation of the electron from the excited dye molecule, necessary to avoid de-excitation, is only possible if the primary acceptor is well-adjusted to the donor. The arrangement of the chromophores in the reaction center must be well optimized for efficient energy storage. This indicates the difficulties in attempts to fabricate non-natural analogues of such systems. To reach the goal of constructing a device according to the given design principles is a great challenge.

3.   SENSORS, MOLECULAR WIRES AND SWITCHING ELEMENTS

3.1 Transducing Signals in Monolayer Assemblies

Several types of supramolecular sensing and switching elements have been made by assembling monolayers[25,26]. Figures 3d-f give some examples.

The problem of transmitting a signal from an excited molecule or group (acting as emitter) through a π-electron system (acting as molecular wire) to a molecule (acting as receiver) is widely discussed[27]. The conduction mechanism considered in the previous section is limited to short distances. Transport of solitons, through a polyene chain connecting a donor and an acceptor and triggered by excitation, has been proposed to enable information transfer over longer distances. Furthermore, soliton transport back and forth was suggested for information storage, since a moving soliton interchanges single and double bonds[28]. One may wonder how to make devices for information processing, based on soliton

811

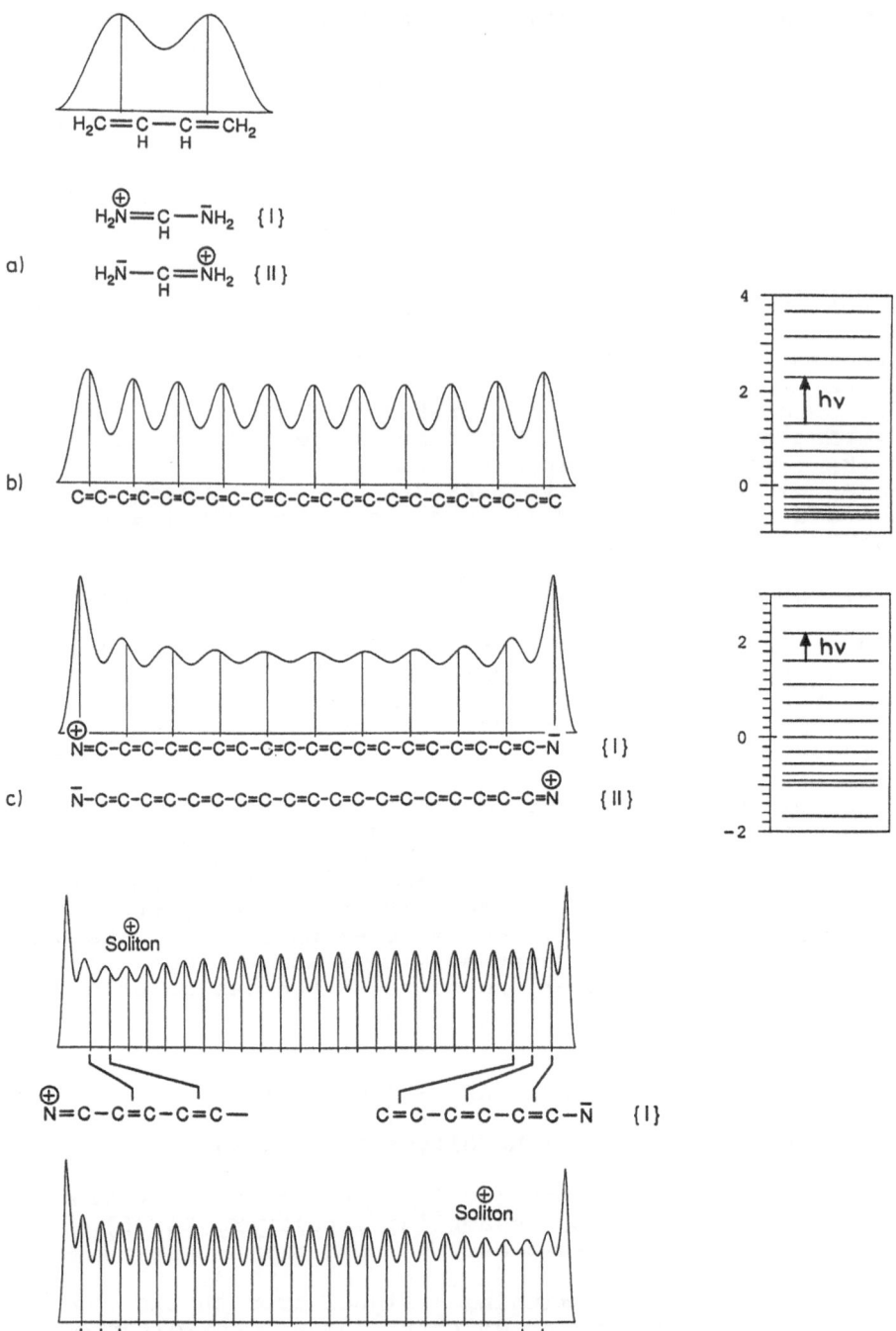

Fig. 4. Geometry of π-electron systems. π-electron
density distribution along molecule representing single or double bond character of each
bond. Additional density peaks on carbon atoms
are not represented in the model. Energy levels
given in units of $h^2/(8\pi^2 md^2) = 1.94$ eV. a)
Butadiene (bond alternation, density maxima at

switching, by assembling appropriate monolayers. A further
study of the processes should be of interest.

3.2 Conditions for Instability of Equal Bonds in Conjugated Chains and Occurence of Solitons

These problems are intimately related to the theory of the
color of organic dyes[29,30], and the basic concepts can be
seen, in the free-electron approximation, for the simple cases
of butadiene and amidinium ion[31] (Fig. 4): The 4 π-electrons
present in both cases occupy the two lowest levels. The charge
density (sum of two sine-squares) has a minimum between two
maxima, indicating the single bond between the two double bonds
in butadiene and the two equal bonds (symbolized by resonating
structures I and II) in the amidinium ion.

The situation is similar for the long-chain analogues,
polyenes and cyanine dyes. Polyenes have alternating bonds: the
π-electron system with equal bonds is unstable since the bonds
with the higher π-electron density tend to shorten, which again
leads to an increase in density until self-consistency between
bond length and density is reached[31,32] (Fig. 4b). A gap
between the energy level of the highest occupied state and the
next state occurs and the absorption maximum converges to 600 nm
as we increase the number of double bonds. Cyanine dyes have
practically equal bonds and can be approximated within the free-
electron model (Fig. 4c). The absorption maximum is 220 nm for
4 π-electrons (Fig. 4a) and shifts by $(4d)^2mc/h = 129$ nm (where
d is the bond length, m the electron mass, c the velocity of
light and h Planck's constant) with each additonal -CH=CH- unit,
independent of the chain length[30,31]. This essential
difference between a polyene and a cyanine – the instability in
the chain of equal bonds in the polyene, resulting in bond
alternation – was the basic concept in the theory of the color
of organic dyes[29-31]. This instability was later considered
in the general case of a one-dimensional metal suffering a
periodic lattice distortion[33].

The simple situation in cyanines (equal bonds, 129 nm
shift) cannot hold for very long chains (an infinite cyanine
chain cannot be distinguished from an infinite polyene chain).
Bond alternation takes place with charge accumulation at one
end. The structures I and II in Fig. 4d describe two equivalent
forms: each form is a push-pull polyene with bond alternation,
while for shorter chains (Fig. 4c), both structures describe one
and the same mesomer. The transition from the first to the
second case occurs above n = 20 where n is the number of π-
electrons, n - 1 the number of atoms in the chain[34]. The bond

---

Fig. 4. double bonds) and amidinium ion (equal bonds,
(cont.) resonance intermediate between structures I and
II, equal density in both bonds). b) Polyene
(bond alternation, energy gap between highest
occupied and lowest unoccupied state). c)
Cyanine, number of π-electrons n = 20 (resonance
intermediate between structures I and II). d)
Cyanine, number of π-electrons n = 24
(structures I and II describe different
(tautomeric) forms with + charge on the left and
right respectively.

lengths are nearly equal at the end where the charge is accumulated and alternate elsewhere along the chain, i.e. a soliton is localized at this end and moves to the other end as we proceed from structure I to structure II.

3.3 Soliton Flip-Flop by External Electric Field

A push-pull polyene may be switched from structure I to II in Fig. 4d, and back by applying an external electric field (the soliton moves from left to right and back). This may be of interest for molecular storage, and then it is important to estimate the threshold field and to find n values for which this field is in an appropriate range[34]. In the case n = 46, the threshold field is 10^6 V/cm: this is in an experimentally well achievable range (fields up to 10^7 V/cm have been applied to monolayers without discharge taking place[35,36]). Such molecules can thus be useful as storage elements.

The data[34] were obtained by applying a very simple model[37] (using a step-potential for electron-phonon coupling in π-electron systems). The model, which has no free parameter, leads to a description of π-electron systems (groundstate geometry and properties of solitons, polarons and excitons), in agreement with more involved standard procedures[38]. By its simplicity and transparency the model is useful for extensions to more complex cases. It will be of interest to investigate the dynamics of signal transfer and the influence of barriers within the cyanine chain that may act as switching elements.

A critical problem in aiming supramolecular information processing, namely addressing single molecules, is discussed in[39,40]. The molecular signal, in the present case the dipole orientation, must also be adequately amplified, e.g. a sufficiently large portion of a liquid crystal must become ordered. Push-pull polyenes can be well organized in monolayers, for instance by inclusion in a monolayer of a functionalized cyclodextrin[41].

4. NATURAL AND ARTIFICIAL CREATIVITY: KNOWLEDGE-GENERATING SYSTEMS

Learning how to obtain biomimetic machines, inspired by reflections on the origin and evolution of life, is a most challenging aim. Even the simplest living structures appear as the result of a highly creative process, comparable with the creative processes in the human mind. Suppose the basic working principle is the same in both cases; we should learn most about the creative process by studying the proposed first steps in the origin of life. At this stage, the situation is less intricate than at later stages of the evolution and in considering creative processes in the brain. This study should help us in seeing general aspects of the creative process, and in finding ways towards artificial creativity.

4.1 Driving Force of the Creative Process

Even the simplest living systems have "knowledge", i.e. the know-how to survive, as a form, in a given environment. Knowledge is a basic property of living systems and is not bound to the presence of a central nervous system. In the evolutionary process living forms become increasingly intricate; they gain knowledge and this is a creative process. Therefore, finding ways to obtain artificial systems that possess knowledge and posses a mechanism to generate knowledge, should be seen as a main goal in attempting biomimetic machines.

The driving force in the origin of life is a diversified microstructure that induces the formation of information-carrying and knowledge-accumulating systems[42,43]. There are two steps; (i) systems appear that carry a coded message; the recipe to reproduce the system. The environment is such that a continuous reproduction takes place; (ii) systems of increasing complexity evolve because a new system, which can be somewhat more complex, may have a selectional advantage: surviving in a hitherto hostile environment which can, from now on, be populated. Gradually, more and more inhospitable regions become populated by increasingly complex systems.

4.2 Modeling First Steps. Aggregate of Molecules: Supramolecular Machine Functioning as Error-filter. Emergence of Translation Machine

How can we model a system according to this proposal? We shall focus on the supramolecular engineering aspects and will not explicitly consider the evolution of the biomachinery because we want to place emphasis on generalities.

Two kinds of elements may join in arbitrary sequence to form a chain. The chain will be copied in the given environment, say via formation of a chain with a complementary sequence. Occasionally, an error may occur in the copying process (i.e. the same kind of element instead of the complementary kind enters the copy at some position). The environment will change periodically between phases where the chains multiply by copying and phases where the chains show specific properties, depending on their sequences. They may, for instance, fold specifically, as indicated in Fig. 5. Different forms have different chances to survive the phase considered, i.e. a selection process takes place where the fittest forms survive the phase, say forms that are hindered from leaving an appropriate region by diffusion.

Particular forms may evolve that aggregate in entering the selection phase. Aggregates, as indicated in Fig. 5, have an increased chance to survive the selection phase since they are prevented from diffusion; most important, aggregates are error filters. If an error has occurred in the copying process, the molecular shape will, in general, be different from the shape of an error-free copy and erroneous copies will not match. A supramolecular machine has evolved: error-free molecules interact and cooperate, forming a functional unit.

The driving force of the process is a periodicity in time, which ensures a continuous change between multiplication and selection phases, and a diversified and compartmentalized microstructure which (i) prevents dispersion of the molecules and thus enables interlocking into an aggregate and (ii) provides a variety of regions where increasingly complex forms can develop.

The emergence of an aggregate forming a supramolecular machine is of particular significance, since biosystems are

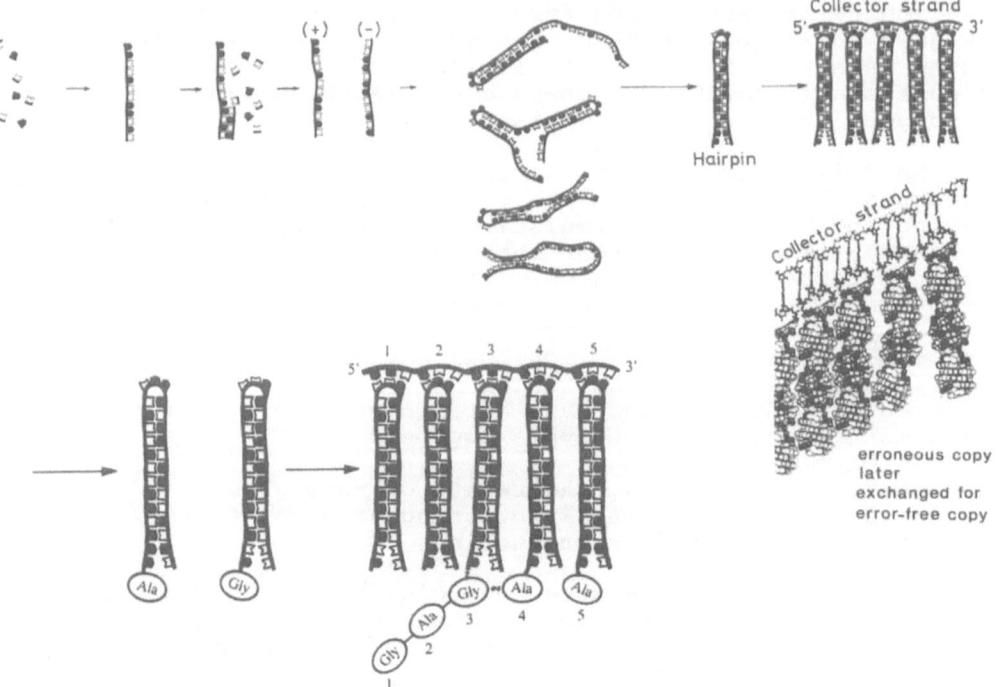

Fig. 5. Evolution of aggregating system and of
translational device.

functional entities on all levels, and the aggregate of
interlocking molecules is considered to be the first supramolecular functional entity. The aggregate has a purpose, but
no processes shaped by this purpose have taken place. The
aggregate is the result of transparent physical processes that
may have happened on a prebiotic planet. The error-filtering
device emerges as a by-product. This elucidates the nature of
creative processes.

A scenario leading to a distinct organizational structure
of evolutionary processes can be described using the same
concept[42]. The formation of an aggregate, acting as catalyst
by connecting monomers to a chain in a specific sequence, can
easily be grasped. This step, the emergence of a translation
machine, is a particularly important breakthrough process
(Fig. 5).

4.3 Knowledge as a Measure of the Creative Process

We restrict ourselves to the basic feature seen in the very
first steps considered (Fig. 5): the process is initiated, and
driven by a specific complex structure in space and time. A
machine emerges that carries *information* (coded message) and
generates *knowledge* (know-how form survives by reproduction).
Information is measured by the number of bits describing the
genetic message while knowledge is measured by the number of
bits of genetic message that are discarded, in the evolutionary
process, from the beginning of the process until the form under
consideration has evolved. The evolutionary pathway is assumed
to be idealized, i.e. a path without detours is considered (any
other path to the same end would require more information to be

816

discarded during the process and thus introduce an ambiguity in defining a measure for evolutionary progress)[42,44]. Knowledge, as defined, is zero until the first self-reproducing system appears and is exposed to selection, and, from then on, knowledge increases in the course of evolution. Generation of knowledge is a creative process. These considerations focus on an essential prerequisite of a creative process: the presence of a complex environmental structure. The evolving system is the response to this structure. The process must be distinguished from the development of structures in a homogeneous system by innate reasons, e.g. the structures forming in cellular automata. These basic aspects of evolutionary processes should stimulate the design of biomimetic machines in many ways.

4.4 Attempts to Mimic Evolutionary Steps

Monolayer assembly may assist in studying the specificity in the process of interlocking molecules in aggregates and its use in designing error filters. A monolayer of a fatty acid is produced on the surface of a dilute solution of two dyes, A and B. Dye A is first adsorbed, and forms a monolayer of close packed molecules and later dye A is gradually exchanged for dye B, which also forms two-dimensional crystals of tightly interlocking molecules. No mixed crystals are formed during the process: the dye A aggregate does not allow incorporation of a dye B molecule and vice versa. An isosbestic point occurs in the reflection spectrum[45,46]. For another remarkable case illustrating segregation from an aggregate in a monolayer see[47].

Attempts were made to construct a translational machine of the kind shown in Fig. 5[48]. A single-stranded oligonucleotide was used as the messenger (collector strand), and oligonucleotides, in hairpin conformation, as adapters that bind to the collector strand by complementary base pairing with the hairpin loop. It could be demonstrated that hairpin loops bind to a collector strand by triplets of complementary base pairs. The collector strand was immobilized (bound to cullulose). Proposed steps are binding amino acids to the hairpins and joining them. The goal is to simulate a critical step in modelling the origin of life, but the principle can be of more general interest in the design of biomimetic machines. The formation of an oligopeptide by polycondensation of the amino acids bound to hairpins can be simulated with the amino acid bound to long-chain alcohols. Oligopeptides are formed by joining the molecules in making a monolayer[49].

4.5 The Prospect of Artificial Creativity

Most challenging, of course, is the aspect of artificial creativity; of using the basic principle of natural creativity for constructing knowledge-generating devices, i.e. machines forming more and more intricate structures (from a great number of transient structures that are generated in parallel processing) by selecting structures according to their rational response to a given complex data set, and by continuously repeating the process, similarly to bio-systems, where structures are selected by their rational response to an external world("internal model of the world").

Artificial creativity (to be distinguished from artificial intelligence) is a fascinating, unexplored future prospect. In

a) Evolution

```
Diversified
microstructure ─────────────────────> Time
                          ∧
                          |
                   Individual
                   interacts with environment

                        Survival and reproduction
                        or extinction?

                        Yes or no?

                             Measuring process
                             contributes to
                             gain of knowledge
                             (creation of new forms)
```

b) Physical measuring process

```
                   Measuring device
                   interacts with environment

                        Amplification of signal
                        or no response?

                        Yes or no?

                             Measuring process
                             contributes to gain of
                             knowledge
                             (creation of new theories)
```

Fig. 6. Evolution as a measuring process and physical measuring process resulting in the increase of knowledge.

the development of artificial creativity it will be important to be inspired by the transparent model steps in the origin of life rather than by the creative processes in the brain, whose analysis is much more difficult. Monolayer assembling can contribute to the development of such machines.

4.6 Bio-evolution as a Measuring Process

A basic aspect of considerations in terms of biomimetic machines is the nature of the measuring process[43]. Bio-evolution is a measuring process: the individual is interacting with the environment. A yes or no decision takes place: to survive and to reproduce in the subsequent multiplication phase or not to survive. Each measurement adds to the knowledge of the evolving system. The result of many measurements is the creation of new forms (Fig. 6).

In the physical measuring process, the measuring device is interacting with the environment. The yes or no decision is reached, in Bohr's words, by an "irreversible act of amplification that brings the measuring process 'to close'", i.e. the recorded result can be described "in plain language".

In the present context, Bohr's human observer is substituted by a knowledge-producing system (natural or artificial). The irreversible act of amplification, that brings the measuring process "to close", corresponds to the decision between elimination, and survival and subsequent multiplication.

The evaluation of physical measurements, the creation of new theories, corresponds to the creation of new forms in bio-evolution.

This remark should indicate that the emergence of knowledge-producing systems is a fundamental physical phenomenon and that biomimetic machines thus play a role in assisting to understand general aspects, in addition to their role in stimulating attempts to design and construct artificial supramolecular machines of increasing complexity.

Acknowledgment

The author is grateful to Professor R.M. Metzger for reading the manuscript and for stimulating discussions.

REFERENCES

1. H. Kuhn, p.245 **in**: "Verhandlungen der Schweizerischen Naturforschenden Ges." (1965)
2. H. Kuhn and D. Möbius, **Angew. Chem.** (Int. edit. in English) 10:620 (1971)
3. D. Möbius, Spectroscopy of Complex Monolayers, **in**: "Langmuir-Blodgett Films", G. Roberts ed. Plenum Publishing Corp. (1990)
4. M. Sugi, **Thin Solid Films**, 152:305 (1987)
5. M. Sugi, Langmuir-Blodgett Films for Molecular Electronics-Recent Trents in Japan p.441 **in**: "Molecular Electronic Devices",F.L. Carter, R.E. Saitkowsi and Y.H. Woltjen, eds., Amsterdam (1988)
6. L.M. Blinov, **Russ. Chem. Rev.**, 52:713 (1983)
7. D. Möbius, ed. Langmuir-Blodgett Films 3, **Thin Solid Films**, 159:1,2; 160:1,2 (1988)
8. K. Fukuda and M. Sugi, ed. Langmuir-Blodgett Films 4, **Thin Solid Films**, 178:1,2; 179:1,2; 180:1,2 (1989)
9. A. Barraud, ed. Report on NATO Advanced Research Workshop on Condenced Systems of Low Dimensionality, **J. Chimie Physique**, 85:991 (1988)
10. D. Möbius and H.Kuhn, **J. Appl. Phys.**, 64:5138 (1988)
11. E.E. Polymeropoulos, D. Möbius and H. Kuhn, **Thin Solid Films**, 68:173 (1980)
12. K. Saito, M. Yoneyama, M. Saito, K. Ikegami, M. Sugi, T. Nakamura, M. Matsumoto and Y. Kawabata, **Thin Solid Films**, 160:133 (1988)
13. H. Schreiber, H. Kuhn and H.D. Försterling, **Ber. Bunsenges, Phys. Chem.**, 91:798 (1987)
14. F. Willig, R. Eichberger, K. Bitterling, W.S. Durfee, W. Stork and M. van der Auweraer, **Ber. Bunsenges. Phys. Chem.**, 91:869 (1987)
15. M. Fujihira and H. Yamada, **Thin Solid Films**, 160:125 (1988)
16. M. Fujihira, M. Sakomura and T. Kamei, **Thin Solid Films**, 180:43 (1989)
17. M. Fujihira and M. Sakomura, **Thin Solid Films**, 179:471 (1989)
18. H. Kuhn, **Pure & Appl. Chem.**, 51:341 (1979); 53:2105 (1981)
19. H. Kuhn, **Phys. Rev. A**, 34:3409 (1986)
20. E.E. Polymeropoulos, D. Möbius and H. Kuhn, **J. Chem. Phys.**, 68:3918 (1978)
21. D. Möbius, Photoelectron Transfer in Organized Assemblies, pp139-159 **in**: "Photochemical Conversion and Storage of

Solar Energy, Part A 1982", J. Rabani, ed., The Weizmann Science Press of Israel (1982)

22. J. Deisenhofer, O. Epp, K. Miki, R. Huber and H. Michel, **J. Mol. Biol.**, 180:385 (1984)

23. W. Holzapfel, U. Finkele, W. Kaiser, D. Oesterhelt, H. Scherr, H.U. Stilz and W. Zinth, **Chem. Phys. Lett.**, 160:1 (1989)

24. E. v.Kitzing and H. Kuhn, **J. Phys. Chem.**, 94:1699 (1990)

25. D. Möbius, **Ber. Bunsenges. Phys. Chem.**, 82:848 (1978);

26. W. Arden and P. Fromherz, **Ber. Bunsenges. Phys. Chem.**, 82:868 (1978)

27. F.L. Carter, A. Schultz and D. Duckworth, Soliton Switching and its Implications for Molecular Electronics, **in:** "Molecular Electronic Devices II", F.L. Carter, ed., Marcel Dekker, New York (1987)

28. S. Roth and H. Bleier, Can Polyacetylene-Solitons be used in Molecular Electroncs? **in:** "Proceedings of the Int. Conf. on Molecular Electronics", Science and Technology, Hawaii (1989)

29. H. Kuhn, **J. Chem. Physics**, 17:1198 (1949)

30 H. Kuhn, The Electron Gas Theory of the Color of Natural and Artificial Dyes **in:** "Progress in the Chemistry of Organic Natural Products", L. Zechmeister, ed., 16:169 (1958); 17:404 (1959)

31. H. Kuhn, **Chimia**, 4:203 (1950)

32. H.D. Fösterling, W. Huber and H. Kuhn, **Int. J. Quantum Chem.**, 1:125 (1967)

33. R.E. Peierls, p.108, "Quantum Theory of Solids", Clarendon, Oxford (1955)

34. C. Kuhn, to be published

35. B. Mann and H. Kuhn, **J. Appl. Phys.**, 42:4398 (1971)

36. E.E. Polymeroupolos and J. Sagiv, **J. Chem. Phys.**, 69:1836 (1978)

37. C. Kuhn, **Phys. Rev. B**, 40:7776 (1989)

38. W.P. Su, J.R. Schrieffer and A.J. Heeger, **Phys. Rev. B**, 22:2099 (1980)

39. H. Kuhn, **Thin Solid Films**, 178:1 (1989)

40. H. Kuhn, Organized Monolayers Building Blocks in Constructing Molecular Devices, p.3 **in:** "Molecular Electronics: Biosensors and Biocomputers", F.T. Hong, ed., Plenum Publ. Corp., New York (1990)

41. A. Barraud, M. Blanchard-Desoe, V. Dentan, I. Ledoux, J.M. Lehn, S. Palacin and J. Zyss, From Molecular to Supramolecular Engineering in Nonlinear Optics: "Mixed Carotenoid-Cyclodextrin Langmuir-Blodgett Films", in print

42. H. Kuhn, J. Waser, **Angew. Chem.**, 20:500 (1981) (Int. Edit. in English)

43. H. Kuhn, **IBM J. Res. Develop.**, 32:37 (1988)

44. H. Kuhn, **Ber. Bunsenges. f. Physik. Chem.**, 80:1209 (1976)

45. U. Lehmann, **Thin Solid Films**, 160:257 (1988)

46. U. Lehmann and H. Kuhn, **Adv. Space Res.**, 4:153 (1984)

47. A. Barraud and M. Vandevyver, p.771 of this volume

48. U. Baumann, U. Lehmann, K. Schwellnus, J.H. van Boom and H. Kuhn, **Eur. J. Biochem.**, 170:267 (1987)

49. K. Fukuda, Y. Shibasaki and H. Nakahara, **Thin Solid Films**, 160:43 (1988)

NATO ADVANCED RESEARCH WORKSHOP ON CONDENSED SYSTEMS OF LOW DIMENSIONALITY
April 23 to 27, 1990, in Marmaris, Turkey

MATERIALS INDEX

quantum, 109, 247, 370
Hall mobility, 98, 143, 147, 411
Hall resistivity, 6
Hartree-Fock theory, 52
HBT, 613, 664, 669
Heisenberg model, 754
HEMT, 145, 242, 640, 643
Heterojunction bipolar phototransistor, 580
Heterojunctions, 3, 45, 100, 261, 647
Highly strained layers, 524
Homoepitaxy, 117
Homogeneous strain relaxation, 454
Hopping transport, 797
Hot electron effects, 236
Hot electrons, 242
Hydrostatic deformation, 582
Hydrostatic pressure, 145, 631

Impact ionization, 629
Impurity scattering, 286, 350
  elastic, 379
In-plane conductivity, 702
Incommensurability, 732
Inelastic light scattering, 4
  resonant, 12
Inelastic scattering, 238
Inelastic tunneling processes, 299
Infrared bleaching, 317
Infrared singularity, 287
Inter-Landau level excitations, 3, 18
Inter-Landau-level transitions, 135
Interband magnetophoto-conductivity, 523
Interband transitions, 218
Intercalation, 678
Intercalation compounds, 695
Intercalation reactions, 709
Intercalation stage, 696
Intercalation synthesis, 697
Interdiffusion, 471, 572
Interface modes, 500
Interface roughness, 197 251, 263, 285, 456, 488, 644
Interface states, 576
Interfacial energy, 561
Interionic potentials, 721
Intermolecular conduction, 760
Intersubband excitations, 36
Intersubband relaxation scattering, 324
  time, 324
Intersubband scattering, 317
Intersubband transitions, 12, 143

Intervalley scattering, 189, 191, 203, 664
Intrasubband excitations, 36
Intrasubband transition, 12
Intralayer forces, 679
Inversion layer, 252
Iodination, 772
Ionic conductor, 690
Ionization lifetime, 110
Ionized donors, 387
Ionized impurity scattering, 263
Island nucleation three-dimensional, 451
Isotopic separation, 704

JFET, 613

$k.p$ model, 145, 202
Kane model, 510, 544, 584
Knowledge-generating systems, 814
Kosterlitz-Thouless phase transition, 59
Kramer's degeneracy, 74
Kronig-Penney model, 74, 585

Lamellar structure, 695
Lanczos (recursion) method, 223
Landau damping, 15, 38, 280
Landau level, 7, 53, 85, 99, 204, 296, 343, 594, 762
  fan, 206, 308
Landau quantum numbers, 525
Landau splitting, 168
Landauer formula, 269, 370
Langmuir-Blodgett films, 784, 795
Lasers
  broad area, 608
  GRINSCH, 604, 614
  quantum well, 74, 579, 623
    strained, 405, 653
  semiconductor, 52
Lattice matching, 146
Lattice mismatch, 471, 495, 509, 533
Layered cathode, 691
Layered structure, 776
LEED, 473
Level splitting, 526
Lewis acids, 697
Lifetime broadening, 66, 361
Light scattering
  inelastic, 3
Liquid phase epitaxy, 435, 607
Lithography, 124
  electron beam, 336, 349
Local density approximation, 201
Localization, 7 ,574, 680

energies, 182
phenomena, 283
Long chain compounds, 771
Luminescence excitation, 5
  spectra, 57
Luminescence, 85, 116
  intensity, 61
Luttinger Hamiltonian, 220, 625
Luttinger parameters, 544, 586

Macroscopic strain effects, 430
Magnetic depopulation, 342
Magnetic polaron, 175
Magnetic quantum limit, 8
Magnetic susceptibility, 713,
  736
Magnetoabsorption, 98
Magnetocapacitance, 293
Magnetoconductance, 293
Magnetodispersion, 135
Magnetoexcitation spectra, 79,
  88
Magnetoluminescence, 52
Magnetooptical measurements,
  149, 208
Magnetooptics, 133
Magnetoreflectivity, 204
Magnetoresistance, 6, 147, 357,
  763
  negative, 237
Magnetoroton minimum, 18
Magnetotransmission, 99, 210
Many-body effects, 70
Many-particle interactions, 4
Mass anisotropy, 204
Matthiessen's rule, 267
Maxwell-Garnett theory, 418
MESFET, 242, 613
Metal-insulator transition,
  285, 295, 388
Metastable alloys, 441
Microelectronic devices, 638
Microscopic strain energy, 435
Microwave performance, 145
Migration enhanced epitaxy,
  404, 449, 535
Miniband, 309, 518
  dispersion, 210
MISFET, 613
Misfit defects, 474
Misfit dislocations, 562
Mismatch dislocations, 438
Mismatch epitaxy, 141
Misoriented substrates, 645
Mixed valence compounds, 771
Mixed valency, 729
Metalorganic chemical vapor
  deposition, 119, 409,
  435, 527, 579, 614
  flow modulation, 141
MODFET, 404, 580

strained, 664
Modulation doping, 4, 36, 317,
  347,
Molecular beam epitaxy, 73,
  104, 124, 147, 166, 191,
  241, 247, 318, 402, 435,
  472, 512, 548, 579, 628,
  636, 669
  atomic layer, 535
  modulated, 534
Molecular electronics, 776,
  779, 805
Molecular field approximation,
  105
Molecular intercalation
  compounds, 716
Molecular orbital calculations,
  788
Molecular wire, 782
Monomolecular layers, 807
MOS structures, 283, 328
MOSFET, 242, 247
Multi-sandwich structure, 678
Multiple scattering, 285

n-i-p-i structures, 141
n-p-n bipolar transistors, 623
Narrow channel, 335
Narrow constrictions, 262
Narrow-gap semiconductors, 141
Nonlinear optical coefficients,
  798
Nonohmic scattering, 236
Nonradiative processes, 125,
  610
Nonradiative recombination, 187
Nonstoichiometry, 729
Nontetragonal distortion, 467

Octahedral coordination, 679
Off-oriented substrates, 453
Ohm's law, 235
Onsager problem, 802
Optical absorption, 5
Optical anisotropy, 417
Optical emission, 5
Optical gain, 626
Optical matrix element, 573
Optical modulators, 120, 641
Optical nonlinearity, 141
Optical switches, 620
Optoelectronic circuits, 623
  integrated, 613
Optoelectronic devices, 73,
  437, 638
Optoelectronics, 141
Order-disorder transition, 565
Order-disorder reactions, 720
Ordering, 442
Organic molecules, 686
Organic rectifiers, 780

Oscillator strength, 55, 95, 124, 185, 221, 226, 319
Out-of-plane percolation, 775

Pauli-blocking, 58
Peierls gap, 799
Persistent photoconductivity, 48
Phase separation, 143
Phase-change switch, 780
Phonon
  dispersion, 477, 504
  drag, 331
  emission, 328
  replica, 186
Phonons, hot, 62
Photoabsorption, 217
Photocreated carriers, 95
Photodiodes, 407
Photoemission, 153
Photoinduced electron pump, 809
Photoluminescence 9, 28, 63, 73, 123, 126, 141, 152, 168, 181, 193, 300, 509, 533, 643
  excitation spectroscopy, 74, 86, 116, 217, 509
  time resolved, 171
Photon funnel, 809
Photoreflectance spectroscopy, 547
Piezoelectric effect, 263
Piezoelectric interaction, 333
Piezoelectric scattering, 63
Piezoelectricity, 116
Piezoreflectance, 427
Plasmon, 35
  excitations, 288
  dispersion, 37
Polarization dependence, 135
Polarons, 800
Pseudomorphic growth, 146, 527
Pseudopotential, 201, 558, 570, 629
Pump and probe experiments, 194

Quantized conductance, 350, 387
Quantum ballistic transport, 341, 387
Quantum confined Stark effect, 112
Quantum confinement, 109, 132, 154
Quantum dots, 109, 124
Quantum efficiency, 609, 625, 632
Quantum oscillations, 339
Quantum phase coherence, 275
Quantum point contacts, 336, 349, 371
Quantum size effect, 40, 580

Quantum wells, 3, 36, 57, 73, 109, 124, 146, 167, 181, 201, 217, 293, 318, 513, 533, 549, 637, 644
  strained layer, 523, 599
Quantum wire 36, 109, 123, 347, 369, 387
Quasi-confined modes, 487, 497
Quasi-electrons, 27
Quasi-holes, 27
Quasi-particle spin, 753

Raman scattering, 35, 167
  micro-, 37
  resonantly enhanced, 317
Raman spectroscopy, 476, 481, 495, 538
Random-phase approximation, 16, 35
Random telegraph noise, 347
Rapid thermal anneal, 671
RBS channeling, 467
Recombination, 8, 34, 86
Reflectance difference spectroscopy, 410
Residual impurities, 150, 284
Resistance quantization, 239, 359
Resonant enhancement, 21, 36, 124
Resonant structure, 364
Resonant tunneling, 293, 317, 381, 640
  structures, 181
RHEED, 534
  oscillations, 403
  pattern, 451
Rigid band approximation, 681
Ring oscillator, 650
Roton structure, 18

Saw-tooth potential, 159
Scanning tunneling microscope, 382, 779
Scattering time, 192, 283
Schottky barriers, 635
Screening, 31, 58
  charge, 12
Segregation, 471
Selective optical excitation, 87
Self-aligned devices, 650
Self-assembling monolayers, 784
Self-consistent potential, 392
Self-organization, 804
Semi-amphiphilic salts, 772
Shallow levels, 660
Sheet carrier density, 46
Shubnikov-de Haas oscillations, 36, 142, 249, 343, 523
Singlet-triplet mixing, 751